AF770602

Developments in High-Speed-Vehicle Propulsion Systems

Developments in High-Speed-Vehicle Propulsion Systems

Edited by
S.N.B. Murthy
Purdue University
West Lafayette, Indiana

E. T. Curran
Wright Laboratory
Wright-Patterson AFB, Ohio

Volume 165
PROGRESS IN
ASTRONAUTICS AND AERONAUTICS

Paul Zarchan, Editor-in-Chief
Charles Stark Draper Laboratory, Inc.
Cambridge, Massachusetts

Published by the
American Institute of Aeronautics and Astronautics, Inc.
1801 Alexander Bell Drive, Reston, Virginia 20191-4344

Copyright © 1996 by the American Institute of Aeronautics and Astronautics, Inc. Printed in the United States of America. All rights reserved. Reproduction or translation of any part of this work beyond that permitted by Sections 107 and 108 of the U.S. Copyright Law without the permission of the copyright owner is unlawful. The code following this statement indicates the copyright owner's consent that copies of articles in this volume may be made for personal or internal use, on condition that the copier pay the per-copy fee ($2.00) plus the per-page fee ($0.50) through the Copyright Clearance Center, Inc., 222 Rosewood Drive, Danvers, Massachusetts 01923. This consent does not extend to other kinds of copying, for which permission requests should be addressed to the publisher. Users should employ the following code when reporting copying from the volume to the Copyright Clearance Center:

1-56347-176-0/96 $2.00 + .50

Data and information appearing in this book are for informational purposes only. AIAA is not responsible for any injury or damage resulting from use or reliance, nor does AIAA warrant that use or reliance will be free from privately owned rights.

ISBN 1-56347-176-0

Progress in Astronautics and Aeronautics

Editor-in-Chief
Paul Zarchan
Charles Stark Draper Laboratory, Inc.

Editorial Board

John J. Bertin
U.S. Air Force Academy

Leroy S. Fletcher
Texas A&M University

Richard G. Bradley
Lockheed Martin Fort Worth Company

Allen E. Fuhs
Carmel, California

William Brandon
MITRE Corporation

Ira D. Jacobsen
Embry-Riddle Aeronautical University

Clarence B. Cohen
Redondo Beach, California

John L. Junkins
Texas A&M University

Luigi De Luca
Politechnico di Milano, Italy

Pradip M. Sagdeo
University of Michigan

Martin Summerfield
Lawrenceville, New Jersey

Table of Contents

Preface

The main accomplishments in the development of high-speed vehicle propulsion in the past 15 years have been the significant advances in a number of foundational technologies, including methods of analysis and prediction, that form the basis for the practical realization of hypersonic cruise and accelerator launch vehicles. A survey of some of the developments in engine design, energy analysis, testing facilities, and mixing processes in scramjet combustors was presented in *High-Speed Flight Propulsion Systems*, Volume 137 of this Progress Series (1991). Since that publication, various members of the technical community have expressed an opinion that other major developments in this area have progressed sufficiently that it would be very helpful, both to new entrants into the filed and practicing engineers, if such developments could be captured in additional volumes.

The area of scramjet propulsion has received considerable attention since the 1950s, and there exists a sizable knowledge base that should be fully utilized in future research and development efforts. Much additional work has been done in recent years. Accordingly, it is proposed to issue a further volume on various aspects of scramjet propulsion, including overall system approaches and propulsion–vehicle integration, in the near future.

Other developments in low-speed propulsion schemes for high-speed vehicles, turboramjets, air collection, separation and storage schemes, and pulse detonation engines form the subject matter of the current volume. Each of these belongs in the spectrum of combined cycle engines that are effective in high-speed flight over different regimes. In addition, the methodology for evolving hypersonic vehicle and propulsion systems to meet the requirements of a stated mission, under given conditions of propulsion performance and industrial and material-structural capability, is gradually becoming clearer and is included in the volume. These subjects have been studied and advanced in several countries, and the authorship of the articles bears testimony to the keen international interest in hypersonic flight. It is to be hoped that such interest can be sustained by continued publications and periodic meetings dedicated to hypersonic technologies.

We have been greatly assisted in the editorial task by the reviewers of the various articles, and we are indebted to them. In the type of subjects covered in this volume, the authors work under a variety of constraints and predominately on their own personal time. We believe the authors and the reviewers have accomplished the task of producing in a timely and clear manner, a comprehensive account of the science, engineering, and art that has been pursued in these interesting technologies over approximately four decades. Such contributions should be of immense value in the future. We have much enjoyed our association with each of the authors and reviewers.

At the beginning of discussions on these volumes, Professor M. Summerfield was the Progress Series Editor-in-Chief. Later, Professor A. Richard Seebass took over the office. Recently, Dr. Paul Zarchan has become the Editor-in-Chief. We are deeply appreciative of their valuable encouragement in this effort.

The Editorial Department of AIAA, with J. Godette, C. Kalmin, and now R. Williams and K. Watters has been most helpful in the preparation of the volume. They have been thorough, patient, and unyielding in quality publication. We have enjoyed working with them and thank them for their immensely cooperative efforts.

S.N.B. Murthy
E.T. Curran
January 1996

Introduction

E. T. CURRAN*
Wright Laboratory, Wright-Patterson Air Force Base, Ohio 45433

Prolegomenon

This volume is the second in a short series, within the larger AIAA "Progress in Astronautics" series devoted to high-speed flight propulsion.

Since the first volume on high-speed propulsion[1] was published, the aerospace industry has been severely downsized, and, in general, the associated research and development base has been reduced and refocused on near-term projects. The drive to pursue large hypersonic projects has also diminished. In the United States the National Aerospace Plane (NASP) program was abandoned and replaced by a flight-test effort named Hypersonic System Technology Program (HySTP), which was aimed at demonstrating supersonic combustion ramjet (scramjet) performance at flight Mach numbers approaching Mach 15. In turn, because of budget constraints, the HySTP program was canceled and redirected to a Mach 8, hydrocarbon-based-fuel technology engine demonstration program named Hypersonic Technology Program (HyTech).

In Europe, following the early cancelation of the British air-breathing Hotol vehicle, the German hypersonics effort was severely curtailed. However, in Russia, despite severe economic problems, there have been several notable achievements in hypersonic propulsion technology with emphasis on scramjet engine development and flight-test experimentation utilizing the "Kholod" hypersonic flight testbed. Although scramjet

---

©This paper is declared a work of the U.S. Government and is not subject to copyright protection in the United States.

*Director, Aero Propulsion and Power Directorate.

technology is the principal subject of the next volume in this series, it may be significant to point out here that a second-generation flight-test effort is under way in Russia, which, if successfully prosecuted, should yield scramjet flight data up to speeds corresponding to approximately Mach 15. This flight-test effort is one segment of the "Orel" program, which is a broad research and development task investigating reusable space transportation system technologies[2].

Despite the overall downturn in hypersonic vehicle projects, there have been impressive achievements in the last few years in the technologies associated with turboramjets and other combined cycle engines. The technologies of these engines, including their thermal management and integration with the vehicle, are the principal concerns of this present volume.

In the Introduction to Ref. 1, the potential performance of both conventional and combined cycle engine systems was discussed using the framework of Fig. 1. It was noted that, for good acceleration performance (based on payload capability), an engine should possess high specific impulse

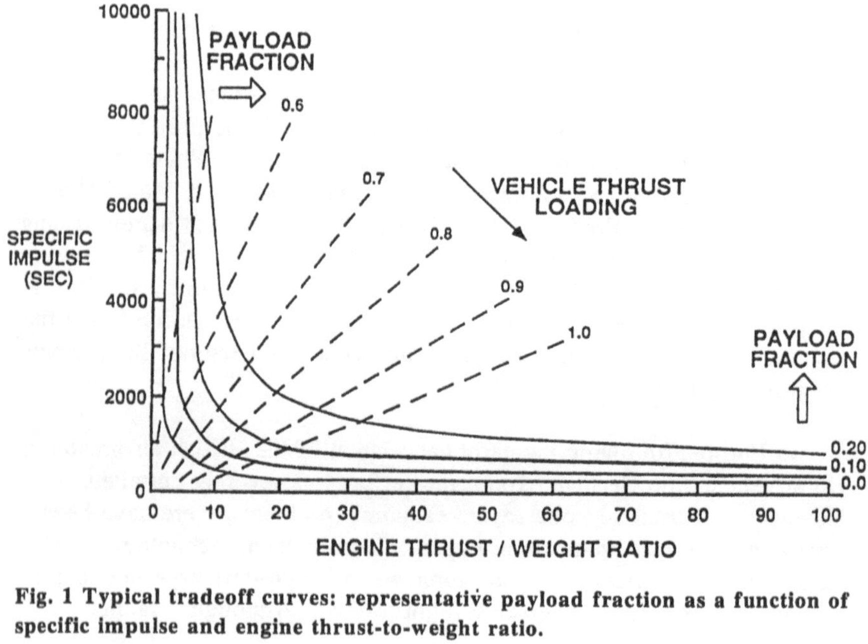

Fig. 1 Typical tradeoff curves: representative payload fraction as a function of specific impulse and engine thrust-to-weight ratio.

and high installed thrust-to-weight ratio and should be able to operate over a wide Mach number range with few mode changes and minimum variable geometry requirements. Furthermore, it was pointed out that there was no single existing class of engine that simultaneously possessed high specific impulse and high installed thrust-to-weight ratio. It was also anticipated that an engine type possessing specific impulse and thrust-to-weight ratio, intermediate between the turboaccelerator class and the rocket engine class, might offer performance superior to its progenitors. It was also simplistically postulated that such a combined cycle engine might be devised by pursuing two general streams of development. The first of these was to improve the thrust-to-weight ratio of the turboaccelerator class of engine—possibly with some reduction of the basic specific impulse of that class of engines. The second approach was to improve the capability of the rocket class of engine by improving the level of specific impulse—possibly with some reduction in the thrust-to-weight ratio. These two approaches provide a convenient, although by no means unique, structure for our further discussion.

Improving Turboaccelerator Capability

A primary and continuing approach to improving the thrust-to-weight ratio of the turboaccelerator class of engine is to concentrate on the technological development of the basic turbojet/turbofan class by increasing turbine temperature, improving internal efficiencies, and reducing structural weight. Although such incremental improvements may seem at first sight to yield small improvements in accelerator performance, it should be noted that in the United States the current integrated high-performance turbine engine technology (IHPTET) initiative aims at doubling the thrust-to-weight ratio of high specific thrust turbojet engines shortly after the year 2000. Sustained progress is being made under this initiative and is well documented in a recent paper by Hill[3]. Obviously such technological advances can be applied to afterburning turbojets, turbofans, and variable cycle engines. Although the variable cycle engine (VCE) may at first sight appear too complex for a combined cycle engine, its wide operating range and potential advantages for vehicle integration are significant. The editors of this volume therefore have included an article, detailing the evolution of the engine from its early beginnings, from J. Johnson of the General Electric Company. This most informative article (see Chapter 3) supplements other papers concerning VCEs in the current literature (see, for example, Refs. 4–6).

In addition to providing technologies for the development of conventional turboengines and variable cycle engines, the IHPTET program also provides improved component technologies for combined cycle engines and, in particular, for the engines being investigated under the joint USAF/NASA High Mach Turbine Engine (HiMATE) program. In this effort, component developments for both variable cycle turboramjets and air turborocket/ramjet engine concepts are currently being pursued. Ultimately, it is hoped to proceed to an engine demonstration phase.

Turboramjet Engines

There has been continuing interest in increasing the flight speed of turbine-powered vehicles. The development of turboramjet (TRJ) systems was a logical step in this process. A unique version of this engine appeared in the successful Nord Aviation Griffon 02 aircraft in France. The TRJ was also the powerplant of choice for the Republic XF-103 fighter in the United States. The XF-103 engine configuration, developed in the 1950s, was a tandem arrangement of turbojet and afterburner: for ramjet engine operation, the airflow bypassed the turbine engine through diverter values and was fed to the afterburner duct, which then functioned as a ramjet combustor. The primary new engineering problems of mode transition from turbojet to ramjet operation, and reverse transition, were investigated, as well as the problems associated with the turbojet installation during ramjet mode operation. Similarly, mode transition problems were also investigated in France in the 1950s[7].

In recent years an extensive investigation of turboramjet configurations and their relative performances has been carried out at Central Institute of Aviation Motors (CIAM) in Russia, and also a comprehensive experimental evaluation of the engineering aspects of those engines to speeds of about Mach 4.5 has been made. These developments are well documented in this volume in a foundational paper in Chapter 5 by V. Sosounov et al.; this chapter discusses the various types of turboramjets that have evolved since the 1950s, including the mechanical incorporation of turbofan and variable cycle elements into the basic turboramjet configuration. Note also that an interesting classification of turboramjet configurations is given, together with a discussion of their relative merits. Another key area discussed in this chapter is the subject of fuels and, in particular, the endothermic decomposition of hydrocarbon-class fuels. Attention was drawn to these fuels in the late 1960s[8] and work has continued sporadically in developing such fuels. These endothermic fuels possess enhanced cooling characteristics and decompose into low-molecular-weight vapors: the resulting high-work potential of such components can be usefully employed in doing work in the engine cycles before actual combustion. Also, of course, the relatively high density of such fuels enables the use of smaller tank volumes compared, for instance, with hydrogen. Similarly, it is possible to generate hydrogen from the higher-density hydrocarbon fuels, by various processes, and potentially gain increased vehicle effectiveness. A recent review of endothermic fuels was given by Ianovski[9] and additional discussion is given in Ref. 10.

Significant and comprehensive research and development have also been under way in Germany on a turboramjet propulsion system related to the Sänger Space Transportation System, the reference concept for the German

Hypersonic Technology Program. This turboramjet investigation was aimed at the speed range from takeoff to Mach 7, with liquid hydrogen as fuel. This work is discussed in this volume in Chapter 4 by F. Heitmeir et al. In addition to discussing the typical problems of turboramjet engines, it is a noteworthy chapter because of its thorough discussion of the key areas of intake and nozzle design, engine subsystems, thermal management, and engine controls. The German work has also been well documented in the recent literature: an extensive treatment of ram combustor development is given by Voss[11]; nozzle studies are covered by Berens[12]; performance optimization is treated by Bareis and Braig[13]; and performance simulation by Esch et al.[14]

Another example of VCE/turboramjet development is the engine program being pursued in Japan[15,16] under the super/hypersonic transport propulsion system research (HYPR) project started in 1989. This combined cycle engine uses a turboaccelerator for the initial speed range Mach 0–3 and a ramjet engine for the speed range Mach 2–5.5. Extensive component analysis, design, and development have been carried out on this engine, including intake, turbomachinery, primary combustor, ramjet combustor, exhaust nozzle, and control system. The next planned step is the fabrication and test of the combined cycle demonstrator engine in an altitude test facility (ATF) exploring turboengine performance to Mach 3, mode transition between Mach 2.5 and 3.0, and finally a ramjet mode test at Mach 3. Development of the basic turboengine designated HYPR 90-T is in progress, as reported in Ref. 15, which also notes that the first test was carried out in December 1994. Researchers plan to carry out the first test of the combined cycle engine demonstrator (HYPR 90-C) in 1997 with the ATF test to follow in 1998. In addition to the preceding references, many papers concerning the extensive investigations supporting the HYPR 90-T propulsion system development are given in Ref. 17.

The Russian, German, Japanese, and U.S. investigations of turboramjets have created a diverse database covering key aspects of engine design, mode transition, integration, and control. It is anticipated that experience with hydrocarbon, endothermic, and cryogenic fuels will be accumulated. The next logical step in the evolution of this engine is the ground test and verification of full engine function, mode transition(s), and ramjet operation, of a suitable demonstration engine. What resources can be found for such developments remain to be seen. The further question of the suitability of this class of engine for a space launcher role will depend largely on the thrust-to-weight and fuel specific impulse performances actually achieved by installed engines. Meanwhile, other turboaccelerator concepts such as the air turborocket engine, including the hydrogen-expander engine, continue to be pursued.

Air Turborocket Engines

A schematic diagram of the air turborocket (ATR) engine is shown in Fig. 2. The key feature of this engine is that the turbine is fed from a rocket-type gas generator. The turbine exhaust products and/or additional fuel are burned downstream in a ramjet type of combustor chamber. For this reason the ATR is often referred to as an air turboramjet. Several studies of the ATR concept have been made since World War II: early developments in the United States are discussed in Ref. 18. An interesting but brief report on some British work is given in Ref. 19. In this latter article, various engine configurations are discussed, including the use of a two-row tip-turbine driven fan; note that four engine designs are shown stemming from work performed at the National Gas Turbine Establishment by A. R. Howell. Other British work by Lombard and Keenan at Rolls Royce is summarized in Ref. 20. In principle, the ATR is capable of high Mach number operation, depending on the stamina of the fan, which will probably be operated in a windmilling mode at the higher speeds. A recent discussion of some of the complexities of ATR engine operation is given in Ref. 21. As noted earlier, there is some continuing effort on ATR concepts in the U.S. HiMATE program.

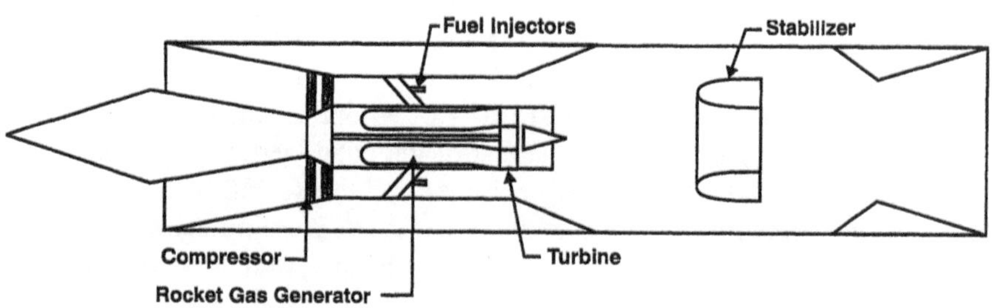

Fig. 2 Schematic of an air turborocket engine.

Hydrogen-fueled Turboexpander Engine

The evolution of the hydrogen-expander engine from early concepts to a definitive form in the Pratt and Whitney Model 304 engine is extensively described by Sloop[18]. A schematic of this engine is shown in Fig. 3, and it is noteworthy that this engine required a large solitary heat exchanger, a 12-stage hydrogen expansion turbine, and a reduction gear to enable fan-turbine matching. Recent studies of this type of engine configuration, reported in Ref. 22, concluded that no weight reduction, compared with a conventional turbojet cycle, was achieved and also that the development risk and cost were anticipated to be higher than for a comparable turbojet. However, a current turboexpander engine that is of considerable interest is the Japanese Air Turborocket Expander Program (ATREX) engine shown schematically in Fig.

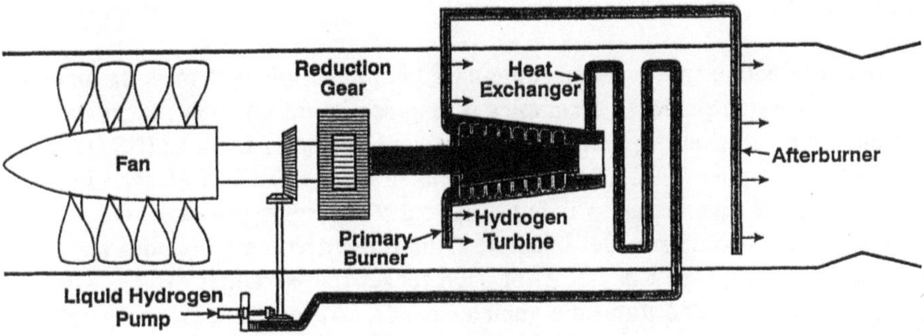

Fig. 3 Schematic of a hydrogen-expander engine.

4. Compared with the Model 304 type of configuration, this engine utilizes heat exchanger elements extracting heat from the inlet stream (precooler) as well as from the combustion chamber (including wall cooling). Also a compact tip-turbine driven fan is used, and no reduction gear is required. With this configuration, a high performing, high thrust-to-weight ratio engine appears possible. The development of this ATREX engine is described in this present volume in Chapter 6 by Tanatsugu. The precooler in the ATREX engine is an important contributor to engine performance; however, as with all hydrogen precoolers, concerns exist regarding icing hazards. The development status of the precooler for ATREX is further discussed in Ref. 23, to which the interested reader is referred. It is anticipated that the evolution of various ATREX engine models will yield a new authoritative baseline for high-speed engine performance.

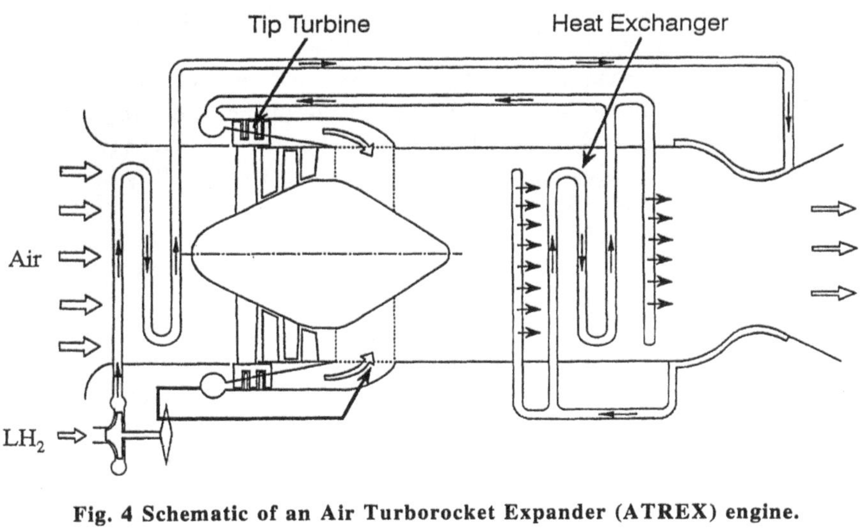

Fig. 4 Schematic of an Air Turborocket Expander (ATREX) engine.

Ejector-based Combined Cycle Engines

An alternative to a turbomachine-based compression system is the use of a jet compression process. Examples of engines based on such processes are the classic ejector ramjet (ERJ), the supercharged ejector ramjet (SERJ), and the ejector scramjet (ESCRJ); these systems are described in Ref. 24. In many cases the ejector pump is fed by a rocket-type gas generator; the pump may also be utilized as a fuel injector. This present volume includes an analysis by Billig (see Chapter 2) of such rocket-ram-scramjet cycles, with particular emphasis on transonic vehicle speeds. At low speeds, compared with turboengine operation, the ejector compression devices yield low specific impulse. Consequently, for those applications that require extended flight operation at low speeds, such ejector concepts fare poorly. However, for those applications where acceleration to a ramjet/scramjet takeover speed is the dominant requirement, such ejector systems can offer relatively simple and effective performance. Typically, aerospace plane concepts have sought both high payload capability and aircraft-like airfield operations; these two objectives were not easily reconciled in previous air-breathing single-stage-to-orbit (SSTO) concepts. However, if modest low-speed performance is acceptable, then the ejector ramjet is an attractive acceleration system that has been receiving increasing attention as one element of the so-called rocket-based combined cycle (RBCC) propulsion system. For example, the ejector ramjet/scramjet principle is also the basis of the "Strutjet" RBCC concept discussed in Refs. 25 and 26.

Liquid Air Cycle Engines and Related Concepts

The introduction of liquid hydrogen as an aerospace vehicle fuel, which possesses outstanding cryogenic heat sink capacity and intrinsic high-work capability, offered new opportunities for the synthesis of novel aeropropulsion systems. An early concept (1950/1960) was the liquid air cycle engine (LACE) introduced by the Marquardt Company. In this air-breathing concept, the intake air was liquified in an LH_2/air heat exchanger and pumped, in the liquid phase, to the combustion chamber to be burned with the LH_2 fuel stream. As shown in Fig. 5, the liquification was performed in two heat exchanger units: a precooler and a condenser. The fundamental limitation of this "basic" LACE system was that the LH_2 flow required for intake air liquification was far in excess of the amount needed for stoichiometric combustion of that airflow. An early unclassified analysis by Jeffs and Beeton, which appeared in Ref. 27, clearly pointed out the excess fuel problem of the basic LACE concept (or in their terminology "straight" LACE). The evolution of the basic LACE development to more efficient engine cycles such as Super LACE, in the United States, is well documented by Escher[28]. The main development thrust was, of course, aimed at more

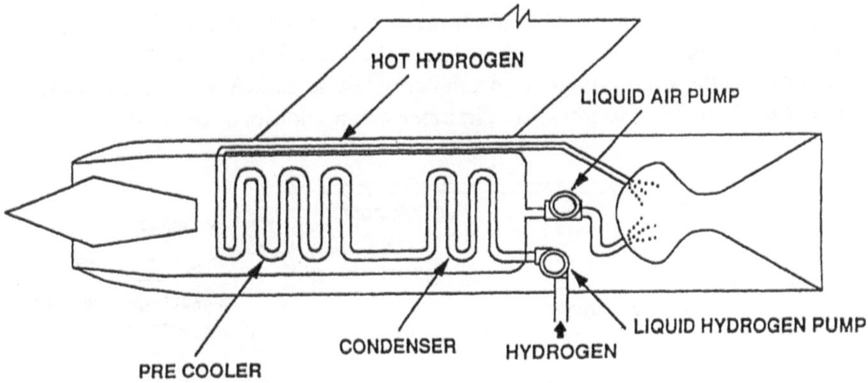

Fig. 5 Basic Liquid air cycle engine.

efficient utilization of the LH_2 cooling potential to enable "weaker" operation of the cycle. The interested reader is directed to Ref. 29, which covers some engineering details of heat exchanger performance. Billig also has a valuable technical discussion in Ref. 30. In recent years there has been a dedicated effort in Japan to develop a LACE concept for SSTO application. This work is well documented in Refs. 31–33. In the latter reference, the progress of engine development from "simple" LACE to more effective LACE concepts is detailed; the work is similar to the developmental activities pursued earlier in the United States, but many additional innovations have been introduced and key hardware components successfully demonstrated.

In Russia, Rudakov and Balepin[34] have proposed a most interesting engine concept. Although similar to LACE, in this engine the "depth" of intake air cooling stops, by design, short of actually liquifying the air. This engine cycle is shown schematically in Fig. 6 and is termed an air turborocket with deep air cooling (ATRDC). The deep cooling permits the engine to be made of conventional materials, and development of such a high-speed engine should require less demanding facility capabilities. It was speculated for many years that an ATRDC system was the basis of the Hotol propulsion system. This speculation was confirmed in 1993 in Ref. 35, which presents a schematic flow diagram for the RB545 Hotol engine; a good discussion of the tradeoff between performance and weight is also given in this article.

In the search for more efficient LACE-based concepts, two approaches are evident. The first is to use cooling/liquification of only a fraction of the intake air, with that portion typically being used to supply the oxidizer flow to the rocket primary element of an ejector ramjet. The second approach is to utilize the total cryogenic propellant flow of a combined engine installation. For example, Ref. 34 shows a LACE plus ATR and also an ATRDC plus ramjet engine installation where the total cryogenic fuel flow to both engines is used for liquification, or deep cooling, of the intake airflow. In a similar vein as shown in Refs. 33 and 36, it is possible in a basic LOX/LH_2 rocket engine to use the total cryogenic propellant flow to liquify air, which is then pumped to the rocket combustion chamber as an additional oxidant.

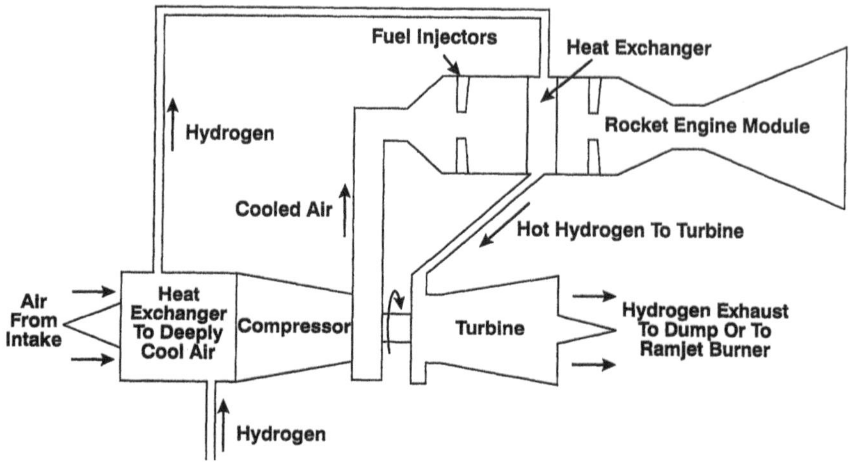

Fig. 6 Schematic cycle of air turborocket with deep cooling (ATRDC).

In summary, the basic LACE invention has spawned a wide spectrum of cryogenically cooled concepts ranging from simple precooling of a turboaccelerator engine through the deep cooling ATRDC to total liquification of the intake air as in basic LACE. The integration of heat exchanger elements in the propellant flow circuits leads to a high level of thermodynamic integration of the engine components while maintaining some flexibility of engine element installation on the vehicle. See, for example, the ATRDC concept, where the rocket chamber may be located apart from the turbocompressor element. This family of engine concepts potentially offers high specific impulse and attractive thrust-to-weight ratios. However, on the downside, the heat exchanger related problems of weight, effectiveness, contamination, and reliability are significant development hurdles, although steady progress is being made (see, for example, Refs. 23 and 37).

Similarly the size and weight penalties associated with hydrogen-driven turbines are significant; once again progress is being made using alternatives to conventional practices, such as tip drives. Overall, the promise of engine performance enhancement will spur continuing efforts to develop reliable engine concepts. This development process is, however, complicated by the essential thermodynamic integration of the vehicle cooling requirements with the engine subsystem.

The early studies of basic LACE concepts in the United States[38], and United Kingdom[27], and more recently in Japan, were closely linked to the concept of air collection and storage in-flight. In that regard, it is interesting to recall that in the early 1960s, in the United States, two distinct approaches to SSTO aerospace plane propulsion were pursued. One approach was the use of a liquid-hydrogen-fueled scramjet system combined with a low-speed accelerator engine and a liquid rocket engine for orbital injection. The other approach was to utilize in-flight air collection to acquire in the ascent phase the necessary oxidizer for use in a subsequent rocket propulsion phase. In this case the cooling capacity of the liquid hydrogen fuel was used to condense the collected air. From a performance viewpoint, it was important to remove as much nitrogen from the collected air as possible, resulting in an oxygen-enriched "air stream". Consequently, this propulsion system was termed the Air Collection and Enrichment System (ACES). Considerable progress in engineering demonstration of the components of the ACES system was made, and this development is well documented in Chapter 7 by J. Leingang et al. in the present volume. One of the more impressive system demonstrations was the successful operation of a flight-type rotating air separator based on the invention by Nau and Campbell[39]. Further extensions of this work have been made in recent years to reduce the amount of cryogenic fueling required by the vehicle so that launch readiness could be improved. This approach led initially to a dual fuel (liquid hydrogen/hydrocarbon) system where only sufficient liquid hydrogen was stored to liquify the collected air. The hydrocarbon fuel was used both for precollection and postcollection propulsion. Later a noncryogenic, all-hydrocarbon-fuel vehicle that used both water and fuel as the basic heat sink, combined with a number of heat pump circuits, was studied. In addition to the studies of ACES systems in the United States, several other countries have explored air collection systems; a significant article on Russian activities written by the Balepin is included in Chapter 8 of this volume (see also Ref. 40). This chapter places emphasis on the vortex-tube (VT) separator (an early discussion of this concept is also found in Ref. 27), which appears to offer significant advantages over the rotary separator discussed by Leingang et al. The VT forms the basis for the flight liquid oxygen (FLOX) plant analyzed by Balepin and Tjurikov.[41] The reader will find the discussion of several variants of FLOX plants and their integration into the vehicle of considerable interest.

The continuing evolution of ACES/FLOX plant concepts and the engineering validation of their performance potentials remain significant challenges. The attractive payload improvements offered by such concepts have been counterbalanced by the engineering challenges of compact heat exchanger performance, lightweight separators, and system complexities. Additional drawbacks include heat exchanger contamination, safety, and reliability. However, it is to be hoped that further exploratory development of such devices will be pursued, and this promising technology will evolve into sound engineering practice. The high promise of such systems to eliminate the marginal performance and operational inflexibility of current all-rocket SSTO concepts is well argued in Ref. 40.

Unsteady Flow Engines

There has been a continuing interest in nonsteady flow engines since the advent of the pulse-jet-propelled German World War II weapon termed the V-1. This author can bear testimony to the effectiveness of this propulsion system and dates his interest in jet propulsion from being on the receiving end of such weapons. These engines have been discussed by Foa[42] and more recently by Kentfield.[43] In the latter reference, there is an excellent discussion of various classes of pulse-jet combustors that operate in the deflagrative combustion mode, with details of their aspiration behavior, including both valved and nonvalved systems. Engines of this class obviously have already demonstrated their potential for low-cost missile or drone propulsion.

In recent years some effort has been devoted to unsteady flow engines as potential candidates for the low-speed acceleration of high-speed vehicles as a result of the absence of heavy turbomachinery in such engines. There is also a continuing interest in near-constant-volume combustion to improve conventional Brayton cycle engine performance.

Chapters 9 and 10 of this book provide two authorative articles dedicated to the pulse detonation wave engine (PDWE or PDE). These contributions reflect the growing attention paid to this class of engine in the last decade, particularly in the use of modern computational tools now available for the analysis of nonsteady flows. Such studies provide a firm basis from which to proceed to the demonstration phase of various engine designs. Here again, Kentfield also offers a fundamental discussion of detonation wave engines. An interesting historical discussion of various investigations of PDE systems is given by Eidelman et al.[44]

System Integration and Selection

It is clear that the future success of high-speed flight vehicles, and particularly the near-marginal SSTO vehicles, will depend on the functional and aerothermodynamic integration and overall optimization of the engine, propellant, and vehicle structure characteristics into a coherent effective aerodynamic vehicle.

The need for optimal aerodynamic performance is addressed by Gusev in Chapter 1, which deals with the efficient integration of the powerplant with a hypersonic vehicle. The related and absolutely vital technology area for hypersonic flight is that of thermal management of the vehicle. This area is discussed in Chapter 11 by the V. Van Griethuysen et al.; who give particular emphasis to airframe/propulsion/thermal integration.

Chapter 12 by Czysz and Murthy proposes a rational methodology for the complex task of synthesizing launch vehicle design. An effective methodology for identifying high-payoff approaches to combined cycle powered space launchers is, of course, a key requirement.

Closing Remarks

This Introduction has presented the various engine systems discussed in this book in terms of their potential and development status. A large number of propulsion candidates exist, and development resources are now severely constrained. Nevertheless, the technology base for such vehicles must be established and the performance of candidate engines clearly demonstrated. Several such demonstrator engines are emerging, and it is hoped that, either by national or international efforts, a firm engineering baseline for air-breathing launch vehicle propulsion will be established.

References

[1]Murthy, S. N. B., and Curran, E. T. (eds.), High-speed Flight Propulsion Systems, Vol. 137, Progress in Astronautics and Aeronautics, AIAA, Washington, DC, 1991.

[2]Lanshin, A., and Sosounov, V., "Russian Space Agency Research and Development Program for Aerospace Plane Combined Propulsion Systems ('OREL-2-1' R&D)," AIAA Paper 95-6149, Apr. 1995.

[3]Hill, R. J., "The Purpose and Status of IHPTET-1995," Proceedings of the 86th AGARD Symposium of Propulsion and Energetics Panel (Seattle, WA), Sept. 1995.

[4]Habrard, A., "The Variable Cycle Engine," L'Aeronautique et L'Astronautique, No. 141, 1990–92, pp. 52–59.

[5]Brazier, M. E., and Paulson, R. E., "Variable Cycle Engine Concept," Proceedings of the Eleventh International Symposium on Air-breathing Engines, AIAA, Washington, DC, 1993, pp. 684–695, ISABE 93-7065.

[6]Johnson, J. E., "Variable Cycle Engine Concepts," Proceedings of the 86th AGARD Symposium of Propulsion and Energetics Panel (Seattle, WA), Sept. 1995.

[7]Calmon, J., and Menioux, C., "The Transition Phase of the Turbo-Stato-Reactor," Entropie, No. 13, 1967.

[8]Lander, H., and Nixon, A. C., "Endothermic Fuels for Hypersonic Vehicles," Journal of Aircraft, Vol. 8, No. 4, 1971, pp. 200-207.

[9]Ianovski, L. S., "Endothermic Fuels for Hypersonic Aviation," Fuels and Combustion Technology for Advanced Aircraft Engines, AGARD Conference Proceedings 536, AGARD, (Neuilly sur Seine, France), Paper No. 44, pp. 44-1–44-8, 1993.

[10]Favorskii, O. N., and Kurziner, R. I., "Development of Air-breathing Engines for High-Speed Aviation by Combining Advances in Various Areas of Science of Engineering," Teplofizika Vysokikh Temperatur, Vol. 28, No. 4, 1990, pp. 793–803.

[11]Voss, N. H., "Ram Combustor Development Within the German Hypersonics Technology Program," AIAA Paper 95-6030, Apr. 1995.

[12]Berens, T., "Experimental and Numerical Analysis of a Two-Duct Nozzle/Afterbody Model at Supersonic Mach Numbers," AIAA Paper 95-6085, Apr. 1995.

[13]Bareis, B., and Braig, W., "Performance Optimization of a Turboramjet Engine for Hypersonic Flight," Proceedings of the 86th AGARD Symposium of Propulsion and Energetics Panel (Seattle, WA), Sept. 1995 (Paper 18).

[14]Esch, T., Hollmeier, S., and Rick, H., "Design and Off-Design Simulation of Highly Integrated Hypersonic Propulsion Systems," Proceedings of the 86th AGARD Symposium of Propulsion and Energetics Panel (Seattle, WA), Sept. 1995.

[15]Miyagi, H., Miyagawa, H., Monji, T., Kishi, K., Powell, T. H., and Morita, M., "Combined Cycle Engine Research in Japanese HYPR Project," AIAA Paper 95-2751, Jul. 1995.

[16]Kondo, M., Morii, S., and Murashima, K., "Overview of the Japanese National Project 'Super/Hypersonic Transport Propulsion System Program,' " AIAA Paper 95-2445, Jul. 1995.

[17]Anon., Proceedings of the Second International Symposium on Japanese National Project for Super/Hypersonic Transport Propulsion System, Japan Industrial Technology Association, 1995.

[18]Sloop, J. L., "Liquid Hydrogen as a Propulsion Fuel, 1945-1959," NASA SP 4404, 1978.

[19]Fulton, K. T., "High-performance Hybrid Powerplants," The Aeroplane, Vol. 92, March 15, 1957, pp. 378-381.

[20]Anon., "The Turborocket," Flight International, Vol. 29, Oct. 1964, pp. 752–754.

[21]Bussi, G., Colasurdo, G., and Pastrone, D., "Analysis of Air-Turborocket Performance," Journal of Propulsion and Power, Vol. 11, No. 5, 1995, pp. 950–954.

[22]Zellner, B., Sterr, W., and Herrmann, O., "Integration of Turboexpander and Turbo-Ramjet Engines in Hypersonic Vehicles," Transactions of the American Society of Mechanical Engineers, Vol. 116, Jan. 1994, pp. 90–97.

[23]Balepin, V. V., Tanatsugu, N., Sato, T., Mizutani, T., Hamabe, K., and Tomike, J., "Development Study of Precooling for ATREX Engine," Proceedings of the Twelfth International Symposium on Air-breathing Engines, Vol. 1, AIAA, Washington, DC, 1995, pp. 173–182.

[24]Escher, W. J. D., Teeter, R. R., and Rice E. E., "Air-breathing and Rocket Propulsion Synergism: Enabling Measures for Tomorrow's Orbital Transports," AIAA Paper 86-1680a, Jun. 1986, (1995 update of a 1986 paper).

[25]Bulman, M. J., and Siebenhaar, A., "The Strutjet Engine: Exploding the Myths Surrounding High-speed Air-breathing Propulsion," AIAA Paper 95-2475, Jul. 1995.

[26]Siebenhaar, A., and Bulman, M. J., "The Strutjet Engine: The Overlooked Option for Space Launch," AIAA Paper 95-3124, Jul. 1995.

[27]Jeffs, R. A., and Beeton, A. B. P., "Liquid Air Cycle Engines for High-speed Aircraft," Journal of the British Interplanetary Society, Vol. 19, Oct. 1964, pp. 484–490.

[28]Escher, W. J. D., "Cryogenic Hydrogen-Induced Air Liquefaction Technologies," Hypersonic Combined Cycle Propulsion, AGARD Conference Proceedings 479, AGARD, (Neuilly-Sur-Seine, France), pp. 14-1–14-12, 1990.

[29]Ahern, J. E., "Thermal Management of Air-breathing Propulsion Systems," AIAA Paper 92-0514, Reno, Jan. 1992.

[30]Billig, F. S., "Propulsion Systems from Takeoff to High-speed Flight," High-speed Flight Propulsion Systems, edited by S. N. B. Murthy and E. T. Curran, Vol. 137, Progress in Astronautics and Aeronautics, AIAA, Washington, DC, 1991, pp. 21–100.

[31]Kajita, M., Ito, T., Hasegawa, K., and Togawa, M., "Air Condensation Type Air-breathing Propulsion System," International Astronautical Federation, IAF Paper 86-180, Oct. 1986.

[32]Miki, Y., Taguchi, H., and Aoki, H., "Status and Future Planning of LACE Development," AIAA Paper 93-5124, Munich, Germany, Nov. 1993.

[33]Hirakoso, H., Aoki, T., and Ito, T., "A Concept of LACE for Space Plane to Earth Orbit," International Journal of Hydrogen Energy, Vol. 15, No. 7, 1990, pp. 495–505.

[34]Rudakov, A. S., and Balepin, V. V., "Propulsion Systems with Air Precooling for Aerospaceplane," Society of Automotive Engineers, SAE Paper 911182, Dayton, Ohio, Apr. 1991.

[35]Hempsell, M., "HOTOL's Secret Engines Revealed," Spaceflight, Vol. 35, May 1993, pp. 168–172.

[36]Balepin, V., Folomeer, E. A., Galkin, S. M., and Tjurikov, E. V., "Rocket-based Combined Cycles for Vertical Take-off Space Vehicles," AIAA Paper 95-6078, Apr. 1995.

[37]Balepin, V. V., Folomeer, E. A., Galkin, S. M., and Svetlakov, E. A., "Cryogenic Heat Exchangers--Key Technologies for Precooled Turbojet Engines," Society of Automotive Engineers, SAE Paper 911183, Dayton, Ohio, Apr. 1991.

[38]Nau, R. A., "A Comparison of Fixed Wing Reusable Booster Concepts," Society of Automotive Engineers, SAE Paper 670384, May 1967.

[39]Nau, R. A., and Campbell, S. A., "Rotary Separator," United States Patent 3,779,452, Dec. 18, 1973, filed Sept. 22, 1960.

[40]Balepin, V. V., Czysz, P., Maita, M., and Vandenkerckhove, J., "Assessment of SSTO Performance with In-flight LOX Collection," AIAA Paper 95-6047, Apr. 1995.

[41]Balepin, V. V., and Tjurikov, E. V., "Integrated Air Separation and Propulsion System for Aerospace Plane with Atmospheric Oxygen Collection," Society of Automotive Engineers, SAE Paper 920974, Dayton, Ohio, Apr. 1992.

[42]Foa, J. V., Elements of Flight Propulsion, Wiley, London and New York, 1960.

[43]Kentfield, J. A. C., Nonsteady, One-dimensional, Internal, Compressible Flows, Theory and Applications, Oxford Univ. Press, New York and Oxford, 1993.

[44]Eidelman, S., Grossman, W., and Lottati, I., "A Review of Propulsion Applications of the Pulsed Detonation Engine Concept," AIAA Paper 89-2446, Jul. 1989.

Optimal Aerodynamic Shapes of a Hypersonic Vehicle with an Airbreathing Engine

V.N. Gusev
Central Aerohydrodynamic Institute (TsAGI), Zhukovsky, Russia

Nomenclature

C_D	=	drag coefficient
C_L	=	lift coefficient
$\bar{c}$	=	thickness ratio
$\vec{F}(F_x, F_y)$	=	force applied to external vehicle surface not wetted by the stream tube passing through the engine
K	=	lift-to-drag ratio
L	=	length
M	=	Mach number
$\vec{n}$	=	normal
$\vec{P}(P_x, P_y)$	=	force applied to the surface of the vehicle and engine wetted by the stream tube passing through the engine
p	=	pressure
$\vec{R}(P_x, P_y)$	=	total force applied to vehicle
Re	=	Reynolds number
S	=	area
T	=	temperature
$\vec{U}$	=	velocity
x,y	=	Cartesian coordinates
α	=	angle of attack

Copyright © 1995 by the American Institute of Aeronautics and Astronautics, Inc. All rights reserved.

θ	=	surface slope
κ	=	specific heat ratio
μ	=	slope of control section
ν	=	total pressure recovery coefficient
ρ	=	density
χ	=	sweep angle

Subscripts

j	=	jet values
m	=	frontal area
w	=	surface values
∞	=	freestream flow
0	=	stagnation parameters
+	=	intake entry
-	=	nozzle exit
*	=	critical section

Introduction

Further space exploration will depend greatly on the availability of economically feasible space transportation systems. Among alternative systems, concepts using airbreathing propulsion with horizontal takeoff and landing are currently being scrutinized. Airbreathing engines provide various possibilities, especially the use of atmospheric oxygen to sustain burning in engine and high (as compared with liquid rocket engines) specific impulse *I*. The application of airbreathing engines with inlets and high-area-ratio nozzles imposes serious requirements on the configuration of a hypersonic vehicle. Apart from the development of a scramjet and novel advanced materials, there are also many important aerothermodynamic problems to be solved when designing such vehicles[1]. As flight speed increases, airframe-propulsion integration for a hypersonic vehicle with airbreathing propulsion increases in importance. The lower surface of such a vehicle serves as part of the propulsion system, with the forebody being used for flow precompression ahead of the engine inlet, and the afterbody as the nozzle wall; interference is a challenge. Large accelerations of a vehicle using airbreathing propulsion cause high dynamic pressures. Vehicle drag and aerodynamic heating during injection must be at a minimum. The lateral range required during vehicle glide without the propulsion system operating may need to be provided for by a high hypersonic L/D ratio.

Some general principles of integration of a hypersonic vehicle airframe with an airbreathing engine that provide a vehicle with minimum drag or maximum lift-to-drag ratio at hypersonic velocities[2] are formulated in this work. In addition to theoretical investigation results, the test data aimed at the search for the most efficient integration technique of propulsion with the airframe are given[3]. These investigations, conducted on schematized models in hypersonic wind tunnels, identify the contributions of the main components of the configuration with its aerodynamics, lay the basis for a databank of the aerodynamic characteristics of any given vehicle class, and can be used to assess the accuracy level of the computational aerodynamic methods applied at the design stage. Appropriate comparisons are presented here.

Peculiarities of the Flow Over a Hypersonic Vehicle with an Airbreathing Engine

Main Relations

Let us consider a vehicle with an integral configuration, the lift of which, along with the lifting surface, can be generated by gasdynamic thrust orientation of the powerplant (see Fig. 1a).

The total force acting on a vehicle with an integral configuration is usually represented in the form of two components: $\vec{R} = \vec{P} + \vec{F}$, where the force $\vec{P}$ is applied to the surface of the vehicle and engine wetted by the stream tube passing through the powerplant:

$$\vec{P} = \int_{S_\infty} \left[\rho\vec{U}\left(\vec{U}\cdot\vec{n}\right) + p\vec{n}\right] ds - \int_{S} \left[\rho\vec{U}\left(\vec{U}\cdot\vec{n}\right) + p\vec{n}\right] ds + \int_{S_f} \left(p\vec{n} + \frac{\partial\vec{U}}{\partial n}\right) ds$$

and force $\vec{F}$ is applied to the external surface of the vehicle not wetted by this stream tube:

$$\vec{F} = \int_{S_W - S'_W} \left(p\vec{n} + \frac{\partial \vec{U}}{\partial n} \right) ds$$

The main feature of a hypersonic vehicle with an integral configuration is that almost its entire lower surface becomes part of the powerplant: the forward part is used for preliminary compression of the air upstream of the intake entry, and the aft part as a nozzle wall. In this case, the airplane external area S'_W wetted by the stream tube passing through the engine becomes comparable with the total external area of the vehicle S_W.

To simplify analysis, let us assume that the flow is two-dimensional and inviscid, the external shapes of the vehicle are rectilinear $(\theta_1 = const, \theta_1 = const)$, the intake entry area is $S_+ = S_\infty$, and the flow variables are constant at the nozzle exit (see Fig. 1b). In this case, when neglecting the fuel weight in the air that passes through the engine, for the component of forces $\vec{P}$ and $\vec{F}$ along the axes x and y, we obtain the following:

$$P_x = \rho_+ U_+^2 y_+ - \rho_+ U_+ U_- y_+ \cos\delta + p_+ y_+ - p_- y_-$$

$$P_y = \rho_+ U_+ U_- y_+ \sin\delta + p_+ y_+ \text{ctg}\mu_+ + p_- y_- \text{ctg}\mu_-$$

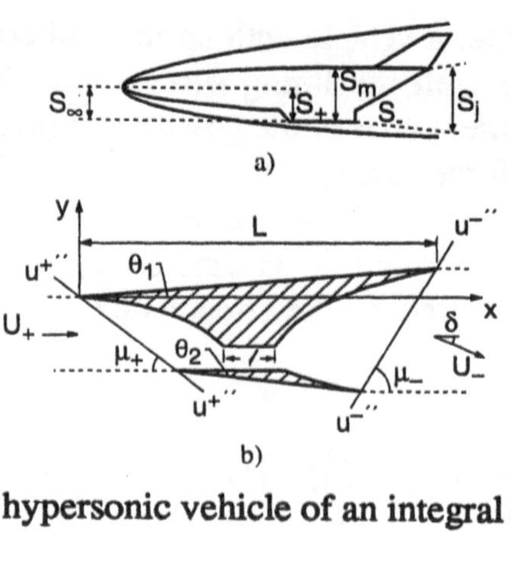

Fig. 1. A hypersonic vehicle of an integral configuration.

$$F_x = p_1 L \,\mathrm{tg}\theta_1 + p_2 \,\mathrm{tg}\theta_2 \left(L - y_+ \mathrm{ctg}\mu_+ - y_- \mathrm{ctg}\mu_-\right)$$

$$F_y = p_2 \left(L - y_+ \mathrm{ctg}\mu_+ - y_- \mathrm{ctg}\mu_-\right) - p_1 L$$

In a nondimensional form, assuming that the energy supply does not cause any variations in the thermodynamic properties of the medium, the preceding equations may be rewritten in the following form:

$$\overline{P}_x = \frac{2P_x}{\rho_+ U_+^2 y_+}$$

$$= 2\left\{1 - \frac{M_-}{M_+}\sqrt{\frac{T_{0-}}{T_{0+}}\left[\frac{1+\frac{\kappa-1}{2}M_+^2}{1+\frac{\kappa-1}{2}M_-^2}\right]}\cos\delta - \frac{1}{\kappa M_+^2}\left(\frac{p_- y_-}{p_+ y_+} - 1\right)\right\} \tag{1a}$$

$$\overline{P}_y = \frac{2P_y}{\rho_+ U_+^2 y_+}$$

$$= 2\left\{\frac{M_-}{M_+}\sqrt{\frac{T_{0-}}{T_{0+}}\left[\left(1+\frac{\kappa-1}{2}M_+^2\right)\sin\delta \Big/ \left(1+\frac{\kappa-1}{2}M_-^2\right)\right]} + \frac{1}{\kappa M_+^2}\left(\frac{p_- y_-}{p_+ y_+}\mathrm{ctg}\mu_- + \mathrm{ctg}\mu_+\right)\right\} \tag{1b}$$

$$\overline{F}_x = \frac{2F_x}{\rho_+ U_+^2 y_+}$$

$$= \frac{2}{\kappa M_+^2}\left[\frac{p_1 L}{p_+ y_+}\mathrm{tg}\theta_1 + \frac{p_2}{p_+}\mathrm{tg}\theta_2\left(\frac{L}{y_+} - \mathrm{ctg}\mu_+ - \frac{y_-}{y_+}\mathrm{ctg}\mu_-\right)\right] \tag{1c}$$

$$\overline{F}_y = \frac{2F_y}{\rho_+ U_+^2 y_+}$$

$$= \frac{2}{\kappa M_+^2}\left[\frac{p_2}{p_+}\left(\frac{L}{y_+} - \mathrm{ctg}\mu_+ - \frac{y_-}{y_+}\mathrm{ctg}\mu_-\right) - \frac{p_1 L}{p_+ y_+}\right] \quad \text{(1d)}$$

Two forms of steady flow over the body with an internal duct - the cross section of which first decreases from the inlet towards the outlet and then increases - are possible: flow with a detached shock wave and subsonic stream in the duct, and flow with a supersonic stream in the duct. In the latter case, the maximum extent of flow contraction in the internal duct will be determined from the conditions of the duct choking when the flow velocity becomes sonic in the minimum cross-sectional area:

$$\overline{y}_* = \frac{y_*}{y_+} = M_+\left(\frac{2}{\kappa+1} + \frac{\kappa-1}{\kappa+1}M_+^2\right)^{\kappa+1/2(\kappa-1)}\left[\frac{p_{0+}}{p_{0*}}\right]$$

In the regulated variant that is, when the duct area is variable $P_{0+} / P_{0*} = 1$; in the unregulated variant, the total pressure recovery coefficient $\nu = P_{0*} / P_{0+}$ in the intake on start varies over the range of $\nu_* \leq \nu \leq 1$. The minimum value of ν_* governs the total pressure recovery coefficient in a direct shock wave,

$$\nu_* = \left(\frac{P_{0*}}{P_{0+}}\right)$$

$$= \left(\frac{2\kappa}{\kappa+1}M_+^2 - \frac{\kappa-1}{\kappa+1}\right)\left[\frac{4\kappa}{(\kappa+1)^2} - \frac{2(\kappa-1)}{(\kappa+1)^2 M_+^2}\right]^{\kappa/(1-\kappa)}$$

$$\cdot\left(1 + \frac{\kappa-1}{2}M_+^2\right)^{\kappa/(1-\kappa)}$$

When M_+ increases infinitely, $\bar{y}_*$ tends to zero when the duct is regulated, that is variable. When the duct is not regulated and $\nu = \nu_*$, value $\bar{y}_*$ remains fixed, equal to

$$\bar{y}_* = (2\kappa)^{1/(\kappa-1)}(\kappa-1)^{1/2}(\kappa+1)^{\kappa+1/[2(1-\kappa)]}$$

In the unregulated case, note that the estimated total pressure loss in the intake due to the direct shock wave is maximum. In the case of an oblique shock intake section, these losses will be much less.

Busemann Biplane

In a supersonic flow, a main feature is that a body of a final thickness suffers a wave drag constituting a considerable part of the total drag. However, in some cases, by application of favorable interference, it is possible to devise systems devoid of wave drag. For example, one system is the Busemann Biplane composed of two airfoils whose plane surfaces are parallel to each other and are at zero angle of attack, their leading and trailing edges being infinitely thin (see Fig. 2a). In this case, the external flow is

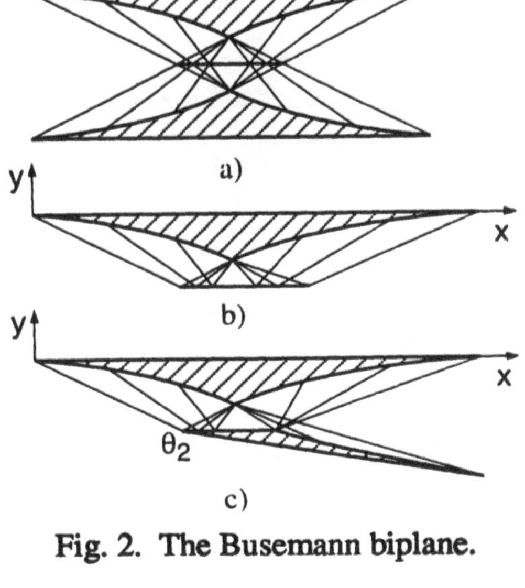

Fig. 2. The Busemann biplane.

equivalent to the flow over a plane plate at zero angle of attack, and the wave drag of the biplane unit will be equal to zero. To eliminate wave drag in the biplane internal duct, the internal flow must be isentropic. In order to achieve this, it is required that the duct shape exclude any possibility of forming envelope compression waves in the flow. In that case, after consecutive compression and expansion of the flow in the duct, the velocity and direction of flow at the outlet will be the same as at the inlet. There is, then, no wave drag of both the external and the internal biplane parts.

Obviously, the above-mentioned flow feature also applies for each part of the Busemann biplane partitioned by a plane located in the plane of symmetry. Each of the parts may be considered as a schematized integral configuration of a hypersonic vehicle when $y_+ = y_-$ and $\theta_1 = \theta_2 = 0$ (see fig. 2b). Let us consider the case in detail.

When energy supply is absent, $T_{0+} = T_{0-}$, $M_- = M_+$, $\mu_+ = \mu_-$, $p_+ = p_- = p_1 = p_2$, and $\delta = 0$, and in accordance with Eqs. (1), we shall have $R_x = R_y = 0$. The force action on the system considered is zero.

When energy supply is present, since then $T_{0-} \neq T_{0+}$, then $M_+ \neq M_-$, $p_+ \neq p_-$, and $\mu_+ \neq \mu_-$. In this case, assuming that $y_+ = y_-$, $\theta_1 = \theta_2 = 0$, $\delta = 0$, and $p_1 = p_2 = p_+$ as before, from Eqs. (1) it follows that

$$\overline{R}_x = 2\left\{1 - \frac{M_-}{M_+}\sqrt{\frac{T_{0-}}{T_{0+}}\left[\frac{1+\frac{\kappa-1}{2}M_+^2}{1+\frac{\kappa-1}{2}M_-^2}\right]} - \frac{1}{\kappa M_+^2}\left(\frac{p_-}{p_+} - 1\right)\right\}$$

$$\overline{R}_y = \frac{2\,\mathrm{ctg}\,\mu_-}{\kappa M_+^2}\left(\frac{p_-}{p_+} - 1\right)$$

Assuming that the energy supply is accomplished in a minimal section of the internal duct, we shall find the Mach number variation in this section depending on the flow heating level

T'_{0*} / T_{0*}:

$$\frac{T'_{0*}}{T_{0*}} = \left(\frac{M'_*}{M_*}\right)\left(\frac{1+\kappa M_*^2}{1+\kappa M_*'^2}\right)\left(\frac{1+\frac{\kappa-1}{2}M_*'^2}{1+\frac{\kappa-1}{2}M_*^2}\right)$$

In the case of an isentropic compression we obtain from the continuity equation the following expression for Mach number M_* in the minimal duct section prior to flow heating:

$$\bar{y}_* = \frac{y_*}{y_+} + \frac{M_+}{M_*}\left(\frac{1+\frac{\kappa-1}{2}M_*^2}{1+\frac{\kappa-1}{2}M_*^2}\right)^{\kappa+1/2[(\kappa-1)]}$$

where the value of $\bar{y}_*$ is found from the condition of the duct activation

$$\bar{y}_* = \frac{M_+}{\nu}\left(\frac{2}{\kappa+1}+\frac{\kappa-1}{\kappa+1}M_+^2\right)^{\kappa+1/[2(\kappa-1)]}$$

The energy supply to a supersonic flow is accompanied by flow retardation $(M'_* < M_*)$, and total pressure losses:

$$\frac{p'_{0*}}{p_{0*}} = \left(\frac{1+\kappa M_*^2}{1+\kappa M_*'^2}\right)\left(\frac{1+\frac{\kappa-1}{2}M_*'^2}{1+\frac{\kappa-1}{2}M_*^2}\right)^{\kappa/(\kappa-1)}$$

The maximum value of T'_{0*} / T_{0*} will be limited by the blockage condition $M'_* = 1$.

In the case of an isentropic flow expansion in the duct exit section, M_- is found from the relation

$$\bar{y}_* = \frac{y_*}{y_+} = \frac{M_-}{M'_*}\left(\frac{1+\frac{\kappa-1}{2}M'^2_*}{1+\frac{\kappa-1}{2}M^2_-}\right)^{\kappa+1/2[(\kappa-1)]}$$

When $T_{0-} \neq T_{0+}$ and $y_- = y_+$, the flow at the duct outlet will be underexpanded as

$$\frac{p_-}{p_+} = \frac{p_{0*}}{p_{0*}}\left(\frac{1+\frac{\kappa-1}{2}M'^2_+}{1+\frac{\kappa-1}{2}M^2_-}\right)^{\kappa/(\kappa-1)} > 1$$

When the duct is not regulated, the amount of flow compression in the duct diffuser is small as a result of the starting limitations. For example, when $\nu = \nu_*$, $M_+ = 10$, and $\kappa = 1.4$, the minimum value of the Mach number in a minimal section is $M_* = 9.01$. When energy is supplied, a considerable underexpansion of the flow in the duct exit section results at $y_+ = y_-$. In this case, the crosswise force $\bar{R}_y$ is larger than the longitudinal force $\bar{R}_x$, and the vector of the force applied to the vehicle will be turned relative to the x axis at a high angle. When $M_+ = 10$ and $\kappa = 1.4$, the nondimensional values of the components of the total force acting on a vehicle as a whole, which depend on T_{0-} / T_{0+}, are as shown in Fig. 3.

At $\nu > \nu_*$ the flow underexpansion level at the nozzle exit may be reduced, which will increase the absolute value of the longitudinal force $|R_x|$, reduce the absolute value of the crosswise force $|R_y|$, and result in decreased rotation around the x axis of the vector of the total force applied to a vehicle. In the preceding case, $M_+ = 10$ and $\kappa = 1.4$, at $M'_* = constant$ the absolute value of the longitudinal force $|\bar{R}_x|$ has its maximum at the final value of T_{0-} / T_{0+}.

In concluding this section, attention may be drawn to one particular point: A majority of the considered hypersonic vehicle

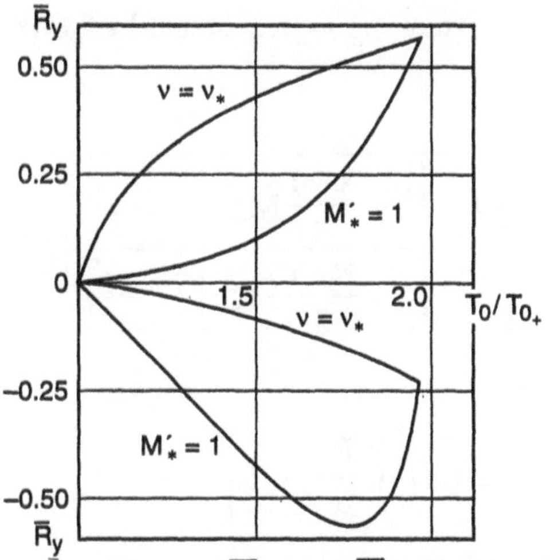

Fig. 3. The relative values of $\overline{R}_x$ and $\overline{R}_y$ vs. the temperature ratio corresponding to the supplied energy T_{0-}/T_{0+} at $y_+ = y_-$.

configurations, developed by using the simplest exact solutions of gasdynamics, is to be used in one regime as a rule. This is the design regime requiring the flow to wet the body that has been used in the preceding analysis for the Busemann Biplane with energy supply. Some peculiarities of the flow over the model in the off-design regime have been considered in Ref. 4.

Busemann Biplane with a Final Inclination of External Shapes

As noted earlier, in the case of energy supply, the flow at the duct outlet at $y_+ = y_-$ always remains underexpanded with respect to the external pressure p_+. An additional expansion of the flow in the duct is required to achieve a pressure at the outlet equal to the static pressure of the freestream flow $(p_+ = p_-)$. This may be realized by means of increasing the vehicle frontal area (see Fig. 2c). In this case, along with increased thrust, there arises additional wave drag.

In the following, for simplicity of the analysis we shall assume one of the external generatrices of the vehicle to be rectilinear and inclined to the freestream flow at the angle of attack θ_2 (see Fig. 2c). In this case, for the components of the total force

$\vec{R}$ on axes x and y, when $p_+ = p_-$, instead of Eqs. (1), we obtain the following:

$$\overline{R}_x = 2\left\{1 - \frac{M_-}{M_+}\sqrt{\frac{T_{0-}}{T_{0+}}\left[\frac{1+\frac{\kappa-1}{2}M_+^2}{1+\frac{\kappa-1}{2}M_-^2}\right]} - \frac{1}{\kappa M_+^2}\left(\frac{y_-}{y_+}-1\right)\left(\frac{p_-}{p_+}-1\right)\right\} \tag{2a}$$

$$\overline{R}_y = \frac{2}{\kappa M_+^2}\left(\frac{L}{y_+} - \mathrm{ctg}\mu_+\right)\left(\frac{p_2}{p_+}-1\right) \tag{2b}$$

In such a case of equal pressures, some flow parameters in the duct, such as M_*, M_*', and ν, remain invariable and are found by using the relations obtained previously. Others are determined from the conditions of an isentropic nozzle expansion:

$$\bar{y}_- = \frac{y_-}{y_+} = \bar{y}_*\left(\frac{M_*'}{M_-}\right)\left[\frac{1+\frac{\kappa-1}{2}M_-^2}{1+\frac{\kappa-1}{2}M_*'^2}\right]^{\kappa+1/[2(\kappa-1)]}$$

$$M_- = \sqrt{\frac{2}{\kappa-1}\left[\left(\frac{p_{0-}}{p_{0+}}\right)^{\kappa-1/\kappa}\left(1+\frac{\kappa-1}{2}M_+^2\right)-1\right]};$$

the relations for the shock wave

$$\frac{p_2}{p_\infty} = \frac{2\kappa}{\kappa+1}\left(M_\infty^2\sin^2\varepsilon - \frac{\kappa-1}{2\kappa}\right)$$

$$\mathrm{tg}\theta_2 = \left[\left(\frac{\kappa+1}{2}\frac{M_\infty^2}{M_\infty^2\sin^2\varepsilon-1}-1\right)\mathrm{tg}\varepsilon\right]^{-1};$$

and the conditions for the characteristic

$$\mathrm{ctg}\mu_+ = \sqrt{M_+^2 - 1}$$

When the length of the internal duct in a minimal section is $\ell = 0$, the vehicle external generatrix inclination will be maximal, $\theta_2 = \theta_{2\,max}$.

As noted earlier, at $\theta_2 \neq 0$ the vehicle has external wave drag. When $\ell = 0$ $(\theta_2 = \theta_{2\,max})$, it is maximal; and when $\ell >> 1$ $(\theta_2 << 1)$, it tends to zero. In the latter case, the longitudinal force acting on a vehicle will be equal to the reactive thrust in the design regime.

$$\bar{R}_x^* = 2\left\{1 - \frac{M_-}{M_+}\sqrt{\frac{T_{0-}}{T_{0+}}\left[\frac{1+\frac{\kappa-1}{2}M_+^2}{1+\frac{\kappa-1}{2}M_-^2}\right]}\right\}$$

Excluding the value of the reactive thrust in the case of pressure equality, $\bar{R}^*$, from the total longitudinal force $\bar{R}_x$, we obtain the following expressions from relations (2) for wave drag C_D and lift C_L of the airframe:

$$C_D = \bar{R}_x - \bar{R}_x^* = \frac{2}{\kappa M_+^2}\left(\frac{y_-}{y_+} - 1\right)\left(\frac{p_2}{p_+} - 1\right)$$

$$C_L = \frac{2}{\kappa M_+^2}\left(\frac{L}{y_+} - \mathrm{ctg}\mu_+\right)\left(\frac{p_2}{p_+} - 1\right)$$

Then the airframe lift-to-drag ratio will be

$$K = \frac{C_L}{C_D} = \frac{L - y^* \cdot \mathrm{ctg}\mu_+}{y_- - y_+} = \mathrm{ctg}\theta_2$$

Such a vehicle with a finite volume will be equivalent in its lifting properties to a zero-thickness plate.

For the case considered earlier, $M_+ = 10$ and $\kappa = 1.4$, Fig. 4 presents the variation in components $\overline{R}_x$ and $\overline{R}_y$ of the total aerodynamic force as a function of the relative value of the supplied energy T_{0-} / T_{0+} *at* $y_- > y_+$.

In comparison with the case considered earlier, $y_+ = y_-$, the vehicle reactive thrust increases considerably in the pressure equality case. However, because of a wave drag penalty, there is essentially no increment in the absolute value of the total longitudinal force $|R_x|$ for the selected vehicle aspect ratio. As for the cross force, the predicted estimates show that the aerodynamic method of its production using lifting surfaces proves to be more efficient than gasdynamic rotation of the engine thrust vector.

The wave drag of the vehicle with an integral configuration may be reduced by increasing the aspect ratio, L / y_+. In the case $L / y_+ >> 1$, assuming that, $\theta_2 << 1, M_+ >> 1$, and $M_+\theta_+ << 1$ in accordance with the linear theory, it follows that

$$p_2 / p_+ = 1 + \kappa M_+ \theta_2$$

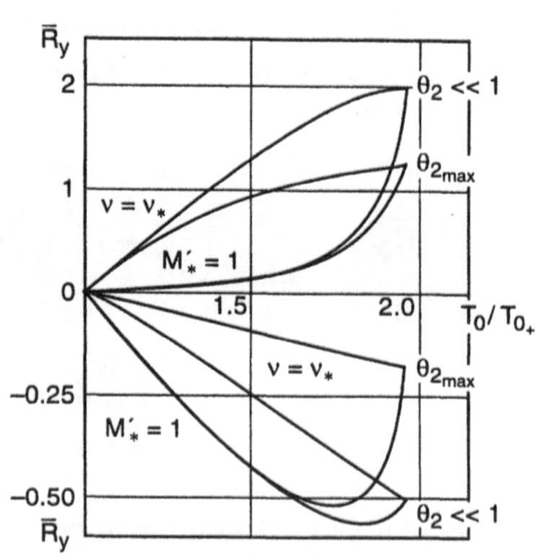

Fig. 4. The relative values of $\overline{R}_x$ and $\overline{R}_y$ vs. the temperature ratio corresponding to the supplied energy T_{0-} / T_{0+} at $y_- > y_+$.

and, for the components of the total force $\vec{R}$ acting on the vehicle, we obtain

$$\overline{R}_x = \overline{R}_x^* + \frac{2\theta_2}{M_+}\left(\frac{y_-}{y_+} - 1\right)$$

$$\overline{R}_y = \frac{2}{M_+}\left(\frac{y_-}{y_+} - 1\right)$$

The limiting values of the components $\overline{R}_x$ and $\overline{R}_y$ as $\theta_2 \to 0$ are shown in Fig. 4. In this case, the vehicle wave drag tends to zero, and the longitudinal force acting on the vehicle will be equal to the reacting thrust of the design regime $\overline{R}_x^*$.

Hypersonic Rule of Areas

Peculiarities of Hypersonic Flow Over Thin Blunt Bodies

At hypersonic velocities near the surface of a thin blunt body with the characteristic size of bluntness **d**, a high-entropy layer forms that includes the streamlines that have passed through the front of the detached shock wave at high angles of inclination (Fig. 5). For the flow parameters in this layer, the boundary of which is given by, $\left(y_\delta / L\right) \sim (d / L)^{\kappa/(\kappa+1)}$, the following estimations[5] are valid:

$$\frac{u_\delta}{U_\infty} \sim 1; \quad \frac{v_\delta}{U_\infty} \sim \frac{w_\delta}{U_\infty} \sim \left(\frac{d}{L}\right)^{\kappa/\kappa+1}; \quad \frac{p_\delta}{\rho_\infty U_\infty^2} \sim \left(\frac{d}{L}\right)^{2\kappa/\kappa+1};$$

$$\frac{\rho_\delta}{\rho_\infty} \sim \left(\frac{d}{L}\right)^{2/\kappa+1}$$

The applicability of these estimations follows from the equation of continuity in the entropy layer:

$$\rho_\infty U_\infty d^2 \sim \rho_\delta u_\delta \left(y_\delta^2 - y_w^2\right)$$

When the area of the entropy layer cross section does not exceed the area of the body cross section in its order of magnitude, from the last correlation it follows that

$$\left(\frac{d}{L}\right)^2 \sim \left(\frac{d}{L}\right)^{2/(\kappa+1)} \left(\frac{y_\delta}{L}\right)$$

$$\left(\frac{y_\delta}{L}\right) \sim \left(\frac{d}{L}\right)^{\kappa/(\kappa+1)}$$

This confirms the applicability of the aforementioned estimations.

Based on the estimations from the equations of motion we next can write for the pressure drop in the radial y and circumferential φ directions,

$$\frac{\Delta p_\delta}{\rho_\infty U_\infty^2} \sim \rho_\delta u_\delta \frac{\partial u_\delta}{\partial x_\delta} \Delta y_\delta \sim \left(\frac{d}{L}\right)^2$$

$$\frac{\Delta p_\delta}{\rho_\infty U_\infty^2} \sim \rho_\delta u_\delta \frac{\partial w_\delta}{\partial x_\delta} y_\delta \Delta \varphi_\delta \sim \left(\frac{d}{L}\right)^2$$

Hence, it follows that when a relative error is

$$\frac{\Delta p_\delta}{p_\delta} = \left(\frac{d}{L}\right)^{2/(\kappa+1)}$$

we can consider the pressure to be constant in the entropy layer.

In this case, if the body is fully embedded inside a high-entropy layer (see Fig. 5a), it is possible to neglect the pressure

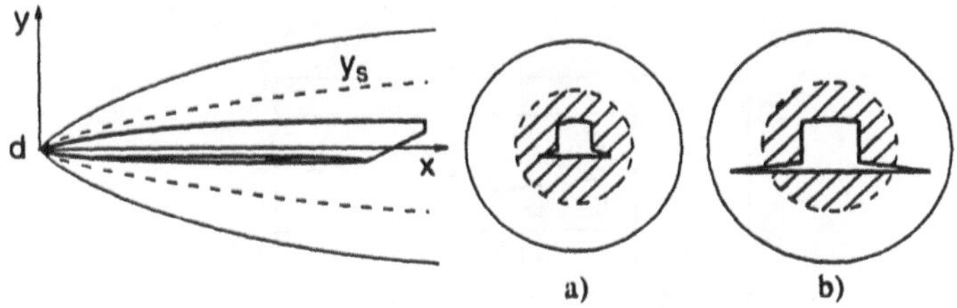

Fig. 5. The hypersonic flow over a thin blunt body.

variation at each section and to consider the compressed gas layer near a shock wave to be axisymmetric. Assuming that the body drag does not exceed in its order of magnitude the bluntness drag, a hypersonic rule of areas has been formulated in Ref. 50. According to this rule, the external drag of nonaxisymmetric blunt bodies will be identical when drag forces of the bluntness F_0 are equal and the laws of the variation of the cross-sectional areas $S(x)$ are the same:

$$R_x = F_x = F_0 + \int_0^l pS(x)\,dx$$

The area rule is satisfied for some classes of bodies and are verified experimentally in Refs. 6 and 7. As an example, drag coefficients C_D of the blunt cone and three-dimensional bodies equivalent to it with the same law of variation of cross-sectional area as for the cone are presented in similarity coordinates in Fig. 6; C_{Dcon} is the sharp-nosed cone drag coefficient, C_{D0} is the bluntness drag coefficient. Up to the length at which the body drag does not exceed (with respect to the order of magnitude) the drag of bluntness, the experimental values of the drag coefficient of the tested bodies fall on a common curve.

Early deviation of experimental values of the drag coefficient of a trihedral prism from the common dependence is necessary because for this body the condition according to which wetted bodies should be entirely within a high-entropy layer is violated. In such a case, when individual parts of the wetted bodies fall outside the limits of the entropy layer (see Fig. 5b) and, in some other cases, outside the limits of the shock-wave surface, the

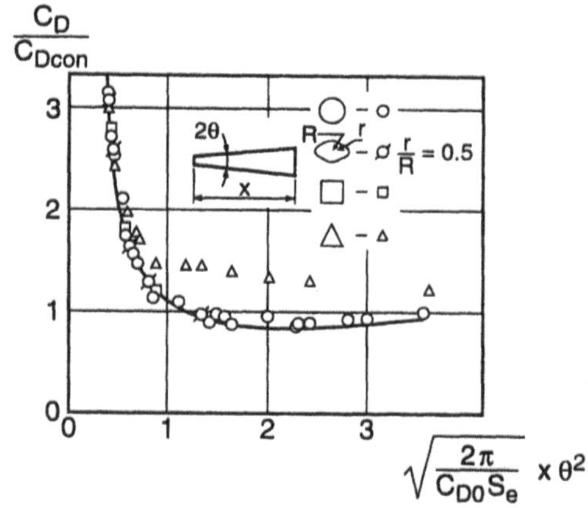

Fig. 6. The drag coefficient of the three-dimensional blunt bodies with the same law of variation of the cross-section areas.

hypersonic area rule has been generalized (see Ref. 8). According to this generalization, the drag forces acting on thin blunt bodies will be equal when the following conditions are satisfied:

1. Drag forces of the body and its bluntness are equal within an order of magnitude.
2. Values of the bluntness drag are equal.
3. Individual components of the wetted body that fall outside the limits of a high-entropy layer and shock-wave surface are the same. And,
4. the laws of the variation of the cross-sectional areas that are inside a high-entropy layer are the same.

Conditions for the Applicability of the Hypersonic Area Rule to Hypersonic Vehicles with an Airbreathing Engine

The characteristic feature of a hypersonic vehicle with an airbreathing engine is that an internal duct is available in this vehicle. However, for such vehicles, it is not possible to apply the hypersonic area rule for thin blunt bodies due to the presence of the duct. Let us formulate the additional conditions to be satisfied.

As noted earlier, the total force acting on a hypersonic vehicle with an airbreathing engine consists of two components, one of which, force $\vec{P}$, is applied to the surfaces of the vehicle and

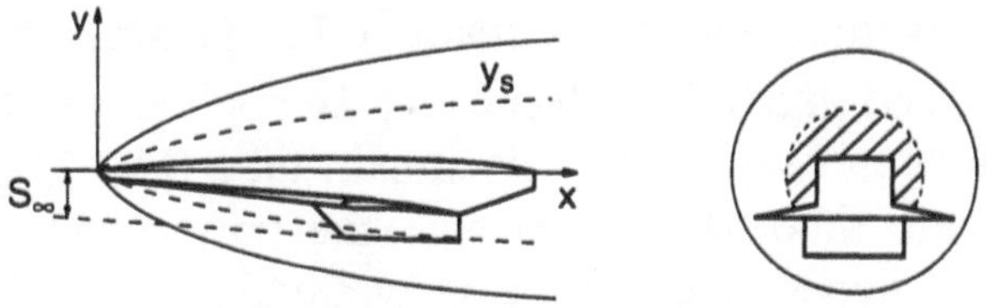

Fig. 7. The hypersonic flow over a thin blunt body with an internal duct.

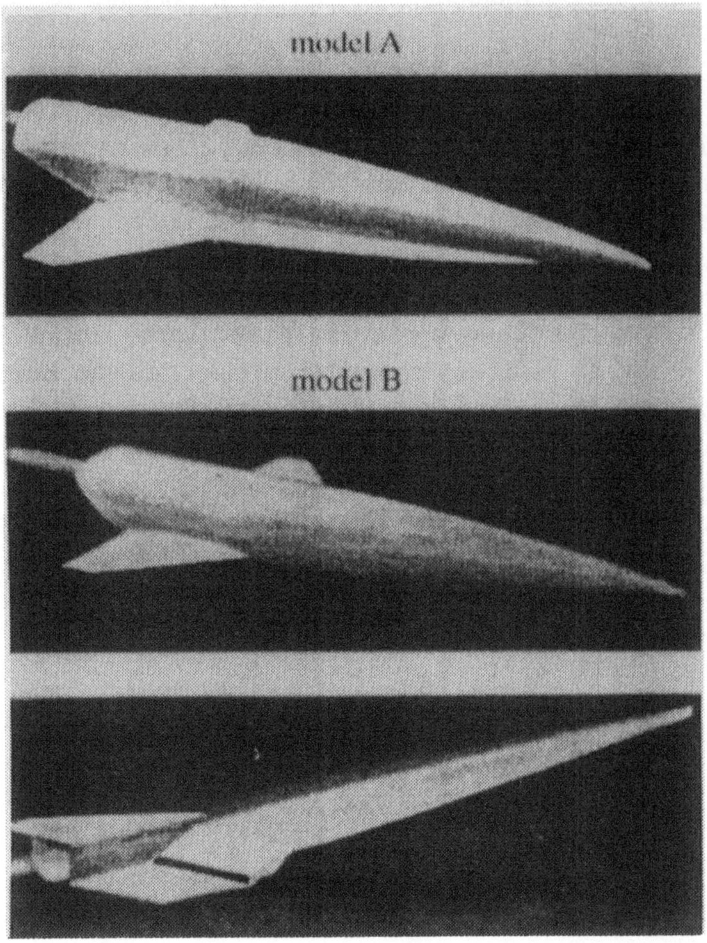

Fig. 8. Two schematized models of a hypersonic vehicle investigated in experiments.

the engine wetted by the stream tube passing through the powerplant, and the other, force $\vec{F}$, is applied to the external vehicle surface not wetted by this stream tube (Fig. 7). Assuming that the stream tube passing through the powerplant remains the same when a hypersonic vehicle is wetted, then, in this case, in accordance with the equation of motion, force $\vec{P}$ does not vary, and for the external vehicle surface not wetted by this stream tube the hypersonic area rule can be used. Therefore, it is necessary to impose the condition of flow constancy in the stream tube passing through the engine in addition to the other conditions in order to be able to apply the hypersonic area rule for hypersonic vehicles with an internal duct.

The conditions for the applicability of the hypersonic area rule for vehicles with an airbreathing engine have been investigated experimentally[3]. The tests have been carried out on two schematized models of a hypersonic vehicle with an airbreathing engine in the hypersonic TsAGI T-121 wind tunnel (Fig. 8). The models differ in body configuration, retaining the same vehicle volume, length, and law of variation of cross-sectional areas. The model scheme and the laws of cross-sectional area distributions are given in Fig. 9. The lower model surfaces are the same, their forward parts being plane with an angle of wing setting relative to the datum line of 3 deg, and the removable engine nacelles are modeled either by a duct having a constant rectangular cross section or by a similar duct having a two-stage air intake with

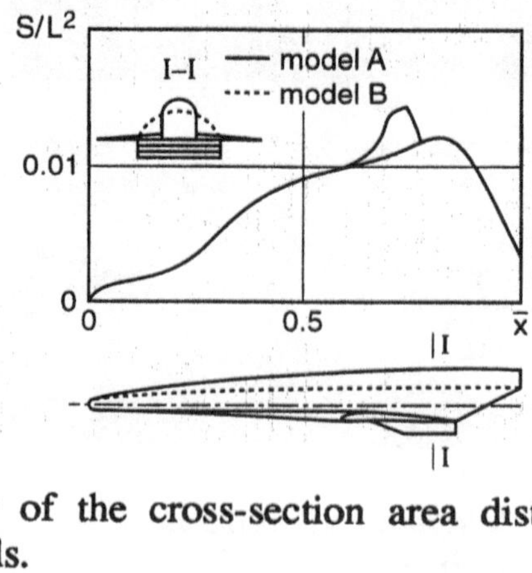

Fig. 9. The law of the cross-section area distributions of the schematized models.

ramps inclined at 6 deg and 24 deg. The low baseline trapezoidal planform wing has a thickness ratio $\bar{c} = 0.04$ and a leading-edge sweep of $\chi = 55\ deg$. The body of model A, whose nacelle width is greater than the body width, has some extensions to ensure a uniform flowfield upstream of the air intake and in the nozzle.

Figure 10 presents variations in the drag coefficient C_D and the lift coefficient C_L as functions of the angle of attack obtained in the tests for nacelle-free models at $M = 8$ and $Re = 3.6 \times 10^6$.

With zero lift at $\alpha = 0$, the drag of the tested models is the same because the requirements to follow the hypersonic area rule, formulated earlier, has been satisfied. This is evident from the results of the calculation of the distribution of Mach numbers using Euler's equation [9] in the plane of the cross-section of model A at $M_\infty = 10$ and $\alpha = 0$ (see Fig. 11). Despite the complex shape of the hypersonic vehicle, the shock wave and boundary of the entropy layer in the model cross section appeared to be close to axisymmetric, and the model wing, which falls outside the limits of the boundary of the high-entropy layer, was the same. Note that the hypersonic area rule also remains valid when viscosity has a considerable effect, despite the fact that this rule was formulated for

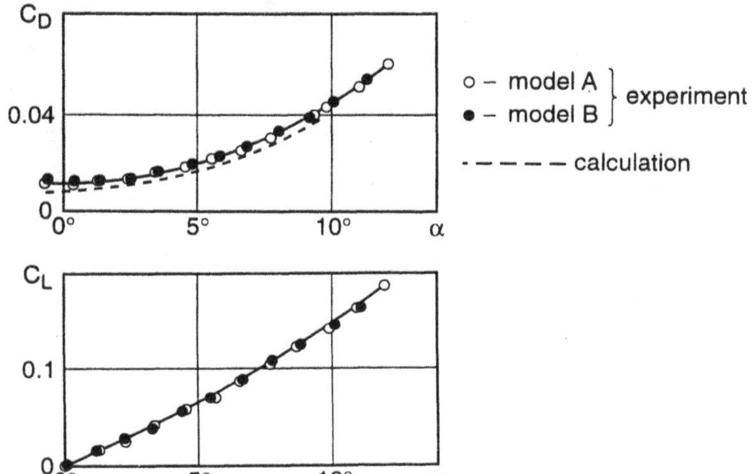

Fig. 10. The drag coefficient C_D and lift C_L vs. the angle of attack α for the nacelle-free model.

hypersonic nonviscous flows. For example, under the test conditions, the friction drag of the model without nacelle at low angles of attack was a considerable part of the total drag. The drag coefficient C_D obtained by Euler's equations [9] and presented in Fig. 10 attest to this.

When the angle of attack is increased, the axisymmetry of the flow is broken (see Fig. 11). However, the drag and lift of the models that have been tested also remain the same in this case. This is because, when $M_\infty \alpha >> 1$, the aerodynamic forces acting on the body will be determined only on the windward side as the pressure on this part behind the strong shock wave will exceed manifold the pressure on the shadow side. Next, the design values of the relative contribution of drag C'_D and lift C'_L of the lower surface into the total balance of drag C_D and lift C_L are given in

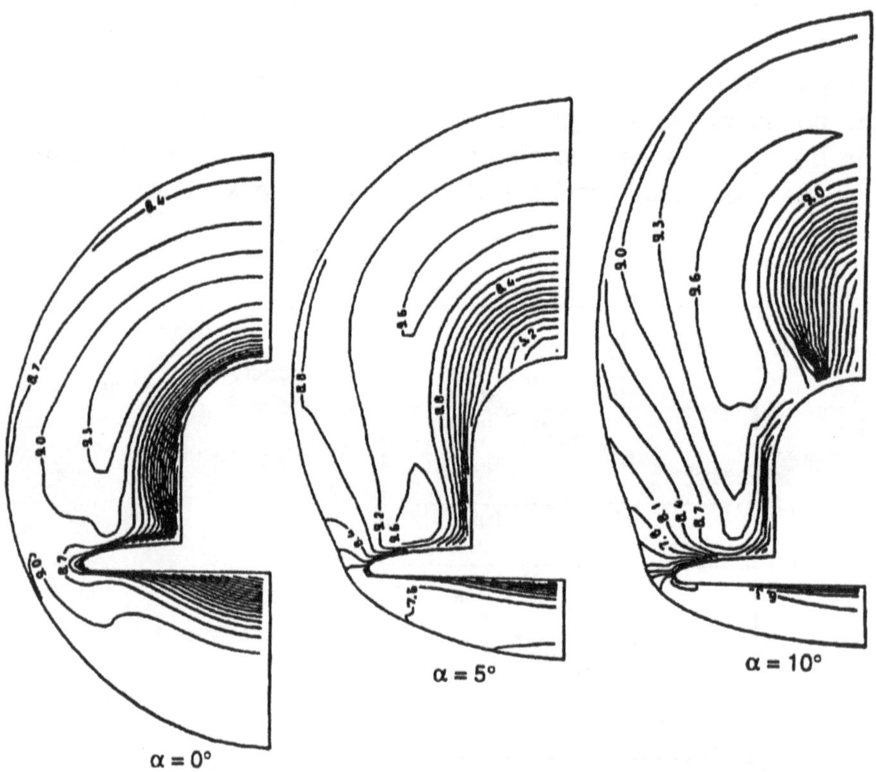

Fig. 11. The distribution of mach number in the plane of the cross section of the model A.

Table 1

α°	0	5	10	15
C'_D / C_D	0.425	0.847	0.953	0.982
C'_L / C_L	∞	1.286	1.058	1.021

Table 1, for model A[9]. When $C'_D \sim C_D$ and $C'_L \sim C_L$, the aerodynamic forces acting on the tested models with the same lower surface become equal. This conclusion was verified in the experiment.

As noted earlier, when the body with internal ducts is wetted, the hypersonic area rule must satisfy an additional requirement: flow constancy in the stream tube passing through the duct. This is a rather strong restriction as the stream in the duct depends on many factors that are specified at the inlet and outlet. When the flow about the body is complex, it is difficult to satisfy this requirement, and during the investigations it was not verified, probably resulting in discrepancies of aerodynamic characteristics for the models with nacelles (Fig. 12). For meeting this

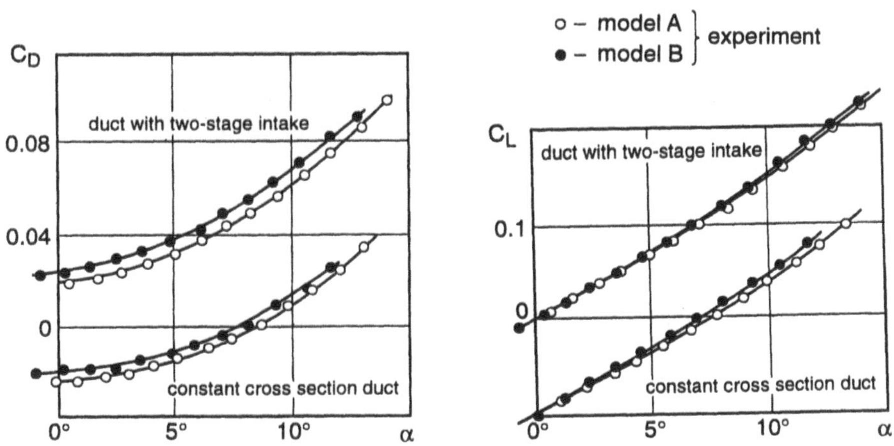

Fig. 12. The drag coefficient C_D and lift C_L vs. the angle of attack α for the models with nacelles.

requirement, it is necessary to measure the parameters of the stream at the inlet and outlet of the duct.

The requirements established earlier to apply the hypersonic area rule for vehicles of integral configuration, that have been determined at $M = 8$, also remain the same when the Mach number changes. This is illustrated by the experimentally observed dependence of the drag coefficient C_D on the Mach number at $\alpha = 0$ (see Fig. 13).

Integration of a Hypersonic Vehicle Airframe with an Airbreathing Engine

Some General Principles

The analyses considered previously make it possible to formulate some principles for the integration of a hypersonic vehicle airframe with an airbreathing engine[2].

The cross-sectional area of the stream tube in the issuing jet S_j will be larger than S_∞ in the case of an energy supply to the gas passing through the engine duct (Fig. 1a). In this case, for the wave drag penalty to be avoided, it is necessary that the intake entry area be $S_+ = S_\infty$, and $S_m = S_-$ in the vehicle frontal area. If

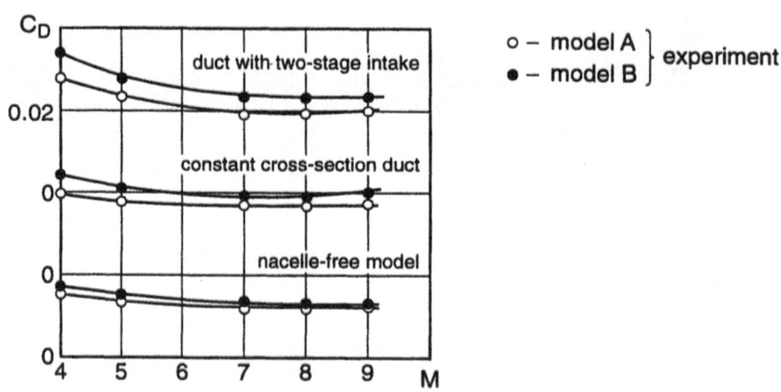

Fig. 13. The drag coefficient C_D vs. Mach number M at the angle of attack $\alpha = 0$.

$S_- = S_j$, the flow pressure at the exit will be equal to the static pressure of the freestream flow, and the engine thrust will be maximum. When there is an off-design flow issuing from the nozzle $\left(S_j \neq S_-\right)$, the thrust losses at $S_\infty = constant$ can be compensated for in part by reducing the vehicle wave drag as a result of reducing its frontal area.

The requirements for the selection of the law of variation of the cross-sectional area of an integral configuration vehicle may be different. It is possible to use the hypersonic area rule for minimizing vehicle drag. In compliance with this rule, the law of variation of the cross-sectional areas for the vehicle must coincide with that for an equivalent blunt body of minimum wave drag so that the drag of the vehicle could be minimized.

The variational problem for the shape of the minimum drag body, when geometrical restrictions are specified and the class of varying surfaces is not constrained, is complex. Therefore, in a simplified approach to shape definition, approximative formulas are used for pressure variation, as a rule, and the class of surfaces is constrained. For example, in the case when the pressure on the body surface is specified by the modified Newton formula, the solution of this problem for bodies with a specified relative thickness gives $y \sim \chi^{3/4}$. The result, obtained within the framework of applicability of the law of two-dimensional cross sections, is close to this. The minimum drag is realized within the class of power-law bodies specified relative thickness when the power exponent is 0.71.

In the integral configuration, the law of distribution of the cross-sectional areas in the vehicle nose part with the chosen drag of body bluntness is close to that for a body of revolution with the generatrix $y \sim \chi^{3/4}$.

For lifting vehicles, the requirements for the choice of the law of variation of the cross-sectional areas will be different. For example, when the maximum *L/D* ratio is realized, the requirements will include, as before, retention of the condition $S_+ = S_\infty$, and since the aerodynamic method of producing lift by wing lifting surfaces is more effective as compared with the nozzle jet rotation, the law of variation of the cross-sectional areas in the range

$(S_m - S_+)$ must be identical to the corresponding law for a plane plate. A schematized hypersonic vehicle of an integral configuration satisfying these requirements is shown in Fig. 14. In its lifting properties, it will be equivalent in an inviscid flow to a zero-thickness plate when the engine thrust is maximum.

Viscosity Effect

When a hypersonic vehicle with an airbreathing engine that performs a long-duration flight in the during the acceleration motion is optimized, it is necessary to take into account the effect of viscosity. At low angles of attack, the friction drag for such vehicles may account for a considerable part of the total value. This has already been noted earlier in the analysis of the experimental data (see Fig. 10).

As the design investigations show, when the values of the coefficient of friction are practically realistic and correspond to high Reynolds number values, inclusion of the boundary-layer effect slightly changes the shape of the optimal body of revolution. This has been verified from the experimental data on the drag of various bodies at zero angle of attack[10]. They showed that at Mach numbers $2 < M < 4$ and Reynolds numbers $14 \times 10^6 < Re < 20 \times 10^6$, the body of revolution with the generatrix $y \sim \chi^{3/4}$ is the optimum with respect to drag considering the bodies that have been tested. Figure 15 exemplifies this showing the dependence of the optimal exponent of power m, for the bodies of revolution of different aspect ratios λ on the Mach

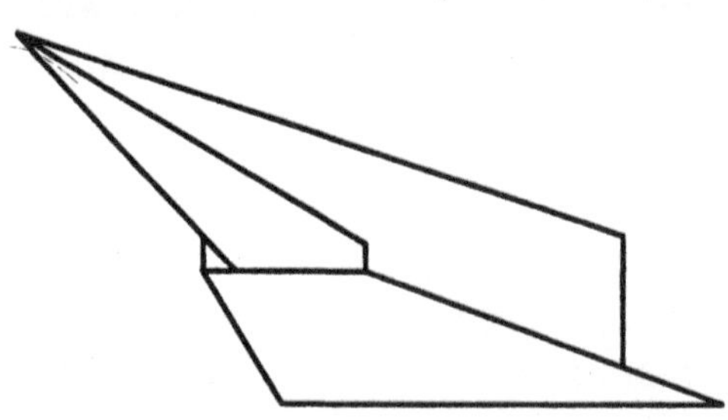

Fig. 14. A schematized hypersonic vehicle of an integral configuration with maximum L/D-ratio.

number M. Therefore, when the drag of a hypersonic vehicle is minimized, the choice of the law of distribution of the cross-sectional areas $y \sim \chi^{3/4}$ in the nose up to the intake entry remains valid with viscosity as well.

When Reynolds numbers are lower, e.g., in the regimes of strong viscous interaction when the boundary-layer thickness can be compared with the thickness of the thin body, its friction drag becomes comparable with the wave drag. Therefore, the effect of the thin body shape on its aerodynamic characteristics at low Reynolds numbers will be less as compared with the similar effect in the nonviscous gas. For example, in these regimes, a slightly blunt-nosed cone will have the same effect with respect to the drag as a sharp-nosed cone, that, at a fixed value of frontal area S_m, has a minimum at a finite value of the half-angle θ. The experimental data obtained in the vacuum wind tunnel VAT-102 TsAGI at $M = 5.15$ and $Re = 10^3$ verify the conclusion (see Fig. 16).

Heat Transfer to Some Elements of a Hypersonic Vehicle with an Airbreathing Engine

The peculiarities of the configuration of a hypersonic vehicle with an airbreathing engine and the trajectories of launch into an orbit and descent result in a considerable increase of thermal loads applied to individual components. High values of heat fluxes in

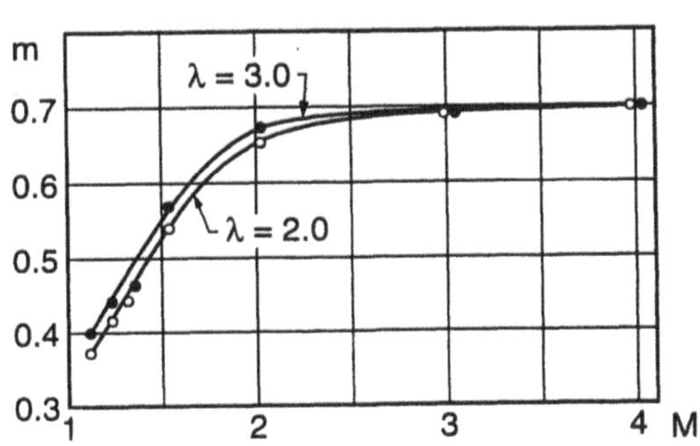

Fig. 15. The dependence of the optimal power exponent m for bodies of revolution, with different aspect ratio λ, on Mach number M.

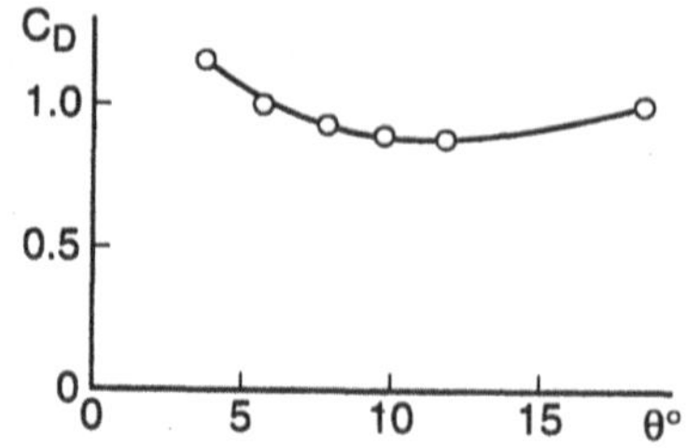

Fig. 16. The drag of a sharp-nosed cone with a fixed frontal area.

such areas can be caused by laminar-turbulent transition, flow separation, the interaction of shock waves and the interaction of shock waves with the boundary layer, the influence of real gas properties and catalytic surfaces, etc. In some cases, the levels of these heat fluxes are so high that, currently, we are only researching for ways to lower them.

The influence of different physical processes and shapes of bluntness on the local and total heat transfer to the leading edges of a hypersonic vehicle has been studied in Ref. 11. Figure 17 illustrates this for such a thermally stressed point, showing the dependence of the Stanton number St on the Reynolds number Re_0 at different recombination rates of atoms of oxygen K_{wO} and nitrogen K_{wN} ($K_{wO} = K_{wN} = \infty$ is an ideally catalytic surface, and $K_{wO} = K_{wN} = 0$ is ideally noncatalytic). Figure 18 presents equilibrium temperatures T_{we} corresponding to heat fluxes for a critical (stagnation) point on the forebody with a bluntness radius of $R = 10\ cm$.

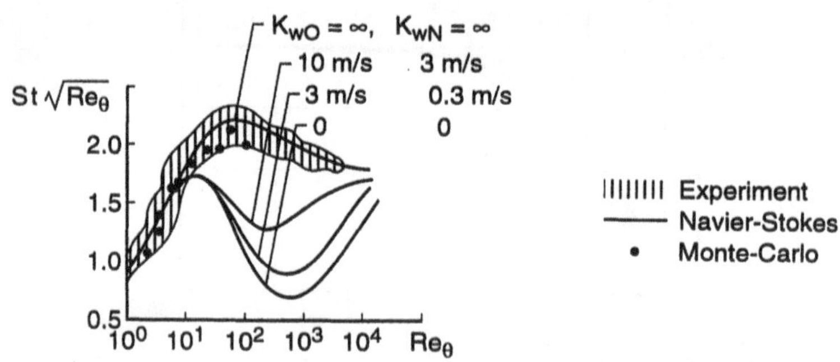

Fig. 17. Stanton number for a critical point at different recombination rates of atoms of oxygen and nitrogen.

The use of noncatalytic materials for thermal protection lowers the equilibrium temperature of the vehicle fuselage nose. However, its value still remains high.

Cooling gas injection is an effective means to lower the equilibrium temperature. The values of T_{we} for two values of the relative rate of injected air, $\overline{\rho v} = 0.1$ and 0.2, are presented in Fig. 18. It is possible to realize a larger effect by injection of a foreign gas of low molecular weight. However, the boundary-layer buildup caused by injection can deteriorate the air intake operation conditions and bring the effect of bluntness radius reduction to nil. In this case, thermal protection by convection, when the coolant is circulated between the external shell and the structure, is more attractive.

When we choose the shapes of thermally stressed leading edges of the fuselage and the wing of a hypersonic vehicle, we should remember that the total heat flux q_{Σ} to the wetted area of the leading edge of a fixed frontal surface varies nonmonotonically as a function of the relative length L / y_m. There is no obvious drop in q_{Σ} as L / y_m decreases, which results from a reduced heat flux at the critical point when the curvature radius increases. The result obtained in the case of a higher length ratio L / y_m is

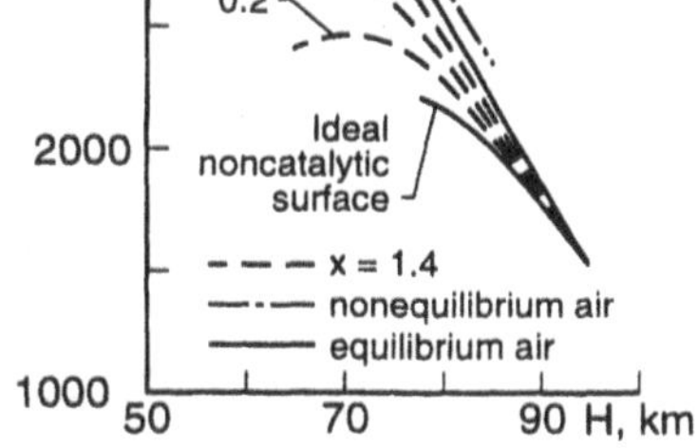

Fig. 18. Temperature for a critical point with the bluntness radius R = 10 cm.

more interesting: despite a higher local heat flux at the critical point, the total heat flux to the body will be smaller. The maximum values of heat flux and drag are observed at small bluntness values. In Fig. 19 we can see how a distribution of local heat fluxes q through the body occurs in this case.

The part of a jet-engine orbiter with the highest heat load is the leading edge of the air intake, whose bluntness radius must be fairly small. The interference between the incident and forward shocks leads to the formation of narrow pressure and heat flux peaks at the blunt edge of the air intake. For laminar flow and relatively high Reynolds numbers, the experimental values of the maximum increase in the heat flux $\bar{q}_{max} = q_{max} / q_0$ on the surface of the leading edge of the air intake are given in Fig. 20 (continuous curve: $M_\infty = 6.5,\ Re_\infty = 10^5 - 10^6$)[12]. In the given classification, the adopted Roman numerals in this figure denote the various types of shock-wave interference. The largest intensification of the heat transfer corresponds to type IV interference, when the oblique shock impinges on the extent of the forward shock perpendicular to the freestream velocity. Under the experimental conditions, the heat flux in the zone of impingement of the high-enthalpy jet on the leading edge of the air intake is six times greater than the heat flux q_0 on the edge stagnation line for flow without interference.

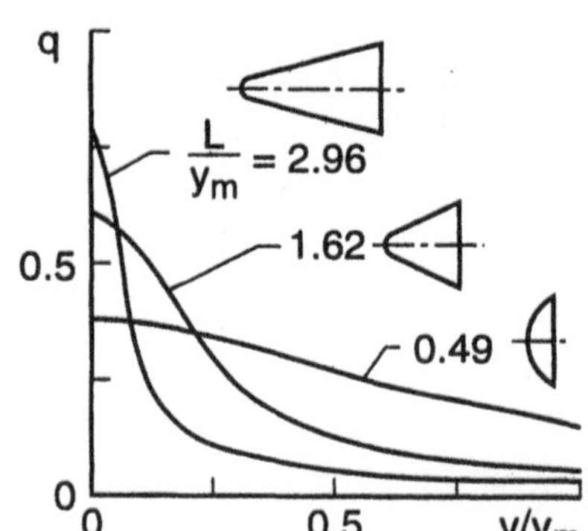

Fig. 19. The distribution of local heat fluxes q along a body.

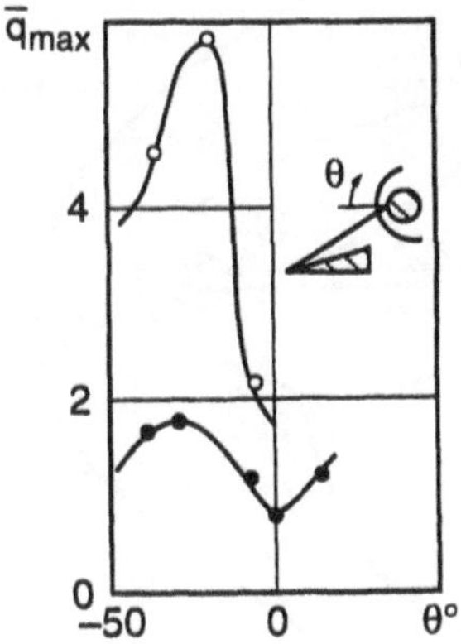

Fig. 20. Cowl heat flux increase due to shock-shock interference.

With a decrease in the Reynolds number, when the shock layer becomes completely viscous, as a result of which the shocks and the boundaries of the mixing layers are heavily smeared out, the extremal values of the heat transfer to the leading edge of the air intake decrease. This information[13] is provided for $M_\infty = 6.5$ and $Re_\infty \sim 10^3$ (broken curve) in Fig. 20. From this figure it also follows that the type IV interference characterized by maximum values of $\bar{q}_{max}$, which occurs at high Reynolds numbers, is not realized at low Reynolds numbers.

In addition to the locations already mentioned, a number of areas of intense heat exchange can appear as a result of the interaction of shock waves and flow separations when a hypersonic vehicle with an airbreathing engine is wetted in flight. For example, this situation can occur during the descent of the vehicle with a closed engine nacelle under the fuselage. For this case, Fig. 21 presents a flow pattern in the cross section of model A in the engine nacelle area, that has been obtained during an experiment[14] by a "laser knife" (1, fuselage; 2, wing; 3, engine nacelle). The experiments have been performed at $M = 5$ and low angle of attack, $\alpha = 5$ *deg*. We can see the nose shock wave (4), shock wave (5) induced by a closed engine nacelle, shock wave (6) near the wing, and internal shock waves (7) over the wing. Flow separations are also shown. Near side surfaces of an engine nacelle, separation areas with a line of separation S_1, and the line of reattachment R_1 appear. A separation area that starts on the line S_2 and finishes on the fuselage (line R_2) appears over the wing. At high Reynolds

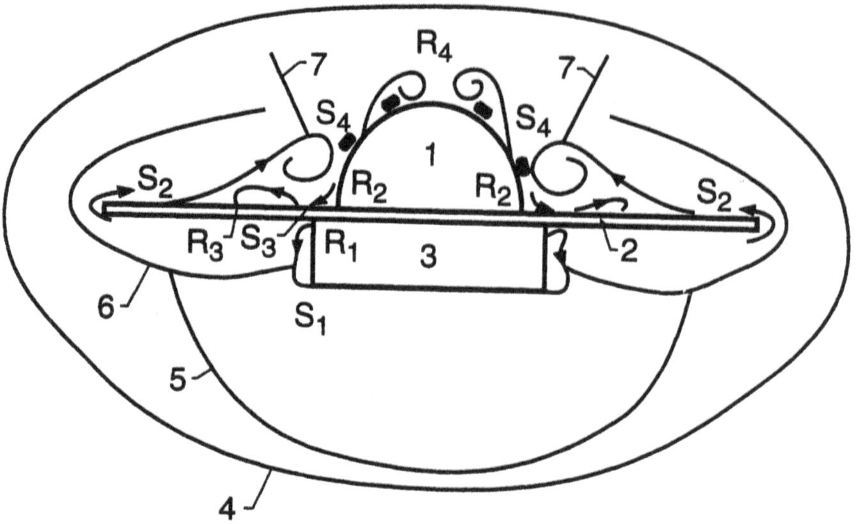

Fig. 21. The flowfield in the engine nacelle cross-section of the model A.

numbers, a secondary separation (lines S_3, R_3) occurs on the wing. A separation area (lines S_4, R_4) originates over the fuselage as well. On some lines of reattachment (e.g., line R_2) heat exchange can increase considerably at high Reynolds numbers, resulting in the temperature near this line approaching the temperature of the lower wing surface.

Conclusions

The present investigations, aiming at an efficient integration of a powerplant with a hypersonic vehicle airframe, make it possible to estimate the influences of individual components of a vehicle on its aerodynamics and to identify the conditions for a favorable interference to optimize vehicle shapes. It is obvious that the integration principles formulated herein do not encompass the whole multiplicity of the phenomena inherent in the atmospheric flight of a hypersonic vehicle. Still, despite this fact, the results obtained are believed to be rather useful for the solution of the problems of the integration of an airplane with a powerplant at hypersonic velocities.

References

[1]Gusev, V. N., "Aerospace Aerothermodynamics," The TsAGI Journal, Vol. 1, No. 1, 1994.

[2]Gusev, V. N., "Hypersonic Vehicle Airframe-Air-Breathing Propulsion Integration," Uchenye Zapiski TsAGI, Vol. 22, No. 5, 1991 (in Russian).

[3]Gusev, V. N., Blagoveshchensky, N., and Zadonsky, S., "The Airbreathing Engine," AIAA Paper 93-5034, 1993.

[4]Stalker, R. J., "Waves and Thermodynamics in High Mach Number Propulsive Ducts," High-Speed Flight Propulsion Systems, edited by S.N.B. Murthy and E.T. Curran, Vol. 137, Progress in Astronautics and Aeronautics, AIAA, Washington, DC, 1991, pp. xx-xx.

[5]Ladyzhensky, M. D., "The Hypersonic Rule of Areas," Inzhenerny Zhurnal, No. 1, 1961 (in Russian).

[6]Gusev, V. N., and Kryukova, S. G., "A Hypersonic Helium Flow Over Very Blunted Bodies," Inzhenerny Zhurnal, Vol. 5, No. 2, 1965 (in Russian).

[7]Krasovsky, V. N., "The Experimental Investigation of Blunt Bodies in a Hypersonic Helium Flow," Inzhenerny Zhurnal, Vol. 5, No. 2, 1965 (in Russian).

[8]Ladyzhensky, M. D., "Generalization of the Hypersonic Rule of Areas," AN USSR, OTN, No. 3, 1961 (in Russian).

[9]Golubinsky, A. A., Kosykh, A. P., Savin, I. V., and Chelysheva, I. F., "Numerical Supersonic Three-Dimensional Flow Over Ideal Configurations of Aerospace Vehicles by Perfect Gas and Equilibrium-Dissociating Air," Proceedings of Reports of TsAGI's Workshop-School "Orbiter Aerothermodynamics", Part 2, 1992 (in Russian).

[10]Aeromechanics of a Supersonic Flow Over Bodies of Revolution of the Power Shape, edited by G. L. Grodzovsky, Mashinostroyelie, Moscow, 1975 (in Russian).

[11]Botin, A. V., Gusev, V. N., and Provorotov, V. P., "Hypersonic Flow Over Blunt Edges at Low Reynolds Numbers," Prikladnaya Mekhanika I Tekhnicheskaya Fizika, No. 4, 1989 (in Russian).

[12]Wieting, A. R., and Holden, M. S., "Experimental Shock-Wave Interference Heating on Cylinder at Mach Numbers 6 and 8," AIAA Journal, Vol. 27, Nov. 1989.

[13]Botkin, A. V., "Investigations of the Interference of Incident Shock Wave with a Shock Layer on the Edge at Low Reynolds Numbers," MZhG, RAN, No. 1, 1993 (in Russian).

[14]Borovoy, V. Ya., "Problems of Heat Exchange and Thermal Protection of the External Surface of Advanced Aerospace Vehicles," Proceedings of Reports of TsAGI's Workshop-School "Orbiter Aerothermodynamics," Part 1, 1992.

Low-Speed Operation of an Integrated Rocket-Ram-Scramjet for a Transatmospheric Accelerator

Frederick S. Billig*
Johns Hopkins University, Laurel, Maryland

Nomenclature

A	= cross-sectional area
A_i	= projected inlet (Fig. 2) or reference inlet area
$\tilde{A}_{ij}$	= cross-sectional area of underexpanded strut injector
$\tilde{A}_4$	= cross-sectional area of constricted inlet airflow
C_T	= thrust coefficient, T/q_0A_i with cowl-lip-to-vehicle trailing-edge force accounting
C_{TFS}	= thrust coefficient with freestream-to-freestream force accounting
C_1, C_2	= coefficients in thrust, specific impulse relationships
D	= vehicle drag, lb_f
D_I	= drag interval
ER_{ij}	= equivalence ratio of engine injectants
ER_0	= overall engine equivalence ratio
$\mathcal{F}$	= stream thrust, $PA(1+\gamma M^2)$
I_{eff}	= vehicle specific impulse, $(T-D)/W_p$, $lb_f s/lb_m$
I_f	= engine specific impulse, $lb_f s/lb_m$
ℓ	= length of strut injectors
ℓ_c	= length of mixing and combustion zone
M	= Mach number
$\mathcal{M}$	= molecular weight
O/F	= oxidizer/fuel ratio, W_{ijO2}/W_{ijH2}
$(O/F)_s$	= stoichiometers O/F for hydrogen, $16/2.016 = 7.9365$
p	= pressure, $lb_f/in.^2$

Copyright © 1995 by the American Institute of Aeronautics and Astronautics, Inc. All rights reserved.

*Chief Scientist, Associate Dept Supv Fellow, AIAA.

q = dynamic pressure, $lb_f/in.^2$
r = nozzle nonequilibrium index
T = temperature, °R, or thrust, lb_f
T_{FS} = thrust in freestream-to-freestream accounting system
T_R = internal thrust of strut injectors, lb_f
u = velocity, ft/s
u_m = takeoff velocity, ft/s
W_a = airflow rate, lb_m/s
W_p, or W_{ij} = total propellant flow rate, lb_m/s
W_{O2} = oxygen flow rate, lb_m/s
W_{RH2} = hydrogen flow rate in strut rocket
W_{RJH2} = hydrogen flow rate from ram-scramjet injectors
W_{BBH2} = hydrogen flow rate from base burning injectors
w = distance between struts
X = axial coordinate, mole fraction
$Y_{2,4}$, etc. = longitudinal coordinate, lateral heights, mass fraction
Z = altitude, ft
α = angle of attack, deg
α_1 or α_c = cowl angle
α_2 = body flare angle (Fig. 2)
β = bypass ratio, main duct airflow rate/rocket injector flow rate, W_a/W_p or W_a/W_{ij}
ρ = density, lb_m/ft^3
τ_w = shearing stress

Subscripts

e = projected area of vehicle at trailing edge
i = inlet reference plane (Fig. 2), ith specie
ij = conditions in exit plane of rocket injector
t = total conditions
0 = conditions in freestream
2 = conditions in plane of cowl lip
4 = conditions at end of isolator duct in discharge plane of rocket
4′ = conditions following normal shock in inlet compression process
5 = conditions at end of mixing and combustion zone
6 = conditions in plane of nozzle throat (Fig. 2) or in plane of cowl trailing edge, Fig. 14
7 = conditions in nozzle exit (Fig. 2) or in nozzle expansion of main propulsion duct where $P_7 = P_0$ for Fig. 14
8 = conditions in nozzle at $P_3 = P_0$ for Fig. 2 or in nozzle expansion of main propulsion duct in plane of vehicle trailing edge for Fig. 14
9 = conditions in inlet plane of boundary-layer bleed duct, Fig. 14
10 = conditions in exit plane of boundary-layer bleed duct, Fig. 14
11 = conditions in expansion of bleed duct flow at beginning of heat release, Fig. 14
12 = conditions in expansion bleed duct flow in plane of vehicle trailing edge, Fig. 14

Superscripts

$\tilde{}$ = conditions corresponding to underexpanded rocket injector

$'$ = conditions following normal shock

$*$ = conditions at sonic point

I. Introduction

THE renewed interest in incorporating air-breathing propulsion in transatmospheric accelerator systems has created a need to identify combined cycles with optimal performance. This paper presents a disciplined approach to the analysis of rocket-ram-scramjet cycles with emphasis on the characteristics of the vehicle at transonic speeds. It is argued that the combined effects of the low density of the primary propellant, hydrogen, and the need for high engine efficiencies at hypersonic speeds lead to a very small engine duct in a single-stage-to-orbit vehicle. This leads to a requirement for a high specific thrust as opposed to a high specific impulse cycle at low speed.

The attractive potential of the ram-scramjet (RSJ) as an alternative to all-rocket-powered propulsion systems for a transatmospheric accelerator has been recognized for more than 30 years.[1,2] Unfortunately, the "pure" RSJ is incapable of providing accelerative force at Mach numbers below 2, and its efficiency as an accelerator is below that of liquid oxygen (LOX) augmented system at Mach numbers above about 18. Pure in the context of this discussion means operation of the RSJ wherein the only propellant is fuel, e.g., hydrogen. In the LOX-augmented system, the RSJ injectors are in effect integrated internal H_2–O_2 rockets. To circumvent these limitations, the pure RSJ can be the propulsion system for the second stage of a three-stage vehicle system with one of a variety of first-stage systems and a rocket third stage. Alternatively, the third stage could be eliminated, and the performance penalty of the pure RSJ would have to be accepted. Another possibility would be to combine rotating machinery with the RSJ to provide thrust at low speed in a single-stage system: principal candidates are an augmented turbojet or an air turborocket. The problem with either of these systems is the penalty for carrying the weight of these extra components over the broad range of Mach numbers where they cannot be used. Moreover, it is very difficult to provide a duct of sufficient size to house the rotating machinery required to yield high net force specific impulse I_{eff} and to effect mode changes.

For single-stage-to-orbit systems or for multistage system (e.g., Pegasus), wherein the first stage is a subsonic aircraft, the rocket-ram-scramjet (RRSJ) is a viable alternative to all of the preceding concepts. The RRSJ is a totally integrated rocket-ram-scramjet. In the bipropellant variant, hydrogen is the fuel and oxygen the oxidizer. Some, if not all, of the injectors are capable of operating over a wide range of oxidizer/fuel (*O*/*F*) weight ratios including zero flow when functioning as a pure RSJ. In another approach to reducing fuel tank volume, a tripropellant system can be used wherein a second higher-density fuel is added and expanded during low-speed flight. Specific impulse penalties are incurred but are more than compensated in overall vehicle performance because of the enormous reduction in fuel tank volume.

Others have discussed the viability and potential of the RRSJ,[3–5] but there is little information in the open literature showing concepts and the details of the design of practical RRSJ vehicles. This paper is intended to partially fill that void and will focus on the low-speed problems where integration must be highly tailored if good performance is to be realized. The first section of the paper is an engine analysis that has been simplified by the assumption of an ideal working fluid to enable the reader to readily verify the results. The paper concludes with a real gas integral analysis of the performance and operation of a recently patented engine configuration.

A. Modeling of Rocket-Ram-Scramjet Engines

At low speed, the rocket serves as the primary driver in an ejector cycle. It is used to raise the total pressure in the "secondary" airstream, thus acting as the compressor in a psuedo-Brayton cycle. The additional fuel for the combustion in the Brayton cycle is added in a separate set of injectors or by operating the primary fuel rich. Schematic illustrations of the two types of systems are shown in Fig. 1. Each has advantages and disadvantages. In type A, the rocket motor is operated stoichiometrically, or lean, to prevent additional combustion in the ejector-mixer section. Maximum cycle performance is obtained when the rocket is stoichiometric ($ER_{ij} = 1$), but the rocket nozzle throat is difficult to cool. "Choking" in the mixing duct is relatively easy to prevent with proper sizing of the rocket and mixing duct exit. Diffusing the flow before combustion leads to minimum Rayleigh losses in the combustor but adds weight, length, and friction losses to the overall system. As shown, the flow is subsonic at the combustor exit and is accelerated in a converging-diverging nozzle.

In type B, the rocket motor operates fuel rich, and the mixing and combustion take place simultaneously. Rayleigh losses are higher, but the system is shorter and simpler and weighs less. Additionally, the rocket nozzle heat transfer is lower because of the rich operation ($ER_{ij} > 1$). As shown, there is no physical throat in the system, and indeed in this configuration the flow surface where the local

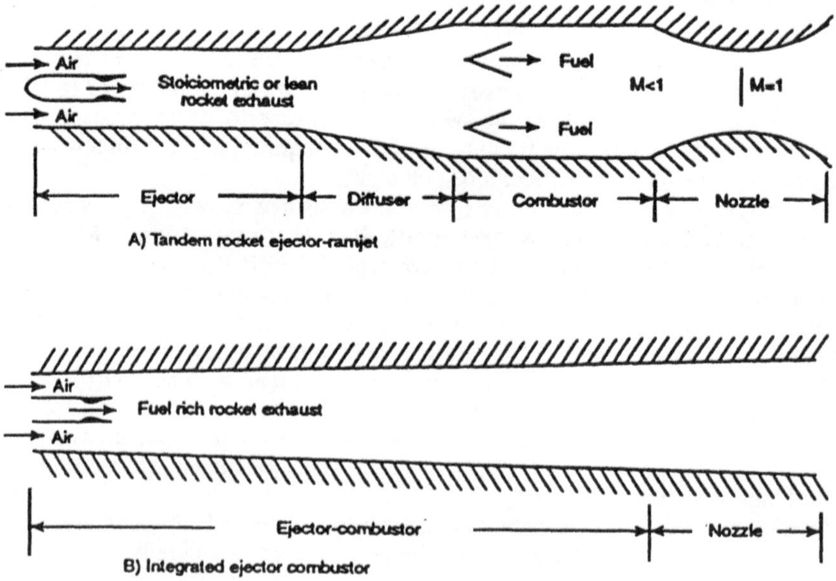

Fig. 1 Schematic illustrations of ejector ramjets.

passes through a physical throat with an area A_6. An exit condition of interest is at the point in the nozzle exhaust where the flow cross-sectional area A_7 is equal to the maximum cross-sectional area upstream. In the case shown $A_7 = A_5$, but in other cases $A_7 = A_4 + A_{ij}$. Another reference condition in the nozzle flow is the point where the pressure is equal to that in the undisturbed freestream; i.e., $P_8 = P_0$ and $A = A_8$.

Some of the arguments regarding optimal design will be based on the net forces on the engine. Consequently, the geometry of the external body must also be defined. Here it is assumed that the forces on the external surface of the inlet cowl can be computed assuming wedge flow at an angle α_1, with respect to the freestream. Similarly, α_2 is the effective wedge angle of the flared afterbody.

As shown, the areas A_4 and A_{ij} are the physical areas of the air duct and rocket exhaust. This is the appropriate representation of flow areas in the equations that will be developed for the case of matched pressure; i.e., $P_{ij} = P_4$. If the rocket would be operated either underexpanded ($P_{ij} < P_4$) or overexpanded ($P_{ij} > P_4$), the flow would adjust to a different matched pressure downstream of the rocket nozzle exit plane. In these cases, the flow areas at the initiation of mixing and combustion are not the physical areas, and a minor modification to the modeling should be made. For this discussion, P_{tij} and/or A_{ij} will be adjusted to yield $P_{ij} = P_4$.

B. Engine Analysis with Ideal Gases

The objective of the cycle calculation is to specify the areas and flow conditions that yield optimum performance under the assumption of complete mixing and combustion. Once these conditions are obtained and the sensitivity to changes from the optimum are determined, additional modeling, perhaps supported by new experiments, can be introduced to redefine optima. The revised models would include losses in the mixing and combustion. To enable the reader to readily check the calculations and to obtain a familiarity with the procedure that ultimately will be done on the computer, a number of simplifying assumptions will be introduced. They are the following:

1) The gas is treated as ideal gas at station 4 with $\gamma = 1.4$ and as a gas mixture at stations ij, 5, 6, 7, and 8 with $\gamma = 1.2$.

2) The molecular weights are $\mathcal{M}_4 = 28.965$ and $\mathcal{M}_{ij} = 7.35$; $\mathcal{M}_5$–$\mathcal{M}_8 = 16.3$. (These conditions correspond approximately to operating the rocket at $ER_{ij} = 3$.)

3) The bypass ratio $\beta = 2.2$.

4) The total temperatures are $T_{t4} = 450°\text{R}$, $T_{tij} = 4500°\text{R}$, and $T_{t5} = 4700°\text{R}$.

5) The total pressures are $P_{tij} = 50$ atm and $P_{t4} = 0.5$ atm.

It is convenient to first examine the problem by analyzing the flow from the plane of injection to the engine exit wherein the entrance conditions at 4 and ij are treated parametrically. In the first exercise, the effects of varying M_4 between 0.2 and 1.0 and M_5 between 0.8 and 1.0 are examined.

Conditions at 4 are obtained from the isentropic flow relationships with $\gamma = 1.4$. These, in turn, set the rocket nozzle exit conditions, which are obtained with $\gamma_{ij} = 1.2$:

$$T_4 = T_{t4}\left(1 + \frac{\gamma_4 - 1}{2}M_4^2\right)^{-1} = 450/(1 + 0.2\,M_4^2) \tag{1}$$

flow is sonic can be positioned in accordance with the flight conditions. In a one-dimensional sense this surface is often referred to as the "thermal throat." Control of this engine cycle is an issue, especially if premature choking occurs.

Whereas weight and volume must ultimately be examined to determine the viability of an ejector system for propulsion of a transatmospheric accelerator, a number of important issues regarding the aerothermodynamics of the flow paths can be examined by the use of propulsion-cycle analysis. These include 1) optimization of the ratio of secondary to primary flow rate (i.e., the bypass ratio β); 2) optimization of, and constraints on, the fuel/oxidizer (or fuel/air) equivalence ratio ER_{ij} of the primary flow; 3) matching the Match number of the secondary flow at the plane of injection with the inlet characteristics; 4) assessing the effects of the total pressure of the primary P_{tij} on engine performance; 5) assessing the tradeoffs between the use of a physical vs a "thermal" throat following mixing and combustion; 6) assessing the effects of simultaneous mixing and burning, vs mixing followed by afterburning; and 7) assessing the effects of different modeling of the wall forces in the mixing and combustion zones.

Items 1–3 and 5 will be addressed in this paper together with the introduction of candidate conceptual designs and flight performance comparisons with turbojet-powered systems. A companion paper covering items 4, 6, and 7 is in preparation.

The subsequent analysis will show that the critical design point of the ejector engine lies between Mach = 1–1.4 where the vehicle drag is maximum and inlet-air capture is minimum. The study will begin by examining two geometries of RRSJ engines operating at $M_0 = 1.25$. The first are axisymmetric geometries that represent podded engines. They are convenient geometries to address design variables when integration of the engine with the rest of the vehicle is not a consideration. Figure 2 represents a sectional view along the axis of symmetry of the simplest ejector system, a rocket in the duct with a normal shock inlet. The second are planar geometries that can be used to examine design variables as they are affected by integration issues. In this case, the sketch of Fig. 2 could be bifurcated along the centerline, and the top half would represent a sectional view of a planar engine of arbitrary width. Area ratios would simply be equal to height ratios; e.g., $A_4/A_0 = h_4/h_0$, etc.

In this general model, A_0 is the cross-sectional area of the flow in the freestream that is ingested into the engine. The projected cross-sectional area of the inlet in the plane of the cowl lip is A_i. In the plane of injection, the primary (i.e., rocket) has a flow area A_{ij}, and the secondary (i.e., ramjet) has a flow area A_4. Mixing and combustion take place downstream of the injection plane and are completed at flow area A_5. When conditions at 5 are subsonic; i.e., $M_5 < 1$, then the flow

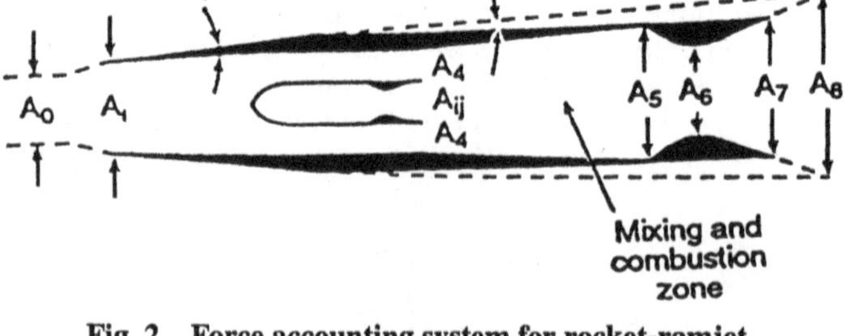

Fig. 2 Force accounting system for rocket-ramjet.

$$P_4 = P_{t4}\left(1+\frac{\gamma_4-1}{2}M_4^2\right)^{-\frac{\gamma_4}{\gamma_4-1}} = 0.5/(1+0.2M_4^2)^{3.5} \tag{2}$$

$$(M_{ij})^2 = \left(\frac{2}{\gamma_{ij}}-1\right)\left[\left(\frac{P_{t_{ij}}}{P_{ij}}\right)^{\frac{\gamma_{ij}-1}{\gamma_{ij}}}-1\right] = 10\left[\left(\frac{50}{P_4}\right)^{\frac{1}{6}}-1\right] \tag{3}$$

$$T_{ij} = T_{tij}\left(1+\frac{\gamma_{ij}-1}{2}M_{ij}^2\right)^{-1} = 4500/(1+0.1M_{ij}^2) \tag{4}$$

$$\frac{(\rho UA)_4}{(\rho UA)_{ij}} = \beta = 2.2; \qquad \frac{A_4}{A_{ij}} = 2.2\,\frac{P_{ij}}{P_4}\,\frac{T_4}{T_{ij}}\,\frac{M_{ij}}{M_4}\,\frac{M_{ij}}{M_4}\,\frac{a_{ij}}{a_4} \tag{5}$$

$$\frac{A_4}{A_{ij}} = 2.2\left(\frac{T_4\gamma_{ij}\quad_{ij}}{T_{ij}\gamma_4\quad_4}\right)^{\frac{1}{2}}\frac{M_{ij}}{M_4} = 1.02602\left(\frac{T_4}{T_{ij}}\right)^{\frac{1}{2}}\frac{M_{ij}}{M_4} \tag{6}$$

$$\frac{\Im_4}{\Im_{ij}} = \frac{A_4P_4(1+\gamma_4M_4^2)}{A_{ij}P_{ij}(1+\gamma_{ij}M_{ij}^2)} \tag{7}$$

Table 1 lists key properties at stations 4 and ij for the range $0.2 \le M_4 \le 1$. Figure 3 shows the large decrease in duct to injector exit area ratio from >8 at $M_4 = 0.2$ to about 1.7 at $M_4 = 1$. The subsequent discussion will show the large impact of A_4/A_{ij} on vehicle design, which dictates the need for high values of M_4.

Conditions at 5 are obtained from simultaneous solution of the continuity and momentum equations:

Continuity equation:

$$(\rho UA)_4 + (\rho UA)_{ij} = (\rho UA)_5 = \left[PAM\left(\frac{\gamma\mathcal{M}}{T}\right)^{\frac{1}{2}}\right]_5 \tag{8}$$

with

$$\beta = (\rho UA)_4/(\rho UA)_{ij} = 2.2$$

$$\frac{P_5A_5}{P_{ij}A_{ij}} = 3.2\frac{M_{ij}}{M_5}\left(\frac{\gamma_{ij}\mathcal{M}_{ij}T_5}{\gamma_5\mathcal{M}_5T_{ij}}\right)^{\frac{1}{2}} = 2.1488\frac{M_{ij}}{M_5}\left(\frac{T_5}{T_{ij}}\right)^{\frac{1}{2}} \tag{9}$$

Table 1 Flow properties at station 4 and ij

M_4	T_4, °R	P_4, atm	M_{ij}	T_{ij}, °R	A_4/A_{ij}	$\Im_4/\Im_{ij}$	P_{t4}/P_4
0.2	446.40	0.4862	3.412	2079.0	8.1109	0.57215	1.0283
0.4	436.04	0.4478	3.456	2050.7	4.0877	0.32632	1.1166
0.6	419.77	0.3920	3.526	2005.7	2.7584	0.26061	1.2755
0.8	398.94	0.3280	3.621	1947.0	2.1022	0.23818	1.5243
1.0	375.00	0.2641	3.737	1878.0	1.7133	0.23155	1.8929

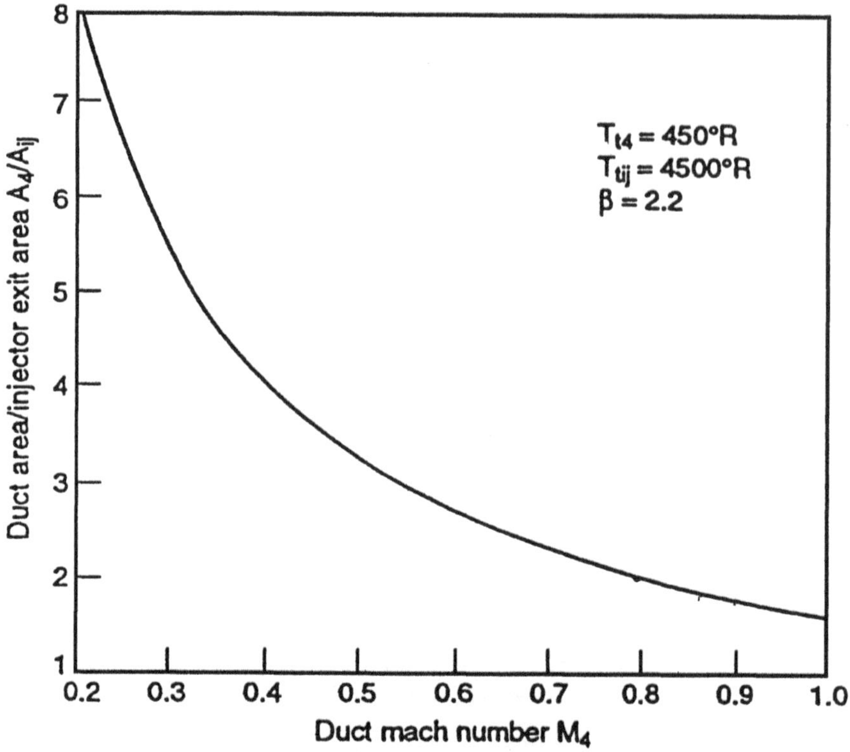

Fig. 3 Secondary to primary area ratio for matched pressure ejector.

Momentum equation:

$$[AP(1+\gamma M^2)]_4 + [AP(1+\gamma M^2)]_{ij} + \int P\,\mathrm{d}A - \tau_w \mathrm{d}A_w = [AP(1+\gamma M^2)]_5 \tag{10}$$

For $\tau_w = 0$ and assuming an area weighted average pressure,

$$\int P\,\mathrm{d}A = \bar{P}[A_5 - (A + A_{ij})] = \frac{P_5 + P_4}{2}[A_5 - (A_4 + A_{ij})] \tag{11}$$

$$\frac{A_5}{A_{ij}} = \frac{-(H) + \sqrt{(H)^2 - 4(K)}}{2} \tag{12}$$

where

$$H = \left[(E) - (G) - \left(\frac{A_4}{A_{ij}} + 1\right)\right] \tag{13}$$

$$E = \frac{P_5}{P_{ij}}\,\frac{A_5}{A_{ij}} \tag{14}$$

$$G = 2(1 + \gamma M_{ij}^2)[D - (1 + C)] \tag{15}$$

$$C = \frac{A_4 P_4 (1 + \gamma_4 M_4^2)}{A_{ij} P_{ij} (1 + \gamma_{ij} M_{ij}^2)} \tag{16}$$

$$D = \frac{P_5 A_5 (1 + \gamma_5 M_5^2)}{P_{ij} A_{ij} (1 + \gamma_{ij} M_{ij}^2)} \tag{17}$$

and

$$K = (E)\left(\frac{A_4}{A_{ij}} + 1\right) \tag{18}$$

Solutions to Eq. (12) are obtained for selected values of M_5 where E is obtained from continuity, Eq. (9):

$$\frac{P_{t_5}}{P_{t_4}} = \left(\frac{P_{t_5}}{P_5} \Big/ \frac{P_{t_4}}{P_4}\right) \frac{P_5}{P_{ij}}$$

$$= \left[1 + \frac{\gamma_5 - 1}{2} M_5^2\right]^{\frac{\gamma_5}{\gamma_5 - 1}} \Big/ \frac{P_5}{P_{ij}} \frac{\{1 + [(\gamma_4 - 1)M_4^2/2]\}^{\gamma_4/\gamma_4 - 1}}{\gamma_4 - 1}$$

$$= [1 + 0.1M_5^2]^6 \frac{P_5}{P_{ij}} / [1 + 0.2M_4^2]^{3.5} \tag{19}$$

$$T_5 = T_{t5}\left(1 + \frac{\gamma_5 - 1}{2} M_5^2\right)^{-1} = T_{t5}(1 + 0.1M_5^2)^{-1} \tag{20}$$

$$\frac{P_5}{P*} = \left(\frac{1.1}{1 + 0.1M_5^2}\right)^6 \tag{21}$$

$$\frac{A_5}{A*} = \frac{1}{M_5}\left[\frac{2 + (\gamma_5 - 1)M_5^2}{\gamma_5 + 1}\right]^{\frac{\gamma_5 + 1}{2(\gamma_5 - 1)}} = \frac{1}{M_5}\left[\frac{2 + 0.2M_5^2}{2.2}\right]^{5.5} \tag{22}$$

Table 2 lists the pertinent isentropic flow relationships at station 5 for $M_5 = 0.8, 0.9,$ and 1.0.

The area and pressure ratios at station 6 can then be found from

$$\frac{A_6}{A_{ij}} = \frac{A_5}{A_{ij}} \Big/ \left[\frac{A}{A*}\right]_5 \tag{23}$$

Table 2 Isentropic flow parameters at station 5

M_5	T_5, °R	$(A_0/A_0)_5$	$(P_0/P_0)_5$	$(P_t/P)_5$
0.8	4417.3	1.0409	1.2210	1.4509
0.9	4347.8	1.0096	1.1102	1.5957
1.0	4272.7	1.0	1.0	1.7716

Table 3 Key pressure and area ratios at stations 5 and 6

M_5	M_4	$(P_5A_5/P_{ij}A_{ij})$	A_5/A_{ij}	P_5/P_{ij}	P_{t5}/P_{t4}	$A_5/(A_4+A_{ij})$	A_6/A_{ij}	P_6/P_{ij}
0.8	0.2	13.359	9.1786	1.4554	2.0535	1.0074	8.8179	1.1920
	0.4	13.545	7.7079	1.7573	2.2834	1.5150	7.4050	1.4392
	0.6	14.055	6.9108	2.0338	2.3135	1.8288	6.6393	1.6698
	0.8	14.650	6.1747	2.3726	2.2584	1.9904	5.9321	1.9432
	1.0	15.394	5.5446	2.7764	2.1281	2.0435	5.3267	2.2739
0.9	0.2	11.781	8.8508	1.3311	2.0656	0.9715	8.7667	1.1990
	0.4	12.015	7.7224	1.5559	2.2235	1.5179	7.6490	1.4015
	0.6	12.395	6.8817	1.8012	2.2534	1.8310	6.8163	1.6224
	0.8	12.919	6.1802	2.0903	2.1882	1.9922	6.1214	1.8828
	1.0	13.576	5.5630	2.4404	2.0572	2.0505	5.5101	2.1982
1.0	0.2	10.511	8.7378	1.2029	2.0724	0.9590	8.7378	1.2029
	0.4	10.719	7.8275	1.3694	2.1727	1.5385	7.8275	1.3694
	0.6	11.059	7.0870	1.5605	2.1674	1.8856	7.0870	1.5605
	0.8	11.516	6.4256	1.7922	2.0830	2.0713	6.4256	1.7922
	1.0	12.112	5.8098	2.0848	1.9512	2.1412	5.8098	2.0848

$$\frac{P_6}{P_{ij}} = \frac{P_5}{P_{ij}} \Big/ \left[\frac{P}{P^*}\right]_5 \tag{24}$$

Table 3 lists the pertinent pressure and area ratios at stations 5 and 6 for M_4 values of 0.2, 0.4, 06, 0.8, and 1.0. Figure 4 is a plot of the mixing and burning duct area ratio vs duct entrance Mach number for three values of the duct exit Mach number. Duct area ratios increase from nearly constant area at entrance Mach numbers of 0.2 by a factor of 2 as $M_4 \to 1$. Total pressure rises as a result of the ejector pumping as shown in Fig. 5. Peak values lie between 2.18 at $M_5 = 1$ and 2.32 at $M_5 = 0.8$. These maxima occur in the $M_4 = 0.5$–0.6 region. The decreases in P_{t5}/P_{t4} with increasing M_4 and M_5 can be attributed to heat release occurring at higher average Much numbers.

Up to this point in the discussion, it has not been necessary to specify the freestream Mach number M_0. However, the choice of $\beta = 2.2$ and, to a lesser degree, T_{tij} and T_{t0} were based on values corresponding to near-optimal vehicle performance at $M_0 = 1.25$. The subsequent results and discussion will show that external drag is an important element in the design process. Vehicle drag is near maximum at $M_0 = 1.25$. For most applications, optimization of the vehicle at $M_0 = 1.25$ and the acceptance of some off-design loss in performance at $M_0 \neq 1.25$ yield the best overall design.

II. Inert Assumption

To proceed with the analysis, it is necessary to model the pressure recovery of the inlet. For this study the following relationship was adopted:

$$\frac{P_{t4}}{P_{t0}} = 1 - 0.0291M_0 - 0.0206M_0^2 \tag{25}$$

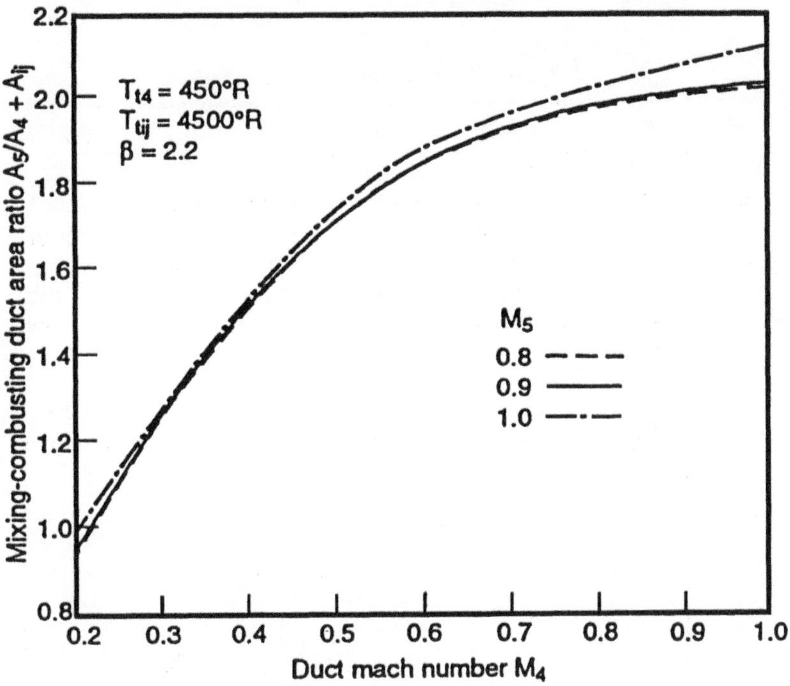

Fig. 4 Mixing duct area ratios as function of entrance and exit Mach number.

At $M_0 = 1.25$, $P_{t4}/P_{t0} = 0.93144$, whereas the loss across a single, normal shock at $M_0 = 1.25$ would be 0.98706. Thus, for near-critical operation of a simple, normal shock inlet, the internal loss is $1 - 0.93144/0.98706 = 0.05635$.

The pressure ratio P_0/P_4 can be written as

$$\frac{P_0}{P_4} = \frac{P_{t4}}{P_4}\frac{P_{t0}}{P_{t4}}\frac{P_0}{P_{t0}} = \frac{P_{t0}}{P_{t4}}\left[\frac{\left(2+(\gamma-1)M_4^2\right)}{\left(2+(\gamma-1)M_0^2\right)}\right]^{\frac{\gamma}{\gamma-1}} = 1.07362\left[\frac{2+0.4M_4^2}{2.625}\right]^{3.5} \tag{26}$$

From continuity between stations 0 and 4 and $T_{t_4} = T_{t_0}$,

$$\left[PAM\left(\frac{\gamma\mathcal{M}}{T_t}\right)^{\frac{1}{2}}\left(\frac{T_t}{T}\right)^{\frac{1}{2}}\right]_0 = \left[PAM\left(\frac{\gamma\mathcal{M}}{T_t}\right)^{\frac{1}{2}}\left(\frac{T_t}{T}\right)^{\frac{1}{2}}\right]_4 \tag{27}$$

$$\frac{A_0P_0}{A_4P_4} = \frac{M_4}{M_0}\left[\frac{2+(\gamma-1)M_4^2}{2+(\gamma-1)M_0^2}\right]^{\frac{1}{2}} \tag{28}$$

$$\frac{\Im_0}{\Im_4} = \frac{A_0P_0\left(1+\gamma M_0^2\right)}{A_4P_4\left(1+\gamma M_4^2\right)}; \qquad \frac{\Im_0}{\Im_{ij}} = \frac{\Im_0}{\Im_4}\frac{\Im_4}{\Im_{ij}} \tag{29}$$

Table 4 lists important flow parameters at stations 0, 4, and ij as a function of M_4 at $M_0 = 1.25$.

Table 4 Flow parameters at station 8

M_4	A_0P_0/A_4P_4	$\Im_0/\Im_4$	$\Im_0/\Im_{ij}$	P_0/P_4	A_0/A_4	A_0/A_{ij}
0.2	0.14021	0.42322	0.24215	0.42620	0.32898	2.6683
0.4	0.28375	0.73893	0.24113	0.46279	0.61313	2.5063
0.6	0.43380	0.91937	0.23960	0.52867	0.82055	2.2634
0.8	0.59331	0.99746	0.23758	0.63181	0.93906	1.9741
1.0	0.76495	1.01594	0.23524	0.78458	0.97498	1.6704

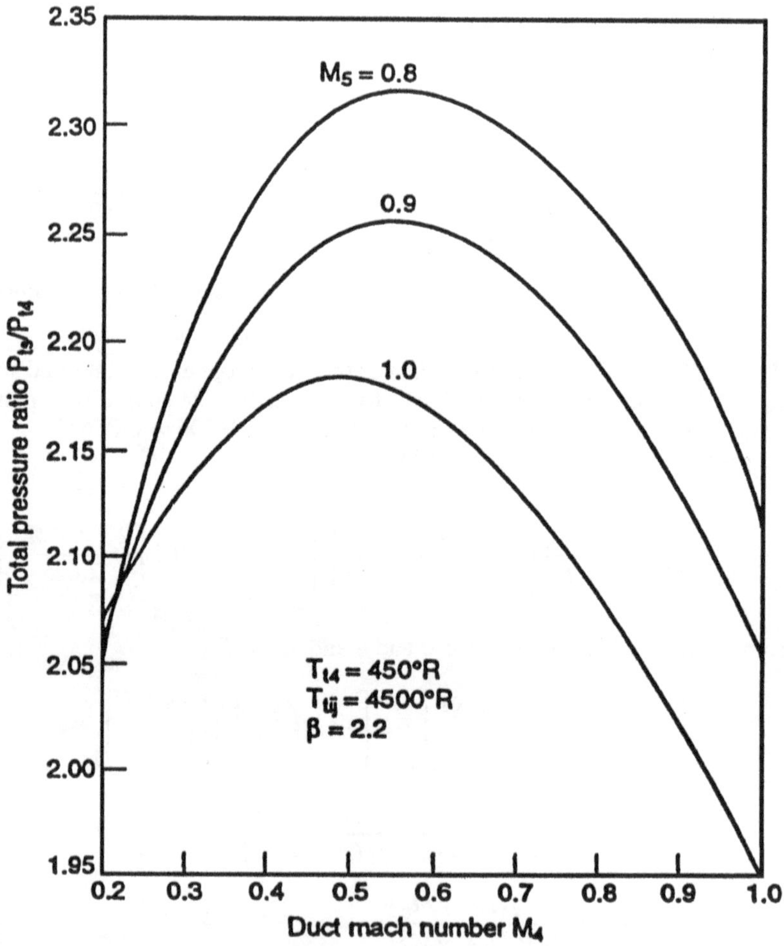

Fig. 5 **Mixing duct total pressure rise as function of entrance and exit Mach numbers.**

III. Nozzle Analysis

Assuming isentropic flow between stations 6 and 8 with $\gamma = 1.2$, the flowing relationships hold, viz.,

$$\frac{P_8}{P_{t8}} = \frac{P_8}{P_0}\frac{P_0}{P_{t0}}\frac{P_{t0}}{P_{t4}}\frac{P_{t4}}{P_{t5}}\frac{P_{t5}}{P_{t8}} = 1 \cdot 0.38606 \cdot 1.0736\frac{P_{t4}}{P_{t5}} \cdot 1 = 0.41448\frac{P_{t5}}{P_{t5}} \quad (30)$$

$$M_8^2 = 10\left[\left(\frac{P_{t_4}}{P_8}\right)^{\frac{1}{6}} - 1\right] \quad (31)$$

$$\frac{A_8}{A_6} = \frac{1}{M_8}\left[\frac{2 + (\gamma_8 - 1)M_8^2}{\gamma_8 + 1}\right]^{\frac{\gamma_8+1}{2(\gamma_8-1)}}$$

$$= \frac{1}{M_8}\left[\frac{2 + 0.2M_8^2}{2.2}\right]^{5.5} \quad (32)$$

$$\frac{A_8 P_8(1 + \gamma M_8^2)}{A_{ij} P_{ij}(1 + \gamma M_{ij}^2)} = \frac{A_8}{A_6}\frac{A_6}{A_{ij}}\frac{P_8}{P_{t_8}}\frac{P_{t_4}}{P_6}\frac{P_6}{P_{ij}}\frac{(1 + \gamma M_8^2)}{(1 + \gamma M_{ij}^2)} \quad (33)$$

$$\frac{\Im_8}{\Im_{ij}} = \frac{A_8 P_8(1 + \gamma_8 M_8^2)}{A_{ij} P_{ij}(1 + \gamma_{ij} M_{ij}^2)} \quad (34)$$

With expansion to ambient pressure, i.e., $P_8 = P_0$, Eqs. (30–34) are solved to yield the values for the flow parameters at station 8, which are listed in Table 5.

IV. Internal Engine Performance

The specific impulse of the rocket ejector is obtained from

$$I_{SP_{ij}} = \frac{\Im_{ij}}{w_{ij}} = \left[\frac{(PA(1 + \gamma M^2)}{PAM(\gamma g/RT)^{1/2}}\right]_{ij}$$

$$= \left\{\frac{(1 + \gamma M^2)T^{1/2}}{M \times (1.2 \times 32.174 \times 7.35/1543.324)^{1/2}}\right\}_{ij}$$

$$= 2.3321\left[\frac{(1 + 1.2M_{ij}^2)T_{ij}^{1/2}}{M_{ij}}\right] \quad (35)$$

Note that the term $P_0 A_{ij}$ does appear in the equation for $I_{SP_{ij}}$ because it is an external force and the rocket is totally enclosed in the overall engine. The specific impulse of the rocket ejector is given in Table 6 for values of M_4.

The specific impulse and thrust coefficients of the overall RRSJ engine cycle can now be computed from

$$I_{SP} = \left[\frac{\Im_8 - \Im_0 - P_0(A_8 - A_0)}{\Im_{ij}}\right] I_{SP_{ij}} \quad (36)$$

Table 5 Flow parameters at station 8

M_5	M_4	(P_8/P_{t8})	M_8	A_8/A_6	A_8/A_{ij}	$\Im_8/\Im_{ij}$	A_8/A_0	$[P_0(A_s - A_0)/\Im_{ij}]$
0.8	0.2	0.20184	1.748	1.4578	12.943	1.7196	4.851	0.29253
	0.4	0.18152	1.814	1.5596	11.549	1.7251	4.608	0.27294
	0.6	0.17916	1.822	1.5722	10.438	1.7276	4.612	0.27148
	0.8	0.18352	1.807	1.5499	9.194	1.7073	4.657	0.27260
	1.0	0.19476	1.770	1.4974	7.976	1.6773	4.775	0.27859
0.9	0.2	0.20066	1.752	1.4731	12.914	1.7220	4.840	0.29170
	0.4	0.18641	1.797	1.5354	11.744	1.7281	4.686	0.27883
	0.6	0.18393	1.806	1.5485	10.555	1.7225	4.717	0.27537
	0.8	0.18942	1.787	1.5212	9.312	1.6989	4.846	0.27705
	1.0	0.20148	1.749	1.4691	8.095	1.6705	4.833	0.28384
1.0	0.2	0.20000	1.754	1.4758	12.895	1.7225	4.833	0.29116
	0.4	0.19077	1.783	1.5155	11.863	1.7239	4.746	0.28330
	0.6	0.19123	1.782	1.5141	10.730	1.7143	4.741	0.28118
	0.8	0.19898	1.757	1.4798	9.509	1.6889	4.817	0.28449
	1.0	1.21242	1.716	1.4267	8.289	1.6603	4.962	0.29241

and

$$C_T = \frac{T}{q_0 A_0} = \left(\frac{\Im_8 - \Im_0 - P_0(A_8 - A_0)}{\Im_{ij}}\right) \times \left[\left(\frac{P_{ij}}{P_4}\right)\left(\frac{P_4}{P_0}\right)\left(\frac{P_0}{q_0}\right)\left(\frac{A_{ij}}{A_4}\right)\left(\frac{A_4}{A_0}\right)(1 + \gamma_{ij} M_{ij}^2)\right] \quad (37)$$

Table 7 lists the thrust coefficients and specific impulses for an ejector ramjet operating at $M_0 = 1.25$ for various values of M_5 and M_4. Several additional conclusions can be drawn from Table 7.

1) For $M_0 = 1.25$, an important design condition, the maximum engine specific impulse occurs at about $M_4 = 0.6$ with a small decrease between $M_4 = 0.6$ and 0.8.

2) The thrust augmentation, which is defined as $[\Im_8 - \Im_0 - P_0(A_8 - A_0) - \Im_{ij}]/\Im_{ij}$ for the cases examined, lies between 13 and 22%.

3) Engine specific impulses for subsonic end states (i.e., $M_5 < 1$) can be somewhat higher than for $M_5 = 1$, but the differences are small and suggest that the

Table 6 Specific impulse of rocket ejector

M_4	M_{ij}	$I_{sp_{ij}}$, lb$_f$s/lb$_m$
0.2	3.412	466.84
0.4	3.456	468.84
0.6	3.526	471.85
0.8	3.621	475.86
1.0	3.737	480.57

Table 7 Specific impulse and thrust coefficients for an RRSJ at $M_0 = 1.25$

M_5	M_4	$\frac{\Im_8 - \Im_0 - P_0(A_8 - A_0)}{\Im_{ij}}$	I_{sp}, $lb_j s/lb_m$	$C_t = (T/q_0 A_0)$
0.8	0.2	1.1838	552.7	14.248
	0.4	1.2110	567.8	14.637
	0.6	1.2165	574.0	14.797
	0.8	1.1971	569.7	14.684
	1.0	1.1635	559.1	14.414
0.9	0.2	1.1882	554.0	14.300
	0.4	1.2082	566.5	14.602
	0.6	1.2075	569.8	14.688
	0.8	1.1843	563.6	14.527
	1.0	1.1518	553.5	14.268
1.0	0.2	1.1892	555.2	14.313
	0.4	1.1995	562.4	14.498
	0.6	1.1935	563.1	14.517
	0.8	1.1668	555.2	14.313
	1.0	1.127	544.4	14.032

simplicity of eliminating a physical throat may lead to a more optimal engine design.

The internal thrust of the engine, however, is but one element of the design consideration. The external forces on the ducting required to produce the M_4, M_5, and M_8 values and, in turn, the geometric implications regarding mixing must be examined. To study these elements, axisymmetric and planar geometries of air-augmented engines similar to that Fig. 2 operated at $M_0 = 1.25$ will be examined. For this comparison, identical rocket motor ejectors will be enclosed in different ramjet ducts. All engines operate with $\beta = 2.2$; thus all have the same A_0, and all are assumed to be operating critically; i.e., the normal shock is located in the cowl lip plane.

Figure 6 shows schematic illustrations of four axisymmetric engine geometries at $M_0 = 1.25$; Figs. 6a and 6b are for $M_4 = 0.2$ and $M_5 = 0.8$ and 1.0, respectively; Figs. 6c and 6d are for $M_4 = 0.8$ and $M_5 = 0.8$ and 1.0 respectively. Key dimensions are given in Table 8. External and internal cowl lip angles of 10 and 7.5 deg, respectively, were chosen, and the mixing-combustion duct lengths were arbitrarily set as $\ell/[r_{ij} + r_4] = 9$. Layouts of the entire engine including the exit nozzle were made to determine the body flare angles and surface areas. The corresponding body flare angles are $\alpha_2 = 0.82$ and 1.80 deg for the $M_4 = 0.2$ and 0.8 cases, respectively. At the scale of the drawing, the presence of the physical throat for cases a and c is barely discernible. Indeed, in an actual engine cycle, whether the flow would "recognize" the throat and thereby be constrained to Mach 1 at this location is debatable.

V. External Drag Consideration

The external drag of the vehicle is composed of wave drag on the cowl and body flow plus friction of the external surfaces. Base drag, when it exists, on these simple geometric shapes is included in the internal thrust of the engine.

Table 8 Dimensions for sketches shown in Fig. 6 and for axisymmetric configuration

Sketches in Fig. 6								
M_4	M_5	A_4/A_0	A_{ij}/A_0	A_{ij}/A_0[a]	A_5/A_0	A_6/A_0	A_8/A_0	$(A_4 + A_{ij})/A_0$,
0.2	0.8	3.040	0.37477	0.02911	3.4399	3.3047	4.8508	3.4148
	1.0	3.040	0.37477	0.02911	3.2743	3.2743	4.8327	3.4148
0.8	0.8	1.0648	0.50656	0.02911	3.1278	3.0050	4.6574	1.5714
	1.0	1.0648	0.50656	0.02911	3.2550	3.2550	4.81167	1.5714
Axisymmetric configuration								
M_4	M_5	d_4/d_0[b]	d_{ij}/d_0	d_{ij}/d_0[a]	d_5/d_0	d_6/d_0	d_8/d_0	$(d_4 + d_{ij})/d_0$,
0.2	0.8	1.7436	0.61218	0.17061	1.8546	1.8179	2.2024	1.8479
	1.0	1.7436	0.61218	0.17061	1.8096	1.8096	2.1983	1.8479
0.8	.08	1.0319	0.71173	0.17061	1.7686	1.7335	2.1581	1.2535
	1.0	1.0319	0.71173	0.17061	1.8042	1.8042	2.1946	1.2535

[a]Refrence dimension, which is not used in sketches.

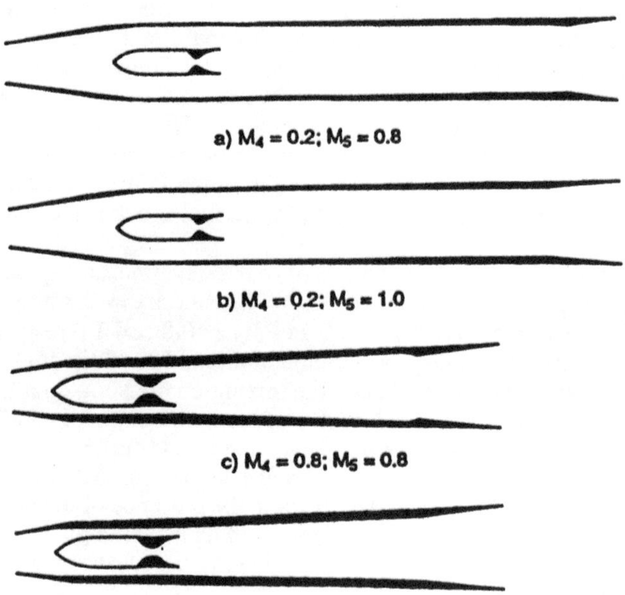

a) $M_4 = 0.2$; $M_5 = 0.8$

b) $M_4 = 0.2$; $M_5 = 1.0$

c) $M_4 = 0.8$; $M_5 = 0.8$

d) $M_4 = 0.8$; $M_5 = 1.0$

Fig. 6 Rocket-ramjet engine configuration, $M_0 = 1.25$.

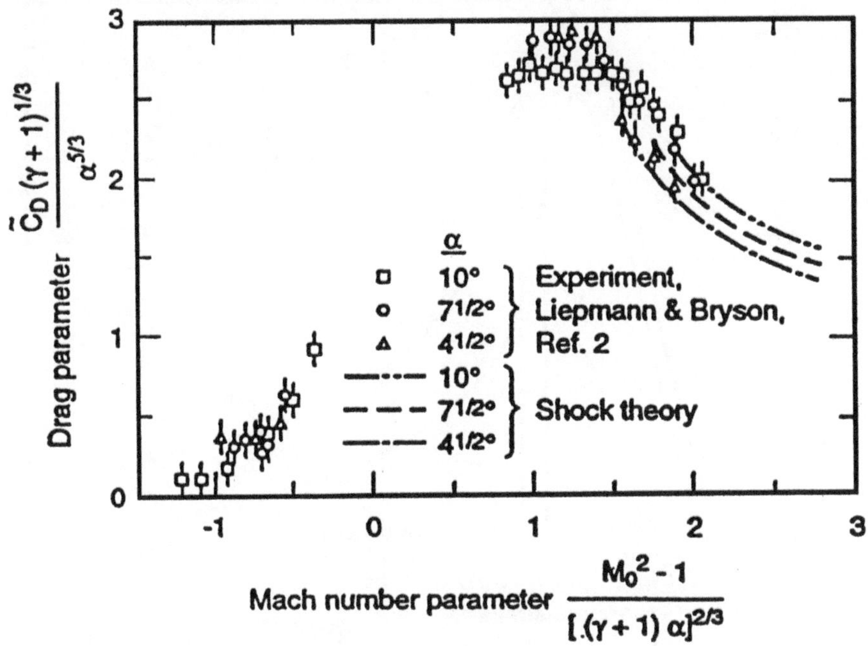

Fig. 7 Drag on wedge-shaped nose in transonic flow.

With $\alpha_1 = 10$ deg and α_2 and the projected area defined, the external wave drag can be calculated. Here it is assumed that the flow over the cowl is "wedgelike" and that drags can be obtained from transonic similarity data. Figure 7, taken from shapiro,[6] is based on the experiments of Leipmann and Bryson[7] at low M_0 and simple shock theory for higher M_0. The drag coefficient C_D shown is based on planform area. Drag coefficients based on projected areas for $\delta = 7.5$ and 10 deg and $\gamma = 1.4$ with flight Mach number as the abscissa are shown in Fig. 8. The friction drag on the cowl and flare can also be obtained for a prescribed friction coefficient. For these calculations $C_{Df} = 0.002$ is assumed.

The engine external drags are obtained from the following relationships:

Cowl–wave drag:

$$C_{D\text{cowl}} = \left[\frac{\bar{P}_{\text{cowl}}}{P_0} - 1\right]\frac{P_0}{q_0} \tag{38}$$

$$D_{\text{cowl}} = C_{D\text{cowl}} q_0 [A_4 + A_{ij} - A_0] \tag{39}$$

$$\frac{D_{\text{cowl}}}{\Im_{ij}} = \frac{C_{D_{\text{cowl}}}(q_0/P_{ij})\left[(A_4 + A_{ij} - A_0)/A_{ij}\right]}{\left[1 + \gamma_{ij}M_{ij}^2\right]} \tag{40}$$

For $M_0 = 1.25$ and $\alpha_1 = 10$ deg, $C_{D_{\text{cowl}}} = 0.675$ (Fig. 8), $[q/P]_0 = 1.09375$, and $(P_4/P_{ij}) = 1$:

$$\frac{D_{\text{cowl}}}{\Im_{ij}} = 0.7383\frac{P_0}{P_4} \times \left[\frac{A_4}{A_{ij}} + 1 - \frac{A_0}{A_{ij}}\right] \Big/ \left[1 + \gamma_{ij}M_{ij}^2\right] \tag{41}$$

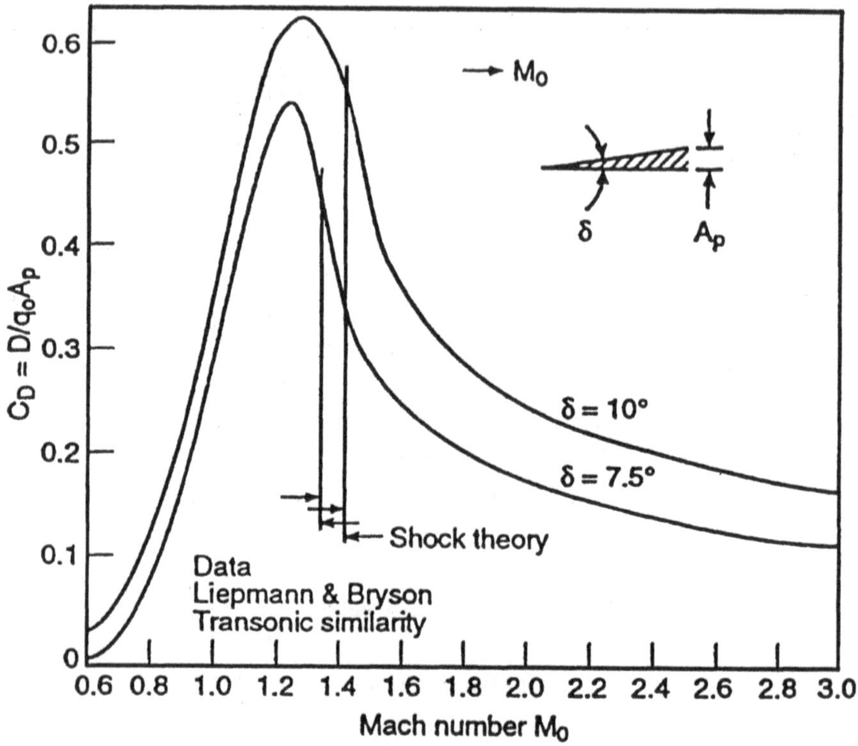

Fig. 8 Cowl pressure drag.

Flare–wave drag:

$$C_{D_{\text{flare}}} = \frac{\left[(\bar{P}_{\text{flare}}/P_0) - 1\right]}{q_0} P_0 \tag{42}$$

$$\frac{D_{\text{flare}}}{\Im_{ij}} = \frac{1.09375\ C_{D\text{flare}} P_0 (A_8 - A_4 - A_{ij})/P_4}{\left[1 + \gamma_{ij} M_{ij}^2\right]} \tag{43}$$

Engine external-friction drag

$$D_f = C_{d_f} q_0 A_{\text{surface}} = C_{D_f} \left[\frac{q}{P}\right]_0 P_0 A_S \tag{44}$$

where

$$A_{\text{surface}} \simeq \pi \ell_{\text{eng}} \left[\frac{d_0 + d_8}{2}\right] \tag{45}$$

$$\frac{D_f}{\Im_{ij}} = \frac{0.00218 A_S P_0 / A_{ij} P_4}{\left[1 + \gamma_{ij} M_{ij}^2\right]} \tag{46}$$

VI. Engine Total Drag

The total engine drag is composed of the cowl drag, the wave drag on the flared body, and the friction drag on the external surfaces.

$$D_{\text{total}} = D_{\text{cowl}} + D_{\text{flare}} + D_{\text{friction}} \tag{47}$$

Whereas the drag on the engine is an important element in determining the optimal values of M_4 and M_5, the effects of overall vehicle drag can be even more important when the engine is relatively small; i.e., A_V/A_0, is large, where A_V is the cross-sectional area of the vehicle:

$$\frac{D_v}{\Im_{ij}} = C_{D_v}\frac{A_V}{A_0}\frac{q_0}{P_0}\frac{P_0}{P_4}\frac{A_0}{A_{ij}}/[1+\gamma_{ij}M_{ij}^2] \tag{48}$$

Table 9 lists the wave, friction, and total drags normalized to the injector exit-stream thrust. Calculations are for $M_5 = 1.0$; the differences for $M_5 = 0.8$ and 0.9 are trivial.

Two additional specific impulse values can now be defined. They are

$$I_{\text{eff}} = \frac{T - D_{\text{total}}}{\dot{w}_f} \tag{49}$$

and

$$\tilde{I}_{\text{eff}} = \frac{T - D_{\text{total}} - D_{\text{vehicle}}}{\dot{w}_f} \tag{50}$$

The vehicle drag can either be treated parametrically or a particular case can be examined. For the value shown in Table 10 and Fig. 5, $\tilde{I}_{\text{eff}}$ is based on a vehicle that has a projected frontal area $A_v = 7A_0$ and a drag coefficient based on that frontal area = 0.65. Thus, the effective vehicle drag coefficient based on A_0 is 4.55. The corresponding net force coefficients are

$$\tilde{C}_F = \frac{\Im_8 - \Im_0 - D_{\text{total}} - D_{\text{vehicle}}}{q_0 A_0} \tag{51}$$

$$C_F = \frac{\Im_8 - \Im_0 - P_0(A_8 - A_0) - D_{\text{total}}}{q_0 A_0} \tag{52}$$

With the aid of Figs. 6 and 9 and Tables 9 and 10, some additional conclusions and comments can be made.

Table 9 Engine drag at $M_0 = 1.25$ and $M_5 = 1.0$

M_4	α_2	C_{d_s}	A_s/A_0	$D_{\text{cowl}}/\Im_{ij}$	$D_{\text{flare}}/\Im_{ij}$	$D_{\text{friction}}/\Im_{ij}$	$D_v/\Im_{ij}$
0.2	0.82	0.0394	24.03	0.13541	0.0046	0.0040	0.14401
0.4	1.15	0.0561	20.55	0.05752	0.0121	0.0034	0.07302
0.6	1.47	0.0723	17.07	0.03666	0.0178	0.0030	0.05746
0.8	1.80	0.0906	13.59	0.03145	0.0233	0.0022	0.05695
1.0	2.13	0.1089	10.11	0.03402	0.0282	0.0016	0.06382

Table 10 Effective specific impulses and net thrust coefficients

M_5	M_4	$D_{vehicle}/\Im_{ij}$	I_{eff}, $lb_f s/lb_m$	$\tilde{I}_{eff}$, $lb_f s/lb_m$	C_F	$\tilde{C}_F$
0.8	0.2	0.5401	485.5	309.0	12.514	7.964
	0.4	0.5378	533.6	357.1	13.754	9.204
	0.6	0.5344	546.9	370.4	14.098	9.548
	0.8	0.5299	542.6	366.1	13.985	9.435
	1.0	0.5247	528.4	351.9	13.624	9.074
0.9	0.2	0.5401	486.9	310.6	12.567	8.017
	0.4	0.5378	532.3	355.8	13.719	9.169
	0.6	0.5344	542.7	366.2	13.989	9.439
	0.8	0.5299	536.5	360.0	13.828	9.278
	1.0	0.5247	522.8	346.3	13.477	8.927
1.0	0.2	0.5401	488.0	311.5	12.580	8.030
	0.4	0.5378	528.2	351.7	13.615	9.065
	0.6	0.5344	536.0	359.5	13.818	9.268
	0.8	0.5299	528.1	351.6	13.614	9.064
	1.0	0.5247	513.7	337.2	13.241	8.691

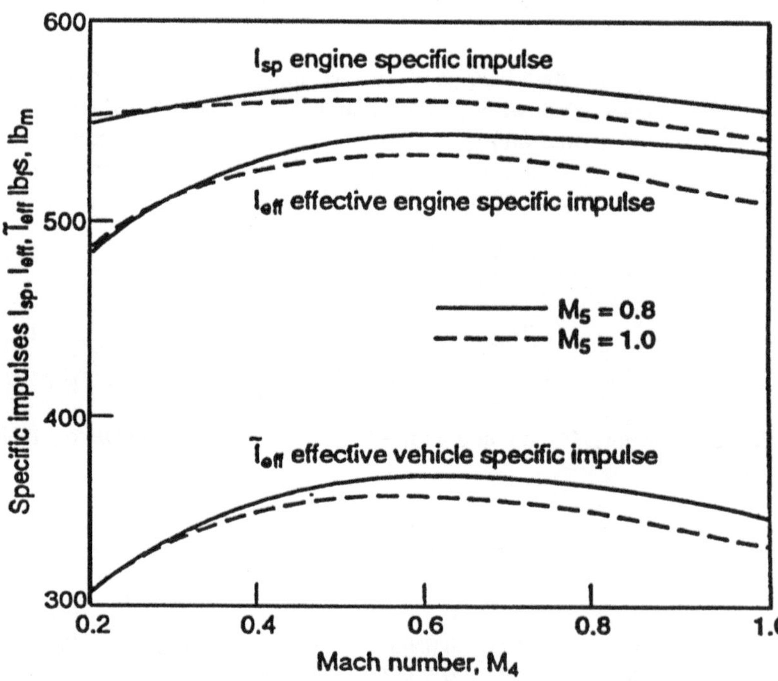

Fig. 9 Mach number M_4 in secondary stream at plane of injector.

1) The inclusion of engine and vehicle drag tends to strengthen the arguments for the desirability of operating at $M_4 > 0.6$ and $M_8 = 1$.

2) When the constraint of a fixed $A_0 = A_i$ is adopted, the principal geometric differences in the configuration are the much larger frontal portions of the engines and the greater length of the $M_4 = 0.2$ designs. The length argument is a consequence of the assumption of a mixing and combustion length that scales linearly with $d_4 = d_{ij}$. Considerable attention, both experimental and analytical, must be directed toward resolution of this issue.

3) The thrust coefficients of the RRSJ systems when operated at a relatively low β values are very high compared with all-air-breathing propulsion items. (If a turbojet having the same $A_0 - A_i$ and A_v would be substituted for the ducted rocket, it would not produce thrust greater than drag at $M_0 = 1.25$.)

4) The benefits that accrue with higher M_4 are much more pronounced when a fixed A_4 or $A_4 + A_{ij}$ is the imposed constraint. With this constraint, the engine becomes relatively smaller as M_4 is reduced, and the impact of external drag becomes overwhelming. For applications that require operation over a wide range of Mach numbers, a variable geometry inlet will be required. In most concepts, $A_0 = A_i$ is the principal variant, and A_4 has some fixed maximum value and a limited range of variation. The engine is operated at maximum A_4 at transonic speeds and is reduced, with limits, at higher speeds.

In a subsequent publication, the results for simple constant-specific heat calculations will be presented that cover the optimization of β, the effects of varying P_{tij} and of modifying the model for the internal pressure distribution, and the assessment of processes where mixing precedes combustion. The remainder of this paper will discuss the evolution of a design approach for a rocket based combined cycle engine (RBCC) engine. Performance estimates for this engine over the low-speed range will be based on thermochemical equilibrium at principal stations in the flow. These results support the contention that the simple ideal gas calculations provide excellent guidance for engine design and good approximations of performance. This RRSJ engine, designed by the Johns Hopkins University Applied Physics Laboratory (JHU/APL), is the propulsion system for an air-breathing transatmospheric accelerator. Its genera were a ram-scramjet aerospace vehicle proposed to the U.S. Air Force in 1960[8] and an augmented rocket proposed to double the range or payload of the U.S. Navy's sea-launched ballistic missiles.[9,10] Recent U.S. patents[11,12] have disclosed important features of the designs that are presented herein. During the intervening 30–35 years JHU/APL has enjoyed the support of the U.S. Navy, NASA, and the U.S. Air Force in developing the enabling technology that establishes the veracity of the engine design.

It is convenient to subdivide the description of the engine design, its performance, and its operability into four speed ranges; low-, low-mid, high-mid, and high-speed ranges. In the low-speed range, M_0=0–3, the engine evolves from an air-augmented rocket to a subsonic combustion ramjet. In the low-mid-speed range, M_0=3–7, the engine operates as a dual-mode ramjet transitioning to a scramjet. In the high-mid-speed range, M_0=7–15, the engine is a scramjet. From M_0=15 to orbit, the high-speed range, the engine transitions from a scramjet back to a rocket. In the parlance of some authors, the engine in the high-speed range is referred to as a LOX-augmented scramjet.

VII. Evolution of a Design Approach—Conceptual Design

A 400,000-lb class gross weight cargo carrier to space will be used to exemplify the conceptual design. Emphasis in this study is on the characteristics of the propulsion flow path, not on the specifics of the vehicle. Only the general geometric features of the vehicle, needed to quantify the performance, will be discussed. Consequently, some of the results will be presented in a parametric form, which will permit their use on a variety of vehicle designs and/or climb trajectories. Figure 10 is an enlarged view of the midsection of the vehicle showing the main engine ducts. It is the midsection of the vehicle that is 160 ft in length and has a midbody width of 32.4 ft. The 307-in. long engine cowl translates vertically with maximum and minimum duct heights of 21 and 5 in., respectively, to provide the varying inlet contraction ratio that is required to obtain good engine performance over a broad Mach number range. The 114-in. long cowl lip section flap rotates to provide modulation of the captured air mass flow, variable cowl wave compression, and inlet closure during final access to orbit, in orbit, and on re-entry. The cowl leading edge is in its maximum outboard position when the cowl lip section is parallel to the 11.5-deg compression surface. The corresponding maximum capture area of 210.3 ft^2 is used as the inlet reference area A_i. The 97-in.-long cowl trailing edge rotates to prevent overexpansion of the internal flow and a corresponding large base drag at very low flight speeds.

With the cowl in the outboard position, the projected area of the engine in the plane of the trailing edge is $A_e = 452$ ft^2 at an angle of attack of 0 deg and 395.7 ft^2 at $\alpha = 2$ deg. In the inboard position, the projected areas reduce to 414.4 and 357.2 ft^2 at $\alpha = 0$ and 2-deg respectively. The corresponding vehicle reference expansion ratios A_e/A_i are 2.15/1.88 at $\alpha = 0$ deg and 1.97/1.70 at $\alpha = 2$ deg for the outboard and inboard cowl positions, respectively.

A boundary-layer bleed duct lies inboard of the main propulsion duct. The louvered closeoff door seals this duct at M_0 less than about 1.05 and at high flight Mach number when the flow is no longer needed to reduce or eliminate base drag. A controllable flap at the exit of the duct is used to regulate flow into the duct.

Figure 11 shows more detail of the geometry of the internal engine. Internal compression is obtained from the cowl lip when the lip is rotated outward while the

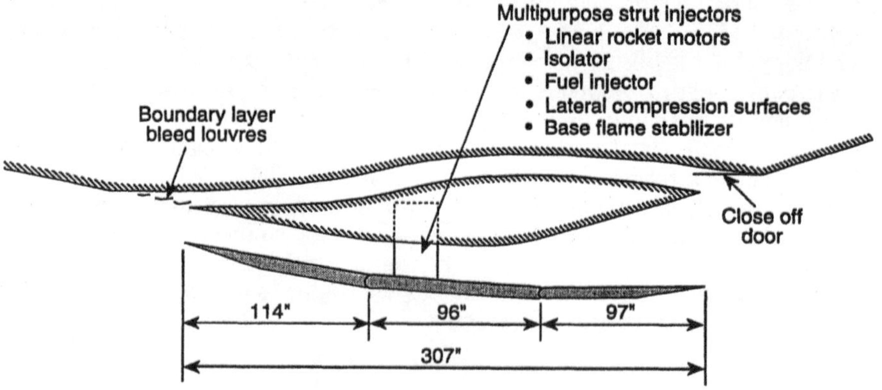

Fig. 10 Midsection of transatmospheric accelerator.

cowl translates inboard. Compression in the lateral plane is provided by the multipurpose retractable instream injectors. In addition to the compression on the 10-deg one-half angle wedge leading edges, the strut injectors serve as insolation ducts to prevent combustion-induced disturbances from disrupting the flow in the inlet. The aft portion of the injectors are segmented linear rocket motors. At low flight Mach numbers, when the engine is operating in the air-augmented mode, these injectors function as rockets. At higher Mach numbers, oxygen is not supplied, and they serve as tangential hydrogen injectors. The vertical segmentation provides the dual capability to reduce rocket thrust as the engine transitions from the air-augmented mode to the ramjet mode and to turn off a portion of the injectors when the cowl translates inboard. The engine is subdivided into six, 50-in.-wide engine modules. Each module houses nine, 0.933-in.-thick, 25-in.-long injectors spaced 5.00 in. on center. The side walls of each bay also contain rocket/hydrogen injectors having a width one-half that of the instream injectors. The resulting internal contraction is 15%, which permits self-starting of the inlet at about Mach 2 for vehicle angles of attack typical of flight. Each isolator duct has a width of 4.17 in. and a length of 22.69 in., which results in a $\ell/w = 5.43$. Additional rocket/hydrogen injectors are located in the cowl and engine body in the plane of the exit of the instream injectors. The instream injectors are fully retracted from the stream at high Mach number to reduce internal shear, drag, and heat transfer and to assure survivability.

Transverse hole hydrogen injectors are located upstream of the rocket motors in the instream struts, in the side walls, and on the internal sides of the cowl and body. Hole injectors are also located on the surface of the expansion nozzle downstream of the exhaust of the boundary-layer duct. These downstream injectors supply the fuel for base burning of boundary-layer bleed air.

Engine operability and performance are dependent on the geometry of the vehicle forebody, which serves as the external compression surface of the engine inlet. The planform of the vehicle has a spatular shaped forebody that blends into a near-to-planar compression surface. The compression surface is composed of a 351-in.-long plane having a wedge angle of 4 deg. and a curved surface to provide near-to-isentropic compression that terminates in a 11.5-deg, 98.95-in.-long wedge surface.

VIII. Low-Speed System Design, Performance, and Operability

One of the complexities in the design of the engine for a transatmospheric accelerator is the interdependence of the engine and airframe, thrust and drag, weight and flight path. The approach taken herein is the design of an engine wherein the flow rates of the propellants can be modulated to accommodate a broad range of vehicle drags, weights, and angles of attack on a variety of flight paths.

To initiate the design, calculations of engine performance are carried out to determine optional fuel/oxidizer flow rates for a broad range of input parameters. For the low-speed cycle, the more important parameters are the inlet total pressure recovery, P_{t4}/P_{t0}, the bypass ratio β, the equivalence ratio of the injected propellants ER_{ij}, and the overall area ratio of the mixing and combustion process $A_5/(\tilde{A}_4 + \tilde{A}_{ij})$.

The perfect gas calculations have provided guidance in the selection of key features of the conceptual design and in limiting the breadth of parameters that

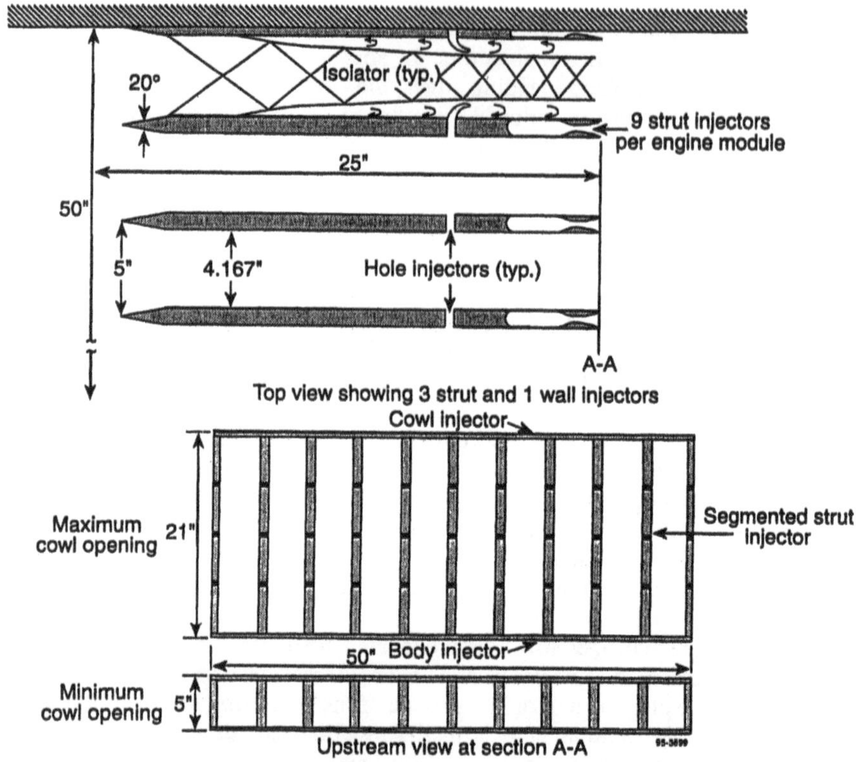

Fig. 11 Detailed view of strut injectors.

needed to be examined to optimize the scheduling of propellant flow rates. The addition of a converging section downstream of the combustion, which thereby permits subsonic Mach numbers at the end of the mixing and combustion zone M_5, yields only small gains in engine specific impulse. These gains could only be realized with the addition of a variable area converging-diverging nozzle and are not sufficient to justify the additional weight, complexity, and cost. Consequently, this engine uses the concept of the "thermal throat" to set $M_5 = 1$ during the low-speed and low-mid-speed modes of engine operation. Precise control of the injection, mixing, and combustion, together with rotation of the aft cowl flap, is needed to optimally locate the thermal throat. Less than optimal control can be tolerated; i.e., relatively slower rates of the mixing and combustion yield $M_5 > 1$ and an associated loss in performance. At $M_0 \leq 2$ when the engine inlet is unstarted, relatively faster mixing and combustion rates would cause inlet spillage, $M_5 = 1$ and $A_5 < A_{5\text{opt}}$, and there is a loss in thrust. At $M_0 > 2$, the inlet is started, and higher than optimal combustion rates would cause premature engine choking and an inlet unstart with a larger loss in thrust. Experimental and more rigorous analytical studies will be required to verify the existing design or to guide a redesign of the injector/combustor to assure the desired control of the thermal throat.

Many other studies, e.g. Ref. 13, have established guidelines for the efficiency of the inlet compression process required to yield competitive performance for air-breathing cycles. This engine uses the translation of the cowl and the rotation of the cowl lip to obtain the guideline performance. A relatively simple analysis was used in this study to assess the inlet performance. Verification of the results must await more complex computational fluid dynamics solutions of the flowfields and/or wind-tunnel tests of the inlets. The analysis used herein began with inviscid calculations of the shock losses and turning on the forebody assuming planar flow for $M_0 \geq 1.125$. For $M_0 < 1.125$ the flow on the forebody is governed by the internal engine flow, which is set by the ejector pumping of the strut rockets and the rotation of the cowl lip. Maximum engine specific impulse was the criterion for selecting the Mach number in the cowl lip plane for $0 \leq M_0 \leq 1.125$. At all flight conditions the cowl lip rotation is set to minimize cowl drag. For $1.25 \leq M_0 \leq 2.00$ the flow is supersonic in the cowl lip plane; i.e., $M_2 \geq 1.0$, the inlet is unstarted, and the cowl is rotated to maintain "shock-on-lip" inlet operation. In this speed range, the Mach number downstream of the terminal normal shock is less than 0.8, and the cowl lip is rotated to provide an increasing area in the upstream portion of the internal duct. At $M_0 > 2$ the inlet is started, and the forebody flow is unaffected by the internal flow. Figure 12 shows the Mach number and pressure ratio in the cowl lip plane as a function of M_0 for angles of attack of 0 and 2 deg. The higher pressure ratio and corresponding lower Mach number at $\alpha = 2$ deg is a result of the stronger shock on the spatular leading edge.

With flow conditions defined at the cowl lip and the inlet started ($M_0 > 2$), calculations were made with the cowl lip aligned with the flow and at various incidence angles to the flow. Solutions were found for properties of the flow following the shock train (Fig. 11) using integral methods and stream-thrust averaging. Figure 13 compares results of the calculated inviscid pressure recoveries for an unstarted inlet with those for a started inlet with no cowl compression and a started inlet with cowl compression for $\alpha = 0$ and 2 deg. The curves marked "Engine Design" are the modeled values of pressure recovery that should adequately account for the viscous losses and nonoptimal wave cancellation not included in the inviscid analysis.

The models are the following: for $\alpha = 0$ deg,

$$\frac{P'_{t4}}{P_{t0}} = 0.96 - 0.02586 M_0^2 \tag{53}$$

and for $\alpha = 2$ deg,

$$\frac{P'_{t4}}{P_{t0}} = 0.96 - 0.02272 M_0^2 \tag{54}$$

where station 4′ corresponds to mean flow conditions at the downstream end of the isolator duct. Note that Eq. (54) is a modification of Eq. (25) that reflects more realistic modeling as $M_0 \to 0$.

The inlet starts at a Mach number somewhat greater than 2, and the cowl forward flap begins to rotate above Mach 2.25. Note that the boundary-larger bleed duct opens at a Mach number just greater than 1, and so forebody viscous losses do not degrade the pressure recovery in the main engine flow path at $M_0 > 1$. Equations (54) and (55) were used for P'_{t4} in the cycle analysis. Complementary calculations made with lower values of pressure recovery at $M_0 = 3$ showed that large losses in performance would accrue if the inlet had to be operated unstarted.

Cross-sectional areas, duct heights, length of the mixing and combustion zone, and cowl flap angle are depicted in Fig. 14. The lower inset shows the air-capture

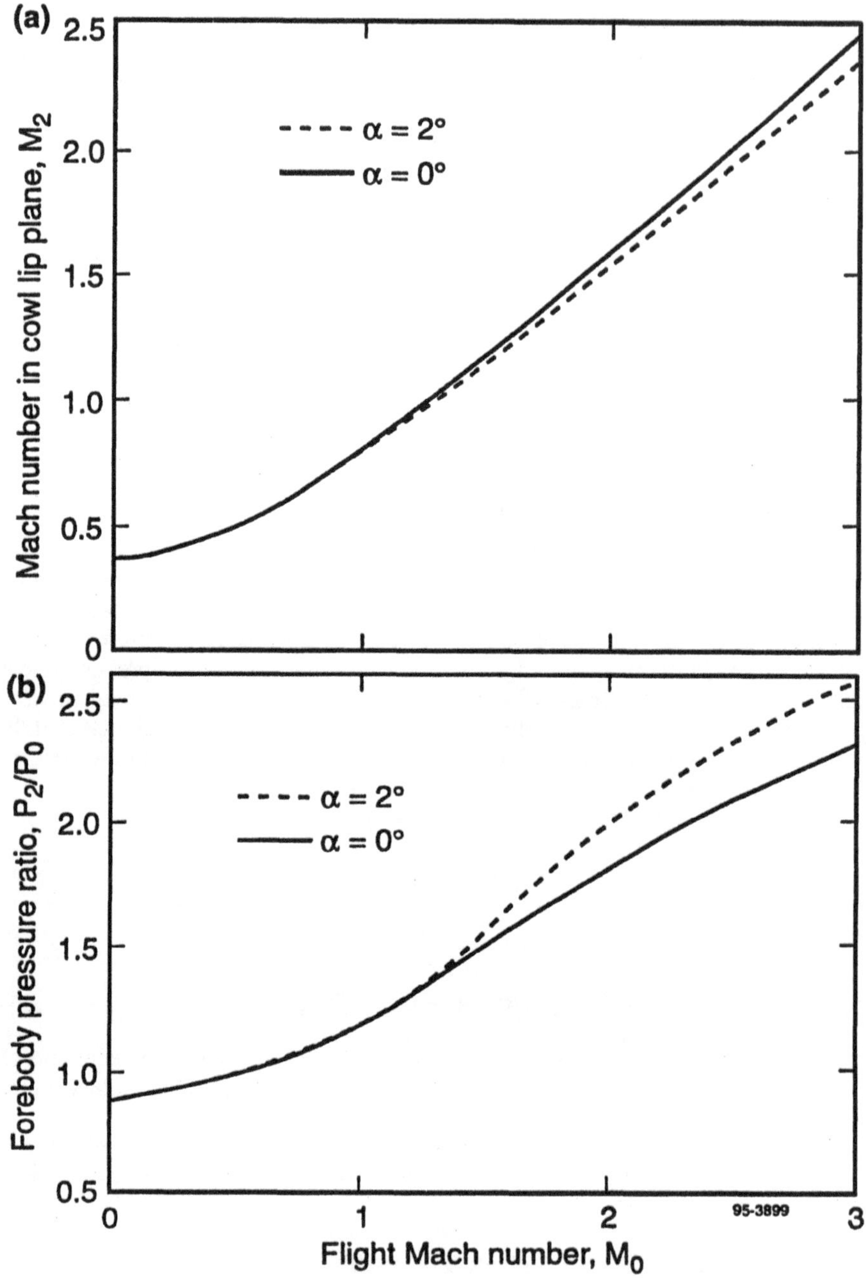

Fig. 12 Flow properties in plane of cowl lip.

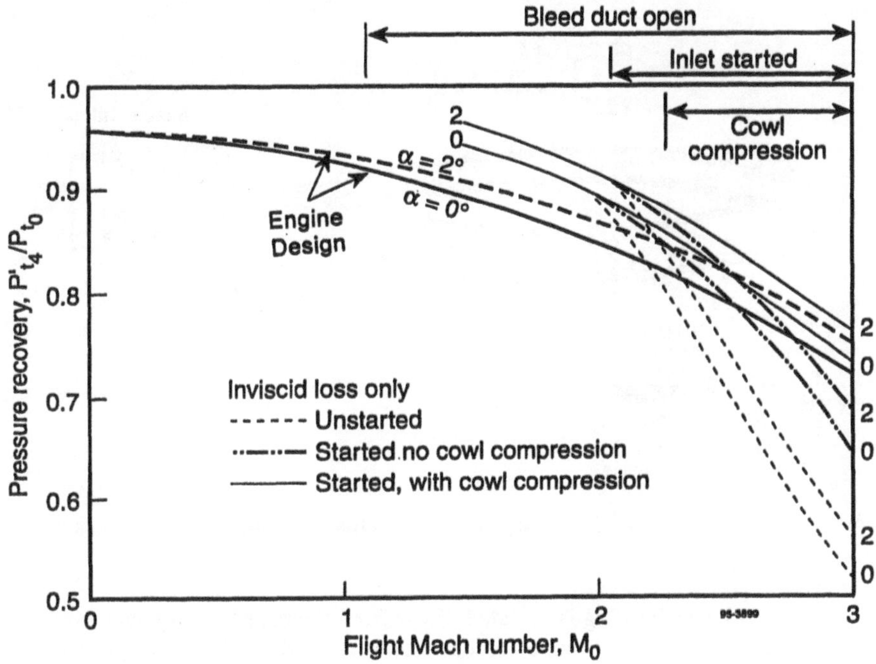

Fig. 13 Inlet pressure recovery.

streamtubes in the plane of the vehicle leading edge. Flow into the bleed duct has a cross-sectional area in the freestream of A_{OBL}; the cross section of the main engine airflow is A_0. With the bleed duct closed, A_0 extends to the vehicle leading edge. The capture height and cross-sectional area of the stream entering the main engine in the cowl lip plane are Y_2 and A_2. The term A, is the cross-sectional area of the bleed duct flow in a plane upstream of the bleed duct doors. The terms Y_4 and A_4 are the height and area of the isolator duct, and A_{ij} is the cross-sectional area of the rocket injectors in the exit plane. When $P_{ij} > P_4$, the rocket nozzle expands to an area $\tilde{A}_{ij}$, and the inlet flow, in general, contracts to $\tilde{A}_4$ such that $\tilde{P}_{ij} = \tilde{P}_4$.

There also are cases where some of the available rocket injectors are not in use and $\tilde{A}_{ij}$ is less than the total A_{ij}, so that $\tilde{A}_4$ can be slightly greater than A_4 even though $P_{ij} > P_4$. Mixing and combustion are completed at an area A_5, in length ℓ_c with $M_5 = 1$. The terms A_6 and Y_6 are the area and duct height in the plane of the cowl trailing edge. The nozzle expansion zone extends from A_5 to A_7 where $P_7 = P_0$. Further expansion of the main engine flow is prevented by the base burning. In this simplified analysis, mixing of the two streams is not considered, and so $A_7 = A_8$, which is the area of the main engine flow in the plane of the vehicle trailing edge. At $M_0 < 1.05$ the cowl aft flap is rotated towards the body such that $P_6 = P_0$, which prevents overexpansion of the main engine exhaust and negates the need for base burning. For $M_0 > 1.05$, even with the flap rotated outboard, $P_6 > P_0$, and so base burning is required to fill the base and to prevent P_7 from dropping below P_0. The cross-sectional area of the bleed duct flow at the sonic discharge is A_{10} where $M_{10} = 1$. This flow expands to an area A_{11} where

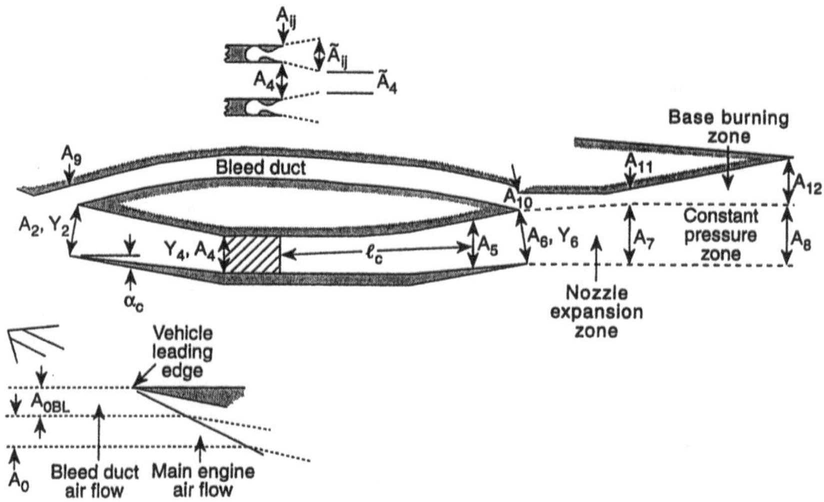

Fig. 14 Stations in the engine flowfield.

$P_{11} = P_7$, and base burning is initiated. Flow from the base burning duct expands from sonic flow at A_{10} to supersonic conditions at A_{11} where the pressure is equal to ambient, i.e., $P_{11} = P_7 = P_0$. The base burning zone maintains constant pressure and near-to-constant velocity and has an increasing speed of sound. For stoichiometric heat release in the base burning zone, the rise in sound speed lowers the Mach number so that the Mach number at A_{12} is subsonic. Entrainment of combustion products from the main engine exhaust and from the unburned air passing around the vehicle causes the base flow to pass through a "viscous throat" downstream of the vehicle trailing edge. The base burning zone extends to an area A_{12} in the trailing-edge plane of the vehicle. For $M_0 > 1$ the sum $A_8 + A_{12}$ is equal to the projected cross-sectional area of the vehicle, which with the cowl fully extended is 65088 and 56981 in.^{2} at $\alpha = 0$ and 2-deg, respectively.

To analyze the flow in the base burning zone, the total pressure at the downstream end of the duct is the average of the static pressure and total pressure following a normal shock in the cowl lip plane. These pressure recoveries are considerably lower than those in the main engine airflow. For example, at $M_0 = 3$ the bleed duct recoveries are 0.4492 and 0.4624 as compared with 0.7273 and 0.7555 from Eqs. (2) and (3) for $\alpha = 0$ and 2 deg respectively.

Figure 15 shows the orientation of the cowl lip relative to the vehicle axis, and Fig. 16 shows the corresponding height of the main engine inlet duct in the plane of the cowl lip. At $M_0 = 0$ the cowl lip is turned very slightly away from the final forebody compression ramp, thus producing a slightly decreasing area in the initial duct. As M_0 increases, the cowl turns toward the compressions surface, reaching a maximum angle of about 15.15 deg at $M_0 = 1.25$ and minimum duct entrance height of slightly less than 14 in. As M_0 increases above 1.25, the cowl rotates outboard, reaching an angle of about 8.8 deg and duct entrance height of about 21 in. at $M_0 = 3$. In the low-mid-speed range, the cowl would continually rotate outboard until it would be parallel to the vehicle axis, $\alpha_c = 11.5$ deg.

Figure 17 shows the translation of the cowl at $M_0 > 2.3$ required to provide the desired inlet contraction. At $M_0 = 3$ the cowl has translated 5.59 and 5.20 in. at

$\alpha = 0$ and 2 deg, respectively. At $M_0 > 3$ the cowl continues to translate inboard until a minimum duct height of 5.0 in. is reached.

To limit the number of discrete points in the engine performance analysis and to provide guidance in design of a vehicle using this engine, a reference trajectory is specified. Modeling is then introduced to provide engine performance for flight on alternative trajectories. Trajectories presented in Ref. 14 for velocities >1200 ft/s were adopted. For velocities < 1200 ft/s a new trajectory was introduced to limit load factors at takeoff and to provide a continuous derivate du/dz at 1200 ft/s. The modeling for this low-speed portion of the climb out is as follows:

$$Z = 1.2276 \times 10^{-4}(u - u_m)^3 + 9.5939 \times 10^{-8}(u - u_m)^4 - 2.435 \times 10^{-10}(u - u_m)^5 \tag{55}$$

For $1200 \leq u \leq 6500$ ft/s,

$$u = 500 + 7.23 \times 10^{-3} Z + 8.95 \times 10^{-7} Z^2 \tag{56}$$

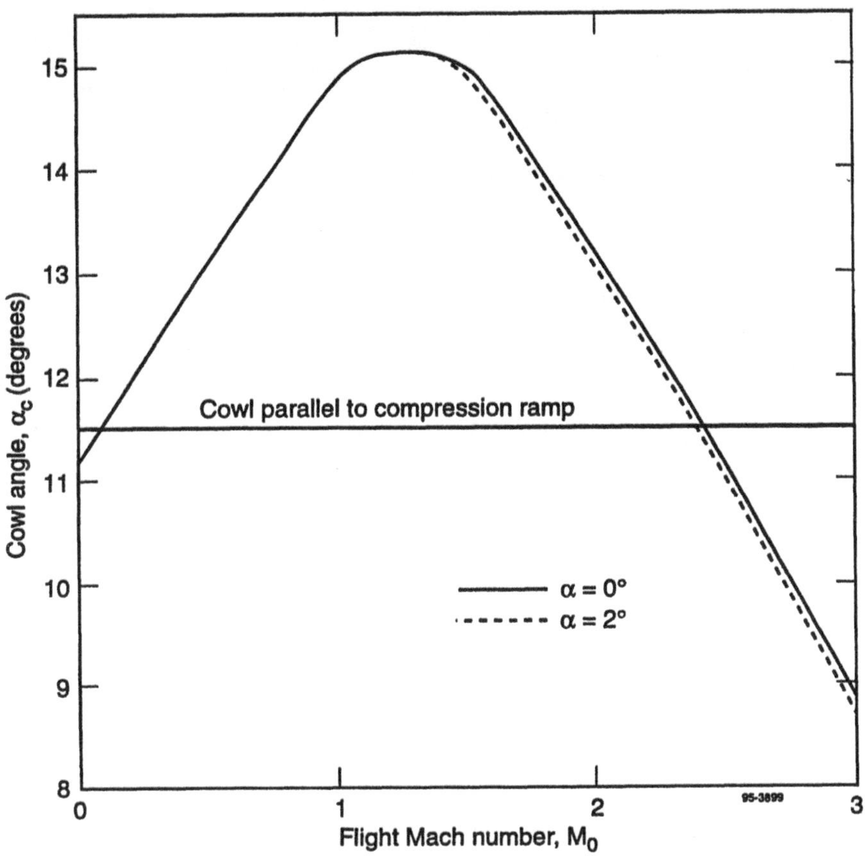

Fig. 15 Cowl lip orientation to vehicle axis.

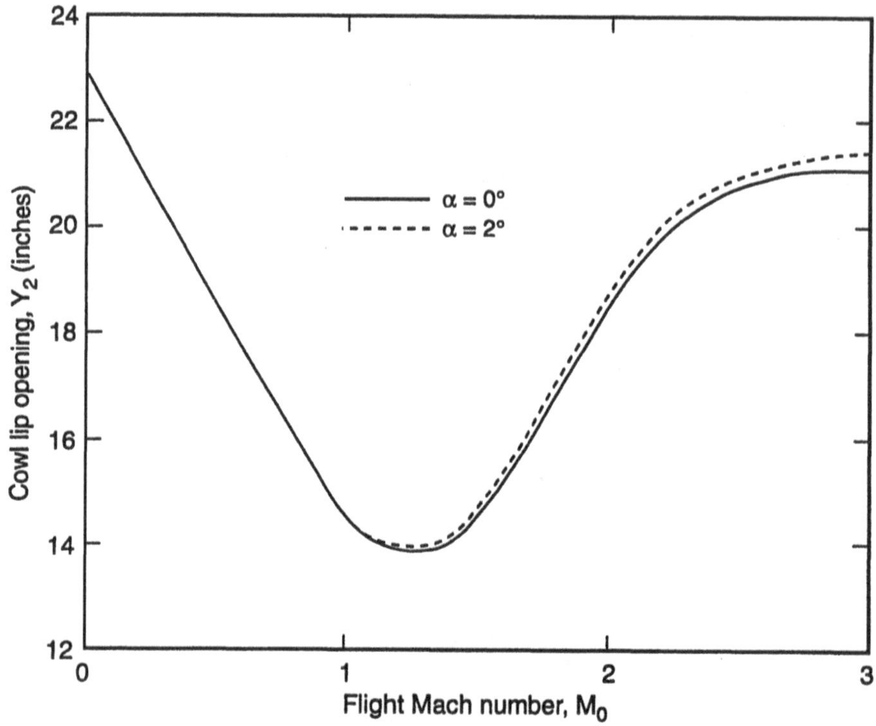

Fig. 16 Inlet duct height in plane of cowl lip.

For $6500 \leq u \leq 14{,}000$ ft/s,

$$\rho = 0.003046 \exp[4.748 \times 10^{-5}(77{,}396 - Z)] \tag{57}$$

For $14{,}000 \leq u \leq u_2$ where u_2 is the velocity required for an unpowered climb on an elliptic transfer orbit, the low trajectory is modeled as

$$u = -39{,}256 + 0.6647Z - 1.648 \times 10^{-6}Z^2 \tag{58}$$

and the high trajectory is modeled as

$$u = -4490 + 0.2059Z - 3.465 \times 10^{-7}Z^2 \tag{59}$$

Figure 18 shows the altitude–velocity relationship for the climbout trajectories.

With the altitude defined as a function of velocity, engine cycle calculations were made at $M_0 = 0, 0.5$, and 1.0 and at intervals of 0.25 between $M_0 = 1$ and 3. At $M_0 = 3$, $Z = 47,950$ ft and $u = 2914.1$ ft/s; thus only Eqs. (56) and (57) are pertinent for the low-speed range cycle analysis.

A key specification in the design study is the selection of the chamber pressure P_{tij}, equivalence ratio ER_{ij}, and area ratio A_{ij}/A^* of the rocket injectors. The principal design considerations are 1) high heat transfer rates in the throat (this problem is exacerbated as ER_{ij} approaches 1 and at high Pt_{ij}); 2) adequate thrust on runway and at transonic speeds (high acceleration is needed to limit runaway length, and the thrust/drag ratio transonically must remain high, i.e., $T/D > 2$); 3) blockage of exhaust jet, which restricts the engine airflow during unstarted inlet

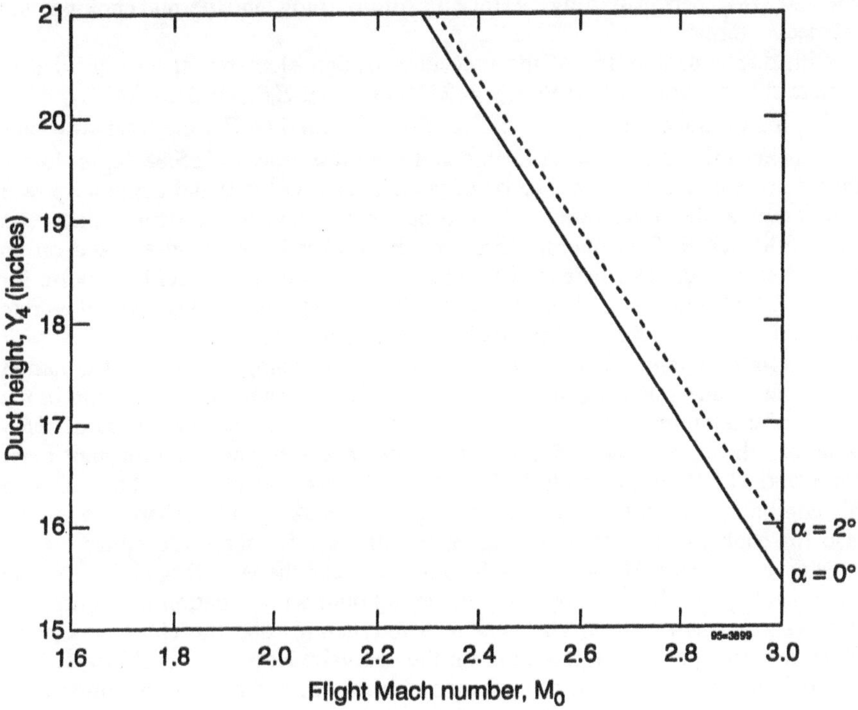

Fig. 17 Cowl translation.

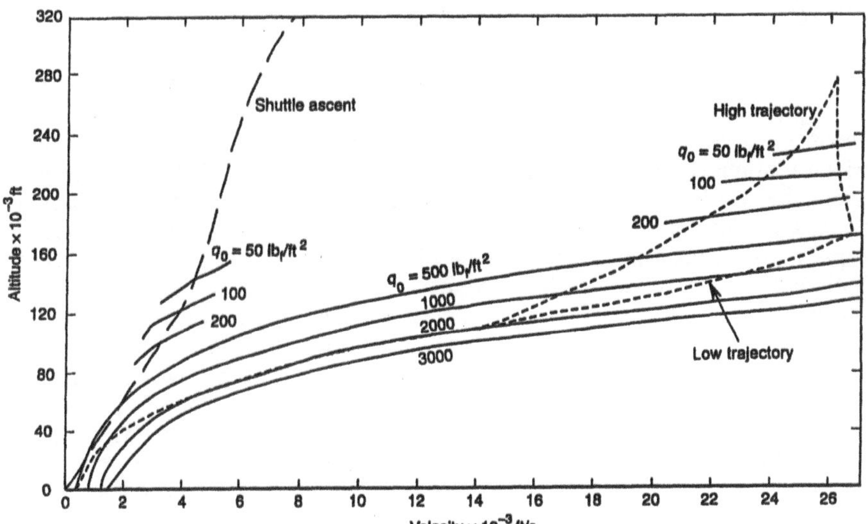

Fig. 18 Typical trajectories for single stage-to-orbit vehicles and shuttle.

operation or causes unstart with started inlet (P_{ij} must be matched with P_4 at selected M_0 to permit starting); 4) mixing, afterburning, and thermal choking; and 5) injector thrust.

With due consideration of the preceding design elements, it was decided to operate the injectors at a fixed $P_{tij} = 2250$ psia and $ER_{ij} = 2.0$ with $A_{ij}/A^* = 13.9$. This produces an $M_{ij} = 3.623$ at $P_{ij} = 16$ lb$_f$/in.^{2} For the total strut plus wall rocket injector, $A_{ij} = 1195$ in.^{2} and the mass flow is 755.88 lb$_m$/s. In the absence of a quantitative evaluation of vehicle heating loads and auxiliary power requirements, the temperature of the propellants entering the strut rockets was set at 500 deg R. The internal thrust of the rocket injectors was based on an exhaust stream thrust efficiency of 0.98. The corresponding specific impulse for $P_{tij} = 2250$ lb$_f$/in.^{2} at $ER_{ij} = 2$ is $453 \times 0.98 = 443.94$ lb$_f$s/lb$_m$. At a flow rate of 755.82 lb$_m$/s, the thrust of the rocket injectors is 335,539 lb$_f$.

The engine cycle calculations were carried out using the JHU/APL ramjet performance analysis program (RJPA)[15]. Several thousand cases were run in the course of examining the effects on engine performance and operability of ER_{ij} variation, the bypass ratio β (mass flow into main engine duct/injectant mass flow), and M_4. From this matrix of solutions, 44 cases were selected to represent this engine operation in the low-speed regime. For $M_0 \leq 1.25$, there is a unique case for each M_0 and α combination; for $M_0 > 1.50$, there are either two or three cases for each M_0 and α combination because the vehicle specific impulse $I_{\text{eff}} = I_f(1 - D/T)$ varies by a significant amount with variation in the drag D. The drag depends on both the vehicle aerodynamics and the climb trajectory, which are not specified at this point in the analysis. However, for $M_0 > 1.50$, a selected case that is typical of operation at each M_0 and α combination is used to illustrate typical flow properties and engine geometries. Flow properties in the freestream and entering the combustor are listed for these cases in Table 11. The RJPA program calculates thrust on a freestream-to-freestream force accounting system:

$$T_{\text{FS}} = \eta_N[(1-r)\mathcal{F}_{8\text{EQ}} + r\mathcal{F}_{8\text{FZ}}] - \mathcal{F}_0 - P_0(A_8 - A_0) \tag{60}$$

where η_N is the exhaust nozzle efficiency, which includes losses as a result of friction, nonuniformity, and divergence. The terms $\mathcal{F}_{8\text{EQ}}$ and $\mathcal{F}_{8\text{FZ}}$ are, respectively, the stream thrusts at station 8 for equilibrium and frozen chemical flow in the nozzle. For the low-speed engine cycle, $\eta_N = 0.98$ was assumed because the expansions are confined by the base burning flow. At low M_0, maintaining equilibrium thermochemistry is generally not difficult, and so $r = 1$ was assumed. To account for losses as a result of incomplete mixing, the engine thrust, not including the rocket injector thrust, was decreased by 5%. The skin friction in the combustor is computed from $\tau_w = C_f A_4 q_4$; the product $C_f A_4 = 39$ in.^{2} was used herein. Freestream-to-freestream force accounting requires a calculation of the inlet and exhaust nozzle additive forces.[16] For these transatmospheric accelerator configurations it is customary to use a cowl-lip-to-vehicle-tail (i.e., cowl-to-freestream) accounting system. The thrust from RJPA must therefore be adjusted by the forebody force, which, for $M_0 \neq 0$, is obtained from

$$C_T = C_{\text{TFS}} + 2\frac{A_0}{A_i}\left(1 - \frac{M_2^2}{M_0^2}\frac{P_2}{P_0}\frac{A_2}{A_0}\right) \tag{61}$$

Table 11 Flow properties in freestream and entering combustor

Case	M_0	α	P_0, lb/in.^{2}	T_0, °R	U_0, ft/s	M_4'	P_4', lb/in.^{2}	T_4', °R	U_4', ft/s	P_4'/P_0	P_{t4}'/P_{t0}
1	0.50	0–2	14.696	518.7	0.0	0.580	11.222	485.6	627.9	0.764	0.960
2	0.50	0	14.682	518.6	559.1	0.677	14.012	518.8	559.1	0.954	0.9535
	0.50	2	14.682	518.6	559.1	0.677	14.024	518.8	559.0	0.955	0.9543
3	1.0	0	7.635	457.9	1051.8	0.800	8.865	487.8	868.0	1.191	0.9341
4	1.0	2	7.635	457.9	1051.8	0.800	8.895	487.8	868.0	1.165	0.9373
5	1.25	0	5.341	427.9	1271.3	0.800	8.365	499.2	877.9	1.566	0.9196
6	1.25	2	5.341	427.9	1271.3	0.800	8.41	499.2	877.9	1.575	0.9245
7	1.50	0	4.465	413.5	1500.0	0.800	9.731	533.8	907.3	2.179	0.9018
10	1.50	2	4.465	413.5	1500.0	0.800	9.808	533.8	907.3	2.197	0.9089
17	1.75	2	3.799	401.0	1723.6	0.800	11.871	576.4	941.9	3.124	0.8904
20	2.00	0	3.273	390.0	1942.8	0.800	14.358	626.2	980.8	4.387	0.8566
23	2.00	2	3.273	390.0	1942.8	0.800	14.69	626.2	980.8	4.488	0.8691
25	2.25	0	2.804	390.0	2185.6	0.759	18.502	708.0	987.6	6.598	0.8291
28	2.25	2	2.804	390.0	2185.6	0.800	18.119	700.0	1035.3	6.462	0.8450
30	2.50	0	2.429	390.0	2428.4	0.700	24.168	803.2	966.8	9.950	0.7984
32	2.50	2	2.429	390.0	2428.4	0.736	23.943	795.7	1013.3	9.857	0.8180
37	2.75	0	2.122	390.0	2671.3	0.667	30.711	902.3	975.5	14.473	0.7645
39	2.78	2	2.122	390.0	2671.3	0.692	30.988	896.9	1009.1	41.603	0.7882
40	3.00	0	1.866	390.0	2914.1	0.637	38.686	1009.3	983.6	20.731	0.7273
43	3.00	2	1.866	390.0	2914.1	0.659	39.452	1004.3	1015.1	21.141	0.7555

where $C_T = (T/q_0 A_i)$ and $A_i = 30,283.2$ in.^{2}. For $M_0 = 0$, defining $C_T = T/P_0 A_i$, the thrust adjustment, for $M_0 = 0$, is given by

$$C_T = C_{\mathrm{TFS}} - \frac{\gamma P_2 M_2^2}{P_0 A_i} \tag{62}$$

The exhaust nozzle additive force is assumed equal to zero because the nozzle expansion flow is turned toward the afterbody of the vehicle.

The gauge force as a result of base burning is zero because the pressure in the combustion zone is equal to ambient. Combustion efficiency in the base burning zone is assumed to be 0.95.

With the engine operating in the unstarted mode (i.e., $M_0 \leq 2$), a matrix of M_4', ER_{ij}, and w_{ij} combinations are examined. For each M_4' there is a unique value of P_4'. For this M_4' and P_4' combination, the degree of underexpansion of the rocket injectors, $\tilde{A}_{ij}/A_{ij}$, is defined, and $\tilde{A}_{ij}$ as a function of w_{ij} is therefore specified. With $\tilde{A}_{ij}$ determined, the remainder of the duct area is $\tilde{A}_4$, which then specifies w_4. Once the engine is started, the rocket injector is overexpanded ($P_4 > P_{ij}$), and the engine geometry sets M_4, P_4, and w_4. For $0 \leq M_0 < 1.5$, the engine cycle calculations showed that the highest specific impulse resulted with $ER_{ij} = 2$ and with $M_4' = 0.58$ at $M_0 = 0$, and then M_4' decrease to 0.50 at $M_0 = 0.5$ and increases monotonically with M_0. To provide less sensitivity to premature choking the maximum M_4' was set to 0.8. Accordingly, for $1 \leq M_0 \leq 2$, $M_4 = 0.8$. For $M_0 > 2$, M_4' decreases slowly with increasing M_0. Figure 19 shows M_4' and P_4' as a function of M_0. This figure supports the foregoing arguments about the selection of the strut rocket nozzle area ratio that leads to overexpansion to permit inlet starting at $M_0 > 2$. Additional flow properties in the freestream and entering the combustor are listed in Table A1 of Appendix A.

The results of the engine cycle calculations showed that for $M_0 \le 1.25$ there was no advantage to injecting supplemental hydrogen into the combustor. At all thrust levels, I_{eff} was maximum at $ER_{ij} = 2$. The mixing and afterburning of the entrained air from the inlet reduced the overall equivalence ratio to approximately 1.3; thus the overall engine is operating fuel rich. For $M_0 < 1.0$, A_5 and A_7, the respective cross-sectional areas of the flow at the combustor exit and at the nozzle expansion ratio where $P_7 = P_0$, are both less than the maximum area $A_{6\text{MAX}} = 17{,}400$ in.^{2}, that can be provided with the aft cowl flap rotated and translated to its more outboard position. For $M_0 > 1.0$, $A_7 < 17{,}400$ in.^{2}, which necessitates the use of base burning to avoid overexpansion of the main engine exhaust flow. At $M_0 = 1.5$ the optimum ER_{ij} is 2.7 for all thrust levels that yield

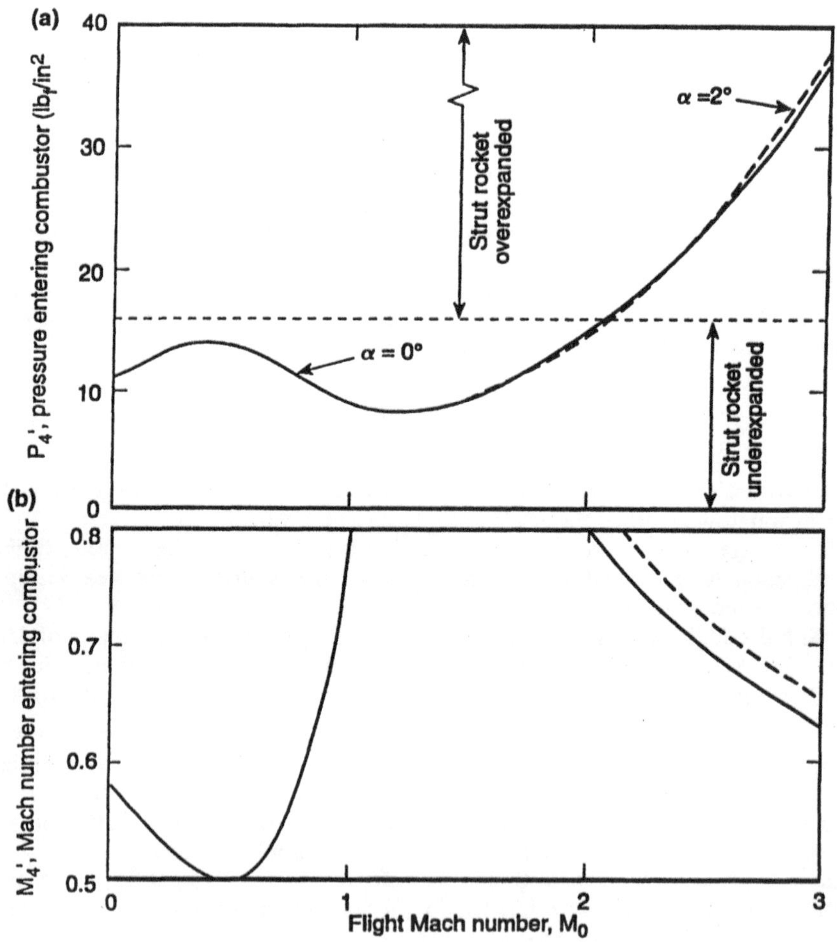

Fig. 19 Flow properties entering combustor.

maximum I_{eff}. At the lowest drag level of interest the thermal throat is located at the cowl trailing edge. At higher M_0 this introduces an additional constraint on the ER_{ij} at which the engine is allowed to operate. This constraint can be more lucidly explained by describing engine operation at $M_0 = 1.75$ where the optimum ER_{ij} varies with thrust level and the maximum area of the thermal throat constrains the optimum ER_{ij}.

Figure 20 shows the important flow parameters that coincide with matched pressure, $P_4 = P_{ij}$ at $M_0 = 1.75$ and $\alpha = 0$ deg. The area of the underexpanded jet $\tilde{A}_{ij}$ increases monotonically with increasing propellant flow w_{ij} and with decreasing ER_{ij} at a specified w_{ij}. The portion of the duct area that can accommodate the airflow $\tilde{A}_4$ and corresponding mass flow w_4 vary in opposition to $\tilde{A}_{ij}$. The resulting bypass ratio β shown in Fig. 20d decreases with increasing w_{ij} and decreases slightly with ER_{ij} at a specified w_{ij}. The engine thrust and area of the thermal throat corresponding to conditions of interest in Fig. 20 are shown in Fig. 21. As the propellant flow at a constant ER_{ij} is reduced, the required cross-sectional area of the thermal throat increases until it reaches the maximum available in the particular engine geometry. The constraint at low flow ratio is more limiting as ER_{ij} increases. The engine efficiency increases slightly at a constant propellant flow rate at this M_0 because the thrust is higher. Thus, at low thrust levels some loss in thrust must be accepted. At $M_0 > 1.75$ this constraint has a larger impact on I_{eff}, which requires attention in the selection of ER_{ij}. Nonetheless, if completion of the mixing and combustion process could not be controlled to meet the 17,400 in.^{2} limit, losses in thrust as a result of moving the thermal throat upstream to a smaller area would not be excessive.

Results at $M_0 = 2.00$, 2.25, and 2.50 showed similar trends to those at $M_0 = 1.75$. Optimum ER_{ij} values varied between 2 and 3 for thrust levels of interest. However, at $M_0 = 2.50$ the total propellant flow rates decrease rapidly as the engine transitions from air-augmented rocket to a ramjet. At $M_0 = 2.75$, ER_{ij}

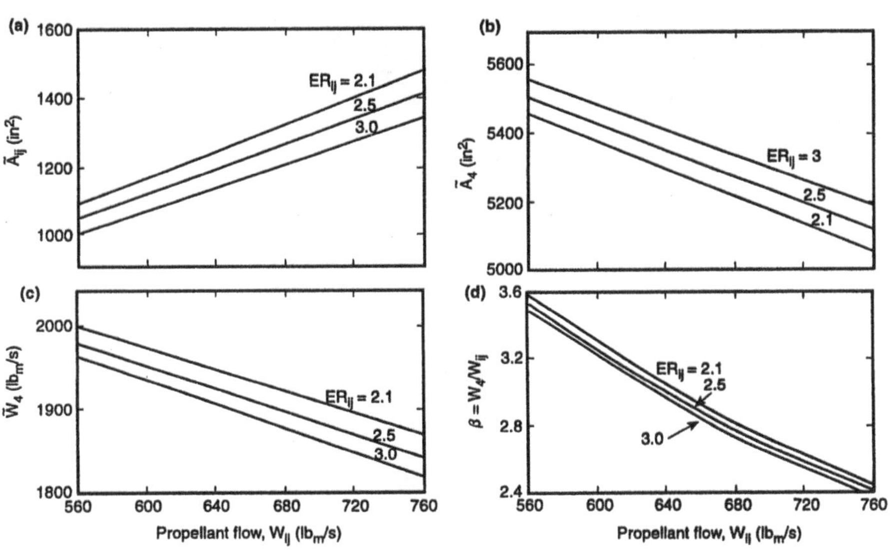

Fig. 20 Parameters for pressure matched under expanded injectors ($M_0 = 1.75$, $\alpha = 0$ deg, $M_4^{\ddot{E}} = 0.8$).

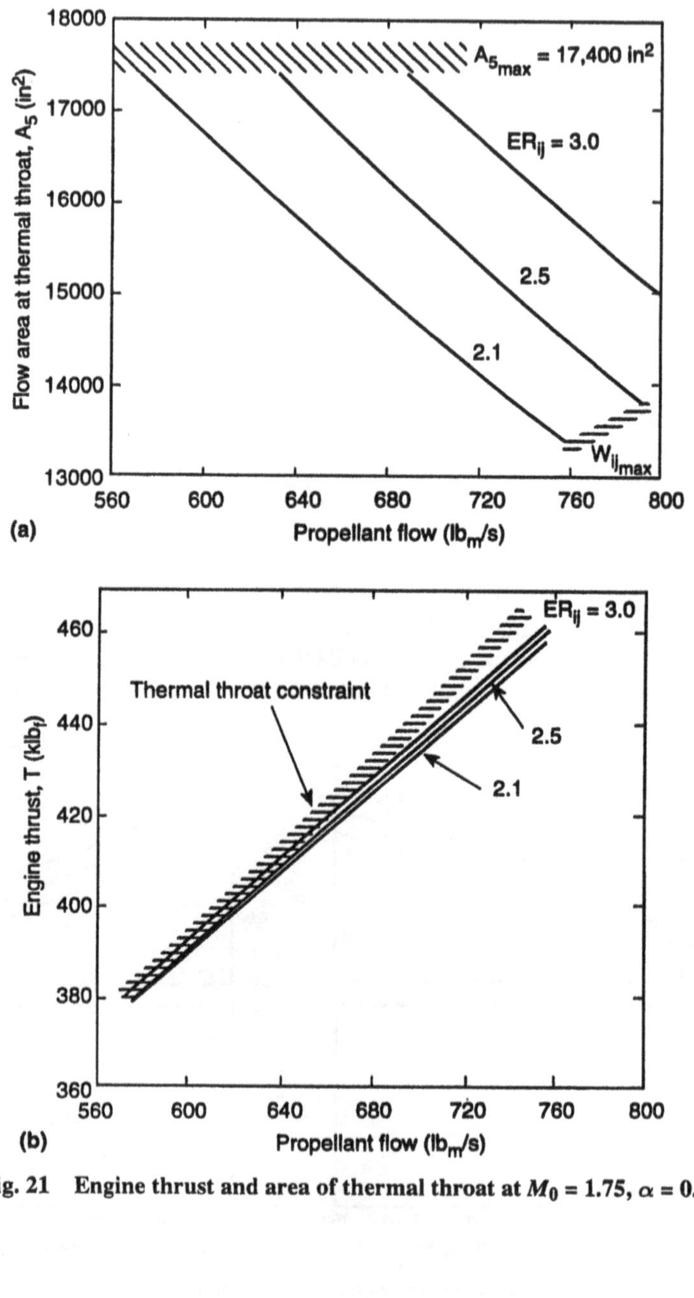

Fig. 21 Engine thrust and area of thermal throat at $M_0 = 1.75$, $\alpha = 0$.

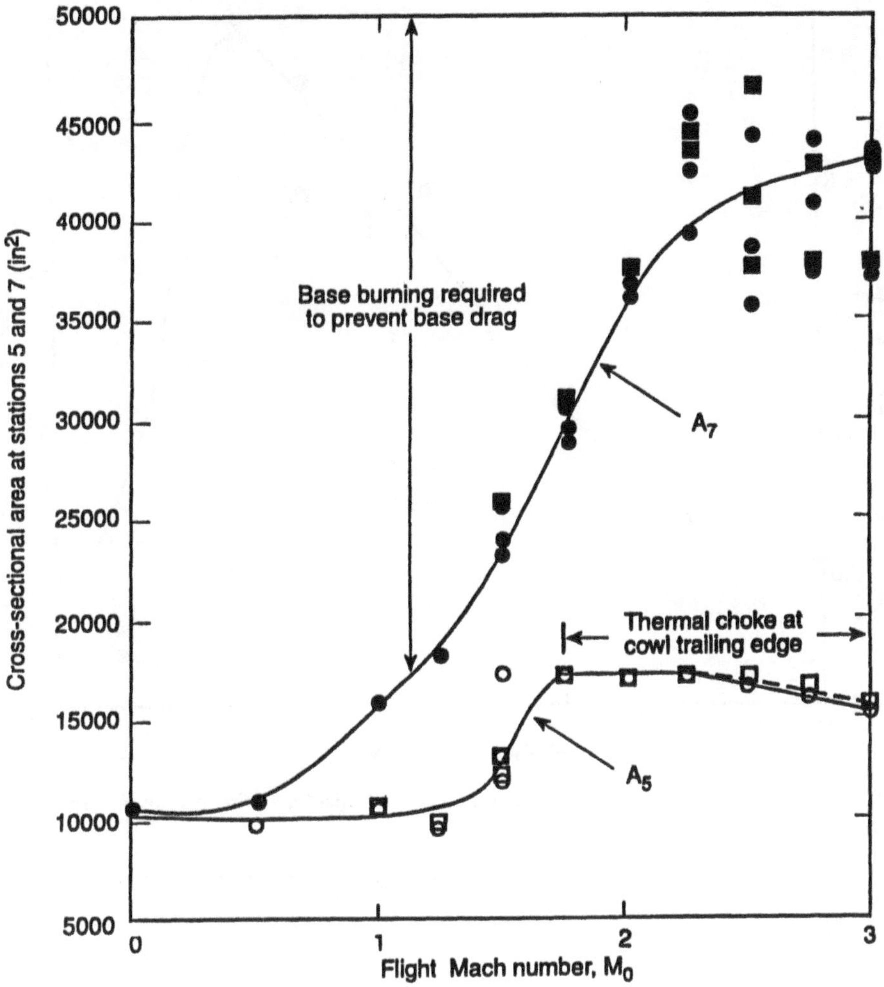

Fig. 22 Flow areas at thermal throat and in nozzle expansion at $p_7 = p_0$.

increases from 4.65 to 10.00. For the selected cases at $M_0 = 3$, only a minimal amount of rocket thrust is desired, and this occurs only for high thrust levels.

The areas of the thermal throat A_5 and nozzle expansion at $P_7 = P_0$ are shown in Fig. 22. Data for all 44 cases listed in Tables A2 and A3 of Appendix A are shown. The scatter in A_7 values at $M_0 \geq 2.25$ is a result of the broad range of ER_{ij} values at different thrust levels. A single curve is drawn through the A_7 points to depict the general trend of a rapidly rising A_7 with increasing M_0. The aft cowl flap is turned toward the body at $M_0 < 1.12$ because $A_7 < 17{,}400$ in.^{2} with $P_7 = P_0$. The required area of the thermal throat rises abruptly at about $M_0 = 1.4$, and it reaches the limiting value of 17,400 in.^{2} at $M_0 = 1.75$. The decrease in A_5 from 17,400 in.^{2} at $M_0 = 2.25$ to 15,839 in.^{2} at $M_0 = 3$ is a result of cowl translation toward the body. Outward turning of the cowl was not considered to avoid increased vehicle drag. Base burning is required to avoid base drag at $M_0 > 1.12$. The area required to be filled decreases rapidly from 47,688 and 39,581 in.^{2} at $\alpha = 0$ and

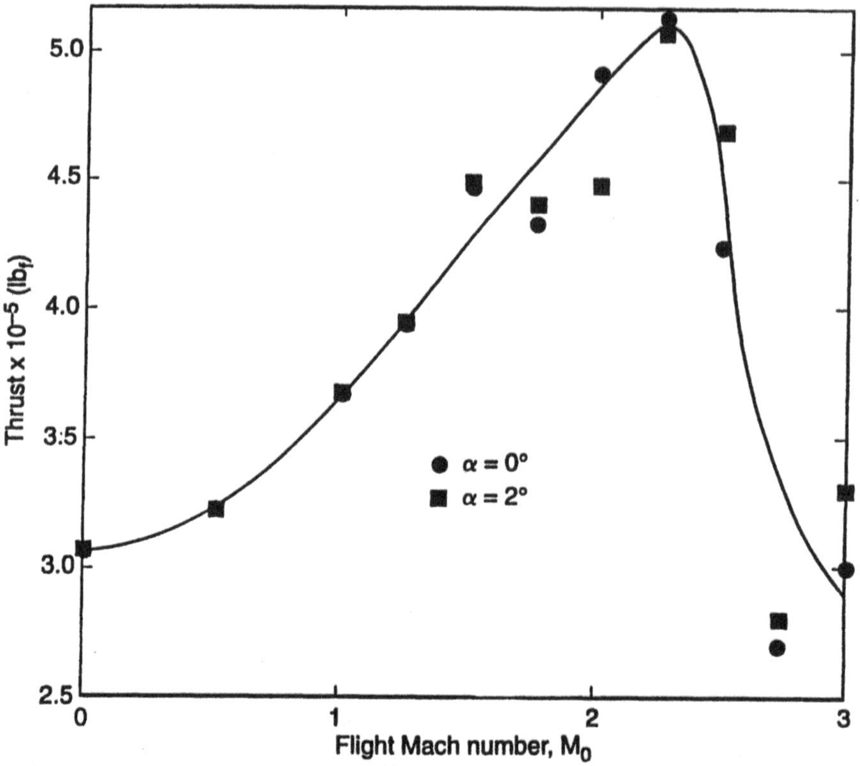

Fig. 23 Engine thrust at maximum I_{eff} reference trajectory with nominal vehicle drag.

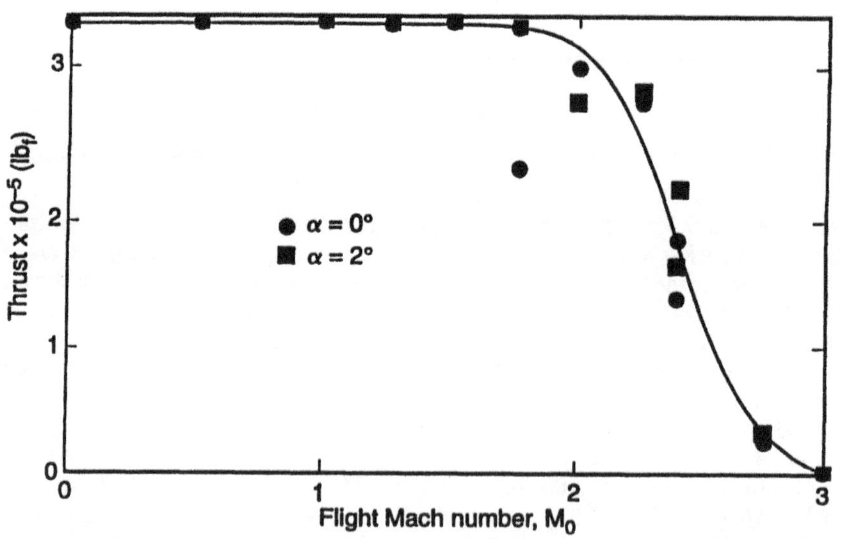

Fig. 24 Rocket motor thrust at maximum I_{eff}; reference trajectory with nominal vehicle drag.

2 deg, respectively, at $M_0 = 1.12$ to about 19,950 and 11,970 in.^{2} at $\alpha = 0$ and 2 deg, respectively, at $M_0 = 3$. The large decrease is a result of the correspondingly large increase in A_7 and the modest decrease in vehicle projected area to 63,192 and 55,217 in.^{2} at $\alpha = 0$ and 2 deg, respectively, at $M_0 = 3$.

Table A2 lists key performance parameters for the 44 cases of interest. At $M_0 \geq 1.50$, where more than one case is listed for each combination of M_0 and α, the cases selected for listing flow properties in Table A1 are indicated by the asterisk. These cases would represent maximum I_{eff} for a vehicle flying the reference trajectory with a nominal drag of 115,000 lb_f for $1.25 \leq M_0 \leq 3.0$. The corresponding values of engine thrust are shown in Fig. 23, and a representative curve is drawn through the data. The thrust increases from $M_0 = 0$, where the air augmentation is trivial, to a maximum of approximately 513,000 lb_f at $M_0 = 2.25$ where the contribution of ram-air-augmentation is nearly equal to that of the rocket injectors. The rocket injectors are gradually shut down as the speed increases from $M_0 = 2$ to 3 as shown in Fig. 24. The decrease in engine thrust as M_0 increases from 2.25 to 3.00 is a result of the combined effect of the reduced rocket thrust and the optimization of I_f and I_{eff} at lower overall equivalence ratio ER_0. Engine specific impulses increase gradually from about 405 $lb_f s/lb_m$ at $M_0 = 0$ to about 775 $lb_f s/lb_m$ at $M_0 = 2.25$ and then climb rapidly to greater than 3000 $lb_f s/lb_m$ at $M_0 = 3.00$ as shown in Fig. 25.

Airflow rates in the main engine and the boundary-layer duct are shown in Fig. 26. As the speed increases from $M_0 = 0$ to about 0.6, airflow increases from 1326 to about 1490 lb_m/s and then dips to a minimum of 1280–1290 lb_m/s at $M_0 = 1.25$. From this point airflow increases monotonically to 2776 and 3011 lb_m/s for $\alpha = 0$ and 2 deg, respectively, at $M_0 = 3$. The bleed duct doors open at about $M_0 = 1$, and the bleed duct flow rate jumps to a maximum of 954 and 786

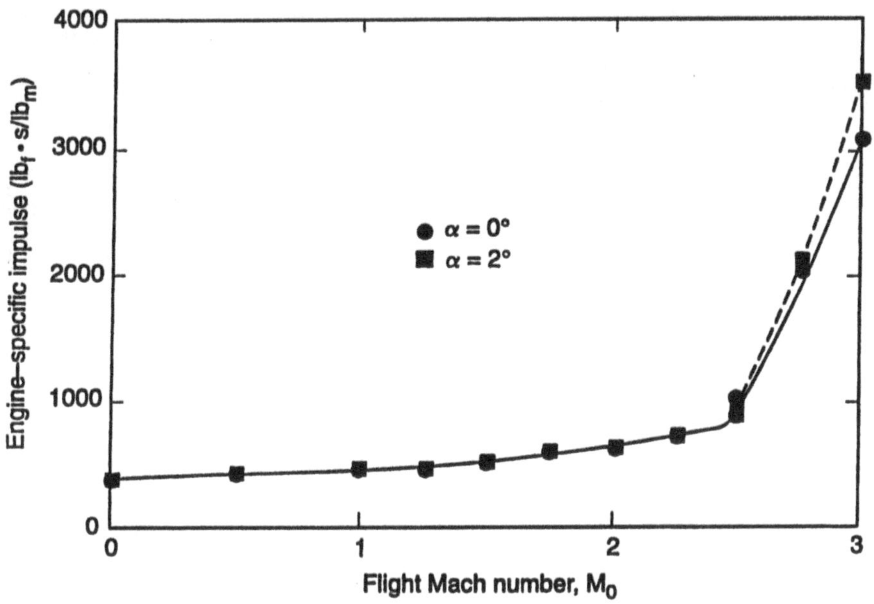

Fig. 25 Engine specific impulse at maximum I_{eff}; reference trajectory with nominal vehicle drag.

lb_m/s at $\alpha = 0$ and 2 deg, respectively. As M_0 increases, bleed flow requirements generally decrease. Values at $M_0 = 3$ are 313 and 190 lb_m/s at $\alpha = 0$ and 2 deg, respectively. The airflows at $\alpha = 2$ deg are a result of the lower projected area of the vehicle and correspondingly lower base filling requirement for the base bleed combustion.

Engine bypass ratio β, main duct flow/total injectant flow from the rocket injectors, is shown in Fig. 27 for the selected cases corresponding to maximum I_{eff}. At $M_0 = 0$, $\beta = 1.75$ and then increases gradually, reaching 3.18–3.81 at $M_0 = 2$. Then β rises rapidly as flow rates to the rocket motors are decreased, and the engine transitions from a ducted rocket to a subsonic combustion ramjet. At $M_0 = 3$, I_{eff} maximizes at $ER = 1$ for $\alpha = 0$ deg and at $ER = 0.875$ for $\alpha = 2$ deg. As shown in Table A2, I_{eff} is relatively sensitive to ER_{ij} and therefore β, and so reoptimization would be required for vehicle drag values other than the nominal values assumed herein.

The overall engine equivalence ratio as a function of flight Mach number is shown in Fig. 28. For $0 < M_0 \leq 1.25$, ER_0 is approximately 1.3. The rise in ER_0 at $M_0 = 1.5$ and 1.75 is a result of the addition of auxiliary hydrogen while the rocket injectors are still operating at very high flow rates. At $M_0 \approx 2.5$ the overall ER_0 is about stoichiometric and then decreases to about 0.8 at $M_0 = 3$. Mixing of the fuel-rich engine exhaust with the bleed duct air would reduce the fuel flow requirements to the base burning stream, but at this stage in the analysis, no fuel saving credit was assumed. The confining of the base bleed burning zone between the hot main engine exhaust and the vehicle body should provide an adequate ignition source as well as prevent undesired overexpansion of either gas stream.

Figure 29 shows that the oxygen mass fraction of the total propellant decreases from 0.7987 at $M_0 = 0$ to about 0.72 at $M_0 = 2.5$, with a local minimum at $M_0 = 1.75$, and then decreases to about 0.1 at $M_0 = 3$ as the rocket motors are shut down. More precise values are given in Table A2.

Pending the specification of vehicle drag values, representative values of drag of 78,000 lb_f at $M_0 = 0.5$, 89, 500 lb_f at $M_0 = 1.0$, and 115,000 lb_f at $M_0 =$ 1.25–3.00 were assumed to obtain the I_{eff} values listed in Table A2 and shown on

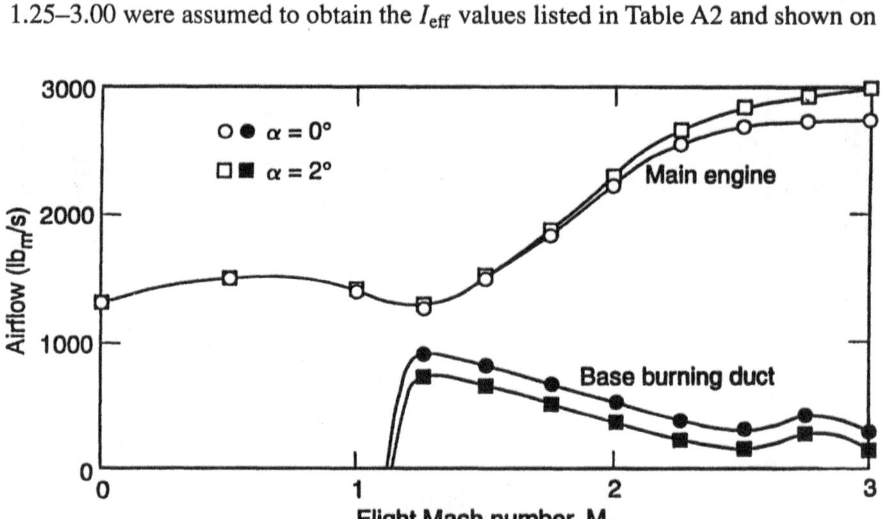

Fig. 26 Airflow rates to main engine and bleed duct.

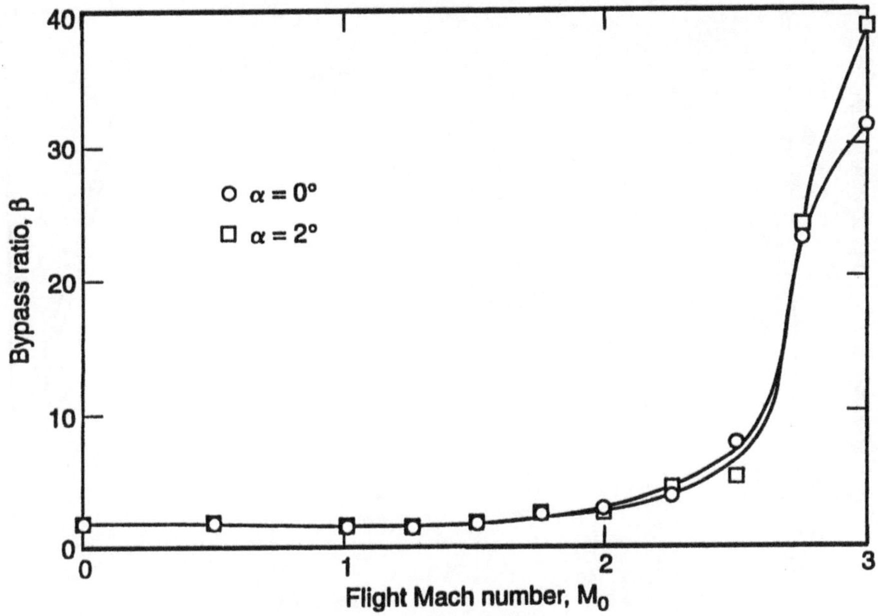

Fig. 27 Bypass ratio in main engine duct at maximum I_{eff}.

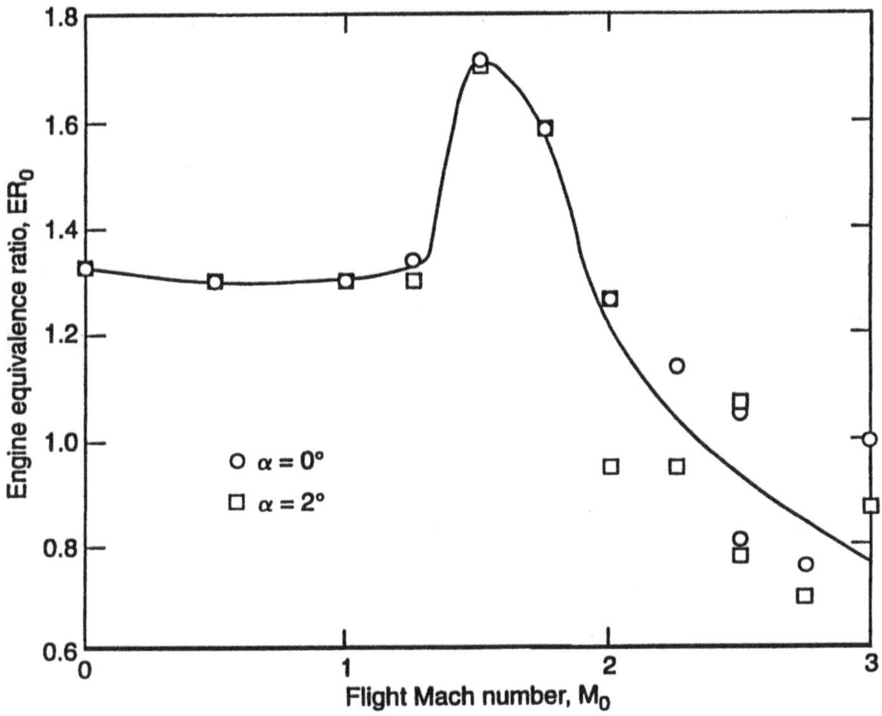

Fig. 28 Engine equivalence ratio; reference trajectory with nominal drag.

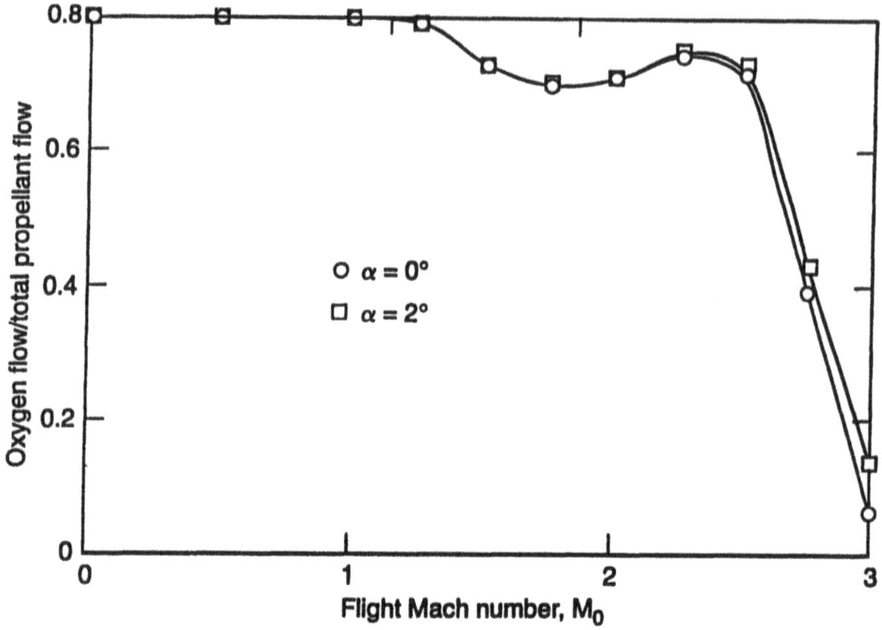

Fig. 29 Oxygen mass fraction.

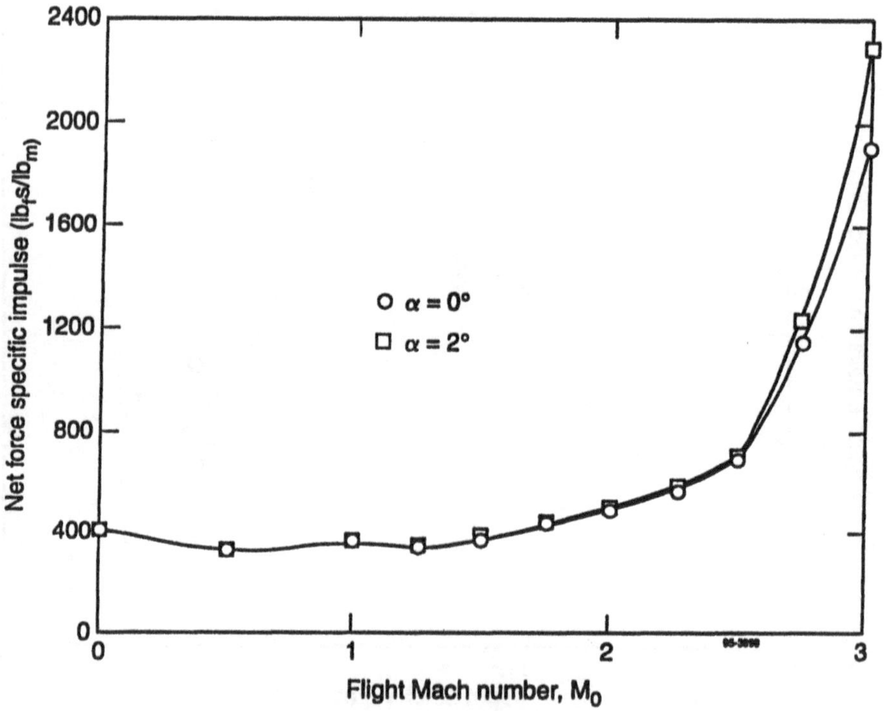

Fig. 30 Net force specific impulse I_{eff}; reference trajectory with nominal drag.

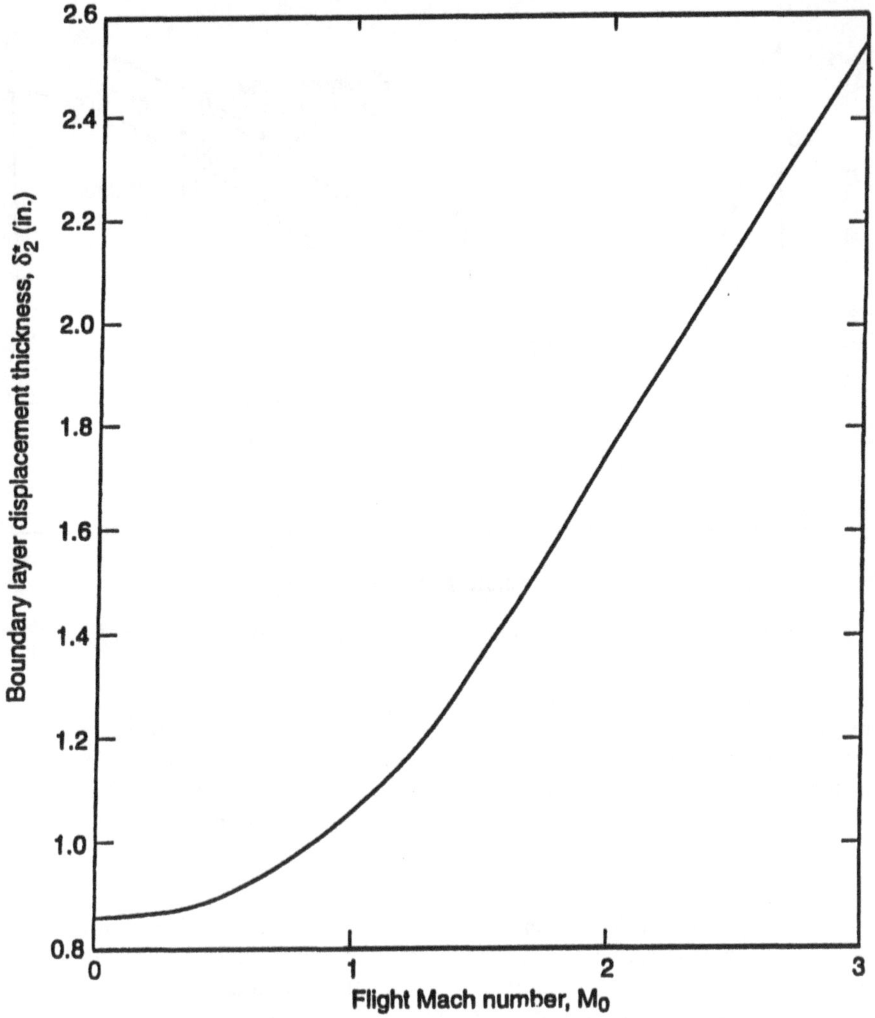

Fig. 31 Boundary-layer displacement thickness in plane of cowl lip.

Fig. 30. The trend of the I_{eff} values with M_0 are similar to those for I_f shown in Fig. 25. The steep rise in I_{eff} corresponds to the reduction in rocket motor thrust. For trajectory calculations using the I_f, T, and W_{O2}/W_p values given in Table A2, and with the vehicle drag specified, the procedure described in Appendix B should be followed.

Boundary-layer displacement thicknesses were computed using the method of Ref. 17 for flight on the reference trajectory. Results at the cowl lip, station 2, are shown in Fig. 31. The δ^* values increase from about 0.85 in. at low speed to 2.55 in. at $M_0 = 3$. These displacement heights together with the flow properties in the plane of the cowl lip were used to calculate the areas at the entrance of the base burning that are listed in Table A3.

The total capture ratio and that of the main engine duct are shown in Fig. 32. The so-called transonic "pinch" where the thrust–drag margin is minimal is a

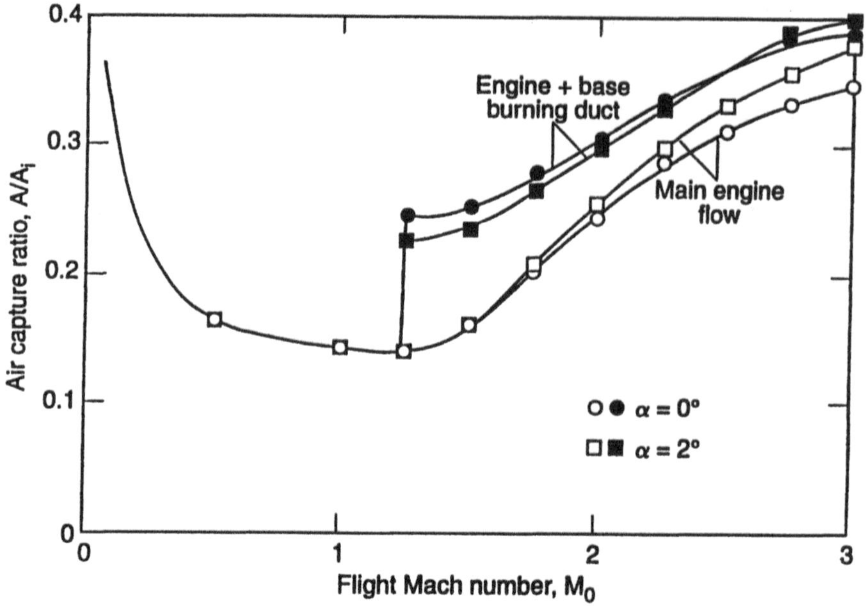

Fig. 32 Air capture ratios; maximum I_{eff} at nominal trajectory.

direct result of the minimum in the air capture that occurs simultaneously with the maximum drag coefficient of the vehicle. Air capture of the base burning duct is the difference between corresponding curves at the same α. Base burning flows decrease from 954 and 786.4 lb_m/s for $\alpha = 0$ and 2 deg respectively, at $M_0 = 1.25$ to 313.2 and 189.6 lb_m/s for $\alpha = 0$ and 2 deg, respectively, at $M_0 = 3.0$.

IX. Conclusions

This transatmospheric accelerator engine concept is the combination of more than 35 years of research and development of air-augmented rockets and dual-mode subsonic combustion ramjets at the Johns Hopkins University Applied Physics Laboratory. The principal features of this engine concept are the following:

1) The engine introduces multipurpose instream struts that contain rocket motors and adjunct discrete hole hydrogen fuel injectors. The struts also serve as lateral compression surfaces and inlet combustor isolators. When the rocket motors are not in use, the bases of the struts function as base flame stabilizers. The characteristic cross-stream dimension becomes the injector spacing instead of the cowl body separation, thereby greatly reducing the required length of the isolator and combustor.

2) The engine uses side wall injectors, which are tantamount to bisected instream strut injectors that provide the same functions as the strut and, in conjunction with cowl and body side wall slot injectors, supply a film of coolant along the combustor walls when needed.

3) The engine has a translating cowl with rotatable leading and trailing cowl lip sections. Translation provides the variable inlet contraction ratio that is mandatory for good overall engine performance over the entire Mach 0.25 operating

range. Rotation of the leading-edge section permits variable internal turning of the captured airflow, shook-on-shoulder operation over part of the flight regime, prevention of subsonic spillage at low Mach numbers, and minimization of cowl wave drag. Rotation of the cowl trailing-edge section permits matching of the internal and external pressures at $M \leq 1$ to eliminate the need for base burning in this speed range and matching pressures between the main engine and bleed duct exhaust during engine operation at $1 \leq M \leq 3.2$.

4) The engine has a base bleed duct system that is used to provide air for base burning to minimize or to eliminate base drag. This system is considerably more efficient than the alternative concept of external buring, which requires flameholding from the engine cowl.

5) The engine operates the linear rocket motors at a constant total pressure and oxidizer/fuel ratio to simplify the mechanical design and propellant control system.

6) The engine has segmented linear rocket injectors to provide thrust to ensure a large thrust–drag margin. Takeoff can be accomplished on relatively short runways, and the climb and acceleration through the transonic pinch can follow a trajectory that minimizes vehicle-induced drag. The thrust-to-weight ratio of the vehicle is lower than all other combined cycle engines, which should lead to the lowest empty weight vehicle.

The same engine concept could also be used in vertical launch vehicle configurations. One of several possible design would cut the vehicle shown in Fig. 1 along the longitudinal axis and make it to a corresponding mirror image. Significant modifications to other features of the vehicle, e.g., landing gear, wings, etc., would be required that thrust in excess of weight would be forthcoming.

Appendix A

Table A1 lists flow properties in the freestream and in the flow entering the combustor for the 20 cases selected to describe the operation of the transatmospheric accelerator engine from take-off to Mach 3. Tables A2 and A3 list data for the entire 43 cases which are discussed in the paper. Table A2 lists points at the selected Mach numbers at angles of attack of 0 and 2 deg. Force, force coefficients, specific impulses, and flow rates are tabulated. Table A3 lists the corresponding cross sectional areas at various flow stations and a few important ratios of areas.

Appendix B: Procedure for Determining Engine Performance Given *M*, *α*, *Z*, and *D*

For adjacent Mach numbers interpolate between cases in Table A2 at $\alpha = 0$ and 2 deg. Assume I_f varies linearly with thrust. Next, interpolate or extrapolate for α, and then adjust for altitude interpolation between cases in Table A2; i.e.,

$$I_{\text{eff}} = I_f\left(1 - \frac{D}{T}\right) = (C_1 + C_2T)\left(1 - \frac{D}{T}\right) \tag{B1}$$

Setting the derivative = 0

$$\frac{\partial I_{\text{eff}}}{\partial T} = 0 = (C_1 + C_2T)\frac{D}{T^2} + C_2\left(1 - \frac{D}{T}\right) = 0 \tag{B2}$$

Table A1 Flow properties in freestream and entering combustor of the conceptual design of a transatmospheric accelerator engine

Case	M_0	α	P_0, lb/in.^{2}	T_0, °R	U_0, ft/s	M_4'	P_4', lb/in.^{2}	T_4', °R	U_4', ft/s	P_4'/P_0	P_{t4}'/P_{t0}
1	0.50	0–2	14.696	518.7	0.0	0.580	11.222	485.6	627.9	0.764	0.960
2	0.50	0	14.682	518.6	559.1	0.677	14.012	518.8	559.1	0.954	0.9535
	0.50	2	14.682	518.6	559.1	0.677	14.024	518.8	559.0	0.955	0.9543
3	1.0	0	7.635	457.9	1051.8	0.800	8.865	487.8	868.0	1.191	0.9341
4	1.0	2	7.635	457.9	1051.8	0.800	8.895	487.8	868.0	1.165	0.9373
5	1.25	0	5.341	427.9	1271.3	0.800	8.365	499.2	877.9	1.566	0.9196
6	1.25	2	5.341	427.9	1271.3	0.800	8.41	499.2	877.9	1.575	0.9245
7	1.50	0	4.465	413.5	1500.0	0.800	9.731	533.8	907.3	2.179	0.9018
10	1.50	2	4.465	413.5	1500.0	0.800	9.808	533.8	907.3	2.197	0.9089
17	1.75	2	3.799	401.0	1723.6	0.800	11.871	576.4	941.9	3.124	0.8904
20	2.00	0	3.273	390.0	1942.8	0.800	14.358	626.2	980.8	4.387	0.8566
23	2.00	2	3.273	390.0	1942.8	0.800	14.69	626.2	980.8	4.488	0.8691
25	2.25	0	2.804	390.0	2185.6	0.759	18.502	708.0	987.6	6.598	0.8291
28	2.25	2	2.804	390.0	2185.6	0.800	18.119	700.0	1035.3	6.462	0.8450
30	2.50	0	2.429	390.0	2428.4	0.700	24.168	803.2	966.8	9.950	0.7984
32	2.50	2	2.429	390.0	2428.4	0.736	23.943	795.7	1013.3	9.857	0.8180
37	2.75	0	2.122	390.0	2671.3	0.667	30.711	902.3	975.5	14.473	0.7645
39	2.78	2	2.122	390.0	2671.3	0.692	30.988	896.9	1009.1	41.603	0.7882
40	3.00	0	1.866	390.0	2914.1	0.637	38.686	1009.3	983.6	20.731	0.7273
43	3.00	2	1.866	390.0	2914.1	0.659	39.452	1004.3	1015.1	21.141	0.7555

$$T^2 = \frac{-C_1 D}{C_2} \tag{B3}$$

$$I_f = C_1 + C_2 T \tag{B4}$$

$$(C_1 + C_2 T)D - C_2 T D + C_2 T^2 = 0 \tag{B5}$$

$$C_2 = \frac{I_{f1} - I_{f2}}{T_1 - T_2} \tag{B6}$$

The coefficients C_1 and C_2 are listed in Table B1; and the flow rate of the propellants is

$$W_p = \frac{T}{I_f} \tag{B7}$$

Note that the bounds for the drag interval are based on

$$D = -T_1^2 \frac{C_2}{C_1} \tag{B8}$$

and

$$D = \frac{I_{f2} - I_{f1}}{(I_{f2}/T_2) - (I_{f1}/T_1)} \tag{B9}$$

Table A2 Illustrative performance parameters for a conceptual design of a transtmospheric accelerator engine

Case	M_0	Z, ft	α, deg	D_{I}, lb_f	ER_0	β	ER_{ij}	T, lb_r	C_r^+	I_r, $lb_r s/lb_m$	T_R, lb_r	W_p, lb_m/s	W_a, lb_m/s	W_{o2}, lb_m/s	W_{RJH2}, lb_m/s	W_{BB}, lb_m/s	W_{BBH}, lb_m/s	I_{EFF}, $lb_r/s/lb_m$
1	0	0	0–2	All	1.3257	1.7545	2.00	306,020	0.6876+	404.9	335,539	755.82	1326.4	603.7	0	0	0	404.9
2	0.5	26	0–2	All	1.3013	1.9621	2.00	322,980	4.1829	427.3	335,539	755.82	1483.0	603.7	0	0	0	324.1
3	1.0	17,050	0	All	1.3024	1.8476	2.00	368,502	2.2643	487.4	335,539	755.82	1396.4	603.7	0	0	0	369.0
4	1.0	17,050	2	All	1.3004	1.8558	2.00	368,814	2.2662	488.0	335,539	755.82	1402.7	603.7	0	0	0	369.6
5	1.25	25,510	0	All	1.3413	1.6937	2.00	390,510	2.1937	497.4	335,539	785.13	1280.1	603.7	0	954.0	29.30	350.9
6	1.25	25,510	2	All	1.3004	1.7060	2.00	391,876	2.2014	502.4	335,539	779.67	1289.4	603.7	0	786.4	24.15	355.0
7*2	1.50	29,540	0	> 84,924	1.7112	1.8620	2.70	446,540	2.0831	535.1	335,539	834.54	1506.5	603.7	53.24	829.8	25.48	397.3
8	1.50	29,540	0	80,450	1.6537	2.0389	2.70	423,050	1.9736	541.7	313,457	780.99	1541.0	564.0	49.74	819.6	25.17	394.4
9	1.50	29,540	0	< 69,040	1.4390	2.8244	2.70	345,030	1.5935	566.7	242,573	608.86	1651.8	436.4	38.49	780.1	23.96	377.8
10*	1.50	29,540	2	> 101,510	1.7050	1.8806	2.70	449,440	2.0966	541.8	335,539	829.49	1521.6	603.7	53.24	665.3	20.43	403.2
11	1.50	29,540	2	87,610–96.380	1.6475	2.0589	2.70	426,440	1.9902	549.8	335,539	775.93	1556.1	563.9	49.74	665.0	20.11	401.5
12	1.50	29,540	2	< 71,830	1.4384	2.8265	2.70	349,770	1.6316	575.4	244,229	607.83	1664.5	439.4	38.76	616.4	18.93	386.2
14	1.75	33,060	0	65,060–108000	1.2960	3.0510	2.50	407,390	1.6405	622.9	267,509	654.00	1931.0	481.3	30.32	687.0	21.10	447.1
15	1.75	33,060	0	< 55,160	1.0484	3.4259	2.10	377,630	1.5086	631.6	250,919	539.17	1955.8	451.5	5.69	725.2	22.27	439.3
16	1.75	33,060	2	> 157,740	1.9053	2.1626	3.25	507,540	2.0438	585.0	335,184	866.86	1838.2	603.1	94.18	549.0	16.86	452.4
17*	1.75	33,060	2	97,450–137,450	1.5920	2.7718	3.00	442,280	1.7530	619.5	281,533	713.99	1934.7	506.3	63.83	520.7	15.99	458.4
18	1.75	33,060	0	< 84,260	1.0482	3.4268	2.10	281,520	1.5368	643.6	253,742	592.82	1976.3	456.0	5.74	525.0	16.12	449.6
19	2.00	36.210	0	> 165,170	1.5607	2.8903	3.00	523,700	1.8736	657.4	314,487	796.63	2253.5	565.5	70.60	551.2	16.93	513.0
20*	2.00	36,210	0	83,840–155,600	1.2693	3.1845	2.50	493,270	1.7647	675.0	301,746	730.78	2273.4	542.9	34.20	549.7	16.88	517.6
21	2.00	36.210	0	< 73,730	0.9508	3.8068	2.00	434,030	1.5804	691.7	270,848	627.53	2322.6	487.3	0	567.7	17.43	508.4

Table A2—*continued*

Case	M_0	Z, ft	α, deg	D_I, lb_f	ER_0	β	ER_{ij}	T, lb_r	C_r^+	I_r, $lb_r s/lb_m$	T_R, lb_r	W_p, lb_m/s	W_a, lb_m/s	W_{o2}, lb_m/s	W_{RJH2}, lb_m/s	W_{BB}, lb_m/s	W_{BBH}, lb_m/s	I_{EFF}, $lb_r/s/lb_m$
22	2.00	36,210	2	> 141, 650	1.5615	2.8874	3.00	540,570	1.9333	668.2	321,586	809.05	2302.1	578.6	72.91	384.4	11.75	526.0
23*	2.00	36,210	2	< 117, 830	0.9501	3.8123	2.00	449,650	1.6349	708.1	276,486	635.06	2374.3	497.4	0	401.0	12.26	527.0
24	2.25	39,440	0	> 192, 840	1.4381	3.4041	3.00	564,900	1.8640	731.3	307,079	772,43	2591.2	552.4	69.60	365.8	11.23	584.4
25*	2.25	39,440	0	120,680–175,730	1.1388	3.9255	2.50	514,800	1.6986	765.0	279,003	672.98	2591.2	501.8	31.63	419.4	12.88	594.1
26	2.25	39,440	0	< 100, 290	0.8863	5.0125	2.20	427,850	1.4117	804.5	224,967	531.81	2591.2	404.8	10.20	484.0	14.86	588.3
27	2.25	39,440	2	< 157, 290	1.1995	3.5609	2.50	573,010	1.8907	751.3	319,326	762.69	2690.2	574.5	36.20	234.1	7.19	600.5
28*	2.25	39,440	2	< 142, 080	0.9867	4.1580	2.20	511,990	1.6894	782.0	281,560	654.69	2690.2	506.6	12.77	250.3	7.69	606.4
29	2.50	42,440	0	> 130, 280	1.0537	5.7893	3.00	453,230	1.3984	957.0	188,231	473.59	2701.9	338.7	42.68	224.4	6.89	714.2
30*	2.50	42,440	0	102,300–106360	0.8192	7.7196	2.80	369,990	1.1415	1027.9	143,801	359.94	2701.9	258.7	26.08	323.7	9.94	708.4
31	2.50	42,440	0	< 91, 950	0.7328	9.0062	2.80	332,450	1.0257	1067.7	123,260	311.36	2701.9	221.8	22.35	369.9	11.36	698.4
32*	2.50	42,440	2	> 137, 520	1.0737	5.1330	2.80	515,010	1.5890	916.7	228,727	561,84	2857.5	411.5	41.48	167.4	5.14	712.0
33	2.50	42,440	2	> 107, 800	0.7874	7.1439	2.50	403,740	1.2457	988.8	169,070	408.32	2857.5	304.2	19.16	271.0	8.32	707.3
34	2.50	42,440	2	> 99, 160	0.7172	8.1644	2.50	367,730	1.3456	1020.7	147,934	360.27	2857.5	266.2	16.77	334.5	10.27	701.5
35	2.75	45,260	0	> 202, 380	1.0520	9.5878	4.80	389,620	1.1370	1311.1	99,403	297.18	2750.6	178.8	63.09	331.4	10.18	924.1
36	2.75	45,260	0	174,040	0.8750	14.5360	6.07	324,060	0.9457	1609.9	59,581	201.30	2750.6	107.2	55.02	392.9	12.07	1038.6
37*	2.75	45,260	0	< 174, 040	0.7637	23.1122	10.00	271,420	0.7917	2049.5	29,269	132.43	2750.6	52.6	53.08	437.3	13.42	1181.1
38	2.75	45,260	2	< 248, 540	0.8750	11.7383	4.65	368,904	1.0766	1434.1	87,802	257.24	2941.2	158.0	52.75	218.4	6.71	1003.6
39*	2.75	45,260	2	< 189, 070	0.7000	24.176	9.12	280,641	0.8190	2142.5	31,459	130.99	2941.2	56.6	50.77	304.2	9.35	1264.6
40*	3.00	47,950	0	> 81, 170	1.0000	31.5910	106.80	304,270	0.8449	3121.3	3,400	97.5	2776	17.6	70.23	313.2	9.61	1941.6
41	3.00	47,950	0	49,810–77,110	0.9698	35.3430	∞	288,970	0.8024	3263.0	0	88.56	2776	0	78.54	326.5	10.02	1964.4
42	3.00	47,950	0	< 43, 250	0.7500	45.7027	∞	241,900	0.6717	3286.4	0	73.61	2776	0	60.74	419.3	12.87	1562.4
43*	3.00	47,950	2	< 76, 970	0.8750	38.8580	57.00	331,570	0.9207	3562.9	6,041	94.8	3011	10.9	77.49	189.6	5.82	2327.2
44	3.00	47,950	2	< 58, 150	0.7000	48.9671	∞	264,310	0.7339	3781.4	0	69.9	3011	0	61.49	273.6	8.40	2136.1

Table A3 Flow areas and area ratio for the conceptual design of atransatmospheric accelerator engine

Case	A_0, $in.^2$,	a_2, $in.^2$,	A_4, $in.^2$,	$\bar{A}_4$, $in.^2$,	$\bar{A}_{ij}$ $in.^2$	A_5 $in.^2$	A_7 $in.^2$	$A_8 + A_{12}$, $in.^2$,	A_{OBL}, $in.^2$,	A_9, $in.^2$,	A_{10}, $in.^2$,	A_{11}, $in.^2$,	A_{12}, $in.^2$,	$A_0 + AA_{OBL}$,/ $A_0 + A_{OBL}/A_i$,/	A_0/A_i, A_0/A_i,	A_{BL}/A_i
1	—	6875	5355	4875	1675	9,497	10,639	10,639	0	0	0	0	0	—	—	—
2	4,997	5368	5355	5238	1312	9,860	11,133	11,133	0	0	0	0	0	0.1650	—	—
3	4,247	4328	5355	4721	1829	10,740	15,927	15,927	0	0	0	0	0	0.1402	—	—
4	4,252	4333	5355	4726	1824	10,944	15,994	15,994	0	0	0	0	0	0.1404	—	—
5	4,304	4153	5355	4643	1907	9,711	18,469	65,088	3206	3417	4051	4055	46,599	0.2480	0.1420	0.1059
6	4,336	4184	5355	4650	1900	9,777	18,563	56,981	2644	2875	3340	3343	38,418	0.2305	0.1432	0.0873
7a	4,961	4383	5355	4858	1692	12,115	23,504	65,088	2732	2777	3072	3281	41,554	0.2540	0.1638	0.0902
8	5,074	4482	5355	4970	1580	13,171	24,014	65,088	2699	2747	3034	3241	41,044	0.2567	0.1676	0.0891
9	5,439	4805	5355	5327	1223	17,400	25,992	65,088	2569	2632	2888	3085	39,066	0.2644	0.1796	0.0848
10*	5,010	4408	5355	4869	1681	12,215	23,647	56,981	2191	2291	2467	2633	33,334	0.2378	0.1654	0.0724
11	5,124	4508	5355	4979	1571	13,280	24,168	56,981	2157	2260	2428	2592	32,813	0.2404	0.1692	0.0712
12	5,481	4823	5355	5326	1224	17,400	26,098	56,981	2030	2149	2285	2440	30,883	0.2480	0.1810	0.0670
13*	6,255	5056	5355	5323	1228	17,400	30,967	65,088	2244	2228	2123	2606	34,091	0.2807	0.2065	0.0741
14	6,307	5098	5355	5367+[b]	1183	17,400	30,921	65,088	2244	2228	2126	2609	34,137	0.2824	0.2083	0.0741
15	6,388	5163	5355	5436+	1114	17,400	29,100	65,088	2369	2328	2242	2751	35,988	0.2892	0.2109	0.0783
16	6,052	4783	5355	5054	1496	14,413	29,750	56,981	1807	1842	1694	2081	27,231	0.2595	0.1998	0.0597
17*	6,370	5034	5355	5319	1231	17,400	31,154	56,981	1714	1769	1607	1974	25,827	0.2670	0.2103	0.0567
18	6,506	5142	5355	5434+	1117	17,400	30,936	56,981	1729	1780	1621	1990	26,045	0.2719	0.2148	0.571
19	7,371	5471	5355	5345	1205	17,400	37,116	65,088	1803	1809	1458	2099	27,942	0.3029	0.2434	0.595
20	7,436	5519	5355	5393	1158	17,400	37,190	65,088	1798	1806	1454	2094	27,868	0.3049	0.2455	0.0594
21	7,597	5639	5393	5509+	1041	17,400	36,282	65,088	1857	1849	1502	2162	28,766	0.3122	0.2509	0.0613

Table A3—*continued*

Case	A_0, $in.^2$,	a_2, $in.^2$,	A_4, $in.^2$,	$\bar{A}_4$, $in.^2$,	$\bar{A}_{ij}$ $in.^2$	A_5 $in.^2$	A_7 $in.^2$	$A_8 + A_{12}$, $in.^2$,	A_{OBL}, $in.^2$,	A_9, $in.^2$,	A_{10}, $in.^2$,	A_{11}, $in.^2$,	A_{12}, $in.^2$,	$A_0 + A_{OBL}/A_i$,	A_0/A_i,	A_{BL}/A_i
22	7,530	5353	5393	5338	1213	17,400	37,582	56,981	1257	1365	998	1451	19,399	0.2902	0.2487	0.0415
23*	7,782	5532	5393	5405+	1045	17,400	36,744	56,981	1314	1405	1041	1513	20,237	0.3004	0.2570	0.0434
24	8,794	6156	5393	5355	—	17,400	45,727	65,088	1241	1402	882	1485	19,331	0.3314	0.2904	0.0410
25*	8,794	6156	5393	5355	—	17,400	42,897	65,088	1423	1529	1012	1703	22,161	0.3374	0.2904	0.0470
26	8,794	6156	5393	5355	—	17,400	39,483	65,088	1642	1683	1168	1965	25,575	0.3446	0.2904	0.0542
27	9,130	6061	5393	5355	—	17,400	44,693	56,981	825	1081	540	936	12,888	0.3287	0.3015	0.0272
28*	9,130	6061	5393	5355	—	17,400	43,845	56,981	883	1119	584	1001	13,136	0.3306	0.3015	0.0291
29	9,527	6315	4954	4954	—	16,928	44,559	64,555	791	1095	489	984	12,422	0.3407	0.3146	0.0261
30*	9,527	6315	4954	4954	—	16,928	39,063	64,555	1141	1327	725	1419	17,918	0.3523	0.3146	0.0377
31	9,527	6315	4954	4954	—	16,928	36,504	64,555	1304	1435	829	1621	20,477	0.3577	0.3146	0.0431
32*	10,076	6273	4999	4954	—	16,982	47,304	56,507	590	937	360	720	9,203	0.3522	0.3327	0.0195
33	10,076	6273	4999	4954	—	16,982	41,605	56,507	956	1165	547	1165	14,902	0.3643	0.3327	0.0316
34	10,076	6273	4999	4954	—	16,982	38,112	56,507	1395	1438	718	1439	18,395	0.3788	0.3327	0.0461
35	10,095	6322	4420	4420	—	16,301	44,553	63,847	1216	1412	701	1585	19,294	0.3735	0.3334	0.0401
36	10,095	6322	4420	4420	—	16,301	40,968	63,847	1442	1554	832	1879	22,879	0.3810	0.3334	0.0476
37*	10,095	6322	4420	4420	—	16,301	37,734	63,847	1604	1656	925	2091	25,428	0.3863	0.3334	0.0529
38	10,789	6328	4499	4499	—	16,393	43,239	55,843	805	1121	440	1023	12,604	0.3828	0.3562	0.0265
39*	10,789	6328	4499	4499	—	16,393	38,276	55,843	1117	1306	613	1426	17,567	0.3931	0.3562	0.0369
40*	10,619	6348	3928	3928	—	15,722	43,956	63,192	1243	1433	636	1641	19,236	0.3917	0.3507	0.0410
41	10,619	6348	3928	3928	—	15,722	43,143	63,192	1249	1463	663	1711	20,063	0.3919	0.3507	0.0412
42	10,619	6348	3928	3928	—	15,722	37,499	63,192	1604	1678	851	2197	25,757	0.4036	0.3507	0.0529
43*	11,518	6584	4028	4028	—	15,839	43,848	55,217	725	1129	374	982	11,389	0.4042	0.3803	0.0239
44	11,518	6584	4028	4028	—	15,839	38,466	55,217	1046	1314	539	1417	16,751	0.4149	—	0.0346

[b]The symbol + denotes $\bar{A}_4 > A_4$ expansion into injector base.

Also note that for intermediate thrust levels in some instances there is a range of vehicle drag values over which no interpolation is required, e.g., at $M_0 = 2.0$, $Z =$ 36,210 ft, $\alpha = 0$ deg for $83{,}840 < D < 155{,}600$ lb$_f$ (case 20, Table A2).

For example, given $M_0 = 1.8$, $Z = 33{,}060$ ft, $D = 90{,}000$ lb$_f$, $\alpha = 1$ deg find T, T_R, I_F, W_p and W_{o2}/W_p. Entering Table 2 at case 14 $T = 407{,}390$ lb$_f$, $T_R =$ 267,509 lb$_f$, $I_f = 622.9$ lb$_f$s/lb$_m$, $W_p = 654.0$ lb$_m$/s and $W_{02}/W_p = 481.3/654 =$ 0.7359. Reentering Table 2 at Case 20.

$M_0 = 2$, $Z = 36{,}210$ ft, $\alpha = 0°$, $T = 493{,}270$ lb$_f$, $T_R = 301{,}746$ lb$_f$, $I_f =$ 673.0 lb$_f$s/lb$_m$, $W_p = 730.78$ lb$_m$, $W_{02}/W_p = 542.9/730.78 = 0.7249$.

Using linear interpolation for $M_0 = 1.8$ with a $\alpha = 0°$, and $D = 90{,}000$ lb, yields $T = 424{,}566$ lb$_f$, $T_R = 274{,}356$ lb$_f$, $I_f = 633.22$ lb$_f$s/lb$_m$, $W_p = 670.38$ lb$_m$ and $W_{02}/W_p = 0.7337$ at $M_0 = 1.75$ and the procedures given in Eqs. B1–B9 for $D = 90{,}000$ lb$_f$

Using C_1 and C_2 from Table A2, Cases 17–18

$$T = \left(\frac{-C_1 D}{C_2}\right)^{1/2} = \left(\frac{-794.5 \times 90{,}000}{-3.958 \times 10^{-4}}\right)^{1/2} = 425{,}040 \text{ lb}_f$$

$$T_R = \frac{425{,}040 - 381{,}520}{442{,}280 - 381{,}520}(281{,}533 - 253{,}472) + 253{,}472$$

$$= 0.7162(28061) + 253{,}472 = 273{,}571 \text{ lb}_f$$

$$I_f = 794.5 - 3.958 \times 10^{-4} \times 425{,}040 = 626.27 \text{ lb}_f\text{s/lb}_m$$

$$W_p = \frac{425{,}040}{626.27} = 678.69 \text{ lb}_m\text{/s and}$$

$$\frac{W_{02}}{W_p} = [0.7162(506.3 - 456) + 456]/678.69 = 0.7250$$

For $M_0 = 2.00$, $Z = 36{,}210$ ft, $\alpha = 2$ deg , and $D = 90{,}000$ lb$_f$ interpolation is not required so the values for Case 23 are used, viz., $T = 449{,}650$ lb$_f$, $T_R = 276{,}48$ lb$_f$, $I_f = 708.1$ lb$_f$s/lb$_m$, $W_p = 635.06$ lb$_m$/s and $W_{02}/W_p =$ 497.4/635.06 = 0.7832.

Linear interpolation between $M_0 = 1.75$ and 2.80 at $\alpha = 2°$ for $M_0 = 1.8$, and $D = 90{,}000$ lb$_f$ yields $T = 429{,}962$ lb$_f$, $T_R = 274{,}154$ lb$_f$, $I_f = 642.64$ lb$_f$s/lb$_m$, $W_p = 669.06$ lb$_m$/s and $W_{02}/W_p = 0.7366$

Finally linear interpolation between $\alpha = 0°$ and $2°$ for $M_0 = 1.8$, $\alpha = 1°$ yields $T = 427{,}264$ lb$_f$, $T_R = 274{,}255$ lb$_f$, $I_f = 637.93$ lb$_f$s/lb$_m$, $W_p = 669.77$ lb$_m$/s and $W_{02}/W_p = 0.7352$.

The final adjustment takes into account the altitude change from that of the example to that on the reference trajectory at the same M_0. For the reference trajectory at $M_0 = 1.8$, $Z = 33{,}640$ ft, and thus the adjustment must be made for $Z = 32{,}000$ ft in the following manner. The rocket thrust remains constant, 274,255 lb$_f$ in this case; the remainder of the thrust is adjusted by the ratio of dynamic pressure, i.e.,

$$(T - T_R)_z = (T - T_R)_{z\text{REF}} \frac{P}{P_{\text{REF}}} = (427{,}274 - 274{,}255)\frac{3.990}{3.692} = 165{,}370 \text{ lb}_f$$

$$T = 165{,}370 + 274{,}255 = 439{,}625 \text{ lb}_f$$

Table B1 Interpolation parameters for a conceptual design af a transatmospheric accelerator engine

Cases	C_1	C_2
7, 8	660.7	-2.814^{-4}
8, 9	667.2	-3.203^{-4}
10, 11	699.9	-3.517^{-4}
11, 12	691.9	-3.305^{-4}
13, 14	829.3	-5.066^{-4}
14, 15	741.3	-2.906^{-4}
16, 17	849.5	-5.202^{-4}
17, 18	794.5	-3.958^{-4}
19, 20	960.3	-5.783^{-4}
20, 21	813.2	-2.802^{-4}
22, 23	905.4	-4.389^{-4}
24, 25	1110.4	-6.710^{-4}
25, 26	999.2	-4.550^{-4}
27, 28	1051.3	-5.036^{-4}
29, 30	1343.1	-8.519^{-4}
30, 31	1420.8	-1.062^{-4}
32, 33	1250.6	-6484^{-4}
33, 34	1315.8	-8.994^{-4}
35, 36	2947.4	-4.184^{-4}
36, 37	4459.8	-8.880^{-3}
38, 39	4395.0	-8.026^{-3}
40, 41	4256.8	-3.732^{-3}
41, 42	3891.4	-2.294^{-3}
43, 44	4640.0	-3.249^{-3}

The specific impulse is assumed to be invariant with Z, and therefore the propellant flow is

$$W_p = \frac{439,625}{637.93} = 689.14 \text{ lb}_\text{m}/\text{s}$$

but the weight flow of oxygen is held constant, thus

$$\left(\frac{W_{02}}{W_p}\right)_z = \frac{(W_{02}/W_P)_{\text{ZREF}} \cdot W_{P\text{ZREF}}}{W_{P_z}} = 0.7352 \times \frac{669.77}{689.14} = 0.7145$$

Finally, the net force specific impulse at this point is

$$I_\text{eff} = 637.93\left(1 - \frac{90,000}{439,625}\right) = 507.33 \text{ lb}_\text{f}\text{s}/\text{lb}_\text{m}$$

References

[1]Ferri, A., "Possible Directions of Future Research in Airbreathing Engines," *Fourth AGARD Colloquium—Combustion and Propulsion,* Pergamon, 1961. Oxford UK pp. 3–15.

[2]Dugger, G. L., Billig, F. S., and Avery, W. H., "Hypersonic Propulsion Studies at Applied Physics Laboratory," Johns Hopkins Univ./Applied Physics Lab., Laurel, MD JHU/APL TG 405, June 1961.

[3]Escher, W. J. D., and Flornes, B. J., et al., "A Study of Composite Propulsion Systems for Advanced Launch Vehicle Applications," Marquardt Corp. Final Rept., NASA Contract NAS7-377, Van Nuys CA. April 1967.

[4]Kramer, P., "Analyse Des Antriebs—Und Einsatzpotentials Luftatmender Kombinationsantriebe Für Ballistische Raumtransporter," Institut für Raumfahrtsysteme, Universität Stuttgart, 1987.

[5]Bendot, J. G., "Composite Propulsion Systems for an Advanced, Reusable Launch Vehicle Application," Proceedings of the 2nd International Symposium on Airbreathing Engines, Sheffield, England, UK, March 1974. The Royal Aeronautical Society London UK, Paper 9.

[6]Shapiro, A. H., *The Dynamics and Thermodynamics of Compressible Fluid Flow,* Vol. 2, Ronald, New York, 1954.

[7]Leipmann, H. W., and Bryson, A. E., Jr., "Transonic Flow Past Wedge Sections," *Journal of Aeronautical Science,* Vol. 17, No. 12, 1950, p. 745.

[8]Gibson, R. E., "Proposed Program for Research on Hypersonic Propulsion to Mr. Weldon Worth, Wright Patterson Air Force Base Ohio," Johns Hopkins Univ./Applied Physics Lab., Laurel MD JHU/APL AC-5165, Dec. 1960.

[9]Walker, R. K., Deklau, B., and Dugger, G. L., "Results of the Recent Tests on Rocket Thrust Augmentation," Minutes of the 31st Bumblebee Propulsion Panel Meeting, Johns Hopkins Univ./Applied Physics Lab., Laurel, MD JHU/APL 63–53, June 1964.

[10]Perini, L. L., Walker, R. E., and Dugger, G. L., "Preliminary Study of Air Augmentation of Rocket Thrust," Johns Hopkins Univ./Applied Physics Lab., Laurel MD JHU/APL TG 545, Jan. 1964.

[11]Billig, F. S., and Van Wie, D. M., "Translating Cowl Inlet with Retractable Propellant Injection Struts," U.S. Patent 5, 214, 914, June 1, 1993.

[12]Billig, F. S., "Propellant Utilization System," U.S. Patent 5, 135, 184, Aug. 4, 1994.

[13]Billig, F. S., and Van Wie, D. M., "Efficiency Parameters for Inlets Operating at Hypersonic Speeds," *Eighth International Symposium on Air Breathing Engines, AIAA,* Washington, DC, 1987. pp. 118–130.

[14]Billig, F. S., "Propulsion Systems from Takeoff to High Speed Flight," *High Speed Flight Propulsion Systems,* Progress in Astronautics and Aeronautics, Vol 137 AIAA, Washington, DC, 1991. pp. 21–100.

[15]Pandolfini, P. P., and Friedman, M. A., "Instructions for Using Ramjet Performance Analysis (RJPA) IBM-PC Version 1.24," Johns Hopkins Univ./Applied Physics Lab., Laurel MD JHU/APL AL-92-P175, June 1992.

[16]Billig, F. S., and Sullins, G. A., "A Generalized Method of Force Accounting," 24th JANNAF Combustion Meeting, Monterey, CA, Oct. 1987.

[17]Persh, J., "A Theoretical Investigation of Turbulent Boundary Layer Flow with Heat Transfer at Supersonic and Hypersonic Speeds," U.S. Navel Ordnance Lab., NAVORD Rept. 3854, White Oak, MD, May 1955.

[18]Billig, F. S., "Research on Supersonic Combiustion," *Journal of Propulsion and Power,* Vol. 9, No. 4, 1993, pp. 499–514.

Variable Cycle Engine Developments at General Electric - 1955–1995

J.E. Johnson
General Electric Aircraft Engines, Evendale, Ohio 45215–6301

I. Introduction

The following three sections contain discussions and descriptions of variable cycle gas turbine engine concepts that address the propulsion needs of mixed mission (subsonic/supersonic) capable aircraft systems.

Section II, Variable Cycle Engines - The Next Step in Propulsion Evolution?, was written in 1976 and presented at the AIAA/SAE 12th Propulsion Conference in Palo Alto, California. This paper discusses the evolution of mixed mission aircraft propulsion from the afterburning turbojet through several novel variable cycle engine concepts that had been evaluated by General Electric in the 1960–1975 time period. Also included are discussions on how basic engine operation affects installation losses associated with inlet spillage and afterbody closure drag levels.

Section III, Variable Cycle Engine Concepts was prepared for a 1995 AGARD meeting. This paper revisits and adds detail to the variable cycle engine (VCE) concepts described in the first paper and then focuses on VCE technology developments and tests that started in 1975 and culminated in the flight test of the YF120 advanced tactical fighter engine (ATFE) in the 1990 time period.

Section IV provides a General Electric perspective on high-Mach capable turbo engines. Several variable cycle propulsion systems capable of Mach 4-6 flight speeds are described along with discussions on nacelle concepts, fuel types, and high-Mach system cooling considerations.

The three sections in this chapter provides an historical record on the development of variable cycle engine concepts that have taken place during the past 40 years, and some guidance for future developments.

Copyright © 1995 by the American Institute of Aeronautics and Astronautics, Inc. All rights reserved.

*Manager, Advanced Studies.

II. Variable Cycle Engines: The Next Step in Propulsion Evolution?

A. Introduction

The development of the gas turbine engine from the early test stand devices of 40 years ago to today's efficient propulsion systems ranks as one of the major technological achievements of the era. Although the basic principles of the gas turbine have been known since the 18th century, it was not until the 1930s that all of the various facets needed to develop an operational jet engine became available. Propulsion pioneers such as Sir Frank Whittle in England and Pabst Von Ohain in Germany, succeeded in designing and demonstrating the world's first jet engines. First generation versions of their engine concepts were built during World War II but had only limited use.

Whereas metal property advancements and design innovations were needed to make even these early engines feasible, the real heart of the machines was the compression system. Two different approaches were taken in developing the compressors for these early engines. In England, the centrifugal compressor concept was pursued, whereas in Germany the axial flow system was adopted. Although both are still in use today the axial flow approach has been the one that has proven to be the most versatile.

Growth of the axial flow turbojet engine since 1945 has been phenomenal. Thrust levels have grown from the 1000–2000 lb class into the 50–60,000 lb class of the turbofans for the wide body transports. In 1970 the GE4 SST engine demonstrated over 70,000 lb of thrust. Future large transport engines could be developed to even higher thrust levels.

Basic engine operational flexibility has been greatly improved by the addition of technology features, such as variable compressor stators, dual rotor compression systems, efficient afterburner concepts and variable exhaust nozzle features. Turbofan developments (both low– and high-bypass versions), air-cooled turbines, better materials, and advanced design practices and manufacturing processes have all been added to the jet engine technology list. Collectively, these advancements have produced a family of propulsion devices that cover the Mach number range from low-subsonic speeds to Mach 3^+ flight.

B. Present Engine Capabilities for Mixed Subsonic/Supersonic Mission Applications

The General Electric Company has been instrumental in all the noted developments and is one of the leaders in propulsion innovation and practical application of new concepts. In line with this, the General Electric Company, in conjunction with the Air Force, Navy, and NASA, has been vigorously studying what the next major step in mixed mission propulsion should be, namely, the variable cycle engine.

First, to determine what a variable cycle engine should be capable of doing let us examine what present engine types can and cannot do.

1. Turbojet Cycle

Figure 1 shows a familiar schematic of an afterburning turbojet engine. This afterburning concept was the first variable cycle feature used in jet engines; the afterburner is being used to produce the augmented thrust levels necessary for short takeoff, acceleration, and supersonic flight. This augmentation feature resulted in a smaller, lighter weight engine than would have resulted from simply scaling up a dry engine. The General Electric Company is the leader in this type of engine development having designed and built the first afterburning turbojet, the J47, in the late 1940s, followed by the first Mach 2.0 engine, the J79, and then the Mach 3 class J93 and GE4 augmented turbojet engines.

A major technology advancement was made in the J79 compressor by adding variable stator geometry in the front section of the compressor. This permitted stable engine operation at the relatively high-flow rates needed for supersonic flight while also providing a means for the simple single rotor system to attain higher pressure ratios than before. The variable stator or concept was further refined for the J93 and GE4 compressors by making the rear stators variable. This allowed the compressor to be high flowed in the 65–85% corrected speed region that corresponds to Mach 2–3$^+$ flight conditions.

Off-design flow potential. As stated at the beginning of this chapter, the compression system is the key element in jet engine operation. It must be capable of stall-free operation while supplying enough airflow for efficient supersonic flight. The amount of airflow that a compressor can deliver at reduced corrected speeds (increased flight Mach number) is a function of design pressure ratio, compressor stator schedule, and operating revolutions per minute. (rpm). Fig. 2 illustrates some generalized off-design flow trends for two levels of design pressure ratio. Rotor speed is held constant for both. As can be seen, the lower design pressure ratio compressor has an advantage in supersonic flow potential. Fig. 3 indicates the value in having rear variable stators for supersonic flight. At Mach 2$^+$ flight conditions, over a 20% increase in operating airflow is made available at the same rotor speed. Increasing rpm relative to sea level static conditions is another way to increase off-design flow levels, but mechanical considerations of stress and weight provide an upper limit as does the increase in turbine temperature required to drive the compressor. The effect of relative rpm on flow potential is shown in Fig. 4. By using variable stator compressor geometry coupled with a reasonable amount of relative rpm increase, the turbojet engine can be tailored to match the maximum flow demands of a supersonic inlet system.

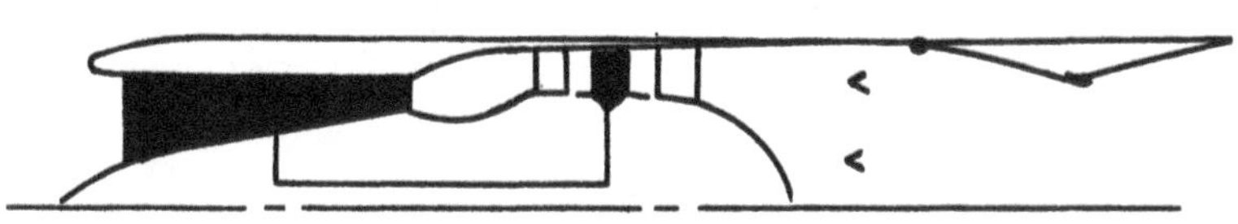

Fig. 1 Afterburning turbojet schematic.

Turbojet potentials. For supersonic flight (Mach 1.5–3.5), the turbojet cycle remains the best choice over all other candidate systems. Technology advancements in turbine inlet temperatures, cooling technologies, augmentors, nozzle, and materials will produce the type of improvements shown in Fig. 5 relative to the J79 engine. Mixed mission propulsion systems, however, must be evaluated at more than maximum power and supersonic flight conditions, and it is in the alternate subsonic legs that the turbojet cycle has proven to be inferior to the turbofan cycle. This has been the major reason for selecting augmented turbofans in the bypass $0.3–1.0^{+}$ class for the latest Mach 2 class mixed mission fighter and attack aircraft - TF30 for F111 and F14, F100 for F15 and F16, F404 for F18, and bypass 2 class F101 for the B1 bomber. As turbine inlet temperatures are increased to improve supersonic performance, subsonic specific fuel consumption levels for turbojets get progressively higher. Increased turbine temperatures also increase dry thrust levels. Whereas this is advantageous for dry maneuvers and dry combat ceiling potentials, it now becomes more difficult to throttle back to dry cruise power settings without having to drastically reduce airflow and, thus, incur the possibility of encountering large spill drag losses and significant afterbody boattail drag losses. Throttle dependent installation losses for some of the present mixed mission aircraft have proven to be much higher than those predicted during initial design studies. A viable solution to these installation losses is one of the main themes of present variable cycle engine studies.

Variable exhaust nozzle scheduling for dry operation. Proper utilization of the variable jet nozzle can help produce an improved airflow – thrust relationship for turbojet engines. By opening the exhaust nozzle, compressor

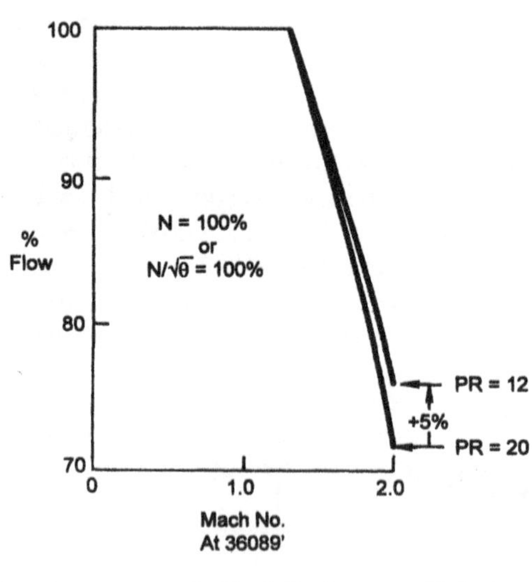

Fig. 2 Turbojet flow trends; design PR effect on off-design flow potential.

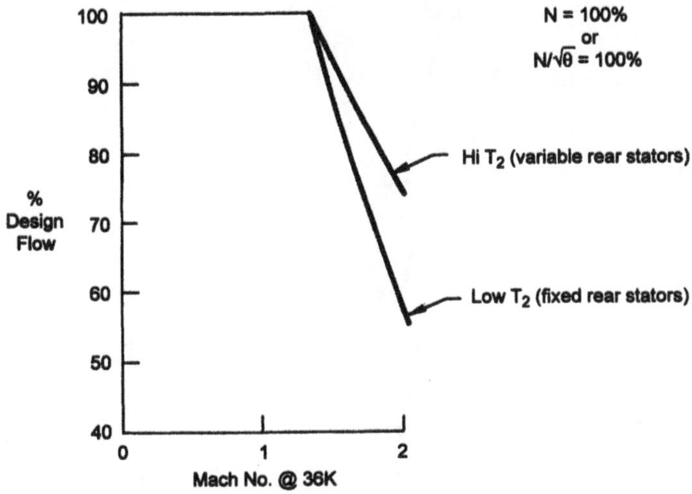

Fig. 3 Turbojet flow trends; rear variable stator effects on off-design flow potential.

flow can be maintained while turbine inlet temperature and compressor operating line (pressure ratio) are reduced. This is depicted in Fig. 6. Depending on the design temperature level, compressor efficiency island shapes, and pressure drops (compressor and turbine exit Mach numbers increase), this technique will permit airflow to be maintained down to the 80% dry thrust region while still maintaining good specific fuel consumption (SFC) levels.

Variable area turbine. The addition of a variable geometry turbine vane system has been studied many times for turbojet engines. This variable geometry feature, when used in conjunction with a variable exhaust nozzle, permits some unique engine operating modes. Within limits, compressor flow and pressure ratio can be maintained while temperature is decreased. This permits a more optimum combination of cycle pressure ratio and turbine temperature during part power operation and can result in 6–8% subsonic SFC improvement for very high-temperature, low-pressure ratio turbojet cycles. Overall, this feature has potential for improving the mixed mission performance of very high-temperature (2600–3000$^{+\circ}$F) turbojet engines.

Turbojet cycle conclusions. Although the turbojet cycle has not been considered for the present mixed mission aircraft fleet, it should not be neglected in future systems studies. Its lightweight and simple, low-cost, easily maintainable design make it attractive from many standpoints of life cycle cost. Variable geometry features such as compressor stators and exhaust systems are mandatory, and afterburners are likely for fastest acceleration and maximum maneuver capability both subsonically and supersonically. For specialized missions using high-temperature cycles without afterburners a variable area

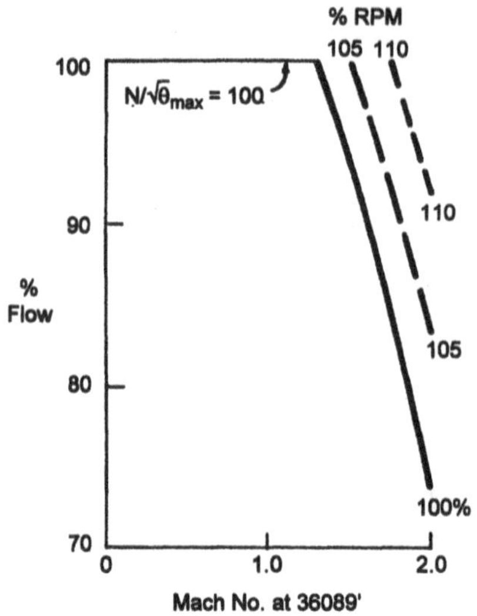

Fig. 4 Turbojet flow trends; increased RPM effects on off-design flow potential.

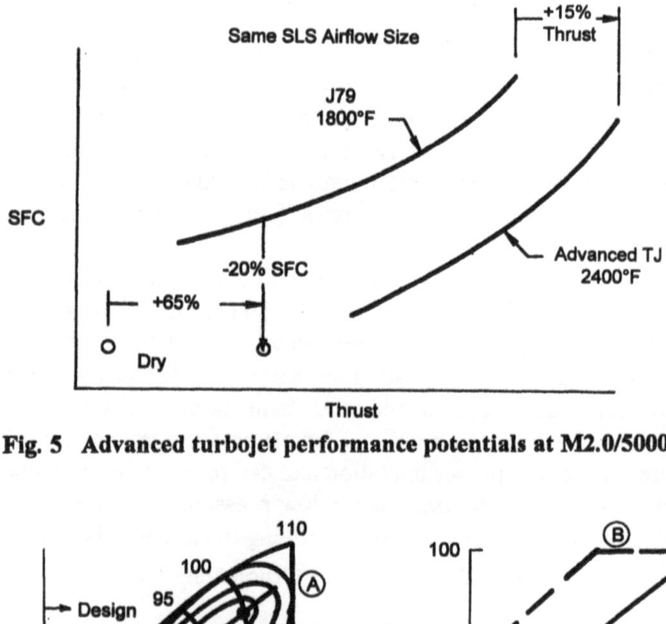

Fig. 5 Advanced turbojet performance potentials at M2.0/50000 ft.

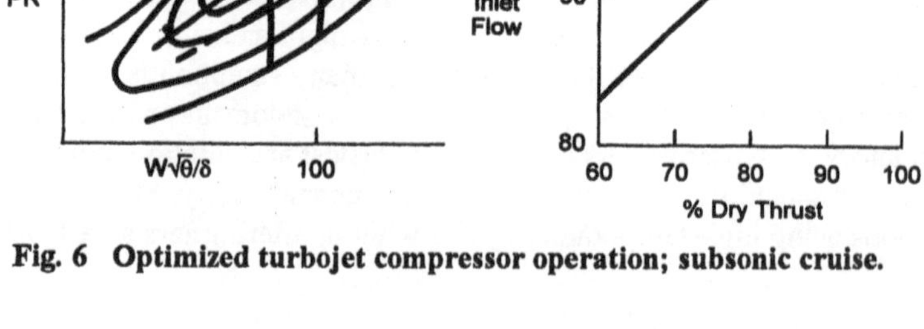

Fig. 6 Optimized turbojet compressor operation; subsonic cruise.

Table 1 Turbojet Suitability for Mixed Mission Application

Pro:

1) Best supersonic performance at high augmentation
2) Excellent potential for maximum power inlet matching
3) Highest specific thrust potential/smallest engine
4) Lightweight
5) Simple
6) Lowest acquisition cost

Con:

1) Poorest subsonic SFCs
2) Potential part power spillage - drag problem

Note:

Turbojet should be evaluated in new aircraft studies to see if cost, weight, and simplicity factors overcome subsonic performance losses in life cycle cost context

turbine would prove cost effective. Table 1 summarizes the pros and cons of the turbojet engine for mixed mission applications.

2. Turbofan Cycle

Selection of augmented turbofans for mixed missions over advanced turbojets during the 1960s and early 1970s was primarily justified by the desire to improve subsonic performance. The higher turbine temperatures being developed permitted cycle pressure ratios to be raised into the 20^+ category with high thrust/weight potential while still having sufficient power left to drive a bypass system. Supersonic SFCs, though not nearly as good as those of advanced turbojets, were still sufficient to meet the missions dictated by the military. Whereas the resultant aircraft have mixed mission capability, most cannot maintain supersonic flight for any appreciable amount of time and are really subsonic machines with supersonic capability. This is partly engine and partly airframe related since supersonic turbofan SFCs on afterburning are very high and aircraft lift/drags are very low (3–5) especially in relationship to supersonic transport (SST) levels of 8–10.

Turbofan types. Two distinct types of turbofans have been considered for mixed mission applications. One, shown schematically in Fig. 7a is a mixed flow afterburning turbofan and the other, shown in Fig. 7b, is the separated flow duct burning turbofan. Only the mixed flow version is presently in service (used for all new military fighters and bombers as well as the Russian SST aircraft).

Mixed flow selection. For current systems and for the new ones in full scale development, the mixed flow concept has been selected over the separated flow duct burner for several reasons.

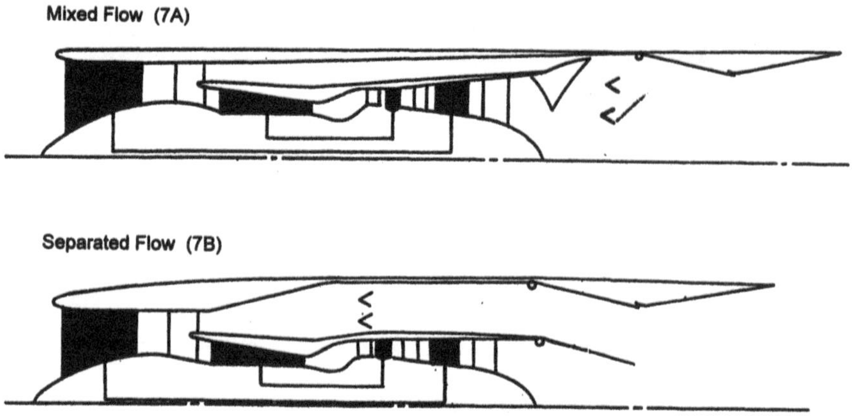

Fig. 7 Augmented turbofan schematics.

1) Mixed flow augmented turbofans maximize specific thrust since all fan air is burned in the augmentor (a second augmentor for the core stream would have to be added to the separated flow engine to approach the mixed cycle potential).

2) Mixed flow offers significant dry subsonic SFC improvement over separated flow.

3) Basic simplicity of the mixed flow afterburner and single nozzle over the duct burner and two throat nozzle is an important asset.

Mixed flow turbofan cycle operation. Several problems have occurred with actual use of the turbofan cycle in mixed mission aircraft. First, the turbofan's ability to maintain high levels of inlet flow for supersonic flight usually do not compare favorably with existing turbojet engines. This is shown in Fig. 8. As can be seen, the J79 swallows a much higher percentage of its design flow at Mach 2.0 than the band depicting a family of existing turbofan engines. This is due to the basic operating principle and present cycle matching philosophy of the mixed flow turbofan. For mixed flow systems, a static pressure match must be maintained in the exhaust system where the cold-duct and hot-core streams join. As a turbofan is flown to higher Mach numbers, the operating bypass ratio increases due to the basic aerodynamic characteristics of the fan and compressor. This trend is shown in Fig. 9. Also, limiting core rpms are reached, and limiting compressor discharge and turbine inlet temperatures are encountered. These operating characteristics result in an inability to maintain the required static pressure balance at full fan speed; therefore, fan speed (and flow) must be reduced. The flight Mach number at which this occurs can be adjusted somewhat by the manner in which the basic cycle is matched: cycle match turbine temperature, high-pressure and low-pressure turbine areas, core match

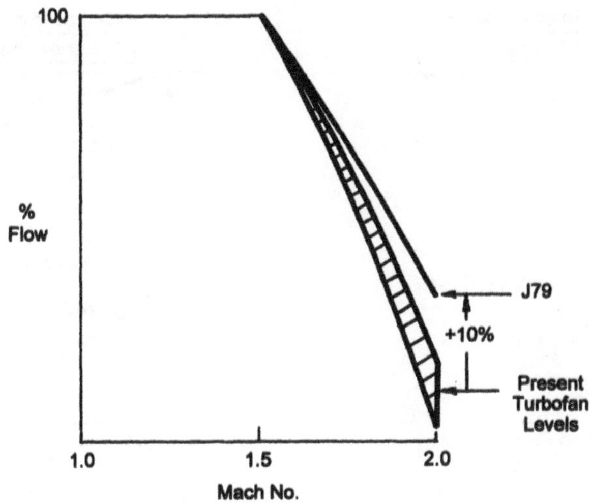

Fig. 8 Flow – Mach number comparisons: turbojet - turbofan.

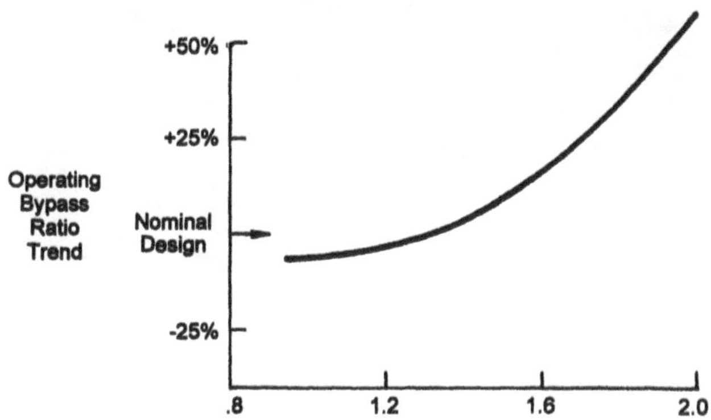

Fig. 9 Typical operating bypass ratio increase with flight Mach number; (present technology mixed flow turbofan trends).

rpm, etc. For typical turbofans, favoring of the high-Mach region normally sacrifices some engine operating in the low-Mach region.

Although the mixed flow turbofan has improved the subsonic performance of military fighter and fighter/bomber aircraft, some of its subsonic cycle advantage has disappeared in throttle-dependent installation drag losses, i.e., inlet-engine airflow mismatch (spillage) and afterbody boattail angle drags due to jet nozzle closure. These losses are shown schematically in Fig. 10. Fig. 11 illustrates a typical airflow – nozzle area – dry thrust relationship for a mixed flow fan cycle at a subsonic cruise condition. Thrust is varied by reducing fan speed and turbine temperature in the sequence dictated by the dry power engine control mode, - i.e., core speed, exhaust nozzle setting, duct Mach number, etc.,

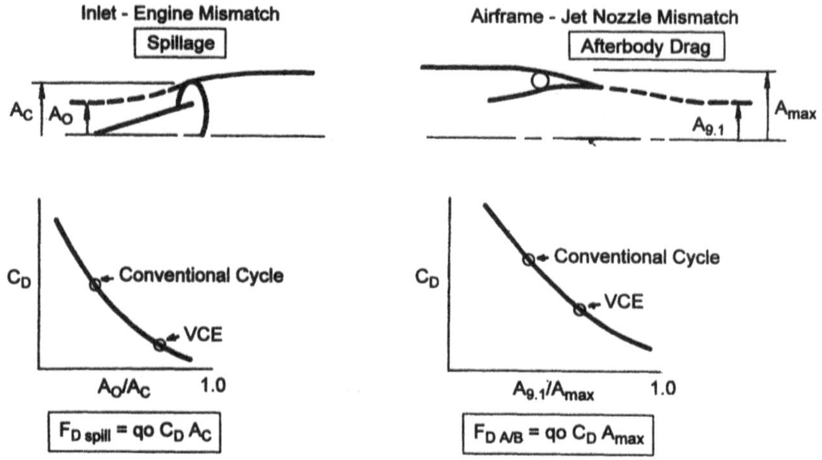

Fig. 10 Part power installation losses.

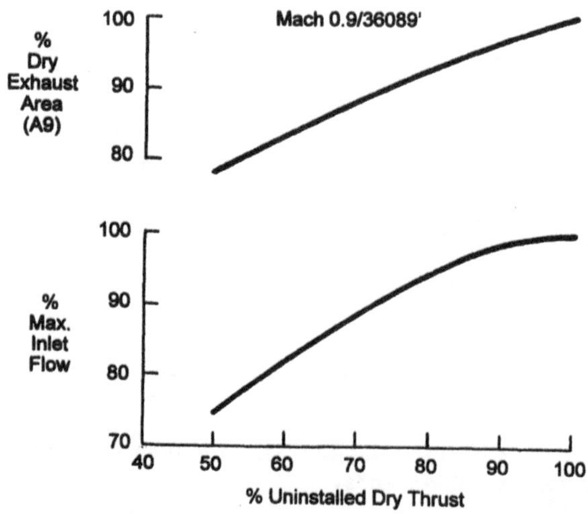

Fig. 11 Inlet and exhaust matching potentials of a typical mixed flow fan.

and the static pressure balance requirement of the mixed flow cycle. It is possible to adjust the airflow decay rate somewhat by scheduling the jet nozzle in a different manner, but this is limited by SFC increases.

Using these inlet and afterbody parameters coupled with spillage and boattail drag characteristics for Mach 2-2.5 fighter systems results in the performance loss buildup shown in Fig. 12. The uninstalled SFCs undergo a large increase in level due to addition of spillage drag and nozzle related drags. The losses shown in Fig. 12 are a strong function of the particular aircraft design, especially the maximum Mach number. Figure 13 indicates the effect of the inlet design on the performance loss buildup due to inlet spillage alone. As can be seen, the more

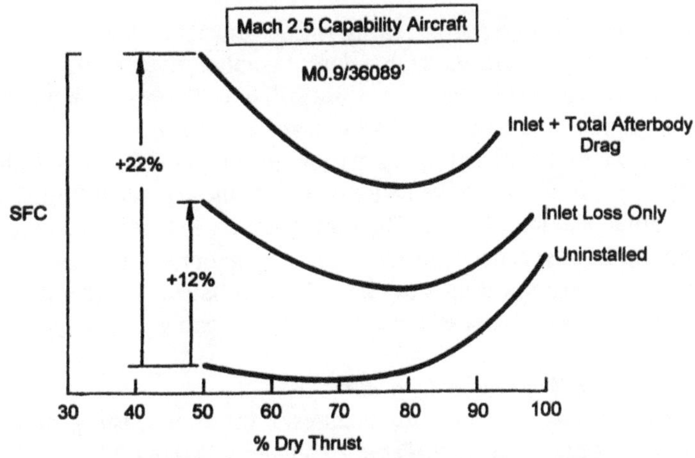

Fig. 12 Installation losses for mixed flow turbofan.

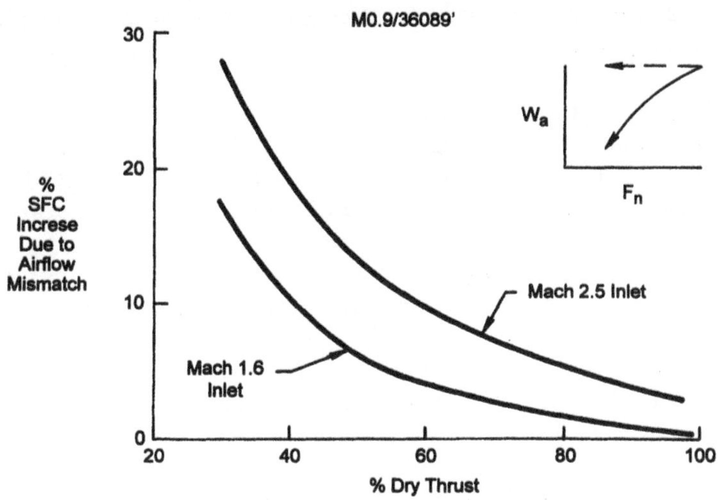

Fig. 13 Effect of inlet spillage on subsonic SFC.

simplified lower design Mach number inlet is less sensitive to inlet–engine airflow mismatch during part power subsonic flight. Afterbody drags for buried installations are more difficult to sort out due to their obvious interactions with the basic aircraft itself. Higher Mach designs require larger exhaust areas to be built into the nozzles to produce maximum thrust potentials and, therefore, result in relatively more nozzle closure (and drag) during part power dry operation.

To utilize the above data some knowledge of approximate thrust match is needed. For modern fighter and fighter bomber aircraft, take-off thrust/gross weight ratios are around 1.0, implying that the engines have been grossly oversized to meet the high power setting and sustained "g" levels needed to make the aircraft highly maneuverable during combat. Even at high-altitude Mach 0.8–0.9 cruise conditions the engines still have to operate in the 40–60% dry thrust regions where large installation losses can occur.

As pointed out in the preceding discussion, the mixed flow afterburning turbofan has several distinct advantages over both the separated flow duct burning turbofan and the turbojet. The static pressure balance needed for the mixed exhaust, however, does introduce some operational restrictions that the other two mixed mission candidates do not have or have to a lesser degree. (Flow matching at part power is a problem for all existing engine types.)

3. Summary of Existing Engine Concepts

Let us now consider the suitability features of the three existing engine types for use in mixed mission aircraft. Table 2 contains a listing of the important parameters that need to be considered and how each engine type compares on a scale of 1–3. Whereas it is obvious from this chart that none of the three fixed cycle concepts addresses all of the important mission aspects, the mixed flow afterburning turbofan cycle has been selected as the best present powerplant for mixed mission aircraft.

C. Variable Cycle Engine Studies

This lack of one clearly superior engine type has led to the various variable cycle engine studies conducted over the past 10–15 years. The General Electric Company has been very active in these past VCE studies and is presently involved in designing and testing several new promising VCE concepts. Let us now examine some of the early VCE types that were basically attempts at combining the desirable features of the turbojet and turbofan cycles in one package.

Table 3 lists chronologically the various VCE concepts studied by the General Electric Company. Let us examine them one at a time and discuss their basic operating principles.

Table 2 Cycle Suitability for Mixed Mission Application

Parameter	Turbojet	Mixed flow TF	Separated flow TF
Supersonic SFC	1	2	3
Subsonic SFC	3	1	2
Specific thrust (sizing)	1	2	3
High Mach flow potential	1	3	2
Part power spillage drag	(none adequately address this problem)		
Thrust/weight at sizing point	1	2	3
Simplicity	1	2	3
Development and acquisition cost	1	2	3

Ranking 1 = best, 3 = worst.

Table 3 Mixed Mission VCE History at General Electric

1960 – 1962 VAPCOM	0 – 1 bypass ratio swing
1960 – 1968 Flex Cycle	concentric turbojets
1970 – 1973 TACE	in-line turbofan – turbojet
1973 – 1974 MOBY	three spool double-bypass duct burning TF
1974 – 1976 Flow matching concepts	single- and double-bypass VCEs

1. Variable Pumping Compressor

The concept of the variable pumping compressor (VAPCOM) was conceived by several engineers at the Aero Propulsion Laboratory at U.S. Air Force Wright Field around 1960. It was the first approach at trying to combine the best features of the turbojet and turbofan into one system. The concept, shown schematically in Fig. 14 works in the following manner: During maximum power and supersonic cruise operation the engine is run as a dual rotor turbojet engine with almost all of the front compressor flow being passed through the core compressor. The outer bypass duct is closed with only a small leakage flow allowed to pass through the duct and outer nozzle. For subsonic flight, the outer duct and nozzle are opened, and the core compressor is low flowed by closing all of the compressor stators. This effectively increases the bypass ratio from 0 to 1.0 and converts the cycle from a turbojet to a separated flow turbofan. Variable turbine geometry is required on both the high- and low-pressure turbines. Whereas in theory the engine should have improved mixed mission performance, in practice the losses associated with the core flow modulation negated much of the higher operating bypass benefits. Also weight and complexity factors detracted from the overall value of the concept. Table 4 summarizes the operating principles and problems associated with the concept.

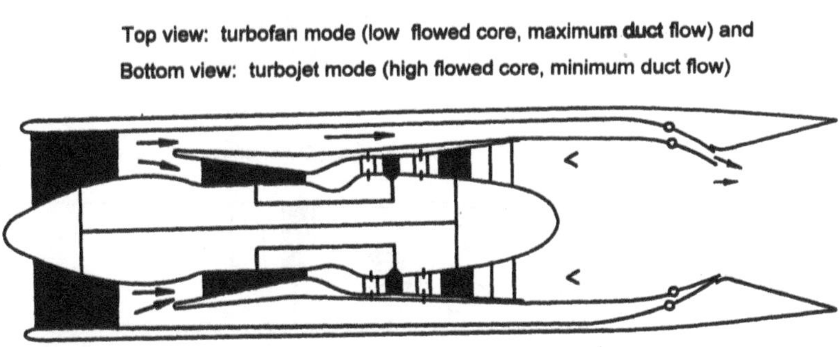

Fig. 14 Variable pumping compressor (VAPCOM) engine schematic.

Table 4 Variable Pumping Compressor

Vary bypass from 0–1 by compressor stator closure
Variable turbines required
Variable stator fan also used for bypass control and flow-speed optimization
Problems
Core compressor performance with wide flow swings
Overall P/P trend goes in wrong direction
Low for subsonic
High for supersonic
Relatively complex, heavy (core sized for full fan flow)

2. *Flex Cycle*

This engine concept was invented and patented in the 1960s by the General Electric Company. It too was conceived as a system that combined the turbojet–turbofan features in one package but in an entirely different manner from the VAPCOM. Figure 15 illustrates the concentric turbojet/turbofan principle of the flex cycle engine. The basic flowpath illustrated shows a front fan followed by a conventional core engine. A second burner is located in the bypass duct. The unusual aspect of the engine lies in the turbines that power the front fan. As can be seen, two separate turbine systems are located on the fan shaft. The first fan turbine located behind the core turbine supplies part of the energy needed to drive the fan. The large aft turbine – the critical component of the engine – supplies the remaining fan energy. The energy split is a function of the cycle parameters and operating flight condition.

The turbojet mode of the engine occurs when the outer burner is on. For low-flight Mach numbers both the outer and inner burners are on for maximum performance. As flight Mach number increases, the core engine can be effectively slowed down to minimize the compressor discharge temperatures, while the front fan is run at full speed to maximize supersonic flow. The outer burner and aft turbine supply most of the energy for supersonic flight. The

Outer burner on, concentric TJ mode, maximum power; Outer burner off, mixed flow TF mode.

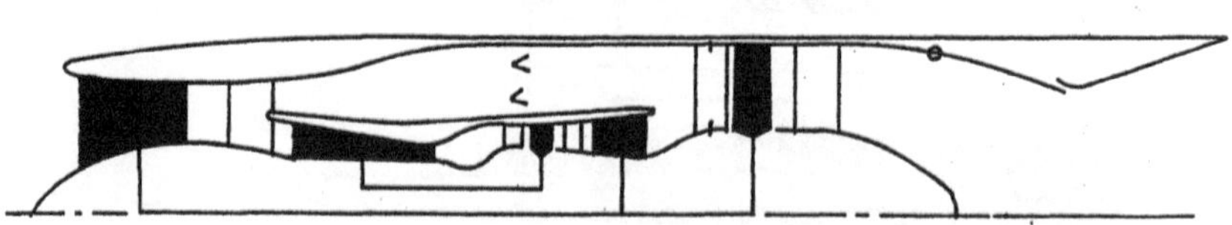

Fig. 15 Flex cycle schematic.

ability to keep the fan at full speed while the core is slowed down permits the cycle to have a very high overall pressure ratio for subsonic flight while not incurring high compressor discharge temperatures during supersonic flight.

The turbofan mode of the engine results when the outer burner is shut off and the main burner alone produces energy for the entire compression system. A static pressure balance occurs in front of the aft turbine, and basically mixed flow exhaust properties result.

Even with the outer burner off, the basic cycle operation requires that the aft turbine still supplies a relatively high percentage of the horsepower needed to drive the fan. It is this dry mode of operation where radical swings in turbine energy, corrected speed, and flow function occur. In past GE work in the 1960s, these turbine parameters and their accompanying wide exit swirl swings and losses have proven to reduce the apparent subsonic advantage of the system. Also, the engine is relatively low in supersonic specific thrust unless configured with an afterburner. In the two-burner configuration shown it is already very complex and heavy, and the addition of another burner system would only add more complexity. Table 5 summarizes the operating characteristics and problems associated with this concept

3. Turbo Augmented Cycle Design

A true marriage of the turbofan and turbojet occur with the turbo augmented cycle engine (TACE). As shown in Fig. 16, this engine is composed of two engines in series. A complete turbofan engine is included for use during subsonic flight. In this mode the bypass and core streams of the turbofan mix together and exhaust through a common nozzle. The aft turbojet is not in operation for normal subsonic cruise. For maximum power and supersonic

Table 5 Flex Cycle

Originally studied in early M 3.0 SST work (called Composite Cycle)
Two modes of operation
Concentric turbojets with outer burner on
Mixed flow turbofan with outer burner off
Only VCE that has overall PR trend go in right direction
High for subsoic
Reduced for supersonic (core slows down)
Problems
Poor aft turbine aerodynamic performance
Low specific thrust unless A/B is added
Complex, heavy (weight increase offset cycle improvements)
Cost

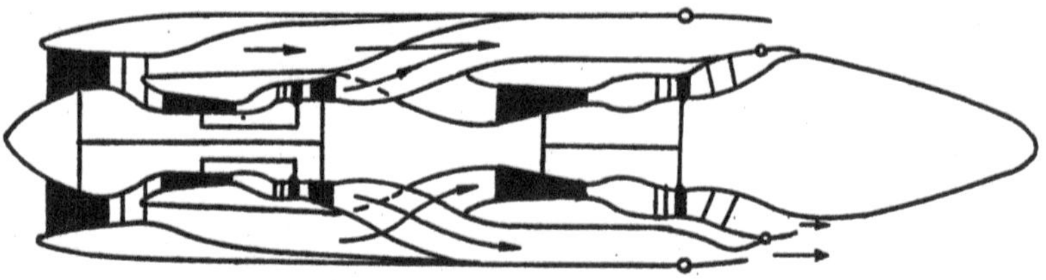

Fig. 16 Turbo augmented cycle engine schematic.

flight, the turbofan bypass flow is diverted into a duct that feeds the aft turbojet engine. This supercharged mode of operation results in the turbojet becoming a very efficient augmentor for the fan duct flow.

Although the system is derived from two well-proven engine concepts it does have several drawbacks. One is the prohibitive overall weight of the total propulsion system. With the weight comes complexity and cost. Another is the lack of specific thrust potential for nonaugmented versions of the engine. Addition of an afterburner to the turbojet and possibly a duct burner to the mixed exhaust of the front turbofan would make the system competitive on a thrust basis but would magnify the overall maintenance, complexity, and cost problems. Table 6 summarizes the characteristics of the TACE concept.

4. Turbojet/Turbofan Combined Concept Summary

The preceding three concepts all had shortcomings in trying to combine a turbojet and a turbofan in one acceptable package. In some, components proved to have extreme aerodynamic or mechanical problems, and in others the weight and complexity of the system more than overcame whatever cycle advantage was initially projected. Although a true turbojet/turbofan mode change is

Table 6 TACE

Basically 2 engines
Turbofan for subsonic - TJ shut off
Turbojet used as an augmentor for bypass flow
Supercharged by fan
Problems
Unacceptable weight and configuration complexity
Low specific thrust at high T2 levels unless A/B is added
Core flow from TF might need an augmentor also
Cost

probably not achievable in a practical, simple, viable system, present VCE studies are indicating that advancements can be made over today's existing engines.

Of the three variable cycle engines thus far discussed only the VAPCOM exhibited any potential for part power airflow tailoring needed to reduce throttle dependent installation drags. Therefore, when installation drag levels became of increased concern in the 1972–1973 time period some of the basic fundamentals of this system were re-evaluated and a new VCE concept was evolved.

5. Modulating Bypass VCE Concept

The modulating bypass (MOBY) cycle was invented by General Electric in the 1973 time period in response to the Air Force's request for engine concepts that addressed the problems of throttle-dependent installation losses. Figure 17 illustrates the initial cycle concept, which is fundamentally a separated flow duct burning turbofan cycle configured as a three-spool engine. One of the unique aspects of the configuration is found in the fan system. As can be seen, the fan is divided into two sections, each being on a separate shaft with a bypass duct located between them. Downstream of the second fan section is a conventional core engine and second bypass duct. A duct burner is located in this duct to provide thrust for acceleration and supersonic flight.

For maximum power and supersonic flight, the engine runs as a standard duct burning turbofan. Almost all of the front fan section flow is passed through the second fan section and then divided between the second bypass duct and core. A small leakage flow goes into the outer bypass duct and out the outer nozzle.

For part power subsonic conditions, the front fan operating mode is set to match the inlet requirement. This airflow match is held while thrust is varied by reducing the second and third spool speeds. Turbine temperature is reduced at an optimum rate by utilizing the variable turbine systems. Matched inlet flow can be maintained down to 50% dry thrust without any sacrifice in uninstalled SFC. Additional installed advantages occur since constant airflow thrust modulation results in more open nozzle settings that reduce aft end closure drag.

Top view: susbsonic, part power (maximum outer duct flow); and
Bottom view: maximum power mode (minimum outer duct flow).

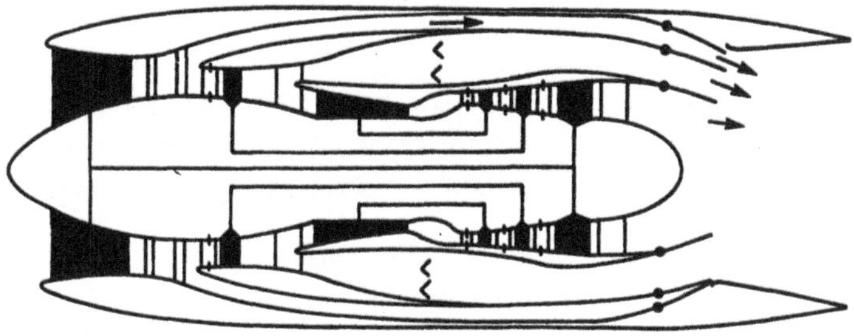

Fig. 17 MOBY schematic, three spool double bypass.

Although this concept did effectively work on installation losses, its overall complexity (three rotors, three variable nozzles, three variable turbines and two bypass ducts) may detract from its serious consideration as a practical propulsion system. Also, since the basic concept was that of a separated flow turbofan, the resultant specific thrust potential was not as high as competitive mixed flow fan systems. Table 7 summarizes the MOBY concept.

6. Present VCE Studies (Through 1975)

Whereas many of the VCE concepts discussed thus far in the chapter have certain very attractive individual capabilities, none appear to demonstrate an overall improvement large enough to replace the mixed flow turbofan as the best mixed mission propulsion system, especially when complexity, risk, and cost factors are also included. As pointed out, however, the mixed flow turbofan does have some serious deficiencies in the areas of maximum and part power flow matching.

The General Electric Company is presently studying two new types of variable cycle engines that were invented at GE in 1974–1975. They are called single and double bypass VCEs. Both retain the best features of the mixed flow cycle — maximum augmented specific thrust, good uninstalled subsonic SFCs, low IR potentials — while incorporating variable geometry components that offer various degrees of flow matching potential.

The single bypass VCE contains variable geometry features that adjust internal engine control areas to improve engine operation at both maximum and part power flight conditions. During supersonic flight they allow fan speed to be maintained at higher levels to maximize thrust and improve supersonic inlet matching. For part power subsonic flight, these variable geometry features permit independent fan speed – core speed relationships to be attained. This allows the development of a wide band of constant airflow thrust modulation

Table 7 Modulating Bypass VCE Concept

Original double bypass		
Conceived as inlet matching concept to minimize installation losses		
Concept operation		
	Maximum power mode:	runs as a standard duct burning turbofan
	Part power subsonic:	second spool and core rpms reduced, T4 drops
		inlet stays matched, afterbody losses reduced
Problems	Weight and complexity (3 nozzles, 3 spools)	
	Relatively low specific thrust levels	
	Cost	

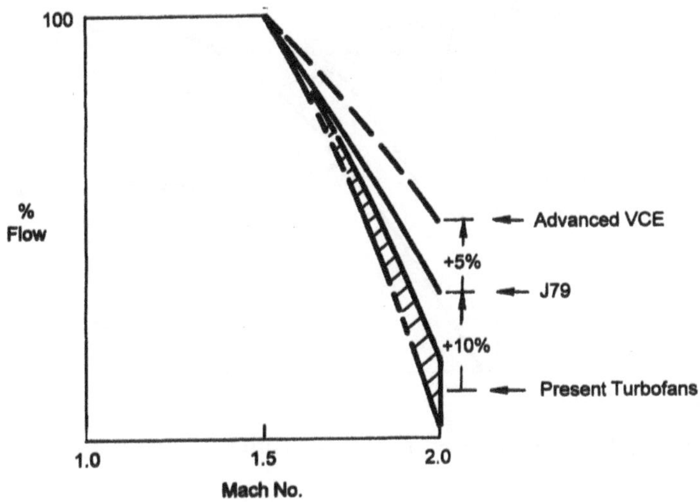

Fig. 18 Improved maximum power airflow potentials for advanced VCEs.

that results in lower installation losses and improved subsonic performance. Double bypass VCEs contain these same basic features, while also incorporating a unique two section fan that has flow modulating capabilities. A second engine bypass duct is located between the front and rear fan blocks. These two additional features allow higher bypass ratios to be attained during part power subsonic flight while also extending the constant airflow, thrust modulation potentials down to 50% of maximum dry thrust. This results in even more subsonic SFC improvement potential.

These new concepts offer the type of maximum power airflow levels shown in Fig. 18. Near optimum maximum power inlet matching at most flight conditions will be possible whereas part power inlet matching will be improved as shown in Fig. 19. For part power subsonic flight the new concepts will practically eliminate throttle-dependent inlet drag while also reducing aft end closure losses. The installed subsonic SFC improvement potentials resulting from lower throttle-dependent drag levels will allow the most basic cycle parameters, fan pressure ratio and bypass ratio, to be selected to favor maximum power and supersonic flight. The end result should be a better balance of performance potentials for mixed mission aircraft.

The mixed mission variable cycle engines presently under study by General Electric for the U.S. Air Force, U.S. Navy, and NASA emphasize simple, straightforward, practical hardware with small weight and cost penalties relative to conventional mixed flow turbofans. In addition to flow matching, these same VCE concepts lend themselves to low-noise, nonaugmented take-off operation for NASA SCAR studies while also offering flow enhancement features for use in V/STOL and supercirculation systems. Table 8 is a general description of the resultant concepts.

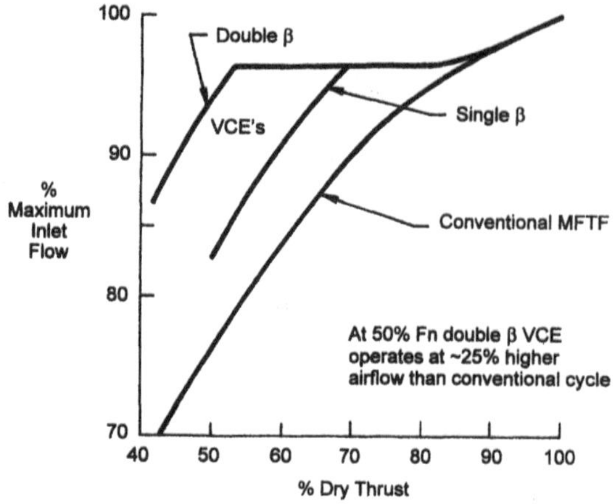

Fig. 19 VCE part power flow matching potentials; same cycle parameters.

D. Final Conclusions (Written in 1976)

If new mixed mission aircraft encounter the high levels of throttle-dependent drag associated with some of today's existing aircraft then a new type of propulsion system is definitely needed. This applies to both military aircraft as well as the second generation SST under study by NASA. The present family of variable cycle engines under study at General Electric contain the basic features to eliminate or greatly reduce these losses while still retaining superior uninstalled performance.

Overall, these new VCE concepts offer the potential of being developed into the versatile powerplants needed for the next generation of mixed mission aircraft.

Table 8 Current VCE Concepts (1976 Reference)

Engine configuration:
Single or double bypass
Only two rotors
Fan flow modulated with IGV change
Static pressure balance control system, simplified exhaust
One variable jet nozzle
Results:
Retains best features of mixed flow TF
Improves supersonic flow potentials
Eliminates part power inlet drag
Improves aft end closure drag
Produces low noise thrust for second generation SST aircraft

III. Variable Cycle Engine Concepts

A. Introduction

General Electric Aircraft Engines (GEAE) has been involved in both the study and development of variable cycle engine concepts for over 30 years. During this time GEAE, working closely with the advanced technology centers in government, has been seeking an affordable engine concept that could combine the attributes of a high-turbine-temperature turbojet, i.e., high dry specific thrust and low-maximum power specific fuel consumption (SFC), with those of a turbofan, i.e., low part power SFC with low exhaust gas temperature. Also sought were engine concepts that could further improve part power SFC levels by reducing inlet spillage and nozzle closure drag by retaining high-operating airflows at reduced power settings.

Section III.B of this paper contains descriptions of several of the novel engine concepts that have been evaluated during the last three decades, starting with the Air Force defined VAPCOM and concluding with a general description of the YF120 type VCE that ran in the YF22 and YF23 ATF prototype aircraft. Section III.C contains descriptions of the family of YJ101 derived VCE concept demonstrators that were jointly funded by the Air Force, Navy, and NASA during the 1975-1981 time period. Section III.D describes the final evolution of the F120 type VCE and Sect. III.E briefly addresses the current status of VCE development.

B. Concept Descriptions

The following paragraphs describe, in chronological order, the major types of VCEs evaluated by GEAE during the past 30+ years.

1. VAPCOM - The Original Variable Cycle

A major change in fighter and fighter/bomber propulsion began in the 1960 time period when augmented turbofans began to replace augmented turbojets as the preferred propulsion concept due to their improved subsonic cruise SFC potentials. These initial turbofan systems, however, were not without their own problems. The lower part power SFCs were accompanied with very high-maximum afterburning SFCs, relatively more complex augmentors, and huge exhaust nozzle sizes that required very large nozzle throat area (A8) variations between maximum dry and max A/B power settings.

The VAPCOM engine concept was conceived by a group of engineers at Wright Patterson Propulsion Lab in this 1960 time period as a possible alternative to the augmented turbofan. They envisioned an engine that could combine the improved lower power SFCs of a turbofan with the maximum power SFCs of a very high-temperature turbojet that required little or no augmentation to produce equivalent maximum power relative to these first generation augmented fans. This was to be accomplished by using variable stator fan, compressor, and turbine systems in a unique manner that actually changed the engine operating mode from that of a pure turbojet, essentially a

- all variable stator fan and compressor system,
- variable HP and LP turbine vane systems, and
- twin variable A8 nozzle system with common A9

Top view: minimum bypass/maximum T41/maximum power mode, BPR ~ 0, PR_{OA} = maximum and
Bottom view: maximum bypass/part power mode, BPR ~ 1.0, PR_{OA} ~ 0.75 maximum.

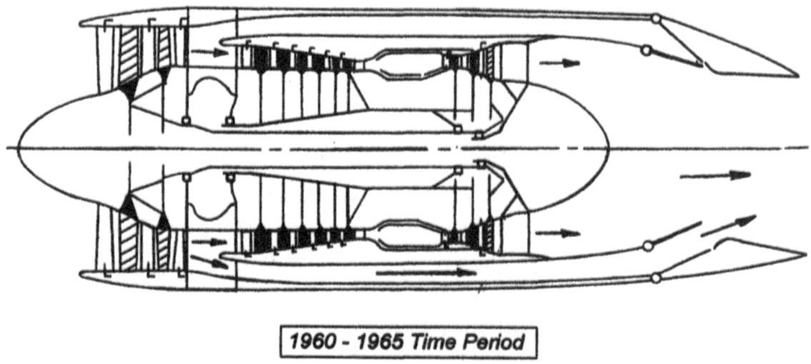

Fig. 20 VAPCOM engine schematic.

near zero bypass ratio dual rotor turbojet, to a bypass 1.0 class dual rotor separated flow turbofan.

Fig. 20 contains an engine schematic of the original VAPCOM engine. As shown, the concept had a dual spool arrangement with a bypass duct that was only used during the turbofan mode of operation. The exhaust system had two variable nozzle throat areas, and for this schematic illustration, a common expansion area. A two-stage fan with all variable stators was driven by a single-stage fan turbine that incorporated a variable area nozzle diaphragm. The core was comprised of an all variable stator six stage compressor, a high-delta T combustor and a two-stage HP turbine that had a variable area stage one-vane system. Turbine temperatures in excess of 3000°F were envisioned for this extremely advanced engine. Afterburning was also considered, but this was not the preferred approach to achieving high-thrust levels. The turbojet mode, with high turbine inlet temperature, was the high specific thrust mode of choice since this would produce maximum combat SFC levels lower than J79 type augmented turbojets and nearly 50% lower than the TF30 class of augmented turbofan. Also, exhaust nozzle areas would be minimized since thrust was being created with both increased nozzle pressure and temperature and not by just temperature alone, as is the case of augmented engines.

For this high specific thrust mode, the core stator systems were full open to allow all of the fan discharge flow to be accepted by the core engine. Fan speed and fan stator schedules coupled with turbine temperature usage and turbine vane control were optimized to maximize engine thrust at all flight conditions. For these high-power modes only the core stream A8 system was utilized as depicted in the top view of Fig. 20. For part power operation the core stators were systematically closed, and fan stators and fan operating line were adjusted to build up the operating bypass ratio from the 0^+ level at maximum power to the 1.0 class at part power. The variable area turbine vane systems were now being utilized to allow core speed, core operating line, and turbine temperature to be optimized to achieve minimum possible fuel flow at each part power setting. Both A8 systems were utilized for part power performance optimization.

Another aspect of performance improvement potential was made possible with the unique spool speed/flow control offered by the VAPCOM. Throttle-dependent spillage drag losses and exhaust system boattail drags (nozzle closure) can be minimized if fan airflow can be maintained as thrust is reduced. In both turbojets and conventional turbofans, only limited capabilities are available for part power airflow tailoring before large penalties are incurred in basic uninstalled performance. The variable stator VAPCOM components and the twin A8 exhaust system allowed high levels of fan flow to be maintained to relatively low-power settings.

As studies progressed several basic problems became apparent with the VAPCOM approach to providing both a turbojet and turbofan operating mode. First, the all variable fan and compressor components did not achieve their target efficiency goals. Also the variable area turbine systems needed in both the HP and LP turbines proved to be complex and relatively high-loss devices. Most importantly, however, was the fundamental problem associated with achieving the turbofan operating mode. To obtain a 0–1 bypass swing the core stator system had to be low flowed, which results in an accompanying loss of core compressor pressure potential (similar to reducing speed to reduce flow). Also, a reduction in fan operating line was also required to achieve the final increment of bypass increase. This further reduced the overall cycle pressure ratio so that when the full conversion from 0–1 bypass was achieved over 25% of the 0-bypass mode cycle pressure potential had been lost, negating a major position of the cycle mode change advantage. This loss, in conjunction with the lower efficiency levels and relative complexity of all variable stator fan, compressor, and turbine systems, lead to an abandonment of this concept in the 1965 time period.

Even though an engine did not result from the VAPCOM program, the basic research conducted on the variable stator fan, compressor, and turbine systems provided valuable knowledge about how to design and utilize these innovative engine features. Also, high-temperature rise combustor technology and advancements in basic heat transfer and turbine cooling were integral parts of the overall technology programs that addressed the needs of the VAPCOM concept. This early research in very high-temperature cooled turbine technology helped form the technology base for General Electric's very successful high-temperature turbine systems.

The author of this article is greatly indebted to the VAPCOM program. His initial assignment at GE involved hand matching of all of the VAPCOM variable geometry components and manually calculating the performance potentials of this machine. Having been introduced to this industry with a turbomachinery concept of nearly infinite flexibility has made the study and definition of ensuing variable cycle concepts relatively easy.

2. Composite Cycle: Turbojet and Turbofan Modes

Another attempt at combining both a turbojet and turbofan mode in one turbomachinery design was the composite cycle concept defined by a General Electric engineer in the early 1960s. His original concept was examined during the initial supersonic transport program whereas the version depicted in Fig. 21 was evaluated in several high-speed advanced bomber and fighter studies in the 1965–1970 time period.

As shown in Fig. 21, this concept had a two-spool design that incorporated a conventional fan and core system and an unconventional bypass duct and low-pressure confluent flow turbine system. The bypass duct contained a second main burner type combustor that fed into a variable area turbine vane system and the tip section of this confluent flow turbine rotor. The hub and tip portion of this turbine plus, for the version shown, three additional conventional fan turbine stages supply the power needed to drive the fan. In the maximum power, or turbojet, mode both the core burner and the bypass burner are on, producing a concentric turbojet operating mode. The flow streams entering the tip and hub of the confluent flow turbine are constrained to have equal static pressure levels and a reasonably homogenized flow exits from the confluent flow turbine for flight speeds from sea level to static to Mach 2^+. For maximum thrust operation beyond Mach 2, the core system can be phased out and the power required to operate the fan can be fully derived from the tip section of the confluent flow turbine. This ability to phase out the core allowed very high overall cycle pressure ratios to be examined for an engine that could also operate in the Mach 3^+ flight regime since core engine speed and temperatures (compressor discharge, T3, and high-pressure turbine inlet temperature T41)

- Two spool design for Mach 3^+/mixed mission applications and
- Unique LP spool turbine system architecture;
  the key component is the aft confluent flow turbine

Top view: turbofan mode for part power, outer burner off and
Bottom view: concentric turbojet maximum power mode, outer burner on.

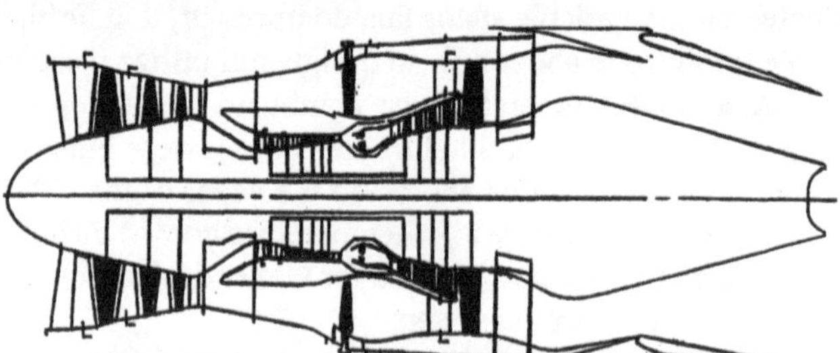

1965 - 1970 Time Period

Fig. 21 Composite cycle engine schematic.

- **combined turbofan and turbojet architecture and**
- **twin variable A8 plug nozzle system**

Top view: turbofan mode only, aft TJ off for low power and
Bottom view: combined mode, aft TJ on for maximum power.

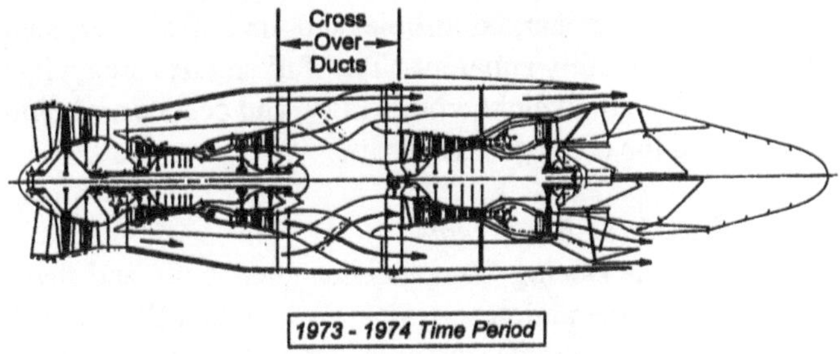

Fig. 22 TACE configuration.

could be minimized with little loss in thrust potential. For part power operation, the outer burner was shut off, and the engine cycle converted into a mixed flow type turbofan. Special demands were placed on the confluent flow turbine tip section during this outer burner off mode. Very large changes occurred in corrected turbine speed, turbine flow function, and turbine energy function that caused untenable swings in turbine exit swirl and high system losses. These losses (verified in a turbine rig) offset the potential turbofan mode advantages and the concept was abandoned. There were some later studies that examined variable pitch confluent flow turbine rotor designs to help handle the swirl problem, but those designs proved to be too heavy and complex to pursue beyond the concept study phase.

Interest in VCEs lay dormant at General Electric from 1970 to 1973 when a series of both in-house and government funded studies were begun to see if a variable cycle engine could be defined that offered distinct advantages over a conventional turbofan while also being both practical and affordable.

3. Combined Turbofan and Turbojet Revisited

Another attempt at defining an engine with two basic operating modes was made in early 1973. Figure 22 illustrates the basic architecture of this engine concept, called a turbo augmented cycle engine. This engine was basically two separate engines connected by a unique set of crossover ducting. The front engine shown is a conventional dual rotor turbofan. The aft engine is a single rotor turbojet that acts as a bypass stream augmentor for the turbofan during high-thrust production.

Two operating modes are shown in Figure 22. In the top view the aft turbojet is shut off and the core and duct streams of the turbofan are mixed in the common duct surrounding the aft turbojet and are exhausted through the outer nozzle. In the bottom view the aft turbojet is on and is supercharged by the bypass stream from the turbofan. In this mode the turbojet adds efficient augmentation to the bypass stream. The hot turbofan core discharge is bypassed around the aft turbojet and exhausted through the outer nozzle.

This concept offered very good part power performance with the aft jet shut off and exceptional maximum power SFC with the aft jet on due to the efficient augmentation that the supercharged turbojet supplied. However, some very obvious drawbacks are readily apparent. Two full engines are really required to make this concept. Length, weight, complexity, and cost overwhelmed the cycle advantages and the concept was dropped.

4. Series/Parallel Mode VCE Concepts

In the early 1970s The Boeing Company designed, built, and successfully tested a unique flow augmentation series/parallel mode VCE concept demonstrator by adding an annular inverter valve (AIV) to a JT3D commercial engine. General Electric evaluated the merits of this concept in the 1973–1974 time period and also looked at several variations on the basic scheme to try to overcome some of the deficiencies found.

Full span annular inverter valve VCE concept. Figure 23 illustrates the conceptual layout of an annular inverter valve added to the fan of a YJ101 engine. This valve permits two distinct modes of engine operation. In the series mode all of the front fan stage discharge flow is fed into the rear fan section and passed on to the core engine and bypass duct, just as in a conventional engine. In the parallel mode all of the front fan discharge flow is diverted around the rear fan section and is then exhausted through a separate outer nozzle system. When the AIV diverts stage-one flow around the rest of the engine, it also opens up an ambient air delivery duct for the rear section of the fan. This auxiliary inlet flow is compressed by the aft fan stages and then divided in a conventional

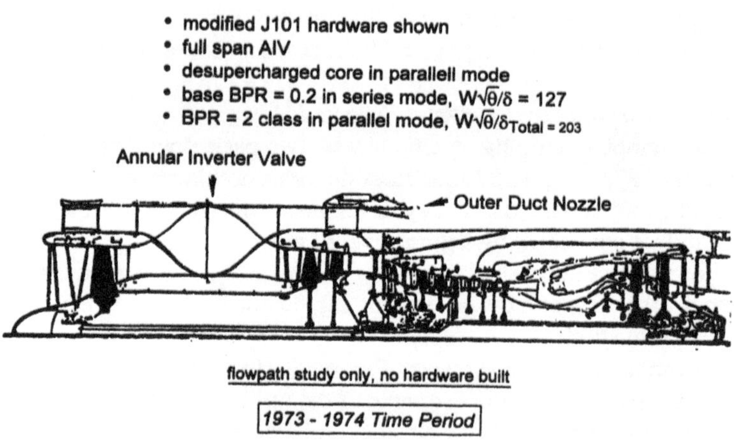

Fig. 23 Annular inverter valve VCE - series/parallel modes.

manner between the core and inner bypass duct. For the YJ101 fan used in this study, total inlet mass flow is increased about 60% and the operating bypass ratio increased from a bypass 0.2 class to a bypass 2 class. Because of the loss of stage-one fan supercharge, overall cycle pressure ratio in the parallel mode is about 40% lower than in the series mode.

This loss of core supercharge results in diminished core energy, and for a given parallel mode thrust class requirement, results in a core size penalty that amplifies the weight problems associated with this concept. Also, the loss of overall pressure ratio that results during parallel mode operation reduces the potential cycle benefit derived from the increased operating bypass ratio. A special oversized inlet system would also be needed to accommodate the increased fan total flow levels that result during the parallel mode of operation.

Part span AIV — constant core supercharge. Figs. 24 and 25 illustrate a modified inverter valve design that retains core supercharging during the series-parallel parallel mode change for a separated flow duct burning turbofan and a mixed flow afterburning turbofan. The splittered fan designs retain full core supercharge while the hinged valve opens and closes the inlet to the rear fan tip section. With this approach core size is minimized for a given thrust size engine; however, the amount of fan flow enhancement is greatly reduced since only the tip portion of the fan rear section is opened to ambient air during the parallel high-flow mode.

These concepts obviously add weight, length, and cost to a basic turbofan engine and were found to have little improvement to offer in fighter, fighter/bomber, and supersonic transport aircraft systems during evaluations conducted in the 1973–1975 timespan. The series/parallel mode concept reappears occasionally, most recently in the tandem fan STOVL concept. Core size penalties and system weight increases again proved to be unacceptable.

- separated flow two-spool duct burning turbofan
- constant core supercharge design

Top view: parallel mode, max bypass ratio for part power and
Bottom view: series mode, max fan PR/minimum bypass mode for maximum power.

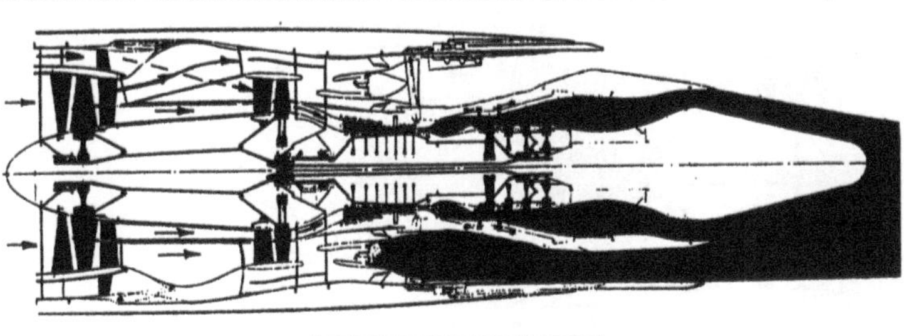

1973 - 1975 Time Period

Fig. 24 Series/parallel variable cycle concept, separated flow version.

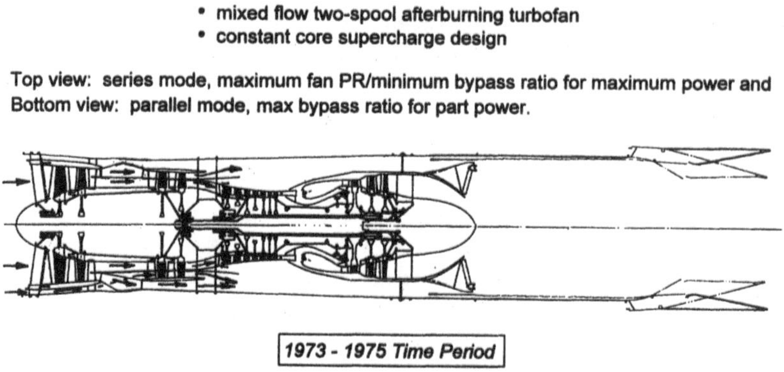

Fig. 25 Series/parallel variable cycle concept, mixed flow version.

5. Three Spool Modulating Bypass Ratio VCE-MOBY

In the 1973 time period attention was starting to be focused on throttle-dependent inlet and afterbody drag levels for the turbofan powered fighter and bomber systems coming on line. Also, NASA had restarted research on efficient low-noise capable propulsion concepts for a Mach 2.2–2.7 supersonic transport.

To address these needs, General Electric defined a very versatile (and very complex) three-spool duct burning engine that allowed nearly independent rotor speed and operating line control for use in optimizing installed part power performance. The modulating bypass engine, shown in Fig. 26 had three spools, three variable area turbines, three variable nozzle throats, two variable stator fan systems, and two bypass ducts, one of which contained a high temperature rise

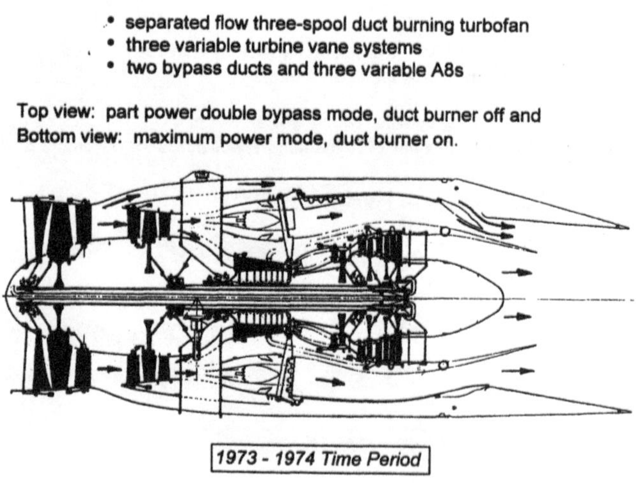

Fig. 26 MOBY engine configuration.

duct burner. The core compressor was conventional, having only a variable inlet guide vane. For maximum thrust production all of the front fan flow was passed through the second fan to maximize fan pressure ratio. As in the VAPCOM system, there was essentially no flow in the outer bypass duct during this maximum dry and maximum augmented operating mode. The discharge flow from the second fan was divided between the core and the inner bypass duct that contained the duct burner. Full temperature rise in this augmentor produced the maximum thrust rating for the engine. The bottom view of Fig. 26 illustrates this mode of operation.

The unique part power operating mode of this cycle concept is shown in the top view of Fig. 26. Front fan flow could be maintained to very low thrust levels by systematically phasing out the core system and intermediate spool pressure rise as core fuel flow was reduced. (The duct burner was not on during part power operation at subsonic flight speeds.) Use was made of the intermediate, and low-pressure turbine variable stators to optimize core speed and intermediate spool speed during part power operation. The high-pressure turbine variable vane system was used to control core operating line. Front fan and second fan operating lines were controlled by utilizing the independent A8s associated with each bypass duct. Finally, the second fan IGV could be scheduled, in conjunction with second spool speed, to regulate the buildup of flow in the outer bypass duct to further improve part power performance by optimizing operating bypass ratio.

This VCE did do something useful relative to a conventional turbofan. All of the variable geometry allowed both optimum uninstalled and, more importantly, optimum installed performance to be achieved, especially at low-power settings. Also, the ability to hold full fan flow at reduced thrust provided an ability to reduce exhaust gas velocity, the prime ingredient in jet noise production and, therefore, tailor noise at reduced power settings, such as community cut back and approach. As with the other VCEs discussed so far, the concept was too complex to be pursued in the form shown in Fig. 26, but many valuable lessons were learned about the attributes that a VCE should have to offer a significant advantage over a conventional propulsion system.

6. *Initial Dual Cycle Single Bypass VCE*

The studies conducted on the MOBY concept provided insight into what each variable geometry feature contributed to overall engine operational flexibility and performance improvement potential. Fig. 27 illustrates how one of the MOBY engine operating modes could be added to a conventional dual rotor mixed flow turbofan engine. The bottom view shows the basic engine mode of operation when the bypass flow and low-pressure turbine flows are joined and exit together through a single exhaust system. A constraint in this operating mode is the requirement for a balanced static pressure to be maintained at the point where the two streams first come together. This constraint, which affects rotor speed/T41 relationships, can be removed by introducing a diverter valve

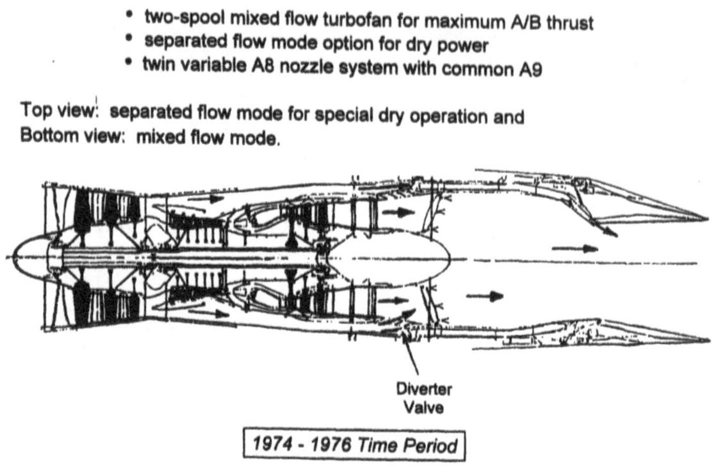

Fig. 27 Initial dual cycle single-bypass VCE configuration example.

system that transfers the bypass flow around the basic exhaust stream as a secondary exhaust stream is opened. The mixed flow engine concept has now been converted into a separated flow turbofan concept with two independently variable nozzle area systems that allow spool speed/T41 flexibilities for fine tuning maximum dry thrust and part power fan airflow scheduling.

This form of single bypass VCE retained the advantage of a mixed flow afterburning turbofan relative to the separated flow duct burning MOBY concept, i.e., all of the fan flow is augmented to provide maximum thrust, while also having the rotor speed flexibility offered by a separated flow turbofan with two variable area nozzles. However, it did not have as much fan pressure ratio control and bypass ratio adjustment as the three spool double bypass arrangement MOBY concept.

7. *Initial Double Bypass VCE Configurations with Three Stream Exhaust*

*1*2 Fan stage split.* Figure 28 shows the first attempt at combining some of the additional flexibilities of the MOBY concept to the mixed flow afterburning VCE shown in Fig. 27.

Two spools were retained, but, as shown in Fig. 28, the three-stage fan was divided into two sections, a single-stage front block and a two stage rear block. Variable stators were added to the rear fan block as were a second bypass duct and third A8 system. Also included was a variable area vane set for the low-pressure turbine. As shown in the bottom view of Fig. 28, the engine retained a basic mixed flow afterburning mode where all of the stage one fan flow was accepted by the two-stage rear block. (The outer bypass duct has essentially no flow during maximum power production, just as in the VAPCOM and MOBY cycle concepts.) The fan flow is then divided between the core inlet and inner bypass duct. Basic mixed flow engine operation produces a static pressure balance between the core and inner bypass duct discharge flows that now mix and burn in the afterburner and then are exhausted through a single A8 nozzle.

For maximum dry power the engine can be operated in a mixed flow mode or, with the diverter valve and second A8 system, in the separated flow mode previously discussed for the preceding VCE concept. The unique split fan architecture and bypass duct/nozzle arrangement for the VCE concept shown in Fig. 28 allows a part dry power mode of operation that comes close to having the fan pressure ratio/fan flow/bypass ratio control offered by the more complex three spool MOBY concept. The variable fan stators of the rear fan block (an original VAPCOM feature) allow nearly the same flexibility as a third spool, especially when variable turbine vanes are also used in the low-pressure turbine.

2∗1 Fan stage split. Fig. 29 illustrates another version of this VCE concept. Instead of dividing the fan into a 1∗2 stage arrangement the fan was split into a 2∗1 configuration. Maximum power single bypass mode performance (no flow in the outer bypass duct) was identical between the two, but part power performance, outer duct sizing, and total A8 demands were impacted.

The simplicity of two spools plus the mixed flow afterburning maximum power mode made either of these initial double bypass VCE concepts more attractive than the MOBY concept; however, the three stream nozzle still added an unacceptable level of complexity and cost.

8. *Single and Double Bypass VCE Simplification*

Rear variable area bypass injector. As already discussed, mixed flow turbofans have a cycle balancing constraint of requiring equal static pressure levels in the core and bypass duct exit streams where the streams first come

- 1∗2 fan stge split,
- two-spool mixed flow turbofan with separated flow double bypass mode for part power
- three variable A8 nozzles with common A9

Top view: double-bypass separated flow mode for part power and
Bottom view: single-bypass mixed flow mode for maximum power.

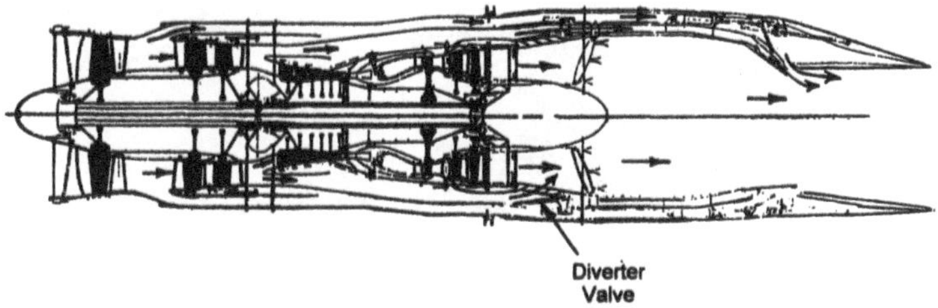

1974 - 1976 Time Period

Fig. 28 Initial double-bypass VCE configuration.

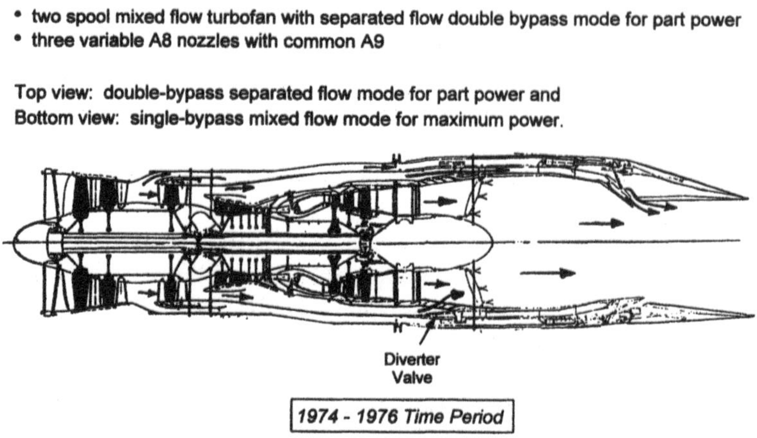

Fig. 29 Initial 2*1 fan stage split configuration, three stream exhaust.

together. The diverter valve used in the preceding single- and double-bypass VCEs effectively eliminated this cycle operational constraint but did introduce the complexity of an additional rear duct and exhaust system. In the 1976 time period an approach to simplifying this was defined. This new concept called variable area bypass injector (VABI), is shown in Fig. 30. Instead of closing off the duct entry into the primary exhaust stream, as was done by the diverter valve, this concept varied the areas at the static pressure balance plane, which allows bypass duct and core discharge total pressure levels to be optimized for both max and part power operation. Although not supplying the same level of fan pressure ratio, speed, and turbine temperature control as a fully separated flow mode, most of the cycle advantages were retained and the exhaust system

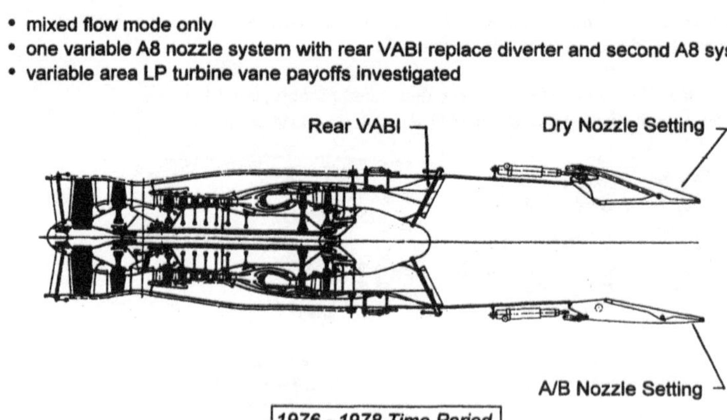

Fig. 30 Improved dual cycle single-bypass VCE configuration.

was greatly simplified with this single-bypass VCE only requiring one variable A8. The variable area low-pressure turbine system added additional spool speed optimization.

Front and rear VABI. Shortly after defining the cycle tailoring potentials of the rear VABI, it became apparent that this same principle could be applied to the inner and outer bypass flows of the double-bypass VCE, resulting in a much simplified version of this cycle concept. Fig. 31 illustrates a potential double-bypass VCE that has both a rear VABI and a first generation front VABI. Note that there is now only one common bypass duct and one conventional exhaust nozzle. This improved mixed flow afterburning double-bypass VCE had a major augmented thrust advantage over the separated flow duct burning MOBY VCE while retaining most of the part power performance optimization potential with one less spool and two fewer exhaust nozzle systems.

The variable area low-pressure turbine vane system shown in Figs. 30 and 31 were found to add an improved level of rotor speed control for optimizing low power performance for this double-bypass VCE concept.

9. Core Fan Architecture for Additional Simplicity

As already mentioned variable cycle engines were also being investigated under a NASA sponsored supersonic transport propulsion study. Several of the VCE concepts described in this paper were analyzed. The flow holding and part power cycle tailoring of the MOBY concept appeared to have potential during these 1973–1974 studies. Therefore, as the basic double-bypass VCE concept was evolved in the 1974–1976 time period, it too was evaluated and found to have merit from both a noise and a performance standpoint. A specialized version of the double-bypass VCE emerged from these studies. Fig. 32 illustrates the features of this VCE and also includes an acoustic nozzle concept that provided stage two noise goal levels. The basic engine is quite similar to

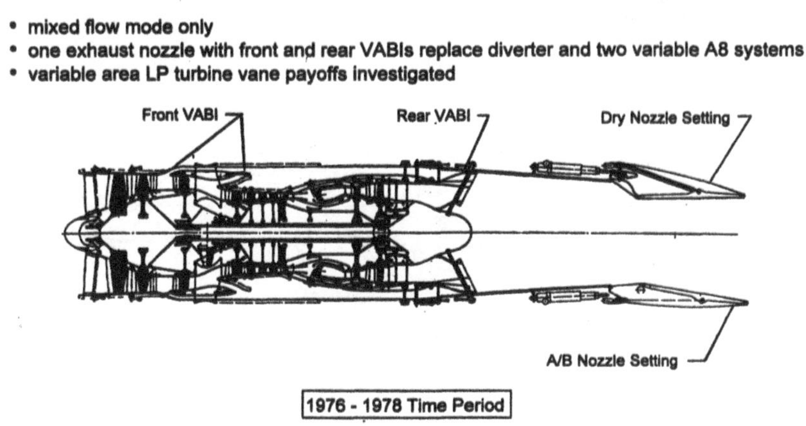

Fig. 31 Improved double-bypass VCE configuration.

the military version shown earlier in Figure 31. Closer inspection shows that the third fan stage is really attached to the high-pressure spool rather than being on the low- pressure spool as was the case in Fig. 31. This unique core engine architecture is called a core fan since the extended tip section of stage one of the core functions is an additional fan stage relative to providing higher bypass duct pressure levels when the engine operates in its single- bypass mode.

With the help of the front and rear VABIs, this core fan version operates in a similar manner to the initial double- bypass concept discussed earlier. A distinct configurational advantage results from the core fan architecture in that a favorable work split occurs between the high-pressure (HPT) and low-pressure turbines (LPT). Since more of the compression work load is given to the HPT, a reduced inlet temperature results for the LPT. This saves a large amount of cooling air and simplifies the LPT design. Also, the LPT loading is minimized since part of the overall fan work is now supplied by the HPT.

C. VCE Concept Demonstrators Based on YJ101 Hardware

The rapid definition of attractive VCE concepts in this 1974–1975 time period attracted the interest of both General Electric management and the advanced technology centers in government. This mutual interest coupled with an unsolicited VCE planning brief started the most successful, cost effective concept demonstration program in jet engine history. Fortunately, for this program several YJ101 engines, initially produced for the YF17 lightweight fighter, were available for use in a series of proof-of-concept VCE demonstrators that were sponsored by the Air Force, Navy, and NASA during the 1975–1981 time period.

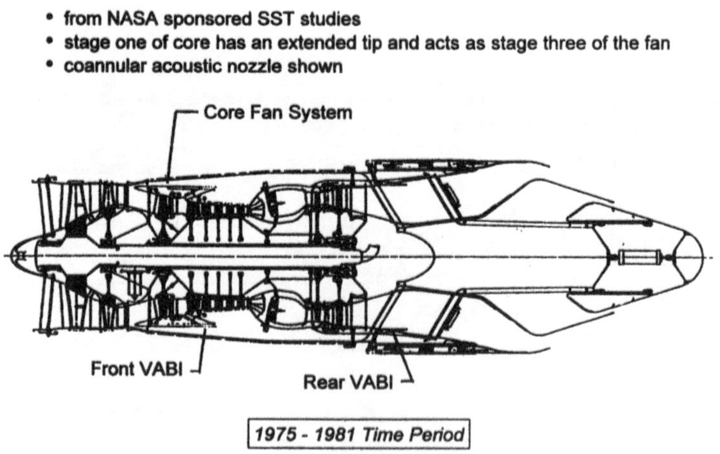

Fig. 32 Double-bypass VCE with core fan architecture.

1. Air Force Sponsored 1*2 Split Fan VCE

The first double-bypass VCE demonstrator was sponsored by the Air Force and is shown in Fig. 33. As can be seen, it has the basic architecture of the original double-bypass VCE shown in Fig. 28. The three-stage YJ101 fan was divided in a 1*2 fan stage arrangement, and an outer bypass duct and rear diverter valve were added. A three stream, three variable A8 nozzle system completed this initial VCE that was tested in the 1975–1976 time period. Fig. 34 illustrates the four unique engine concepts that were tested with this one basic configuration, these being 1) baseline mixed flow turbofan,2) separated flow mode–single-bypass VCE with two variableA8s, 3) mixed flow mode – double-bypass VCE with two variable A8s , and 4) separated flow mode – double-bypass VCE with three variable A8s. Each version ran as predicted with each increased level of engine complexity offering both better uninstalled part power SFC plus improved airflow holding that would provide additional performance improvement potentials from reduced inlet spillage drag.

*2. Navy Sponsored 2*1 Split Fan + Rear VABI*

A second build of this double-bypass VCE was sponsored by the Navy and is shown in Fig. 35. A 2*1 fan stage split was selected for this build, and the first rear VABI system was added. Also, a fully variable low- pressure turbine nozzle system was designed, built, and tested on this two stream nozzle concept demonstrator. The following tests were conducted in the 1976–1977 time period: 1) first 2*1 fan split test, 2) first drop chute rear VABI test, 3) first augmented VCE testing, 4) first variable LPT test, and 5) fixed Cycle, Single Bypass, Double Bypass mode testing.

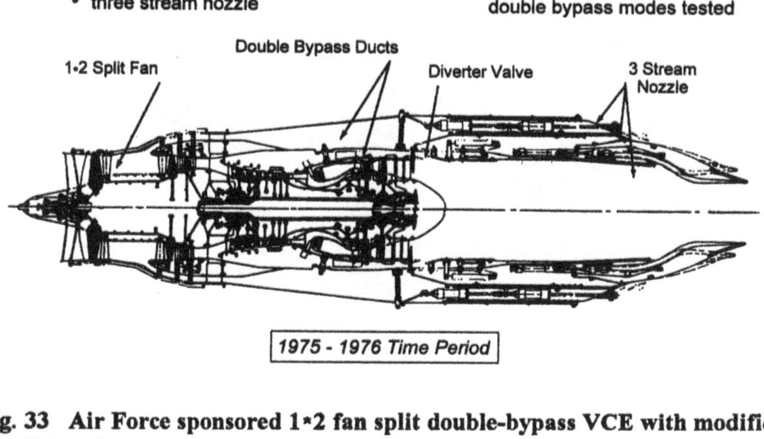

Fig. 33 Air Force sponsored 1*2 fan split double-bypass VCE with modified YJ101 hardware.

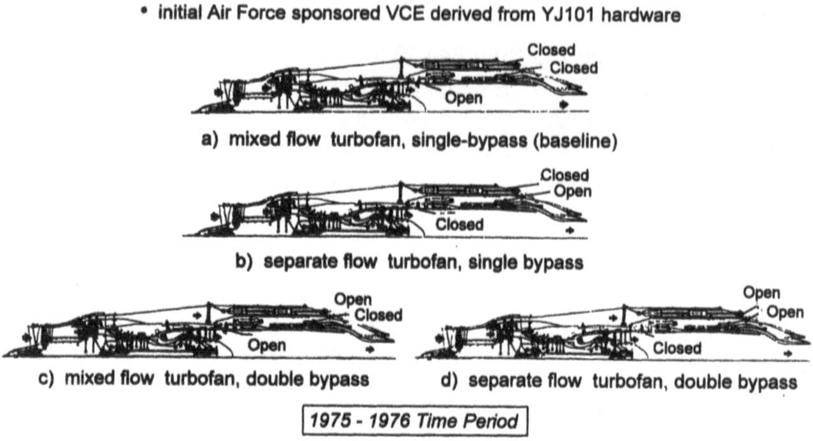

Fig. 34 Test mode flexibility of 1*2 fan split VCE concept demonstrator.

3. NASA Sponsored Front VABI and Acoustic Nozzle

NASA took the basic Navy VCE vehicle and added the worlds first front VABI system. The resultant concept demonstration now required only one variable area exhaust nozzle. Fig. 36 contains both single- and double- bypass mode geometry settings for this engine. The engine ran extremely well, and everything performed as predicted.

As was mentioned earlier, the double-bypass VCE concept was being evaluated in the ongoing NASA sponsored SST studies. A very simple stage two noise capable acoustic nozzle concept had been evolved for this type engine, and a YJ101-sized version was built and added to the NASA sponsored

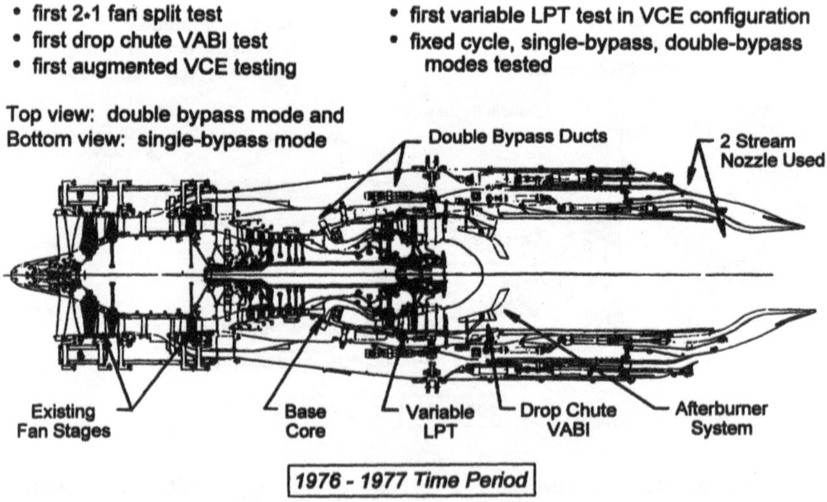

Fig. 35 2*1 fan split double-bypass VCE with modified YJ101 hardware.

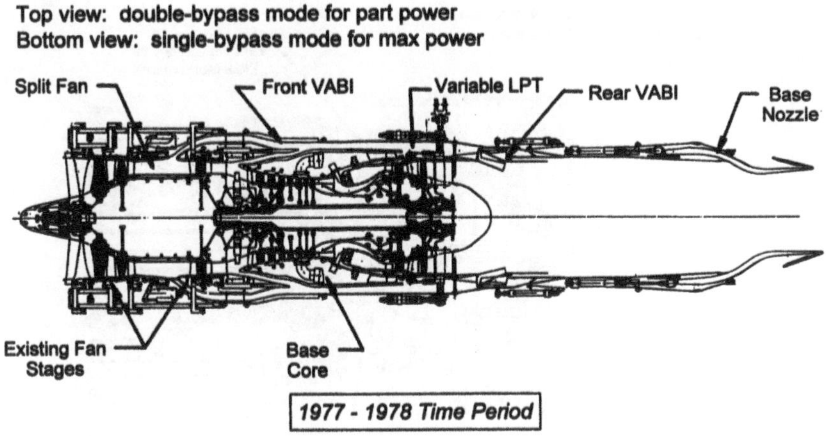

Fig. 36 NASA front VABI test demonstrator single stream exhaust derived from Navy 2∗1 J101 based VCE concept demonstrator.

front VABI VCE concept demonstrator. Fig. 37 shows the complete engine test configuration that produced very successful acoustic test results.

4. NASA Sponsored Core Fan Concept Demonstrator

One last YJ101 VCE concept demonstrator configuration was defined, built, and tested in the 1979–1981 time period and is shown in Fig. 38. Stage three of the YJ101 fan was removed, and a core fan stage was added to the seven-stage YJ101 compressor. This unique compression system architecture was successfully demonstrated and closed out the highly successful YJ101- based VCE concept demonstration testing. (Several additional VCE tests were conducted in the 1982–1984 time period as part of the Air Force and Navy

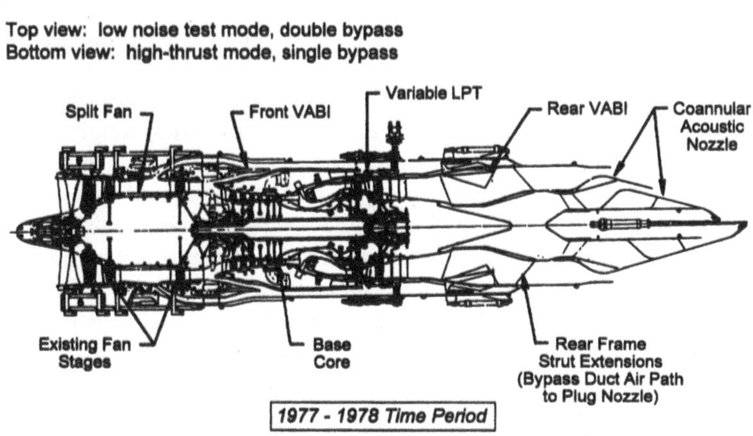

Fig. 37 NASA acoustic test VCE (Navy 2∗1 fan with acoustic nozzle), initial double-bypass VCE noise test configuration.

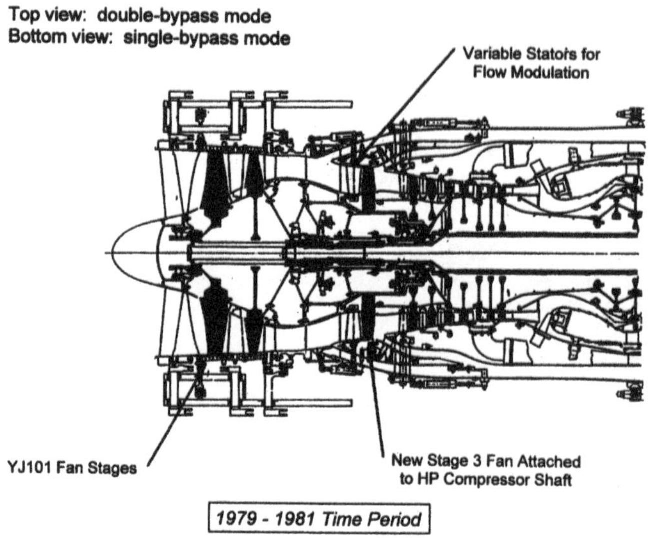

Fig. 38 NASA core driven fan stage VCE with modified YF101 hardware plus new core fan.

sponsored GE23 ATEGG/JTDE program that was based on an upgraded F110 class core and a new advanced technology counter rotation low-pressure spool.)

D. ATFE Configuration Definition

Lessons learned from these demonstrations were built into the VCE cycle and configuration studies that were evolving during the pre-ATF aircraft and mission studies conducted in the 1978–1982 time period. Fig. 39 contains a typical cross section of one of these double-bypass VCE study engines that was latter evolved into the F120 type VCE.

The final architecture selected for the XF, YF, and F120 engines is illustrated in the configuration schematic shown in Fig. 40. Both the XF120 and YF120 and the proposed F120 had two-stage front fan systems and five-stage core compressors. Both compression components were driven by single-stage turbines. Stage-one of the core compressor had an extended tip (core fan stage) that served as a third stage for the fan system. Counter rotating spools plus the optimum turbine energy split made possible by the core fan architecture greatly simplified the design of the low-pressure turbine system. As shown in Fig. 40, the ATFE concepts retained the single- and double-bypass modes that were initially tested on the YJ101 VCE concept demonstrators. The successful test stand results from the XF120 lead the way to an even more successful series of proof-of-concept flight testing of the YF120 in both the YF22 and YF23 ATF prototype aircraft. The worlds first flight VCEs performed as predicted and powered both aircraft to their maximum attained supercruise flight speeds.

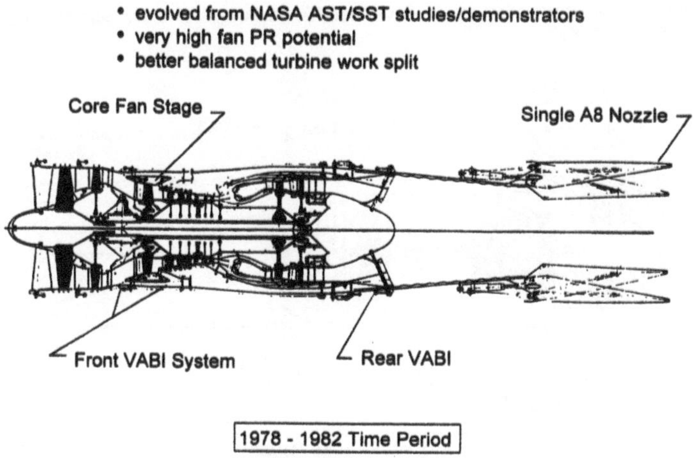

Fig. 39 Double-bypass VCE with core fan architecture.

E. Continuing VCE Developments

Fundamental VCE development work continues to be an integral part of General Electric's ATEGG and JTDE programs. More advanced versions of the basic YF120 VCE were defined and tested during the XTC/E45 ATEGG and JTDE program that ran during the 1984–1991 time period. A new type of VCE was defined in the 1991–1992 time period and is now being jointly pursued with Allison Advanced Development Company under the joint XTC/E76 ATEGG and JTDE programs. The performance and architectural simplicity of this new engine will allow the full phase 2 (IHPTET) goals to be achieved by 1998.

IV. High-Mach Capability Turbo Engines: A 1995 GEAE Perspective

A. Introduction

Ever increasing flight speed capability was one of the major goals during the early development of jet powered flight. Mach 1, then Mach 2, then Mach 3 systems were defined, developed, and tested by the mid 1960s time period, only a short 25 years after the first flights of the original jet powered aircraft. These initial high-Mach systems were for military use only, but in the late 1960s interest in commercial supersonic transports increased dramatically with development programs being initiated in the U.S., England/France, and Russia. The U.S. program was canceled before a flight vehicle was developed, however, several prototype engines were produced and tested. Continued military interest in even higher flight speeds lead to a series of Mach 4–8 capable aircraft/propulsion studies with many new air-breathing engine concepts being defined to cope with the unique demands of these higher Mach numbers. Various forms of ramjets were evaluated along with supersonic combustion ramjets or scramjets for the Mach 6–8 flight speed regions. These studies

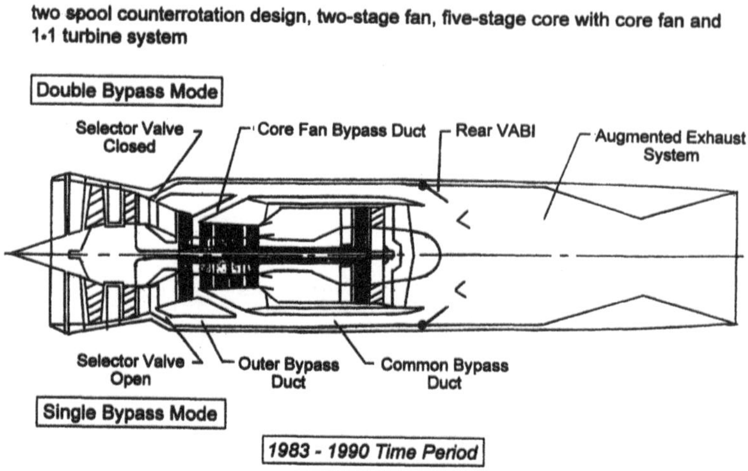

Fig. 40 XF, YF, and F120 variable cycle configuration schematic.

continued through the 1960s and into the early 1970s, but the technology problems associated with very high-flight speeds, primarily the high inlet temperatures (1150°F at Mach 4, 2500°F at Mach 6) and the overall system cost increment for going from Mach 3^+ to Mach $4\text{–}6^+$ could not be reduced to acceptable levels.

These studies lay dormant until the 1984–1985 time period when Mach 4–6 air-breathing propulsion systems were re-examined. The work associated with this most recent round of studies has been used as a technical source for the Mach 4–6 propulsion systems described in Sec. IV.C.

B. GE Mach 2–3^+ Engine Technology Development

General Electric has been involved with the development of Mach 2^+ capable turbomachinery since the mid 1950s. A key technology item found in all of GE's high-Mach engines is the use of variable stator compressor geometry. This engine feature provided early single rotor turbojets such as the J79 and J93 shown in Fig. 41 with stall free operation throughout their flight regimes without resorting to inefficient interstage bleed or dual rotor systems. The GE4 SST demonstrator engine shown in Fig. 42 utilized this same basic compressor stage matching philosophy. Each succeeding engine generation has also pushed the state of the art in materials, aerodynamics, manufacturing processes, basic structural design, and starting with the J93, turbine cooling technologies. In particular, efficient turbine blade and vane cooling with compressor air has helped revolutionize the jet propulsion industry, unlocking tremendous energy potentials offered by increased turbine temperature levels.

The Mach-3 class J93 is the highest flight speed engine that General Electric has built and tested. Current production engines like the F404 and F110 mixed flow afterburning turbofans shown in Fig. 43 have Mach 2 class capabilities.

J79 Engine Cross Section

- 1955 technology single rotor augmented turbojet
  - seventeen stage compressor, three stage turbine
- Mach 2$^+$ flight speeds

- $W\sqrt{\theta}/\delta$ = 170 lbs/sec
- PR = 13.5
- Turbine temp. = 1830°F
- SLS Aug. thrust = 17900 lbs
- FN/Wt = 4.7

J93 Engine Cross Section

- Early 1960's technology single rotor augmented turbojet
  - eleven stage compressor, two stage turbine
- Sustained Mach 3 flight in B70 aircraft

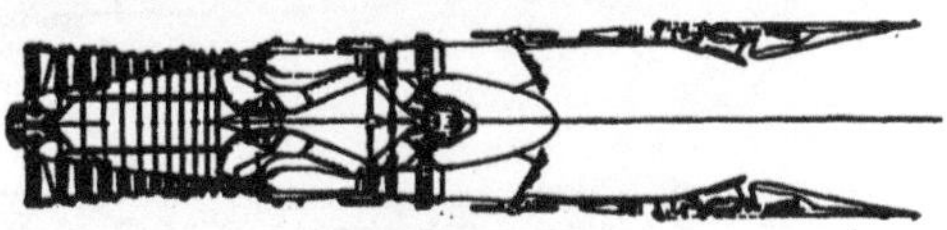

- $W\sqrt{\theta}/\delta$ = 264 lbs/sec
- PR = 8.7
- Turbine temp. = 2100°F
- SLS Aug. thrust = 28000 lbs
- FN/Wt = 5.4

Fig. 41 Mach 2–3$^+$ class turbojets.

These dual rotor engines utilize variable stator systems in their core compressors and have turbine temperature in the 2600–2700°F class.

C. Renewed Interest in Mach 4–6 Propulsion

A resurgence of interest in high-Mach air breathing propulsion took place in the 1984–1985 with particular focus being given to Mach 4–6 capabilities. These new Mach 4–6 engines were to be based on the aero/mechanical/material improvements developed since the 1960–1970 time period with additional projected technology improvements coming from the integrated high performance turbine engine technology initiative. IHPTET is a phased, goal related technology improvement initiative that will, by the end of phase III,

GE4/J5 Engine Cross Section

- Late 1960's technology single rotor augmented turbojet
  - nine stage compressor, two stage turbine
- 1800 engine test hours, 200 at simulated M2.7 conditions

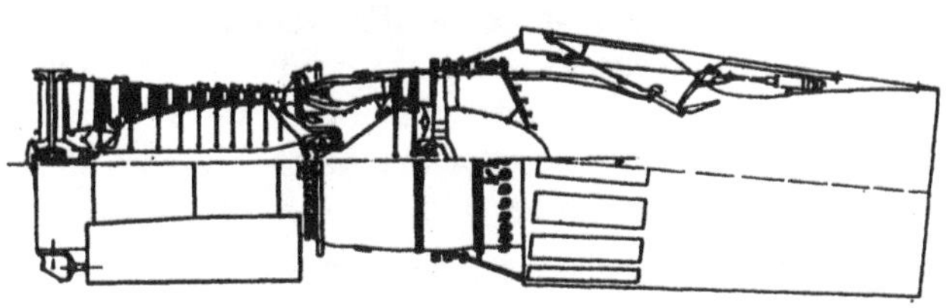

	GE4/J5 Demo Augmented	GE4/J5 Production (Proposed) Dry
$W\sqrt{\theta}/\delta$	633	900
PR	12.3	12.3
T41 - °F	2200°	2400
SLS thrust - Lbs	69900 test with A/B	< 60000
Nozzle type	As shown	Acoustic plug

Fig. 42 Mach 2.7 capability turbojet for SST.

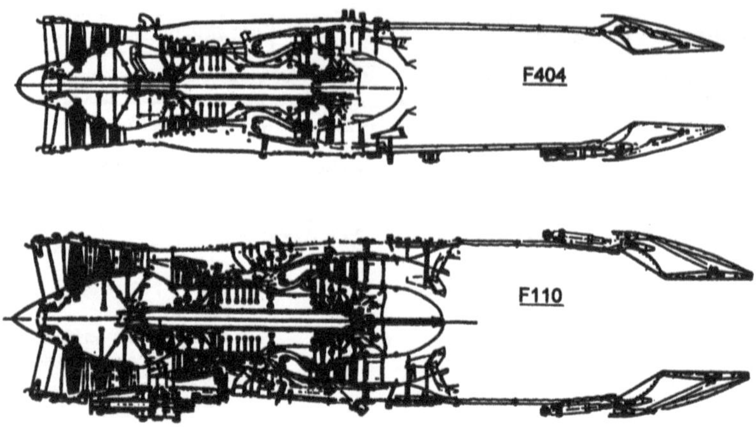

Fig. 43 Mach 2 class mixed flow afterburning turbofans.

provide a doubling of engine capabilities relative to today's thrust/weight, with comparably aggressive improvement goals in fuel burned, operating engine temperatures (both compressor discharge T3 and turbine inlet temperature T41), and other major aspects of engine design and operation. Phase I of the initiative has been completed, and the target for completing phase II is 1997–1998. Phase III technologies will be demonstrated in the 2000–2005 time period.

Fig. 44 indicates both current and projected flight regimes for air-breathing engines. Most of today's engines operate in the dotted region that extends into the Mach 2–2.2 capability region. New SST studies are currently using an upper speed limit of Mach 2.4. The B70 and recently retired from service SR71 pushed air-breathing flight speeds into the Mach 3–3.2$^+$ band, and at least one Russian fighter/interceptor is supposed to be capable of Mach 3.5 class speed. The Mach 4–6 region shown in Fig. 44 would be filled with high-Mach surveillance or interceptor military systems or possibly a superfast SST if economic factors are favorable. Another possible application for this flight zone would be a launch aircraft of a two-stage-to-orbit system with this mother ship being used to attain Mach 4–6 speeds before the orbiter was separated.

The same problems that were encountered in the 1960–1970 time period are still present as flight speeds are increased above Mach 3. Figure 45 illustrates the dramatic buildup of ram temperature with increased flight speed. As can be seen, Mach 4 flight will produce a ram temperature around 1150–1200°F. This engine inlet temperature level approaches current maximum exit temperatures from compressors and so any compression work added will result in T3 levels exceeding current limits. As was already discussed, the IHPTET initiative has set aggressive goals for raising this engine technology barrier, and these new technologies will be needed if the compressor system is still being used at these flight speeds. As will be discussed later, most high-Mach systems convert from

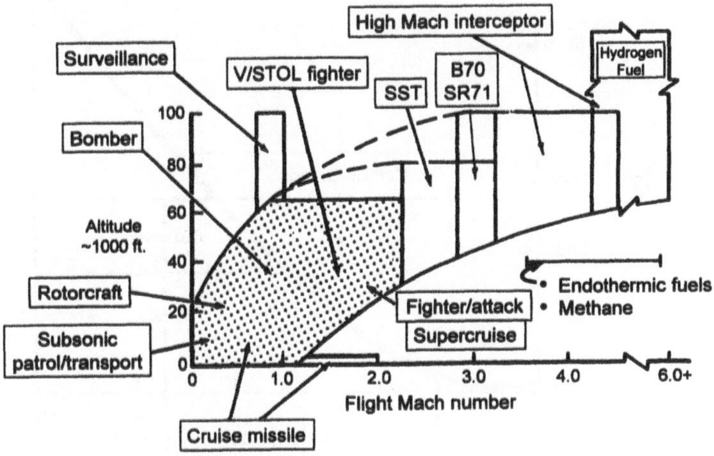

Fig. 44 Flight regimes of the 21st century.

turbomachinery operation to ram burner operation at flight speeds under Mach 4 so that this particular problem may not be a major concern. As flight speeds enter the Mach 5–6 region, ram temperatures increase from 1800°F to over 2500°F. New materials become mandatory, and their general classes are noted in Fig. 45. Not all engine parts can be made from these materials, and special cooling techniques coupled with high fuel heat sink capabilities will be required.

D. Mach 4–6 Propulsion System Descriptions

1. Generalized High-Mach Propulsion Discussion

Figure 46 contains a sea level takeoff to Mach 8 propulsion corridor as defined by a minimum and maximum level of dynamic pressure q at each Mach number. The upper bound is set by a minimum $q = 200$ psf to keep combustion viable, whereas a somewhat arbitrary lower bound is formed by a $q = 2000$ psf level to keep propulsion and airframe weight reasonable. As indicated, turbomachinery associated with turbojets and turbofans can be operated into the Mach 3–3.5 region (and possibly higher as will be discussed later). Attention must be paid to basic cycle parameter selections, particularly with respect to overall cycle pressure ratio in order not to exceed T3 limits during turbomachinery operation. Effective ramjet operation can be started in the Mach 3^+ region and continued into the Mach 5–6 flight regime. For air-breathing flight speeds beyond Mach 6, a scramjet system must be used because of the impracticability of diffusing the inlet flow down to subsonic speeds needed for ramjet combustion operation. The fuel change zones noted in Fig. 46 are a function of fuel thermal stability and fuel heat sink capacity that will be needed for cooling critical engine and airframe parts. Special jet propulsion fuels can be

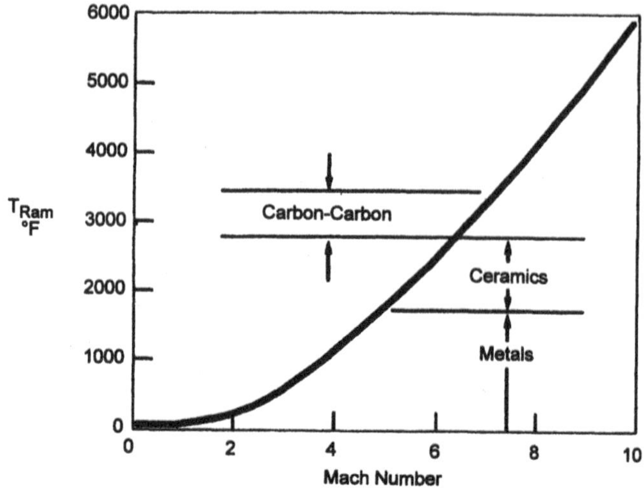

Fig. 45 Ram temperature, °F.

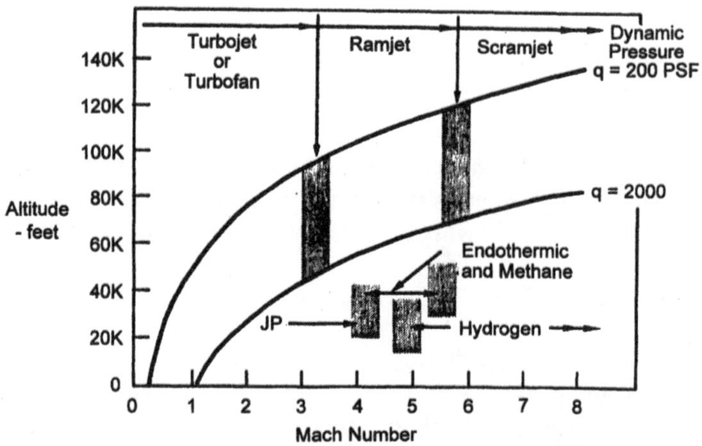

Fig. 46 Air-breathing propulsion corridor.

Table 9 Fuel Characteristics

Fuel type	Maximum Mach	Heating value, BTU/lb	Storage temp, °F	Heat sink, BTU/lb at 1200°F	Density, lb/ft^3
JP7	$3.0 - 4.0^+$	18,930	70	824	48.6
Methane	5.0 – 6.0	21,500	-259	1,244	23.4
Hydrogen	To orbit	51,800	-423	5,855	4.42
MCH (endothermic)	$4 - 5.0^+$	18,800/ 19,500	70	871/ 1,811	43.5

used into the Mach 4 region whereas methane and endothermic fuels such as methylcyclohexane (MCH) could be used into the Mach 5–5.5 region. Hydrogen fuel must be used for flight speeds beyond Mach 6. Pertinent fuel properties for these fuels are shown in Table 9. As noted, JP7 and MCH have conventional storage temperature values, heating values, and densities. The heat sink potential for MCH is much greater, however, and this becomes a critical factor in fuel suitability for flight speeds greater than 4.0. Methane and hydrogen are cryogenic fuels and require insulated tanks. Their densities are lower, particularly hydrogens, and this requires larger tank volumes. Their Btu per pound heating values are higher, especially in the case of hydrogen, which partially offsets these volume penalties on a total fuel energy basis. Hydrogen heat sink is by far the highest, which makes it the only candidate for Mach 6 and above usage.

Figure 47 contains schematics of six Mach 4–6 propulsion candidates that have been studied during the past 10 years. They range in concept from a simple single- rotor turbojet with near stoichiometric turbine temperature levels to an external oxidizer air turboramjet that uses a fuel-rich rocket gas mixture to power a turbine that drives a multistage fan. The flow from this fan mixes with the fuel-rich turbine exit flow, ignites, and is exited through a high area ratio nozzle to make thrust. Discussions on each engine are contained in Secs. IV.D.5–IV.D.9 of this chapter.

2. Nacelle Concepts

Engine/airframe integration is a critical design consideration at any flight speed, but it becomes a dominant factor for high-Mach systems. Each of the engine concepts shown in Fig. 47 would require a tailored nacelle designed to properly blend in with the basic aircraft. Fig. 48 illustrates notional nacelle

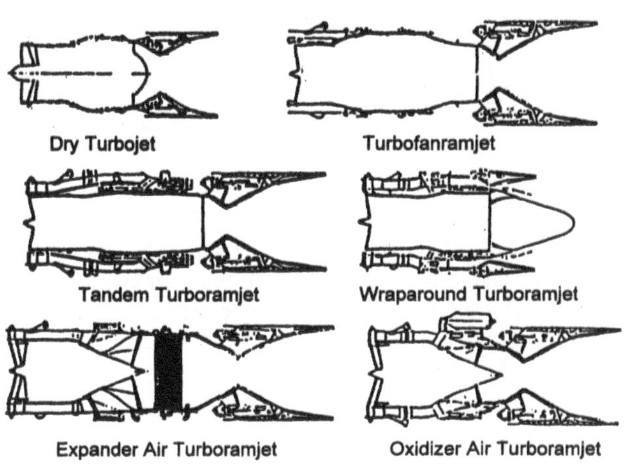

Fig. 47 Mach 4–6 engine concepts shown in equal thrust size.

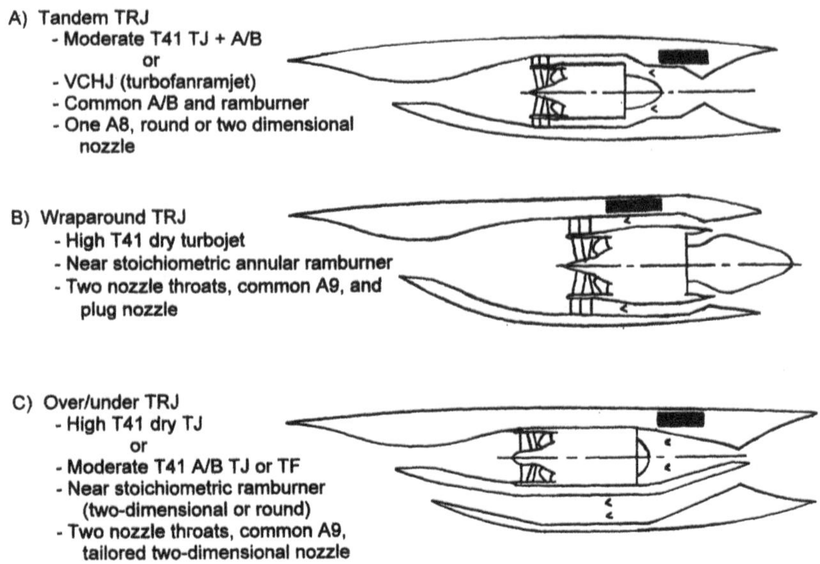

Fig. 48 Turboramjet nacelle schematics.

concepts for these various types. Single-throat engine types, such as the single rotor turbojet, tandem turboramjet (TRJ), or the turbofan ramjet, would fall under nacelle Concept A. Concept B would be required for the wraparound TRJ and a special over/under nacelle such as concept C would be used for a system that had the ramjet separated from the basic engine.

3. Engine Airflow Characteristics

Another key parameter for high-Mach systems relates to defining an engine airflow schedule that provides adequate thrust production to meet aircraft needs while also being compatible with basic inlet off-design flow scheduling characteristics. Fig. 49 contains an engine corrected flow schedule for a reference 400-lb/s study class engine that adequately provided thrust for several different high Mach system studies. Specific airframe-inlet-engine matching studies would obviously result in some tailoring of this notional schedule. Also turboramjet systems with two exhaust nozzle throats such as the wraparound TRJ or an over/under TRJ could use their auxiliary ram ducts to accept inlet spillage flow during low-flight speeds where a Mach 4–6 inlet design provides more airflow than the turbomachinery can swallow. This transonic flow mismatch is a drag producing problem that can be reduced if the flow can be passed through the ram ducts of these engines. More acceleration thrust could be produced if this flow is burned.

4. Exhaust Nozzle Requirements

High-Mach systems not only require that attention be paid to inlet requirements but also place great demands on exhaust system designs. Fig. 50

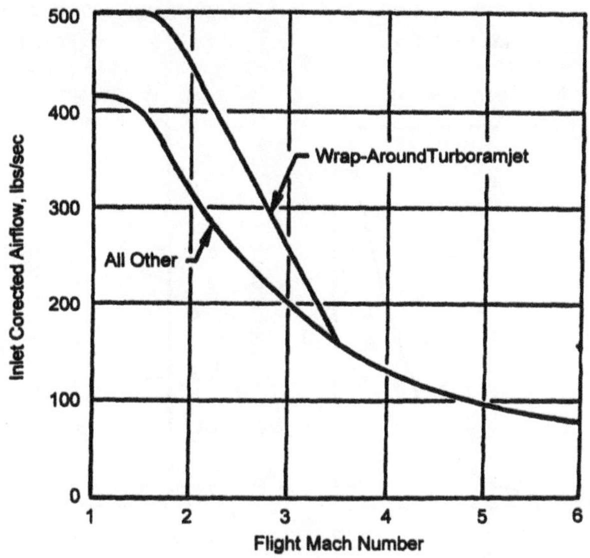

Fig. 49 Engine inlet flow characteristic.

contains typical nozzle requirement in terms of nozzle pressure ratio and relative nozzle expansion area for Mach 0–6 engines. As can be seen, P8/P0 levels of 15–20 at Mach 2 turn into 200–600 as flight speeds go into the Mach 4–6 region. Also, to attain high thrust coefficient levels, which are critical to good high-Mach performance, a final nozzle expansion area 6–10 times that needed at sea level static will be required. For Mach 4–6 systems this means that nozzle weight and size will play a key role in the overall propulsion system design.

The inlet and nozzle comments in the preceding sections apply to all of the high-Mach concepts shown in Fig. 47. Each concept will now be discussed separately.

5. Mach 4⁺ Dry Turbojet Concept

Fig. 51 illustrates the basic architecture of a very high-turbine temperature single-rotor turbojet capable of operating at Mach 4 flight conditions. Variable stator geometry is included in the compressor system to maximize high flight speed performance. Since there is only one burner in this engine, the compression system must be kept on line in order to get flow into the combustor. Depending on the design pressure ratio of the engine, very high compressor exit temperatures T3 could result. Reference to Fig. 52 will show the impact of design pressure ratio on T3 as flight speeds are increased along a representative climb and acceleration path. Current technology limits of 1200°F are soon reached even with a 4:1 pressure ratio system that is not a very efficient low flight speed machine. Even adding 200–400° to current limits would just start to give an 8:1 pressure ratio design Mach 4 capability. Also, combining

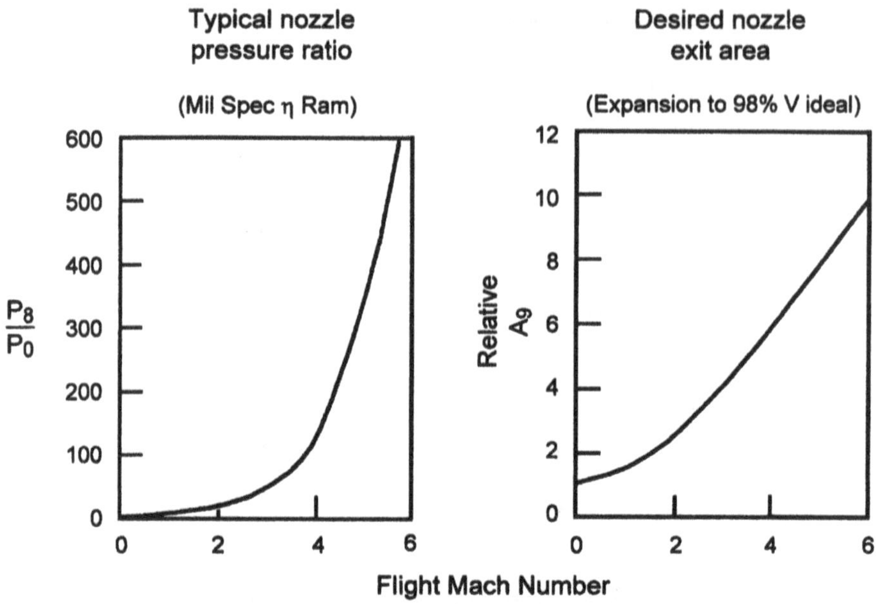

Fig. 50 High Mach nozzle requirements.

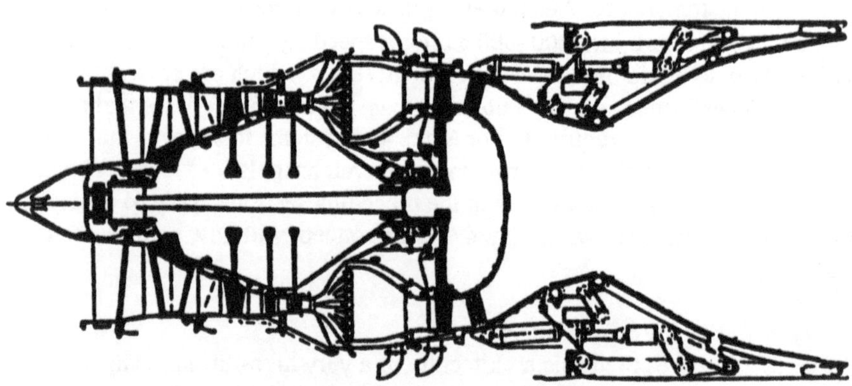

Fig. 51 Single rotor dry turbojet.

these extreme T3 levels with stoichiometric turbine temperatures presents a very challenging cooling problem for the turbine system and some sort of fuel/air heat exchanger would be needed to make a viable man rated system. Whereas this single rotor dry turbojet concept is basically very simple, it does require very advanced materials and cooling technologies to make a realistic Mach 4 system. Higher flight speeds soon become very impractical for this type of engine.

6. Wrap Around Turboramjet Concept

The basic dry turbojet referred to in the preceding section can be converted into either a wraparound turboramjet as shown in Fig. 53 or into an over/under

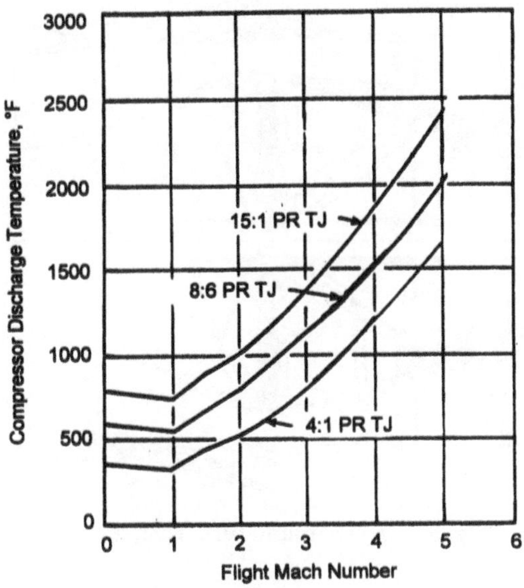

Fig. 52 Compressor discharge temperature at maximum power.

configuration as shown in concept C in Fig. 48. For this concept the turbojet would be kept on line into the Mach 3–3.5 flight speed region, where it would be phased out of operation by reducing turbine temperature and closing down the exhaust nozzle throat located just downstream of the rear frame. By shutting down the turbomachinery in the Mach 3–3.5 region, compressor discharge temperatures do not become a problem, and a higher pressure ratio, more low speed efficient compression system can be utilized. As the turbomachinery is taken off line, the ram burner is brought up to full thrust production levels, and a smooth transition from turbojet mode to ram burner mode is attained. As was noted in an earlier section, a turboramjet with either a wraparound or over/under burner design can utilize the ram duct for inlet flow tailoring at transonic speeds. Also, as soon as sufficient duct pressure is attained, the ram burner can be lit to supply additional transonic thrust. This dual flowpath concept requires a twin nozzle throat design. A common A9 plug nozzle works well for a wraparound design. Relatively high-turbine temperatures would be needed for a dry turbojet, since there is no afterburner downstream of the turbine to provide additional exhaust energy. Also, an auxiliary power system driven by a ram air turbine will be needed to produce energy for aircraft and engine systems while the turbojet is shut off.

7. Tandem Turboramjet Concept

Fig. 54 illustrates the basic configuration of a tandem turboramjet. A single-rotor low/moderate T41 turbojet engine is shown with an afterburner being incorporated downstream of the turbine frame. For takeoff through Mach 3^+,the

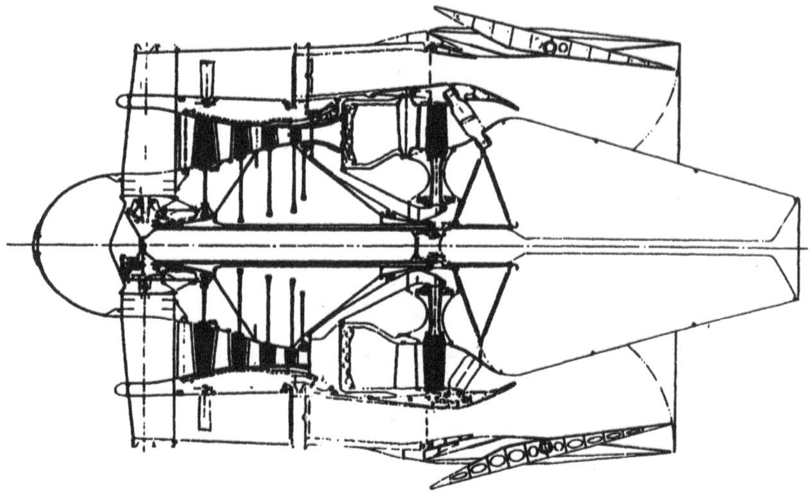

Fig. 53 Wraparound turboramjet.

engine operates as a normal afterburning turbojet similar to a J93 engine. At flight speeds around Mach 3 turbine discharge pressure and ram pressure levels begin to approach each other, as indicated in the sketch at the bottom of Fig. 54. By proper adjustment of a VABI, it would be possible to begin mixing unburned ram air with turbine discharge air and burning the mixed flow in the afterburner. The turbojet would be gradually phased out, shut off, and motored above a flight speed of Mach 3^+, and the auxiliary bypass duct shown would pass the inlet flow around the engine and direct it into the afterburner that now has become a ram burner. Smooth thrust transitions into and out of these turbojet-on – turbojet-off modes should be possible. It would also be possible to utilize a mixed flow afterburning turbofan in place of an afterburning turbojet. The same engine operating philosophies would apply. Again, an auxiliary power system driven by a ram air turbine will be needed to supply engine and airframe power at flight speeds above Mach 3.

8. Air Turboramjet Concept

An entirely different high-Mach engine concept is shown in Fig. 55 and at the bottom of Fig. 47. Instead of a combustor system located behind the compressor of a turbojet or inside the core of a turbofan, an air turboramjet relies on either an external oxidizer/fuel rich rocket gas mixture or a heated (warm) hydrogen or methane gas stream for energy to power a small multistaged turbine located within the inner flowpath. No inlet air is used in this turbine system. Bypass duct air is mixed with the fuel rich/fuel only discharge from the turbine, and burning takes place in a manner similar to an afterburner in a conventional engine. In the case of the oxidizer air turbormjet (ATR) sufficient oxygen is mixed with the fuel to attain a low/moderate temperature level for the gas entering the turbine. For the expander version a heat exchanger located either in

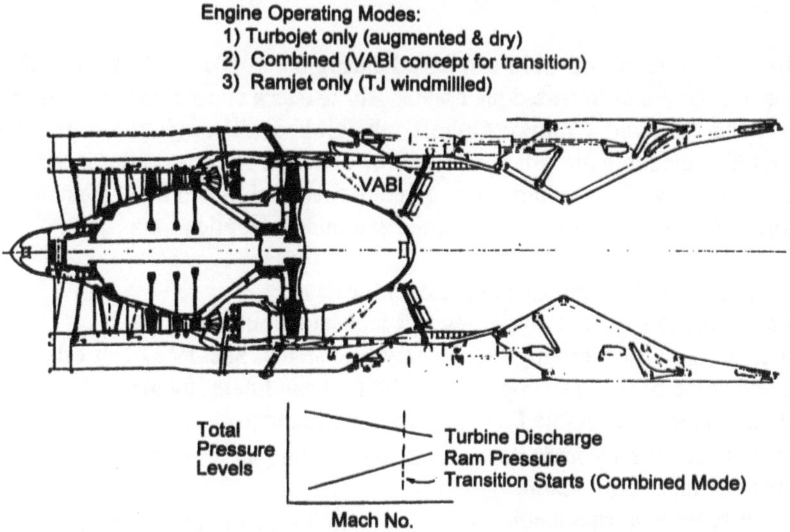

Fig. 54 Tandem turboramjet.

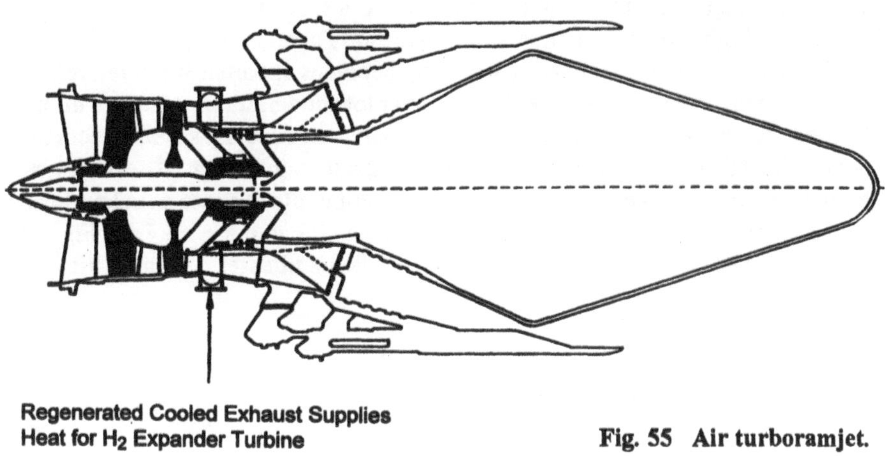

Fig. 55 Air turboramjet.

the plug of an AT, as shown in Fig. 55 or in a full annulus heat exchanger, as whown in Fig. 47, will be needed to heat up the cold hydrogen or methane to an acceptable level for energy extraction purposes. Turbine inlet temperatures must be kept compatible with nonair cooled turbine blade designs because of the fuel rich nature of the turbine gas stream.

Maximum power performance of a hydrogen expander ATR is similar to that of a turboramjet or turbofan ramjet engine. Part power performance will suffer greatly, as shown in Fig. 56, since the ATR turbine system can only support a relatively low-- ratio system, around 5 for a hydrogen fueled system and around 2 for a methane fueled version. Any ATR that is based on a oxidizer system would not be range competitive with the engines shown in this figure.

The ATR concept was built and tested during the 1950s. Both Aerojet, the rocket company that invented the concept for use as a simple, low-cost missile engine, and Pratt and Whitney Aircraft, who looked at the system for use in a man-rated supersonic aircraft, evaluated the concept. Recent studies have relooked at the various versions discussed in this section and have concluded that the concept is best suited for high-speed missile applications

E. Fuels Impact/System Cooling Considerations

The importance of fuel properties and fuel selection cannot be overemphasized for these high Mach engine systems. Sec. IV.D.1 briefly reviewed some of the key properties of the fuel candidates available for high Mach use. Thermally stable fuel properties are mandatory, and available fuel heat sink is of primary importance relative to having an effective system cooling concept and an overall viable system.

The difference in fuel heating value noted in Table 9 convert into the specific fuel consumption deltas shown in Figs. 57 and 58. As can be seen, hydrogen fuel has by far the lowest SFC values, but keep in mind that hydrogen has only 1/10 the density of a JP or endothermic fuel. Not much thrust difference is noted at low flight speeds, but at higher Mach conditions, see Fig. 58, hydrogen fuel provides a higher burning temperature that results in an increase in net thrust.

The importance of fuel heat sink quickly becomes apparent when reference is made to the ram temperature vs flight speed plot shown in Fig. 45. Cool air must be made for purging critical engine and airframe areas, such as sumps, actuators, control interfaces, basic engine flowpath parts, and engine bays. The only way to make cool air is to divert some of the hot engine inlet air into a fuel/air heat exchanger system and use the available fuel heat sink to make cool air. Fig. 59 shows a high-Mach cooling system schematic that addresses cooling critical

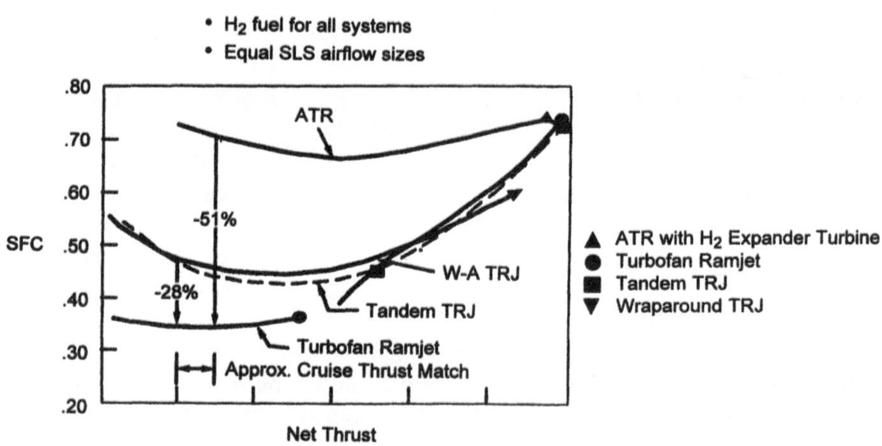

Fig. 56 Subsonic performance comparison

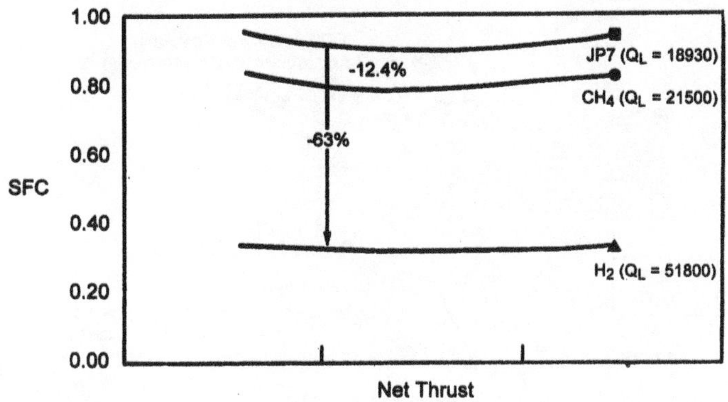

Fig. 57 Impact of fuel on performance at 0.9/50,000 ft, basic cycle parameters held constant.

engine parts as well as using cooled ram air in an energy generating system. The schematic shows ram air extraction ports, heat exchangers, cooling air pumps (to replace piping and heat exchanger pressure losses), and a common compressor discharge air/ram air turbine plus generator device for producing power for the aircraft and engine during very high-Mach operation. These are all system-critical components, and attention must be given to every one of them in the definition and design of a high- Mach system. Similar equipment will be needed for any of the high-Mach propulsion concepts discussed in this chapter.

F. Final Comments

The preceding sections serve only as an introduction to the configurations, operations, and system needs of a family of Mach 4–6 propulsion systems that have been evolved over the past 30–40 years. Selection of one system over any

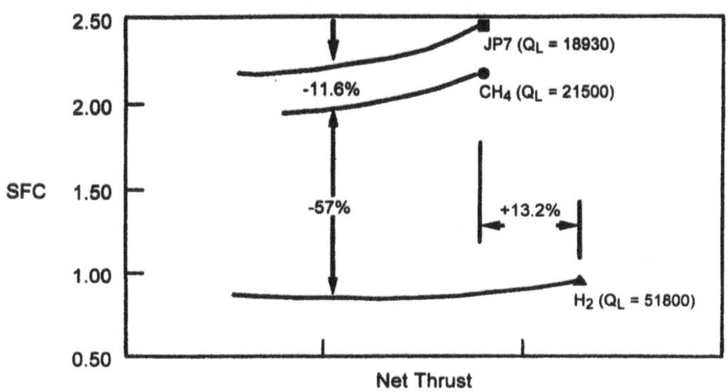

Fig. 58 Impact of fuel on performance at Mach 4/70,000 ft, basic cycle parameters held constant.

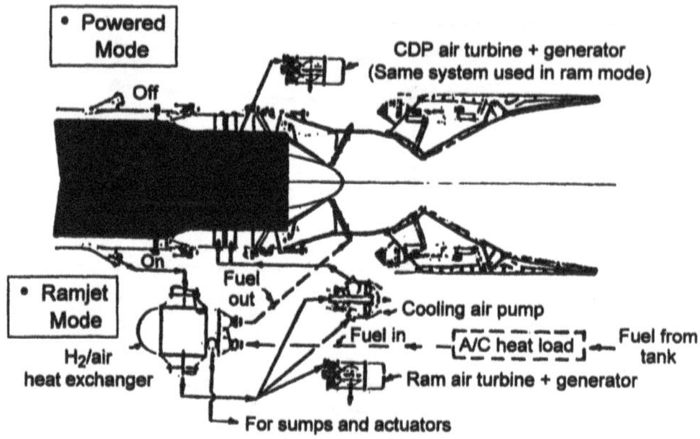

Fig. 59 High Mach cooling system schematic.

of the others will require a series of detailed airframe engine inlet nozzle cooling system-fuel selection-etc., trade studies. Also, the technologies associated with key high-Mach materials, high heat sink fuels and related heat exchangers, and basic high-Mach engine aero/mechanical/cooling designs must be developed before one or more of these engine concepts could be made ready to fill a Mach 4–6 military or commercial need.

Turboramjets and Installation

Franz J. Heitmeir * and Roland Lederer †
MTU Motoren- und Turbinen-Union München GmbH, Munich, Germany

and

Norbert H. Voss ‡ and Norbert C. Bissinger § and Otfrid W. Herrmann ¶
Deutsche Aerospace AG, Ottobrunn, Germany

I. Introduction

As early as the beginning of spaceflight investigations, the potential of airbreathing propulsion systems as well as the operational and economic advantages of a reusable aircraftlike space vehicle were recognized. However, despite the merits of such spaceplanes, none of the concepts was ever realized because a great number of the key technologies required were not available at the dawn of spaceflights. Even today, many of technologies still need to be developed further. Existing space transport systems are based on expandable launchers or, at best, on partly reusable systems such as the Space Shuttle or the Buran. The necessary thrust is produced by rocket engines. Since the operation of such systems is very costly, engineers were looking for less expensive means of gaining access to space. In recent years, the investigation efforts in the field of single- or two-stage-to-orbit vehicles with airbreathing propulsion systems were intensified in various countries.

To lay the technological foundations in Germany, the Federal Ministry of Research and Technology initiated a hypersonic technology readiness program in 1988 based on a reference concept of a two-stage, fully reusable space transport system. Simultaneously, a basic research initiative was started by the Deutsche Forschungsgesellschaft (DFG) in the same field. These programs are

Copyright © 1994 by the American Institute of Aeronautics and Astronautics, Inc. All rights reserved.

***Manager, Hypersonic Propulsion Systems.**
†Manager, Hypersonic Nozzles.
‡Manager, Hypersonic Ram Combustors.
§Manager, Hypersonic Inlets.
¶Manager, Engine/Airframe Integration.

intended to establish a national hypersonic technology basis with the aim of expanding these programs and cooperating on an international level.

This paper highlights some of the technological challenges associated with an airbreathing propulsion system for hypersonic applications and also shows what is being done and what is planned within the German Hypersonics Technology Programme to solve these challenging problems. First, various possible airbreathing propulsion system concepts are discussed, a short glance at the an selected reference concept serving as a guideline for technology development is given. Furthermore, the highly integrated work necessary to deal with the high complexity of such space transportation systems is described. Afterward the challenges resulting from the various engine components are discussed. For an optimized overall system, engine/airframe integration is one of the key design aspects. Finally, the paper deals with system aspects such as secondary power supply, thermal management, and control. It is necessary to consider these systems very early because they have a decisive influence on the design of the entire propulsion system.

II. Propulsion Systems

According to the principle of Newton (1643-1727), thrust can be generated in different ways. Consequently, there are different types of jet engines, classified by their working principles (Fig. 1) or by the regimes of increasing flight speed:

1) For subsonic cruising speeds up to Mach 0.6, the turboprop is the best choice because of its excellent propulsive efficiency (bypass ratio, BPR > 40).

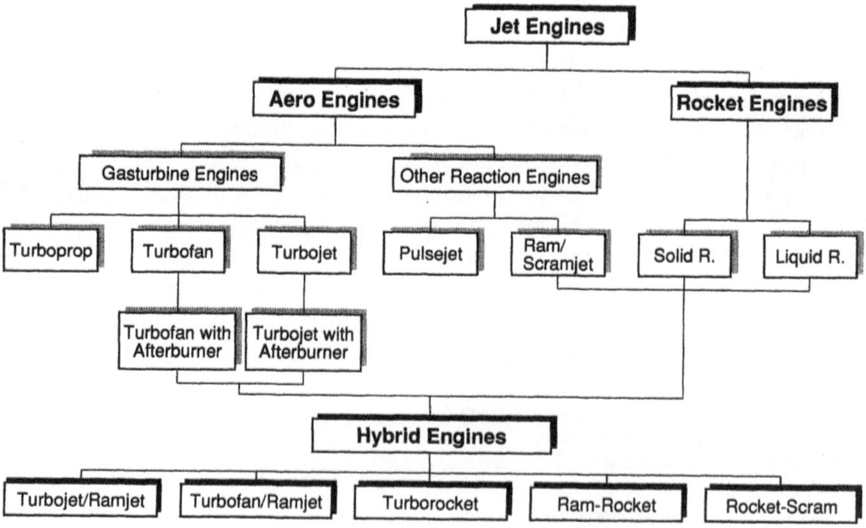

Fig. 1 Propulsion system classification.

2) Higher subsonic flight speeds up to Mach 0.9 are covered by turbofans with bypass ratios of about 6.

3) The lower supersonic flight regimes are typical applications for turbofans with smaller bypass ratios (BPR < 1) and medium pressure ratios. Their specific fuel consumption is better than that of a turbojet because of its higher mass flow. Alternatively, a turbojet with higher pressure ratio could be installed, which offers some advantages with respect to its thermodynamic cycle.

4) Pure turbojets with medium compression ratios are dominant in the upper supersonic regime (Mach < 3).

5) Lower hypersonic speeds (Mach > 3) may be covered by a low-compression turbojet with very high turbine entry temperatures and with reheat; greater advantages are offered by a hybrid (combination) engine with a small, low-compression gas turbine (turbojet or turbofan) contained within or parallel to a subsonic ramjet combustor. Increased ram entry temperatures require extremely high combustion temperatures, otherwise the thrust per air mass flow would be too low.

6) The upper hypersonic regime (Mach > 6) could be covered by scramjets, a ram duct with supersonic combustion. In this case, turbocomponents would become obsolete since the multishock precompression inlet already provides sufficiently high pressure ratios.

Even higher Mach numbers can be achieved with rocket engines. A rocket engine operates on the jet propulsion principle and carries its fuel and an oxidizer to burn with the fuel either in the rocket itself or aboard the vehicle that the rocket propels. Unlike ramjets, pulsejets, and gas turbine engines, a rocket engine is not an airbreathing engine; therefore, it is completely independent of the outside atmosphere and can even be operated in the airlessness of outer space. Hybrid versions of ramjets and rocket engines are also being considered as potential power units for either single- or two-stage-to-orbit systems.

Based on the different jet engines, as indicated in Fig. 1, many different propulsion system concepts are possible. There is a strong dependence on the required Mach number range and flight trajectory. Therefore, as a guideline for the orientation of the development effort within the German Hypersonics Technology Programme, a reference concept was devised. This two-stage-to-orbit (Sänger) concept provides for an advanced space transport system that is fully reusable and features a highly integrated airbreathing propulsion system.

The Sänger spaceplane reference concept is outlined in Fig. 2. This vehicle accelerates to Mach 6.8, where stage separation is initiated. The first stage returns to the home base, whereas the second stage is orbited. Detailed information is given in Ref. 1. Fig. 2 also shows that the payload is only a very small fraction of the total launch mass of the vehicle. In general, this is true for all spaceplane concepts and clearly indicates that all of these concepts are very weight sensitive.

From an engineering point of view, it is recognized that the achievement of sustained hypersonic flight will depend largely on the successful development

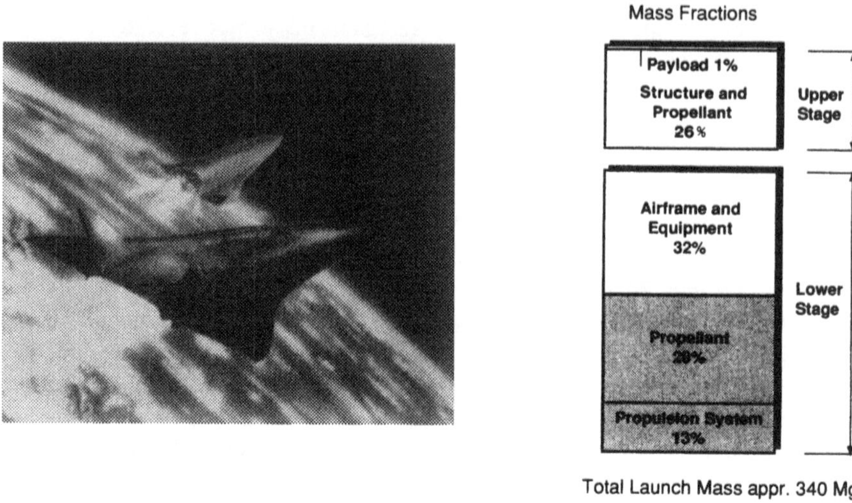

Fig. 2 Two-stage-to-orbit reference concept Sänger.

of an airbreathing propulsion system. First, the propulsion system has to fulfill the thrust requirement along the whole flight path. Furthermore, high thrust-to-weight ratios, low fuel consumption, and high reliability of the system are mandatory. To realize this, basic key technologies must be developed. As a result of the highly integrated nature of airframe and propulsion systems, a great deal of the propulsion work can be done only iteratively in conjunction with the airframe design.

A. Selection of Reference Propulsion System

The propulsion unit is integrated at the lower aft end of the vehicle, thus taking advantage of the precompression achieved by the aircraft forebody, which improves the air inlet performance, and using the aft end of the vehicle as the thrust nozzle expansion base (see Sec. V).

Numerous engine cycles are proposed in the literature for such a type of propulsion system. For a closer investigation of typical questions associated with hybrid propulsion systems, six engine concepts were set up (see Fig. 3).

1) Turbojet/ramjet in tandem configuration with the ramjet air being ducted around the turbojet core engine. Valve-type splitter mechanisms divert the air stream either through or around the turbojet. The combustion chamber and nozzle are common for the turbojet and ramjet modes. A pure turbojet with a low-pressure ratio was chosen as the core engine.

2) Turbojet/ramjet in parallel configuration. Here, too, a splitter mechanism permits the air to be ducted either through the turbojet or the ramjet engine. This

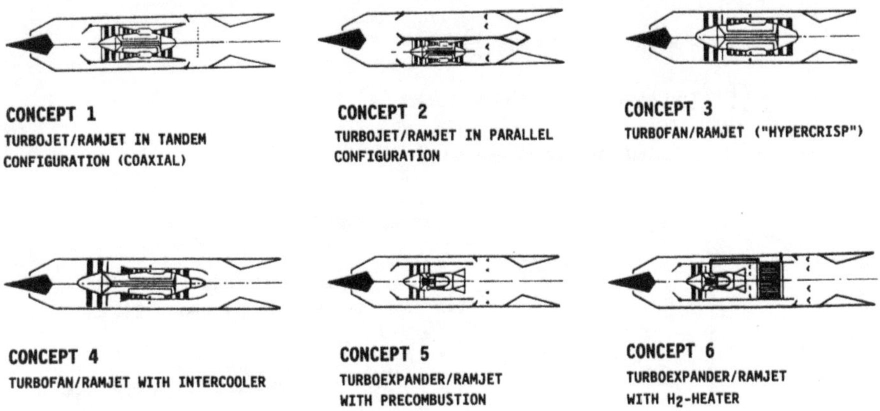

Fig. 3 Engine concepts for high-speed propulsion.

design requires a dual thrust nozzle. For this concept, a low-bypass turbofan was chosen as the core engine. In the flight regime between Mach 2 and 4, the turbojet and ramjet are operated simultaneously to take the greatest advantage of the possible intake flow capacity.

3) Turbofan/ramjet in tandem configuration. Here large switchover valves are avoided. Instead common air ducting is used in the turbojet and ramjet modes. To reduce the windmilling drag during ramjet operation and to minimize the number of cooled stages (at high flight Mach numbers), a two-stage, counterrotating, variable-pitch fan was chosen.

4) This concept is similar to concept 3. However, an air-cooler is installed in front of the high-pressure compressor. This allows the turbofan to be operated at high speeds up to very high Mach numbers. A rather conventional fan was selected, and air ducting during ramjet operation is similar to concept 1.

5) Turborocket/ramjet in tandem configuration. A gas generator burning fuel rich hydrogen and oxygen drives the turbine which, in turn, drives a fan. Air ducting for the different operational modes is identical with concept 1.

6) Turboexpander/ramjet in tandem configuration whereby, similar to concept 5, a turbine drives a fan. In this case, however, the turbine is driven by pure hydrogen gas heated in the ramjet combustion chamber. Air ducting for the different operational modes is similar to either concept 1 or 3.

Cycle analysis calculations for all six propulsion concepts have been carried out. The methodology of performance calculation is given in Ref. 2. The resulting associated installed net thrust values at discrete points of the trajectory are shown in Fig. 4.

The thrust requirement at Mach 1.2, which results from the corresponding aerodynamic drag and a 0.1-g minimum acceleration capability, determines the sizing of the turbojets and, consequently, in all concepts (except for concept 2) also the ramjet size.

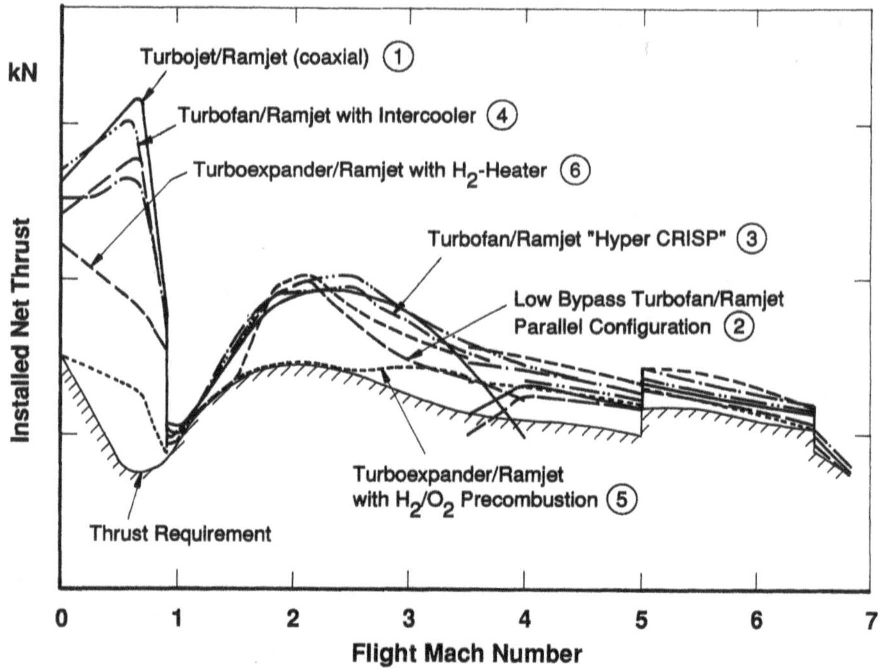

Fig. 4 Installed net thrust vs flight Mach number for different engine concepts.

During takeoff and subsonic climb, the installed net thrust significantly exceeds the minimum thrust requirement: hence, engine operation with reduced reheat or without reheat would be possible except for concepts 5 and 6. In the mid supersonic regime again, all concepts (except concept 5) are characterized by rather high excess thrust. Beyond Mach 3, concept 1 reveals the typical thrust breakdown in the turbomode. In the other concepts, measures were taken to avoid this decrease in thrust or, at least, to shift it to higher flight Mach numbers: simultaneous turbojet and ramjet operation between Mach 2 and Mach 4 in concept 2; a variable pitch fan in concept 3, which allows a very effective control of engine bypass ratio; in concept 4, the engine core flow is precooled by liquid hydrogen in the intercooler before the air enters the high-pressure compressor; and progressive overfueling (overstoichiometric combustion) is used in concepts 5 and 6.

In the ramjet mode, the thrust levels of all concepts vary only slightly.

At Mach 5 - 6.5, the thrust level is increased to make up for the increased aerodynamic drag as a consequence of an almost 90-deg flight turn.

The ramjet mode starts with stoichiometric combustion of hydrogen and air. From Mach 5 onward, the combustion has to be overstoichiometric (excess hydrogen) to achieve the required thrust level.

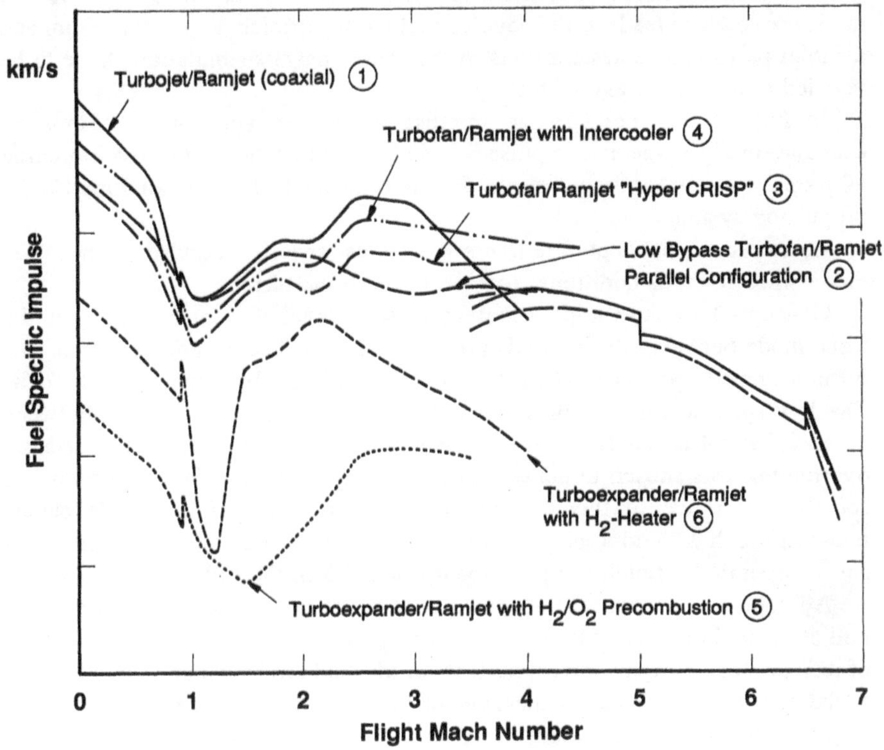

Fig. 5 Fuel specific impulse vs flight Mach number.

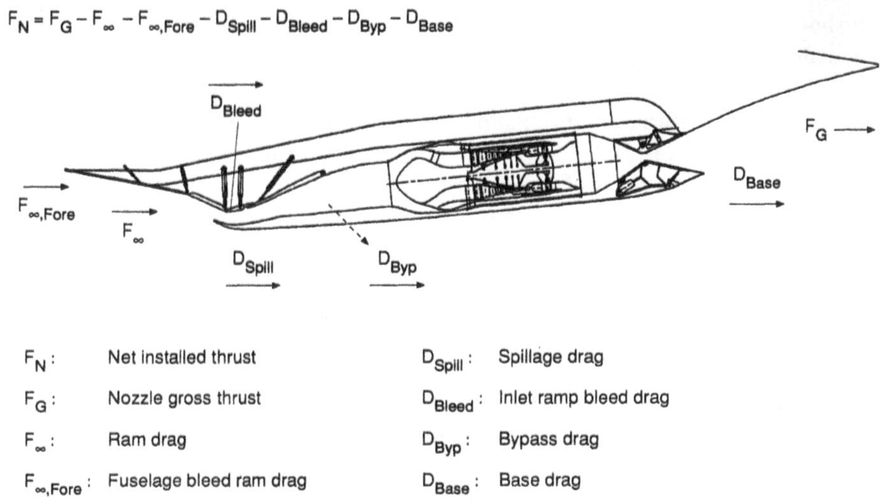

Fig. 6 Propulsion system force accounting.

The fact that the required minimum thrust and the actual thrust in the ramjet mode are so close leads to the conclusion that the efficiency, pressure drop, and air inlet performance assumptions in the cycle analysis simulations have to be verified at an early stage of testing.

In Fig. 5, the fuel specific impulse of the six concepts is compared. Calculation of the specific impulse is based on the installed net thrust, including all propulsion-related drag portions. Figure 6 shows the force accounting for the propulsion system.

As expected, concept 5 delivers the lowest specific impulse in the turbo mode since it needs additional oxygen to drive the turbine.

Concepts 1 - 4 show approximately the same level of specific impulse in the turbo mode between Mach 0 and approximately 3.5. The differences reflect the different nozzle pressure ratios. Concept 6 is inferior by approximately 20%. The low specific impulse of concept 6 in the transonic regime and between Mach 2.5 and 4 is caused by heavily overstoichiometric operation. In this case, overfueling was chosen to achieve a fairly compact engine despite the low air specific net thrust of this concept. Switchover to the ramjet mode occurs between Mach 3.5 and 4 except for concept 4. The intercooled turbofan/ramjet can be operated efficiently up to about Mach 4.5 in the turbo mode.

All concepts, except for concept 2, deliver the same specific impulse in the ramjet mode. For concept 2, the values are approximately 10% lower as a result of inferior nozzle expansion capabilities of the dual thrust nozzle. Beyond Mach 5, the specific impulse drops because of progressive overfueling.

From the performance analysis, it can be concluded that: the turbojet/ramjet in tandem configuration delivers the best thrust and fuel consumption values at maximum reheat; the turbojet/ramjet in parallel configuration is inferior in the ramjet mode as a result of the limited performance of the dual nozzle; the turbofan/ramjet in tandem configuration is superior during subsonic flight and with respect to air ducting for the different operational modes; and the turbofan/ramjet with intercooling offers advantages in terms of thrust and fuel consumption; moreover, it has a higher switchover Mach number (4.5).

In a final comparison and assessment of all of these findings, concept 1 was chosen as the reference propulsion system to serve as a baseline for further investigations and is aimed at providing a solid basis for tackling the necessary technologies in the field of hypersonic propulsion.[3-5]

The selected propulsion system, outlined in Fig. 7, consists of a rectangular inlet, a turbine engine featuring front and rear closure mechanisms to protect the rotating machinery during ramjet operation at extreme temperatures, an afterburner/ram combustor representing the only source of propulsion beyond approximately Mach 3.5, and a rectangular nozzle. The vehicle afterbody is designed for use as a single expansion ramp nozzle. The entire assembly is characterized by a high degree of airframe/powerplant integration.

Fig. 7 also shows a typical temperature distribution along the gas flow path for turbo- and ram-mode operations. This indicates that materials and structures of the engine components have to meet exacting requirements, especially with a view to the fact that all of them must be very lightweight structures.

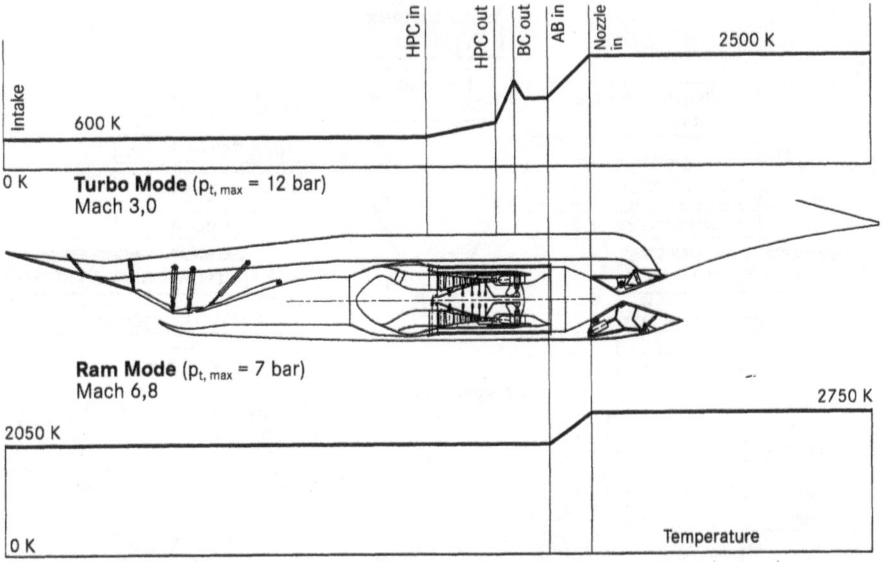

Fig. 7 Temperature distribution along flow path of high-speed engine. (HPC = high pressure compressor; BC = combustion chamber; AB = after burner)

The design and size of the propulsion system depend largely on the trajectory defined for the mission. It is imperative to identify clearly critical flight conditions with respect to engine components to allow optimization of the entire propulsion system. For this reason, an initial set of spot points for different engine modes, angles of attack, altitudes, flight Mach numbers, power settings, etc., must be calculated. After this, mission analysis is carried out by using the above-mentioned points and by scaling the engine or parts of the engine as necessary. This results in a new flight trajectory and new requirements for the propulsion system design. This procedure is described in greater detail in Sec. III.

III. Integrated Team Work

As indicated in the previous sector, the development of an airbreathing space transport system requires a technological standard that is at the frontier of the state of the art. However, the development approach has to be such that only a very close cooperation between all disciplines involved will ensure the necessary progress. Therefore, apart from being highly skilled, the engineers must also be excellent team workers.

Fig. 8 shows the various specialist areas involved as well as their interdependences. The necessity of the above-mentioned close cooperation is obvious. For example, to develop a vehicle configuration, aerothermodynamics,

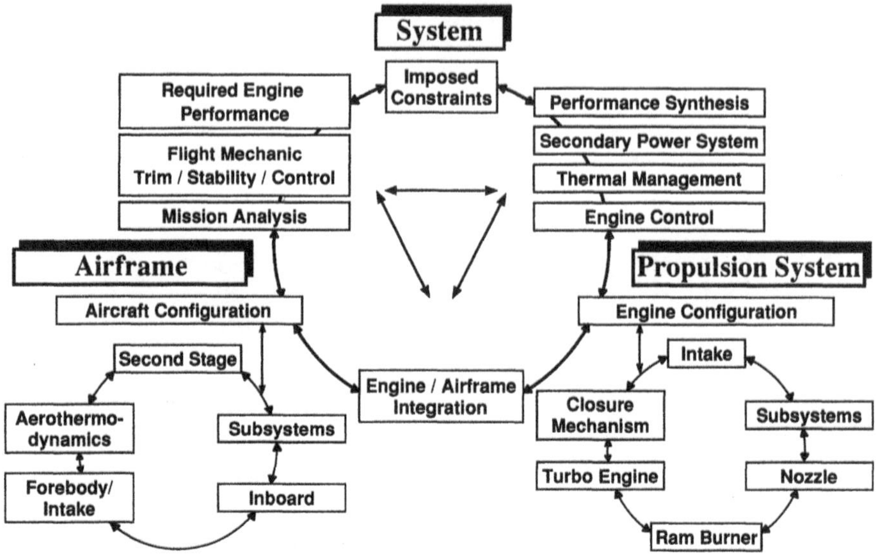

Fig. 8 Integrated teamwork.

inboard requirements, and subsystems must be investigated, and a first rough estimate of the propulsion system behavior is necessary. On the basis of these data, a mission analysis is performed to allow statements to be made with respect to stability, control, and required engine performance. Furthermore, constraints imposed on the system, such as costs and economic viability, safety aspects, operating scenarios, and environmental conditions must be taken into consideration.

After this a new loop within the propulsion system development starts. Based on the new thrust requirement, the performance calculations are carried out again, leading to a revised engine design. For example, it might become necessary to change the size or design of some components or even of the entire engine. With this input, all of the component characteristics, including precompression, installation losses, intake characteristics, turbine engine characteristics, combustion chamber characteristics, nozzle characteristics, base drags, etc., are reviewed. These data are included in new calculations aimed at meeting the performance requirements. Finally, this results in a revised propulsion system design. Arriving at this design, system aspects such as engine control, secondary power demand, and thermal management as well as restrictions imposed by engine/aircraft integration must not be forgotten within the iterative process. The inboard configuration of the airframe must also be taken into account since it has to accommodate subsystem hardware and tanks. Thereafter, more precise performance data are available, and the loop for mission analysis described earlier is entered again until a feasible configuration is obtained. In the following, some aspects within this development process are discussed.

IV. Engine Components

A key area for successful development of a hypersonic vehicle is the design of the engine components. In this section the technological challenges associated with the component design are discussed briefly.

The operating conditions of the propulsion system are shown in Fig. 7. High temperatures up to 2800 K call for actively cooled structures with an efficient cooling system. At the same time, the need for overall weight reduction using lightweight structures must be borne in mind. Furthermore, the engine components cannot be designed individually without considering their interaction. For example, the size of the intake is determined by the thrust required at maximum flight speed. This, however, has a severe impact on the turbine engine sizing. The turbine engine is designed for the transonic regime at approximately Mach 1.2. In this flight regime, the intake usually delivers a mass flow in excess of engine demands. The excess mass must be spilled around the intake cowl, thus producing an additional spillage drag (see Sec. V). The higher the thrust requirements at Mach 6.8, the greater the intake capture area, and, in addition, the greater the size of the turbine engine necessary to overcome the additional drag. However, because of the larger engine bay, the thrust demand at Mach 6.8 rises again, and the loop must be entered again. A detailed discussion of the interaction of component size and thrust, fuel consumption, and engine weight is presented in Ref. 6.

A. Intake

The intake is one of the most critical parts of an airbreathing propulsion system. Its design must be such that it delivers air to the engine at the desired mass flow rate and flow conditions for all flight Mach numbers. This delivery must be accompanied by as little losses, drag, weight, and complexity as possible. This task is often very difficult because the intake is at the interface between the vehicle and engine. Thus, many factors, such as vehicle configuration and its overall performance affecting the design of an intake, are outside the responsibility of the intake engineer,[7-8]. Therefore, the engineers responsible for external aerodynamics, intake design, vehicle stability, and vehicle performance must find a compromise for the final forebody/intake configuration. Fig. 9 summarizes the requirements and constraints to be considered in the intake design.

Forebody. The task of an intake is to convert the kinetic energy of the incoming flow into pressure energy. At high supersonic Mach numbers, precompression of air because of a forebody can be utilized in addition to the compression inside the intake (see Sec V.). Since increased precompression results in increased pressures on the lower side of the fuselage, the lift and momentum balance of the whole vehicle are affected. Furthermore, these

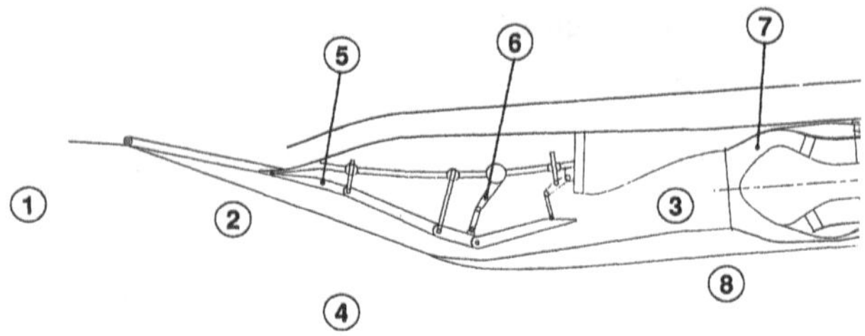

1. External Flow: Strong Shock Waves; High Temperature
2. Shock-Boundary Layer Interaction; Shock-Shock Interaction
3. Internal Flow: Transition Laminar/Turbulent; Corner Flow; Boundary Layer (Bleed)
4. Transient Effects
5. Structural Requirements: High Temperature; High Pressure; Great Variability
6. Intrinsic Kinematic and Control
7. Homogeneous Flow at Engine Face Required; Optimum Pressure Recovery
8. Minimum Drag

Fig. 9 Requirements and constraints for the intake.

pressures influence the boundary layer on the lower fuselage side and, thus, the quality of the flow entering the intake. For example, a pressure increase on the upper side of the forebody can induce crossflows on the lower fuselage side and boundary-layer air can be accumulated on the centerplane. Fig. 10 shows the development of the flows for two forebodies with two different crossplane shape distributions. The increased boundary-layer height on the centerplane can be recognized clearly. In addition, the shock standoff distance from the body is different for the two forebody geometries. The first result determines the selection of a boundary-layer diverter, and the second influences the maximum intake area.

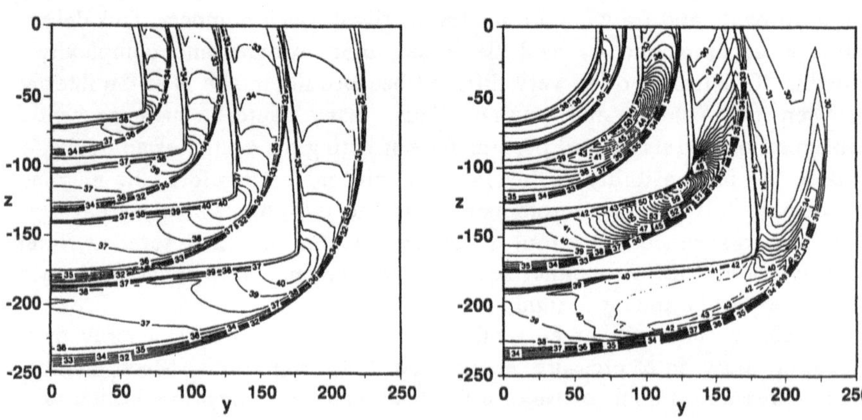

Flat Forebody, Ma=15, α=0°

Transition Forebody, Ma=15, α=0°

Fig. 10 Longitudinal variation of mass flow.

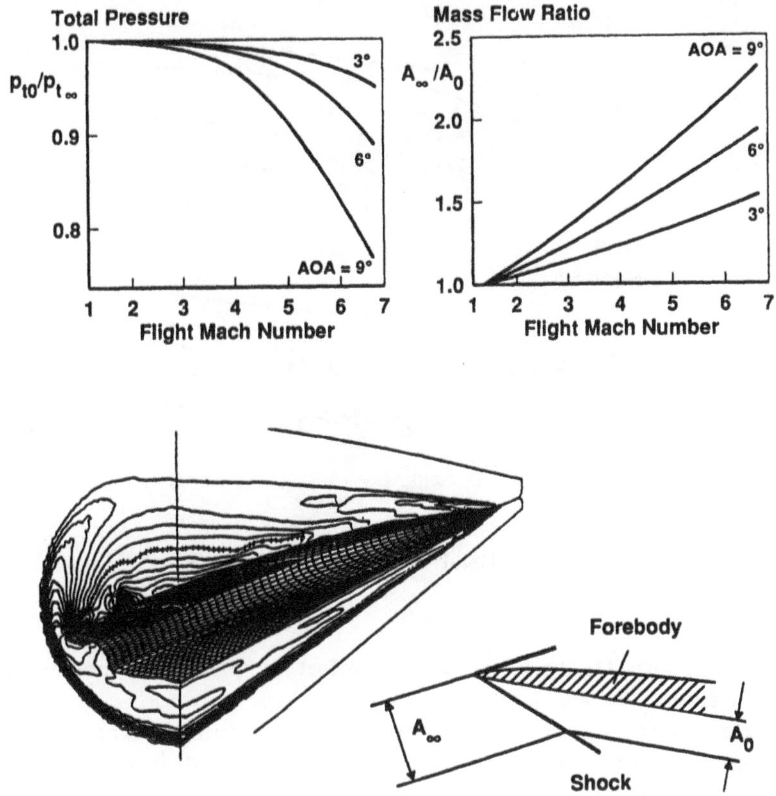

Fig. 11 Effect of forebody precompression on mass flow ratio and local total pressure.

During early development stages, the flow around the forebody can be determined by numerical methods. Full Navier-Stokes calculations are usually too expensive and time-consuming. Parabolized Navier-Stokes calculations are used somewhat less frequently. However, usually Euler calculations in connection with boundary-layer calculations are sufficient, especially if the forebody is designed such that the flow at the intake station is uniform. For a slender forebody with a flat lower side, Euler calculations produce results similar to those plotted in Fig. 11. As can be seen, the precompression is a function of flight Mach number and angle of attack.

Whereas the results in Fig. 11 have been obtained only for ideal gas flows, at even higher Mach numbers real gas effects and radiation will have to be included in those calculations. In addition, the bluntness of the tip of the forebody will become important because its flow can be dominated by effects such as entropy swallowing. However, for long slender forebodies, these effects do not seem to be too important. One problem of all numerical calculation methods still is the lack of validated transition and turbulence models. For

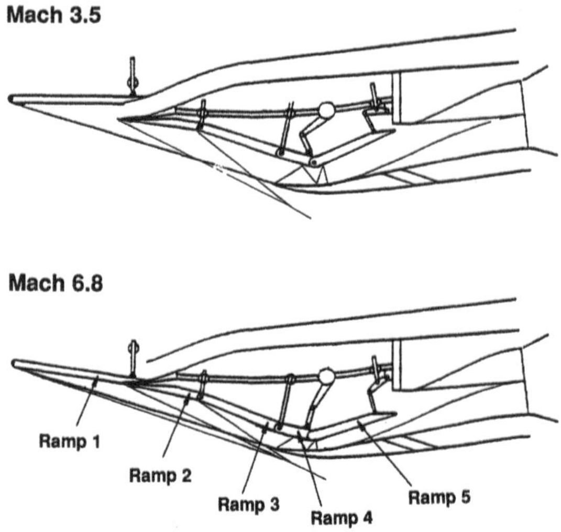

Fig. 12 Highly variable intake concept.

engineering purposes, this lack can be offset by parametric studies of the forebody flows.

Intake selection. For a Mach 6+ vehicle, several intakes and intake arrangements, each of which has its advantages and disadvantages, seem to be possible. Two-dimensional circular or half-circular intakes could be used. One intake could supply air to one or several engines. That is, the engines could be arranged in individual or multiple engine pods either on the wing or below the fuselage. The selection of an intake type depends on its performance but also on other items, such as intake/engine matching, structure mass and complexity, engine nozzle/base integration, landing gear arrangement and installation drag, etc. (see Sec. III). For the Sänger concept, consideration of all of these aspects led to the selection of a two-dimensional mixed compressor intake (see Fig. 12). At subsonic and low supersonic flight Mach numbers, it is used as a purely external compression intake. At high supersonic and hypersonic flight Mach numbers, its mixed compression mode is utilized. The variability of the intake geometry is accomplished using movable ramps. The first ramp is a dual-position ramp. During the turbopropulsion mode, this ramp is set parallel to the fuselage and is, therefore, not aerodynamically active. The fuselage boundary layer is removed by a diverter. The intake boundary layer is removed by wall bleed, where necessary. In the ram mode, the first ramp is moved such that it closes the diverter opening. Thus it produces one additional shock contributing to the compression of the intake flow. Boundary-layer bleed is shut off.

Drag and pressure recovery. During the design of such an intake, a compromise between low drag and high performance must be found. This compromise depends strongly on the mission of the vehicle, e.g., whether the vehicle is mainly an accelerating or cruising vehicle. In addition, it also determines the ratio of external to internal compression of the intake, which cannot be selected independently. For example, external compression is governed by the high total-pressure recovery and low-distortion demand of the turboengine in the lower Mach number range. If the mixed compression is also used during the turbo mode, the external ramp schedules have to be selected such that the mixed compression produces high-pressure recoveries, too. Compromises in one or all of the flight regimes are inevitable and can be found only after several propulsion and vehicle performance calculations. These calculations require tradeoffs between spillage and dumping of intake air. At the same time, their aim must be an intake design with as little variability, complexity, and weight as possible.

Aerodynamics. The use of shocks for flow compression inevitably leads to shock/boundary-layer interactions. These interactions are associated with total pressure losses. Strong shocks can produce boundary-layer separations that can produce unstarts of the intake. Unsteady shock movements and flow instabilities can reduce propulsion performance. Bleed through porous walls, a common means to solve the problems with these shock/boundary-layer interactions at intermediate flight Mach numbers, is no longer feasible at high flight Mach numbers because of the high total temperatures and the difficulty of manufacturing cooled structures of sufficient strength with bleed holes. In addition, bleed flows are always accompanied by momentum losses. The same holds true if the fuselage boundary-layer air is taken onboard by a diverter channel. To minimize these losses, it is desirable to swallow both the fuselage and the intake boundary-layer air at high flight Mach numbers. This, however, aggravates drastically the problems connected with shock/boundary-layer interactions. So far, the only way to solve these problems appears to be reduction in the strength of the intake shocks. This can be accomplished by the use of smaller flow deflection/ramp angles. To obtain the same flow compression, more shocks are needed, which result in more shock/boundary-layer interactions. Other consequences are increased intake length and weight. To the authors' knowledge, there is no project for which this concept has proved to be feasible.

Intake Control. A control system for a mixed compression intake must include at least three control loops. The ramps must be controlled according to schedule depending on Mach number and vehicle attitude. The position of the terminal shock must be controlled, and the start/unstart situation of the internal compression must be checked and corrected if necessary. As control parameters, the static wall pressures inside the diffuser duct seem to be sufficient. They must be derived from model tests and/or intake flow calculations. The requirements imposed on the air data system are very stringent, because for a high-performance intake the Mach number and angle of attack must be known within the limits of $+\Delta Ma = 0.05$ and $-\Delta\alpha = 0.1$ deg. Larger deviations without an intake control system reaction may lead to intake unstart.

B. Turbine Engine

The size and mass of the turbomachine depend mainly on the airframe thrust requirement at transonic acceleration. In hypersonic flight, the overall aerodynamic drag of the vehicle has several major components; the largest, wave drag. The wave drag coefficient reaches its highest value in the transonic speed range. To aggravate the situation, the airbreathing propulsion system experiences thrust losses in the transonic regime as a result of inlet efficiency losses, spillage, flow distortion, nozzle losses, etc. (see Sec. V).

Sufficient thrust must also be ensured in the subsonic and transition regimes to ramjet operation at approximately Mach 3.5. To design an optimized powerplant that satisfies all of these criteria, an appropriate core cycle has to be matched with an inlet, ramburner, and nozzle, thus ensuring adequate performance not only in the turbo mode but also in the ramjet mode.

The main parameters affecting the turbine engine thermodynamic cycle are as follows:

1) Overall pressure ratio. This pressure ratio determines the engine rundown characteristics and, as a result, the performance at transition Mach numbers.

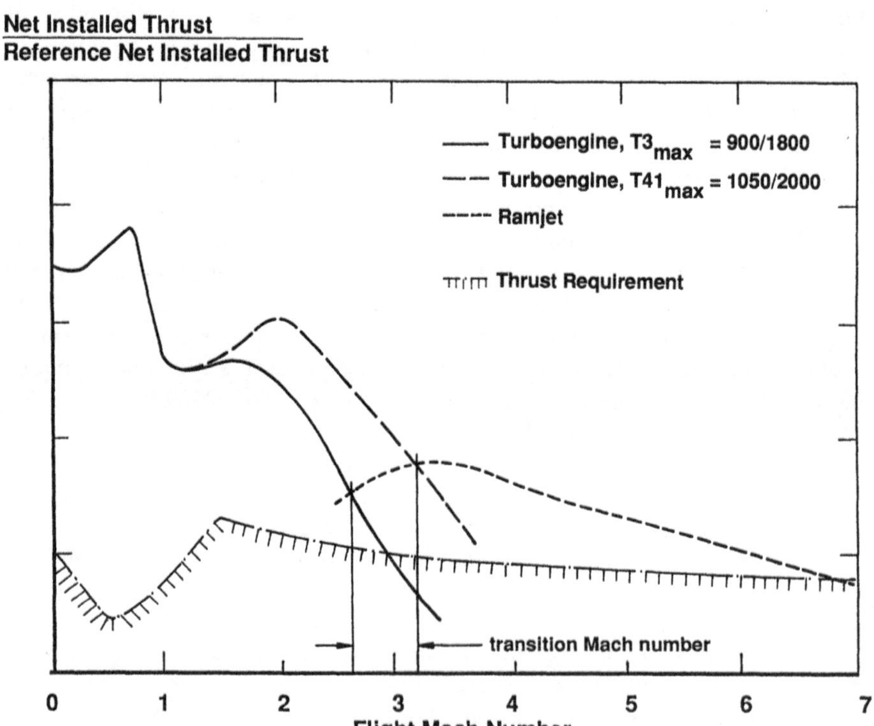

Fig. 13 Influence of internal temperatures on thrust.

Windtunnel testing and computational fluid dynamics. For proof of concept and determination of the internal performance of intakes, wind-tunnel model testing has been used extensively in the past. Deficiencies in flow modelization have been of minor importance during those tests. Because of restricted flight Mach numbers with low total flow temperatures, wind tunnels for small- and full-scale intake testing were technically and economically feasible.

This situation is different for the simulation of high-enthalpy hypersonic flight Mach numbers. For small-scale intake model tests, "cold" wind tunnels are available. The total temperatures in these wind-tunnels are just high enough to prevent condensation of oxygen or nitrogen. Because of the reduced size of the models and the low temperature, the wind-tunnel flow Reynolds numbers are nearly identical to the flight Reynolds numbers. Fuselage/intake integration aspects can be simulated only very roughly by using flat plates to produce boundary layers comparable to fuselage boundary layers.

Larger intake models can be tested in combustion-heated wind-tunnels. In these wind tunnels, freejet testing inside an evacuated test cabin is being performed. Total temperatures of up to $T_t \approx 1100$ K are reached by burning butane, methane, or hydrogen. Water vapor produced during combustion may form water or even ice. The maximum Mach number achievable in those wind tunnels is $Ma \leq 8$. Although the wind-tunnel nozzles have exit diameters of about 1m testing of reduced-scale intakes only is possible.

Intake tests in high-enthalpy wind tunnels are known from literature only for self-starting small-scale intake models. The test times of these existing blowdown facilities are much too short to allow starting of the internal compression of an intake model. In addition, new instrumentation or even strategies, e.g., for the measurement of mass flow, must be developed. Nonintrusive measurement systems developed for these wind tunnels are seldom practical for application during testing of realistic intakes. Because the heat transfer in the intake structure cannot be simulated during these tests, it is very difficult to assess the influence of the intake wall temperature on the flow.

The best tool currently available for the analysis and diagnostics of intake flows is computational fluid dynamics (CFD). Apart from the known deficiencies, such as the lack of transition and turbulence models, intake flow calculations have to deal with confined flows that require very dense grids to be able to resolve all flow details very accurately. For example, very small errors in the shock angles of the internal compression can accumulate and falsify the results at the intake exit (engine face). Convergence of a flow solution on these dense grids requires extensive computer time. Possibly this is the reason why the literature presents mostly two-dimensional calculation results.

A practicable approach toward the development of a hypersonic intake seems to be small-scale model testing under servicelike flow conditions for the verification of the design methods and CFD.

Obviously, this work must be accompanied by basic research of the flow physical problems relevant to intake flows. With these proven methods, the full-scale intake can be designed and its performance predicted. Final proof of concept must be obtained during flight tests with an experimental vehicle.

Compressor aerodynamic off-design operation results from rising inlet temperatures with rising flight Mach number while the turbine engine internal temperatures are at their limits.

2) Turbine entry temperatures that depend strongly on the technological standard of the turbine vanes and blades.

To illustrate the influence of an advanced technological standard, Fig. 13 shows the thrust for two different turbine engine cycles along a certain flight path as an example. In the baseline cycle, the maximum compressor exit temperature (T3) is 900 K, and the maximum stator outlet temperature (T41) is 1800 K. In the second cycle, the limits are set to T3 = 1050 K and T41 = 2000 K. The engine size and the cooling mass flow remained unchanged.

The changed cycle parameters result in higher thrust for Mach numbers in excess of approximately 1.5. Up to this Mach number, engine performance is restricted by the compressor speed rather than the temperature, thus producing the same thrust characteristic.

In the baseline cycle, the injected fuel must be reduced beyond approximately Mach 1.5 as a result of the limited turbine entry temperatures. At about Mach 2.5, the compressor runs down because of its limited exit temperatures. Both effects result in the steep decrease in thrust shown in Fig. 13.

With an engine design that allows higher compressor exit temperatures and turbine entry temperatures, rundown occurs at higher flight Mach numbers (see Fig. 14), which means higher thrust of the turbine engine in this flight regime.

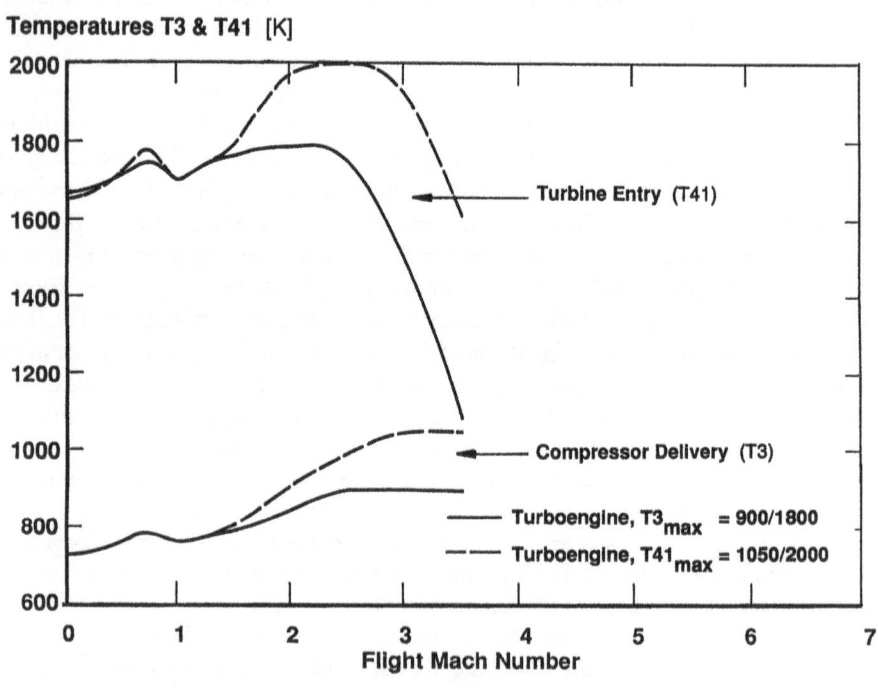

Fig. 14 Engine internal temperatures.

This is a favorable effect since transition from the turbo mode to the ramjet mode can be shifted to a flight Mach number where the ramburner operates with maximum thrust, compare Fig. 13. To find a compromise with respect to the best design, all factors, such as cycle temperatures, thrust, engine size, and operating life, must be weighed carefully for the entire flight path.

From the technological point of view, it is all but easy to raise the cycle temperatures of the engines. The operating life of a turbine blade depends strongly on its metal temperature. Figure 15 shows that, for standard materials, an increase of 50 K in metal temperature reduces the lifetime by a factor of 10.

There are several possibilities of realizing higher cycle temperatures without any penalties regarding the operating life. The first is to increase turbine blade cooling so that, despite the higher gas temperatures, the metal temperature remains unchanged. The second is to introduce thermal barrier coatings on the blade surfaces to reduce heat flux. A third possibility consists in the use of new materials capable of withstanding higher temperatures. The advantages and disadvantages are discussed subsequently.

Fig. 16 illustrates the effects of increased blade cooling. It can be seen that the amount of cooling air needed is very sensitive to the permissible metal temperature and to the compressor exit temperature which nearly equals the temperature of the cooling air for the high-pressure turbine blades. The higher the permissible metal temperature, the less coolant flow is necessary. This

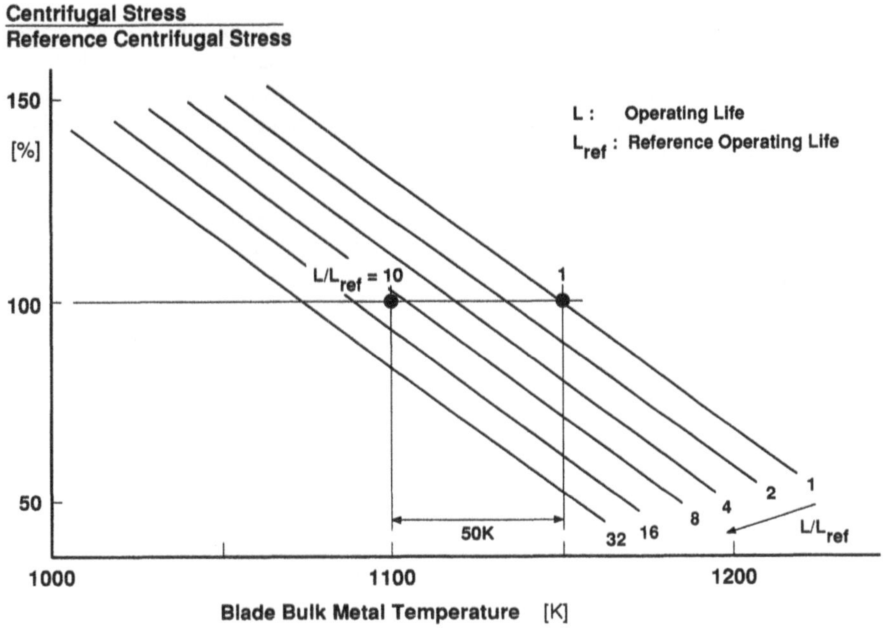

Fig. 15 Operating lifetime at given stress and thermal loading.

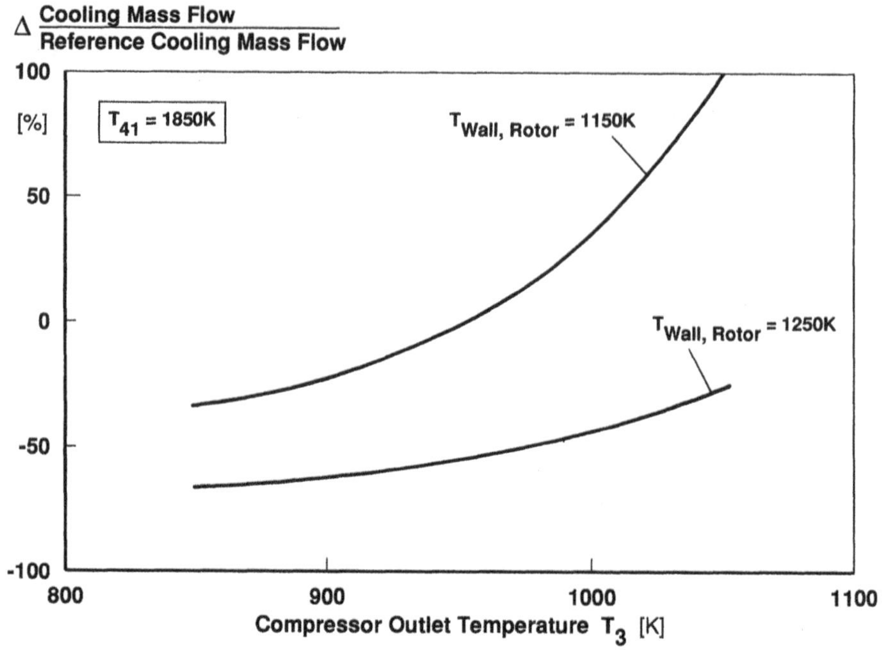

Fig. 16 Varying cooling bleeds: high-pressure rotor cooling air.

results in better specific fuel consumption for the turbine engine and vice versa. However, of more significance is that with higher temperatures, the rundown of the turbine engine occurs at higher Mach numbers. Introducing these findings in an overall balance shows that, without changing the permissible blade surface temperature, the thrust gain from increased cycle temperatures is more than outweighed by the penalty attributable to the additional cooling air demands. Thus, it becomes obvious that advanced thermal barrier coatings and/or advanced structures and materials must be taken into account as a means of increasing engine performance.

C. Combustion Chamber

An additional key area for the successful development of a hypersonic vehicle is the ram combustor design. As mentioned earlier, high temperatures up to 2800 K call for actively cooled structures with an efficient cooling system. At the same time, the combustion chamber must be a lightweight structure to reduce overall weight. To meet these requirements, advanced cooling methods and new materials, such as ceramic-matrix composites and metal-matrix composites, are presently being developed. Fig. 17 shows the requirements and constraints for the combustion chamber.

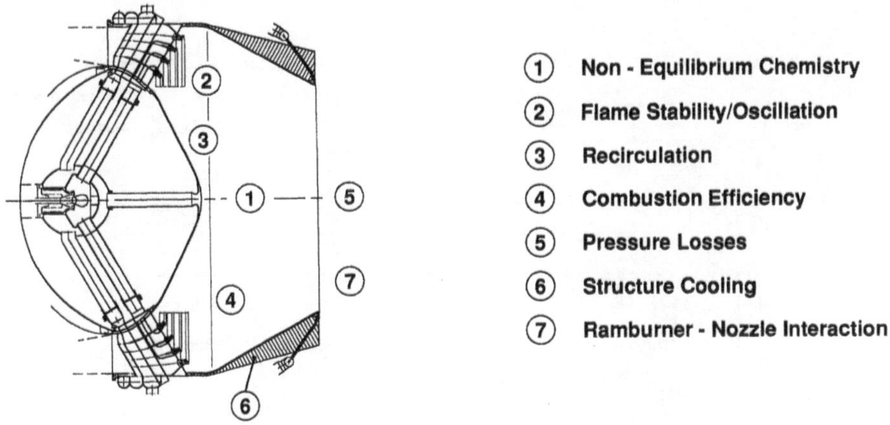

Fig. 17 Requirements and constraints for combustion chamber.

The main objective in the German Hypersonic Technology Programme for the ram combustor technology development is to lay the technological foundations for a cryogenic hydrogen ramjet combustor by using the following methods:

1) Numerical analysis to gain understanding of the and combustion flow phenomena in the combustor to reduce the number of development tests and, consequently, to be able to reduce development cost;

2) Structural design to obtain a combustor with adequate structural strength and an efficient cooling system; and

3) Subscale testing of a combustor with an internal diameter of 13 in (330 mm) to identify performance capabilities and deficiencies, detect combustion instabilities, and assist the development and verification of CFD computer codes.

In the reference propulsion system, the ramjet combustor is located downstream of the turbojet with the airflow entering through an annular duct around the turbojet. Consequently, the ram combustor concept is configured as a sudden dump with an inward-facing step. The resulting test configuration is depicted in Fig. 18: vitiator, combustor, and thrust nozzle are assembled on a rig, which allows the thrust (exit impulse) to be measured with a high degree of accuracy.

Early fuel injector configurations consisted of 24 fuel injector pins with one sonic injection orifice each. Injection was directed either downstream or normal to the flow direction, thus producing a swirl in the hydrogen flow. The latter was found to have a favorable effect on combustion efficiency because the residence time of the fuel is obviously longer.

Fig. 19 shows some results in terms of combustion efficiency vs combustor length. As expected, the efficiency drops with reduced combustor length. One important result of the first test series is that, with fuel-lean mixtures, combustion efficiency is high (>98%) even for short combustors. However, when approaching stoichiomertric mixture ratios, combustion efficiency drops sharply. More important, combustion instabilities were identified at

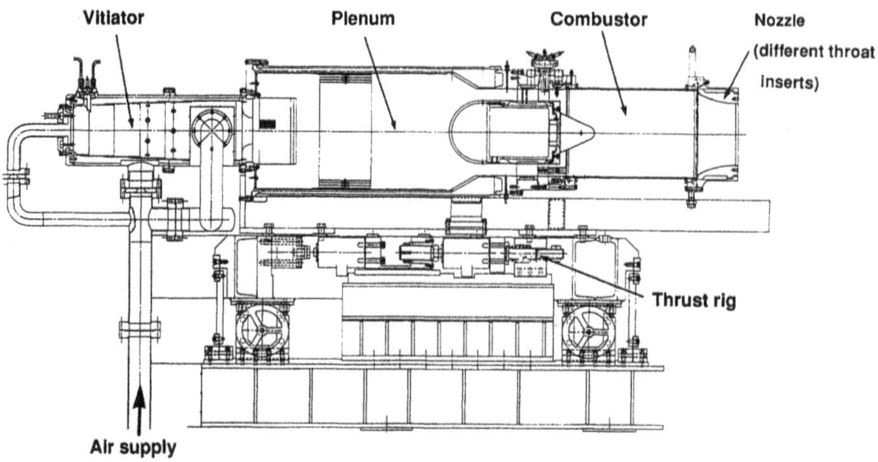

Fig. 18 Combustion chamber test facility

stoichiometric combustion with about 4000 Hz and approximately 22% pressure amplitude.

In an attempt to explain this behavior, it was determined that temperature stratifications, resulting from more than 24 discrete points of energy release, were causing a thrust deficit, which the common method of ramjet test data reduction interprets as a loss of combustion efficiency.

As a consequence, a fuel injection device was designed, which would allow a more homogeneous distribution of hydrogen in the air stream. Fig. 20 shows two vanes that are installed to direct airflow toward the axis. At each vane base, hydrogen is injected normal to the air stream through injection holes ranging from 0.4 to 0.9 mm in diameter.

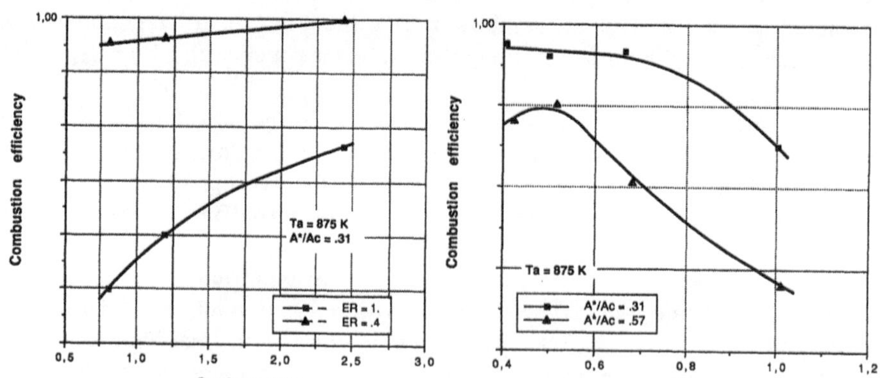

Fig. 19 Selected test results for combustion efficiency.

Although the level of combustion efficiency was somewhat higher (as illustrated in Fig. 20), it still is not satisfactory at stoichiometric combustion conditions. Two-dimensional Navier-Stokes calculations were carried out. They indicated (see Fig. 20) a massive streak of low-temperature exhaust originating from the annulus between the two vanes leading to temperature stratification and, thus, lower combustion efficiency.

An attempt is currently being made at increasing the number of vanes, thus ensuring a more homogeneous hydrogen distribution.

As far as combustion instability is concerned, the vanes had a favorable influence on the pressure amplitude, which dropped to about 3.5%.

For performance evaluation test runs, the combustor wall is lined with silicon rubber ablative material. However, to obtain first indications on design specifics, a cryogenically cooled chamber was designed, built, and tested. It consists of spiral cooling tubes with an electroplated nickel outer shell providing the pressure vessel function. The combustor was tested 18 times at full heat load; the results with respect to structural integrity are excellent.

To reduce the cost of the combustor, manufacturing methods have been investigated that could replace the relatively expensive electroplating procedure. For future chambers, metal powder spray methods, such as vacuum plasma spray or high-velocity oxygen fuel method, are recommended.

One final remark on the method of air heating that is applied in the Deutsche Aerospace (DASA) facility follows: Commonly, a suitable fuel (e.g., propane or hydrogen) is burnt with air to produce the required simulated flight stagnation temperature. Gaseous oxygen is added to make up for the loss of reacted atmospheric oxygen. This method is appropriate for low to medium stagnation temperatures (e.g., simulated Mach < 4.5). However, at higher stagnation temperatures, greater amounts of reaction products (water steam and/or carbon dioxide) are in the gas stream. They may affect the combustor and nozzle reaction kinetics, thus rendering the experimental performance data questionable.

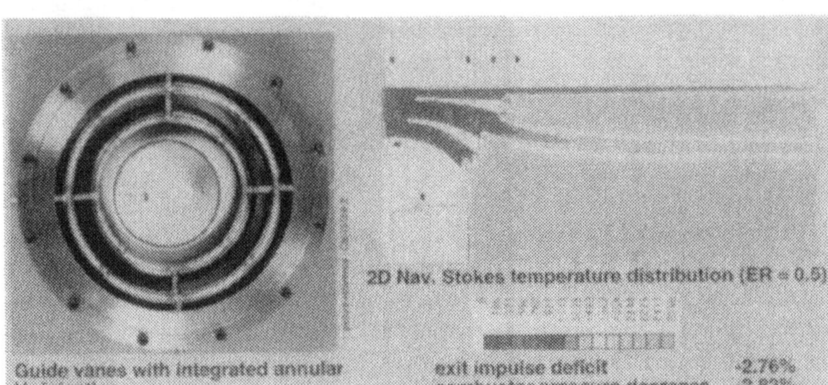

Fig. 20 Two-dimensional Navier-Stokes flowfield calculation.

Fig. 21 Ram combustor test facility at DASA, Ottobrunn.

This drawback is significantly minimized when the "air" is created synthetically by thermal decomposition of laughing gas (N_2O) into its constituents. The fundamental advantage is the exothermic character of this reaction, since laughing gas exhibits a positive heat of formation. Thermal decomposition is achieved either by burning laughing gas with hydrogen at a fairly high oxidator/fuel ratio or by a hydrogen/air pilot flame. The final balance with respect to atmospheric airlike composition is obtained by adding nitrogen or air to the decomposition products. Using this method, the amount of vitiation products is reduced by a factor of almost 3 as compared with common vitiation methods.

Figure 21 shows such a laughing gas reactor on the thrust rig together with a cryogenically cooled combustor. This setup made test conditions corresponding to flight Mach numbers of almost 7 possible.

D. Exhaust Nozzle Technology

The nozzle exhaust system is an essential component in any airbreathing propulsion system,[9]. In hypersonics, the nozzle becomes even more important because of large area variations, strong weight constraints, and close interaction with the vehicle. The nozzle, as part of the propulsion system of the reference concept Sänger, is shown in Fig. 7.

The nozzle development process comprises three major tasks:

1) To maximize propulsion performance, the expansion process of the nozzle flow should be completed with minimum internal energy losses since the

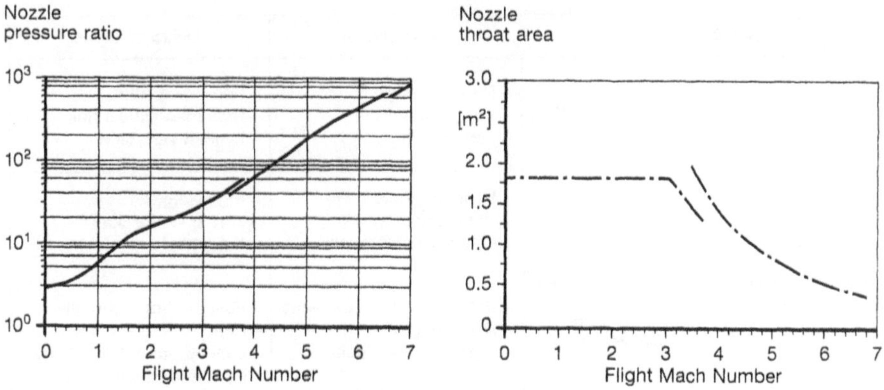

Fig. 22 **Nozzle pressure ratio and nozzle throat areas vs flight Mach number.**

internal efficiency is reflected directly by engine performance. Furthermore, the additional drag caused by nozzle integration must be within an acceptable range determining the installed efficieny.

2) To minimize propulsion weight and because of cooling requirements, the aerodynamic design must be transformed into a lightweight, actively cooled structural design.

3) A sophisticated testing methodology has to be applied to guarantee a straightforward, cost-effective approach toward verification of the aerodynamic and structural design concept.

Nozzles for hypersonic applications have to operate over a wide range of flight Mach numbers (e.g., for Sänger: Mach 0-7). This consequently leads to a wide range of nozzle working conditions.

Throughout the flight envelope, the nozzle has to transform pressure ratios (nozzle entry/ambient) from about 3 (takeoff) to about 800 (maximum Mach) into thrust (see Fig. 22).

Furthermore, as a result of intake flow characteristics, the nozzle needs a throat variation on the order of 1:4. These are exacting requirements, resulting in a variable geometry design and favoring the use of the rear end of the vehicle as nozzle surface, compare Sec. V.

A large number of nozzle types have been investigated during several decades of rocket and jet propulsion development,[10]. However, the nozzle operating requirements described earlier as well as integration constraints limit the number of types suitable for hypersonic vehicle applications. A survey with a brief description of advantages and disadvantages concerning hypersonic application is given in Fig. 23 for each nozzle configuration of interest.

For the reference propulsion system, a single-ramp expansion nozzle was selected. Fig. 24 provides insight into the complexity of this hypersonic nozzle, and Fig. 25 shows the great variations in geometry required along the flight path.

Nozzle Configuration		Advantages	Disadvantages
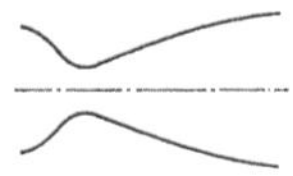	**Axi-symmetric C-D**	**– Proven design** **– Low aero risk** **– Low weight**	**– Poor integration** **– High mechanical risk** **– Limited variability**
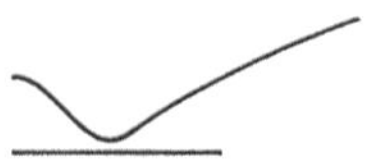	**SERN**	**– Best integration** **– High variability** (throat and exit areas)	**– High weight** **– Thrust vector bias** **– High mechanical risk**
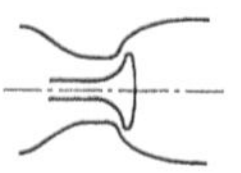	**Expansion-Deflection** (ED)	**– Broad performance range** **– Short length** **– High throat variability**	**– Plug cooling difficulties** **– High loading on plug** **– Limited boattail flexibility** **– Poor integration**
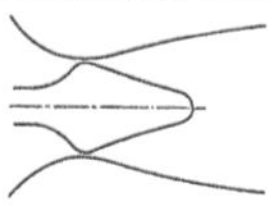	**Plug**	**– Good high Mach performance** **– Low weight**	**– Plug cooling difficulties** **– Under-expansion problems** **– Poor integration**

Fig. 23 Characterization of different nozzle types.

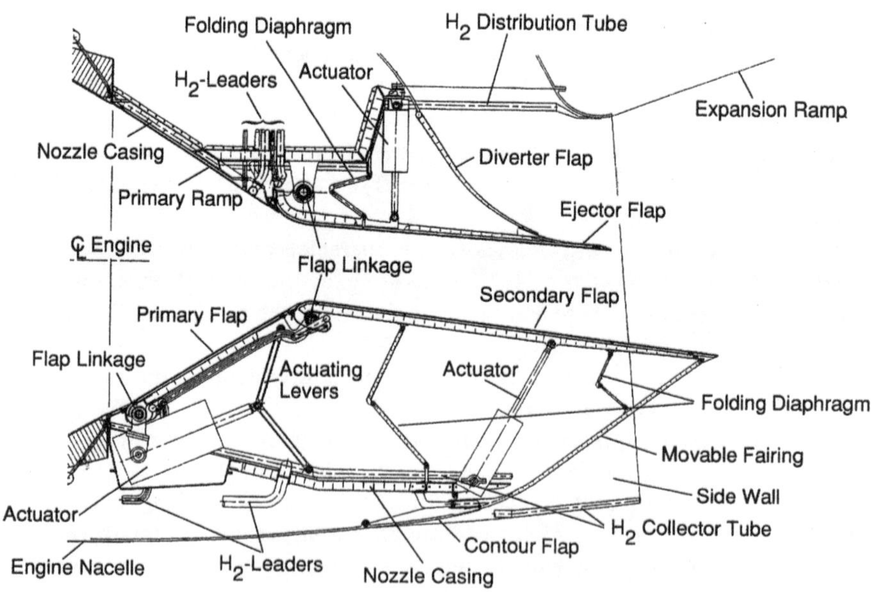

Fig. 24 Two-dimensional nozzle concept.

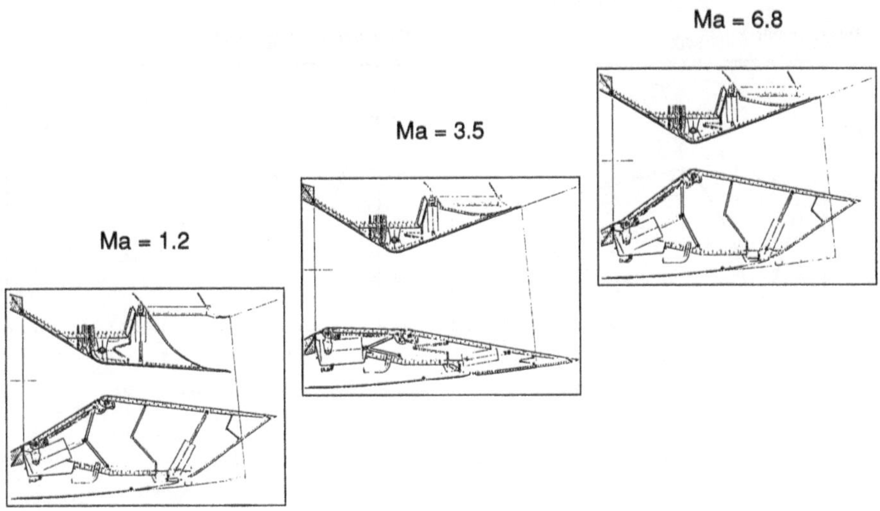

Fig. 25 Typical nozzle throat area variation.

Because of the high temperatures, the walls as well as the movable flaps must be cooled with liquid hydrogen, this means that all the problems associated with the hydrogen supply, especially to movable walls, need to be solved. Furthermore, the flaps must be designed properly with regard to pressure balancing. The great pressure variation along the nozzle flaps calls for different pressure chambers, vented with cooling air and separated by folding diaphragms. However, despite this, the temperatures in the actuator compartments are so high that only sophisticated actuator thermal management can guarantee reliable operation. Moreover, the need for forebody boundary-layer injection into the nozzle requires another movable system, doubling the above-mentioned problems.

The typical resulting thrust coefficient $C_{FGI\text{-}x}$ over the flight range is shown in Fig. 26. Note that the $C_{FGI\text{-}x}$ value shown includes only the axial thrust component. The significant impact at transonic flight Mach numbers is caused by the thrust vector angle variation shown in Fig. 26 (right-hand side). This problem is a matter of nozzle integration and is discussed in greater detail in Sec. V.

To give an idea of the nozzle cooling requirements, the heat transfer coefficients are given in Fig. 27. The hot-gas heat transfer coefficient reaches its maximum at the nozzle throat. As a result of the strong pressure drop, it strongly decreases downstream toward the nozzle exit. In addition, analytical investigations showed that the expansion ramp can be cooled adequately by radiation alone.

To investigate the wide field of nozzle-related aspects, an overall strategy was set up to cover the most important technological challenges within the budget constraints and taking the availability of appropriate test facilities into

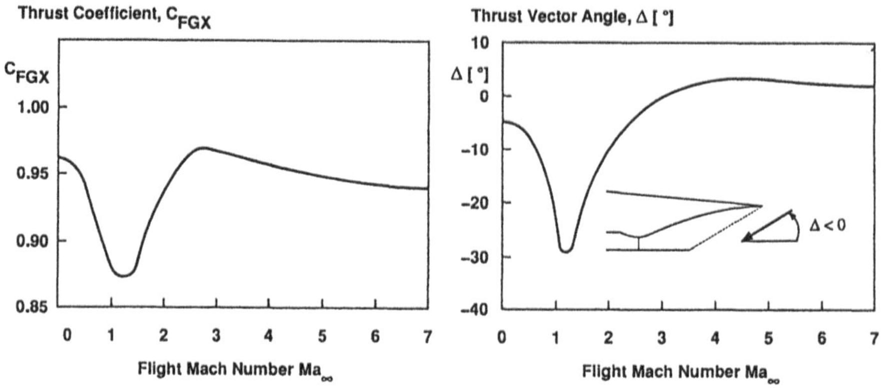

Fig. 26 Nozzle performance over flight range.

account. Figure 28 shows the development philosophy. Four different groups of subscale nozzles were defined. Each group covers a major technological area. Any possible additional aspects were also taken into account to allow comparison with data from other test setups.

With this philosophy in mind, testing comprising the areas of nozzle aerodynamics, high-temperature thermodynamics, cooling systems, structures, materials, and the nozzle afterbody integration was started. Fig. 29 shows some of the subscale nozzles. The most advanced is the so-called technology demonstration nozzle. This nozzle is liquid-hydrogen cooled. The cooling structure is mounted on a load-bearing structure. The uncooled expansion ramp is made from carbon/carbon with an oxidation barrier coating. To simulate the

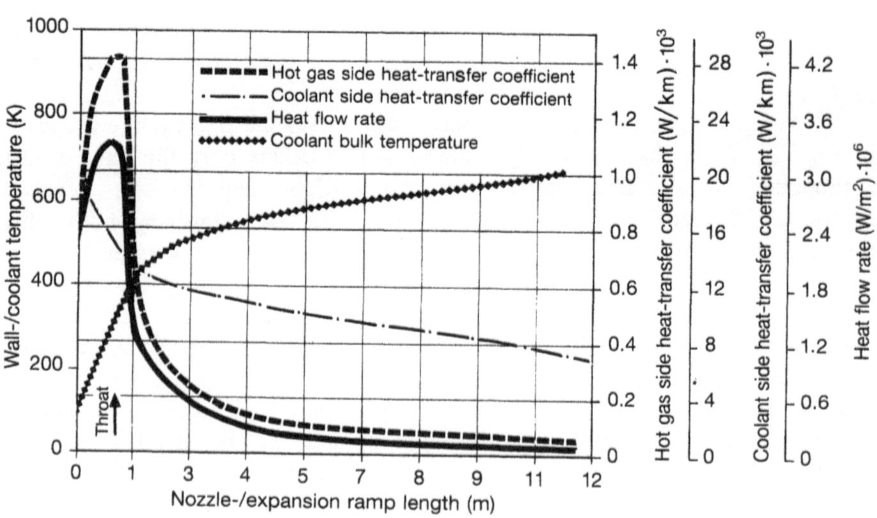

Fig. 27 Rate of heat transfer along wall of hypersonic nozzle.

	Subscale Nozzles (Cold Flow)	**Subscale Nozzles** (Hot Gas Flow)	**Technology Demonstration Nozzle**	**Integration Model**
Aerodynamics	■	□	□	□
High Temperature, Thermodynamics		■	□	
Cooling System		□	■	
Structures, Materials		□	■	
Integration				■

□ Additional Aspects ■ Focus

Fig. 28 Nozzle development philosophy.

Rectangular Nozzle

Plug-type Nozzle (1)

Plug-type Nozzle (2)

Technology Demonstration Nozzle Ready to Test

Fig. 29 Subscale nozzles.

Fig. 30 Technology demonstration nozzle during test.

different flight Mach numbers, the lower flaps of the nozzle are adjustable to vary the nozzle throat area. The hydrogen supply system to the flaps as well as the cooling structure itself were investigated in great detail. A variety of production feasibility tests have been carried out to find the most suitable structure. The same holds true for the sealing technique. This nozzle was tested under servicelike conditions for the first time in March 1993 to verify the cooling technology and the structural concept (see Fig. 30).

V. Engine/Airframe Integration

The literature available on this subject emphasizes that propulsion integration is one of the basic prerequisites for a successful approach toward airbreathing hypersonic flight,[11-15]. However, what does this really mean?

Figure 8 makes an attempt at defining at least the major areas involved. Most of them - as can be seen - represent interaction problem areas (interfaces) between aircraft and engine manufacturers. This means that a very close collaboration between the different parties involved is essential for the development of an overall optimized aircraft/propulsion system (see Sec. III). Therefore, "propulsion integration" stands for the activity of a joint group or team of specialists.

The integration requirements of the airbreathing propulsion system depend on the maximum Mach number and cause various problems. Typical configurations for different speed ranges are highlighted in Fig. 31.

Propulsion systems for the first speed range - from subsonic to about Mach 4 - are characterized by the following aspects: the engines are installed in low-drag nacelles (pods) arranged outside of the disturbed flowfield of the

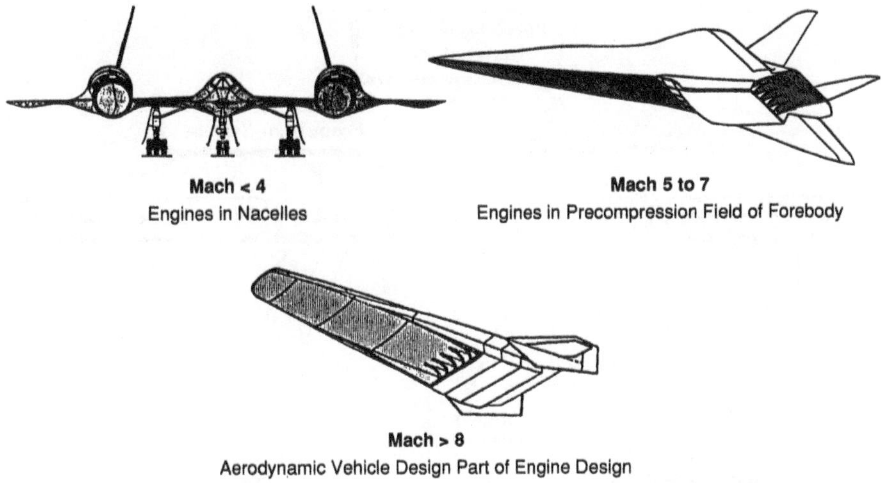

Fig. 31 Propulsion integration for different flight Mach numbers.

aircraft. The propulsion system is designed quite independently of the aircraft system. Intake, engine, and nozzle are optimized with respect to best installed specific fuel consumption. For the second speed range Mach 5-7 the interference between the airframe and propulsion system is already noticeable. The size of the intake and exhaust systems has to be reduced to save weight and limit drag. This is achieved by using precompression and postexpansion effects within the flowfield of the fuselage or wings. Conversely, the forebody and afterbody are still shaped according to aerodynamic aspects, which result in a low-drag design.

For the third speed range above Mach 8, an independent design of the airframe and propulsion system is no longer possible. The forebody of the aircraft is part of both the aerodynamic design, producing lift, drag, and pitching moments, and the propulsion system, acting as a compressing intake. Similarly, the afterbody is used as a large expansion area to produce maximum gross thrust and, in addition, a significant part of the aircraft lift.

From Fig. 31, it can be concluded that the required level of integration increases significantly with higher Mach numbers. This does not only affect the pure aerodynamic integration, but also the thermodynamic integration (heat management), structural integration (weight reduction), and functional integration (subsystems, safety aspects, etc.).

Both, the forebody and the nozzle/afterbody were selected as examples to illustrate the variety of interacting problem areas and interdependent disciplines, which have to be investigated to obtain an optimized propulsion/airframe system.

A. Forebody Design and Fuselage Precompression Effects

In highly integrated propulsion systems with intakes located below the fuselage (belly type), there are many different mechanisms, how the bottom side

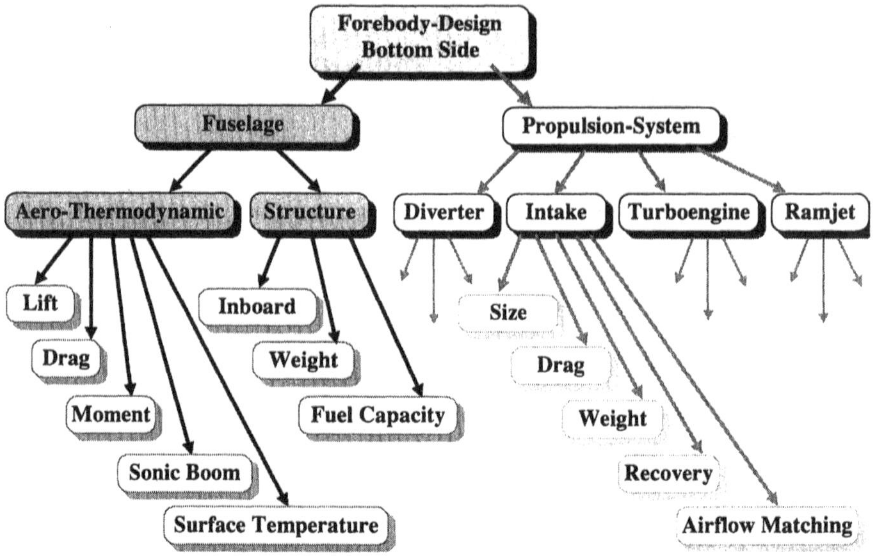

Fig. 32 Influence paths in forebody design.

of the fuselage influences the intake flow while the intake itself influences the flow around the engine cowl, parts of the fuselage and wings, or fins. These interactions between airframe and propulsion system generally are designated as propulsion integration tasks.

The first example deals exclusively with forebody effects on the airframe and propulsion system. Fig. 32 indicates some of the areas influenced by the forebody bottom side shape, starting with the tip of the aircraft and ending (in our bookkeeping definition) at the first ramp of the intake. The effects on the fuselage are aerothermodynamic and structural.

A modified design of the forebody changes lift, drag, and moment coefficients and derivatives, surface temperatures (values and distribution), and sonic boom generation. Structural aspects are weight, inboard configuration, and fuel capacity.

The propulsion system is influenced by the flowfield generated by the forebody in a very complex manner. It affects thrust, fuel consumption, and weight of turbojet and ramjet engines via the intake and boundary-layer diverter system.

In addition, pressure recovery, drag, size, and weight of the intake and boundary-layer diverter system itself are strongly dependent on the entering flowfield. Some of these propulsion-related features are explained in the following.

During high supersonic and, even more, hypersonic operation, the use of the fuselage precompression has several major advantages:

1) The fuselage provides an additional shock that is beneficial for the total pressure recovery of the intake as a result of the reduced local Mach number.

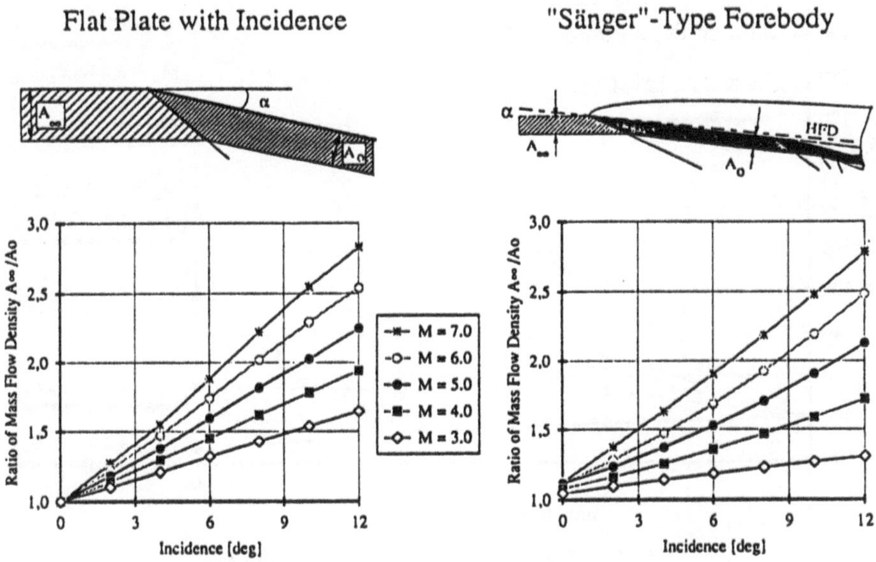

Fig. 33 Precompression of forebody bottom side.

The total pressure losses across the bow shock are relatively small. However, the shape of the fuselage bottom has to be designed such that no flow acceleration between the bow shock and intake occurs.

2) The intake ramps are shorter for the same design (i.e., shock on lip) conditions.

3) The intake mass flow, which is essential for ramjet thrust, is significantly higher for a given intake capture area. Precompression causes an increased airflow density and thus an increased captured mass flow. The ratio of mass flow densities is often expressed by the area of the entering stream tube in the freestream divided by the area at the intake entry plane (A/A_0).

Fig. 33 shows these area ratios as created by a flat plate with incidence (left-hand side) and of a typical Sänger-like forebody. The precompression ratio is increased signficantly by freestream Mach number and aircraft incidence.

The effect of precompression on net thrust is shown in Fig. 34, for a constant intake capture area. It is obvious that during ramjet operation at high speeds, where the mass flow is not limited by the ramjet burner or nozzle throat area, the thrust increases nearly proportional to the airflow. Therefore, this effect is very valuable for an increase in the accelerating force (thrust minus drag) at the ultimate speeds attainable with subsonically burning ramjets. Consequently, high internal pressure (as a result of high precompression) means high structural loads to the components and, therefore, high component weight. For achieving high precompression, a greater aircraft incidence angle is necessary, resulting in higher vehicle drag. A careful tradeoff between all effects is required.

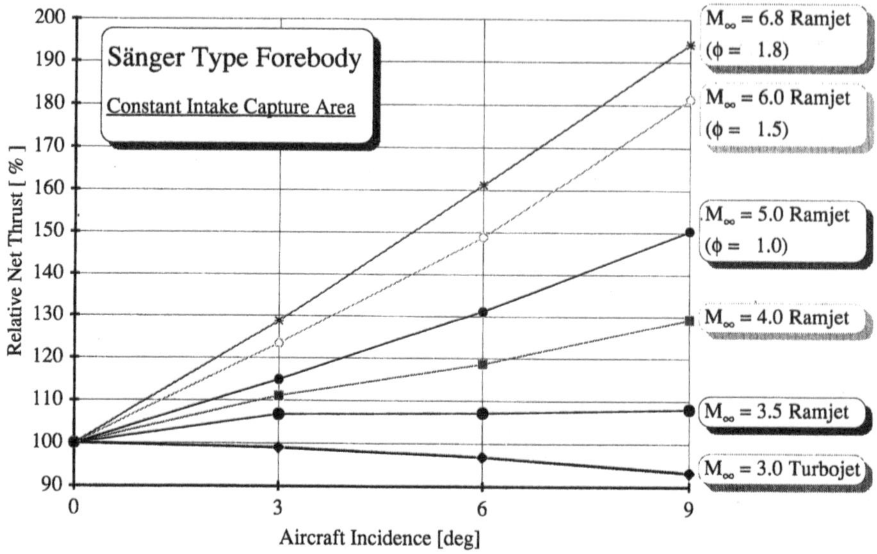

Fig. 34 Precompression effect on net thrust.

The intake/engine airflow mismatch, which is typical for high-speed airbreathing accelerators, is indicated in Fig. 35. Here the various airflows passing through or spilled around the intakes are shown schematically. The numbers indicate the engine mass flow related to the maximum possible mass flow to be swallowed by the given capture area.

The diverter system prevents the boundary layer created by the large aircraft forebody from entering the intake during turboengine operation. In addition, parts of the boundary layer created by intake ramps and side walls are diverted through holes or slots indicated as ramp, throat, and side wall bleed. This allows for low flow distortion at the turboengine face and, consequently, stable and surge-free engine operation.

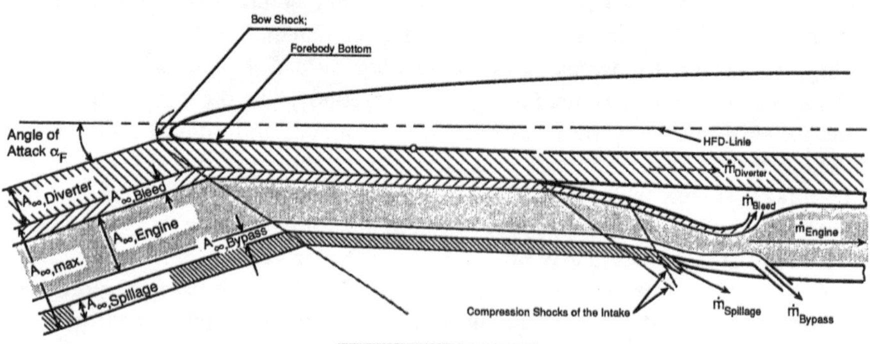

	M_{∞}=1.2	M_{∞}=3.5	M_{∞}=6.8
Engine Flow	40%	60%	100%

Fig. 35 Mass flows along stream tube of intake.

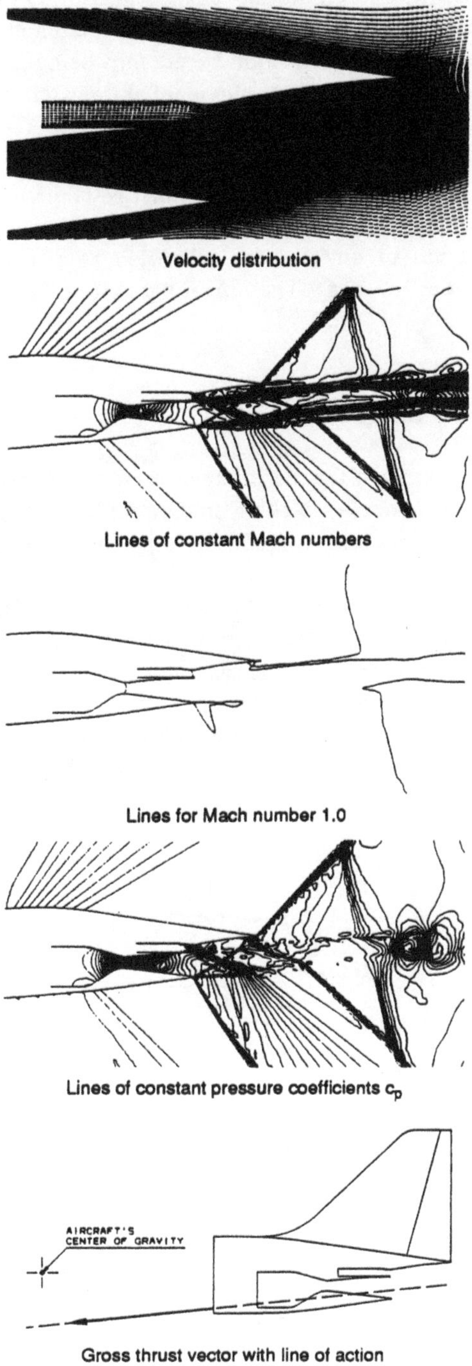

Fig. 36 Example for complex nozzle/afterbody flowfield at off-design (Mach 1.2).

During turboengine operation, the engine mass flow is defined by the compressor flow capacity. Conversely, the intake size (capture area) is defined by the thrust demand at maximum flight speed, thus being greatly oversized for the turboengine demand. This becomes evident at transonic and low supersonic speeds. To reduce the spillage flow, which leads to high pre-entry and cowl lip drag (spillage drag), part of the excess air is dumped through doors at the lower side of the engine cowling (bypass flow). It must be taken into account that the size of the dumpdoors is limited bcause of the structural considerations, and only a limited amount of air may be exhausted at a low angle to prevent excessive door or crossflow drag. The remaining part of the air is spilled around the intake cowl.

B. Nozzle/Afterbody Integration

As an additonal example for the interaction of the airframe and propulsion system, some aspects of nozzle/afterbody integration and the complex interaction with aircraft aerodynamics and weights, flight performance, and mechanics are addressed.

Nozzle systems for combined-cycle hypersonic turbo- ramjet engines have to be operated from the low-speed region (takeoff and landing) with dry or reheat turboengine operation up to the maximum required flight Mach number, which may well exceed Mach 6, with ramjet operation. The scope of nozzle pressure ratios (i.e., internal total to static ambient pressure) ranges from about 2 to 500, or even 1000 depending on the inlet recovery achieved. Therefore, a

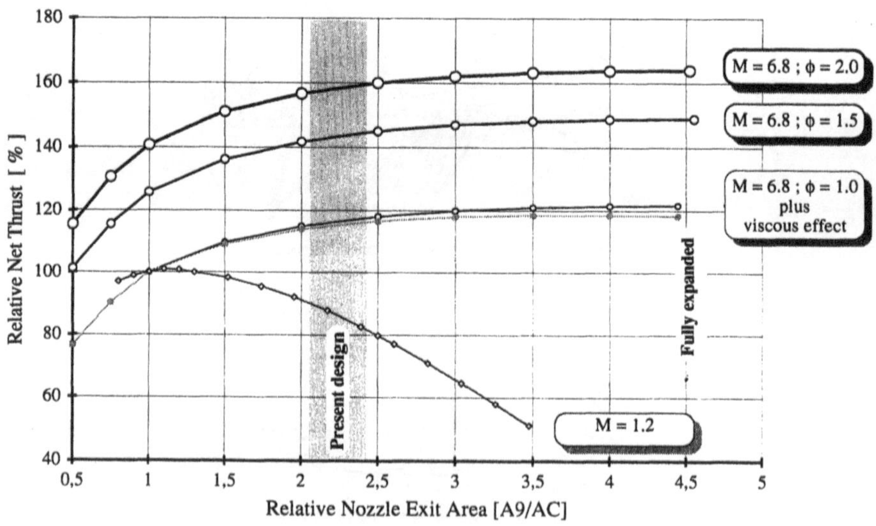

Fig. 37 Effect of nozzle exit area on net thrust.

high variability in nozzle throat area and nozzle exit area is required to obtain adequate internal performance (see Sec. IV.D). The problem arising is the absolute size of the nozzle exit area for high flight Mach numbers required to avoid detrimental underexpansion losses, which lead to undesired internal and/or external losses caused by overexpansion, unfavorable afterbody angles, and flow detachment at base areas in the transonic and low supersonic flight speeds, where the nozzles operate in extreme off-design modes. For axisymmetric and symmetric two-dimensional convergent-divergent nozzles, these demands result in severe technical problems and unacceptably high weights. Nozzles with a single expansion ramp (SER) are attractive solutions for the challenging variation requirements outlined above as a result of their better off-design adaptability. Throat area variation can be achieved with comparably small flaps, which is important with respect to weight, especially in the high-pressure convergent part of the nozzle. The maximum usable nozzle end area is shaped to a large extent by the aircraft fuselage itself.

A typical two-dimensional SER-type nozzle with ejection of forebody boundary-layer air during turboengine operation is shown in Fig. 24.

The flowfield presented in Fig. 36 calculated by the CFD-Euler method is an example of the complex expansion process in the exhaust system during off-design operation for a flight Mach number of 1.2 characterized by low pressures at the expansion ramp and base areas, thus leading to strong drag and downward forces.

In Fig. 37, the relative net thrust is plotted vs nozzle exit area for transonic (Ma = 1.2) and hypersonic (Ma = 6.8). flight Mach numbers. The data were approximated for a two-dimensional SER-nozzle.

For the high-speed case where different fuel/air ratios prevail (ϕ = 1 is stoichiometric), the thrust is increased for larger nozzle exit areas (A_9), but levels off for A_9/A_C of about 3 (A_C = intake capture area), when viscous effects are taken into account. This value again is dependent on the level of forebody precompression achieved for the relevant high-speed case.

The low-speed case shows just the opposite trend despite injection of the total forebody boundary-layer flow diverted in front of the intake. Since the low speed (transonic) case is as thrust-minus-drag critical as the hypersonic case, the optimization of the effective nozzle exit area represents one of the most important tradeoffs. Considering the effects of the nozzle exit area not only on the internal performance but also on external drag, thrust vector direction and longitudinal moments, nozzle weight and cooling requirements, this task becomes even more difficult.

An additional integration aspect is the asymmetry of the SER-nozzle configurations. The gross thrust vector is neither in line with the engine nor with the aircraft axis, but will vary with the configuration and the nozzle operating condition. For a given configuration, the position and angle of the gross-thrust vector were found to depend on nozzle pressure ratio, nozzle mass flows, temperature and fuel/air ratios, flight Mach number, and aircraft incidence angle.

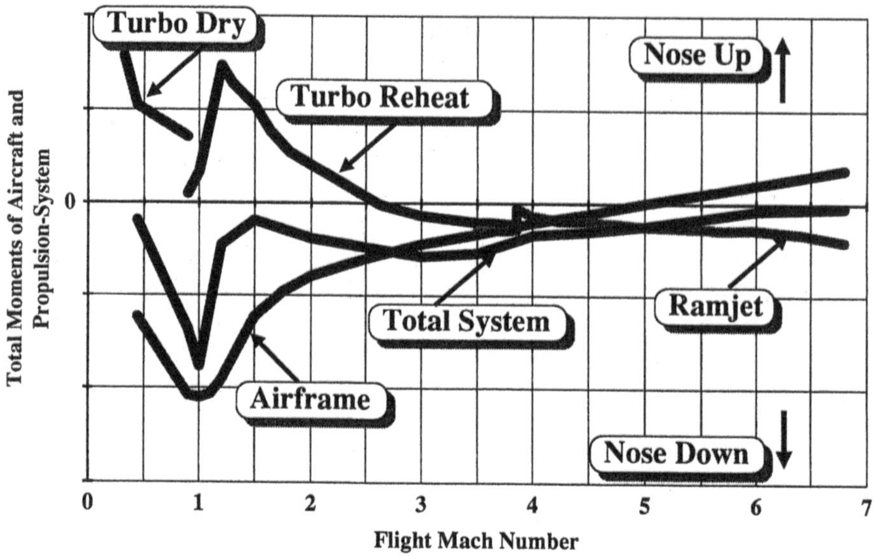

Fig. 38 Longitudinal moments of propulsion system and airframe.

The magnitude of the forces acting on intake and nozzle/afterbody show clearly that the longitudinal moments of the propulsion system together with aerodynamic forces and moments have to be optimized with respect to flight mechanics and performance.

For evaluating the moments of the propulsion system, it is necessary to assess the forces acting as vectors. Because of the asymmetric design of the intake and nozzle and the great distance between the propulsion system and the center of gravity of the vehicle, the resulting moments are in the same order of magnitude as the aerodynamic moments of the aircraft itself.

The resulting moments of the propulsion system for a nominal ascent trajectory are shown in Fig. 38. They are compared with the moments of the airframe (fuselage, wings, and upper stage) and added to the moments of the total system.

With the intake located some meters below the center of gravity, the resulting forces from the swallowed air stream act as nosedown moments for all full-power cases.

For the exhaust system, there are differences between low speeds (transonic and lower supersonic) and high speeds (hypersonic). In the low-speed regime, the exhaust system is operated strongly off-design. The highly excessive expansion rate of the expansion ramp leads to overexpansion of the nozzle flow, Prandtl-Meyer expansion of the surrounding airflow, and flow detachment. This results in low pressures at the nozzle and ramp surfaces. The consequences are a strong downward-directed force and the above indicated nozzle losses. Conversely, it can be seen from Fig. 38 that the noseup moment of this downward-directed force at the aft end helps to compensate for the high aerodynamic nosedown moments of the airframe in the transonic flight regime.

At high speeds, the airframe shows slight nose-up moments, which are compensated for by the nosedown moments of the intake and exhaust system. The small resulting moments of the total system can be trimmed with a relatively small deflection of the flaps, resulting in a favorably small trim drag.

It is one of the aims of integrated engineering (see Sec. III) to minimize the losses created by moment characteristics of the airframe and propulsion system. In the iterative process of establishing the airframe aerodynamics and the propulsion system characteristics, the engineers of the participating disciplines are working closely together. A better understanding of the mechanisms of interaction between aerodynamics and propulsion forces may result in an overall aircraft system that is optimized with respect to performance, and controllability may be achieved.

VI. Subsystem Considerations

A. Secondary Power System

For safe operation of the propulsion system and the vehicle itself, secondary power is needed for the engines and the airframe throughout the entire flight.

In conventional turbine-engine-powered aircraft, all fuel pumps, hydraulic pumps, and electric generators are driven mechanically by the turbine engines. For the Sänger propulsion system, this is not possible since the turbine engine does not operate in all flight phases. An independent secondary power system (SPS) is therefore required, designed as a lightweight system with low fuel consumption. Furthermore, the SPS shall ensure starting of the engines on the ground, and the provision of sufficient energy in the event that no engine is running (emergency power unit), and the supply of cooling fluids for the thermal management system.

Thermal management involves cooling of the propulsion system, air inlet, cockpit, and avionics (see Sec. VI. B). As a result of the high flight Mach numbers, the cooling systems for the propulsion system are of size and complexity unknown to conventional aircraft. Hydrogen fuel is used as a heat sink for the whole vehicle, while compressed and cooled air is used for sealing and secondary cooling purposes. Therefore, the SPS is highly integrated not only with the engine but also with the cooling system.

These constraints resulted in a secondary power system comprising five units, which operate in different flight segments. Each main engine has its own secondary power system. It drives three LH_2 pumps for different hydrogen fuel demands (a small pump for the SPS combustion gas turbine/gas generator fuel supply, for turboengine starting and idling, for ram combustion and cooling during descent; a medium-size pump for turbine engine operation; a large one for the reheated turboengine and ramjet operation during ascent), a cooling air compressor for the cooling air after it leaves the LH_2 cooling air heat exchanger, a hydraulic pump, a generator, and an oil pump for the SPS lubrication system. To reduce fuel consumption, the energy for these components is generated by several sources. Three bleed-air turbines (one for SPS ground start and to drive the hydraulic pump, generator, small LH_2 pump, and lubrication oil pumps during turboengine operation, the other for the turboengine fuel pump), three expansion turbines (one for hydraulic pump, generator, or small LH_2 pump and oil pumps, one for the ram/afterburner fuel pump, one for the cooling air compressor), using heated H_2 from the cooling system, and one combustion gas turbine driven by H_2 and H_2O from the gas generator (for hydraulic pump, generator, small LH_2 pump, and oil pumps, for use during low Mach number descent and as an emergency power unit).

To increase the redundancy and simplify the pipe connections each main engine has a separate SPS. The maximum power demand of one SPS is 1.8 MW. The integrated secondary power system is shown in Fig. 39. For details, see Ref. 15.

B. Thermal Management System

Active cooling is required for most parts of the propulsion system. Present standard metallic materials are limited to temperatures in the range of 1200-1300 K. Materials such as ceramics and ceramic composites, which are still under development, are capable of withstanding temperatures of about 2000 K.

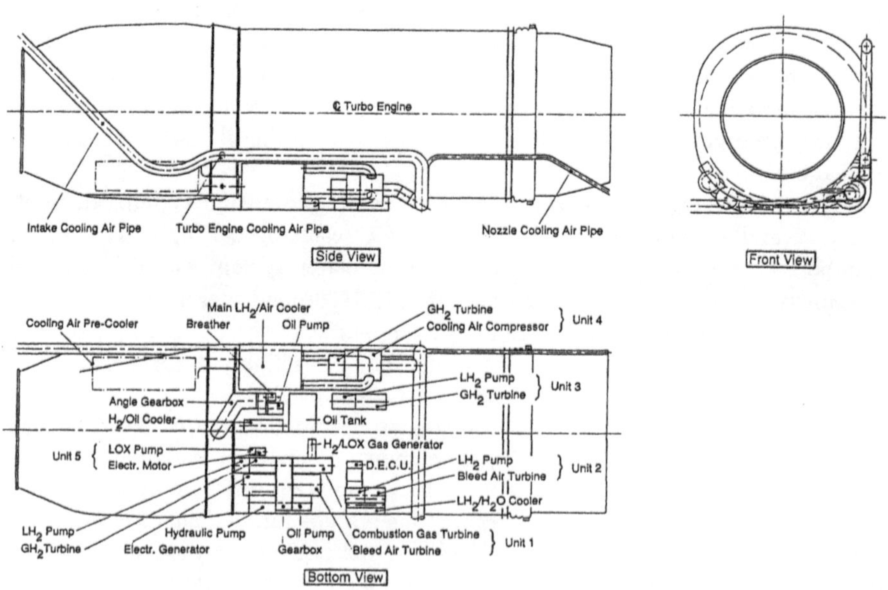

Fig. 39 Integration of secondary power system.

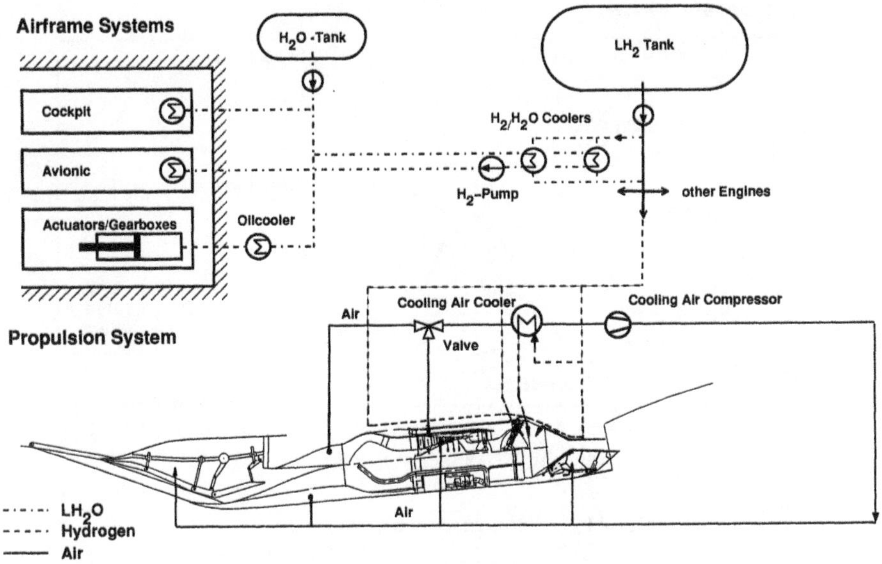

Fig. 40 Thermal management system for high-speed propulsion system.

However, even if these materials are available, a sophisticated design and an efficient cooling system are necessary in any case.

The design of the cooling system to ensure maximum cooling effectiveness combined with minimum pressure losses is a cumbersome task. Hydrogen is used as a direct coolant as well as a heat sink to generate cooling air. Figure 40 shows the thermal management system.

The design of effective cooling systems requires detailed knowledge of heat transfer from the hot gas to the structure and from the structure to the coolant flow,[16-18]. Phenomena similar to those encountered in aerothermodynamics, such as separation, transition, and shock-wave/boundary-layer interaction, must be taken into account. The boundary layer may separate and reattach at some location downstream. At the reattachment point, the boundary layer will be thin, and as a consequence the heat transfer will be high. If the boundary layer is laminar before separation, often there will be a transition to turbulence depending on the pressure gradients. As a result, the heat transfer rate will increase with downstream distance. Turbulence modeling[19] is important in the calculation of heat transfer. Today's methods rely mainly on the use of turbulence models based on the use of Reynolds-averaged Navier-Stokes equations. These models suffer from limited applicability. For the design of effective cooling systems, the scatter in results caused by different turbulence models is unacceptably high. To predict the distribution more accurately, advanced turbulence models with a wide scope of application must be developed.

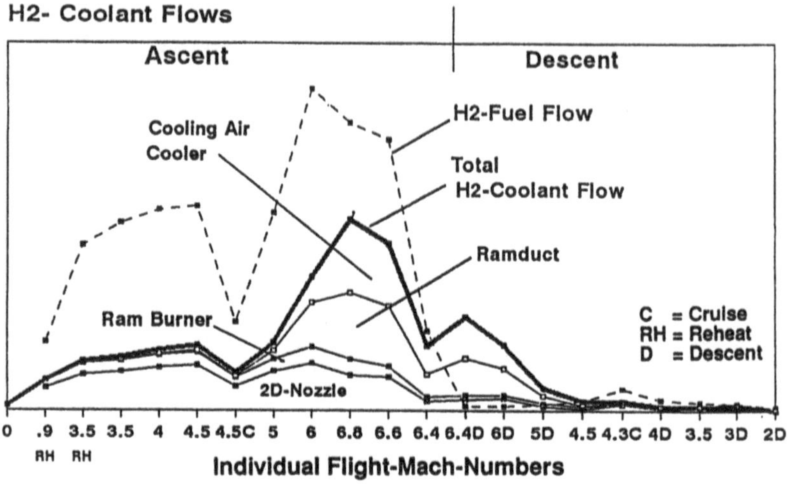

Fig. 41 Hydrogen cooling requirements.

In Fig. 27 variation of the heat transfer coefficient along the nozzle wall is shown. The design of the cooling system must ensure sufficient cooling at all downstream locations. Careful balancing of the cooling passages and local heat transfer rates of the coolant is therefore necessary.

The hydrogen flow necessary for cooling purposes should not exceed the flow required for combustion to avoid penalties with respect to hydrogen tank size, since the structural weight of the vehicle depends largely on tank volume. As indicated in Fig. 2, excess hydrogen adversely affects the payload.

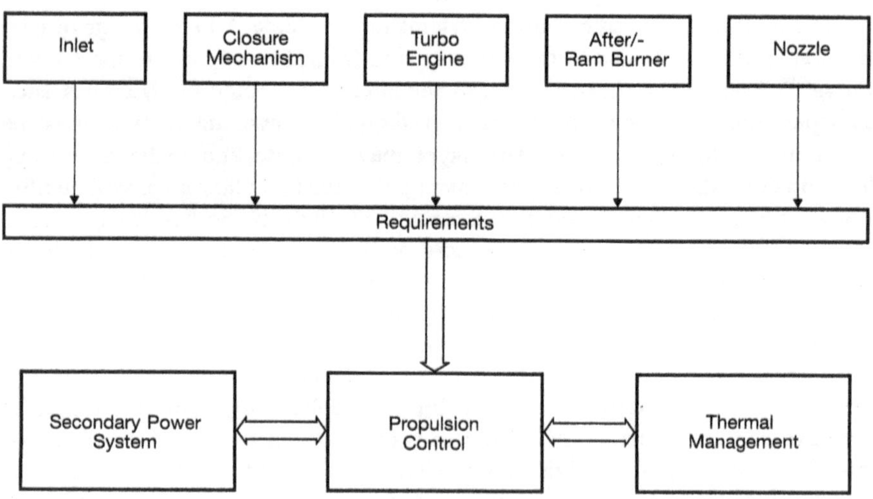

Fig. 42 Schematic of control interfaces.

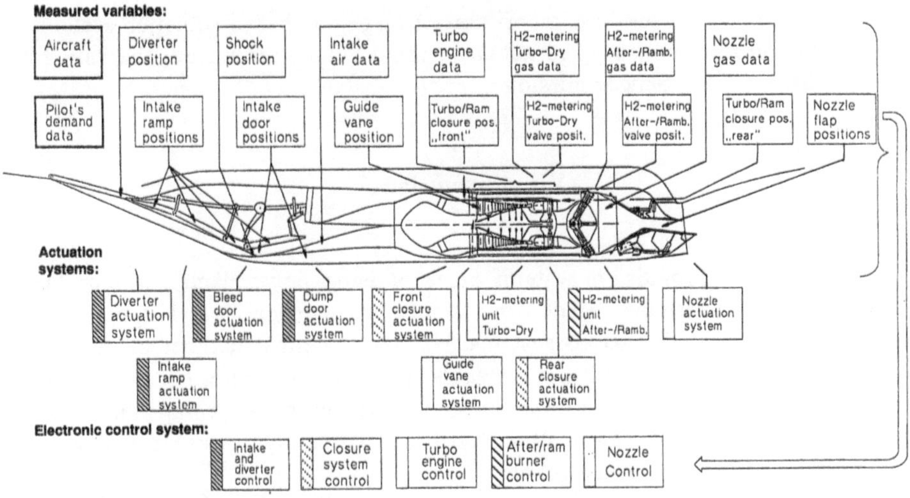

Fig. 43 Control system of highly variable propulsion system.

In the ascent phase, no additional hydrogen for cooling purposes is needed (see Fig. 41). However, as expected, in the first phase of the nearly unpowered descent, the hydrogen consumption for cooling purposes exceeds the hydrogen demand of the engines. Additional investigations are necessary to optimize this flight sector, either by introducing new materials and structures that can withstand higher temperatures and/or by changing the mission profile so that hydrogen consumption of the propulsion system equals the amount of hydrogen needed for cooling purposes.

C. Control System

The wide Mach number range in which the engine has to deliver thrust requires a variable-cycle engine. This has a severe impact on the engine control system. Fig. 42 gives a global survey of the different components and subsystems that must be controlled. These are mainly the intake, the turbine engine and the closure mechanism, the ram combustor, and the nozzle, but also the secondary power system that has to deliver the energy necessary for all pumps, compressors, actuators, etc., and the thermal management system.

The engine control is shown in greater detail in Fig. 43. The control functions are subdivided into groups in accordance with the main propulsion components. The measured engine variables are also listed. They are used as input parameters to the electronic control system of the powerplant.

All major components, including the secondary power system and the thermal management system, have to be constantly monitored and/or controlled by using open- or closed-loop algorithms. A large number of actuation systems then allow correct matching of actual to rated values throughout the various flight phases of the Sänger mission.

VII. Conclusions

Airbreathing propulsion systems are promising powerplants for reusable space transport systems. They offer significant economic and operational advantages over rocket-based systems. However, a great number of the key technologies needed are still not available today. Furthermore, because of the highly integrated design and the close interaction of all components, the development of such an airbreathing system requires highly skilled engineers who are also efficient team workers.

In the design of each engine component as well as the subsystem design, engineers have to face many technological challenges. For flight Mach numbers above 4, nearly all metal structures require cooling. It would be preferable to use advanced materials such as ceramic composites, or carbon-carbon composites, to keep the total weight of the propulsion system at a minimum. For this purpose, advanced materials with high specific strength at high temperatures must be developed.

The overall objective of the German Hypersonic Technology Programme is to make a major step forward in the development of a feasible hypersonic airbreathing propulsion system by a thoroughly set up strategy. In the beginning, component tests for the most challenging elements were performed using sophisticated subscale models, numerical calculations, and advanced measurement techniques.

In the next step, integrated tests were carried out, the results were fed back into the component design as well as into the overall system concept. The next goal is ground testing of a complete subscale propulsion system.

Since ground testing offers only limited possibilities of simulating actual flight conditions, the development process also must be supported by computational fluid dynamic calculation and flight testing.

Acknowledgments

The authors wish to acknowledge the assistance and contributions of colleagues from MTU Motoren- und Turbinen-Union GmbH and Deutsche Aerospace AG. A great deal of this work was funded by the German Ministry of Research and Technology.

VIII. Addendum

Since 1993, the work within the German National Hypersonics Technology Program has progressed, concentrating especially on the experimental exploration of new technologies at the component and engine levels. For example, the combustion chamber and nozzle were built in a larger size with a 50-cm inner combustion chamber diameter, and a nozzle that is fully variable during operation. The connected pipe test facility at DASA-Ottobrunn mentioned in this paper was upgraded to accomodate the larger engine. It was also equipped with a diffuser to simulate altitude conditions. The progress made is described in Refs. 20–28.

References

[1] Kuczera, H., and Hauck, H., "The German Hypersonic Technology Programme, Status Report 1992," IAF Paper 92-0867, Aug. 1992.

[2] Kurzke J., "Calculation of Installation Effects within Performance Computer Programme," AGARD Lecture Series 183, 1992.

[3] Heitmeir, F., Lederer, R., and Herrmann, O., "German Hypersonic Technology Programme, Airbreathing Propulsion Activities," AIAA Paper 92-5057, Dec. 1992.

[4] Heitmeir, F., Schmitt-Thomas, K., and Dietl, U., "Hypersonic Propulsion, Challenges in Design and Materials," Space Course, TU-München, Vol. 2, 1993, pp. 22.1.-22.28.

[5] Krammer, P., Heitmeir, F., Bissinger, N., and Voss, N., "German Hypersonics Technology Programme," AIAA Paper 93-5094, 1993.

[6] Schaber, R., and Heitmeir, F., "Einfluß der Komponentendimensionierung auf die Leistung eines Hyperschallantriebs," DGLR Jahrestagung, Göttingen, Sept. 1993.

[7] Wilson, G. J., and Davis, W. H., "Hypersonic Forebody Performance Sensitivities Based on 3-D Equilibrium Navier-Stokes Calculations," AIAA Paper 88-0370, 1988.

[8] Monnoyer, F., "Hypersonic Configuration Optimization with an Euler/Boundary Layer Coupling Technique," AIAA Paper 93-3116, 1993.

[9] Lederer, R., and Krüger, W., "Nozzle Development as a Key Element for Hypersonics," AIAA Paper 93-5058, 1993.

[10] Dusa, "Exhaust Systems and Turbo-Ramjet Propulsion for Mach 2.0 to 6.0 Aircraft," UTSI Aeropropulsion Short Course, 1988.

[11] Berens, T., "Numerical Investigation of Thrust Vectoring by Injection of Secondary Air into Nozzle Flows," AGARD CP-534, Apr. 1993.

[12] Bissinger, N. C., and Schmitz, D. M., "Design and Wind Tunnel Testing of Intakes for Hypersonic Vehicles," AIAA Paper 93-5042, 1993.

[13] Herrmann, O., and Rick, H., "Propulsion Aspects of Hypersonic Turbo-Ramjet Engines with Special Emphasis on Nozzle/Aftbody Integration," ASME Paper 91-Gt-395, June 1991.

[14] Herrmann, O., "Integration of Turbo-Ramjet Engines for Hypersonic Aircraft," AGARD-FDP Symposium, Oct. 1991.

[15] Trollheden, S., and Streifinger, H., "Secondary Power System Study for the SÄNGER First Stage Vehicle." AIAA Paper 93-5033, 1993.

[16] Patankar, S. V., "Recent Development in Computational Heat Transfer," ASME Journal of Heat Transfer, Vol. 110, Nov. 1988, pp. 1037-1045.

[17] Launder, B. E., "On the Computation of Convective Heat Transfer in Complex Turbulent Flows," ASME Journal of Heat Transfer, Vol. 110, Nov. 1988, pp. 1112-1128.

[18] Pletcher, R. H., "Progress in Turbulent Forced Convection," ASME Journal of Heat Transfer, Vol. 110, Nov. 1988, pp. 1129-1144.

[19] Marvin, J. G., "Turbulence Modelling for Computational Aerodynamics," AIAA Journal. Vol. 21, No. 7, 1987, pp. 941-955.

[20] Kania, P., "The German Hypersonics Technology Programme—Overview," AIAA Paper 95-6005, April 1995.

[21] Kania, P., "Latest Results in Hypersonics Technology, German Hypersonic Technology Program," IAF Paper 94-V.4.547, Oct. 1994.

[22] Heitmeir, F.J., "Combined Cycle Engines for Hypersonic Applications," AGARD Paper PEP-16, Sept. 1995.

[23] Heitmeir, F.J., and Bissinger, N.C., "Development and Test of an Air-Breathing Propulsion System for Hypersonic Speed," AIAA Paper 95-6027, April 1995.

[24] Heitmeir, F.J., and Sacher, P.W., "Ground- and Flight-Testing of an Air-Breathing Propulsion System for Hypersonic Speed," IAF Paper 94.V.5.554, Oct. 1994.

[25] Schaber, R.W., and Heitmeir, F.J., "Einflussentscheidender Triebwerksparameter auf das Leistungs- und Betriebsverhalten eines Hyperschallantriebs," DGLR Jahrestagung, Erlangen, Germany, Oct. 1994.

[26] Schaber, R.W., and Heitmeir, F.J., "Zusammenfassung der wesentlichen Erkenntnisse aus den Studien zu luftatmenden Hyperschallantrieben," DGLR Jahrestagung, Bonn, Sept. 1995.

[27] Voss, N.H., "Ram Combustor Development within the German Hypersonics Technology Program," AIAA Paper 95-6030, April 1995.

[28] Lederer, R., and Krüger, W., "Testing the Hypersonic Technology Demonstration Nozzle, Results from the Test Campaign 1993/94," AIAA Paper 95-6084, April 1995.

Turboramjets: Theoretical and Experimental Research at Central Institute of Aviation Motors

Vladimir A. Sosounov*, Mikhail M.Tskhovrebov†, Valentin I. Solonin‡
Pavel A. Kadjardouzov§ and Vladimir A. Palkin$

Central Institute of Aviation Motors, 111250, Moscow, Russia

Nomenclature

A = area
C = aerodynamic drag coefficient related to the wing area
D = aerodynamic drag
d = humidity
E = mechanical energy
F = thrust
H = altitude
H_e = flight energetic altitude, ratio of vehicle mecanical energy to vehicle weight
H_u= lower heating value,mass heat of combustion
Hv= volumetric heat of combustion
k = adiabatic exponent
L/D= lift-to-drag ratio
M = Mach number
m = bypass ratio
N = revolution per minute

Copyright © by CIAM. Published by the American Institute of Aeronautics and Astronautics, Inc. All rights reserved with permission. Released to AIAA to publish in all forms.

* First Deputy Director, CIAM, 2, Aviamotornaya St.
† Head of Engine Division, CIAM, Aviamotornaya St.
‡ Head of Advanced Engine Department, CIAM, 2, Aviamotornaya St.
§ Senior Researcher, Engine Division, CIAM, 2, Aviamotornaya St.
$ Senior Researcher, Engine Division, CIAM, 2, Aviamotornaya St.

P = pressure
PRC= compressor pressure ratio
q = dynamic pressure
R = gas constant
S = surface area of wing
T = temperature
V = flight speed
v = volume
W = mass airflow, weight
α = air excess coefficient, air-to-fuel equivalent ratio
β = relative airflow in ramjet or turbofan duct
ΔH_{hs} = heat sink capacity of fuel
η = efficiency
ρ = density
σ = intake recovery coefficient
τ = time

Subscripts and Superscripts

a = acceleration, ambient
c = compressor, absolute flow velocity in compressor, or turbine flowpath
col = cooling system, cooling air
cor = corrected parameters
cr = cruise
en = engine
ex = exit
f = fan, fuel
g = turbine inlet
in = intake, inlet
m = mean value
n = nozzle
o = sea level static conditions, optimum, overal
p = airplane
r = ARCC, reheat
rd = RJ outer duct
rj = ramjet engine; ramjet operation mode
s = specific parameters, separate
sum= summary parameter
th = throat cross section
tj = gasturbine operation mode
u = tangential component
v = flight conditions
w = windmilling mode, perforated screen wall
x = ARCC inlet
* = total parameters

I. Introduction

In the second half of the 1950s in connection with the successful launch of the first Earth satellite there was a significant growth in interest in advanced high-supersonic and hypersonic flight technology. (CIAM) began fundamental studies of combined air-breathing engines (ABE) whose operation process is based essentially on evolution of the concept of a bypass engine with reference to flight conditions at high-supersonic speeds. At the same time the foundation of their theory was laid.

Studies have been carried out on a variety of turboramjet engines (TRE) based on a turbojet engine (TJ) and a turbofan engine (TF) as a core, and areas of their expedient application have been determined. For the first time methods of engine configuration selection have been developed, which is essential in the TRE case for optimum matching of gas turbine and ramjet parts. A theory of TRE operation has been developed when switching from one operational mode to another, operation at the ramjet mode at maximum speed of flight under conditions of high kinetic heating, including the modes with turbofan (turbocompressor) windmilling. Theoretical research of turboramjet performance, structures, and application possibilities were carried out before beginning experimental work.

Investigations into real operating conditions and performance of the TRE were carried out in the frame of a development program and test bench studies of several versions of full-scale TRE demonstrators, created on the base of production of TJ, TF, and their units. The performance at switching for various operation modes, the ramjet modes with cooling of the heat loaded components, the operation conditions of rotor supports, etc were studied. For the first time tests were carried out of an experimental TRJ under simulated flight conditions at $M = 3–4$ ($T^*_{in} = 300 \div 600$ °C) for a rather long time (several hours).[1-3]

All of these investigations were intended for the development of a cruise mission flight vehicle (FV) using kerosene. Results obtained and the Russian experience with the use of liquid hydrogen[4,5] open the possibilities for development of TRE for an FV using liquid hydrogen as fuel. This chapter includes the results of both theoretical and experimental studies of TRE being carried out at CIAM. The problems connected with the TRE's development and their application had been studied in other Scientific and Research Institutes and Design Bureaus previously in the USSR, Russia, and other countries. We present the point of view of CIAM scientists on some of the problems.

II. Turboramjet Engines: Types and Performance

A. Variable Cycle Combined Engines

When developing the supersonic or hypersonic propulsion system (PS) the main factor defining the configuration is the wide range of the operating conditions. Engine efficiency in cruise flight is ensured by optimization of thermodynamic parameters, which change considerably depending on flight speed and altitude. On the other hand, a significant requirement of the PS of a high-speed aircraft is providing high thrust at acceleration, which at high-Mach number flight is determined by the airflow.

A typical property of turbojets is the lowering of the PRC and the corrected airflow in the turbocompressor with the rise of flight speed.[6] Specific thrust parameters of the augmented turbojet are draw together with ramjet specific parameters at flight Mach numbers of 3–3.5. At higher flight speed ramjet engines (RJ) give higher specific thrust at equal air-to-fuel coefficients than turbofan engines with afterburners (TJAB). Corrected airflow in RJ with variable inlet and nozzle can be maintained constant in wide range of flight speeds. As a result, the RJ thrust rises with Mach number more rapidly than turbojet thrust. Then at Mach numbers higher than 3–4 the RJ thrust exceeds TJAB thrust, and at this flight-speed range the RJ can be more effective than the TJAB.

Combining a single PS, the TJ and RJ give the possibility of obtaining the ABE class, which is operable and effective up to M = 4–5 and higher. Combined TRE is, in essence, a bypass ABE, in which a high-pressure inner part represents the TJ or TF gas turbine part, and a low-pressure external part represents the RJ (ramjet part). Depending on engine type some of its structural elements can be common for both parts. Usually, in these engines afterburning chambers represent the combustion chambers of RJ too, which are formed by switching out the TJ part by means of special devices and fuel feed to RJ combustion chambers only.

The selection of one or another type of engine is determined by the requirements of the hypersonic aircraft. The change of these requirements can significantly change the PS reasonable design. For example, at a high aircraft thrust-to-takeoff weight ratio and with stringent requirements to the specific weight of an engine, the PS with the air-turborocket engines (ATR) may have certain advantages; however, they have low-fuel efficiency, particularly at low speed flight. More sophisticated turbocompressor engines which use hydrogen high cooling and working capacity (air cooling at the engine entry, vapo-hydrogen turbine) have an extended flight speed range and higher efficiency, but also have higher weight as compared to the rather simple gas generator ATR.

Problems of matching the essentially different operational modes in combination with the more stringent requirements of noise level, sonic boom intensity, and detrimental emissions create a very complicated problem in the development of the multimode hypersonic PS. There is a need in using variable cycle engines (VCE) to match the conflicting ecological and thrust characteristics requirements. Usually, VCE is a bypass engine with variable geometry components.[7,8] Adding the ramjet to such engine types allows the development of a combined variable cycle PS with a very wide range of performance. When a large part of a flight is at subsonic cruise flight level (for instance, the flight over populated territory, subsonic hold, emergency situation in case of supersonic or hypersonic transport aircraft) there are severe requirements for providing for high fuel efficiency at the mentioned modes. In this case, the advantage will be in the PS with the turbojet or turbofan (VCE), whose fuel efficiency at the throttle cruise modes is much higher as compared to the ATR of regenerative type.

B. Turboramjet Engine Types

There are the many types of TRE and their selection depends on the vehicle mission. The multitude of TRE types can be divided according to the principle of using the engine gasturbine part "free power" to press air in ramjet part and on the type of afterburner-ramjet combustion chamber (ARCC), common or separate.[3,6]

1. Turboramjet Engine Without Energy Transfer to the Ramjet Part

The TRE without energy transfer to the ramjet part is just a mechanical combination of turbojet (turbofan) and ramjet engines (Fig. 1). Their thrust performances are obtained by adding together the performances of gasturbine and ramjet parts. Relative airflow in ramjet part is equal to

$$\beta = W_{rj} / (W_{tj} + W_{rj})$$

In the gasturbine operational mode, that is when the ramjet part is switched off ($\beta = 0$), such an engine has TJAB or turbofan engine with afterburner (TFAB) performance, whereas in the ramjet operational mode, when the TJ (TF) is switched off ($\beta = 1$), it has RJ performance. If there is a separate combustor in the ramjet part, then the parts are independent. This allows switching on the ramjet at the transonic flight to increase the total thrust of the engine turboramjet engine with TJ and separate RJ (TRJ_S) and turboramjet engine with TF as a core and duct afterburner ramjet combustion chamber ($TFRJ_S$) in Fig. 1. Thus, in the TRE of this type a simultaneous (parallel) operation of the parts is realized.

For better performance of the engine operating in the ramjet mode, the switched-off part should be shut off by special devices, such as doors, louvres, moving center body, etc.

The drawback of the TRE type with a separate combustion chamber and without energy transfer to the RJ part consists in the requirement of larger dimensions, greater complexity, and higher structure weight, which is necessitated by the parallel arrangement of two combustion chambers with variable nozzles. The most simple, compact, and advanced in this group is the TRJ with separate gasturbine and ramjet ducts (TRJ_S), especially when using "stoichiometric" TJ.

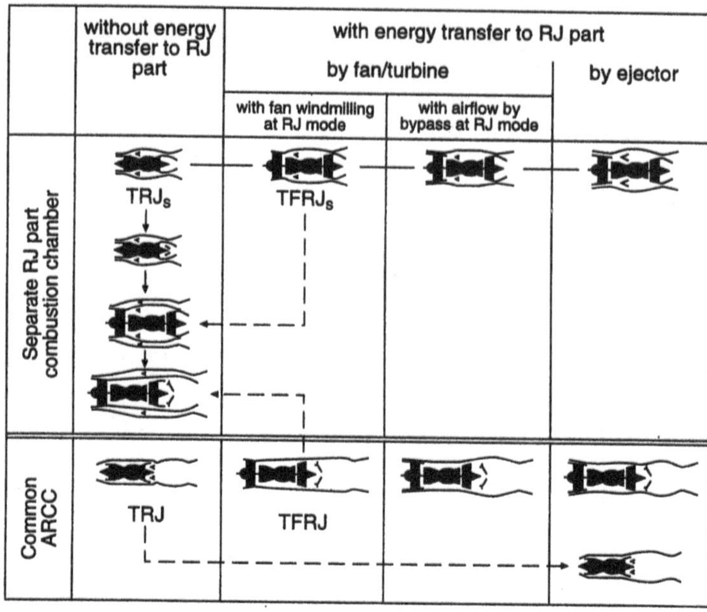

Fig. 1. Classification of TRE

A group of TRE types without energy transfer to the ramjet part at low speed flight includes the TRJ with a common afterburner ramjet combustion consecutive layout of the TJ and ARCC, the ramjet operational mode with the flight speed increase is changed from the TJ mode to the ramjet mode. The TRJ principle feature is impossibility of realizing the simultaneous (joint) operation mode of the parts at low and moderate flight speeds due to the large pressure difference in the TJ and RJ passages, hence, the necessity of shutting off the ramjet duct at the gasturbine operation mode with a special device and special control at the transition mode when switching to RJ takes place.

The necessity of isolating the switched off TJ at high-Mach numbers may dovetail with the influence of kinetic heating on the structure. Hence, if the structure durability and oil system operability are sufficient, then along with switching to the RJ mode, it is possible to transfer the TJ to windmilling or idling and use it for driving the accesories. Usually at this mode only a small part of airflow passes through the TJ part ($\beta \rightarrow 1$), and the thrust performance difference from that of pure RJ operation is rather small. The TRJ has the advantage from the point of view of dimensions, weight, and simplicity.

2. Turboramjet Engine with Energy Transfer to the Ramjet Part

In the TRE with energy transfer to the ramjet part (Fig.1.1), the free power produced by the gasturbine engine is used to increase the pressure of the air supplied to the ramjet. As a result both parts of the engine can be involved in the operation within the whole range of speeds and altitudes of flight. These types are essentially the development of the TRJs and the turboramjet engine with TJ as a core (TRJ) by application of a blade (or jet) compressor to increase air pressure in the ramjet part.

In TRE types with energy transfer by ejector, the active operating fluid is the airflow from the fan or turbine. These TRE types, in principle do not need the shutoff device for the RJ part duct. For effective operation at takeoff and low speed flight conditions in which one can obtain some effect from use of ejection, it is necessary to take some constructive measures. Among these there are lowering the turbine or fan passage exit area for maintaining high active gas parameters, using narrowed mixing chamber and special mixer for lowering the losses, using ARCC with large cross section for loss lowering, and, consequently, using the diffuser with a large area ratio before ARCC. The necessity for these measures, consistant with greater engine complexity, higher weight, and larger dimensions, is conditioned by the low efficiency of the jet compressor when limiting its dimensions. As a result, the TRE types with ejection have no practical advantages over the types with energy transfer to the RJ part by turbofan.

The TRE with energy transfer by fan [turboramjet engine with TF as a core (TFRJ) or $TFRJ_S$] does not differ significantly from augmented turbofans with separate primary and secondary flow turbofan engines with duct afterburner ($TFAB_S$) or with mixing flow TFAB (Fig. 1). Their feature is the transition to the RJ operational mode with turbofan windmilling at high flight speeds ($\beta \rightarrow 1$). In this connection the windmilling turbofan drag presents an important factor. The possibility of multimode operation and the use of the windmilling turbofan to drive the accessories are the important advantages of the $TFRJ_{(s)}$. In the case with rather high design fan pressure ratio, which degrades windmilling performance, or if the TF should be isolated from kinetic heating at the RJ mode operational conditions as a result of reliability requirements, it is necessary to include into the engine structure a passage around the fan for airflow bypass at high-Mach number flight. For the same reasons as in the case of TRJ, this passage should be closed by special device in low speed flight conditions.

The TRE types with energy transfer and separate combustion chamber of RJ part inherently have a compact configuration because of the parallel arrangement of the parts. But these types require closing of the gasturbine passage exit in RJ operation mode to maintain the thrust performance. The TRE types with common ARCC do not need such closing but are longer and need a wider range of moving nozzle flaps. The expediency of using the TRE based on TJ or TF is defined by the vehicle mission. If high fuel efficiency at low speed flight and a good performance at the cruise flight with M = 4–5 are required, the TF as the basic engine is preferable; if the mean requirement is high performance at maximum speed flight, then it is expedient to analyse the possibility of using the TRE types based on the TJ too. It is possible to construct the TFRJ structure with slightly lower length and weight than of the TRJ with equal design thrust.[8]

The main advantages of the TFRJ are effective performance in a wide range of flight conditions, lower noise level at takeoff and low speed flight conditions, and the wide range of thrust variations at constant airflow.

C. Turboramjet Engine Design Parameters and Dimensions

The independent selection of both the gasturbine part and the ramjet part sizes (a fan/compressor inlet area A_f or A_c and an ARCC or ramjet burner cross-sectional area A_x) is a common feature of the combined engines (Fig. 2). In general, sizes are determined as follows.

1) The size of inlet area A_{in} is determined by the design thrust value at the maximum flight speed and is optimized by the influence on the flight range and external drag at the transonic flight speed.

2) The size of the fan (compressor) area A_f (A_c) is determined bearing in mind ecological requirements, namely, by a design thrust at the transonic flight speed (sonic boom limitations) or at takeoff (noise level limitations), for engines with lower dry specific thrust (TFRJ) the subsonic cruise thrust can be the determinating factor.

3) The size of ARCC cross-sectional area A_x is determined by the flight speed at which the transition to the ramjet mode is realized and is optimized taking into consideration matching of TJ (TF) and RJ operation modes.

4) The size of the nozzle exit area A_n is optimized taking into consideration nozzle efficiency at the maximum flight speed, external drag at the trasonic flight speed, and nozzle weight.

Selection of the design bypass ratio is of important role at the development of a multimode engine. The lower the bypass ratio, the closer the TFAB performance is to those of the TJAB. With rising bypass ratio, TFAB properties approacn those of the RJ (Fig. 3). It is possible to base the choice of reasonable bypass ratio value, for example, of the hypersonic transport (HST) airplane engine, on the following considerations. When the design bypass ratio m at given cycle parameters is rising the optimum fan pressure ratio and engine core airflow are decreasing. This decrease relaxes the core airflow steepness influence and leads to an increase of summary engine airflow at supersonic flight speeds. As a result, the relative engine thrust at maximum augmented ratings (F/F_o) increases. Corresponding specific fuel consumption

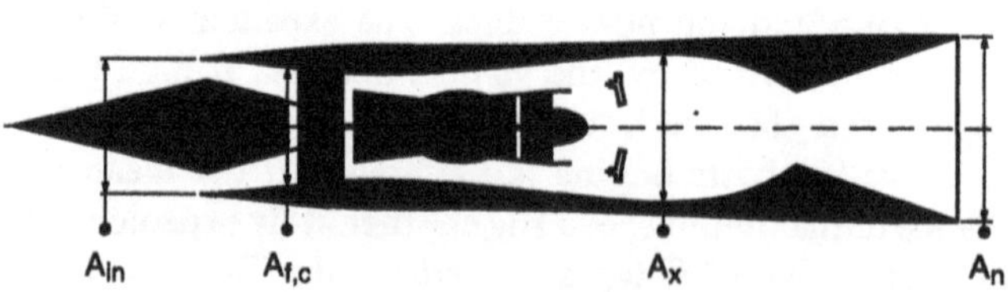

Fig. 2. Characteristical cross-section of propulsion system with TRE

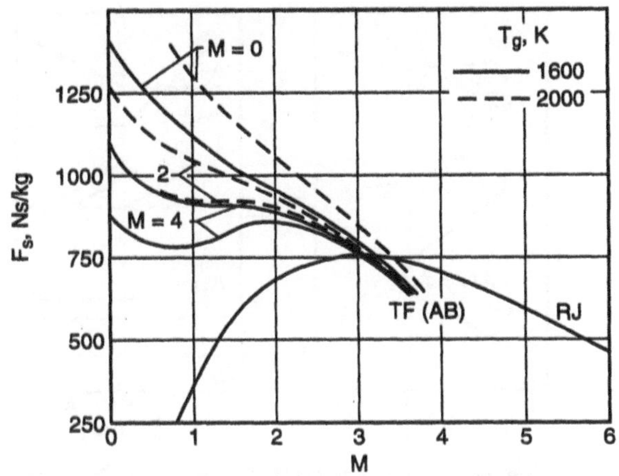

Fig. 3. Specific thrust comparison of both turbofan and ramjet engines

(SFC) also increases, but usually this does not lead to a significant rise of required fuel amount, because of the simultaneous increase of excessive thrust and higher acceleration rate. At high supersonic flight speed (M > 3.5) the engine must run according to the RJ cycle, which can be realized more simply at the larger dimensions of the external low-pressure passage, that is, the higher design bypass ratio. The bypass ratio increase in this case decreases the demand in special mechanization of the passage and in additional bypass duct. The fuel efficiency at subsonic cruise is also improved. For example, corresponding to a bypass ratio of order $m_o = 3$, the fan pressure ratio values consist of $PRC_f = 2{,}2.5$. The high pass capacity of such low-pressure ratio fans in a wide range of rpm is one of the main factors that determines the possibility of creating the TFRJ types of TRE, which can be transferred to RJ operation mode with fan windmilling at high speed flight.

The design bypass ratio increase results in both an engine airflow and its cross dimensions increase. In the case of an aircraft of moderate supersonic flight speeds (M=2,2.5), this is one of the reasons for selection of the low-bypass-ratio values. In the case of high supersonic flight speeds (M > 3) the relative cross dimensions of an inlet and a diverging nozzle rise considerably, whereby a significant increase of the bypass ratio within the range determined by the dimensions of the PS is possible. As for the subsonic cruise part of flight, which is inherent to heavy-weight supersonic vehicles, fuel economy considerations also require the choice of higher bypass ratios but with an upper limit of nearly $m_o = 4$ because of the substantial falling of the engine dry thrust.

Typical TFRJ performances interrelations are shown in Fig. 4, which should be considered when making the choice of design bypass ratio. In this case, the critical factor is the engine dry thrust at subsonic cruise. The requirements for the design takeoff (or transonic) thrust and thrust value at cruise with maximum hypersonic speed are providing for by corresponding

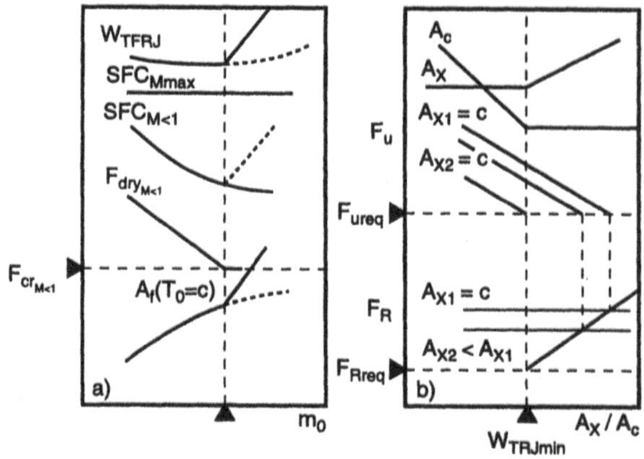

Fig. 4. Definition of design bypass (a) and area (b) ratio in TRE

design dimensions of inlet, fan, and ARCC in considering the whole range of engine parameters. With higher design bypass ratios, a fan dimension increases, but this does not influence fuel efficiency at hypersonic cruise, slightly reflects on engine weight W_{en} and leads to higher fuel efficiency at subsonic cruise. However, subsonic dry thrust decreases. Increasing bypass ratio higher than some limiting value would lead to higher engine dimensions and weight and to engine oversizing relative to other flight conditions (Fig. 4a). In this case, the maximum bypass ratio at which the dry thrust corresponds to required level is the expedient design value. Then, for engines of multimode FV with high supersonic flight speed adoption of relatively high design bypass ratios may be rational by decided. The optimum bypass ratio values depend on the engine cycle parameters, the aircraft aerodynamic characteristics, the extent of flight parts with supersonic and subsonic speed, etc. It is possible to use a similar approach when defining the optimum correlation between the dimensions of RJ and TJ parts of TRJ. One can express his correlation as the ratio of the ARCC section area to the area of compressor inlet. The choice of each of them is to some extent mutually independent and is defined by required thrust values at RJ and TJ (for example, at takeoff) operation modes. At rather low A_X/A_C ratios the engine would be oversized for TJ mode of operation, whereas at the high ratio values it would be overised for the RJ mode (Fig. 4b). At some ratio value both areas and engine weight W_{en} are minimum, and it is possible to consider this value as an optimum.

D. Combined Engine Types Comparison

The analysis of mission performance and advantages of some various combined ABE concepts was carried out for two advanced hypersonic aircraft[9]: the two-stage-to-orbit (TSTO) aerospace plane (ASP) launcher with stage

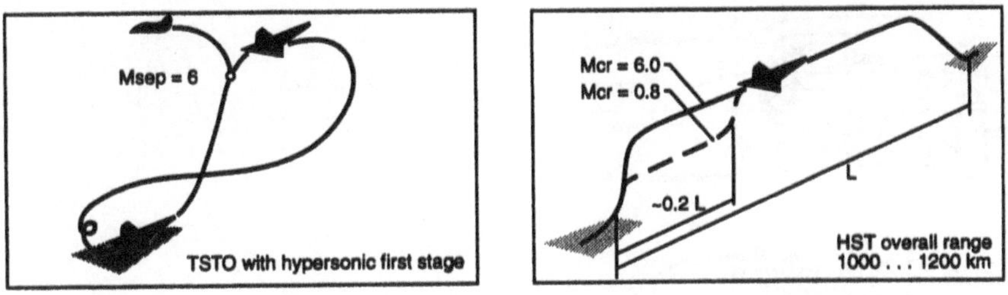

Fig. 5. The flight profiles of TSTO and HST

separation at Mach 6 and the HST with Mach 6 cruise speed (Fig. 5). **Liquid hydrogen fuel** is used. The use of LH_2 allows the consideration of a wide variety engine concepts as it is of great combustion heat, high- cooling capacity, and high-working capacity when expanding in a turbine.[4,6] In the performed comparison, for both hypersonic missions the climb and acceleration trajectory was practically identical and the flight at maximum speed was conducted at an altitude out of the atmospheric ozone maximum concentration area.The efficiency criterion of the TSTO PS was the relative mass of the second stage. For the HST with commercial load mass constant for all variants, the criterion was takeoff weight. For the latter mission airplane (HST), two flight concepts were considered: hypersonic cruise and combined subsonic (about 20% of total range) cruise and hypersonic cruise only. The combined mission reflects the possibility of flight speed reduction over populated territory for sonic boom impact elimination.

1. Engine Concepts

The engine concepts shown in Fig. 6 are discussed.

1) The turboramjet engine consists of TJ and RJ parts and has a common ARCC, as well as the ramjet duct with shutoff device. The TJ has a compressor with a moderate pressure ratio and high inlet turbine temperature (T^*_{gmax} up to 2000 K). A variation of the TRJ with a separate combustion chamber of RJ part (TRJ_s) was also considered. The independence of the TJ and RJ parts in TRJ_s provides parallel operation of the parts at low and moderate flight speed with higher thrust performance.

2) The turbofan-ramjet engine (TFRJ), in essence is a high-temperature, variable geometry turbofan. The selection of a design bypass ratio value at a higher level improves matching of the TF and RJ sizes and transition of the engine to windmilling fan mode at high flight Mach numbers.

3) The regenerative air-turborocket engine (RATR) has transition to the RJ mode at high Mach numbers. Its compressor is driven by a turbine working on gaseous hydrogen, the heating of which is performed in the heat exchanger located in the combustion chamber. Because of heat regeneration, the SFC of the engine is nearly equal to the SFC of the TFRJ at operation with the same

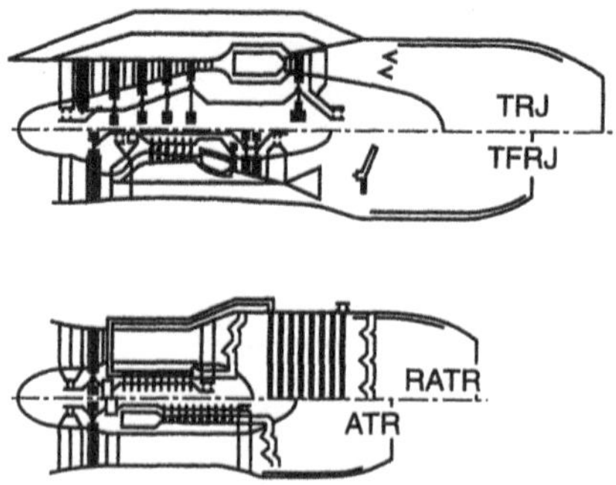

Fig. 6. The engine concepts

value of a, but the RATR has a higher thrust-to-weight ratio because of the low pumping work of a liquid fuel. The turbine inlet temperature is 1200 K.

4) The air-turborocket engine (ATR) has transition to the RJ mode at high Mach number flight. The oxygen/hydrogen burner (gas generator) provides high-pressure fuel-rich flow to the uncooled turbine. This fuel-rich flow is then mixed with the compressor discharge flow and burned in the combustion chamber. The turbine inlet temperature is 1200 K.

2. Engine Performance Comparison

The variable geometry inlet and nozzle performance efficiencies are the same for all engine types. Design fan (compressor) pressure ratio in the TFRJ, RATR, and ATR cases was selected to be nearly the same. For this reason the thrust characteristics of the TFRJ and RATR are nearly similar (Fig. 7a). The SFC of the ATR is more than twice as high compared to other engine types because its use of onboard oxygen. It was determined that all engines would operate on the ramjet mode at flight speed higher than Mach 3.5.

The TFRJ has the highest fuel efficiency at subonic cruise flight. In accordance with its higher propulsive efficiency, its dry thrust SFC is lower about 25–35% in comparison with the TRJ and about 2–3 times in comparison with the RATR (for very high fuel consumption the ATR engine cannot be a competitor at this flight condition) (Fig. 7b). The low dry specific thrust of the TFRJ at the minimum SFC condition, however, may lead to a decrease of subsonic cruise flight speed and altitude or to an increase of engine size and PS weight. It should be noted that the range of a RATR throttling performance is limited by the heat exchanger structure overheating.

For the engines considered, one can recognize some interdependence between the engine thrust-to-weight ratio F/W and SFC at subsonic cruise flight. The ATR has a highest F/W. The RATR has an F/W about 10–15% lower comparised to the ATR because of heat exchanger weight and lower working medium flow in the turbine. The TRJ and TFRJ have an F/W about 1.5–2 times lower than ATF or RATR (Fig. 7c).

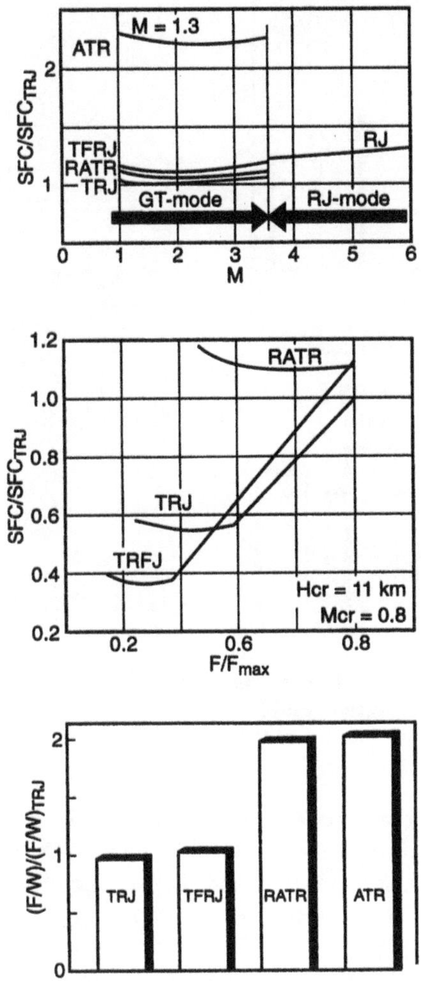

a) relative SFC ($\alpha = 1.2$)

b) subsonic relative SFC

c) relative thrust-to-weight ratio

Fig. 7. Engine performance comparison (fuel - LH_2)

3. Mission Performance Analysis

The results of the mission performance analysis are shown in Fig. 8. The highest second- stage mass corresponds to the airplane-to-orbit system (ATOS) launcher with the RATR PS in accordance with a good engine F/W and a relatively low fuel consumption at maximum thrust rating (Fig. 8a). The second-stage mass lower by only about 5–7% corresponds to the system with the TFRJ or TRJ. The lowest second-stage mass corresponds to the ATR (about 12–14% lower in comparison with turboramjets) because of its worst fuel consumption, which cannot be compensated for by low engine weight (Fig. 8a).

The mission performance results for the HST are substantially influenced by the presence of subsonic cruise. In the case of fully hypersonic

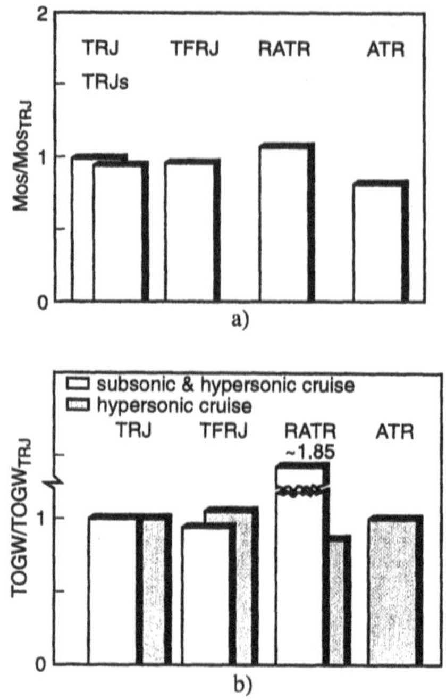

Fig. 8. Mission performance comparison (fuel - LH_2)
a) TSTO second stege relative mass
b) HST relative take-off gross weight

cruise the lowest HST takeoff weight corresponds to the RATR PS. The effectiveness of the other engine types is some lower and nearly equal (TOGW is higher by 18–20 %), (Fig. 8b). If combined subsonic and hypersonic cruise is considered (subsonic part is about 20% of the total range), then the turboramjet engines reveal the lowest takeoff weight (TFRJ has a nearly 5% advantage in comparison with TRJ). Takeoff weight with the RATR is about 1.8 times higher than with turboramjets. Taking into account the takeoff noise limitations, the advantages of the TFRJ are additionally increase. Hence, TRJ and TFRJ could be considered as universal multimode engine types for hypersonic (up to nearly Mach 6) aircraft with different missions.

E. Optimization of Turboramjet Engine Operation Mode Switchibg

At some flight Mach numbers specific parameters of the turbine and ramjet engines become the same. On the other hand, the TJ corrected airflow decreases with the inlet air temperature increase, whereas in the RJ case it is possible to maintain constant corrected airflow. As a result, the RJ thrust increases more rapidly vs Mach number than the augmented TJ type engines.The supersonic vehicle high sensitivity to components weight requires identification of those PS control methods which provide maximum vehicle effectiveness. For accurate resolution of such problems, usually variation methods are used. Different variations of the combined ABE thrust and SFC vs Mach number when operating in gas turbine or ramjet modes result is some flight speed when switching to the ramjet operation mode on an acceleration trajectory at which the maximum efficiency corresponding to some criteria is provided. Usually it is required to provide minimum time expenditure for acceleration, minimum fuel expenditure for the acceleration part of the flight trajectory (for example, for TSTO first stage), and maximum flight range (for example, for HST). Only in particular cases will the optimum for mode-switching flight Mach number (Mo) correspond to the flight speed at which the specific thrust parameters of the TJ and RJ are equal to one another.

It should be mentioned that it is possible to define the appropriate Mach numbers for mode-switching flight from other points of view (for example, proceeding from the minimum dynamic influence on vehicle, from the minimum variable geometry components changing, or from higher life considerations, etc.). The afore mentioned principle criteria, however, are important for TRE working process analysis.

1. Definition of the Optimum Mach Numbers

When obtaining the corresponding equations we will proceed from the following assumptions.[10]

1) Switching the operational mode from TJ to RJ is being accomplished instantaneously.

2) Acceleration trajectory is given.

3) The air-to-fuel ratio at each trajectory point is optimum in accordance with afore- mentioned criteria.

If the thrust vector is directed toward the tangent to flight trajectory, then the acceleration flight time from point (M_1, H_1) to point (M_2, H_2) of the trajectory

$$\tau = \int_{He_1}^{He_2} W_p \, dHe \, / \, [V \cdot (F_{sum} - D_p)]$$

where W_p, V, D_p, and vehicle PS summary thrust F_{sum} vary vs $M(He)$. If $M_1 < Mo_\tau < M_2$, then one can find the optimum value of Mo_τ (for time criteria) by resolving the equation

$$\tau_{min} = \int_{He_1}^{He_o} W_p \, dHe \, / \, [V \cdot (Fsum_{tj} - Dp_{tj})] + \int_{He_o}^{He_2} W_p \, dHe / [V \cdot (Fsum_{rj} - Dp_{rj})] \qquad (1)$$

where the energetic altitude He_0 corresponds to Mo_τ. The minimum value of the right-hand side of Eq. (1) corresponds to the minimum values of the underintegral functions. These values correspond to an engine operating at maximum rating in each point of the trajectory. Because engine thrust when operating in the RJ mode increases vs Mach number more rapidly than in the TJ mode, then minimum value of right-hand side of Eq. (1) corresponds to the value of the variable integration limit He_0, where underintegral functions for the TJ and RJ modes of operation are becoming equal to one another. A geometry interpretation is shown in Fig. 9. The summary area under the curves of underintegral functions, which is proportional to the time spent for acceleration, would be minimum in the case when the He_0 value (or M_0) corresponds to the point of the curves crossing. Transfer to the RJ mode of operation is accomplished at rather high Mach numbers, when the additional inlet drag decreases substantially. Then one can neglect by difference in overall vehicle aerodynamic drag at TJ and RJ operating modes in trajectory point with Mo_τ. Consequently, the Mo_τ value practically corresponds to the condition when the engine thrust values at the TJ and RJ operation modes are becoming equal to one another, that is,

$$F_{tj} = F_{rj}$$

Combined ABE always has some common structure elements for both parts (inlet, ARCC, external casing, etc.) with characteristic cross-sectional area A. Then the equation to define Mo_τ is

$$\frac{Fs_{tj}}{Fs_{rj}} - \frac{(W_{cor}/A)_{rj}}{(W_{cor}/A)_{tj}} = 0 \tag{2}$$

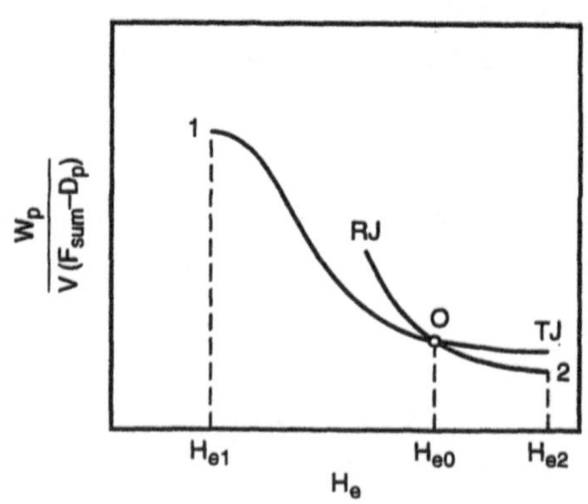

Fig. 9. Definition of optimum flight Mach numbers

Thus, the Mo_t value depends on the gas turbine part cycle parameters, the engine control law, the correlation of both engine parts dimensions (for example, A_x/A_c), but it is independent of flight trajectory and vehicle drag parameters. If with increasing flight Mach number the first part in the left-hand side of Eq. (2) decreases and the second increases, which is inherent to TRE, then the Mo_τ value would be between the Mach numbers, corresponding to equal values of specific thrust M_{Fs} and airflow M_W when operating in the TJ and RJ modes,

$$M_W < Mo_\tau < M_{fs} \quad \text{or} \quad M_W > Mo_\tau > M_{Fs}$$

When increasing A_x/A_c at constant $(W_{cor}/A)_{rj}$ the ratio Fs_{tj}/Fs_{rj} in Eq. (2) increases too, i.e., the Mo_τ value decreases. Acting in the same way as earlier, one can obtain that the optimum fuel Mach number Mo_f of transition to the RJ operating mode at which minimum fuel expenditure is provided corresponding to the flight speed when the relative to excessive thrust SFC values for both operational modes are becoming equal to one another. The equation defining Mo_f looks as follows:

$$\frac{Fs_{rj} - \{ [\, \boldsymbol{d} \cdot U(Mo_f)\,] / (W_{cor}/A)_{rj} \}}{Fs_{tj} - \{ [\, \boldsymbol{d} \cdot U(Mo_f)\,] / (W_{cor}/A)_{tj} \}} = \frac{\alpha_{tj}}{\alpha_{rj}} \tag{3}$$

where the subtrahend values in the left part present the ratio of the vehicle aerodynamic drag to the summary airflow of the whole PS (specific vehicle aerodynamic drag).

The $\boldsymbol{d}$ value is proportional to the vehicle drag coefficient and inversally proportional to the sum of engine section areas A_{sum} for all engines of the PS:

$$\boldsymbol{d} = 0.5 \cdot k \cdot P_o \cdot (C_p / A_{sum}/S) \cdot \sqrt{(T_a/\ T_o)}$$

where A_{sum} is the sum of compressor areas A_i, and S is the wing area. Included in Eq. (3), the flight Mach number function $U(M)$ can be seen to be.

$$U(M) = \frac{M^2 \cdot [1 + 0.5 \cdot (k-1) \cdot M^2 \cdot (1 - \eta_{in})]^{[\,k/(k-1)\,]}}{[1 + 0.5 \cdot (k-1) \cdot M^2]^{[\,0.5 \cdot (k+1)/(k-1)\,]}}$$

where intake efficiency η_{in} is the ratio of airflow kinetic energy after isentropic expansion from behind the intake section to that of the freestream airflow.

Then, unlike $\underline{Mo_\tau}$, the value $\underline{Mo_f}$ also depends on the vehicle aerodynamic perfection and its $(\underline{F}/\underline{W})$ ratio, which is reflected by $\boldsymbol{d}$ value. If air-to-fuel ratios α in both operational modes are equal, then with d value decreasing M_{0f} approaches the value at which specific thrust values on both operating modes are becoming equal. With *d* increasing, M_{0f} approaches the value at which airflow values in both operational modes are becoming equal. Then, depending on cycle parameters, dimensions correlation, and engine control law, with *d* increasing M_{0f} can both increase and decrease. The latter case is more possible.

Using a similar approach the equation defining M_{0L} was obtained, corresponding to the maximum range.[10] In addition to being subjected to the influence of the same factors as M_{0f}, the value M_{0L} depends on the trajectory altitude (dynamic pressure) at the point of transfer to the RJ mode, the wing loading factor, and the range factor (i.e., on vehicle L/D, overall engine efficiency, and flight speed at cruise). With an improvement of vehicle quality at cruise (increasing L/D or engine overall efficiency), M_{0L} approaches M_{0f}, because it is profitable to redistribute the onboard fuel in favor of cruise flight part.

As was shown, the optimum Mach numbers depend on the variation of specific thrust and airflow in both the gas turbine and the ramjet operational modes.

The TRE PS transition to the RJ mode process depends on the correlation between the airflow characteristics of the variable geometry inlet and both combined engine parts. As an illustrative example, in Fig. 10 are shown typical variations of maximum possible airflow at the inlet at constant altitude vs flight Mach number (lines for A_{in1} and A_{in2}, where for this example it is known that $A_{in1}>A_{in2}$), and also the curves of maximum airflow capacity for the TJ (curve T) and RJ (curve R) engine parts. One can see that the real PS airflow variation vs flight speed would be defined by the correlation between the air inlet and engine parts dimensions (i.e., fan or compressor and ARCC cross section areas).

If the engine airflow at the transition to the RJ mode is defined by the ARCC airflow capacity and is not limited by the variable inlet, which is more peculiar to the TRJ with its rather high specific thrust values on TJ operation

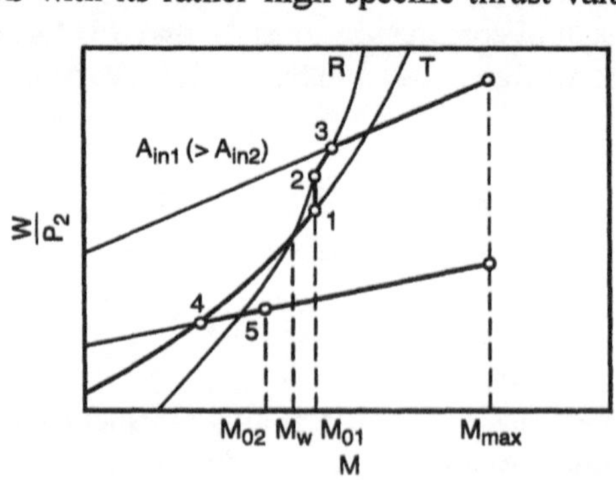

Fig. 10. TRE airflow variation at the transition to the ramjet operation mode

mode, then the suitable governing PS corresponds to maintaining the TJ part maximum rating (curve T) up to point 1 at Mo_2, at which the transition to the RJ mode of operation takes place. Then between points 2 and 3, the airflow corresponds to the ARCC maximum capacity and at higher flight Mach numbers is limited by the maximum inlet airflow capacity. In this trajectory part the best engine performances are provided by a corresponding decrease of the critical sections of the inlet and nozzle.

For the TFRJ type of TRE (with rather low specific thrust in the TJ mode) the transition to the RJ operational mode is to a greater extent characterized by limiting the engine airflow by the inlet flow capacity (line A_{in2} on Fig. 10). In this case, the transition to the RJ mode consists in gradual by slowing the TJ(TF) part in some range of flight Mach number up to the TJ(TF) switching off. In the process of transition, the air-to-fuel ratio in the ARCC should be maintained corresponding to maximum rating. The PS control law in this process is defined by the accordance of the inlet and engine variable geometry elements on the condition of providing the minimum pressure losses and necessary surge margin. The transition begins at flight Mach number, when the engine airflow approaches the limit of the inlet airflow capacity (point 4), and finishes with the TJ part approaching the windmilling mode. In this case the optimum Mach numbers on various energetic criteria coincide.

2. Optimum Mach Numbers for Turboramjet Engine with Turbojet Engine Core

The influence of various factors on optimum Mach numbers analysis was performed for TRJ based on the obtained equations. Minimum time spent for vehicle acceleration corresponds to maintaining minimum air-to-fuel ratio values along the acceleration trajectory.The range of possible Mo_τ is limited by

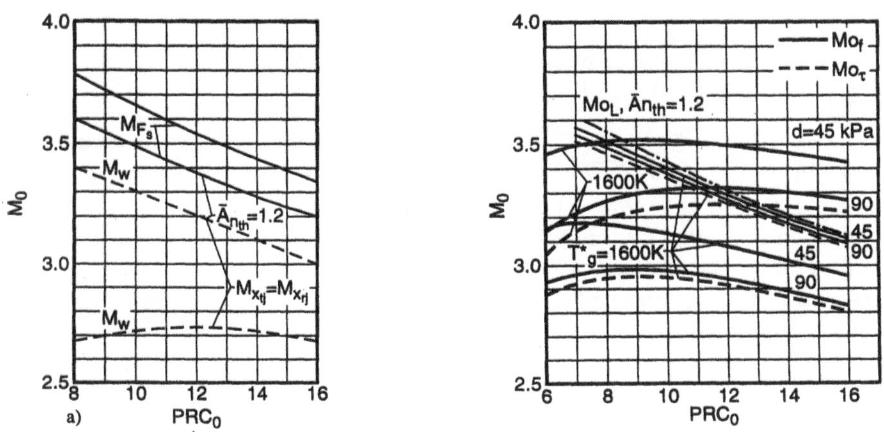

a) M_{F_S} and M_W **b) Mo_f ——, Mo_τ - - - -**

Fig. 11. Mach numbers vs design PRC_0

values of Mo_W and Mo_{Fs}. For the latter value, equal engine nozzle pressure ratio expansion values correspond in both operational modes.

The parameter M_W, at which the airflow values in both operational modes are becoming equal, depends on both engine design cycle parameters along with its control law and both engine parts dimensions correlation. As an example, typical M_{Fs} and M_W variations vs design engine compressor pressure ratio are shown in Fig. 11a. When increasing the compressor work by the nozzle opening, the curves for M_{Fs} and M_W draw together because the specific thrust is decreasing and the airflow is increasing in the gas turbine operational mode. This is illustrated by the curves corresponding to the critical nozzle area increase by 20% (An_{th}=1.2). The mentioned curves in Fig. 11a are correspondingly equal for both the operational modes corrected stream velocity at the ARCC inlet. Increasing the airflow and stream velocity at the ARCC inlet in the RJ operating mode decreases M_W.

To define the engine ratings corresponding to minimum fuel spent at the vehicle acceleration phase, one can use the idea of instant acceleration efficiency as the ratio of vehicle mechanical energy increment to the engines thrust work within an infinitesimal time. Bearing in mind the approximate correlation for the vehicle mechanical energy derivative on time, this equals.[10]

$$\eta_a = dE_p/(F_{sum} \cdot V \cdot d\tau) = 1 - D_p/F_{sum}$$

The η_a efficiency characterizes that part of the useful engines thrust work which passes to the vehicle mechanical energy.

Similarly, defining the instant overall acceleration efficiency η_{oa} as the ratio of vehicle mechanical energy increment to the energy of fuel spent in the PS within infinitesimal time, one can obtain η_{oa} equal to the product of engine overall efficiency and the instant acceleration efficiency,

$$\eta_{oa} = \eta_o \cdot \eta_a = \eta_o \cdot (1 - D_{sum})/F_{sum}$$

Minimum fuel spent on the vehicle acceleration phase corresponds to such an engine control law, when at each acceleration trajectory point, the maximum possible η_{oa} value is provided. Such propulsion governing corresponds to the minimization of SFC related to excessive thrust. These minimum points displace to the right (on lower air-to-fuel ratio values) along the engine rating curve SFC(F) relative to the usual SFC (relative to inner thrust) minimum points. In practically possible cases the optimum, from the fuel spent point of view, engine ratings on supersonic part of acceleration trajectory coincide with the maximum (afterburning) ratings.[10] The TRJ optimum Mach number calculations were performed at constant values of air-to-fuel ratio (α=1.2), ambient air, and turbine inlet gas temperatures. The dependence of optimum Mach number vs design compressor pressure ratio is shown in Fig. 11. The presence of maximum points on the curves for Mo_τ and

M_{0f} is explained by the compressor pressure ratio variation contrary influence on specific thrust and airflow in the TJ operational mode. When increasing the design turbine inlet temperature value, the maximum optimum Mach numbers displace to the higher design compressor pressure ratios, and Mo increases mainly because of the relative decrease of the ARCC dimensions and the corresponding decrease of thrust in the RJ operational mode. If it is possible to increase the compressor work while maintaining its efficiency at the initial level, then the optimum Mach values also rise because of both the airflow and the thrust increasing in the TJ operational mode (Fig. 11a, curves corresponding to the nozzle throat area increasing by 20% $\underline{An_{th}}/\underline{An_0}$=1.2). As follows from the Fig. 11b, at typical values of d=45÷90 kPa, the $\underline{M_{0f}}$ values are close to the $\underline{M_{0\tau}}$ values, the former being some higher than the latter. The curve for $\underline{M_{0L}}$ (dotted with points line) is practically analogous to the corresponding curve for M_{0f}. The dependence of $\underline{M_{0\tau}}$ and $\underline{M_{0f}}$ on the ratio of the ARCC cross-sectional area to the compressor inlet area is shown in Fig. 12. The optimum Mach numbers are decreased with rising of the $\underline{A_x}/\underline{A_c}$ value and are lower than the TJ degeneration Mach numbers.

When the air-to-fuel ratio is given, then the main factor by which one can vary the thrust performance in the RJ operational mode in the PS with variable geometry inlet and nozzle is the airflow. Increasing the flow velocity at the ARCC inlet while maintaining the inlet efficiency leads to the thrust increase on the RJ operational mode and consequently to Mo decrease (Fig. 13), in spite of the engine specific parameters getting somewhat worse. This makes it possible to use the forcing of the TRJ on airflow capacity at the RJ operational mode as a method for lowering the flight Mach number for transition to the RJ operational mode. The TRJ performance analysis has revealed that usually optimum Mach numbers for transition to the ramjet operational mode lie in range 2.5–3.5, and because of higher airflow in the RJ mode, as a rule the optimum Mach number is lower than the TJ degeneration flight Mach number, i.e., at Mo there is a difference in total pressure values behind the TJ and RJ

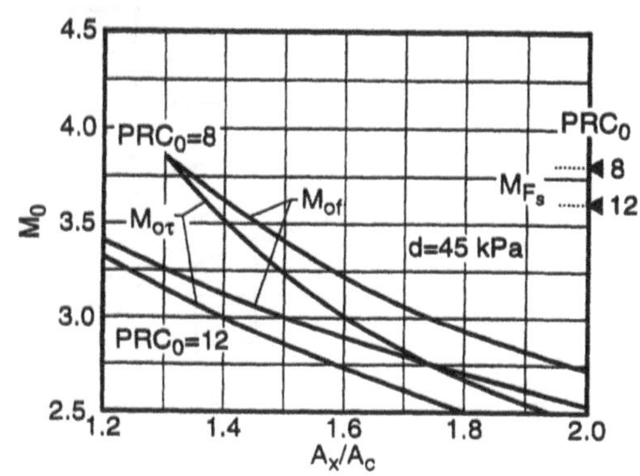

Fig. 12. Optimum Mach numbers vs ARCC/compressor areas ratio

engine parts in favor of the first. In this case adequate control of the propulsion system is needed to provide stable operation in the transition mode.

F. Alternative Fuels for Turboramjet Engines

The choice of fuel candidate for hypersonic FV, i.e., aircraft, airspace plane, or others, depends on two groups of main factors.

1) Some factors depend on the flight vehicle: its designation, that is, conception, range, and size/volume; its maximum Mach number and flight time; and the type of engine, simple cycle or combined.

2) Some factors depend on the fuel: mass heat of combustion H_u, volumetric heat of combustion (or fuel density): $\underline{H_v} = \underline{\rho_f} \cdot \underline{H_u}$, thermostability without coking, heat sink capacity, specific work capacity, and acceptability or feasibility of operation (existing or new infrastructures).

Some factors of the first group were discussed earlier in Sec.II.D. Some possibilities connected with fuel will be examined here.

1. Potential Chemical Fuels: Heating Value and Cooling Capacity

The Mendeleev Table of chemical elements gives us only seven elements with the mass heat of combustion $H_u > 25$ MJ/kg (Fig. 14). Hydrogen, beryllium, and boron have the most mass heat of combustion, and boron, beryllium, carbon, and the metals aluminium and magnesium have the most volumetric heat of combustion. Five elements H, C, B, Al, and Mg, may be practical for use in ABE. These elements can be used as fuels either directly or primarily in chemical compounds or mechanical mixtures with each other and with other components, i.e., liquids, solid propelants, and slushes. Compounds such as the C-H and B-C-H types and LH_2 are usually considered. In the past (in the 1950s) there was an attempt to utilize the boron hybrides (B-H) in

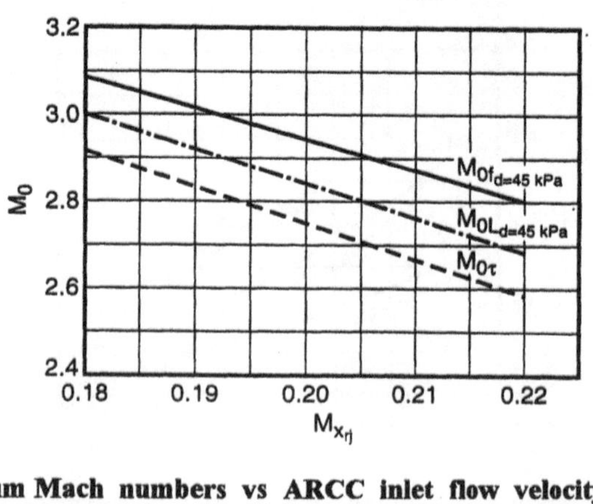

Fig. 13. Optimum Mach numbers vs ARCC inlet flow velocity (ramjet mode)

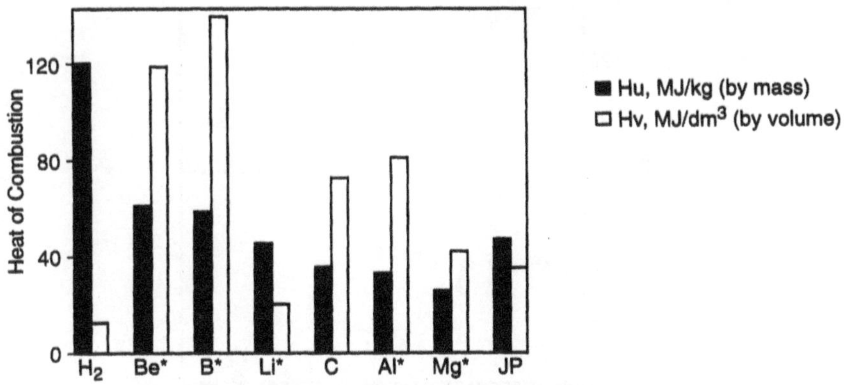

Fig. 14. Basic combustible elements (oxidizer - oxygen; basic temperature - 298.16 K; * - solid oxide) ■ Hu, MJ/kg (by mass) □ Hv, MJ/dm^3 (by volume)

aircraft but the toxicity of these fuels and the viscous boron oxide formation while burning made these attempts unrealizable.

The relative volume of the fuel tanks depends on the size of FV; actually it derives from the geometric similarity of the relative vehicle (and tanks) volume, i.e., the ratio of volume to thrust, or to drag, or to wing area

$$\bar{v} = \underline{v}/\underline{F} \ldots = \underline{v}/\underline{D} \ldots = \underline{v}/\underline{S},$$

which must be proportional to some characteristic size: $\underline{\bar{v}} \sim \underline{L}$. For this reason the mass heat of combustion H_u is preferable for large-size FV and volumetric H_v for small-size FV, so-called low-volume (Fig. 15). Hydrocarbon fuels are suitable for medium-size aircraft. As for large cruise aircraft and aerospace planes, hydrocarbon fuels and liquid hydrogen are applicable for them. The advantage of LH_2 for hypersonic flight with very high Mach number is the higher value of specific impulse because of its exclusive mass heating value and also the highest heat sink and specific work capacity.

The increase in FV flight speed is associated with the necessity of providing the PS and FV structure component firmness due to the rise in heat fluxes and operating process temperatures. This is one of the most important factors for fuel choice.

In Fig. 16 are shown the requirements for engine cooling and fuel candidates for hypersonic flight. Storable high thermostable hydrocarbon fuels T-6 (Russia) and JP-7 (US) can provide flight at $\underline{M} = 3 \div 3.5$. Cryogenic LH_2 ($T \leq 20$ K) is beyond the comparison. Precooled liquid methane ($T \leq 111.5$ K) covers the range of $\underline{M}=4 \div 6$. Endothermic hydrocarbon fuels have rather impressive potential properties, $\underline{M}_{max}=7 \div 8$. The use of hydrogen and other cryogenic fuels is limited only by future FV of large masses and volumes and requires long preparation for flight and solution of many infrastructure

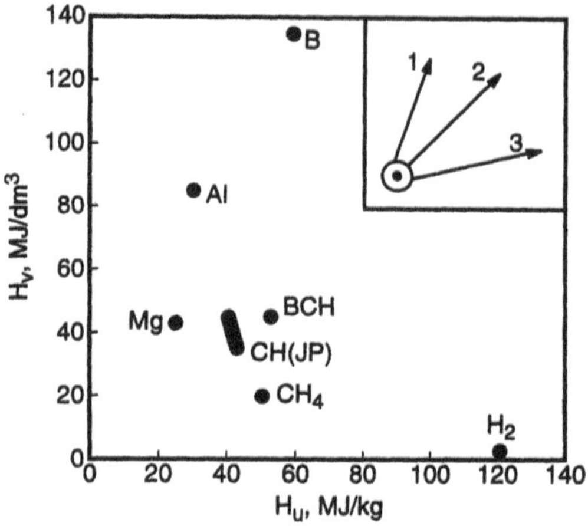

Fig. 15. Potential chemical fuels. Range increase tendency:
1 - Low volume FV (small missiles)
2 - Medium volume FV
3 - Large cruise/aerospace planets

problems. It is advantageous to use liquid hydrocarbon fuels in case of a significant increase in their cooling capacity for some high-speed FV having not very high maximum flight Mach numbers and moderate dimensions. Endothermic fuels (EF) have such properties; they decompose by heating without coking to some gas components: hydrogen, methane, and others. The cooling capacity of such fuels is better than that of methane (Fig. 16) and the density is practically twice as large as that of the density of liquid methane. Endothermic fuels are storable, and so convenient service will be better. Thus, depending on conceptions of the FV using TRE, either LH_2 or EF might be the most suitable fuels. The properties of LH_2 are well known, and so we will discuss some EF peculiarities.

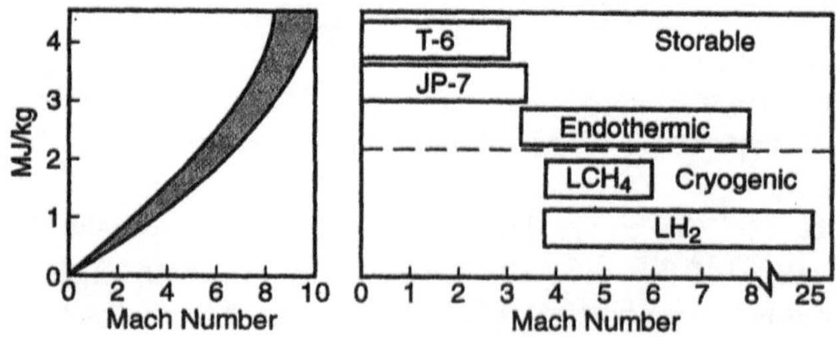

Fig. 16. Cooling requirements and fuel candidates for hypersonic aircraft

2. Endothermic Fuels: Fundamentals

Endothermic thermodecomposition of the hydrocarbon jet fuels is a complicated process including a host of reactions proceeding sequentially and simultaneously with the product formation of a lower molecular mass. The models based on a radical chain mechanism of the thermal decomposition are the most theoretically justified kinetic models of this process.[11] These models take into account a large number of the elementary reactions (up to 2000), however, because of the uncertainty in the selection of the constant numerical values of some reactions, at present there is no point in using such models to determine the transformation degree of the fuels and the products composition of their thermal decomposition as well as the heat of the reaction.

The semiempirical models of thermal decomposition including the molecular reactions which define the yield of the main products of the thermal fuel decomposition and the thermal effect of the reaction are the most convenient in practice. The thermal decomposition of the endothermic fuel may be approximately described by the pseudostoichiometric relation

$$C_nH_m \xrightarrow{k} \gamma_1 \cdot H_2 + \gamma_2 \cdot CH_4 + \gamma_3 \cdot C_2H_6 + \gamma_4 \cdot C_2H_4 + \gamma_5 \cdot C_3H_8 + \gamma_6 \cdot C_3H_6 + \gamma_7 \cdot C_4H_{10} + \gamma_8 \cdot C_4H_8 + \gamma_9 \cdot C_4H_6 + \gamma_{10} \cdot C_5H_{10} + \gamma^* \cdot C_xH_y + Q_{end}$$

where

k = constant of the thermal decomposition brutto-reaction rate,

γ_i = mass yield of i gas component of pyrogas,

C_xH_y = takes into account all of the rest of the condensted reaction products, where x and y are derived from the atomic balance,

Q_{end} = value of the thermal reactions effect and determined by Hess law.

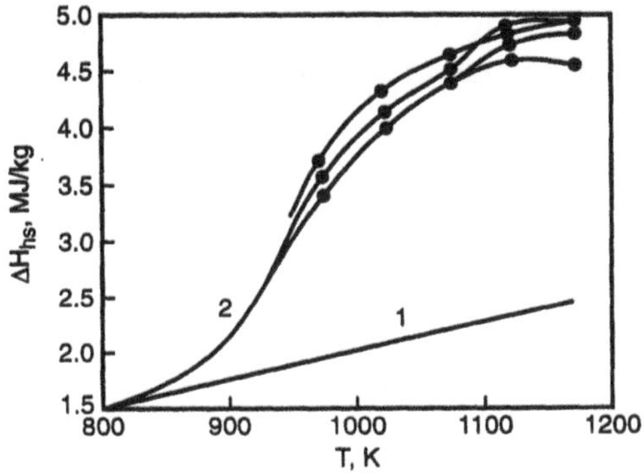

Fig. 17. The experimental dependence of endothermic fuel T-15 heat sink of temperature ($\Delta H_{hs} = 0$ by hardening temperature $T_0 = 213$ K)
1 - physical heat sink
2 - total heat sink with decomposition effect (different samples)

Use of such a type of the model requires the experimental determination of the thermal decomposition rates for each type of the fuel. From CIAM experimental data the yield of each pyrogas component γ_i may be approximated by the polynomials of the form:

$$\gamma_{\underline{i}} = [P/(\underline{a} + \underline{b}P)] \cdot (\alpha\underline{Z} + \beta\underline{Z}^2 + \delta\underline{Z}^3)$$

where
Z = decomposition extent, = 0–1
α, β, δ, a, b = empiric coefficients

The determination of the afore-mentioned decomposition process indices for each endothermic fuel type requires carrying-out a large volume of experimental tests.

3. Endothermic Fuels: Properties

The first publications on endothermic fuels (EF) appeared in the 1960s.[12] At present the problems of EF realization are being studied in some countries, mainly in the United State and Russia. CIAM developed jointly with some organizations of the oil-chemical industry in Russia a new endothermic jetfuel T-15 having rather good properties.[13] This fuel has superior heat sink qualities such U.S. fuels as CRC-1, CRC-2,and Soltrol-170.

The endothermic heat sink effect of fuel decomposition begins at a temperature of ~ 800 K (Fig. 17). At 1000 K decomposition heat sink is equal to physical heat sink of heated fuel without decomposition, the total heat sink is doubled. Common fuel heat sink onboard the aircraft board is characterized as its ratio to common fuel heat onboard, i.e., as the ratio of 1- kg fuel heat sink capacity ΔH_{hs} to the fuel lower mass heat of combustion H_u; $\Delta H_{hs}/H_u$. This parameter gives the real value of fuel cooling capacity. As is shown in Fig. 18, the EF T-15 cooling capacity at T=1000 K is equal to 72% of hydrogen relative cooling capacity, and EF significally surpass methane. At T=1000 K, the heat

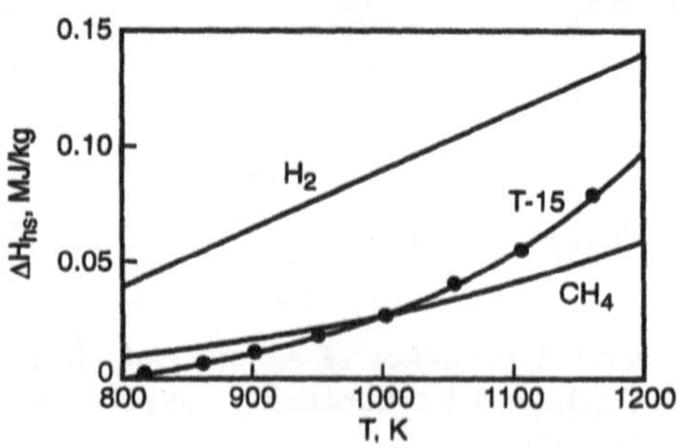

Fig. 18. Relative heat sink of ABE fuels

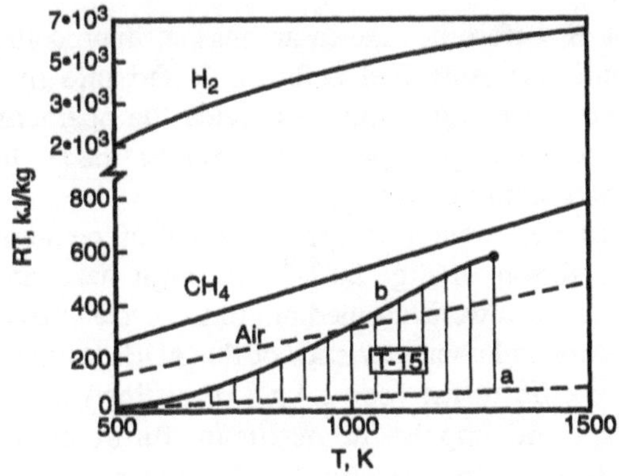

Fig. 19. The specific working capacity of fuel vapours: hydrogen, methane, endothermic fuel T-15 and air (a - without decomposition Z = 0, b - with decomposition, • - decomposition extent Z = 0.85)

sink of EF T-15 reaches 9% of its mass heat of combustion. In this figure the values of heat sink ΔH_{hs} were calculated for T-15 and CH_4 from the smelting point, T_0 =213 K and 89 K, respectively, and for H_2 from the boiling point T_0 = 20 K.

For combined engines the specific work capacity of fuel vapours is of great importance. This property is characterized by the value of RT and the gas constant R grows as the molecular mass of gas become lower. In thermodecomposition of EF pyrogas, components with low molecular mass are formed (see II.F.2), and the gas constant of the pyrogas grows (Fig. 19). The high working capacity of EF pyrogas makes it possible to use it for different purposes, for example, to drive the turbopump unit for engine fuel supply, as was shown for liquid hydrogen in Ref.[4] Such a turbopump unit is especially suitable for turboramjet engines operating a long time in the ramjet mode because there is no source of mechanical energy at that mode.

The use of endothermic fuels requires no special aggregates because the thermodecomposition process can be organized in cooling channels or in heat exchangers with catalytic coated surfaces.

The properties of EF, its cooling and working capacity and rather large density, make this fuel one of the best candidates for use on high-velocity FV having medium free volume for fuel tanks and flying at not very high Mach numbers (5–7).

G. Heat Exchangers for Engines of High-Speed Flight Vehicles

Heat exchangers (HE) of gas turbine and combined engines are intended to raise thermodynamic cycle efficiency or to provide normal temperature operational mode of engines parts. With the help of HE decreasing temperature or liquefaction of air before its use in an engine, intercooling in compressors, gasifying of liquid cryogenic fuel before it is fed into the injectors of the combustion chamber, cooling of air to provide the operability of turbine and other high heat-loaded engine parts, etc. can be made. In some cases HE execute several tasks simultaniously.

As an example, some principal schemes of combined engines using LH_2 and HE are shown in Fig. 20.[4,8] The high heat sink and high work capacity of hydrogen as a working medium in the cycle enable the significantly improved parameters and characteristics of the ABE. Improvements include an increase of the specific thrust (per 1 kg/s of airflow) by 1.5–2.0 times, and increase of the specific impulse at maximum thrust by 1.2÷1.5 times, and increase of thrust-to-weight ratio by 1.5–2.0 times, and an increase of the maximum speed of the gas turbine engine usage by use of the cooling capacity of LH_2 untill M=5÷6. Most of these engines have a ramduct and so are used as an improved core engine in TRE. The use of HE in TRE of conventional type for some purposes can improve its characteristics also.

Liquid methane or endothermic fuels instead of LH_2 can be used for improving engines characteristics by using of HE. The result may be significant but not as high as by using LH_2.

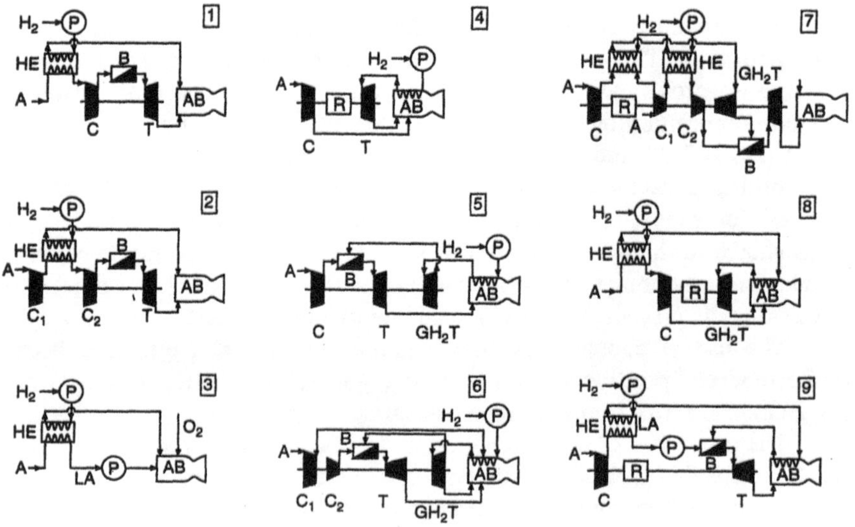

Fig. 20. Some new types of LH_2 air-breathing engines using the hydrogen cooling and working capacities by means of HE system

1 - TJ with precooling; 2 - TJ with intercooling; 3 - LACE; 4 - RATR; 5 - TJ with a gaseous hydrogen turbine GT_2T; 6 - the 5-th schemes based on TF; 7 - TF with the intercooling and GT_2T; 8 - RATR with precooling; 9 - ATR with air liquefaction (A - air; AB - afterburner; B - combustion chamber; LA - liquid air; P - reducing gear)

1. Technical Parameters

In HE for engines of high-speed FV, heat carries that are usually used are the air and the fuels with high heat sink: hydrogen or methane in liquidor gaseous state and endothermic fuel too. Schemes of engines with air-to-air HE for heat regeneration are also considered. The parameters of heat carriers are defined by the HE setting and are varied in wide ranges, temperature in the range from 16–30 K to 1300–2500 K and pressure in the range from one to some hundred atmospheres. Airflow rate through HE can change from several tens to several hundred kilogram per second.

2. Heat-Phisical Process Features

In a series of experiments held by CIAM it was revealed that heat-physical processes in HE of high speed FV engines had some characteristic features. High values of temperature drops between heat carriers are characterized in HE intended for air cooling by cryogenic fuel. Air temperature at HE entrance can be 1300÷2500 K and exit temperature 80÷400 K. In this case, the heat-physical properties of air, i.e., density, specific heat capacity, viscosity, thermal conductivity, etc., are changed several times while the heat carriers moving within the HE ducts. On the other hand, the high temperature drops bring high temperature stresses on the HE structure. That is why, for provision of optimal thermohydraulic and acceptable strength characteristics, it is advisable to make HE several sections so that the flow path cross section of the air duct is decreased down stream from the flow. In such HE a high difference in the temperatures of flows and heat exchange surfaces take place resulting in worsening of the heat emission from cold flow. In this case, the temperature factor (T_{wall}/T_{gas}) value can be equal to 10÷50.

Inlet air contains moisture which contacts the HE walls, is cooled lower than the dew-point temperature and 0°C, freezes, and falls out on the walls as a hoarfrost or ice. Dew-point temperature depends on pressure P and humidity d and can be calculated by

$T_{dp} = 52.9 + 1563.3 \{9.88 - \lg[dP/(d+0.622)]\}$ where T is in Kelvin and P in Pascal.

Humidity of the airflow is defined by humidity of atmospheric air on the flight height without that part of the moisture which has condensed on the surfaces of other HE locating upstream of flow. Condensate which falls out on the HE walls increases the heat resistance between heat carriers and deispite some intensification of heat emission, in general, it results in the worsening of the heat exchange and jamming of ducts raising hydraulic resistance. Active and passive methods to deal with this occurence were developed by CIAM. Active methods consist of preliminary drying of the air before its feeding to the HE; for example, it can be performed by spraying liquid air or nitrogen into the flow to freeze moisture with further sedimention in special swirl chambers. Passive methods consist of using of hydrorepellent coatings.

High-level temperature drops are typical for heat exchanger-gasifiers of cryogenic fuels. A flow of boiling fuel in it is a nonequilibrium mixture of overheated vapour and drops of liquid. Heat emission to such flows are not characterized by high intensity.

For HE of launch mission engines working under conditions of permament change of both flight speed and flight height, an important feature is operation in an unstationary mode which is characterized by change of all input parameters: temperature, pressure, and flow rate.

*3. **Structure and Materials***

In different projects the various HE structure for engines of high-speed FV are proposed. Tube type structures were mainly considered, but one should not put aside the possibility of using of plate HE. In some cases it is possible to use the regenerative and mixture types of HE and HE joined with high heat loaded parts of engines structure, such as combustion chamber or compressor vanes.

The scheme of the annular HE original structure developed in CIAM intended for air cooling at the compressor entrance of a TFAB in flight up to M=5–6 is shown in Fig. 21. In this HE a liquid hydrogen is used as a cold heat carrier. The HE is a tube annular type with 1t, 2d, and 3d ways of LH_2. Tubes are placed perpendicularly to the airflow on evolvents. The length of the tube is 0.5 m, and the diameter is 0.2 mm. Tubes are produced from heatproof Cr-Ni alloy. The number of tubes in one evolvent row is 64: 15+21+28. The quantity of evolvent rows on the circle is 480. The whole heat exchange surface square is 8.5 m^2. Weight-to-flow rate ratio is 141 kg per 1 kg/s. The HE begins to operate on the final part of climb-acceleration trajectory. The application of fuel-air HE using the heat sink of cryogenic fuel allows engine operability in high speed flight conditions (M>3) and even influences the type and parameters of the thermodynamic cycle and, consequently, the engine performance.

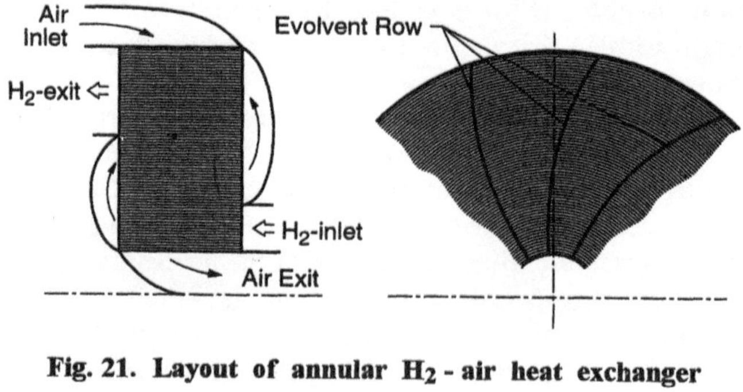

Fig. 21. Layout of annular H_2 - air heat exchanger

4. Methods of Calculation

To calculate the HE characteristics for engines of high-speed FV special computational methods were developed in CIAM.[14]

In developing the mathematical models the following assumptions were adopted: 1) the flow of the heat carrier is homogeneous, 2) the change of humidity exerts a small influence on the heat carrier flow, 3) all of the sublimation heat is given off on the surface of the hoar frost layer, 4) density of hoarfrost is constant in the cross section of its layer but is changing in time, 5) heat resistance of duct walls in the cross direction is absent, and 6) thermal conductivity of the heat carrier, of the elements of HE matrix, and of the layer of hoar frost in the longitudinal direction is not considered.

Calculation algorithms allow the determination of the points of the beginning and the end of the phase transitions in the flows of the heat carriers and the regions of frost appearance on the heat exchange surfaces. This method passed extensive experimental checking which proved its adequacy. With help of the developed methods both the thermal and the hydraulic characteristics of HE, even on the startup modes can be analyzed; the influence of hydraulic irregularity at the entrance on the operation efficiency of the HE can be studied; the regions of hoarfrost condensation on the heat exchange surfaces can be defined; and the thermal stress in tube row bank, etc., can be estimated.

III. CIAM Experimental Turboramjets

A. Experimental Turboramjets and Specialized Test Cell

Large programs both in the investigation of engines for high-speed flights and the use of liquid hydrogen in aviation were carried out in Russia in series of 5-year programs in 1960–1980. In the framework of these programs the investigations of advisable engine concepts, operating process parameters, and control modes of the TRE were conducted. In this period the program of development and experimental investigation of full-scale TRE of various types in real and simulated flight conditions was carried out at CIAM. The experimental TRE were assembled from units of series produced TJ and TF. Within the tests, the flight conditions corresponding to the Mach numbers M=4.0--4.5 were simulated in the test facility.

For simplification in solving of problems of maintanability, service, preparation, manufacturing, and controlling, the unified ram duct with a shutoff device with remote control was used. The ram duct consists of front and back manifolds and four connecting side tubes; replaceable afterturbine nozzles, which provided different exit turbine areas, were used (Fig. 22).

In the course of study of the heat state problems, the ARCC of the TRJ was equipped with a perforated screen along its length, a distribution system of manifold fuel supply, and a separate air cooling system. This cooling system could provide variable parameters of cooling air ($\underline{W}_{col}$, $\underline{P}^*_{col}$, $\underline{T}^*_{col}$) into the cooling duct, i.e., into the space between the external case and the perforated screen of ARCC. The bearing supports of the TRJ were also reconstructed.To provide operability of engine transmission, thermoprotective screens were located over the support cases. The air cooling system with variable parameters ($\underline{W}_{col}$, $\underline{P}^*_{col}$, $\underline{T}^*_{col}$) and the oil system with variable oil feed were used for bearing support cooling.

All of these investigations were intended for the development of a cruise mission FV of relatively low takeoff weight, using kerosene. Obtained results and the Russian experience in area of liquid hydrogen use[4,5], however, open the possibilities for development of TRE for ASP, using liquid hydrogen as fuel.

The main goals of the experimental research were 1) complex study of the operating process, 2) afterburning ramjet combustor operation, 3) engines switching modes and operation stability, 4) windmilling and ramjet operation modes characteristics, 5) passage hydraulic characteristics, 6) power output possibilities, 7) ramjet combustor and nozzle cooling at $\underline{M}$=3.0÷4.5, 8) cooling system impact on operating parameters, and 8) structure heat state and transmission operability.

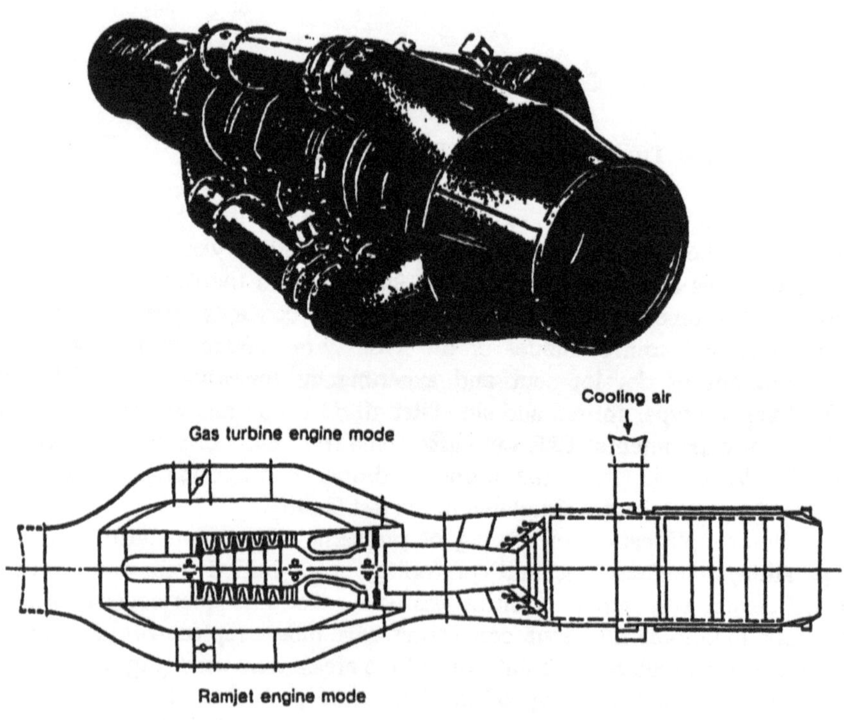

Fig. 22. CIAM experimental turboramjet

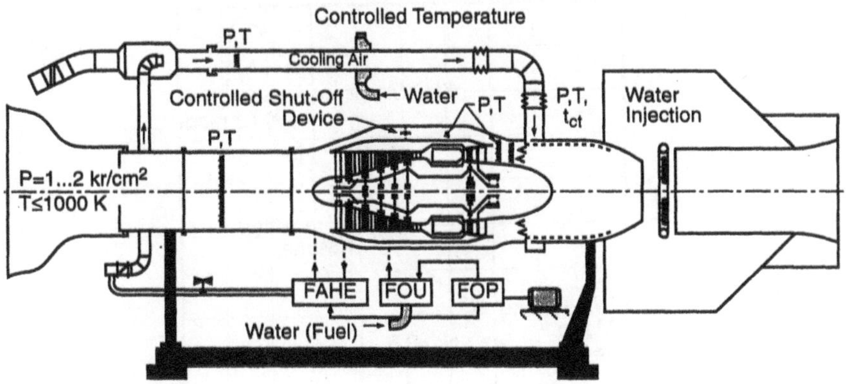

Fig. 23. Layout of CIAM test facility
(FAHE - fuel-air heat exchanger, FOU - fuel-oil unit, FOP - fuel-oil pump)

Tests of experimental turboramjet engines were carried out at the CIAM-branch test facility with a direct connected pipeline which permitted simulation of the flight conditions corresponding to the Mach numbers M=4.0÷4.5 (Fig. 23). The entry air temperature realized at the facility is provided for by the heat exchanger and combustion heaters (with recovery of air normal composition by oxygen replenishment). Air pressure at the entry reached up to 900 kPa. The facility has several hundreds channels for boththe temperature and the pressure measurements in static and dynamic modes. The facility dimensions allow the testing of any current supersonic engines. This cell was also equipped with injector exhaust system. The test facility was equipped with separate controlled air cooling and oil systems.

The range of parameters of airflow at the engine inlet was P^*_{in}=0.1–0.2 MPa, and T^*_{in} = 270–900 K. The variable range of parameters of the cooling air for ARCC cooling was T^*_{col} = 400–900 K, P^*_{col}/P^*_{ab} = 0.95–1.40, and W_{col}/W_x = 0.03– 0.23. The bearing cooling air pressure was varied in range up to 0.5 MPa and oil pressure on entrance of bearing supports was in the range P_{oil} = 0.16–0.22 MPa. The air mass flows through ram duct and ARCC cooling air tube were controlled by the change of shutoff devices damper angles.

During the experimental investigations the following parameters were measured 1) total and static pressures and total temperature at the engine entrance, 2) total/static pressures and total temperature at the compressor entrance, 3) total/static pressures and total temperatures on the compressor and the turbine outlet, 4) total and static pressures in the ARCC and in the ARCC cooling duct, 5) total and static pressures in the ram duct, 6) total and static pressures and temperature at the outlet section of both engine parts ducts, 7) total temperatures in both the ARCC and the cooling air streams, 8) wall temperature along of perforated screen, 9) cooling airflows through both the cooling duct and the bearing supports, 10) temperatures of bearing support

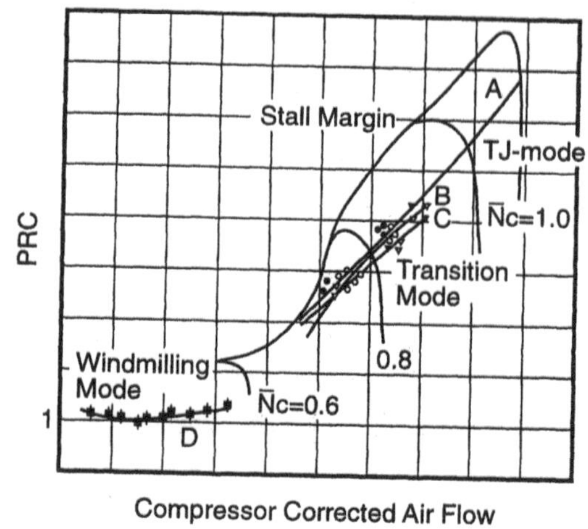

Fig. 24. Compressor map of experimental TRE

cases, 11) cooling air temperatures at the entrance and exit of the supports, 12) oil temperatures at the entrance and exit of the supports, 13) oil flow through bearing supports, and 14) engine rotational speed and some other parameters. The engine continuous operation time at $M = 4 - 4.5$ was about 3 h.

B. The Experimental Investigations of Turboramjet Engines with Turbojet Engine Core Transition to Ramjet Mode

The transition from gas turbine to ramjet mode (also windmilling mode) is the great interest for TRJ. The first tests have been directed just to these modes.[1-3] Some results of these investigations will be discussed.

The rational value of flight speed for transition to the ramjet mode depends on the design operating parameters, the TJ control mode, and the size correlation between typical PS passage areas. The transition Mach number usually is changed in the range 2.5–3.5 (Ref. 3, 6, and 10).

The working lines on the compressor map for different modes of operation are shown in Fig. 24. The gas turbine mode corresponds to the curve A. The joint operation of TJ and RJ is realized at the transition mode. In these conditions, the pressure behind turbine is higher than that at the ram duct entrance. If the shutoff device opens, the control system of the TRJ must respond in such way that stable operations of the inlet and the engine are provided.[1,3,10] The RJ starting mode corresponds to zero airflow through ram duct. The curves B and C (Fig. 24) correspond to this mode for different values of turbine exit duct area (this area for curve C is higher than that for curve B). At the selected turbine exit area, the certain engine nozzle throat area corresponds to this condition. The region located above the curves B and C corresponds to the reverse flow of the gas part from behind the turbine to the entry of engine. Such reverse flow is realized if the engine nozzle throat area An_{th} is less than the value An_{tho} corresponding to the curve $W_{rd} = 0$. With lower values of An_{th} an unstable operation of the TRJ and propulsion system is possible. The points located below the curves $W_{rd} = 0$ correspond to the direct airflow through the ram duct. Such flow is realized if the engine nozzle throat area An_{th} is bigger than the An_{tho}. If An_{th} is increased additionally, there is the increase of airflow through ram duct at practically the same airflow as through the TJ. The experimental points at the joint operation mode are located nearly on the curves B and C.

At the definite values of the corrected rpm the curves W_{rd}=0 cross the curve A and approach the curve of compressor stall. Usually for the TRJ, however, the corrected rpm for the transition to the beginning of the RJ mode located right of the curves crossing point. Therefore, an error of control at the start of the RJ (premature open) of the shutoff device presents danger rather than stable operation of the inlet (Fig. 25). As follows from Fig. 25, forestalling the shutoff device opening leads to the TJ inlet corrected airflow lowering with corresponding consequences. It was confirmed by some testings of TRJ. [1-3]

The rational control of the engine at the transition to the RJ mode process must provide not only stable operation of the propulsion system but also thrust proceeding without significant drops. Investigations have revealed a rather narrow operation interval, inside which PS stable working process with parallel operation of the TJ and RJ parts could be realized. Maintaining the TJ operating parameters with respect to the limitations provides propulsion system reliability at the transition to the RJ mode. The

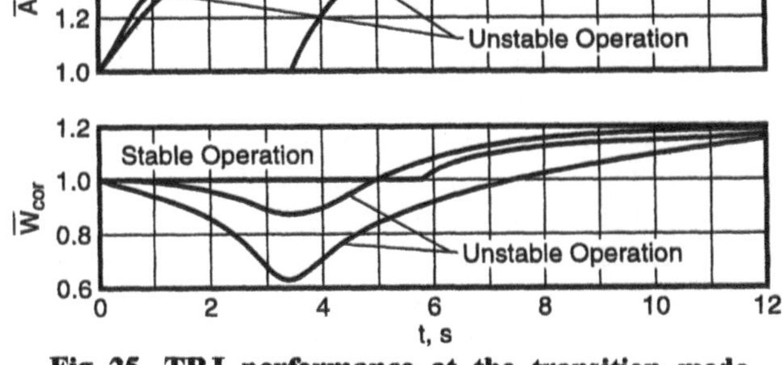

Fig. 25. TRJ performance at the transition mode

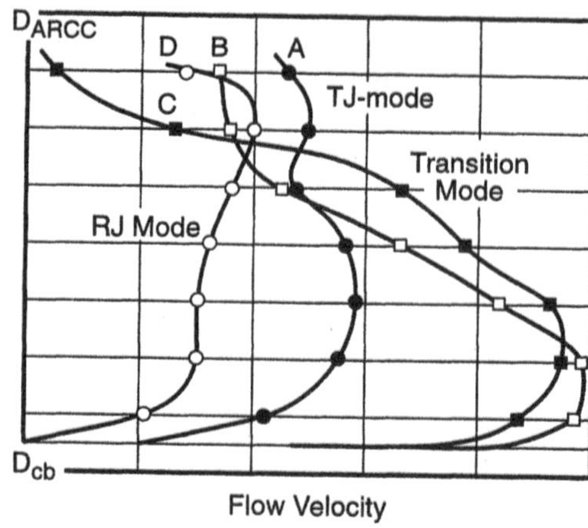

Fig. 26. Distribution of flow velocity on ARCC entrance

transition to the RJ mode is finished with TJ windmilling, which may be used to drive engine accessories.

At the transient mode, the structure of flow at the ARCC entry is strongly changed (Fig. 26). The results of TRJ testing show that the ARCC inlet flow velocity increases at the constant rotation speed approximately in proportion to the increasing engine nozzle area. In this case, a considerable distortion of the flow take places. For example, at the TJ mode the distortion of the flow is not more than 10– 20 % (the curve A in Fig. 26), but at the engine parts joint mode the maximum flow speed can exceed the mean flow velocity about 2.0÷2.4 times (the curves B and C in Fig. 26). Therefore, special measures possibly would be demanded for stable combustion in ARCC at the transition mode.

C. Ramjet Mode of the Turboramjet Engine with Turbojet Engine Core with Windmilling of Turbojet Engine Core

At the RJ mode the TRJ can operate with either opened or closed gas turbine part of engine. The curve A in Fig. 27 corresponds to the windmilling mode of the TJ turbocompressor. The pressure losses in the TJ at windmilling very strongly depends on compressor corrected airflow (the curve A in Fig. 27 with the ram duct closed). At the RJ mode the TRJ pressure losses do not exceed 5÷7 % if the TJ is closed (the curve C in Fig.2.6) or is operated at the windmilling mode (the curve B in Fig. 27). The application of the TJ windmilling mode provides the increase of airflow through the engine as the additional part of airflow is passed through the TJ and gives some possibilities for accessories drive. Expediency of the possibilities, however, depend on the TJ efficiency in the windmilling mode.

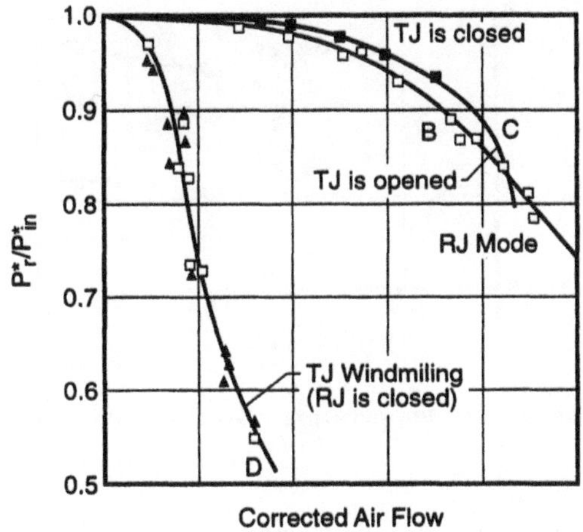

Fig. 27. TRJ pressure losses on the ramjet mode (core windmilling)

At the RJ mode the ARCC inlet flow has rather a little distortion of speed. At the begining of the transition mode, the region of maximum speeds is located in the central part of the ARCC and very strong distortion take place, but at the windmilling mode it moves to the peripherial part of the ARCC (the curve D in Fig. 26). The unevenness of the ARCC inlet flow decreases if the corrected airflow is decreasing too.

D. The Cooling of Afterburner-Ramjet Combustion Chamber

The ARCC and nozzle are two of the most heat loaded units of the TRE operating in the ramjet mode. To provide reliability of these units, it is natural to use an air cooling system in which a part of the air is taken away from the engine flow path to the ARCC cooling duct (i.e. a space between the ARCC case and perforated screen). In an experimental TRJ for cooling of the ARCC, the air cooling system with a perforated screen was used and for cooling the nozzle the slot system was used. Experimental investigations of heat state and operability of high heat loaded TRJ units were carried out at in ramjet mode operation with windmilling of the gas turbine core and simulation of Mach number in the range 2.5–4.

The next tasks of heat state problem were investigated: 1) ARCC combustion performance with the distortion of ARCC inlet total pressure and flow velocities fields, 2) joint operation of the ARCC and its cooling system, 3) heat state of the ARCC structure, and 4) heat state of bearing supports on the ramjet operation mode. Some results of the experimental investigation of full-scale ARCC in the ramjet mode working process and an estimation of cooling system operation influence on engine performance are discussed next.

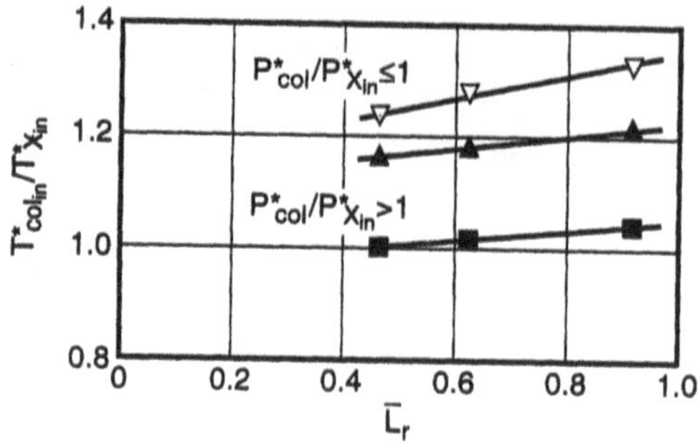

Fig. 28. Cooling air temperature variation along cooling duct

1. Operation of Afterburner-Ramjet Combustion Chamber Cooling System in the Ramjet Mode

The operation of the TRJ is characterized by wide range of ARCC inlet parameter changes with significant distortion of the total inlet pressure fields. The value of the total inlet pressure distortion depends on the operation mode and value of corrected airflow through the engine. In the course of the experimental investigations of the TRJ, the ARCC inlet parameters of flow were characterized by the values T^*_X = 450÷900 K, M_X = 0.1÷0.25. The radial distortion of the ARCC total inlet pressure was 6÷12% with the gas turbine core windmilling and 7÷14% with the closed gas turbine core. Nozzle flaps location and cooling air blowing through the perforated screen faintly influence the shape of the total pressure fields in the nozzle throat section. The inlet total pressure distortion decreases along the ARCC in the modes without burning, and in the throat nozzle section distortion is no more than 2÷3% when combustion takes place. The ARCC ombustion efficiency under these conditions depends mainly on the excess air coefficient and locations in the range typical for afterburners.

The length of the combustion zone significantly influences distribution of the cooling air along the perforated screen of the ARCC. One can characterize combustion zone length indirectly by the static pressure drop on the ARCC screen measured along the ARCC. As experiments have shown, the fuel combustion is completed within 2/3 of the ARCC length.[2,3] The cooling airflow variation and its distribution along the ARCC are defined by very complex processes in nonuniform flow with combustion in the ARCC. The hydraulic characteristic of the ARCC cooling duct depends on the discharge coefficient of holes and duct pressure loss coefficient. Variation of these factors was examined in the course of the investigation[2]. The analysis of the cooling air parameters in the ARCC cooling duct has revealed that the cooling air temperature along cooling duct changes faintly. When the total pressure ratio is $P^*_{col}/P^*_X < 1$ the cooling air temperature at duct exit is 20÷30% higher than the value of duct inlet temperature (Fig. 28).

This temperature variation is explained by the penetration of hot gas into the cooling duct through the holes of the perforated screen at that part of the duct. Here, as well, the change of cooling air temperature is not large. It should be noted, that a small change of T^*_{col} along the ARCC has no practical influence on the hydraulic characteristic of the ARCC cooling duct, and it is possible to use an assumption about the constant value of cooling air temperature along the ARCC cooling duct.

The results of experimental investigations showed a rather weak distribution of cooling air along the ARCC perforated screen depending on a variation of both the cooling air and the ARCC gas parameters in a wide range. The variations of the corrected airflow into the cooling duct and the gas flow in the ARCC was obtained depending on the ratio of total and static pressures in the characteristic cross sections. The results of experimental investigations led to the formulation of a hydraulic characteristic of the cooling duct and ARCC at the joint operation. A base was adopted in this case for the following reasons.

1) A pressure characteristic defining the cross section, which is located at approximately the half length of the ARCC, was adopted; in this section the combustion is mainly completed and the main part of pressure losses is realized.

2) The total air pressure at the cooling duct characteristic cross section is taken to be equal to the mean value between the values for inlet and exit of cooling duct; ARCC total pressure is taken to be equal to the ARCC exit total pressure.

3) One part of the cooling air from the cooling duct blows out through the perforated screen and other part of the cooling air is used for nozzle cooling.

Obtained by means of experiment and calculation, the hydraulic characteristic of the ARCC and its cooling system is shown in Fig. 29 in the form of a ratio of the dependence of the corrected airflow in the cooling duct and gas flow in ARCC to the corresponding total pressures ratio. The curves are given at constant values of corrected flow density in the duct and the ARCC. The obtained hydraulic characteristic advances to form the principles governing the coolant flow. The hydraulic characteristic of the cooling system showed, that when the total pressures ratio is near to 1 mass flows ratio changes are rather small in a wide range of parameter variation both of the ARCC and the cooling duct.

The data has led to the approach for the development of the optimally controlled cooling system of the ARCC. For the development of a controlled air cooling system with the possibility of cooling airflow change in different engine operation modes it is necessary to select the total pressure ratio somewhat higher than 1.

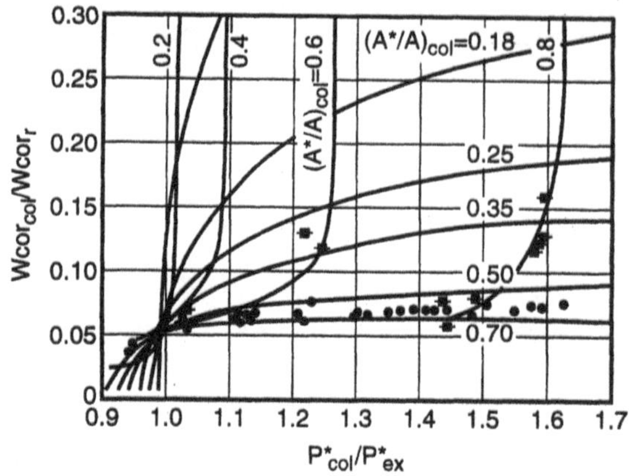

Fig. 29. Hydraulic characteristics of ARCC cooling duct

2. Heat State of Afterburner- Ramjet Combustion Chamber

The heat state of ARCC, which is characterized by the perforated screen temperature T_W, depends on the engine operating mode, flight conditions, parameters, and flow of cooling air. Using the hydraulic characteristic of the air cooling system, one can consider the correlation between relative cooling airflow W_{col}/W_X, mean screen temperature T_{wm} and flight Mach number, keeping in mind, for example, the noncontrolled cooling system of ARCC without conditioning of cooling air ($\underline{P}^*_{colin}=\underline{P}^*_X$, $\underline{T}^*_{colin}=\underline{T}^*_X$). Two possible laws of the PS control at the RJ operating mode are characteristic: 1) constant value of engine inlet corrected airflow, which corresponds to the constant value of the ARCC inlet Mach numbers M_X independent of flight conditions, and 2) decreasing value of inlet engine corrected airflow vs flight Mach number, which corresponds to the decreasing ARCC inlet Mach numbers; in this case the airflow through engine is limited by inlet maximum productivity.

The variation of mean screen temperature and relative cooling airflow are shown in Fig.30 in depending on flight Mach numbers in the ramjet mode where $\alpha = 1.2$. As is shown in Fig. 30, the relative cooling airflow changes weakly although the change of TRJ parameters for these two laws of PS control is significant. At the flight Mach number M=4 the mean screen temperature is T_{wm}=1150÷1230 K and the relative cooling airflow is approximately 12%.

At flight Mach number $\underline{M}$<4, the cooling airflow in uncontrolled cooling system is significantly higher than it is nesessary for optimum cooling. For this reason, the mean screen temperature at lower flight Mach numbers is rather low. For the two laws of PS control the mean screen temperatures are weakly distinguish able. For example, at $\underline{M}$=4 the increase of M_X from 0.13 to 0.18 provides the decrease of T_{wm} of only 80 K. In this case, the cooling airflow decreases by approximately 2%.

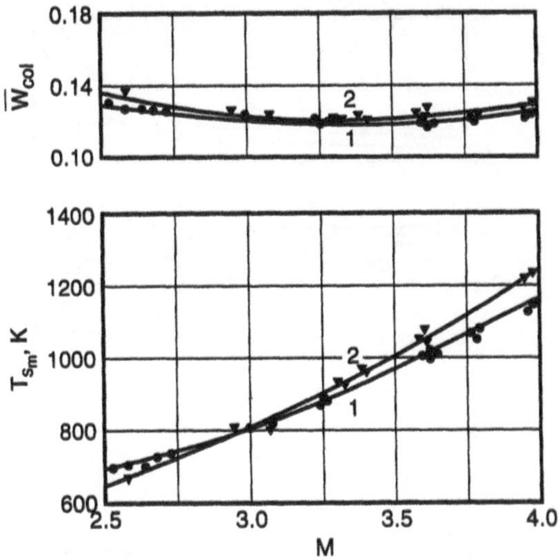

Fig. 30. Relative cooling airflow and mean screen temperature vs flight Mach numbers

1 - $\overline{W}x_{cool}$ = const, 2 - $\overline{W}x_{cool}$ = var

TRJ throttling at flight Mach number M = 4 by decrease of the fuel supply to the ARCC provides the decrease of W_{col} because of lower heat pressure losses. However, the decrease of air cooling flow is lower than necessary for the optimum cooling system. In this case, bigger coolant flow also provides low screen temperature.

The results of full-scale TRJ experimental investigations in simulated flight conditions at Mach numbers M=2.5–4 have shown that in an uncontrolled air cooling system of the ARCC with a perforated screen in nondesign flight conditions (lower flight Mach numbers or engine throttling at the M = 4) the relative cooling airflow is significantly higher than the required value. This leads to thrust performance aggravation in these modes. Improving the performances is possible by choosing rational cooling air parameters and using the controlled cooling airflow system.

The results of TRJ experimental investigations allowed the development of the method to calculate the engine thrust performance characteristics taking into account the influence of cooling airflow. Performance estimations were made according to the two types of ARCC which are used in different types of TRE[6]: ARCC in engine common exit duct (cylindrical ARCC) and ARCC in separate ram duct (coaxial ARCC). In performance definition the nonuniformity of the total pressure and temperature in the ARCC and nozzle and the screen heat state were accounted for.

The cooling system operating at the optimum mode in which the necessary value of cooling airflow is defined by the limit value of screen temperature, was considered. As a measure of degree of cooling airflow

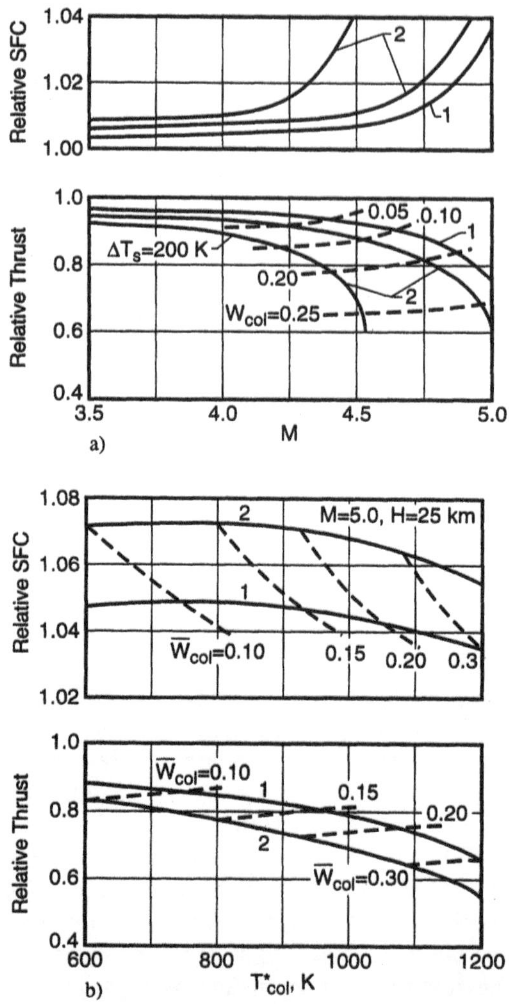

Fig. 31. Relative thrust and SFC variation vs flight Mach numbers (a) and cooling air temperature (b)
1 - common ARCC, 2 - separate ARCC

output influence the ratio of thrust and SFC values to their conditional quantities was used without accounting for cooling airflow output (Fig. 31). The performances are given in dependence with flight Mach number along the trajectory with dynamic pressure q_V=50 kPa and α=1.2 for screen materials with different temperature limits.

The significant increase of cooling airflow takes place at the flight Mach numbers M=4÷5. At the higher flight Mach numbers the necessary value of cooling airflow becomes very high with a significant increase of the loss of performance (Fig. 31a). It is possible to decrease the amount of cooling air at the necessary high flight Mach numbers by use of the conditioning system, which decreases the cooling air temperature. Figure 31b ,as an example, shows the influence of cooling air temperature T^*_{col} on both the thrust and SFC for the two different types of ARCC using the same screen material under the flight conditions M=5, H=25 km. The TRE with common ARCC has higher a thrust value of 7÷8% and lower SFC of 2% than the TRE with separate ram duct because of the larger area of cooling surfaces of the latter. To provide equal values of cooling airflow output in both engine types, the cooling air temperature in the TRE with coaxial ARCC must be lower by 100÷150 K than the other type of ARCC. The SFC variation vs T^*_{col} is rather small. Up to flight Mach numbers M=4÷4.5 there is no need of a conditioning system to ensure ARCC operability. At the higher flight Mach numbers, the use of such system is a necessity. Rational type of cooling air conditioning system depends on the duration of cruise flight at the maximum flight speed, fuel cooling capacity (heavy hydrocarbons, LH_2), etc.

E. Bearing Support Heat State in Ramjet Mode (Windmilling of Core)

The bearings used in the supports of the TRJ gas turbine core define the engine's reliability to a high degree. Reliable and long-term operation of the bearings is possible only when the bearing details temperature is lower on 40÷50°C than the tempering temperature of the bearing's material.[2]

For cooling and lubrication of the aviation engine bearings usually synthetic oils are used. Rising flight speed leads to increase of inlet engine air temperature and heat loads on the units and structure elements of the engine. The heat flow into the engine's supports grows since oil with limited operating temperature is used for cooling. When flight speeds are higher than M=2.5÷3, operability of gas turbine engine supports should be achieved by using special structure measures. One of the most effective methods to ensure bearing supports operability is the use of active thermoprotection. In essence it is the setting up of a protective screen over the support case under which conditioned air from the engine cooling system is supplied. The layout of the

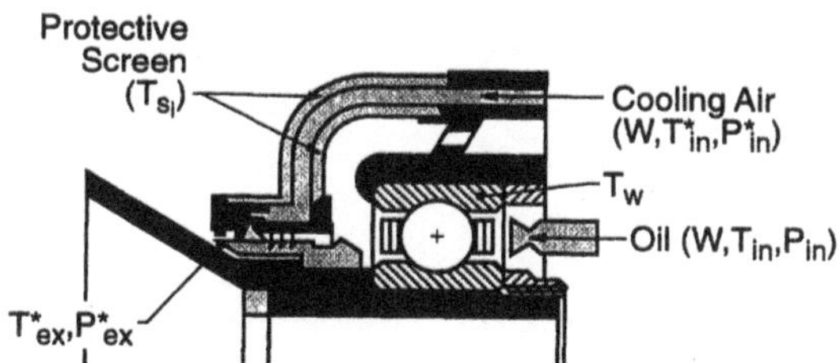

Fig. 32. Layout of bearing unit

experimental TRJ support is shown in Fig. 32. All supports of the experimental TRJ had an autonomously controlled oil supply and a thermoprotective screen under which cooling air with controlled parameters was supplied.

Some results of heat state investigation of the experimental TRJ bearing supports in the ramjet operational mode with windmilling of core are discussed. The summary value of the heat flow into the oil cavity of the supports depends on 1) heat of friction in bearings and seals, 2) heat flux through the details of bearing supports, and 3) heat flux from hot air, which penetrates from engine flow path through seals into the oil cavity.

The experimental investigations of the TRJ supports heat state showed that along with the TRJ transition to the ramjet mode with windmilling of core the redistribution of the relative values of each of the afore mentioned heat components takes place. It has been determined that part of the friction heat in the bearings decreases because of the decrease of both the rotor rpm and the axial forces. Heat flux through the details of supports and heat coming from hot air penetrating into the oil cavity components rise because of the inlet temperature increase and aggravation of the seals operating condition. The total heat flux into the supports, as compared with the gas turbine mode rises (approximately twice), so the support exit oil temperature also rises and reaches its limit at some flight Mach numbers (Fig. 33). Autonomous oil supply to each support of the engine in combination with air cooling supply have allowed the investigation of the influence of both oil and cooling air parameters on TRJ support operability. The experimental data of relative cooling airflow and relative support exit oil temperature depending on relative cooling air pressure are shown in Fig. 34. At the relative pressure ratio $(\underline{P}^{*}_{col}/\underline{P}^{*}_{in})>1.7$ the rising cooling airflow is stopped, and the decrease of support exit oil temperature is also stopped. The cooling air supply under the screen space significantly lower the support exit oil temperature and ensures the engine transmission operability at the flight Mach numbers $\underline{M}>2.5\div3$. The use of the thermoprotective screen, under which the cooling air is supplied, ensures the engine supports operability up to $M = 4\div4.5$.

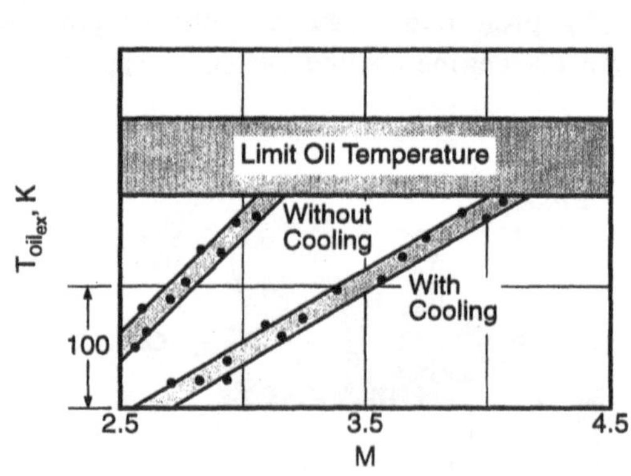

Fig. 33. Support exit oil temperature vs flight Mach numbers

F. Turboramjet Engine with Turbofan Engine in the Ramjet Operation Mode

The operating parameters, engine performances, and control principles of the TRE types based on the TF core (TFRJ and $TFRJ_S$) at the high Mach numbers flight (transition and ramjet operation modes), including some experiments with imitation of these conditions, are considered next.

1. Fan Performance in the Ramjet Mode

To maintain rather low losses in the air intake with flight Mach number increase, the fan corrected airflow should be decreased, as it follows from the equality of airflows in the stream at infinity before the air intake and in the fan at the areas ratio $(f_{in} \cdot Ai_{in}/A_f)$,

$$(W_{fcor}/W_{fcoro}) = \frac{P_o\,(k/R \cdot T_o)}{(W_{fcoro}/A_f)} \; \frac{M \cdot (T^*_a / T_a) \cdot f_{in} A_{in}}{\sigma \cdot (P^*_a/P_a) \cdot A_f}$$

where (W_{fcor}/W_{fcoro}) is the relative corrected airflow in the fan. This tendency is demonstrated particularly when the area ratio becomes constant.

Let us consider the peculiarities of the fan operation at low--rotational speed (n_{cor}/n_{coro}<0.4). The typical fan characteristic at low rpm (curve A) is shown in Fig. 35 (it is simplied the conventional fan stage which reflects the low staged fan operating process). At point 1 the flow angle of attack at the mean radius is of some positive value, the difference of tangential velocity components ΔC_u>0, the fan expended work is positive, and the fan pressure ratio value is near 1. Reducing the counterpressure behind the fan, the operating point is shifted to the right down on the characteristic, and this is followed by stretching the velocities triangles and decreasing the angle of

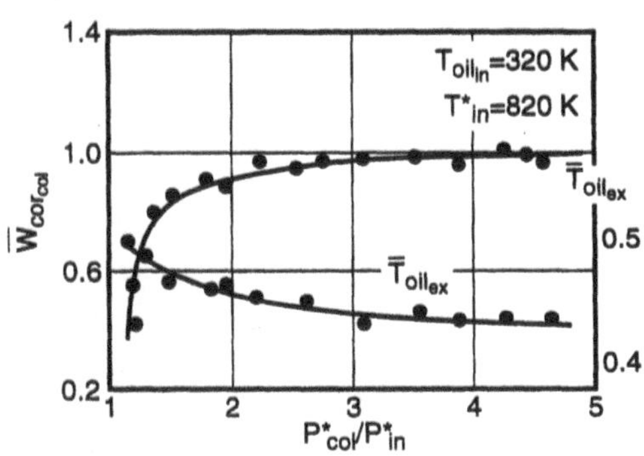

Fig. 34. Relative support exit oil temperature and cooling air flow vs air pressure ratio

attack, the ΔC_u, the expended work, and the PRC_f. Point 2 corresponds to the case when the ΔC_u and expended work values are equal to 0, i.e., the fan operates at the free windmilling mode and presents the hydraulic drag in airflow ($PRC_f<1$). The combination of such points at various rotational speeds gives a line of fan windmilling mode operation (curve B). With further shift of the operation point down along the characteristic (up to exit section choking) the transition to turbine operation mode takes place. At point 3 the ΔC_u and expended work values are negative, and it is possible for the fan to provide some power. The example of zero work (free windmilling) performance lines for the fans with different design pressure ratio values is shown in Fig. 36. The lower the PRC_{f0} is, the less pressure losses there are, and the wider is the range of air flow in the fan in windmilling mode. At design pressure ratio values 2÷2.5 and $(W_{cor}/W_{cor0})<0.4$, the total pressure drop in the windmilling fan is of the same order as the pressure losses in the separate ramjet duct of the TRE.[6] At the RJ operational mode the airflow in the TRE inner part (gas turbine core flow path) is near zero. At these conditions the turbine becomes the consumer of power being produced by the windmilling fan. The turbine resistance to rotation is defined by the so-called ventilation losses, which arise as a result of energy fed to the airflow in the turbine flowpath and also by the turbine disk friction losses. The turbine driving work is proportional to the air density and to rpm in the power of about 2.7. The analysis of a turbofan working in typical RJ operation mode condition has shown that because of low rpm an additional pressure decrease in the fan driving the turbine is usually small.[6]

It is possible to use the windmilling fan for driving engine accessories at the RJ operational mode. The expediency of using the fan to drive the PS accessories at the ramjet mode depends on its efficiency as a turbine, the values of the takeoff power, the thermal condition, and other factors. When loading the windmilling fan with external energy consumers,

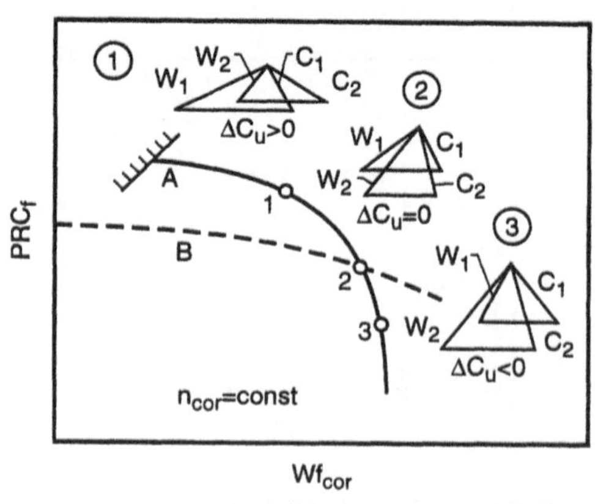

Fig. 35. Fan stage parameters variation along n_{cor} line

the operating points on the fan map fall beneath the zero work line. At the TFRJ ramjet operation mode, however, this deflection is usually insignificant, and it is possible to adopt the zero work line as an operating line for the RJ operational mode. In some cases when the fan design pressure ratio is rather high, it is possible for the TFRJ thrust performance to improve at the RJ operational mode by arranging the bypass duct around the fan case corresponding controlled shutoff device.

The possible use of the windmilling fan to drive the accessories at the RJ operational mode is a significant advantage of the TFRJ and $TFRJ_S$.

2. Turbofan-Ramjet Engines Transition to Ramjet Mode

As an example let us consider the change of parameters and characteristics of turbofan-ramjet engines at the transition to the RJ operational mode (Figs. 37 and 38). Points B and E correspond to the start and the finish of the transition to RJ mode; point W corresponds to the begining of the fan windmilling mode.

The engine transition to the RJ operational mode, in essence, consists of switching off the gas turbine part without engine thrust loss. For the TFRJ and $TFRJ_S$ the gradual transition process is typical and takes place in some range of flight Mach numbers. The transition process is begun at Mach number $\underline{M_B}$ when the engine airflow at maximum rating approaches the inlet maximum airflow capacity (point B in Fig. 37) and finishes at flight Mach number M_E when the whole (or practically the whole) airflow is fed through the outer (ramjet) engine part (β=1). With the flight speed increase in the range of $M_B<M<M_E$ it is necessary to gradually decrease the fan corrected airflow. To lower the TF airflow, it is necessary to decrease the engine rpm. It is expedient to use a rather simple method of rpm lowering by gradually decreasing the turbine inlet temperature up to the engine going to windmilling mode. In this case decreasing the nozzles throat areas or the common nozzle throat area becomes necessary just after cessation of the combustion chamber fuel supply.

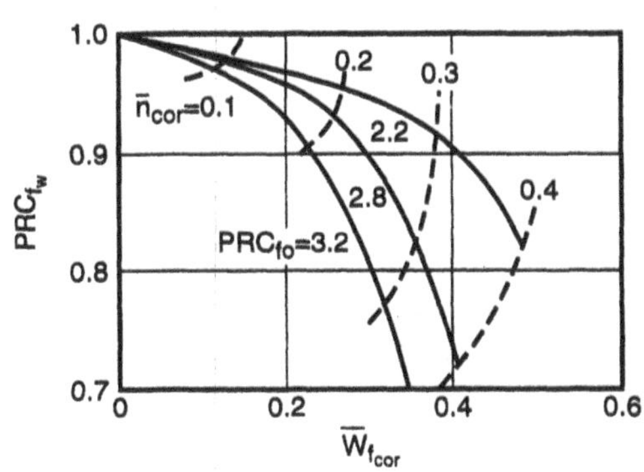

Fig. 36. Fan zero work lines vs design fan pressure ratio

Turbofan-ramjet engine with duct afterburner-ramjet combustion chamber. For selected $TFRJ_S$ (solid lines in Figs. 37 and 38) the inlet maximum airflow capacity is reached at the $\overline{M}$=2.7. Along with further flight speed increase the engine-inlet matching for maintaining σ_{max} is provided by a corresponding decrease of fuel feed to the combustion chamber. The inner and outer nozzles throat areas $\overline{An}_{th1}$ and $\overline{An}_{th2}$ at the beginning are maintained constant. As a result at $\overline{M}>\overline{M}_B$ the temperature ratio $\overline{T^*_g/T^*_{in}}$ and the turbo-fans rpm quickly decrease. Maintaining the maximum $\overline{ARCC}$ flow capacity in conjunction with decreasing of the airflow in the engine gas turbine part leads to the quick increase of the bypass ratio and the β value, i.e., the engine operation mode draws near the RJ mode (Fig. 37). As a result, the overall heat addition in the engine increases because of maintaining the ARCC rating at the minimum air-to-fuel ratio. Decrease of the engine nozzles expanding ratios, however, is found to be the prevailing factor in this case, and as a result the engine specific thrust slightly decreases. For this reason the thrust increase rate is somewhat lower when $\overline{M}>\overline{M}_B$; the SFC increases simultaneously (section B-W in Fig. 38).

After going to windmilling mode (point W at $\overline{M}$=3) the engine governing is provided by the optimum variation of the nozzles throat areas. At the flight conditions corresponding to $\overline{M}_W$ the β value is approximately equal to 0.9, i.e., about 10% of the overall airflow goes through the engine inner part (Fig. 38a). When the flight Mach number is higher than $\overline{M}_W$ it is expedient to decrease the airflow in the engine gas turbine part and its drag by decreasing the $\overline{An}_{th1}$. For this purpose it is possible to use, for example, the inner nozzle central body shifting. The nozzle area decrease which corresponds to the inlet-engine matching with minimum pressure losses and constant outer nozzle throat area is shown in Fig.38a (section W-E in Fig.37).

Along with $\overline{An}_{th1}$ decrease, the turbine expanding ratio and work become lower, which leads to both the fan rpm and the pressure ratio decrease (Fig. 38a). When the flow in the LP turbine becomes under critical, the inner ozzle area decrease provides the HP turbine work decrease along with core rpm and corrected airflow lowering. As a result, the engine airflow redistributes in

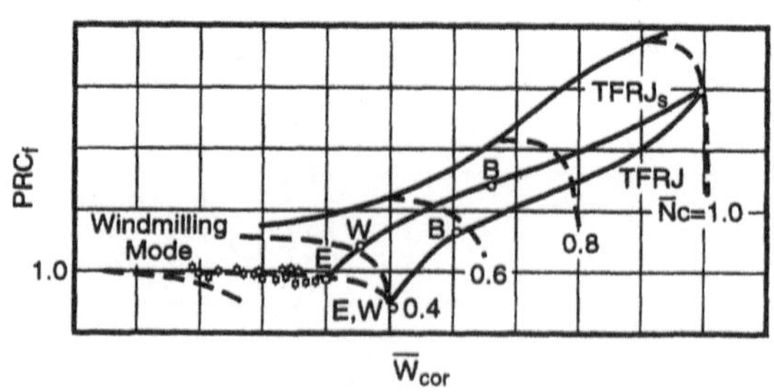

Fig. 37. Fan map

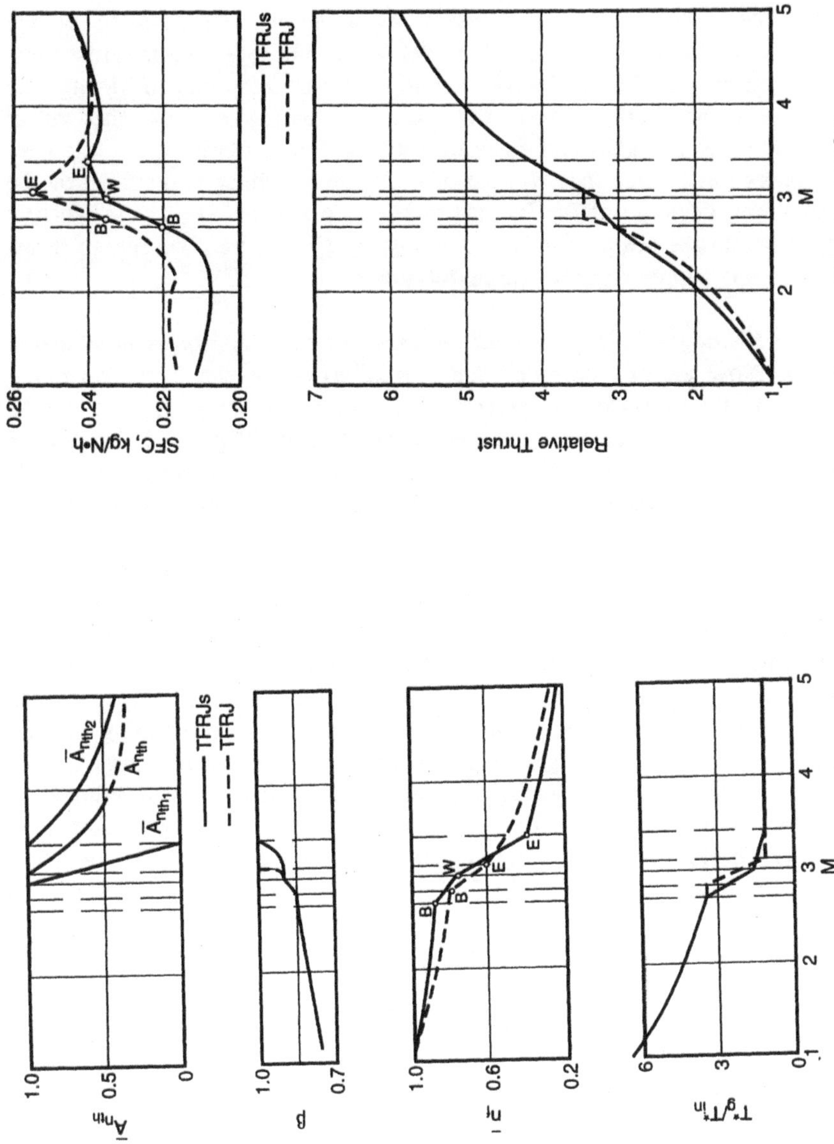

Fig. 38. Characteristics of turbofan-ramjet engines at the transition to ramjet mode ——— **TFRJs** – – – – **TFRJ**

favor of the RJ part with the full gas turbine part shutoff ($\underline{An_{th1}}$=0, b=1). At this moment (M=3.4 in Fig. 38) the transition to the RJ operational mode is being completed. The corresponding operating point on the fan map is found to be practically on the zero work line (point E in Fig. 37).

With lowering of the inner nozzle area, the engine specific thrust increases because of the core losses decrease and the increase of the overall heat addition. For this reason, in the range $\underline{M_W}\div\underline{M_E}$ engine thrust rises, and SFC varies slightly. At the flight conditions corresponding to $\underline{M}=\underline{M_E}$ the TFRJs parameters differ from the pure RJ parameters by the value, corresponding to the additional pressure losses in the windmilling fan. At the RJ operational mode ($\underline{M}>\underline{M_E}$) the inlet-engine matching is provided by the corresponding decrease of the RJ part nozzle throat area $\underline{An_{th2}}$. The fan rpm and pressure losses are gradually decreasing ($\underline{PRC_f}$ >1), and the engine thrust performance is approaching the RJ performance.

Turbofan-ramjet engine with common afterburner-ramjet combustion chamber. Now let us consider some peculiarities of the TFRJ operating process when the transition to the RJ mode is carried out using, as an example, the engine with the same cycle parameters and transonic ($\underline{M}$=1.3) to hypersonic ($\underline{M}$=4÷5) thrust ratio as just discussed. For this condition the area ratio $\underline{A_{in}}/\underline{A_f}$ would be some higher than for the TFRJs engine type (in correlation with transonic specific thrust values).

The TFRJ operating parameters variation peculiarities are conditioned mainly by the mutual influence of gas and airflows at the common ARCC entry. As a result, the core pressure ratio decrease vs flight speed, which is not compensated by the fan duct outlet pressure variation, the LP turbine specific work decreases. Because of the fan duct flow influence the fan operating line on the fan map deflects down from the corresponding line for $TFRJ_S$, and bypass ratio vs flight Mach number rises more rapidly (Figs. 37 and 38). Meanwhile the corrected airflow variation vs flight Mach number is approximately the same as for the $TFRJ_S$. At the higher $\underline{A_{in}}/\underline{A_f}$ ratio in the PS with a TFRJ, the maximum inlet capacity condition is approached at some higher flight Mach number than in $TFRJ_S$ (Fig. 38), and the corresponding fan pressure ratio is nearer to 1. Then, since the transition is begun the TFRJ operation mode is nearer to the RJ mode. The turbine inlet temperature lowering at $\underline{M}>\underline{M_B}$ leads to β value rise up to almost 1 when TFRJ approaches the windmilling mode. The corresponding fan operating point lies practically on the fan zero work line (Fig. 37). Then, in the case of TFRJ, the completion of transition to the RJ mode practically coincides with the moment of ceasing the fuel feed into the combustion chamber. As a result, $\underline{M_E}$ number for TFRJ is lower than for the $TFRJ_S$ (in the example considered, $\underline{M_E}$ = 3.1 against 3.4, respectively, Fig. 38).

Rather fast TFRJ transition to the RJ mode is a consequence of the simultaneous effect of turbine inlet temperature decrease and gas dynamic influence of fan duct flow on the gas turbine part performance.

TFRJ thrust performance variation vs flight Mach number is nearly the same as in the case of $TFRJ_S$. Any difference is connected with fan efficiency decrease and higher fan duct pressure losses. At the RJ operational mode ($\underline{M}>\underline{M}_E$) the inlet-engine matching is provided by the corresponding nozzle area decrease, and this is accompanied by rpm and engine flow path pressure losses decreasing (dotted lines in Figs. 37 and 38).

When analyzing engine parameters variation at transition to RJ mode, it was assumed that the air inlet operates at maximum capacity rating. This approach, although not accounting for possible inlet capacity limitations when $\underline{M}<\underline{M}_{max}$, permits simplifying the analysis and revealing the main correlations. The account of the real variable geometry inlet performance characteristics would not change the engine parameters variation in principle but, in the case of lower inlet capacity, the transition flight Mach number range would be displaced to the lower flight speeds.

The TRE thrust performance differences at the ramjet operational mode and that of the pure RJ are conditioned mainly by the additional pressure losses in the engine flow path. The TRE performance aggravation compared with pure RJ at hypersonic flight speeds is possible when joined with higher airflow or other agent expenditure for engine structure cooling. Some engine thrust performance aggravation is also possible in connection with the part of the airflow energy used for auxiliaries drawing. In this case, the use of the windmilling turbofan would be possible. When the power takeoff is low, the auxiliaries drawing this way would be expedient, in spite of the rather low fan efficiency when running in turbine mode.

IV. Conclusion

Performed theoretical investigations and computational modeling of air-breathing PS for $\underline{M}=4\div6$ have included the classification and performance analysis of the turboramjet engine of various types. The approach for theoretical design of engine configuration was elaborated.

Comparative analysis of the merits of various ABE types being installed on future hypersonic vehicles with cruise and acceleration flight trajectories has revealed the TRJ and TFRJ advantages as universal variable cycle engines for high Mach number flight.

Optimization of specific TRE conditions for transition from TJ or TF to RJ modes along with other investigation results give the approach for governing TRE inflight operation and also give the base for formulating the requirements of the experimental TRE and the problems of their investigations.

In the framework of a wide program of engines for high-speed flight study which was carried out in Russia in 1960-1980, the investigations of advisable engine concepts, operating with the process parameters of the TRE, were conducted. The experimental full-scale TRE of different types assembled from the units of production gas turbine engines were tested at the CIAM special test facility at simulated flight conditions, corresponding up to the Mach numbers M=4.0÷4.5.

The results of theoretical and experimental investigations of TRE have led to the definition of 1) the rational methods of the transition from gas turbine to ramjet modes, 2)the conditions of stable operation of the PS with the TRE at the transition mode, 3) the characteristics and pressure losses at the windmilling mode for the TRJ and TFRJ turbocompressor, 4) the expediency of the application of windmilling mode at the ramjet operation mode, 5) performance of the ARCC at the distortion of the ARCC inlet total pressure fields, 6) conditions of operation of the ARCC with air cooling system and hydraulic characteristics of cooled ARCC, 7) influence of ARCC air cooling on the TRE thrust performance, 8) heat states of the ARCC with perforated screen and engine bearing supports, and 9) possibilities of heat protection of bearing supports by controlled oil and cooling air supplies.

The theoretical and experimental results achieved in the TRE study program, along with the well-known Russian experience of LH_2 use in ABE and FV [4,5] and also in EF study[13] give the initial base for the next steps in developing TRE for future hypersonic planes.

Solving the complicated problems in developing new aviation techniques requires high expenditures. International cooperation in the building up of scientific-technical base for advanced PS is one of the important factors to successfully realize the projects of 21th century air transport.

Acknowledgments

The authors acknowledge their collegues for their contribution during the work and preparation of this paper and thank the Directorate of Central Institute of Aviation Motors (CIAM) for permission for publication. The authors are indebted to CIAM colleagues for the available materials and help of its preparation: L. Yanovski (II.F) and A. Antonov (II.G).

References

[1]Sosounov, V.A., Tskhovrebov, M.M., Solonin, V.I., and Palkin, V.A., "The Study of Experimental Turboramjets", 28th Joint Propulsion Conf., AIAA paper 92-3720, 1992, Nashville, TN, July 6-8, 1992.

[2] Sosounov, V.A., Solonin, V.I., Tskhovrebov, M.M., Kadjardouzov ,P.A, and Palkin, V.A., "The Study of Experimental Turboramjets: Heat State and Cooling Problems", 29th Joint Propulsion Conf., AIAA paper 93-1989, Monterey, CA, June 28-30, 1993.

[3] Sosounov, V.A., and Tskhovrebov, M.M. (eds), Turboramjet Engines for Hypersonic Flight Speed, Aspects of Aviation Science and Technique, Ser. Aviation Engine, Vol. 3, 1993, in Russian.

4 Sosounov, V.A., "Some Aspects of Hydrogen and Other Alternative Fuels for Application in Air-Breathing Engines", Ninth International Symposium on Air Breathing Engines, Athens, Greece, Sept., 1989.
5 Sosounov, V.A., "Study of Propulsion for High Velocity Flight", Tenth International Symposium on Air Breathing Engines, Nottingham, UK, Sept., 1991.
6 Shlaykhtenko, S.M., and Sosounov, V.A. (eds), The Theory of Turbofan Engines, M, Mashinostroenie, 1979, in Russian.
7 Rotmistrov, N.Yu., Tskhovrebov, M.M., "Variable - cycle engines for boosting - cruising vehicles", International Astronautical Federation Congress, IAF-88-249, Bangalore, India, 1988.
8 Kurziner, R.I., Jet Engines for High Supersonic Flight Speed. M, Mashinostroenie, 1989, in Russian.
9 Tskhovrebov, M.M., and Palkin, V.A.,"Combined Engines for Hypersonic Flight", 18th International Congress of Aeronautical Science, ICAS-92-3.4.3., Sept., 21-25, 1992, Beijing, People Republic of China.
10 Tskhovrebov, M.M., About Optimal Mach Numbers of Combined Engine Transition on Ramjet Operation Mode. CIAM, No. 542, Moscow, Russia,1972, in Russian.
11 Zhorov, Yu.M., Modelling of Physical-Chemical Processes of Oil Processing and Oil Chemistry. M, Chemistry, 1978, in Russian.
12 Nixon, A.C., Endothermic Fuels as Heat Sink for Hypersonic Flight. Space Aeronautics, Vol.47, No. 5, 1967, pp. 112-114.
13 Yanovski, L.S., "Endothermic Fuels for Hypersonic Aviation", AGARD Conf., Proceeding 536, Fuels and Combustions Technology for Advanced Aircraft Engines. 81 Symposium, Fuiggi, Italy, May 10-14, 1993, pp. 44.1-44.8.
14 Antonov, A.N., and Chivanov, S.V., "Investigation into Dynamic Characteristics of Aviation Gas Turbine Engine Heat Exchangers" Proceeding of the 1st ICHMT International Numerical Hear Transfer Conference and Software Show, Guildford, Surrey, England, pp. 136-144, July, 22-26, 1991.

DEVELOPMENT STUDY ON AIR TURBORAMJET

Nobuhiro Tanatsugu*
Institute of Space and Astronautical Science, Kanagawa, Japan

Nomenclature

A area, m^2
C_1 correlation coefficient in Eq.15
C_p specific heat at constant pressure, J/(kg K)
C_b blockage ratio
C_x optimum design parameter, W/Pa
d_e equivalent diameter, m
d_i inner diameter of heater tube, m
d_o outer diameter of heater tube, m
d_t ring diameter, m
f friction factor
G mass flow rate, kg/s
h heat transfer coefficient, W/(m^2 K)
i enthalpy, J/kg
L_p length of heater tube, m
l length, m
M mach number
N tube number
N_s number of heater tube set
N_t number of heater tube
N_u Nusselt number
P_r prandtl number
p pressure (total), P_a
Q heat transfer rate, W
q heat flux, W/m^2
Re Reynolds number
T temperature (total), K
u velocity
V valve throttling position, mm
W combustion gas flow rate, kg/s

Copyright © 1994 by the American Institute of Aeronautics and Astronautics Inc. All rights reserved.

*Professor, Space Propulsion Systems Division

Wa air flow rate, kg/s
x distance from heat exchanger front edge, m
γ isentropic exponent
Δp pressure drop, Pa
ρ density, kg/m^3
τ shear stress at wall, Pa
ϕ equivalent ratio
ΔT average temperature difference between coolant of hydrogen and combustion gas, K
ε mass capture ratio
Φ equivalent mixture ratio
η pressure recovery factor

Subscripts
air induced air
d_i inner diameter
d_o outer diameter
f flight condition
g combustion gas
h hydrogen
h_x hydrogen heater
i inner wall
i_s static condition at inlet
o heater outlet
o_s static condition at outlet
out outer wall
p heater tube
r recovery
s static
x distance from heater front edge

I. Introduction

The airbreathing propulsion system offers superior performance potential by its essentially high specific impulse. However, its low thrust-to-weight ratio and high development cost have to be considered when applying it for the space launch system. These airbreather features are preferable to the fully reusable system utilizing wing lift force. Another airbreather problem is that no one propulsion cycle concept is optimum over the entire flight regime from liftoff to orbital velocity. Some combination of two or more propulsion cycles is necessary for space launch vehicle to avoid the large weight and volume penalty caused by employing many separate propulsion systems optimum for each flight regime. This requirement is strongly imposed on the single-stage-to-orbit (SSTO) vehicle. This problem is, however, eliminated in the two-stage-to-orbit (TSTO) vehicle of which the flyback booster operates below approximately Mach 6. In this case, the air turboramjet (ATR) is one of the most preferable candidates as well as the liquid air cycle engine and the ejector ramjet. The TSTO launch system has other benefits from the viewpoints of launch capability, technology level required and development cost. The payload ratio of the TSTO system is essentially more than doubled when compared with the SSTO based on the same technology, and so it must have reduce the vehicle size and weight for the same payload requirements. To improve the payload ratio of the SSTO, we need more advanced technology on the propulsion system and the vehicle structure together with the materials, which surely increase the development cost. In the launch operation of the TSTO system, the operating time of the flyback booster is relatively short compared with the orbiter and so one flyback booster can be used for carrying many orbiters in turn. Such an operational system also reduces the development cost.

The Institute of Space and Astronautical Science (ISAS) has been engaged in the development study of the ATR engine since 1986. This work is now under way in cooperation with Ishikawajima Harima Heavy Industries (IHI) on the turbomachinery and combuster including the heat exchanger installed in it, Kawasaki Heavy Industries on the air intake including the precooler, Sumitomo Heavy Industries on the test facilities, and Mitsubishi Heavy Industries on the flying test bed. The ATR is one of the most prefeable candidates for the propulsion system of a future spaceplane. Our ATR engine system was named "ATREX (Air TurboRamjet of EXpander cycle)" after employing the expander cycle. ATREX is energized by thermal energy extracted regeneratively in both the precooler installed in air intake and the heat exchanger in the combustion chamber. ATREX works in the flight condition from sea-level(SL) static conditions up to Mach 6 at 35 km altitude. It also employs the tip turbine configuration for compactness of turbomachinery. We have tested the ATREX system at SL static conditions using the one-quarter-scale model (ATREX-500) with a fan inlet diameter of 300 mm and an overall length of 2200 mm.

The present paper addresses analysis of the ATR engine to be applied for the TSTO spaceplane, the feasibility test results of the ATREX-500 engine and the ram combuster under SL static conditions, the design and preliminary experiments of the heat exchanger, and the preliminary experiments of the air intake coupling with the precooler in the wind tunnel.

This paper presents the following: an analytical study of the ATR cycle derivatives, the test results of the one-quarter-scale ATREX engine performed under sea-level static conditions, the experimental study of the ATREX combuster, the analytical and experimental studies of the ATREX heat exchanger, and the experimental study of the precooler (intake air cooler).

II. Description of ATR Cycle

The ATR is a combined cycle engine that performs similar to a turbojet engine at subsonic flight and a fan-boosted ramjet at supersonic and hypersonic flight below Mach 6. Thus, the ATR is available to produce effective thrust with high specific impulse from liftoff to Mach 6 without switching hardwear. As shown in Fig.1, there are two primary cycles available for the ATR cycle, the gas generator cycle and the expander cycle. In both cycles, the turbine inlet temperature is decoupled from the incoming air temperature, which increases as the flight Mach number increases, thus the ATR cycle does not reduce thrust capability because of the turbine material limits on hypersonic flight. It is the primary advantageous feature on the ATR cycle. The performance characteristics of the basic ATR cycle

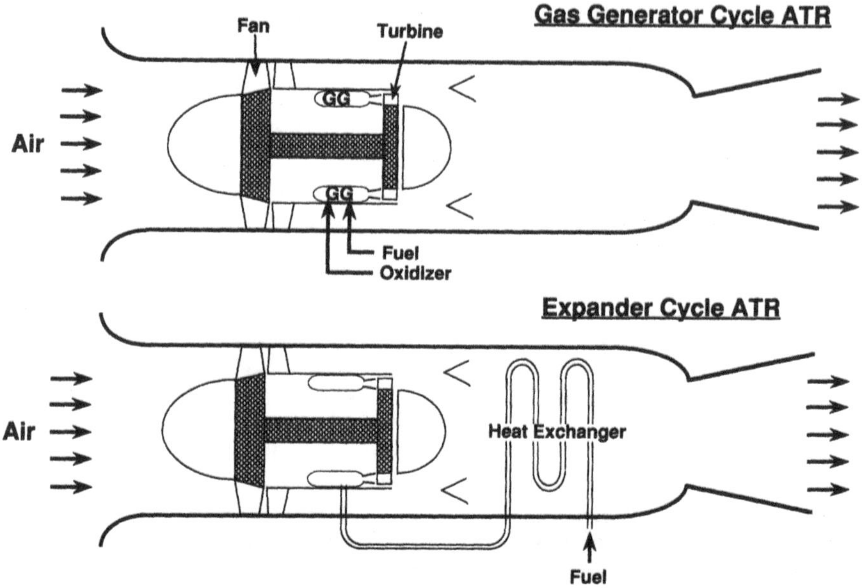

Fig. 1 Basic cycle of air turboramjet (ATR).

engine are discussed in Sec. III for comparison with other derivative ATR cycles (see G-0-type ATR engine).

III. Analytical Evaluations of ATR Cycle Derivatives[1-3]

A. ATR Engine Designs

1) *Engine description and design criteria.* The schematics of the candidate ATR engines are shown in Figs. 3-5 and Fig. 2 is shown for reference. The engines have a two-stage fan with a pressure ratio of 2.5 and adiabatic efficiency of 0.85 at SL static conditions. The fan is driven by a multistage, partial admission, tip turbine. The turbine has an expansion ratio of 6 and adiabatic efficiency of 0.5 at the design point. The main burner is located aft of the fan, and fuel is fed as exhaust gas out of the tip turbine and through one or two fuel injectors inside the burner. Stoichiometric combustion in the main burner was assumed, whenever possible. A mixture ratio other than stoichiometric was utilized when necessary to observe the design criteria discussed later.

Off-design-point performances of the fan and turbine were estimated in the conventional manner. The temperature rise, composition, thermodynamic, and transport properties of the combustion gas were computed assuming equilibrium flow. The rotational speed of the fan was kept constant throughout the entire engine operating regime. This is possible by mainly adjusting the engine exhaust throat area. Thus, corrected rotational speeds, air mass flows, and fan pressure ratios decrease at high flight Mach number.

A precooler is installed at the front of the fan. The precooler's main purposes are reduction of incoming air temperatures to protect the fan blades, and the

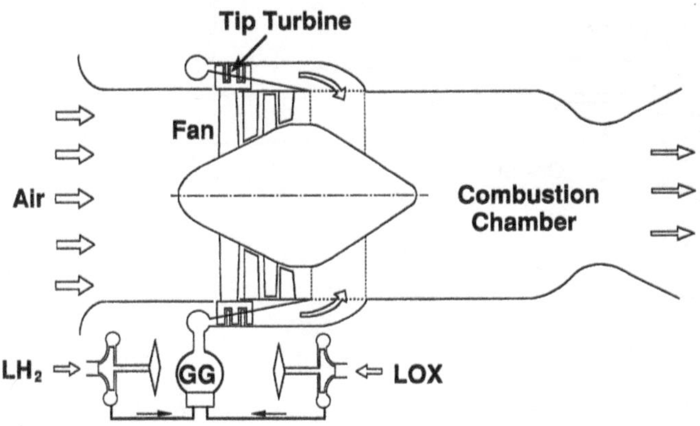

Fig. 2 G-0 engine.

increase of airflow into the engine, which results in the increase of thrust. There are additional effects of precooling that will be discussed subsequently. Calculation of the heat transfer and pressure loss was conducted at each engine operating point by dividing the heat exchanger into a number of small elements. The same procedure was followed for the in-stream hot section heat exchanger. The air intake was assumed variable, and the total pressure recovery was estimated according to MIL-E-5008C. The exhaust nozzle was also assumed variable. The nozzle exit was truncated to avoid excessive area. The following design constraints were assessed for structural integrity of the engine: 1) fan inlet air temperature, <1270 K; 2) precooler metal temperature, <1270 K; 3) turbine inlet gas temperature, < 1470 K. We assumed these limits to be realistic with near-term technology levels.

We assumed carbon-carbon (C-C) composite for tip turbine blade material. C-C composites have very low density (one-fourth of typical metallic turbine blade alloys), a good strength-to-density ratio, as well as high strength at elevated temperatures. The turbine working gas is hydrogen gas or hydrogen plus water vapor, thus vulnerability of C-C composites for oxidizing the atmosphere may not impose problems.

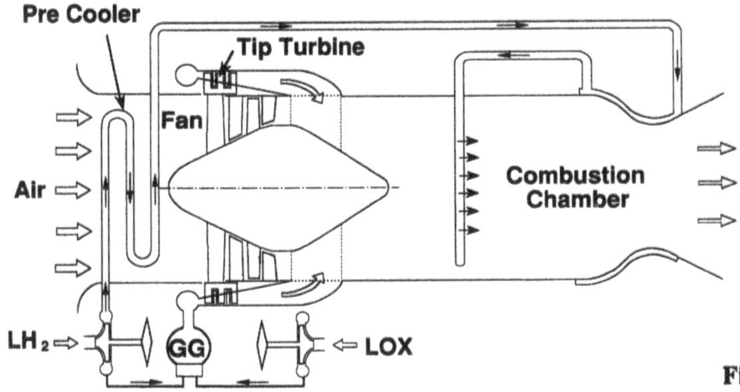

Fig. 3 G-2 engine.

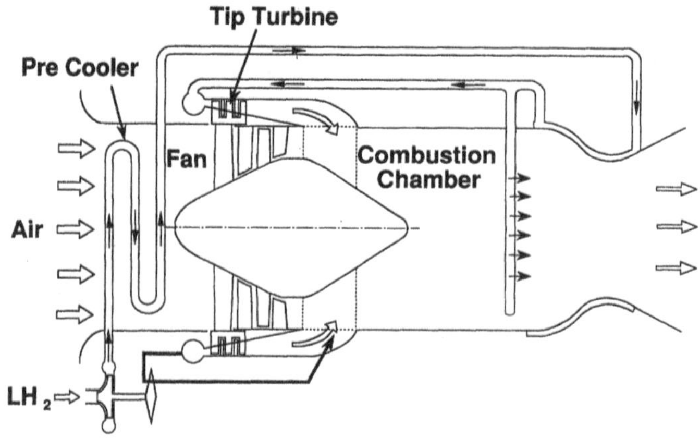

Fig. 4 E-1 engine.

2) *Variation of ATR engines.* Four types of ATR engine were considered here (Figs. 2-5).

G-0 is the basic gas-generator cycle engine without a precooler, considered only for reference purposes. Hence, design criteria 1 and 2 of the previous subsection do not apply for the G-0 engine. An equivalence mixture ratio of 1.0 (stoichiometric combustion) in the main burner was assumed at all operating points. The engine with a near-conventional metallic fan blade without blade cooling becomes inoperable at flight Mach numbers > 4.5.

G-2 is the precooled gas-generator cycle engine. At high flight Mach number, fuel flow rates are increased beyond stoichiometric combustion, and the fan inlet temperature and precooler metal temperature are kept within specified limits. In both the G-0 and G-2 engines, the LH2 and LOX pump driver turbines are driven by hydrogen gas tapped off the regenerative cooling path around the main burner and exhaust duct.

E-1 is the precooled expander cycle engine. Heating of liquid hydrogen is performed only through the precooler and regenerative cooling path around the engine. Thus, when the flight Mach number is low and the incoming air temperature is low, the gas temperature into the fan-driver turbine is too low to produce

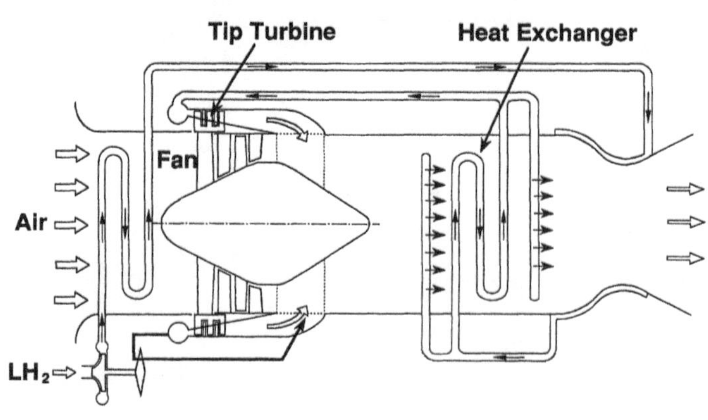

Fig. 5 E-2 engine.

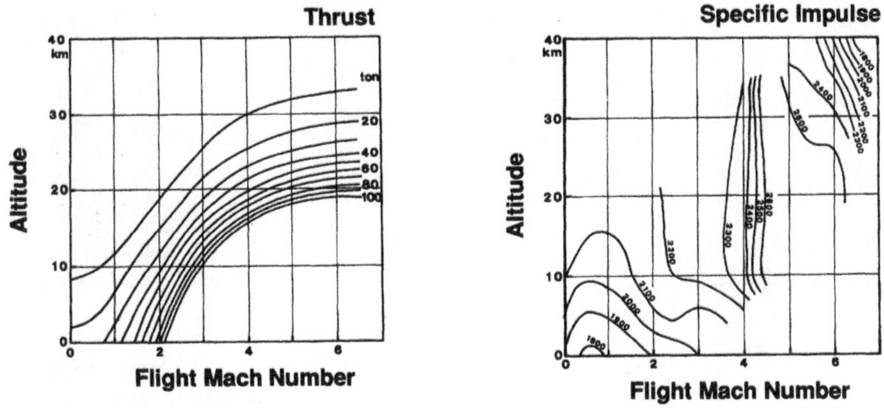

Fig. 6 Maximum thrust contour and specific impulse contour of G-0 engine.

enough energy, if fuel flow is restricted to near-stoichiometric combustion. In this case, the fuel flow is increased beyond stoichiometric, so that turbine driving power is attained.

E-2 is the precooled expander cycle engine with a heat exchanger inside the exhaust system of the engine. The main burner fuel flow is divided into two, one in front of and one behind the heat exchanger. In this way, gas temperatureat the exit of the heat exchanger, as well as metal temperature at the heat exchanger itself, is controlled within the design limit. The high turbine inlet temperature results in high metal temperaturesin the hot section heat exchangers of the E-2 engine.

The use of C-C composite, ceramics, or refractory materials must be assumed for this heat exchanger. Lowering of turbine inlet temperatures is possible with some sacrifice in the turbine design or low flight Mach number performances.

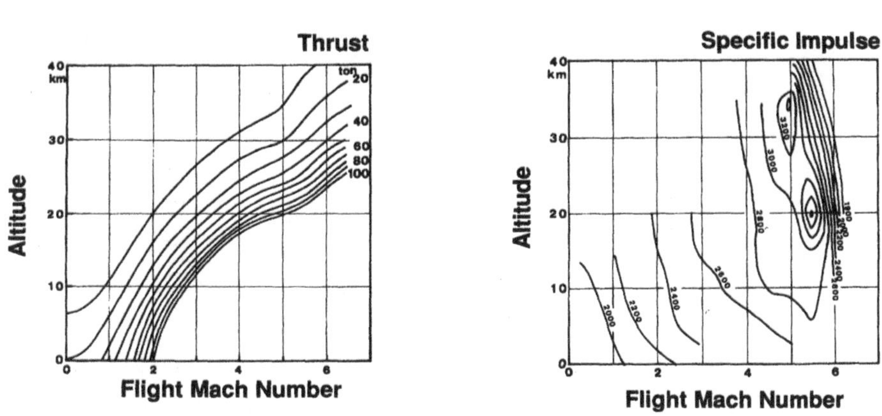

Fig. 7 Maximum thrust contour and specific impulse contour of G-2 engine.

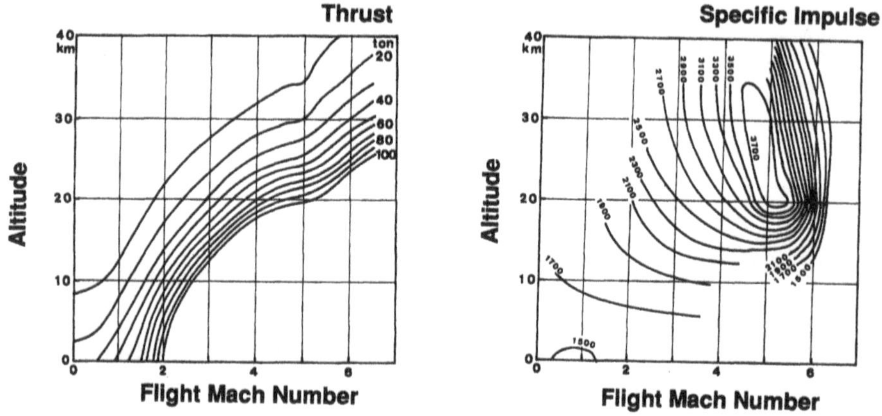

Fig. 8 Maximum thrust contour and specific impulse contour of E-1 engine.

3) *Performance characteristics of the engines.* Figures 6-9 show the constant thrust contours and constant specific impulse curves on flight Mach number-altitude maps. These performances are calculated for the engine with a fan inlet diameter of 1.3 m. In the case of G-0, constant thrust curves follow fairly closely with constant dynamic pressure curves. In the cases of G-2, E-1, and E-2, thrust curves again follow closely with constant dynamic pressure curves from sea-level static up to certain points. Beyond these points, at higher flight Mach number, fuel flow rates of G-2, E-1, and E-2 are increased to observe temperature limit design criteria, thus thrust increases more rapidly than constant dynamic pressure curves.

Among the four engine types, E-1 has the lowest specific impulse at near SL static conditions. This is because fuel flow rates were increased beyond stoichiometric to produce necessary power out of the fanturbine. It is possible that precooling must be suspended at low flight Mach numbers to prevent iceformation

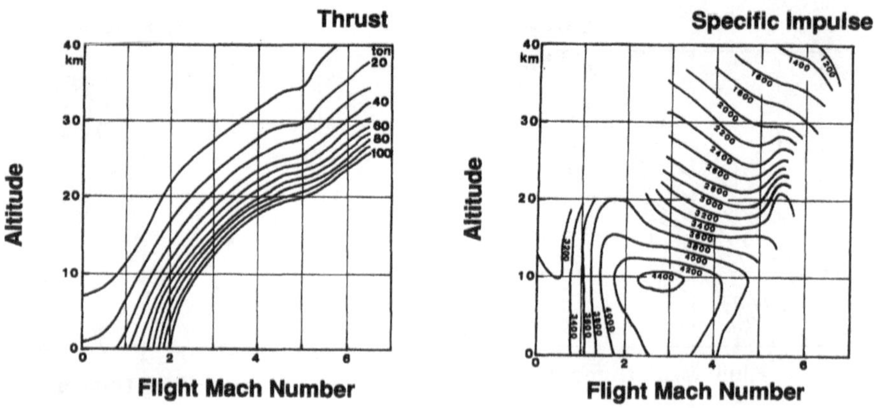

Fig. 9 Maximum thrust contour and specific impulse contour of E-2 engine.

In this case, specific impulse of E-1 is decreased further to 700 or 1000 s near SL static conditions. Effects on other engine types will be smaller.

Comparison of specific impulse characteristics of G-0 and G-2 reveals that at flight Mach number of approximately 4.5-5.5, specific impulse of G-0 and G-2 is much higher than that of G-0. In this operating regime, the precooler performs very effectively as the heat exchanger, so that the performance of G-2 approaches that of an expander cycle engine. At flight Mach numbers greater than 5.0, equivalence ratios have to be increased for cooling purposes. This tends to reduce specific impulse of the precooled engines, although thrust is increased. Figure 10 shows the variations of equivalence ratio on the flight Mach number-altitude map for the E-2 engine, for example. The decreases of specific impulse at high flight Mach number also occur by the dissociation of combustion products for all engine types, and this will impose the limit for effective use of the subsonic combustion ramjet.

In summary, precooling of the ATR engines has the following effects. the first effect is reduction of the fan inlet temperatures, thus allowing operation of the engines at higher flight Mach numbers without cooling the rotating turbomachinery. Second is the decrease of fan driving power and the increase of thermal efficiency, as a result of the intermediate cooling effect on the compression processes. Third is the increase of thrust, with some penalty in specific impulse.

Preliminary design studies to estimate engine weight suggested that the weight of the heat exchanger (precooler or internal hot section heat exchanger) is rather large, as much as 30-40 % of the weight of the basic engine (engine without heat exchangers). Thus, evaluation of the engine type must be done in the overall vehicle system studies, taking thrust, specific impulse, and engine weight into account.

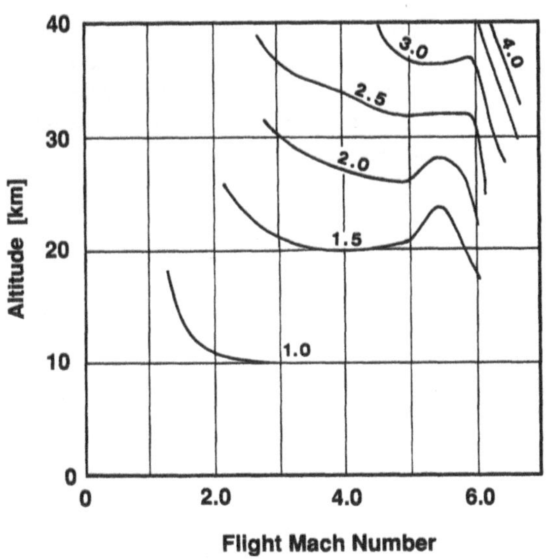

Fig. 10 Variation of equivalence ratio (E-2 type).

B. Flight-Path Analysis of TSTO SpacePlane Powered by ATREX Engine (E-2 ATR)

The performance characteristics of the airbreathing propulsion system depend largely on the flight path on which it is operated. The flight-path analyses by three-dimensional numerical simulations integrating the performance characteristics of the four types of ATR engine mentioned earlier are presented here. The flight simulation was performed for the first stage of the flyback booster and the second stage of the orbiter, respectively, in the TSTO spaceplane to obtain optimization over the entire flight path from liftoff to orbit.

1) *Flight path for flyback booster.* The flight path of the flyback booster was divided into six phases, as follows:
(1) Liftoff phase: The vehicle is acceleratedon the runway by ATR engines without any ground support facility until the vehicle can lift off using its own aerodynamic lift force. The takeoff speed is approximately 150 m/s, which depends on the wingloading at liftoff.
(2) Constant ascent angle phase: The vehicle climbs along the way of the constant ascent angle to transonic speed.
(3) Constant Dynamic Pressure Phase: This is characterized as the main propulsion phase, where the ATR engines produce nearly constant thrust, as mentioned earlier. At the end of this phase, the vehicle reaches a Mach number of 4 to 5 at an altitude above 20 km.
(4) Constant aerodynamic heating phase: Aerodynamic heating becomes a severe problem as vehicle velocity increases. In this phase, the vehicle climbs higher along the way of constant skin temperature (stagnation equilibrium temperature), where aerodynamic heating is in balance with radiation cooling.
(5) Terminal powered flight Phase: To reduce the maneuver loss of the orbiter, It is preferable that the vehicle vertical velocity is as high as possible before staging. From this point of view, the vehicle should climb using lift force and thrust pitch control as long as the ATR engine can work at the terminal phase of powered flight.
(6) Coasting phase: The dynamic pressure at ATR engine burnout is approximately 10-20 kPa, which is considerably high for the pure rocket-powered orbiter. The vehicle must climb using the lift force of the flyback booster until the dynamic pressure decreases to some extent.

The six flight phases of the flyback booster are illustrated in Fig. 11. Throughout all phases the following assumptions were adopted:
1) Initial thrust accelerationis approximately 0.6-1.0 g.
2) The vehicle utilizes the maximum thrust available in the ATR engine as long as thrust acceleration does not reach the upper limit of 3.0 g.
3) The thrust direction is always set to be coincident with the flight direction except for phase 5.

The maneuvering velocity loss does not appear in this thrust control, and ascent angle control is obtained utilizing only the lift force of the flyback booster. The orbiter does not contribute to the ascent angle control since the orbiter's wing is designed for reentry. In phase 5, the thrust is directed toward the body axis to increase the ascent angle.

From the results of flight simulation assuming 8.0 deg of constant ascent angle, 60 kPa of constant dynamic pressure, and 1200 K of stagnation equilibrium temperature, it is revealed that the G-0 engine is available up to the flight Mach

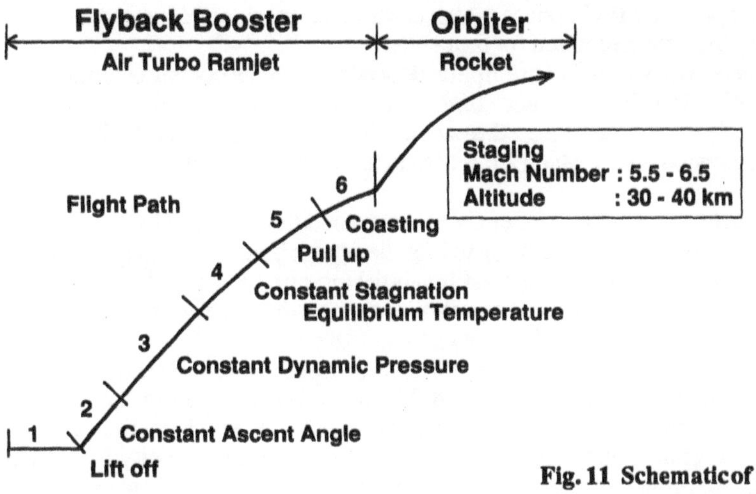

Fig. 11 Schematic of flight path.

number of 5.5 at an altitude of about 30 km, and the G-2, E-1, and E-2 engines, all of which equipped the precooler, up to a Mach number of 6.5 at 35 km. It should be remarked that these values are attained in almost horizontal flight conditions, and the staging velocity decreases when the vertical velocity is required in the terminal powered flight phase.

The required mass ratio of the flyback booster is about 1.2 using the G-0, G-2, and E-1 engines and 1.1 using the E-2 engine, and the specific impulse averaged over powered flight is approximately 2100, 2200, 1900, and 2900 s, respectively. The superior performance of the E-2 engine comes from both the high specific impulse in the lower Mach number and the high thrust in the higher Mach number.

Wing loading is also one important key parameter for the flyback booster powered by an airbreather flying inherently under considerably high dynamic pressure. In the vehicle powered by the E-2 engine with wing loading of 1250 kgf/m^2, the velocity increment of 3.8 km/s is required for the propulsion system, although the actual vehicle velocity reaches only 2.0 km/s. This difference is caused by both the gravity loss of 0.5 km/s and the drag loss of 1.3 km/s. It seems obvious that the higher wing loading is more favorable to reduce the drag loss.

In the TSTO system discussed here, the velocity increment required for the propulsion system of the orbiter is 1.6 times larger than that of the flyback booster, and the specific impulse of the orbiter is less than one-fifth that of the flyback booster. Thus the sizing of the flyback booster does not have a large impact on the entire system. The optimizing point of the entire system focuses rather on how the flyback booster can get the flight condition at the terminal phase powered by the ATR engine to minimize the velocity increment required for the orbiter. In this case, the optimization is performed between the lowest velocity increment required for the orbiter and the least number of ATR engines.

As the result of present examinations, the thrust and specific impulse requirements for the ATR engine are clarified as follows:

Large thrust is required in two parts of the flight regime. One part occurs in the transonic regime to overcome transonic drag divergence. The other occurs in the terminal powered flight phase which contributes considerably to reduction of the velocity increment required for the orbiter.

In general, total fuel consumption depends on engine operating duration as well as specific impulse, also more time is spent in the lower speed region during vehicle acceleration. Therefore, the specific impulse prefers to be higher in the lower speed region below Mach 2 rather than in the higher speed region.

2) *Flight path for orbiter.* Under the staging condition attained by the flyback booster, the velocity increment assigned for the orbiter is approximately 6.0 km/s. As mentioned earlier, the staging condition is the most important problem concerning total system optimization. The preliminary flight-path analyses of the orbiter were conducted based on the "linear tangent guidance law."

The required velocity increment from the staging condition at an altitude of 40-200-km circular orbit was calculated The result is shown in Fig. 12 with the parameters of ascent angle and Mach number at staging. As is evident from Fig. 12, the ascent angle at staging has an intensive influence on the velocity increment required for the orbiter as a result of the large velocity loss on orbiter maneuver. It seems that the staging altitude has little effect on the velocity increment.

The additional velocity requirement of 300 m/s is equivalent to an increases of approximately 10% of orbiter initial weight in the case of the 15-ton-payload-class orbiter. As a result of examination, it is revealed that the ascent angle at staging should be taken more than 15 deg to minimize the velocity increment of the orbiter.

C. Vehicle Design

The TSTO spaceplane powered by the ATREX engine was designed based on the analytical results mentioned in the previous sections. The vehicle weight was estimated by the modified WAATS computer program.[4, 5]

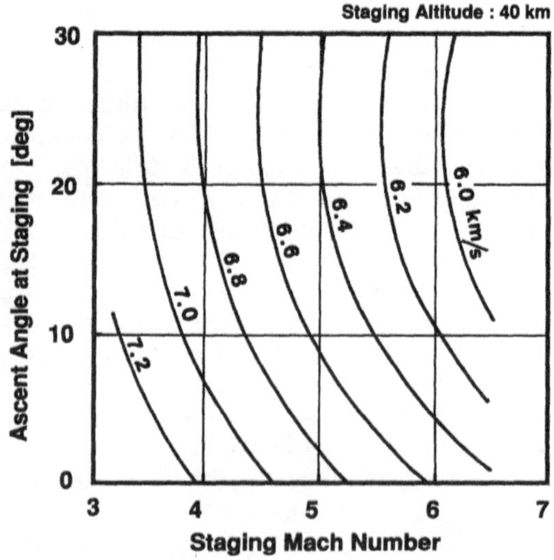

Fig. 12 Velocity increment required for orbiter.

1) *Vehicle definition and design criteria.* The vehicle is the horizontal takeoff, landing TSTO, of which the flyback booster is powered by the ATREX engine (E-2 type ATR engine) using LH_2 as a propellant and the orbiter by LH_2/LOX pure rockets. Both the flyback booster and the orbiter are manned. The payload capability to 200 km circular orbit is 15 tons, and the retrieval capability from orbit is 7.5 tons. This design assumed 0.8 g of initial acceleration, 8.0 deg of initial ascent angle, 60 kPa of constant dynamic pressure, 1200 K of stagnation equilibrium skin temperature, and a maximum Mach number of 6.5 attainable by the ATREX engine. These measurements are consistent with the engine design criteria. The precompression of incoming air by the vehicle fuselage was taken into account. Wing loading of the flyback booster and the orbiter was also assumed to be 1250 and 240 kgf/m^2, respectively.

Cooling of the fuselage is performed by radiation, and so the vehicle skin temperature is controlled below the available temperature of refractory metal. The essential elements are partially cooled utilizing LH_2.

2) *Vehicle specifications.* The vehicle configuration designed is illustrated in Fig. 13. The size and weight of the vehicle are shown in Tables 1 and 2 and Fig. 14. The initial weight of the compared vehicle is 360 tons, and the payload ratio becomes 4.2%, which is considerably high with 1.5% of the present U.S. Shuttle.

3) *Flight properties.* The typical flight properties of the flyback booster phase are shown in Figs. 15-18. The vehicle lifts off at Mach number 0.4 and climbs with a constant ascent angle of 8 deg intersecting the path of constant dynamic pressure of 60 kPa at an approximate altitude of 1 km under the transonic region. From transonic speed to Mach 5, the vehicle is accelerated along the path of constant dynamic pressure and then goes into the pull-up and coasting phase. The maximum velocity of the flyback booster is approximately 6 and the maximum skin temperature 1200 K. The vehicle is exposed an high temperatures beyond 1000 K for about 60 s, as shown in Fig. 16. Staging occurs at an altitude of 50 km and a downrange of 260 km. The duration of powered flight by the ATREX engine is 300 s.

D. Concluding Remarks

Three different cycles (one gas-generator cycle and two expander cycles) of the ATR propulsion system equipped with a precooler (intake air cooler) were analyzed and assessed for application to the flyback booster of the TSTO spaceplane. LH_2 was used as fuel. The engine design criteria were assessed based on the near-term technology level.

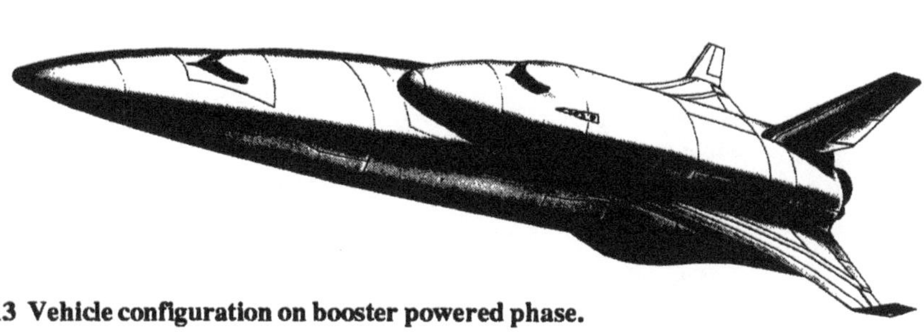

Fig. 13 Vehicle configuration on booster powered phase.

Table 1 Weight distribution of first and second stages

	Flyback booster	Orbiter
Body length, m	63.4	38.4
Wing span, m	24.2	21.8
Wing aera, m^2	290.0	182.0
Wing loading (takeoff), kgf/m^2	1250.0	--
Wing loading (landing), kgf/m^2	370.0	240.0

Precooling of the ATR engines has the following effects. First, it reduces fan inlet temperatures, thus allowing operation of the engines at higher flight Mach number. Second, it decreases fan driving power and increases thermal efficiency. Third, thrust is increased, with some penalty in specific impulse.

Preliminary estimation on engine weight suggested that the weight of the heat exchanger (precooler or heat exchanger) is rather large, as much as 30-40% of the weight of the basic engine without heat exchangers.

The flight-path analyses were performed for the flyback booster and the orbiter, respectively, revealing the optimizing point between them. Atmospheric flight of the flyback booster was analyzed divided into six phases (liftoff, constant ascent, constant aerodynamic pressure, constant aerodynamic heating, terminal powered flight, and coasting phase). The vehicle flies mostly along the way of constant dynamic pressure. Optimization of the entire system primarily depends on the ascent angle at staging, which contributes to reducing the maneuvering velocity

Table 2 Weight specification of first and second stages kg

	Flyback booster	Orbiter
Required mass ratio	1.113	4.370
Aerodynamic surface	4,200	3,200
Body surface	29,200	8,200
Environmental protection system	20,100	10,300
Launch and recovery system	6,900	800
Main propulsion system	34,200	3,200
Miscellaneous	7,400	6,600
Vehicle dry weight	102,000	32,300
Design reserve	5,100	1,600
Crew	600	600
Main propellants	36,500	166,800
Stage gross weight	144,200	201,300
Payload weight	216,300	15,000
Initial weight	360,500	216,300

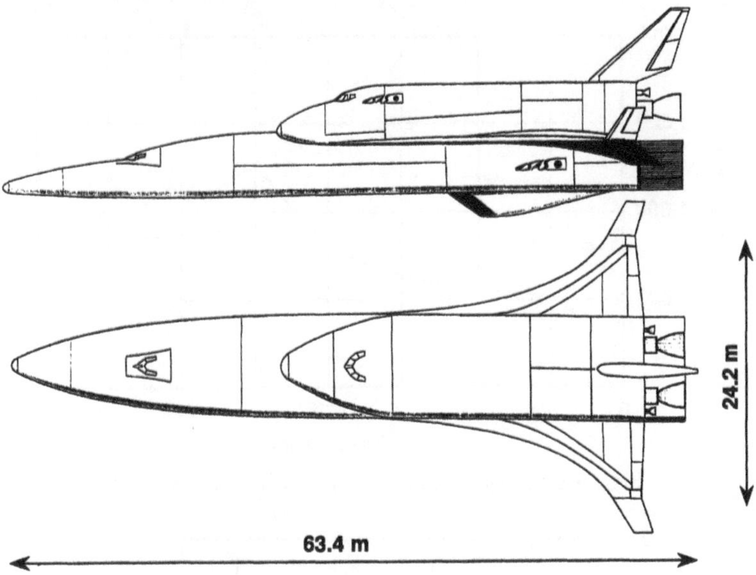

Fig. 14 Dimensions of designed vehicle.

loss of the orbiter and thereby minimizing the velocity increment required for the orbiter. The preferable ascent angle at staging is more than 15 deg. The staging altitude has little effect on this optimization.

Some requirements from the flight analyses are imposed on the performance characteristics of the ATR engine. The ATR engine is required to produce relatively larger thrust in the transonic regime to overcome the relatively large drag force and also in the terminal powered flight phase to climb by high ascent angle at high altitude. The ATR engine prefers to produce higher specific impulse in the lower flight speed below Mach number 2 where relatively longer time is needed for acceleration.

It was revealed from ATR cycle analysis and flight-path analyses that the E-2 ATR engine (ATREX engine) produced the best performance.

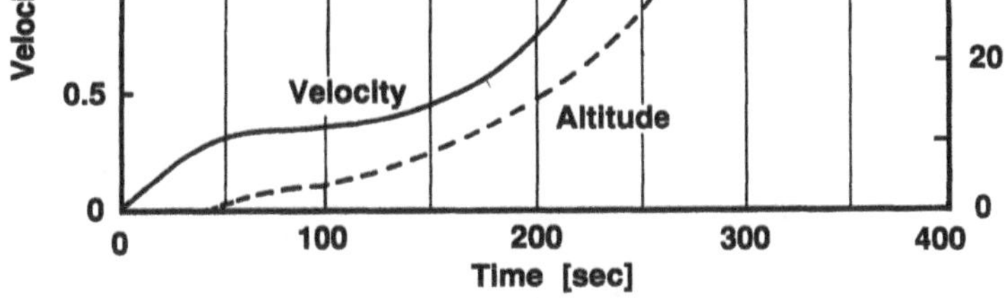

Fig. 15 Trajectory of altitude and velocity.

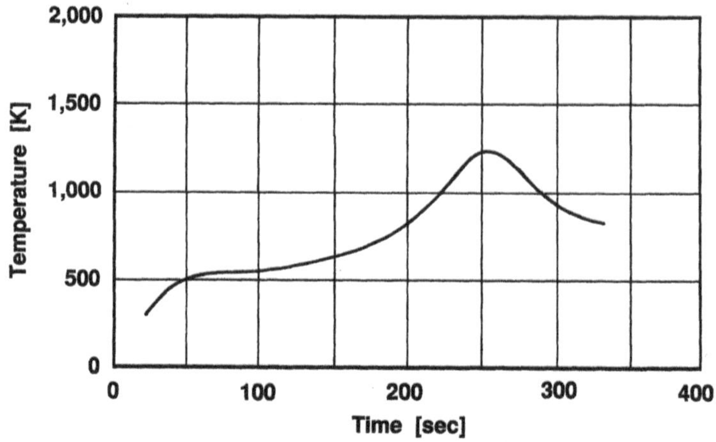

Fig. 16 Profile of stagnation equilibrium temperature.

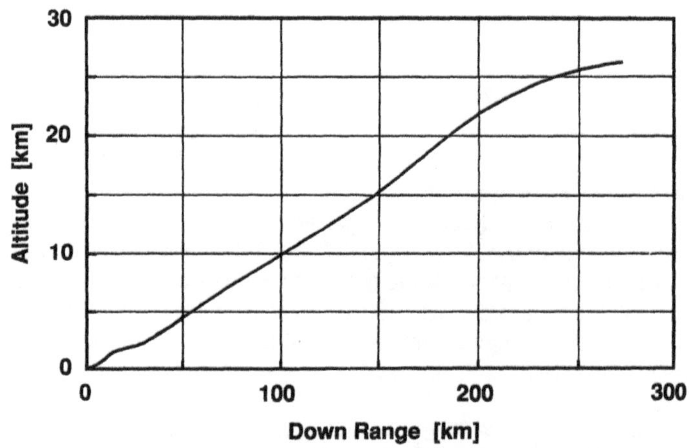

Fig. 17 Profile of altitude and downrange.

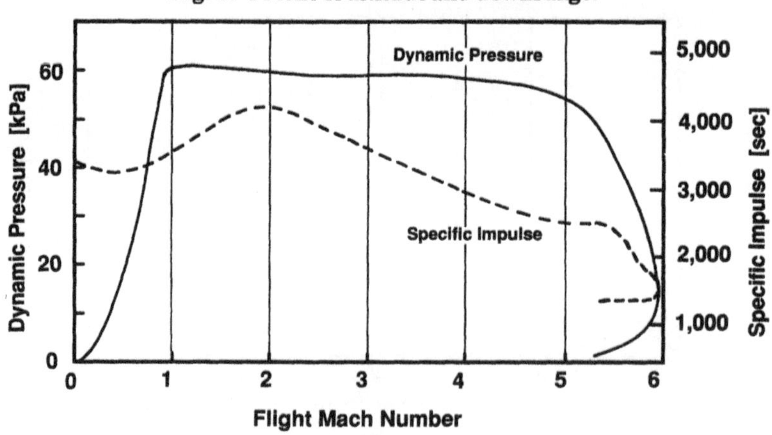

Fig. 18 Dynamic pressure and specific impulse map on Mach number.

IV. Development Study of the ATREX Engine[6]

The development study on the ATREX engine system was made to examine the feasibility of the ATREX engine as a propulsion system for a future spaceplane. This development study primarily verified the performance characteristics and functions of the ATREX system, and the engine system operation as follows:
1) Overall system performance characteristics (thrust and specific impulse)
2) Mechanical characteristics of turbomachinery composed of tip-turbine/fan configuration which is suspended by ceramic bearings (vibration and sealing problems)
3) Combustion characteristics coupling with fan, turbine, and heat exchanger, which are coupled closely in the expander cycle
4) Heat transfer characteristics of the heat exchanger installed in the combustion chamber
5) Engine system operations at startup and shutdown transients and thrust maneuvering operation

A. Development Study Approach

ISAS and IHI have examined in cooperation the various cycle derivatives of the ATR propulsion system as mentioned in the previous section, of which the E-2-type expander cycle ATR with the precooler was selected because of its highest performance. This ATR engine was named "ATREX" after employing the expander cycle.

The development study on the ATREX engine was initiated to assess its feasibility in 1988 based on the three-step verification program as follows (Fig. 19): 1) sea-level static test, 2) hypersonic simulation test with wind tunnel, and 3) actual flight test with a flying test bed.

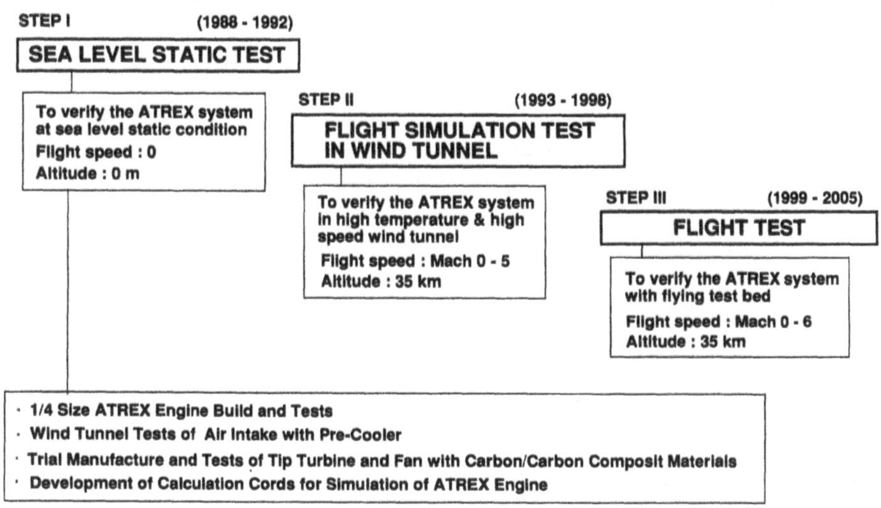

Fig. 19 Approach of development study of ATREX engine.

The first step was initiated in 1988 and accomplished in 1992. The primary objective of the first step was to assess the feasibility of the ATREX engine system by using the sea-level static test. Another study works in the first step was to examine the air intake with the precooler in the usual wind tunnel, to manufacture for trial the tip turbine and the fan blades-disk with the carbon-carbon composite materials, and to develop the calculation cords to simulate the ATREX engine flying in a practical flight path.

B. Configuration of ATREX Engine for Sea-Level Static Test

The flow diagram of the ATREX engine system is shown in Fig. 20 which is same as the E-2-type system mentioned in Sec. II. The ATREX has a two-stage fan driven by three stages of tip turbine. The main burner is located aft of the fan, and the fuel is fed as exhaust gas out of the tip turbine and through two fuel injectors inside the combustion chamber. There are two heat exchangers installed in the engine, one is located in the air intake, the so-called precooler or intake air cooler, and the other in the combustion chamber. The regeneratively cooled chamber wall works also as a heat exchanger.

The tip turbine configuration is employed to eliminate complexity of turbomachinery, thereby improving the thrust-to-weight ratio. The target thrust-to-weight ratio of ATREX engine is estimated at approximately 15 to 20 at SL static conditions.

The one-quarter-model of the practical ATREX engine was built for the SL static test. This subscale engine produces the maximum thrust of approximately 500 kg at SL static, thereby designated the "ATREX-500." The ATREX-500 engine has the same system as the practical ATREX engine except for the following three parts: the precooler is not assembled and the bell mouth is equipped instead of the air intake diffuser; the combustion chamber is cooled by water; and the thrust nozzle is the fixed configuration.

The overall configuration of the ATREX-500 and the close schematics of turbomachinery are shown in Figs. 21 and 22. The turbine consists of three stages

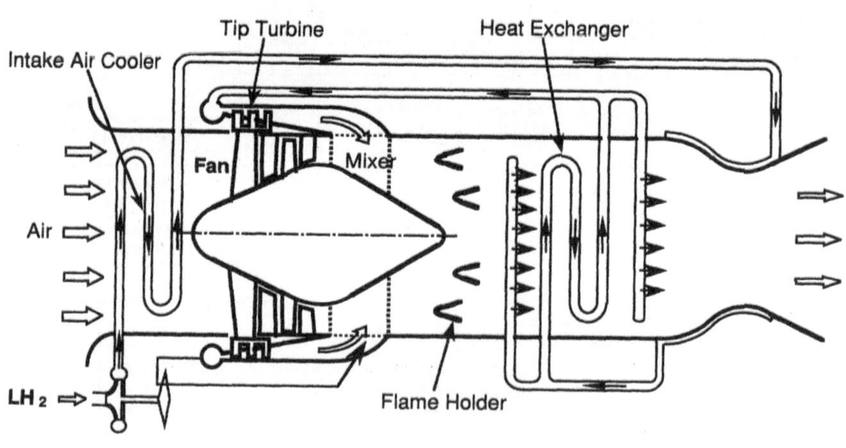

Fig. 20 ATREX engine.

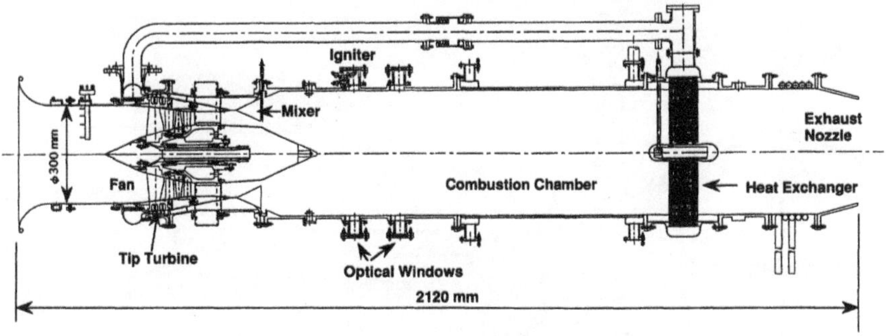

Fig. 21 Schematic of ATREX-500 engine.

and the fan two stages. The turbine rotor blades and ring disk are made as one body and integrated on the peripheral tip of the first stage fan (see Fig. 22). The shroud of the first-stage fan is extended to the fan inlet. A part of the turbine disk is also extended to the same direction. The edges of both ring extensions are coupled by using pin joints, as shown in Fig. 23. This coupling configuration allows the deformations caused on the turbine and the fan assembly by thermal expansion and centrifugal force. In the radial separate configuration similar to this, the centrifugal force worked on the turbine blades appears as the hoop stress of the turbine ring disk, thus only the rotational torque and the axial thrust of turbine are transmitted to the fan blades. Titanium alloy is utilized for the turbine rotor disk and fan rotor blades (Fig. 24).

Two pairs of labyrinth seals are equipped at the inlet and outlet sides of the turbine, respectively (see Fig. 23). The room between the labyrinth seals is purged out by the inert gas to prevent the mixing of hydrogen gas and air flowing through the turbine and the fan, respectively.

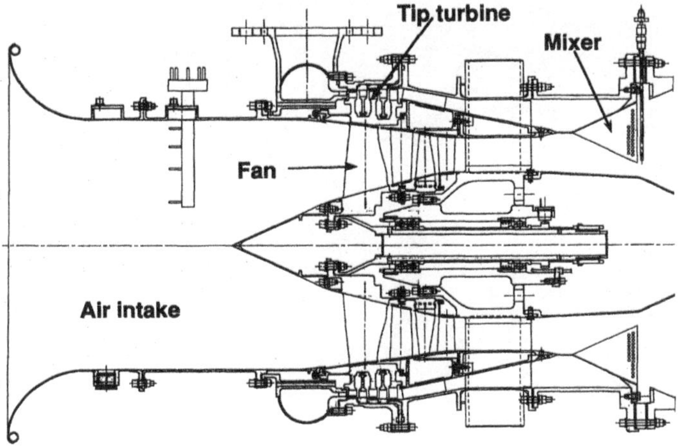

Fig. 22 Turbomachinery of ATREX-500 engine.

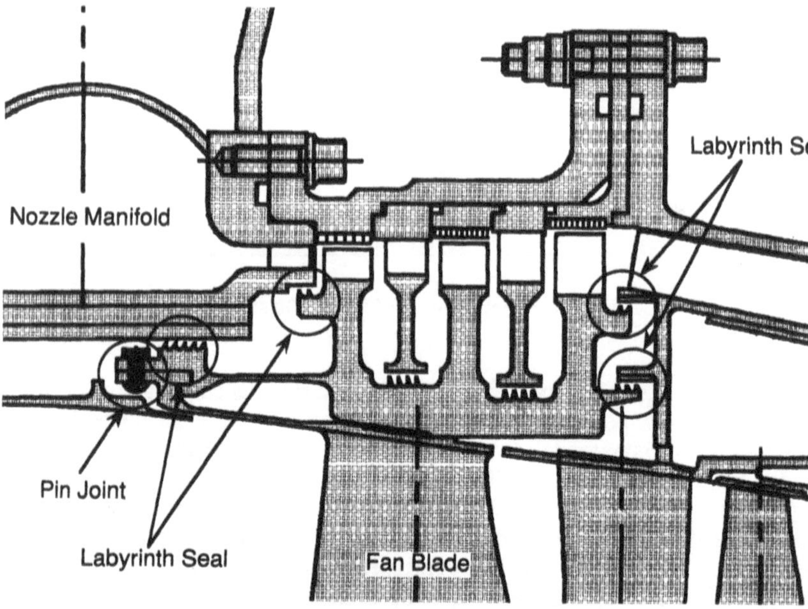

Fig. 23 Installation of tip turbine and labyrinth seals.

Fig. 24 Tip turbine and fan assembly.

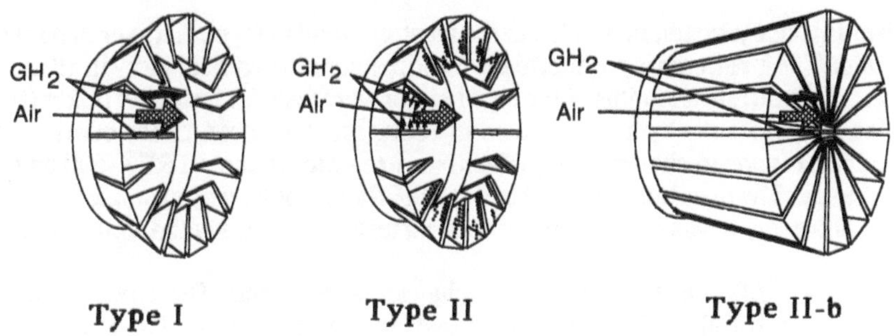

Fig. 25 Mixer configuration of ATREX-500 engine.

The rotation shaft is supported by three ceramic bearings without lubrication and cooling. It can eliminate the complexity on the rotating machinery. The bearing case is supported by the engine outer casing through six struts in which the pipings for purge gas flows and the wirings for measurements are arranged.

The combustion chamber consists of several cylindrical parts and the fixed geometric convergent exhaust nozzle. The chamber length can be changed by adjusting the length of the cylindrical parts between the mixer and the heat exchanger. To see through the combustion flame, two pairs of window with 30-mm-diam quartz glass are equipped on the chamber wall for optical observation and measurement of the combustion process. One touch igniter using gaseous hydrogen, and oxygen is equipped closely downstream of the mixer.

Two mixer types as shown in Fig. 25 were tested to mix the hydrogen gas discharged out of the turbine and the air compressed through the fan. The type I

Fig. 26 Type I mixer integrated in ATREX-500 engine combustor.

mixer injects hydrogen in parallel to the air stream and the type II mixer in perpendicular. The air stream is parallel to the rotation axis of turbomachinery in all mixer types. Two different lengths of the type I mixer were tested. The longer mixer (type Ib) easily guides the hydrogen stream injecting into the center of the combustion chamber. Figure 26 shows the type Ia mixer integrated in the ATREX-500 engine combustor. The usual annular V-shape type of flame holder was examined in early tests; however, it was removed in subsequent tests. The mixer acted well as a flame holder.

Two different types of heat exchanger were tested. The type I heat exchanger is a spiral configuration, as shown in Fig. 27 and 28. The type II heat exchanger is an annular configuration, as shown in Fig. 29 and 30. They are of the shell-tube configuration and integrated by the brazing technique. The dimensional data of the heat exchanger are given in Table 3.

Table 3 Heat exchanger dimensions

	Type I	Type I I
Heat exchage aera, m^2	0.885	1.781
Tube outer diameter, mm	5	5/7 [a]
Tube wall thickness, mm	0.5	0.3
Tube length/channel, m	0.82	1.61
Number of tubes	72	65/58 [a]

[a] Inner/Outer

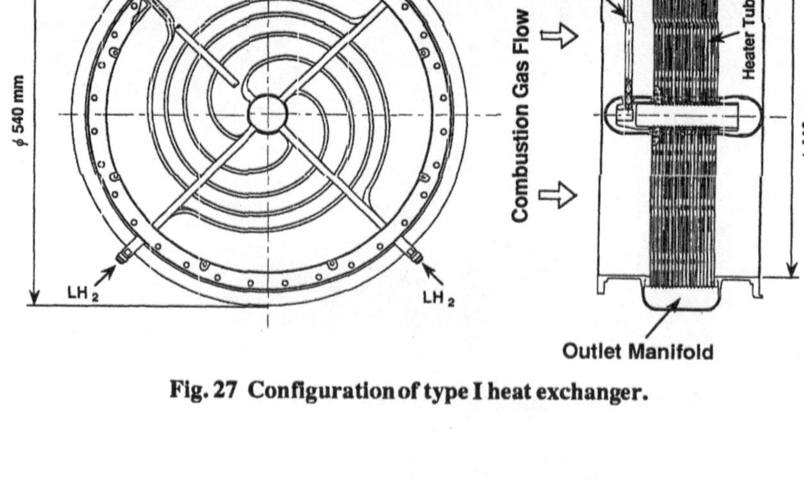

Fig. 27 Configuration of type I heat exchanger.

Fig. 28 Type I heat exchanger installed in ATREX-500 engine.

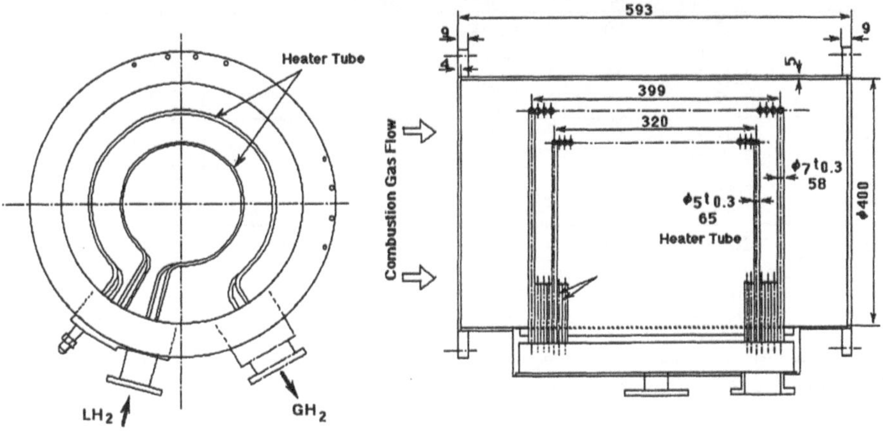

Fig. 29 Configuration of type II heat exchanger.

Table 4 Design performance characteristics of ATREX-500 engine

Thrust, kgf	410
Specific Impulse, s	1360
Rotational Speed, rpm	17,800
Air Flow Rate, kg/s	6.5
Hydrogen Flow Rate, kg/s	0.3
Equivalent Mixture Ratio	1.58
Turbine Inlet Temperature, K	650
Fan inlet diameter, mm	300
Thrust nozzle diameter, mm	330
Combustor diameter, mm	400

Table 5 Summary of test runs

No.	ATREX engine	Date	Duration, s	Maximum rotation speed, rpm	Engine operation	Working fluid: Turbine gas	Working fluid: Heat exchanger
1	1-1	9/24/90	40	8,800	Cold run	GN_2	- -
2	1-2	9/25/90	40	7,000	Cold run	GN_2	- -
3	1-3	9/26/90	20	12,050	Hot run	GH_2	- -
4	1-4	9/28/90	30	17,920	Hot run	GH_2	- -
5	1-5	9/29/90	30	18,020	Hot run	GH_2	- -
6	1-6	10/1/90	60	17,300	Hot run	GH_2	- -
7	2-1	11/21/90	30	12,500	Cold run	GHe	- -
8	2-2	11/21/90	30	13,000	Cold run	GHe	- -
9	2-3	11/22/90	60	17,900	Hot run	GH_2	- -
10	2-4	11/24/90	30	17,600	Hot run	GH_2	- -
11	2-5	11/26/90	60	18,000	Hot run	GH_2	- -
12	3-1	9/18/91	40	17,380	Hot run	GH_2	- -
13	3-2	9/19/91	40	18,440	Hot run	GH_2	- -
14	3-3	9/23/91	15	- -	Cold flow	- -	LH_2
15	3-4	9/24/91	25	12,550	Hot run	GH_2	LH_2
16	3-5	9/26/91	40	18,100	Hot run	GH_2	LH_2
17	3-6	9/27/91	40	18,250	Hot run	GH_2	LH_2
18	4-1	11/5/91	30	14,800	Hot run	LH_2	
19	4-2	11/6/91	40	14,300	Hot run	LH_2	
20	4-3	118/91	45	15,200	Hot run	LH_2	
21	4-4	11/991	60	15,200	Hot run	LH_2	
22	5-1	7/17/92	50	17,700	Hot run	LH_2	
23	5-2	7/20/92	35	17,800	Hot run	LH_2	
24	5-3	7/2292	40	17,500	Hot run	LH_2	
25	5-4	7/23/92	60	18,200	Hot run	LH_2	
26	3-1	10/28/92	24	18,520	Hot run	LH_2	
27	3-2	10/29/92	40	18,210	Hot run	LH_2	
28	3-3	10/31/92	26	18,920	Hot run	LH_2	
29	3-4	10/31/92	50	18,490	Hot run	LH_2	
30	3-5	11/2/92	60	18,440	Hot run	LH_2	

cumulative test duration: 1190 s

The bell mouth is attached to the fan inlet for uniform admission of the air stream. For measurement of the incoming air flowrate, four sets of pitot tubes are integrated in the cylindrical part of the bell mouth.

The design performance characteristics of the ATREX-500 are shown in Table 4.

C. Sea-Level Static Tests

The ATREX-500 engine test was initiated in September 1990 and completed in November 1992 at the Noshiro Testing Center (NTC) of ISAS. Table 5 shows the summary of test runs carried out. The total number of test runs is 30 with acumulative operation time of 1192 s. In addition to the ATREX-500 test runs, various types of mixer and flame holder elements were also examined by firing tests with one-fifth-scale simulation chamber of the ATREX-500.

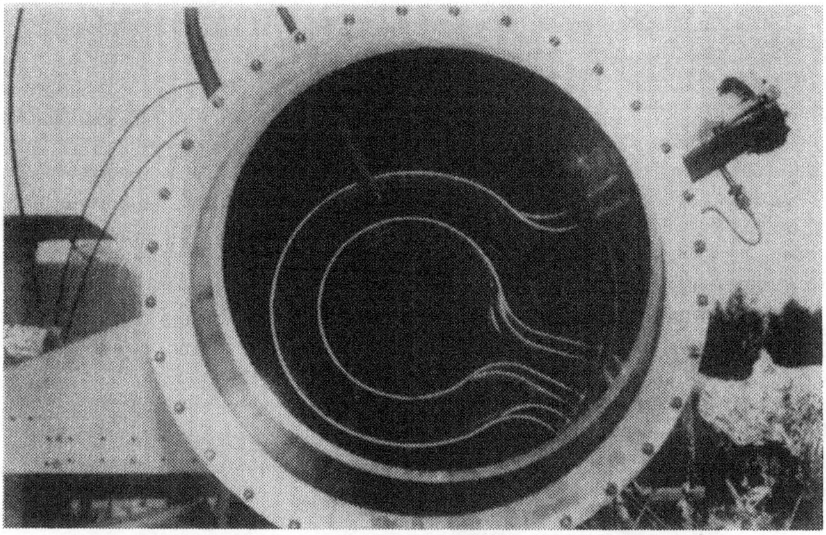

Fig. 30 Type II heat exchanger installed in ATREX-500 engine.

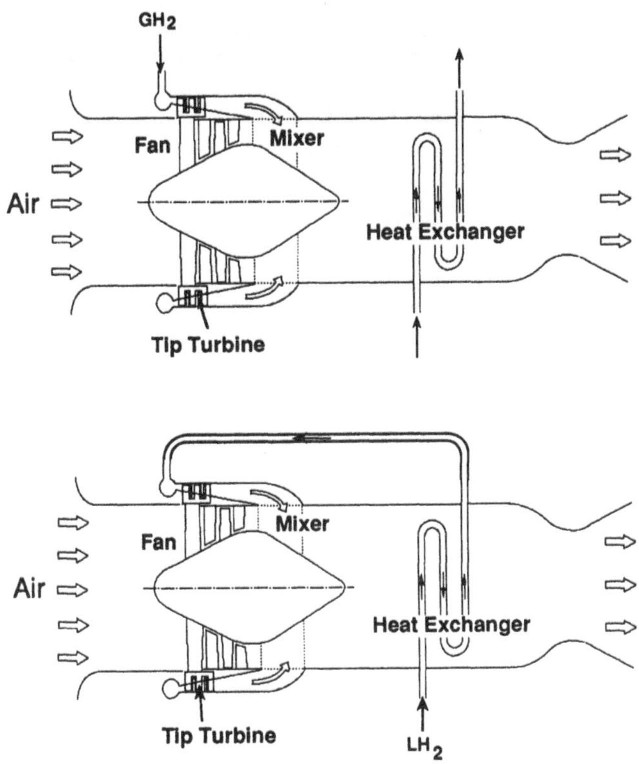

Fig. 31 System configuration of verification tests on ATREX-500 engine.

The tests were planned to be carried out by using the following approach to verify the performance characteristics of the ATREX-500 systems (Fig. 31).
1) Cold run test of turbomachinery: The turbine was driven by inert gas (nitrogen or helium) at room temperature. The rotational vibration of the tip turbine was primarily checked in this test.
2) Hot run test of turbomachinery and combustion chamber: The turbine was driven by hydrogen gas at room temperature and operated at the rated rotational speed. The hydrogen gas injected into the combustor was burned. The matching between the turbomachinery and the combustor was primarily examined in this test.
3) Hot run test including the heat exchanger: The heat exchanger was installed in the combustion chamber to verify its performance alone in the hot run test, the same as in step 2. The hydrogen gas heated by the heat exchanger was discharged out of the ATREX engine without injection into the combustion chamber.
4) Hot run test of overall ATREX system: The heat exchanger was integrated in the combustion chamber. The turbine was driven by the hot hydrogen heated through the heat exchanger. The entire ATREX system was examined in this test.

D. Test Results of ATREX-500 Engine

The test results obtained in the tests carried out using the approach mentioned in the previous section are discussed here.

1) *Tip turbine.* Tip turbine efficiency is shown in Fig. 32 as a function of the velocity ratio defined by the ratio of turbine peripheral velocity to hydrogen gas spouting velocity. The test results are beyond the predicted design value, as indicated in this figure. The velocity ratio was beyond the design value at the lower temperature of hydrogen gas. It leads to the higher efficiency. This is because the spouting gas velocity is slower when the turbine is driven by the colder gas.

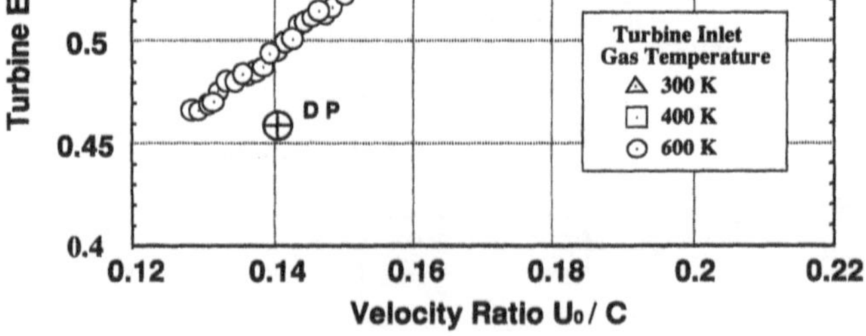

Fig. 32 Turbine efficiency.

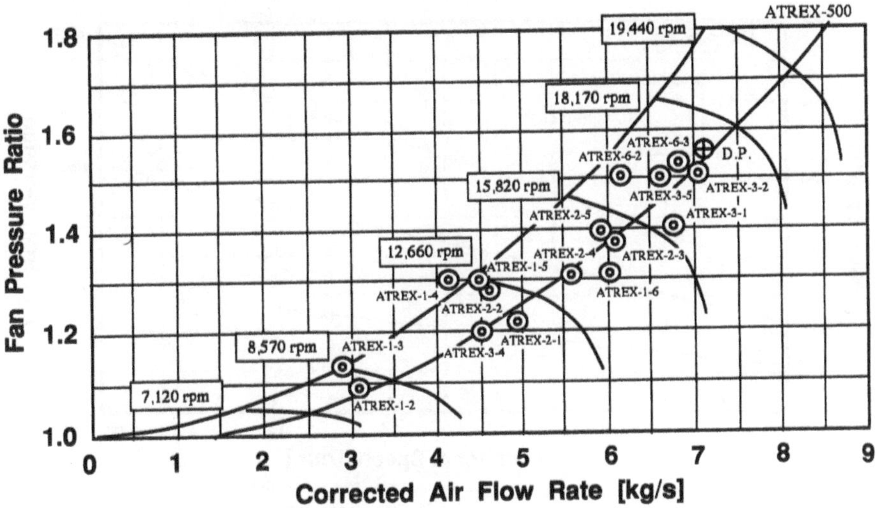

Fig. 33 Fan pressure ratio and flow rate characteristics.

2) *Fan.* The fan pressure ratio/flow rate characteristics are shown in Fig. 33, where the test results are indicated by the dotted circles. The test results nearly fit the predicted characteristic curves. However, the pressure ratio is slightly lower than the predicted value. The fan was designed by scaling the existing fan to half. The primary differences from the original fan are the shroud equipped on the first-stage fan blade and the purged gas flow injected from the labyrinth seals. The shroud surface is convergent from inlet to outlet on the air stream, and, therefore, some inverse flows are generated on the shroud surface by the centrifugal force. The inverse boundary-layer flows seem to spoil the performance characteristics of the fan. To improve fan performance, the gas discharged out of the labyrinth seal located at the fan inlet was injected along the shroud surface to blow out the inverse boundary-layer flows.

3) *Mechanism of turbomachinery.* There are two inherent problems from the tip turbine configuration. One problem is how to suppress the vibration caused by the coupling between the tip turbine and fan blade. The other is how to prevent the turbine gas (hydrogen) from flowing into the air stream through the fan. The coupling structure between the tip turbine and fan blade was changed three times before the vibration was suppressed below an acceptable level. The final coupling structure obtained is shown in Fig. 24. As is evident from the Campbel and spectrum diagrams shown in Figs. 34 and 35, respectively, critical vibration from start to rated rotational speed is not large.

Two pairs of labyrinth seals worked well by adjusting the inert gas supplied between them, respectively. As mentioned, in the fan, the seal gas discharged out of the labyrinth seal at the inlet side of the fan is injected along the fan shroud surface to blow out the inverse boundary-layer flow in the air stream. Some improvement is necessary to reduce the amount of seal gas flow rate when put into practice.

The ceramic bearings worked well: however, their temperature could not reach the steady-state condition in the operation duration as long as 60 s (see Fig.

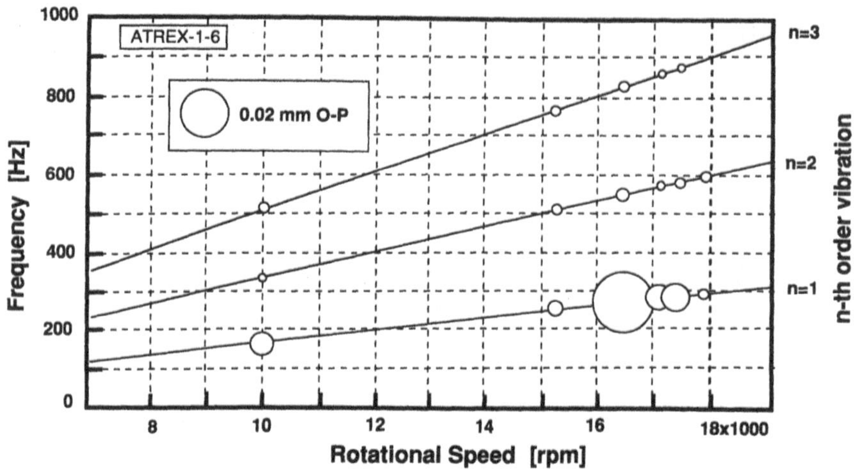

Fig. 34 Campbel diagram of ATREX-500 operation.

36) The ceramic bearings can withstand temperatures up to 700 K. From posttest examination after six startup operations, damage could not be found on the ceramic bearings.

4) *Combustor.* In the early design of the ATREX engine combustor, the annular V-shape type of flame holders were integrated just downstream of the chute-type mixer (type I). As a result of the firing tests, it was determined that the conventional mixer/flame holder configuration was not suitable in the hydrogen/air combustion

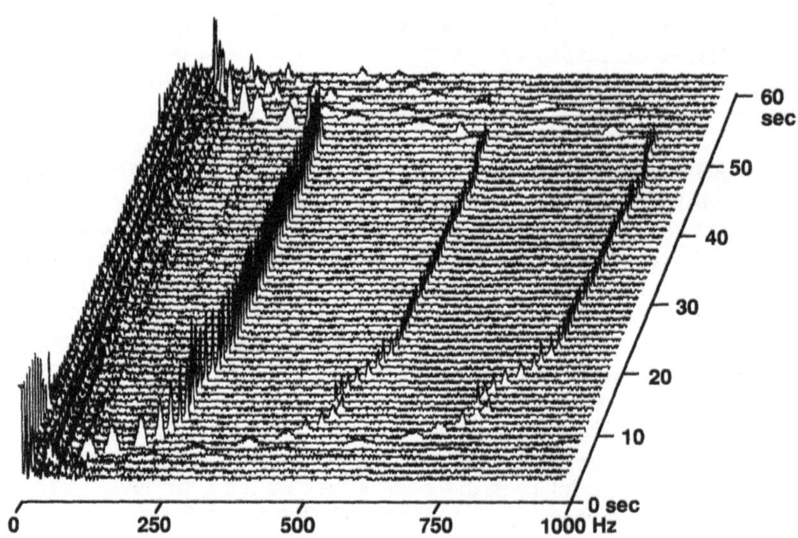

Fig. 35 Spectrum diagram of ATREX-500 operation.

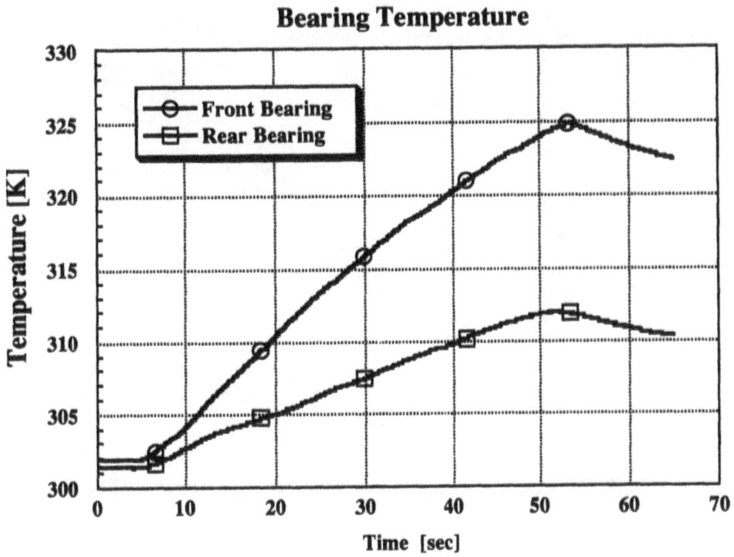

Fig. 36 Temperature rise of ceramic bearings.

system. The flame holder was severely damaged as a result of melting and deforming. To make this problem clear, the simulation tests were carried out using the subscale combustion chamber, as mentioned in the next section. The optical observation of the ultraviolet ray emitted from the OH radical of the combustion process revealed that the flame was hold just downstream of the mixer exit and the flame holder was exposed in the violent combustion flame. Then the conventional flame holders were removed from the ATREX engine. In this cases, the mixer acted well as a flame holder.

It was also revealed from the optical observation that the combustion process was completed at a distance of approximately 1.5 times the chamber diameter downstream of the mixer exit end.

Thrust efficiency, which is defined by the ratio of measured to theoretical-thrust, was from 80 to 100%. Theoretical thrust can be calculated from the air and fuel flow rates and the nozzle throat area assuming perfect combustion. Thus, thrust efficiency includes all losses in the combustion process and thrust nozzle. Thrust efficiency depended on the combustion pressure and the mixture ratio coupled with the mixer configuration.

5) *Heat exchanger.* The performance characteristics of the types I and II heat exchangers are shown in Figs. 37 and 38, respectively, where the heat flux is indicated as a function of the combustion gas flow rate (W) and the average temperature difference (ΔT) between the coolant of hydrogen and the combustion gas. The total heat amount obtained by hydrogen and the hydrogen temperature discharged from the heat exchanger are shown in Fig. 39. The amount of heat exchange was dependent largely on combustion efficiency. It is result of heat transfer from the hot gas to the coolant gas dominated by the hot gas side heat transfer coefficient.

The design and preliminary tests are mentioned in Sec. VI. From the posttest examination, there was no damage in both heat exchanger.

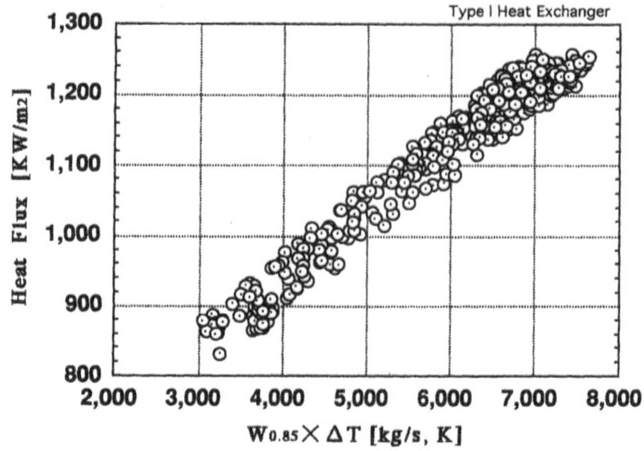

Fig. 37 Heat transfer characteristics of type I heat exchanger.

6) *Engine operation.* Liquid hydrogen was supplied to the ATREX engine from the run tank pressurized at constant. The supply line was chilled adequately to the boiling point of hydrogen before engine startup. The ATREX engine was started up by opening the control valve of the liquid hydrogen. The turbine started to rotate at 500 rpm when the turbine purge and seal gas were supplied, and its rotational speed increased further after the hydrogen control valve opened. The fan airflow rate followed, and the mixture ratio reached the appropriate value for main combustion. At the beginning of startup, the mixture ratio is several times the rated value (see Fig. 40), which is inherent in the expander cycle. It is caused that the airflow driven by the fan delays after the hydrogen flow to drive the turbine. The excessive

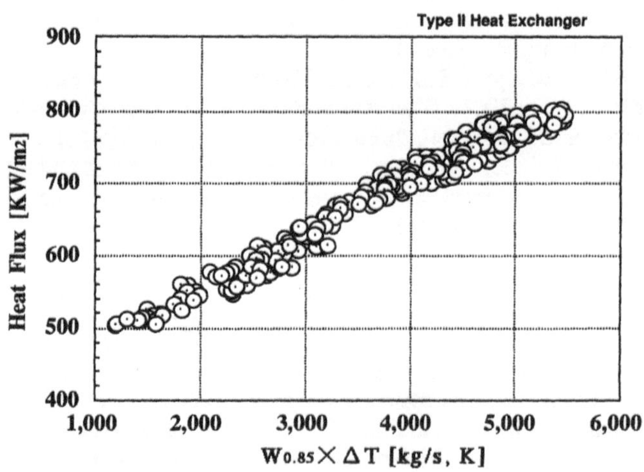

Fig. 38 Heat transfer characteristics of type II heat exchanger.

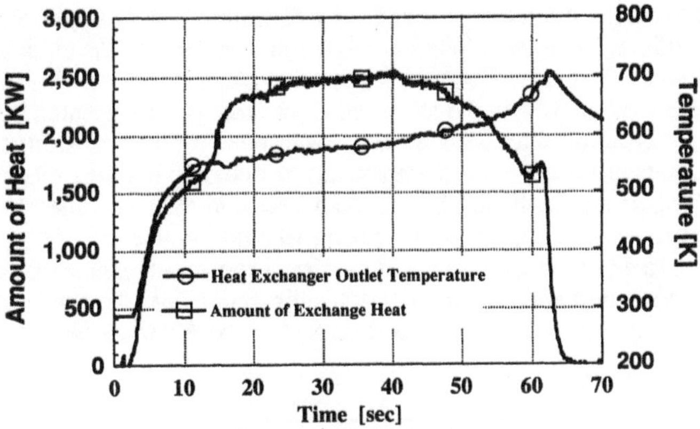

Fig. 39 Total amount of heat exchange.

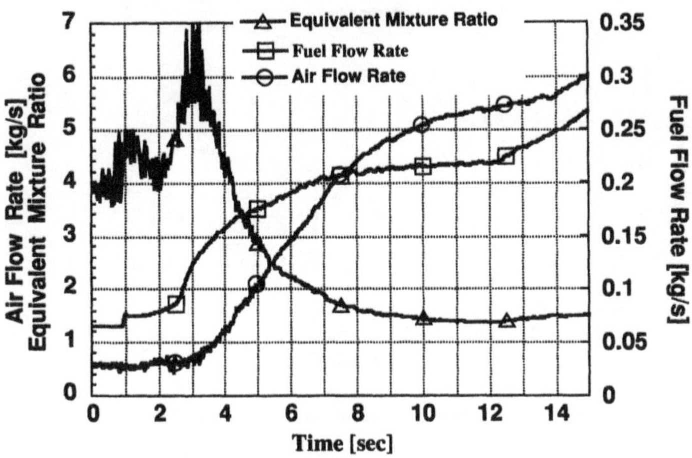

Fig. 40 Fuel and airflow rate, and equivalent mixture ratio at startup transient.

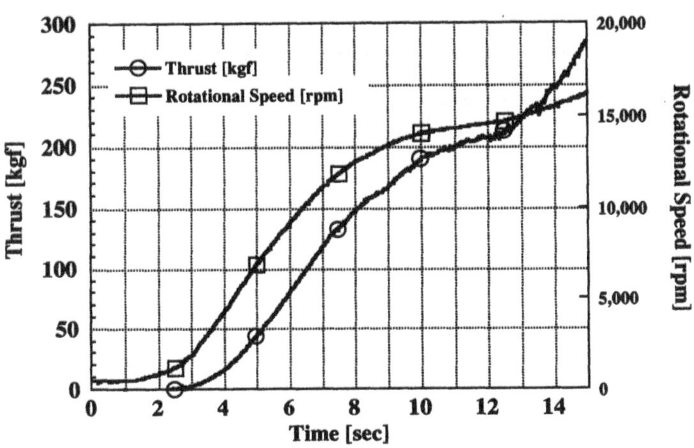

Fig. 41 Thrust and rotational speed at startup transient.

rich hydrogen effects the main ignition, and ignition timing is very important for smooth startup without backfiring. When the ignition was down at the equivalent mixture ratio above 5, the backfire was seen at the fan inlet in an instant. The main ignition also depended largely on the mixer configuration. It was revealed by the firing tests that the type Ib mixer (longer one) was suitable for smooth startup. At the beginning of startup until around 15 s after startup including ignition, all operations were performed automatically by the computer and then switched to the manual operation to increase the rotational speed to the rated value.

Figures 41 to 43 show the thrust and rotational speed, the heat exchanger discharge gas temperature andthe fuel injection temperature, and the combustion gas pressure and temperature at the startup transient of the ATREX-500, respectively.

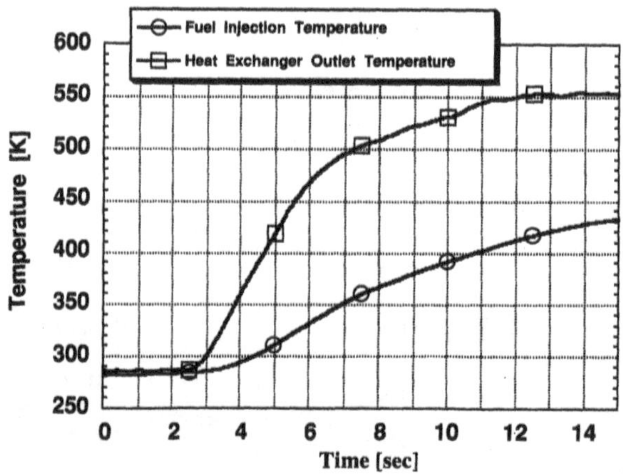

Fig. 42 Heat exchanger discharge gas temperature and fuel injection temperature at startup transient.

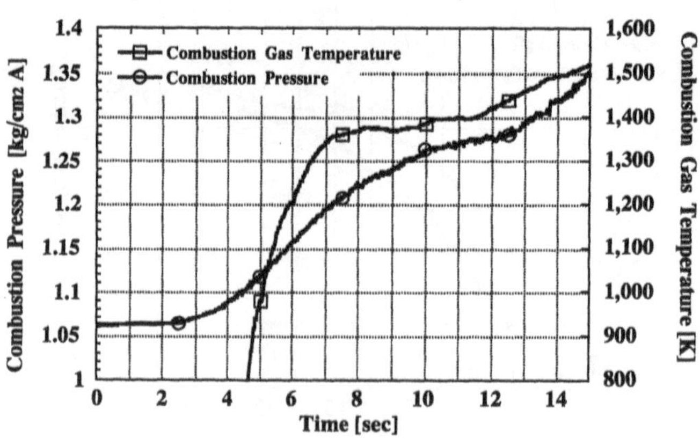

Fig. 43 Combustion pressure and combustion gas temperature at startup transient.

Table 6 Test results in final test run series

	ATREX6-1	ATREX6-2	ATREX6-3-A	ATREX6-3-B	ATREX6-4	Design Value
Date	10/28/92	10/29/92	10/31/92	10/31/92	11/2/92	
Duration, s	24	40	26.5	50	60	
Rotational speed, rpm	18,520	18,210	18,918	18,490	14,190 - 18,440	17,800
Thrust, kgf	435	430	421	383	172 - 423	409
Specific impulse, s	1195	1287	1403	1390	1390 - 994	1360
Air flow rate, kg/s	6.46	6.26	6.74	6.64	5.16 - 7.04	6.5
Hydrogen flow rate, kg/s	0.364	0.334	0.300	0.275	0.173 - 0.304	0.3
Equivalent mixture ratio	1.93	1.83	1.53	1.42	1.15 - 1.48	1.6
Thrust efficiency	1.08	1.08	0.90	0.84	0.62 - 0.85	1.00
Heat exchanger outlet temperature, K	416	442	584	616	663 - 581	650
Heat exchange, kW	2030	2000	2346	2300	1530 - 2370	2800
Type of heat exchanger	Type I x 2	Type I x 2	Type I+Type II	Type I+Type II	Type I+Type II	

The test results obtained in the final test run series are summarized in Table 6. The maximum thrust and specific impulse were 460 kgf and 1400 s, respectively. The performances obtained are slightly lower than the design value due to the lower efficiency in the fan and the combustion.

Figures 44 and 45 show the ATREX-500 setup on the test stand at NTC and a photograph of the exhaust jet flame, respectively. To take a clear picture of the jet flame, the water solution of $NaHCO_3$ was injected into the combustion chamber.

E. Concluding Remarks

Four different cycles of the ATR propulsion system using liquid hydrogen as a fuel were analyzed in Sec. II. The E-2 cycle engine (the expander cycle ATR with the precooler in air intake and the heat exchanger in combustion chamber) was tested at SL static conditions using the one-quarter scale model named "ATREX-500."

Thirty test runs with a cumulative operation duration of 1190 s were carried out from 1990 to 1992. The entire ATREX system worked well with successful results. The maximum specific impulse and the maximum thrust were 1400 s and 460 kgf, respectively. The turbomachinery (tip turbine), the combustor, and the heat exchanger worked well under the design conditions.

V. Preliminary Study on ATREX Engine Combustor[7]

A. Introduction

The purpose of the present study was to find a flame holder concept suitable for ATREX combustors, which is a key component of the engine. In the after-

Fig. 44 ATREX-500 engine setup on test stand.

Fig. 45 Firing test of ATREX-500 engine.

burner-type flame holder used for conventional aircraft jet engines, JP4 liquid fuel is injected perpendicularly into the air stream emerging from the turbine and fan, mixed and vaporized partially, and ignited and held in the flame by the flame holder or "gutter" installed further downstream[8]. In the case of the ATREX engine, the hydrogen fuel has the burning rate and the calorific value around 7.5 and 2.7 times, respectively, the corresponding values of JP4, which permits a much higher combustion intensity to reduce combustion chamber volume. The basic principles that govern the combustor configuration and characteristics however, have not been well clarified as yet. A series of tests performed ona land-based stationary bench for making a comparison of performance between the gutter and the alternative-mixer type flame holder is described. The mixer-type flame holder works with both

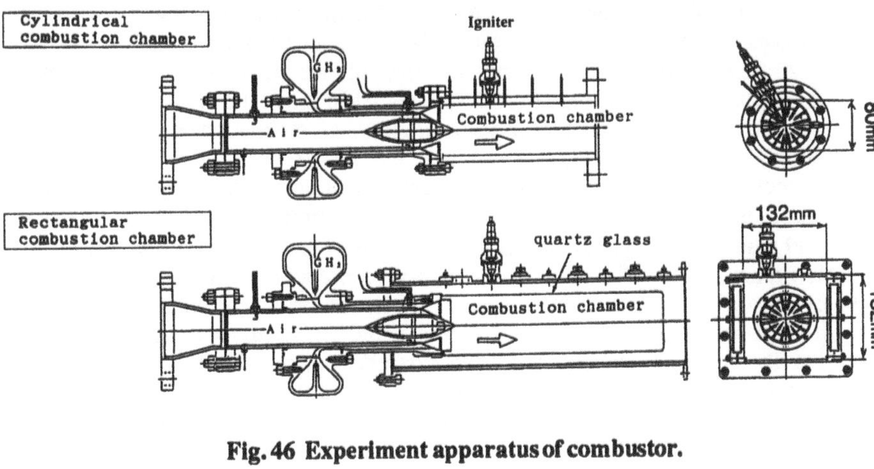

Fig. 46 Experiment apparatus of combustor.

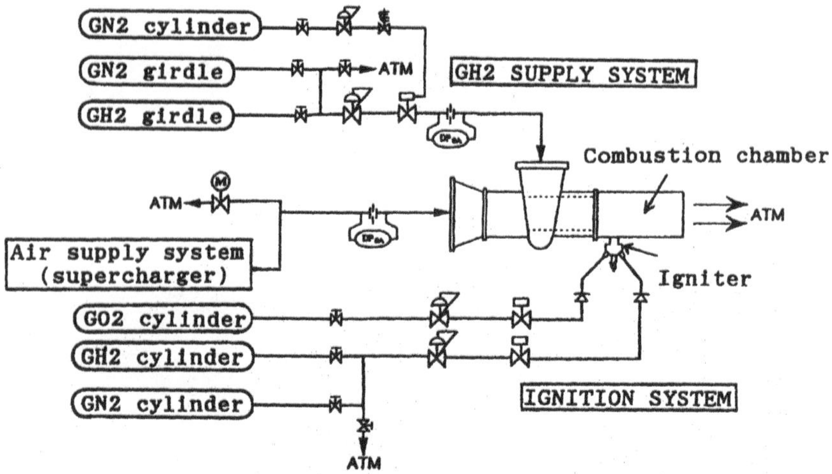

Fig. 47 Flow diagram of test facilities.

functions of flame holding and mixing in one piece, to realize a more compact combustor.

B. Test Facility and Method

The schematics of the test apparatus used is shown in Fig. 46. Two combustion chamber types were tested: cylindrical (80 mm diameter) and rectangular (132 x 132 mm). The rectangular combustion chamber equips the quartz glass windows to allow the optical observation and measurement of the combustion flame to get the flame behavior and the temperature distribution inside the combustion chamber. The two combustion chambers were alternatively tested using the same fuel and air supply systems (left hand side of Fig. 46). The combustion chambers are quite close to one-fifth scale of the ATREX-500 engine combustor. A flow diagram of the test facility is presented in Fig. 47. The atmospheric air is compressed by the turbocharger and supplied into the combustion chamber in the axial direction, and the pressurized hydrogen gas fuel is fed radially inward through a girdle. After mixing with air in the combustion chamber, the fuel is ignited by the gaseous hydrogen/oxygen torch igniter using an electric spark and held in the flame. In the test runs, the first airflow was set at the prescribed rate, and then the hydrogen fuel was fed at a rate to ensure stoichiometry (fuel equivalence ratio = 1). The duration of stable combustion was ensured to be 15 s in each test run of the cylindrical combustion chamber, and 15-80 s in the case of a rectangular combustion chamber. The test conditions are given in Table 7. In all runs, the airflow rate was switched in three steps where the stoichiometric equivalence ratio was targeted, but the setting errors resulted in scatter from 0.84 to 1.21. In tests of the cylindrical combustion chamber, the fuel was supplied gradually, throttling down to determine the lower limit for stable flame holding. The positions of measurement were indicated in Fig. 48 for the cylindrical combustion chamber, and in Fig. 49 for the rectangular one. In the cylindrical combustion chamber case, the flame temperatures were given by the optical fiber temperature sensor set at the chamber exit and the static pressure sensors set along the chamber wall. In the rectangular combustion chamber case, ultraviolet pictures were taken to observe the flame behavior, such as flame detachment from the flame holder, and temperature distribution on the flame holder was mea-

Table 7 Conditions in test

Flameholder	Gutter-type	Mixer-type
Air supply		
Total temperature , T, K	300	
Total pressure, Pa, MPa	0.16 - 0.28	
Flow rate, Wa , kg/s	0.3 - 0.6	
Mixer exit velocity, Va, m/s	60 - 110	
GH_2 supply		
Total temperature , T, K	300	
Total pressure, Pa, MPa	0.11 - 0.23	
Flow rate, Wa, kg/s	0.01 - 0.02	
Mixer exit velocity, Va, m/s	140 - 400	270 - 600
Equivalent mixture ratio	1	

sured by the infrared temperature sensor, and temperature distribution in the flame was measured by the Na D line temperature measurement.

C. Comparison Tests of Various Configuration Flame Holders

1) *Flame holder configurations tested.* The gutter-type of flame holder is shown in Fig. 50, and the mixer type in Fig. 51. With the gutter-type flame holder, the mixer was designed to keep the same static pressure for fuel and air upstream of the folder. The number of lobes is 16.

The gutter flame holder has radial legs. The blockage ratio of the gutter area projected to the duct cross-sectional area is approximately 40%, which is almost equivalent to the afterburners of conventional aircraft jet engines.

With the mixer-type flame holder, the mixer lobe configuration is similar to the gutter type, but the fuel folds are blinded and pierced with 160 1.4-mm-diam

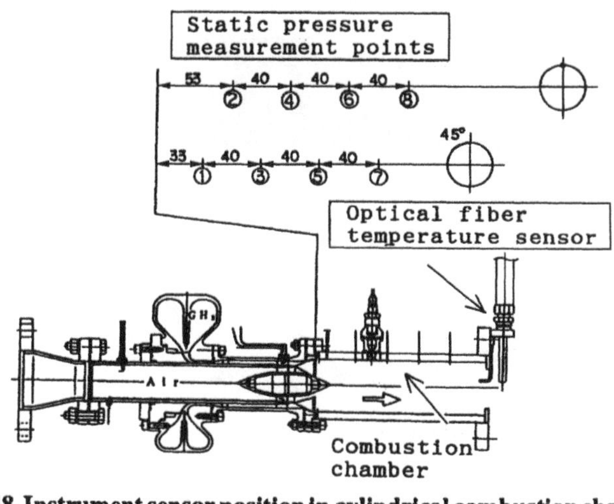

Fig. 48 Instrument sensor position in cylindrical combustion chamber.

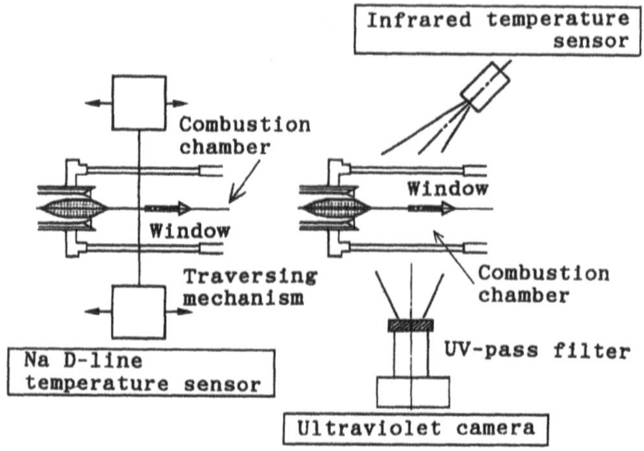

Fig. 49 Instrument sensor position in rectangular combustion chamber.

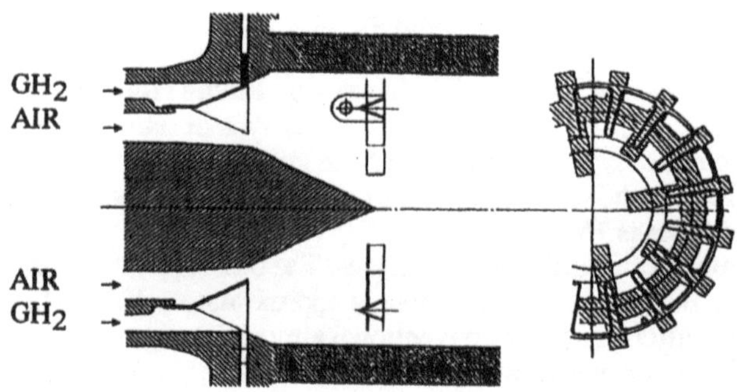

Fig. 50 Gutter-type flame holder.

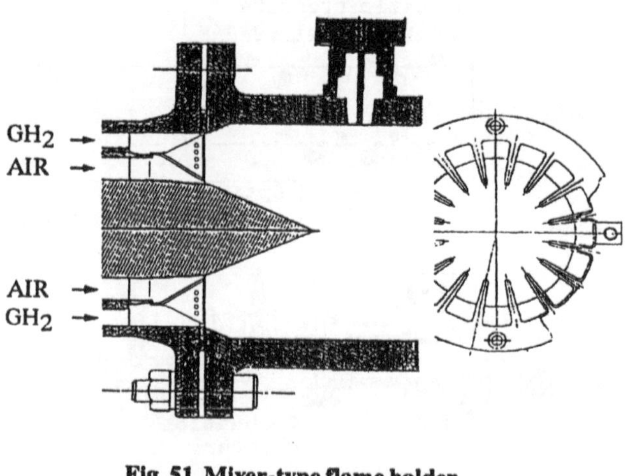

Fig. 51 Mixer-type flame holder.

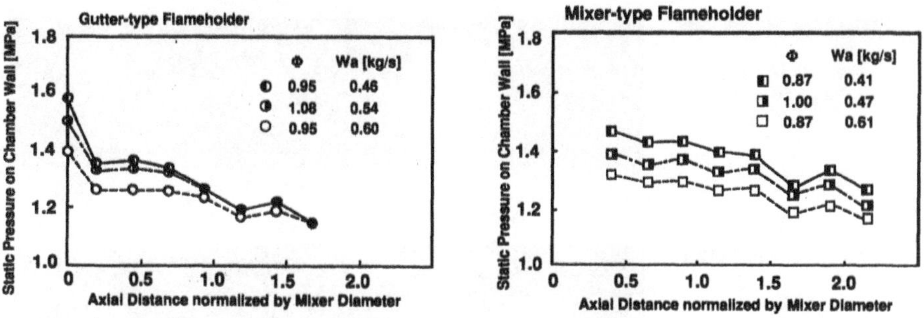

Fig. 52 Static pressure distribution along combustion chamber wall surface.

holes. The holes are arranged to allow fuel injection perpendicularly to the air stream with a view toward enhancing fuel/air mixing and flame holding.

2) *Test results.*

Static pressure along combustion chamber wall. Static pressure distribution along the cylindrical combustion chamber wall is presented in Fig. 52. The axial distance is normalized with the nondimensional ratio of axial distance to combustion chamber diameter. In the gutter type, a sharp pressure drop can be seen immediately downstream of the flame holder and followed by an additional gradual drop. In the mixer type, the static pressure could not be measured in the zone close to the mixer flame holder, but beyond this zone the pressure decreased gradually.

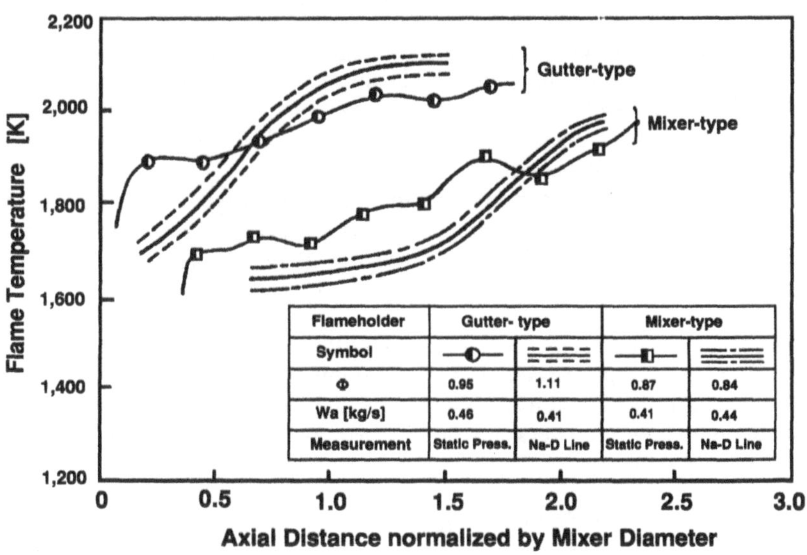

Fig. 53 Flame temperature distribution.

Gutter-type flame holder

ϕ : 0.87
Wa : 0.56 kg/s

Mixer-type flame holder

ϕ : 0.91
Wa : 0.51 kg/s

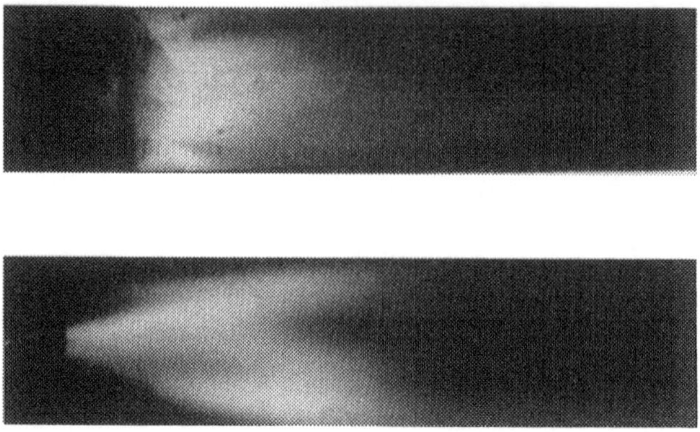

Fig. 54 Ultraviolet photographs of combustion flame.

Flame temperature distribution. The flame temperature distributions of the cylindrical combustion chamber were calculated from the static pressure on the chamber wall, which are plotted in Fig. 53 for both gutter- and mixer-type flame holders. Also shown are the corresponding data obtained in the rectangular chamber, which were taken by using an Na D line temperature measurement. Despite the different configurations in the combustion chamber, the substantially good agreement can be seen between the two data groups. The gutter type is characterized by a sharp temperature rise immediately downstream of the gutter flame holder, which is verified by the violent flame generated directly from the flame holder, as shown in the top portion of Fig. 54. In the mixer-type flame holder, the temperature rise is seen at some distance from the mixer flame holder, which seems to be the result of poor flame holding as is evident from the detached flame shown in the bottom portion of Fig. 54. It should be noted, however, that an unstable pulsating combustion characterizing the flame detachment in the case of liquid fuel afterburners did not occur in the present case. This is caused by the gaseous state of hydrogen fuel that dispenses with the evaporation process in combustion.

Another difference between the gutter and mixer types of flame holder is seen in the flame holder wall temperature. The infrared temperature measurements have shown approximately 1200 K of average wall temperature in the gutter-type flame holder, and approximately 900 K in the mixer type.

D. Test of Improved Mixer Type Flame Holder

The flame tended to detach from the flame holder in the mixer-type flame holder, which results in retarding of the flame temperature rise. This is caused by the small mass flow ratio between the hydrogen fuel and air, which prolongs the time required for mixing.

1) *Improved mixer-type flame holding system.* To promote mixing of hydrogen fuel with air, the mixer was modified as shown in Fig. 55. The improved mixer has the arrangement of 12 fuel folds pierced by 168 holes of 1.4 mm in diameter which provides more fuel pockets at the fuel fold, and ensures better uniformity if the fuel

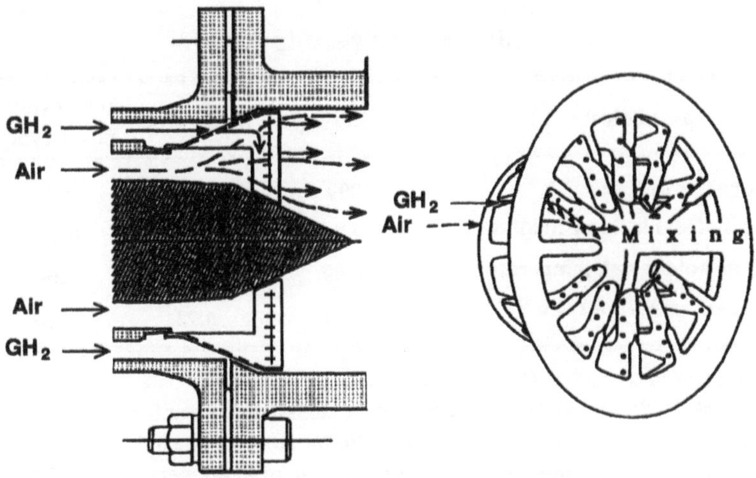

Fig. 55 Improved mixer-type flame holder.

is distributed in the radial direction of the mixer. These modifications improved the mass flow ratio by factors of approximately 2.6 and 1.3 compared with the gutter-type and original mixer type flame holder, respectively.

2) *Test results.* The flame temperature distribution in the axial direction of the rectangular combustion chamber using the improved mixer-type flame holder is shown in Fig. 56, which was taken by Na D line measurement, and the ultraviolet flame photographs in Fig. 57. Table 8 gives the data obtained in the cylindrical combustion chamber. In these figures and table, the results are compared with that taken by the gutter-type and original mixer-type flame holders.

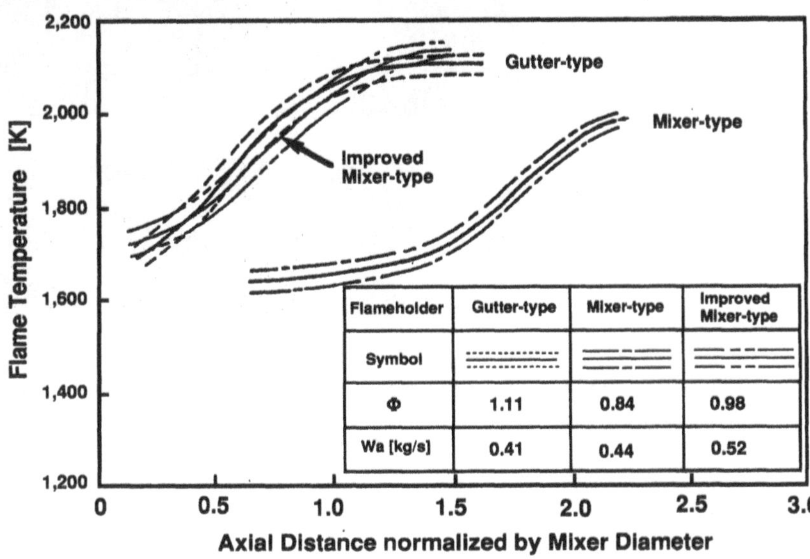

Fig. 56 Flame temperature distribution measured by NaD line instrument.

Table 8 Test results

Flame holder	Gutter type	Mixer type	Improved mixer type
Flame temperature [a], K	1800	1750	2100
Limit of stable flame holding [b]	0.24	0.26	0.12
Flame holder wall temperature [c], K	1200	900	1200
Fuel/air mass flow ratio [d]	0.11	0.22	0.29

[a] Measured by optical fiber temperature sensor located about 20 mm inside chamber wall at combustion chamber exit.
[b] Lower limit of equivalence mixture ratio for stable flame holding.
[c] Maximum wall temperature measured by infrared temperature sensor.
[d] Condition at mixer exit.

Gutter-type flame holder

ϕ : 0.87
Wa : 0.56 kg/s

Mixer-type flame holder

ϕ : 0.91
Wa : 0.51 kg/s

Improved mixer-type flame

ϕ : 0.95
Wa : 0.49 kg/s

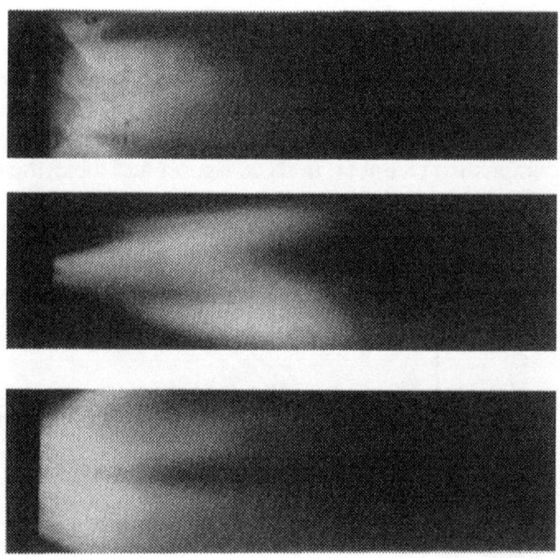

Fig. 57 Ultraviolet photographs of combustion flame (compare improved and original flame holders).

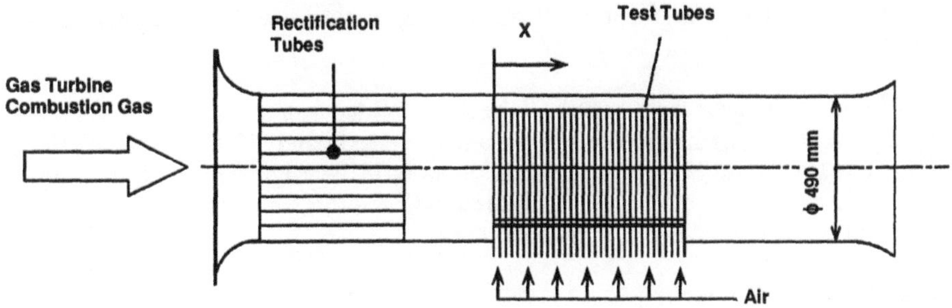

Fig. 58 Experimental apparatus using a gas turbine exhaust jet.

3) *Flame temperature distribution.* As indicated in Table 8, the flame temperature of the improved mixer was higher than that of any other flame holder if comparing the temperature measured by the optical fiber temperature sensor located about 20 mm inside from the duct wall at the cylindrical combustion chamber exit (about 270 mm downstream of the flame holder). This was caused by enhancement of mixing between the fuel and air as expected, that can be realized by increasing the momentum imparted to the fuel and air. In Fig. 56, the flame temperature distribution of the improved mixer-type flame holder in the rectangular chamber is almost equivalent to that of the gutter type. The delay in flame temperature rise, as seen in the original mixer type, is improved substantially. This good performance is verified by the ultraviolet flame photographs in Fig. 57, where the violent flame is seen just downstream of the mixer flame holder similar to the case of the gutter flame holder, with slight flame detachment.

4) *Lower limit of fuel equivalence ratio for stable flame holding.* As shown in Table 8, the promotion of fuel air mixing made by the improved mixer has also reduced the lower limit for the flame extinction, thus extending the allowable range

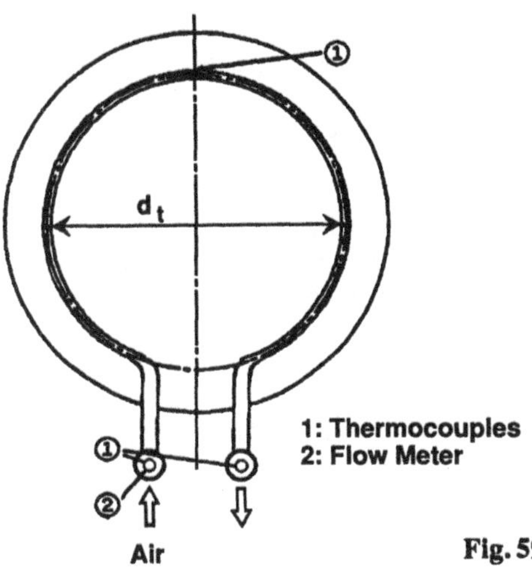

Fig. 59 Cross-sectional view of heater assembly.

of fuel equivalence ratio for stable flame holding compared with the original flame holders.

5) *Flame holder wall temperature.* The flame holder wall temperature measured by the infrared temperature sensor has risen to approximately 1200 K, which is the same level as the gutter type, from 900 K obtained in the original mixer-type flame holder. In practical application, this heating will request some cooling system as used in the conventional gutter flame holders.

E. Concluding Remarks

Comparison tests between the gutter- and mixer-type flame holders were conducted with a view to assessing the characteristics of different flame holding systems for ATREX combustors. The following results were obtained:
1) Improvements given by the mixer-type flame holder are elimination of flame detachment from the flame holder, improvement of flame temperature distribution in the combustion chamber, extension of the allowable range of fuel equivalence ratio for stable flame holding, and diminution of flame instability. The resulting performance is almost equal to that obtained in the gutter-type flame holders.
2) The improved mixer-type flame holder promises the successful practical application to the ATREX engine combustors and the possibility of more compact combustors, while the flame holder wall temperature rise may request some protection measures in practical applications. This type of flame holder can be expected to provide a promising new system.

VI. Development Study on ATREX Engine Heat Exchanger[9]

A. Introduction

In the ATREX engine, the hydrogen fuel is heated by the combustion gas with the heat exchanger installed in the thrust chamber. ATREX performance is dependent largely on the temperature rise of hydrogen to drive the turbine. In the practical ATREX engine, hydrogen fuel is required to be heated from 25 K (liquid state) to 1500 K. The heat exchanger is installed within the combustion chamber and operated at high heat flux and large temperature differences between the combustion gas and hydrogen fuel.

Glickstein and Powell[10] showed that the ordinary expander cycle ATR engine is heavy as a result of the installation of hydrogen heater, gearbox, and a lot of turbine stage. In the ATREX engine, the tip turbine integrated on the peripheral of the fan is employed, and the compact heat exchanger is installed in the combustion chamber to improve the disadvantages of the ordinary system. Thus the heat exchanger is required the light weight and the small pressure losses of combustion gas flow, and in addition to withstand the high heat flux and vibration. It is important to estimate correctly the heat transfer characteristics to develop the engine system successfully.

The preliminary experiments on the heat exchanger to obtain the basic data necessary for design of the ATREX engine system and the practical concept of the heat exchanger. This is based on a numerical analysis using the data obtained in the preliminary experiments, where the heater elements were set within the exhaust flame of the conventional aircraft jet engine fueled by JP4. Following this conceptual design study, the subscale heat exchanger (hydrogen outlet temperature: 650

K) was made for the ATREX-500 subscale engine and tested to verify its heat transfer characteristics.

B. Preliminary Experiments on Heat Transfer Characteristics

When the heat exchanger is installed in the high-speed turbulent stream of combustion gas, and hydrogen of coolant is also supplied into the heat exchanger at high pressure, the heat exchanger assembly is predicted to vibrate without any rigid supports. Thus, the ATREX heater tubes are assembled continuously in rows without any space between them.

The heat transfer characteristics of tube banks in the ordinary concept have been established[11, 12], but those of the ATREX heater concept have never been established.

Turbulence in the combustion gas is important to evaluate the heat transfer characteristics of the ATREX heater concept. The turbulent strength should be measured but it is very difficult to measure in the high temperature gas flowfields. Thus, the influence of turbulent strength is not evaluated here.

1) *Experiments using exhaust gas flow of gas turbine.* A schematic of the experimental apparatus utilizing the exhaust gas flow of the gas turbine is shown in Fig. 58. A gas turbine exhaust gas is used to simulate the ATREX turbulent combustion gas flow. The turbine exhaust gas is rectified upstream of test heater tubes to eliminate the swirling effect on heat transfer. Combustion gas temperature and velocity profiles were measured by thermocouples and pitot tubes, respectively.

The test heater tube is shown in Fig. 59. Ring heater tubes with outer diameters of 4.8 mm were used. These heater tubes are set in rows in which air is fed in place of hydrogen. The airflow rates in every tube were controlled at the same value.

The inlet and outlet air temperature and the tube wall temperature were measured by using thermocouples. Two different ring diameter tube sets were tested to evaluate the effect of the ring diameter on heat transfer. Table 9 shows the experimental conditions.

Experimental results are shown in Fig. 60. It was revealed from these tests that the effect of the ring diameter on heat transfer could be neglected. In ordinary tube banks, the heat transfer coefficients become larger in the tubes set more downstream because of the effect of turbulence generation. However the heat transfer characteristics of the heater tubes tested are similar to those of a flat plate, as established by Katayama et al.[12] Nusselt numbers of the heater tubes tested are ap-

Table 9 Conditions in experiment using a gas turbine exhaust jet

d_o, mm	0.0048
d_t, m	0.18, 0.37
T_g, K	690 - 780
U_g, m/s	48 - 135
Re_{do}	3,400 - 7,500

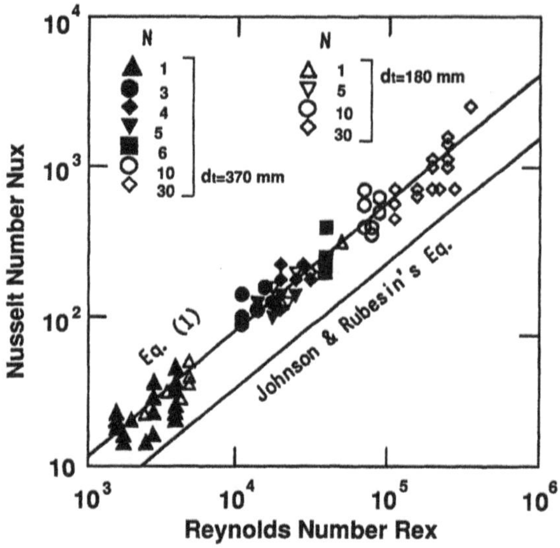

Fig. 60 Experimental results using a gas turbine exhaust jet.

proximately three times larger than that of the flat plate. It is considered that this difference is caused by the turbulence in the combustion gas. An experimental data correlation equation is found to be:

$$Nu_x = 0.038\, Re_x^{0.8}\, Pr_g^{0.67} \tag{1}$$

We are not able to estimate the effect of Prandtl number on heat transfer with some accuracy, and it is assumed that the effect of Prandtl number is similar in a flat plate.

2) *Experiments using the wind tunnel.* The schematic of the experimental apparatus is shown in Fig. 61. This experiment has been done to calculate the heat transfer data in the absence of turbulence.

Heated air was supplied to the test section through a plenum to reduce turbulence. Heater tubes with an outer diameter of 4.8 mm were set in rows. The

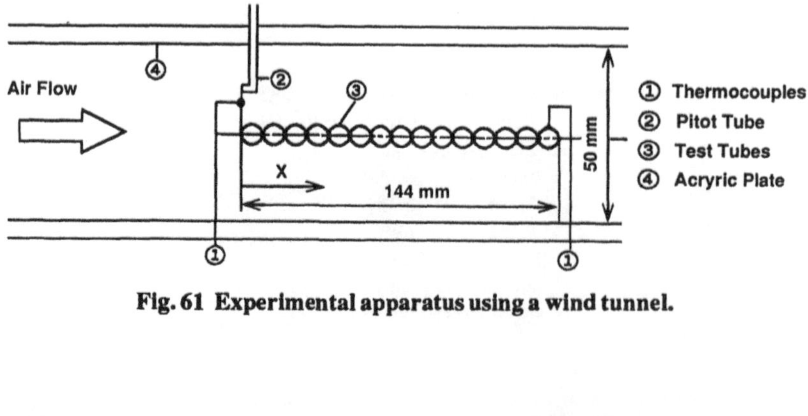

Fig. 61 Experimental apparatus using a wind tunnel.

Table 10 Conditions in experiment using a wind tunnel

d_o, mm	0.0048
N_t,	30
T_g, K	313 - 322
U_g, m/s	78 - 180
Re_{do}	10,240 - 25,000

thermocouples were set on these tube surfaces to measure the surface temperatures. Water was used as a coolant passing through the tubes. The experimental conditions are shown in Table 10.

Experimental results are shown in Fig. 62. Nusselt numbers in this experiment are less than those in the experiments using exhaust gas flow of the gas turbine. It appears that this difference is caused by turbulence in the main flow. An effect of Re_{do} also appears and should be considered in heat transfer characteristics of heater tubes. This effect did not appear in the experiments using a gas turbine. At this experiment, Nusselt numbers are proportional to $x^{0.95}$ and decreasing rates of heat transfer coefficients are smaller than at a flat plate. It appears that some turbulence was generated on the surface of the tubes and it promoted largely the heat transfer. Re_{do} depends on the strength of the generated turbulence. In the experiment using a gas turbine, the main flow is turbulent, and the generated turbulence on the surface may not affect the heat transfer. Consequently, Nusselt number is proportional to $Re_{do}^{0.5}$ and $Re_x^{0.95}$. Thus, an experimental data correlation equation for low turbulence flow in wind-tunnel tests is

$$Nu_x = 2.89 \times 10^{-5} \, Re_x^{1.45} \, Pr_g^{0.67} \sqrt{x/d_o} \qquad (2)$$

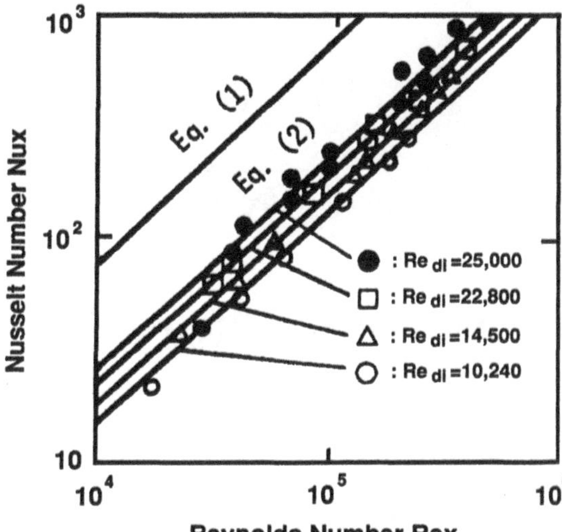

Fig. 62 Experimental results using a wind tunnel.

Only the velocity effect of the main flow was examined, and an effect of Prandtl number could not be evaluated. It is assumed that the effect of Prandtl number is similar in a flat plate.

C. Practical Concept of Heat Exchanger

1) *Optimum design point.* The hydrogen heat exchanger proposed for the practical flight-type ATREX engine requires lighter weight. The ordinary heat exchangers are designed usually to obtain a high heat exchanger effectiveness, but the primary object of the ATREX heater is to attain the required outlet total temperature of hydrogen fuel minimizing the pressure drops in the combustion gas flow and heater weight. Thus, the design parameter $Q_{hx}/(\Delta p_g A_{hx})$ is recommended as a figure of merit. This parameter is able to be substituted as

$$C_x = q_{hx}/\Delta p_g \tag{3}$$

The heat transfer characteristics of the hydrogen heater were analyzed by numerical simulation. The main equations used are as follows:

For heat balance

$$Q_{hx} = G_h\left(i_{h,o} - i_{h,i}\right) \tag{4}$$

$$= h_g\left(T_{g,r} - T_{p,out}\right) \tag{5}$$

$$= h_h\left(T_{p,in} - T_{h,r}\right) \tag{6}$$

For combustion gas

$$Nu_x = 0.038 Re_x^{0.85} Pr_g^{0.67} \tag{7}$$

$$T_{g,r} = T_{g,s}\frac{\gamma_g - 1}{2} Pr_g^{1/3} M_g^2 \tag{8}$$

$$T_g = T_{g,s}\left\{1 + \frac{\gamma_g - 1}{2} M_g^2\right\} \tag{9}$$

$$h_g = C_{p,g}\tau\left[u_g\left\{1 + 5\sqrt{\frac{\tau}{\rho u_g^2}}\left(Pr_g - 1\right) + \ln\left(\frac{1 + 5Pr_g}{6}\right)\right\}\right]^{-1} \tag{10}$$

For hydrogen

$$Nu_{di} = 0.023\, Re_{di}^{0.8} Pr_h^{0.4}\left(T_{p,in}/T_{h,r}\right)^{-0.57} \tag{11}$$

$$T_{h,r} = T_{h,s}\frac{\gamma_h - 1}{2} Pr_h^{1/3} M_h^2 \tag{12}$$

$$T_h = T_{h,s}\left\{1 + \frac{\gamma_h - 1}{2} M_h^2\right\} \tag{13}$$

$$f_{di} = 0.0032 + 0.221 Re_{di}^{-0.237} \tag{14}$$

Equation (1) was used in the flight-type ATREX engine assuming that the combustion gas characteristics are similar to that of the gas turbine combustion gas. Combustion gas pressure loss was calculated by Eq. (9) given by Karman's analogy [13]. Equations (11) and (14) are quoted from Refs. 14 and 15 respectively. Combustion gas and hydrogen fuel were treated as compressible gases. The analytical conditions are shown in Table 11. Mass flow rate in all heater tubes is the same, and the hydrogen velocity discharged out of all heater tubes is below the choking limit.

Analytical results are shown in Fig. 63. A blockage ratio is the ratio of the heater tube projected area to the cross-sectional area of the combustion chamber. The combustion gas pressure loss is reduced as the blockage ratio is reduced. However, the heater length becomes longer for a given heat transfer. A high blockage ratio leads to a compact heater with high pressure loss. The parameter C_X indicates a maximum value when the blockage ratio is about 8% for a given heat transfer.

2) *Practical Heat Exchanger Concept.* The flight-type concept of the heater is shown in Fig. 64. The blockage ratio given by the analysis mentioned earlier re-

Table 11 Conditions for analytical calculation

G_g, kg/s	219
T_g, K	2,440
G_h, kg/s	4.49
$T_{h,o}$, K	1,470
$T_{h,i}$, K	212
$O_{h,o}$, MPa	1.49

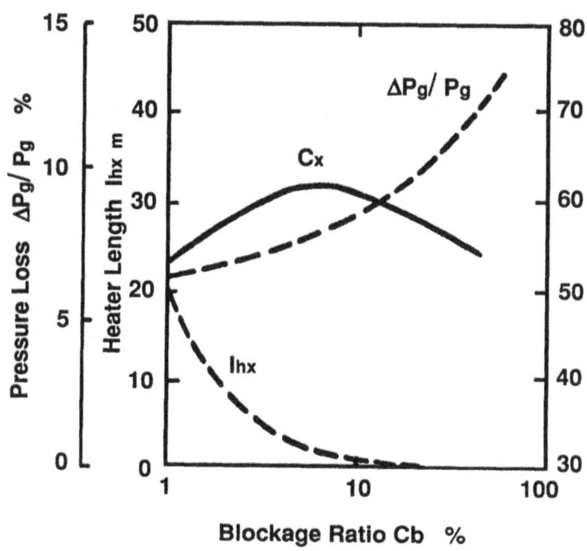

Fig. 63 Optimum design point.

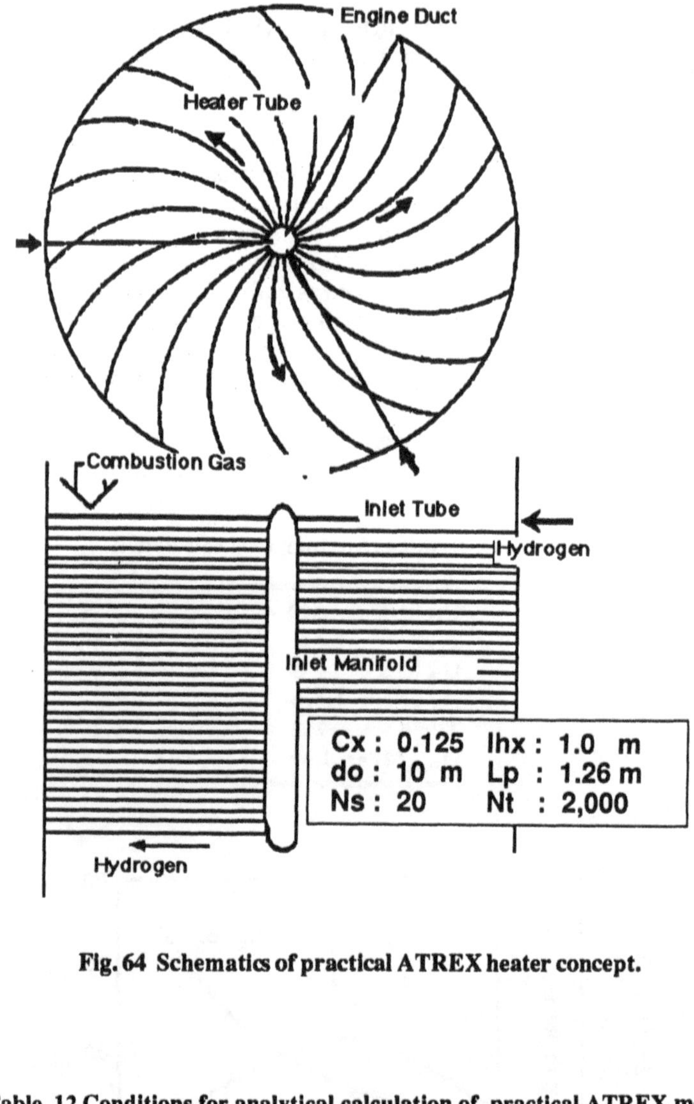

Fig. 64 Schematics of practical ATREX heater concept.

Table 12 Conditions for analytical calculation of practical ATREX model

M_f	0	2	4	5	6	6.5
G_g, kg/s	219					
T_g, K	2440	2560	2720	2710	2440	3010
G_h, kg/s	4.49	8.90	22.8	31.0	31.4	90.2
$T_{h,o}$ K	1470	1470	1470	1470	1470	1120
$T_{h,i}$, K	212	287	588	807	1120	617
$P_{h,o}$, MPa	1.49	2.95	52.9	52.6	34.7	23.7

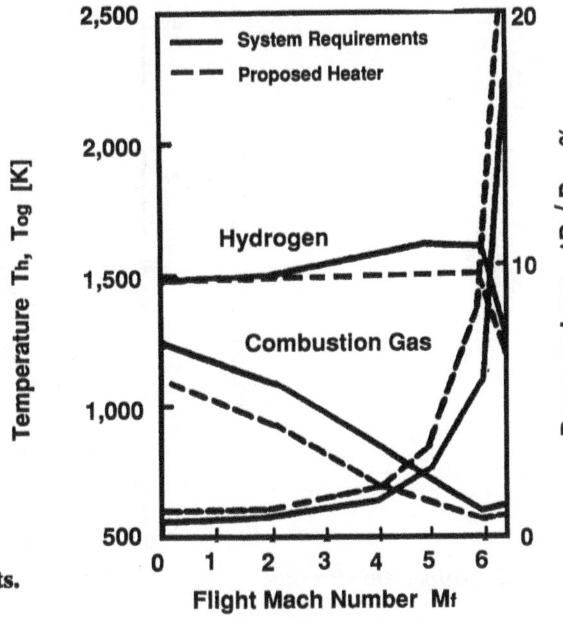

Fig. 65 Analytical results.

quests to have hydrogen inlet manifold in the center of the combustion chamber. Twenty sets of heater tubes are arranged around this manifold. The heater length is less than 1 m, which makes it possible that the hydrogen heater proposed is applied for the ATREX engine.

Calculated outlet total temperatures and combustion gas pressure loss corresponding to the requirements of the ATREX engine are shown in Fig. 65. Table 12 shows the engine system requirements for the heater. The heater proposed here satisfies the system requirements.

3) *Configuration of heat exchanger applied for the ATREX-500.* A schematic of the hydrogen heater to be used for the ATREX-500 engine is shown in Fig. 64. An inlet manifold is set in the center of the combustion chamber. Hydrogen fuel is fed to the manifold through four hydrogen inlet tubes. A twisted plate is inserted within an inlet tube to enhance heat transfer. Hydrogen fuel cools the manifold wall exposed to the stagnation point of the high-temperature combustion gas flow and then flows into the header in the center of the manifold. In the header, the static pressure of hydrogen fuel increases as a result of the diffuser effec,t and then the hydrogen fuel flows into the front edge tubes of the heater. The four sets of heater ubes are joined to the manifold. The heater tubes have a spiral shape to respond to the thermal deformation by high heat flux and has a twisted plate in it. The blockage ratio is about 12%.

4) *Numerical Analysis.* A numerical simulation was carried out for the heater. The main governing equations mentioned earlier were used. The following equation was used as the governing equation for hydrogen because of the twisted plate installed in the heater tubes.

$$\mathrm{Nu}_{de} = 0.023(C_1 \mathrm{Re}_{de})^{0.8} \mathrm{Pr}_h^{0.4} \left(T_{p,in}/T_{h,r}\right)^{-0.57} \tag{15}$$

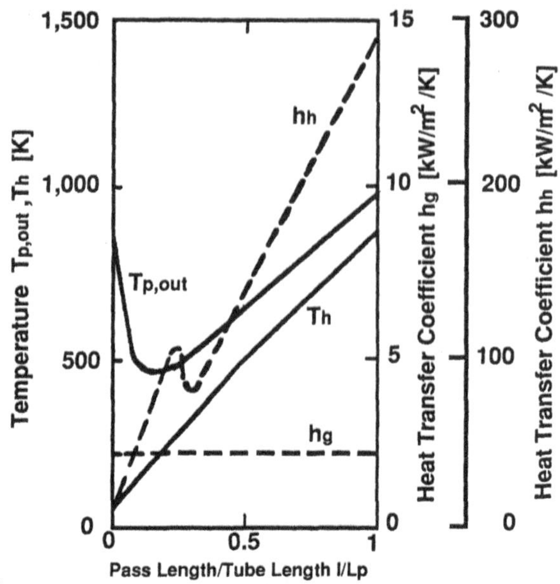

Fig. 66 Temperature Profiles.

It is not right to use Eq. (15) for the tube with any twisted plate. The established analyses [12] show that the Nusselt number is a function of C_1Re, therefore, it is assumed that the effects of the twisted plate appear in the Reynolds number coefficient. Table 13 shows the conditions set in the analyses. To match the discharge conditions of hydrogen with the tip turbine inlet interface, the discharge total pressure of hydrogen is fixed and the inlet total pressure is calculated. The profiles of the hydrogen total temperature, the outer tube wall temperature, and the heat transfer coefficient for the front edge tube are shown in Fig. 66. The tube wall temperature near the inlet is high because the coolant side heat transfer coefficient is relatively lower than the hot gas side. This is caused by the low hydrogen thermal conductivity.

The nondimensional profiles of Mach number and static pressure of the hydrogen fuel in the front heater tube are shown in Fig. 67. The Mach number is set below 0.6 to avoid choking at the heater tube outlet.

Table 14 shows the calculated heater performance. The outlet total temperature is higher than the required temperature of 650 K. The heat transfer characteristics are affected strongly by the forced convection of the combustion gas because the predicted heat transfer coefficient is smaller at the hot gas side than at the coolant side. However, this result depends on the degree of turbulence in the combustion gas and this effect can not be evaluated with confidence. Measurement of convection transfer from the gas flow at high speed and high temperature is quite difficult. Thus, the design was carried out with higher outlet total temperature.

D. Heat Exchanger Experiment on the ATREX-500 Engine

The experimental results of the heat exchanger installed in the ATREX-500 engine are shown in Fig. 68 compared with the Nusselt number correlation obtained in the model experiments.

Table 13 Conditions for analytical calculation

G_g, kg/s	7.41
T_g, K	2380
P_g, MPa	0.14
G_h, kg/s	0.25
$T_{h,o}$, K	650
$T_{h,i}$, K	72
$P_{h,o}$, MPa	0.87
d_e, mm	400
d_o, mm	4.7
d_i, mm	3.8

Table 14 Analytical results

$T_{h,o}$, K	714
$T_{h,os}$, K	658
G_h, kg/s	0.25
$T_{h,i}$, K	40
$P_{h,o}$, MPa	0.87
$P_{h,i}$, MPa	1.96
N_t,	18
N_s,	4
l_{hx}, mm	85
C_x, %	8
A_{hx}, m^2	0.9

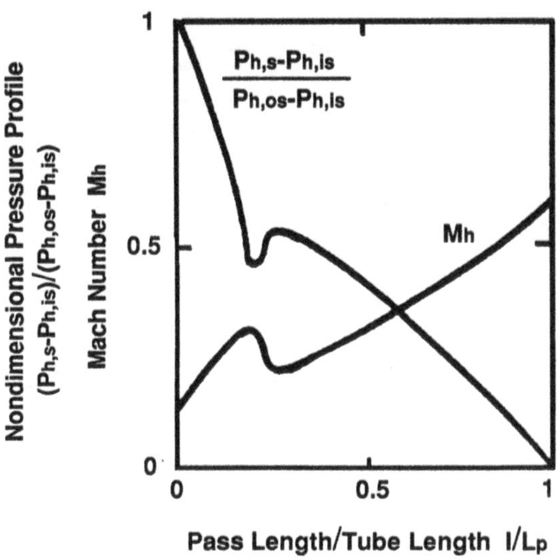

Fig. 67 Pressure profiles.

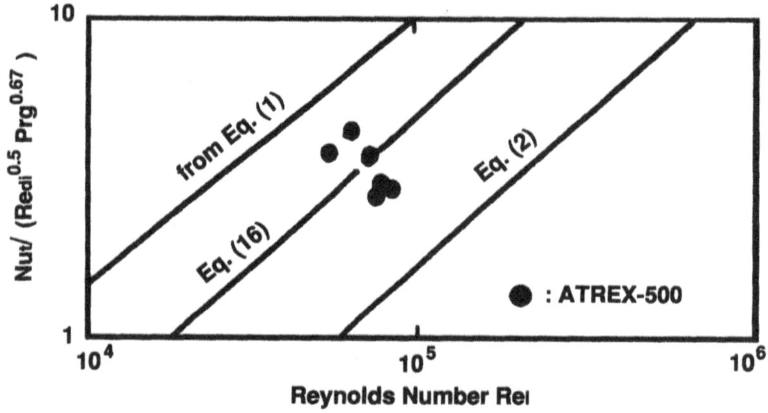

Fig. 68 Experimental results.

Combustion gas temperature is less than 2000 K and has not attained the design point temperature of 2380 K in the ATREX-500 experiments. These Nusselt numbers are average over the heat exchanger length.

Combustion gas temperature and velocity are calculated from the engine thrust efficiency. This efficiency is derived from the hydrogen flow rate, the equivalent mixture ratio, and the combustion efficiency. Nusselt numbers in the ATREX-500 experiments are larger than in the wind-tunnel experiment but less than in the gas turbine experiment. It is concluded that the effect of turbulence is not large in the ATREX-500 experiment. The effect of turbulence in the combustion gas is assumed to appear only in the coefficient of the Nusselt number function,

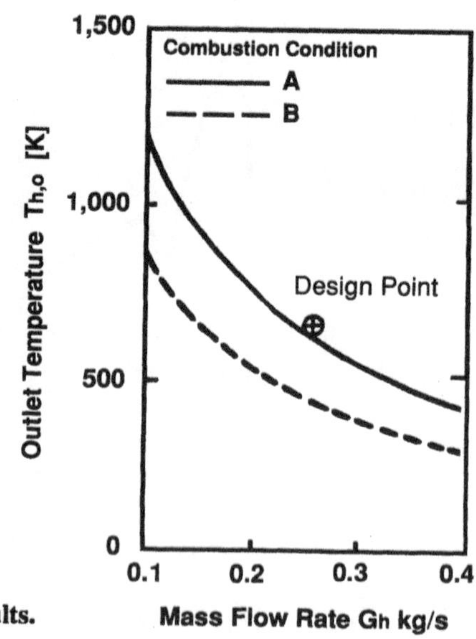

Fig. 69 Analytical results.

Table 15 Conditions on ATREX-500 engine test

Condition	A	B
T_{h}, K	650	290
G_{h}, kg/s	0.25	0.54
ϕ	1.2	2.83
T_{air}, K	290	290

thus the experimental correlation becomes

$$Nu_x = 1.6\times10^{-4}\, Re_x^{1.45}\, Pr_g^{0.67}\, \sqrt{x/d_o} \tag{16}$$

When the combustion gas condition is as shown in Table 15, the outlet hydrogen total temperature calculated by Eq. (15) is shown in Fig. 69. When the mass flow rate of hydrogen fuel is the design value of 0.25 kg/s, it is expected that the outlet total temperature becomes higher than 620 K. This shows that the design requirement to the heat exchanger is satisfied within a 5% error.

The heat exchanger designed here was operated without difficulty and satisfied sufficiently the system requirements. Thus, the heat exchanger configuration design is feasible to apply for the practical ATREX engine.

E. Concluding Remarks

The concept of the heat exchanger applied for the ATREX engine is shown, and the heat exchanger has been manufactured and tested in the ATREX-500 subscale engine.

It was shown by the preliminary tests using an aircraft jet engine flame that the optimum design point existed in the heat exchanger characteristics. The heat exchanger was designed for the practical ATREX engine based on this design point. It was verified by numerical simulation that the system requirements for the practical ATREX engine were satisfied well by the heat exchanger presented herein.

Table 16 Design value of air intake and precooler

Case	Mach 3 model		Mach 2 model	
Air intake				
Entrance area, mm2	4400		2400	
Throat area, mm2	1360		1416	
Pressure recovery, %	65		80	
Precooler (air side)				
Number of precooler	1 section	2 sections	1 section	2 sections
Entrance area, mm2	4800	4800	4800	4800
Entrance Mach number	0.198	0.198	0.221	0.221
Total pressure loss, KPa	58	136	96	Choking
Temperature drop, K	33	60	32	Choking
LN_2				
Temperature rise, K	25.0	47.9	36.5	Choking

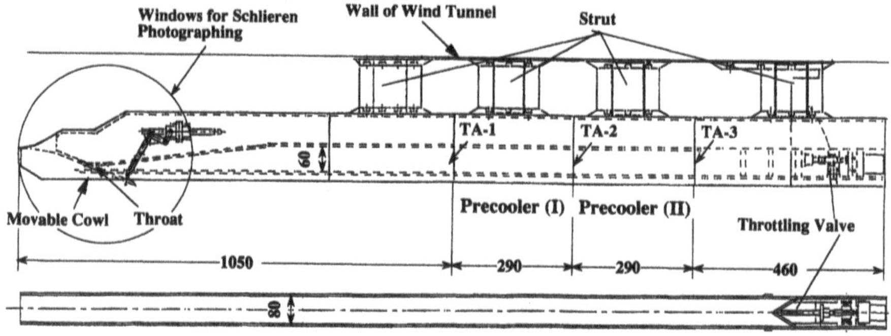

Fig. 70 Air intake model with precooler for wind-tunnel test.

Forced convection effects were examined in a gas turbine and a wind-tunnel experiment, where the former involved turbulence in the hot gas side flow and the latter did not.

On the basis of the concept given by the present study, a subscale heat exchanger for the ATREX-500 engine was made and tested to verify the heat transfer characteristics. There was a slight difference on the estimated forced convection characteristics of the combustion gas because they depend on the degree of turbulence in the combustion gas. However, the heat exchanger was verified to satisfy the design requirements within a 5% error, thus the heat exchanger configuration proposed here may be applied for the practical ATREX engine in the future.

VII. Preliminary Study on ATREX Engine Precooler

A. Introduction

Precooling of the ATR engines has the following effects. First is the reduction of the fan inlet temperatures, thus allowing operation of the engines at higher flight Mach numbers without cooling of the rotating turbomachinery. The second effect is the decrease of fan driving power and the increase of thermal efficiency, as a result of the intermediate cooling effect on the compression processes. Third is the increase of thrust due to the increasing density of incoming air, with some penalty in specific impulse caused by excess fuel consumed for cooling at higher Mach number flight.

It was revealed from the preliminary analysis that it is not suitable to set the precooler in the supersonic region of the airintake as a result of the large pressure losses. Thus, incoming air is cooled primarily in precoolers installed in the subsonic region although the intake wall is cooled in the supersonic region. The wind-tunnel test of the airintake with the precooler was initiated to evaluate its performance, where liquid nitrogen was used instead of liquid hydrogen.

B. Air intake Models and Experimental Apparatus

Two types of fixed geometric airintake models were tested coupling with one precooler configuration. The airflow passage of the airintake and the precooler has a two-dimensional rectangular cross section. A schematic view of a Mach 3 airintake model is shown in Fig. 70. The ramp shape of this model with three staged

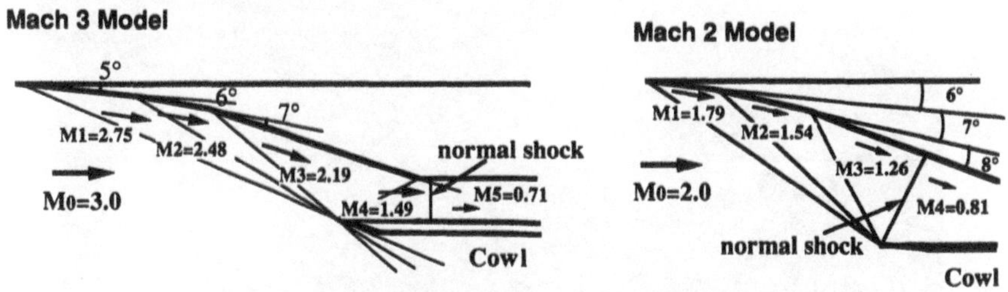

Fig. 71 Design flow pattern in ramp section of air intake.

angles is a mixed compression type in which the external shock waves focus on the cowl lip and make smaller total pressure losses (top, Fig. 71). A part of the incoming air stream is bled from the variable area slits in both the ramp and the throat and also from small holes on the cowl (Fig. 72). This air bleeding allows suppressing growth of the boundary layer, self-starting, and adjusting the normal shock wave at the throat. The rectangular glass windows are placed around the intake throat to observe the air flowfield there. Figure 73 shows the integration of the head of Mach 3 intake model in the wind tunnel.

Two precooler units are installed downstream of the airintake diffuser throat, and the plug-type valve is placed at the tail end of the model to control the back pressure and the position of the shock waves. In each precooler unit, 6 x 15 straight circular tubes are arranged in rows and columns, which are directed perpendicularly to the subsonic air stream. The precooler tube is 5.0 mm in diameter and 0.5 mm thick. The precooler tube row occupies 35% of cross-sectional area in the air stream. The total surface area of each precooler is 0.170 m^2. The circular tube

Fig. 72 Variable area slits in ramp and throat and small holes on cowl for air bleeding.

Fig. 73 Mach 3 intake model set up in the wind tunnel.

allows low total pressure loss for LN_2 flow and low thermal stress. Figure 74 shows the precooler tube assembly.

The model is supported on the top wall of a wind tunnel by four struts through which the inlet and outlet tubes of liquid nitrogen and the instrument lines are arranged. The Mach 2 model is similar to the Mach 3 model except for its external compression and no bleeding at the ramp region (bottom, Fig. 71). Table 16 shows the design value of each model with precoolers. The performancesof the airintakes and flow conditions is evaluated by the mass flow rates, the pressure losses, and the temperature differences across the precooler and also by Schlieren visual observation.

Fig. 74 Tube assembly of precooler.

Liquid nitrogen is supplied to the precooler from the run tank pressurized beyond the supercritical pressure of nitrogen to prevent a drop of heat exchange capability caused by film boiling.

The wind tunnel used has a rectangular cross-sectional flowfield with 0.6 x 0.6 m^2 and a Mach number capability of 1.5 to 4.0 with stagnation temperature around atmospheric temperature.

C. Results and Discussions

Schlieren photographs of flowfields in the Mach 3 intake model are shown in Fig. 75.† When the exhaust plug valve was opened widely to decrease back pressure, the air stream rear the throat was supersonic and many oblique shock waves appeared (supercritical condition, Fig. 75a). In this case, the total pressure loss in the precooler was large due to the higher flow velocity, and the flow separation was observed at the upper wall of the diffuser section. As the back pressure increased by closing the valve, the shock waves moved forward. These normal shock waves were kept at the throat under the design condition (critical condition, Fig. 75b). In this case, total pressure recovery was maximum with small flow spillage. As the valve was closed further, the shock wave rushed out of the cowl together with larger flow spillage (unstart condition, Fig. 75c). Figure 75 shows similar photographs of the Mach 2 model. In the supercritical case (Fig. 76a), the shock wave generated at the cowl lip reflected twice, and the flow separated behind the last shock wave. For the critical case, (Fig. 76b), the normal shock wave was kept just at the cowl lip, and the total pressure recovery became maximum with little spillage, as in the Mach 3 model. In the unstart condition (Fig. 76c), the bow shock wave stood ahead of the model and vibrated violently.

In the case of the Mach 3 model, the temperature drop of the airflow is shown in Fig. 77, where P0 is the stagnation pressure of the incoming airflow and the temperatures were measured at the inlet of precooler I, between precoolers I and II, and at the outlet of precooler II, respectively. It can be seen that the air was cooled from the inlet to outlet of the precoolers by the temperature of 75 K for the airflow rate of about 0.9 kg/s. This value is larger than the design value as a result of the smaller air mass flow through the precoolers. Conversely, the temperature rise of LN2 was 25-35 deg in precooler I and 20 deg in precooler II, respectively, for a total LN2 flow rate of about 1.0 kg/s (Fig. 78) . The total amount of heat transfer rate was around 50 kW. From these results, it was verified that the heat exchange rate between the air and LN2 was almost satisfied with the design value.

In the case of the Mach 3 model, the mass capture ratio (ε) is shown in Fig. 79 with a changing throttling position of the exhaust valve (V), where the back pressure increases with decreasing V. The mass capture ratio (ε) is defined as follows:

$$\varepsilon \equiv \frac{\text{Mass flow rate of air swallowed through throat}}{\text{Mass flow rate of air incoming to intake}} \tag{17}$$

In the case of no cooling, the air intake turned to the unstart condition and ε decreased by about 15% at V = 5. The limit of the start condition was extended by

†Figures 75 and 76 of this paper may be found in the color section.

Color Plates: Schlieren Photographs

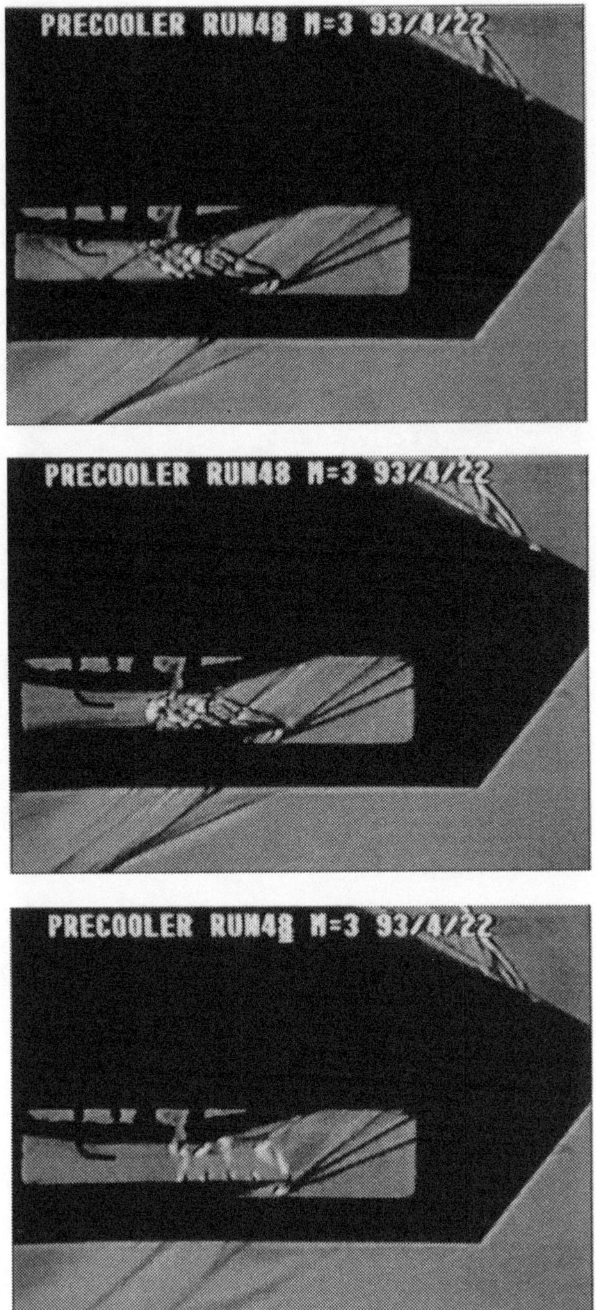

a)

b)

c)

Fig. 75 Schlieren pictures of flow fields in Mach 3 intake model: a) supercritical condition, b) critical condition, c) unstart condition.

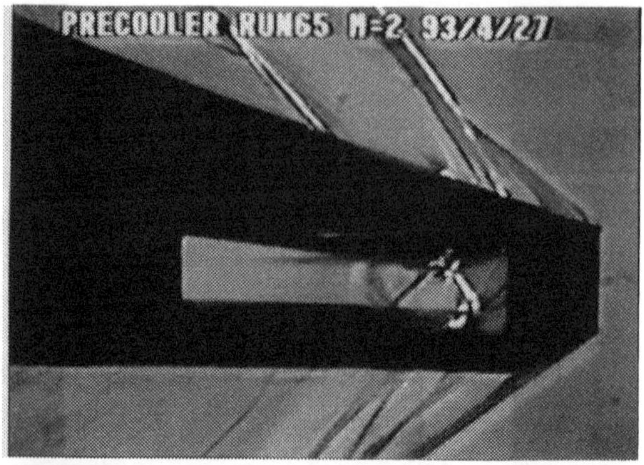

a)

b)

c)

Fig. 76 Schlieren pictures of flow fields in Mach 2 intake model: a) supercritical condition, b) critical condition, c) unstart condition.

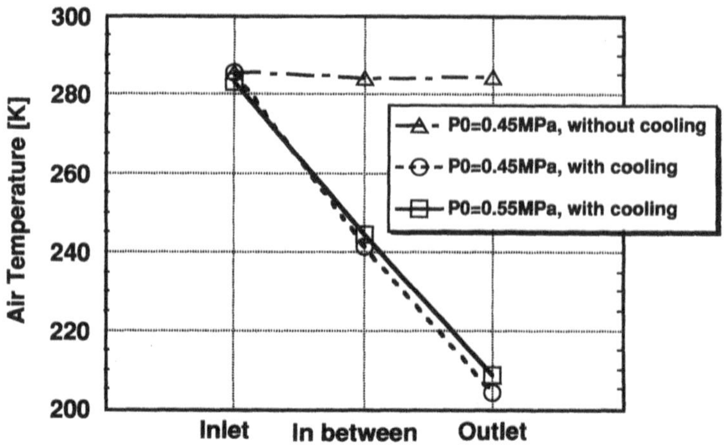

Fig. 77 Temperature drop of air flow (M = 3 model).

cooling, thereby ε did not decrease even at V = 5. Total pressure recovery (η) along the air stream is shown in Fig. 80 as a parameterof valve throttling. The total pressure recovery factor (η) is defined as follows:

$$\eta \equiv \frac{\text{Total pressure at specified point}}{\text{Total pressure of incoming air stream}} \tag{18}$$

At the throat, the total pressure recovery for every valve throttling was the same, except in the case of the unstart condition (V = 5, noncooling). The total pressure recovery upstream and downstream of precooler could increase with an increasing of the back pressure and about 55% pressure recovery was achieved in

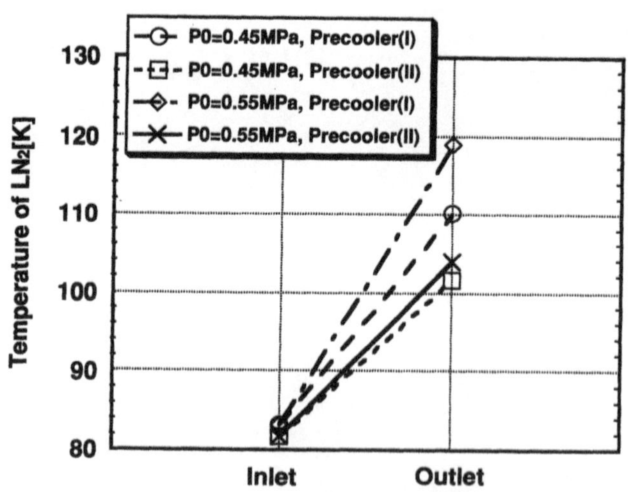

Fig. 78 Temperature rise of liquid nitrogen coolant (M = 3 model).

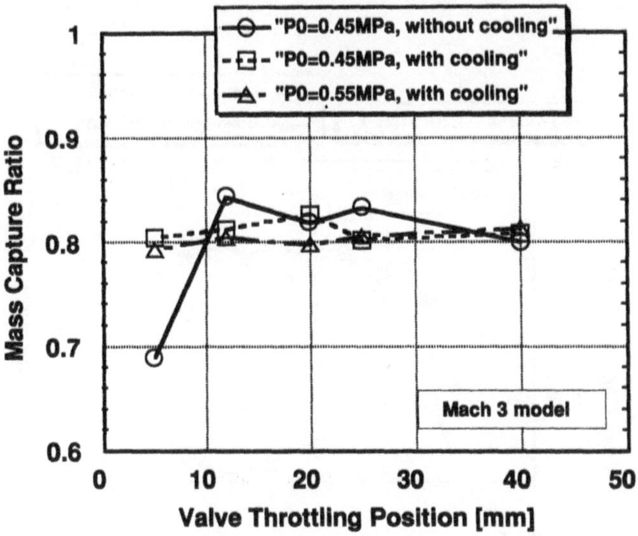

Fig. 79 Mass capture ratio (M = 3 model).

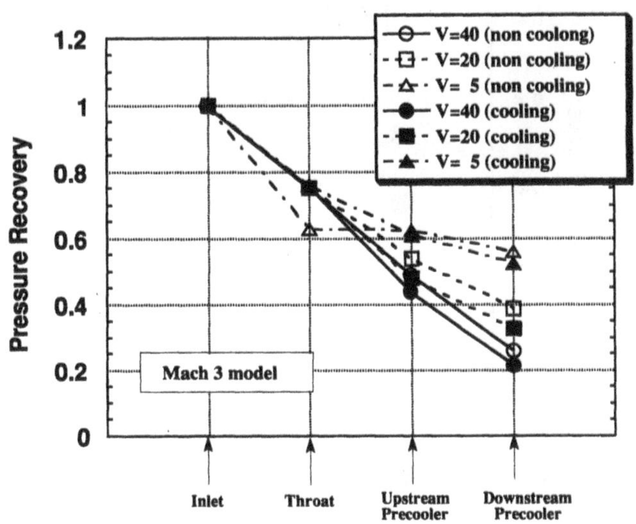

Fig. 80 Pressure recovery factor (M = 3 model, P0 = 4.5 ata).

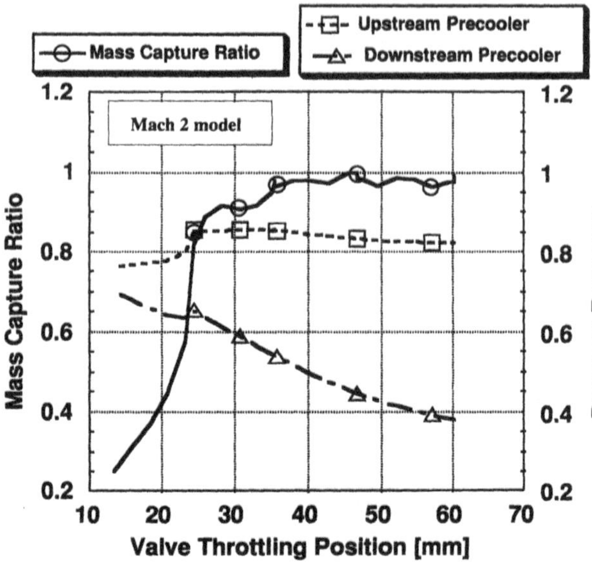

Fig. 81 Mass capture ratio and pressure recovery factor (M = 2 model).

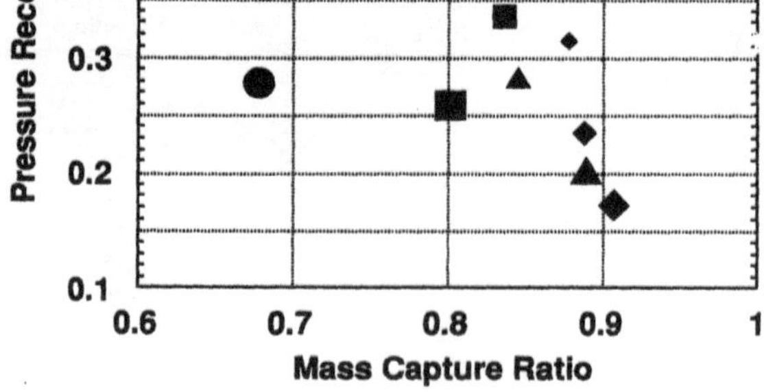

Fig. 82 Mass capture ratio and pressure recovery factor at off-design Mach numbe (M = 3 model).

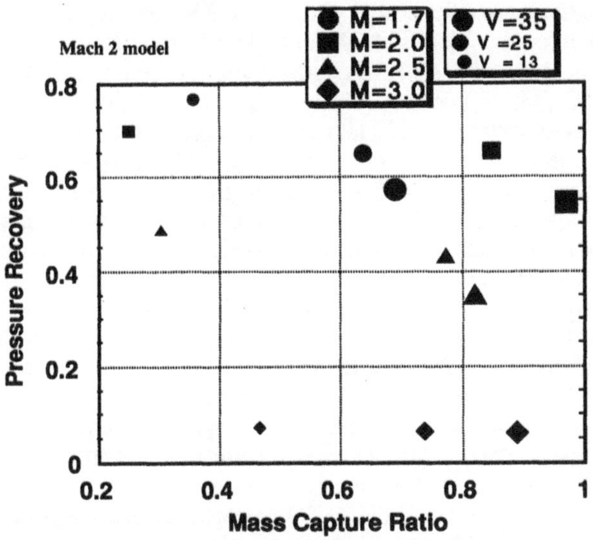

Fig. 83 Mass capture ratio and pressure recovery factor at off-design Mach number (M = 2 model).

this model. In the case of the Mach 2 model, the air mass capture ratio (ε) and the pressure recovery (η) are shown in Fig. 81 as a function of the valve throttling position (V). The intake kept the supercritical condition at the valve throttling position V above 35 and the pressure recovery downstream of the precoolers increased with decreasing V. The unstart condition occurred at approximately V = 25 and ε decreased abruptly.

Figures 82 and 83 show the maps of the air mass capture ratio (ε) and the pressure recovery factor (η) for the Mach 2 and 3 intake models, respectively, in the off-design Mach number condition without cooling. Mach 3 intake shifted to the unstart condition at a Mach number of 2.7 below the design point not depending on the valve throttling position. At Mach 3.3 and 3.5 beyond the design point, the start condition was still constant although the pressure recovery factor reduced slightly (Fig. 82). Conversely, the Mach 2 intake model shifted to the unstart condition at Mach 1.7 and 2.5 in the off-design condition, and so the intake performance fell.

D. Concluding Remarks

These test results made it clear that the heat transfer rate of the precooler installed in the subsonic air stream met the design value, and this precooler type is feasible for assemble in the ATREX engine without large total pressure losses. In the next step, another configuration of the air intake and the precooler will be examined to improve their performance, and especially the variable geometric airintake coupling with the precooler will have to be examined to adapt to the wide Mach number operation from the subsonic to the hypersonic flight up to Mach 6.

VIII. Concluding Summary

Development studies on the expander cycle air turboramjet (ATR) with precooler (intake air cooler) are discussed. This ATR engine is designated

"ATREX." The Institute of Space and Astronautical Science is carrying out development studies in cooperation with Ishikawajima Harima Heavy Industries and Kawasaki Heavy Industries. ATREX engine is going to be applied for the propulsion system of the flyback booster of the two-stage-to-orbit (TSTO) space plane.

Section II described the main feature of the basic ATR cycle.

In Sec III, three derivatives of the ATR cycle (one gas-generator cycle and two expander cycles) equippe with a precooler (intake air cooler) were analyzed and assessed for applying to the flyback booster of the TSTO spaceplane. LH_2 was used as a fuel in the present study. The engine design criteria were assessedbased on the near-term technology level.

Precooling of the ATR engines has the following effects. First, it reduces fan inlet temperatures, thus allowing operation of the engines at higher flight Mach number. Second effect is a decrease of fan driving power and increase of thermal efficiency. Third is the increase of thrust, with some penalty in specific impulse.

Preliminary estimation on engine weight suggested that the weight of the heat exchanger (precooler or heat exchanger) is rather large, as much as 30-40% of the weight of the basic engine without heat exchangers.

The flight-path analyses of the TSTO spaceplane were performed to clarify the requirements for the ATR engine. The flyback booster flies primarily along the way of constant dynamic pressure, which allows the constant thrust of the ATREX engine. Optimization of the entire flight depends primarily on the ascent angle at staging, which contributes to reducing the maneuvering velocity loss of the orbiter and thereby minimizing the velocity increment required for the orbiter. The preferable ascent angle at staging is more than 15 deg. The staging altitude has little effect on this optimization. Some requirements from the flight analyses are imposed on the performance characteristics of the ATR engine. The ATR engine is required to produce relatively larger thrust in the transonic regime to overcome the relatively large drag force and also in the terminal powered flight phase to climb by high ascent angle at high altitude. The ATR engine prefers to produce higher specific impulse in the lower flight speed below Mach number 2, where relatively longer time is spent for acceleration.

It was revealed from the ATR cycle analysis and flight-path analyses that the E-2-type ATR engine produced the best performance. Thus, the E-2-type ATR engine was selected for the present development study. Since then, E-2-type ATR was called "ATREX."

In Sec. IV the one-quarter-scale-model ATREX engine ("ATREX-500") was tested at sea-level static conditions to verify its feasibility. Thirty test runs with a cumulative operation duration of 1190 s were carried out from 1990 to 1992. The entire ATREX system worked well with successful results. The maximum specific impulse and maximum thrust were 1400 s and 460 kgf, respectively. The turbomachinery (tip turbine configuration), combustor, and heat exchanger worked well under the design conditions. The engine operation features at startup and shutdown transient, and the thrust maneuvering were clarified.

In Sec V, comparison tests between the gutter- and mixer-type flame holders were conducted with a view toward assessing the characteristics of different flame-holding systems for the ATREX combustors. The following results were obtained 1) Improvements given by the mixer type flame holder are elimination of flame detachment from the flame holder, improvement of flame temperature distribution in the combustion chamber, extension of the allowable range of fuel equivalence

ratio for stable flame holding, and diminution of flame instability. The resulting performance is almost equal to that obtained in the gutter type flame holders.
2) The improved mixer-type flame holder promises successful practical application to the ATREX engine combustors and the possibility of more compact combustors, whereas the flame holder wall temperature rise may require some protection measures in practical applications. This type of flame holder can be expected to provide a very promising new system.

In Sec VI, it was shown by preliminary tests using an aircraft jet engine flame that the optimum design point existed in the heat exchanger characteristics. The heat exchanger was designed for the practical ATREX engine based on this design point. It was verified by numerical simulation that the system requirements for the practical ATREX engine were satisfied well by the heat exchanger presented herein.

Forced convection effects were examined in a gas turbine and a wind-tunnel experiment, where the former involved turbulence in the hot gas side flow and the latter did not.

On the basis of the concept given by the present study, a subscale heat exchanger for the ATREX-500 engine was made and tested to verify the heat transfer characteristics. There was a slight difference in the estimated forced convection characteristics of the combustion gas because they depend on the degree of turbulence in the combustion gas. However, the heat exchanger was verified to satisfy the design requirement within a 5% error, thus the heat exchanger configuration proposed here may be applied for the practical ATREX engine in the future.

In Sec. VII, the wind-tunnel test of the air intake installed in the precooler was carried out to evaluate its performances, where liquid nitrogen was used instead of liquid hydrogen. Two types of fixed geometric air-intake models were tested coupling with one precooler configuration.

The preliminary test results made it clear that the heat transfer rate of the precooler installed in the subsonic air stream met the design value, and this type of precooler is feasible to be assembled in the ATREX engine without large total pressure losses. In the next step, another configuration of the air intake and the precooler will be examined to improve their performance and especially the variable geometric air-intake coupling with the precooler will have to be examined to adapt to the wide Mach number operation from subsonic to hypersonic flights up to Mach 6.

IX. Future Plan

Trial manufacture of tip turbines made of advanced carbon-carbon composite material is now under way. In the near future, the present titanium alloy tip turbine-fan assembly will be replaced by the carbon-carbon tip turbine-fan, which will improve fan performance by increasing its peripheral speed as a result of the high specific strength of carbon-carbon at high temperature.

Actual flight test of the ATREX engine is now being examined. In this flight test, the flying test bed will be used, which is the similar configuration of the "X-7" project carried out in the United States in the 1950s. In the early phase, the flying test bed is about 13-16 m in length and 1 m in diameter and is going to take off horizontally accelerating up to Mach 0.5 by the ground assist system and then powered by the ATREX engine itself.

Acknowledgments

The author would like to acknowledge Yoshihiro Naruo and Tetsuya Sato of the Institute of Space and Astronautical Science; Tsuyoshi Hiroki, Itaru Rokutannda, Masakazu Obata, Takeshi Kashiwagi, Tomoaki Mizutani, and Masato Oguma of Ishikawajima Harima Heavy Industries; Hiroshi Uchida and Kenji Hamabe of Kawasaki Heavy Industries; and Yoshiyuki Furuya of Sumitomo Heavy Industries for their tremendous contributions to the development of the ATREX engine.

X. Addendum

A. Air Precooler

In the early designs the precooler was assumed to operate above a flight speed of Mach 2 to prevent ice formation on it. The ATREX engine operation has since been changed, following the results of flight analysis, such that the precooler works continuously from liftoff in order to augment the thrust, especially in transonic flight.

One of the changes in engine operation pertains to an increase in air mass flow of the ATREX-500 engine. The engine, with air at a temperature of 160 K after cooling and a pressure recovery factor of 0.9, provides twice the thrust and 1.5 times the specific impulse compared with the nonprecooled engine at sea level static conditions. In the modified engine the air flow rate is increased from 6.7 to 11.5 kg/s and the fan pressure ratio from 1.53 to 2.03 in the same turbomachinery, following precooling of air.

The thrust sensitivity can be shown to be about 1.0% per 1 K of precooling temperature change, and 2.5% for each 1% of pressure recovery change around an air cooling temperature of 160 K and a pressure recovery factor of 0.9. For specific impulse, these figures become 0.4 and 1.4%, respectively. It is seen that the engine performance is affected significantly by the cooldown temperature and pressure losses of the precooler.

In the beginning of the precooler design study, some key targets were set on its performance characteristics and configuration as follows: it must provide precooling to 160 K with an air pressure recovery factor of 0.85 and higher at SL-static conditions; it should be a shell-tube configuration with a tube diameter of 3 mm and a wall thickness of 0.1–0.15 mm; the precooled ATREX engine should be capable of being operated with higher than the stoichiometric mixture ratio of fuel; and an icing-free regime of operation of the precooler should be provided from the SL-static conditions. Based on those requirements a number of precooler configurations were examined, of which two were selected for technology development. These two precoolers are designated as "blin" and "baraban" in Russian, which mean "pancake" and "drum," respectively.

The blin type precooler provides a pressure recovery factor of 0.86 at the design point of 160 K. In the baraban type the factor is 0.94. It is estimated that the higher pressure recovery of the baraban type can provide a large increase of thrust of 20% and specific impulse of 12% compared to the blin type. As the vehicle accelerates, the pressure recovery factor of the precooler is estimated very quickly to become close to 1.0 and, therefore, the degree of air precooling is important rather than the pressure recovery factor. A more sophisticated analysis is necessary for the selection of the planned precooling level at SL-static conditions, taking into consideration the optimization of the total amount of fuel required in the overall mission. This is one of the subjects in the next step of the study, following the verification of the current design in the SL-static test of the precooled ATREX-500, conducted in September and November 1995.

An analytical and design study of air precooling for the ATREX engine showed that the precooler outlet air temperature T = 160 K and the air pressure recovery factor 0.9 are almost optimum at an equivalence ratio of 1.3–1.4 under SL-static conditions.

Under those conditions the thrust and specific impulse of the existing ATREX-500 engine at SL-static conditions are expected to be increased by 2 and 1.5 times, respectively, compared to the nonprecooled one.

B. Precooler Icing

The icing problem in a heat exchanger for atmospheric air cooling was a matter of concern from the beginning of the precooler study. However, icing-free operation of the precooled engines has been examined with a physical model of ice formation generated by the CIAM of Russia and its subsequent verification by them. Two ice-formation mechanisms have been postulated by the CIAM, depending on the vapor partial pressure Pst.

1) Ice formation on a micro level. When the vapor partial pressure is lower than the vapor pressure at the triple point (Ptr = 0.006228 atm), ice is formed, avoiding the liquid phase. This process occurs by crossing the sublimation line in the P-T diagram of water. In this case an ice layer becomes formed at a molecular level. Icing of this kind is not strong and generally is carried over by an air stream.

2) Ice formation on a macro level. When the vapor partial pressure is higher than that at the triple point, and air is chilled below the water dew point, a liquid phase of water can be seen in the air stream. This process occurs by crossing the saturation line and the solidification line in the P-T diagram of water. In this case, the ice formation can be said to occur at the macro level, that is, water droplets settling on the heat exchange surface turn into dense ice. Thus, according to this physical model of ice formation, there is no possibility of ice formation spoiling the performance of a heat exchanger at Pst < Ptr.

These conditions occur at low air humidity, which is found typically in a high altitude, and at an air temperature lower than the water triple point. In other words, no ice is formed on the heat exchanger surface when at least one of the following two conditions is satisfied:

Air pressure < pressure of water triple point (1)
Air temperature < 273 K (2)

Relations (1) and (2) allow limit lines to be drawn for preventing icing on the speed–altitude map of a flight path.

From December to March, a no-icing window was expected around the launch site candidate town, Taiki. Maximum duration of icing formation along the typical flight trajectory was estimated to be as long as 25–40 s in the most humid season.

C. Axisymmetric Variable Geometry Air Intake

Axisymmetric air intakes have been tested using a supersonic wind tunnel at ISAS since 1993, to assist in developing a design method and to establish change of performance characteristics during boundary layer bleeding.

The center spike of the intake is movable for adjusting to the given Mach number. The outer diameter of the test model is 127 mm, which is one-sixth of the ATREX engine for flight test. Air bleeding is arranged by small holes on the center spike and slits on the cowl. The amount of bleeding was fixed during the tests, though it has to be adjusted for the Mach number of the incoming air stream in real flight. The performance decreases abruptly and the inlet shifts to an unstart condition beyond the design Mach number, especially in an intake with increasing internal compression and, therefore, the design point should be set at the maximum flight Mach number. The Mach number is 4.5 for the present design. Type E, F, and G models were designed by means of the method of characteristics.

All of the models were tested below Mach 4, which is the maximum possible in the ISAS wind tunnel. The type E model meets the target value below Mach 4. The type F model values are smaller in lower and higher Mach number ranges and, therefore, type F is estimated to be not useful beyond Mach 4. Although the type G model values are relatively lower over a Mach number range of 1.5 to 3.5, the value is larger at Mach 4 than at Mach 3.5, and thus is considered to be better beyond Mach 4. Mass capture ratio is 70, 33, and 58% at Mach 3.5 for types F, E, and G, respectively.

In the type F model, the mass capture ratio decreases while the maximum pressure recovery factor is nearly constant as bleeding in cowl increases. The startability becomes worse without cowl bleeding and leads to reduced performance. From these results, there arises an optimum amount of bleeding in the cowl. When air bleeding was decreased in the center spike, the performance became worse, not because of decreasing the throat area but because the boundary layer effects become larger over the larger surface area of the center spike.

The pressure recovery factor in each type of air intake model is being improved so as to meet the target of the ATREX engine. The bleeding over the center spike should be made larger to reduce the boundary layer, whereas the bleeding in the cowl needs to be precise both in amount and location.

D. High Loading Ram Combustor

A compact ram combustor is required to reduce substantially the size and weight of the ATREX. A compact ram combustor is accomplished through the use of high, specific combustion loading. The combustor performance is strongly dependent on the mixer and, thus, the improvement of the mizer is one of the well-worn approaches to increasing the combustion loading.

The mixer of the ATREX combustor has two functions, namely mixing and flame holding without the need for any other flame holder and, therefore, the combustor performances in the ATREX engine are dependent only on the mixer configuration. Six different configurations of mixers were tested in subscale tests as well as with CFD analyses. It was found that the mixer with skewed lobes, which cause a swirl flow of hydrogen and air in the mixing process following injection into the combustor, reduces combustion in the chamber. The swirl flow improves the combustion performance by lengthening the residence time for mixing and combustion.

Of these mixers with different skewed angles, the skewed angle of 15 deg gave the best performance. In this case, combustion is completed over a length of 1.3 D from the mixer outlet end compared to 1.8 D of the nonskewed mixer (D being the combustor diameter). From these results it appears that a significant enhancement of specific combustion loading can be realized using mixers with skewed lobes, approximately by a factor of 1.5, compared to the nonskewed mixer. The combustion flame structure was analyzed by means of ultraviolet photography of the OH-radical luminescence, and the intensity distribution was similar to the temperature distribution. The pressure drop through the combustion chamber was not greatly increased by the mixers with skewed lobes compared with a nonskewed mixer.

A swirl mixer with 15 deg of skewed angle is used in the ATREX-500 engine and was proved effective in tests during 1995.

E. Flight Test Plan on ATREX Engine

In the final phase of the ATREX engine development study, it is planned that the overall system performance, functions, and operations of the engine will be proved by means of an actual flight test. The flight test will be performed by using the flying test bed (FTB), which is expected to be powered by the ATREX engine itself after liftoff. The FTB takes off horizontally with the takeoff assist system running on rails up to 0.4–0.5 of Mach number. The takeoff assist system is powered by a conventional turbo jet engine or solid motors in parallel with the ATREX engine. In the early phase of the flight test, the maximum flight speed is planned to be up to Mach number 4.5 while using metal turbomachinery. The final target speed for flight tests is Mach 6, which can be attained with the use of carbon-carbon composite turbomachinery. After reaching the maximum speed planned, the FTB will decelerate by throttling the engine power, then gliding down, and finally landing on the sea by means of a parachute for recovery. The FTB will fly by means of an autonomous and programmed control in addition to the radio guidance system with GPS. The data obtained during the flight will be transmitted by means of telemetry.

Two ATREX engines have been designed for flight tests, one is the metal ATREX, which is available for flight up to Mach 4.5, and the other is the ACC ATREX, which can attain the final goal of Mach 6. They are of the same size as the ATREX-500 developed for SL-static tests. The engine performance is improved in the ACC ATREX primarily through improvement of the turbomachinery with its higher peripheral speed at a higher temperature compared to the metal ATREX. Its performance is also improved by precooling compared to the test results of nonprecooled ATREX-500. The target thrust and specific impulse of the modified ACC ATREX engine with a fan inlet diameter of 0.3 m are estimated to be 1730–1930 kgf and 3150–3300 s at the SL-static condition. A decrease in specific impulse is caused beyond Mach 3.5 because of an increase in the hydrogen flow rate for cooling purposes, although thrust is increased.

The flight test needs approximately 170–180 km in range and 190–260 s in time to reach the target flight speed for the ACC and the metal ATREX powered flight, respectively. The dynamic pressure will be kept at 50 kPa from flight Mach 1 to 5 and reduced gradually beyond Mach 5 in order to reduce the combustion pressure.

Additional information on the air precooler, the variable geometry air intake, the combustor modifications, and the flight-test plan can be found in Refs. 16–18.

References

[1]Tanatsugu N., et al., "A Study on Two-Stage Launcher with Air-Breathing Propulsion", *Proceedings of the AAS-JRS Joint Symposium*, 1985.

[2]Tanatsugu N., et al., "An Analytical Study on Two-Stage Launcher with Separate Ramjet and Rocket Propulsion", *Proceedings of the 15th International Symposium on Space Technology and Science*, 1986.

[3]Tanatsugu N., Inatani Y., Makino T. and Hiroki T., "Analytical Study of Space Plane Powered by Air-Turbo Ramjet with Intake Air Cooler", IAF Paper 87-264, 1987.

[4]Glatt C.R., "WAATS-A Computer Program for Weights Analysis of Advanced Transportation Systems," NASA CR-2420, 1974.

[5]Schoettle U.M., "Massenanalyseeinstufigergefluegelter Raumtransporter," *IRA-83-iB-2*, 1983.

[6]Tanatsugu N., Naruo Y., Sato T., Rokutannda I. and Uchida H., "Development Study on ATREX Engine", IAF-93-S.4.483, 1993

[7]Kashiwagi T., Obata M., Ohkita Y., Tanatsugu N. and Naruo Y., "Test Results of the Hydrogen Fueled Model Combustor for the Air Turbo Ramjet Engine," *Eleventh International Symposium on Air Breathing Engines*, 1993

[8]Sato Y. and Kashiwagi T., "Recent Developments in Jet Engine Afterburner Technology," *AVIATION ENGINEERING*, March-June 1985 (in Japanese).

[9]Tanatsugu N., Oguma M., Mizutani T. and Yano T., "Heat Transfer Characteristics of Hydrogen Heater for Air Turboramjet Engine," *First International Conference on Aerospace Heat Exchanger Technology*, 1993.

[10]Glickstein, M. R., and Powell, T. H., "Advancements in Hydrogen Expander Airbreathing, Engines," AIAA Paper 87-2103, 1987.

[11]Holman, J. P., *Heat Transfer*, 4th McGraw-Hill, Tokyo, 1976.

[12]Katayama, K., et al., *JSME Data Book: Heat Transfer*, 4th JSME, Tokyo, 1986.

[13]Rohsenow, W. M. and Choi, H., *Heat, Mass, and Momentum Transfer*, Prentice-Hall, Englewood Cliffs, NJ, 1961.

[14]Taylor, M. F., "A Method of Prediction Heat Transfer Coefficients in the Cooling Passages of Nerva and Phoebus Rocket Nozzles," NASA TM X-52437, 1968

[15]Inoue, M., et al., *JSME Data Book : Hydraulic Losses in Pipe and Ducts*, JSME, Tokyo, 1980.

[16]Tanatsugu, N., Sato, T., et. al., "Development Study on ATREX Engine," Proceedings of 45th IAF Congress, Jerusalem, Oct. 1994.

[17]Balepin, V.V., and Tanatsugu, N., "Some Considerations of Precooler for ATREX Engine," Proceedings of the Space Transportation Conference, ISAS, Jan. 1995.

[18]Balepin, V.V., Tanatsugu, N., Sato, T., Mizutani, T., Hamabe, K., and Tomike, J., "Development Study of Precooling for ATREX Engine," Proceedings of XII ISABE Symposium, Melbourne, 1995.

In-Flight Oxidizer Collection Systems for Airbreathing Space Boosters

J. L. Leingang*, L. Q. Maurice †, L. R. Carreiro ‡
Wright Laboratory, Wright-Patterson Air Force Base, Ohio 45433

Nomenclature

CR	= collection ratio, kilograms of oxidizer collected per kilogram of hydrogen or water coolant
g_C	= gravitational constant
GW_{ACES}	= gross-takeoff weight of ACES vehicle
GW_{LC}	= gross-takeoff weight of LOX carrying vehicle
I_{SP}	= specific impulse
I'_{SP}	= apparent or effective specific impulse
K	= relative first stage weight factor of a LOX-carrying vehicle compared to an ACES vehicle, defined by eq.(A14)
MR	= mass ratio
MR'	= apparent or effective mass ratio
O/F	= oxidizer to fuel ratio
W_{ACES}	= weight of ACES equipment
W_1	= first stage structure, propulsion and fuel weight of ACES vehicle
W_F	= weight of fuel
W_{F2}	= second stage fuel weight
W_{FC}	= fuel used by ACES during collect
W_{LEA}	= weight of liquid enriched air
W_{OR}	= second stage weight in orbit

This is a work of the U.S. Government and is not subject to copyright protection in the U.S.

*Supervisory Aerospace Engineer, Aero Propulsion and Power Directorate, Associate Fellow AIAA.
†Aerospace Engineer, Aero Propulsion and Power Directorate, Senior Member AIAA.
‡ Chief, Experimental Development Branch, Aero Propulsion and Power Directorate.

W_{OX} = weight of second stage oxidizer
ΔGW = gross-take-off weight difference between a LOX-carrying vehicle and an ACES vehicle, defined by eq. (A19)
ΔV = boost phase velocity increment

I. Outline of Chapter

This chapter discusses the development and application of a propulsion fluid system known as the Air Collection and Enrichment System (ACES). In an airbreathing space launch vehicle, the role of ACES is to acquire, early in the ascent phase, the liquid oxygen that will be used later in a rocket mode once the airbreathing propulsion phase has been completed. The refrigeration capacity of liquid hydrogen is employed to condense incoming air, and fractional distillation is used to extract the oxygen. Earth-to-orbit capability is achieved without carrying liquid oxygen (LOX) from takeoff or relying on scramjets for high Mach propulsion. The ACES concept is introduced, technologies and equipment for generating oxidizer in-flight are discussed, and operating regimes of the process are covered. Discussion then proceeds to vehicle design studies done as part of the original USAF Aerospaceplane program of the 1960s and a description is given of the ACES air separator and heat exchanger developments conducted as part of that program. The chapter then reviews recent comparison studies of ACES-equipped launch vehicles with scramjet-powered vehicles and LOX-carrying vehicles. The concepts of a

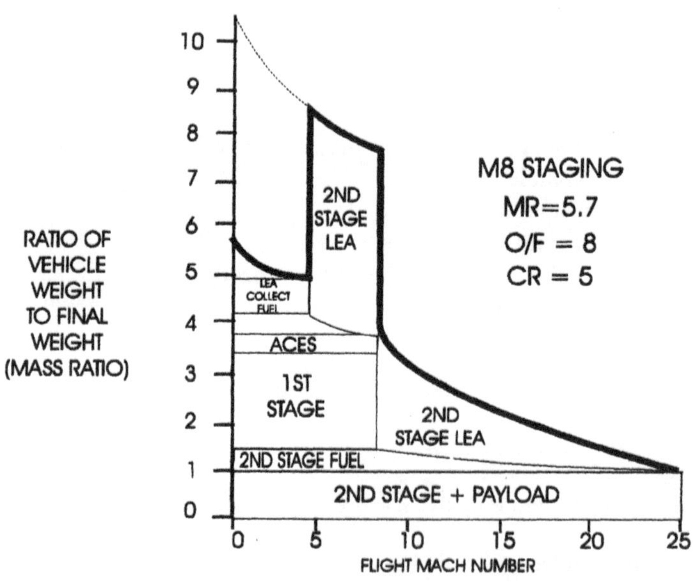

Fig. 1 Vehicle weight history.

dual-fuel vehicle and an all-hydrocarbon vehicle are introduced, and design studies of these vehicles are presented. The chapter concludes with a summary of other recent analytical and experimental work done on the ACES concept.

II. Introduction To Concept of In-Flight Oxidizer Collection

In the late 1950s, interest grew in developing fully reusable orbital launch systems. The prospect that nonrecoverable systems would impose immense costs for space operations was as apparent then as it is now. As is also the case today, there was interest in horizontal takeoff and landing, single-stage vehicles, and maximum use of airbreathing propulsion. The weight penalties for single-stage, horizontal takeoff vehicles having no airbreathing propulsion were known to be prohibitive. In that era, very little was yet known about scramjets and their potential for propelling vehicles to orbit. Only very cursory studies of scramjets' prospects existed before 1960. In an effort to extend airbreathing propulsion specific impulse levels to flight speeds airbreathing function, concepts such as an "air hoarder" were investigated. Such a vehicle would use the refrigerant capacity of the liquid hydrogen fuel to condense and store liquid air for a subsequent rocket mode after airbreathing operation was no longer possible. The limit of airbreathing propulsion was that of the subsonic burning hydrogen fuel ramjet (i.e., about Mach 8). By about 1959, it was concluded that collecting and storing liquid air would not be effective, the inert weight penalty of the nitrogen constituent being much too high. Just prior to 1960, studies were initiated to investigate air collection approaches in which the nitrogen component was removed and expelled from the vehicle while retaining the oxygen and storing it as a liquid. This concept, known as the Air Collection and Enrichment System (ACES), was the subject of extensive Air Force-funded vehicle system study and propulsion hardware experimentation from 1960 through 1967. This work was done as part of the first USAF Aerospaceplane program. The ACES concept employed conventional airbreathing propulsion to approximately Mach 8. Figure 1 shows the weight history of a two-stage ACES vehicle as it proceeds from takeoff to orbit. The vertical axis show the vehicle weight relative to the weight in orbit (weight of second stage structure plus payload is 1.0). Looking at fig. 1 and reading up the left side of the figure, one sees that the vehicle takeoff weight consists of the weight in orbit (second-stage vehicle plus payload), the second stage fuel, the first-stage vehicle and propulsion, the liquid enriched air (LEA, 90% O_2) collect equipment, the fuel for acceleration from Mach 5 to 8, the fuel used during Mach 5 cruise for collecting LEA, and the fuel used to accelerate from takeoff to Mach 5. The vehicle starting weight is approximately 5.5. It accelerates to Mach 5 at about 75,000 ft, the weight falling to about 4.9. Then, during an ensuing cruise phase of about 20 min, LEA is collected and stored in the second stage. During the Mach 5 cruise phase, the vehicle weight grows to the value suggested by the figure, approximately 8.5. After collecting LEA, the vehicle accelerates on ramjet propulsion to the Mach 8 staging speed. The amount of fuel expended to collect LEA is considerably less than the weight of LEA collected. The amount of fuel needed for LEA collection can be easily quantified. For a typical rocket powered second stage, a mass ratio of 5.7 is required to achieve orbit from a Mach 8 starting speed. Assuming a rocket O/F of 8 and the final mass in orbit to be 1.0, then 4.112 units of LEA and 0.588 units of fuel are required by the second stage. The second stage LEA is generated during first stage cruise at a collection ratio *CR* of approximately 5 lbs LEA collected per pound of fuel used for collection. In the example of fig. 1, the amount of fuel required to collect LEA is

$$\frac{O/F}{CR} \times F = \frac{(8)}{5} \times 0.588 = 0.9408 \text{ units of fuel}$$

Note that the amount of fuel required to generate the second-stage oxidizer in-flight is less than a quarter of the oxidizer weight. By not carrying second stage oxidizer at takeoff, a large reduction in takeoff weight is achieved through a smaller vehicle and propulsion as well as much less acceleration fuel. Had the vehicle of this example carried LOX at takeoff, a starting weight exceeding 10 units would have been required.

The apparent effective specific impulse of the second stage is defined by the propellants onboard the vehicle at takeoff, i.e., the second stage fuel (0.588) plus the first-stage fuel used to collect LEA (0.9408). Adding these together gives a second stage apparent mass ratio of 2.5288, and applying the rocket equation gives the following apparent effective specific impulse

$$I'_{SP} = \frac{\Delta V}{g_c \ln MR'} = \frac{(26{,}000-8000)}{(32.17)(\ln 2.5288)} = 603 s$$

Note that the apparent mass ratio is about one-half of a vehicle which carries LOX from takeoff, and the apparent effective I_{SP} is nearly double.

Use of a scramjet engine is an alternative to be considered. If the fuel used to collect LEA in the previous example were instead carried in the second-stage, in addition to the second stage fuel, and used in a scramjet operating to Mach 20, the following would be obtained. Some of the total 1.5281 units of fuel would have to be allocated to LOX and fuel for the final rocket acceleration beyond Mach 20. This amounts to 0.528 units of LOX and 0.066 units of fuel, leaving 0.9341 units of fuel available for scramjet acceleration from Mach 8 to Mach 20. For this example, an effective I_{SP} of 809 s would have to be delivered by the scramjet for the vehicle to be no heavier at takeoff than the ACES vehicle. Achieving such a higher I_{SP} level over this broad Mach number range is a large challenge for the scramjet. The foregoing is a brief and only approximate illustration of how an ACES vehicle can be very much lighter than a LOX carrying vehicle using a rocket second stage and how very challenging it would be for a vehicle with a second-stage scramjet to be launch weight competitive. Later, in Sec.V, more detailed design studies of these three concepts are presented and reinforce the general conclusions just reached.

In Appendix A, material from Ref. 1 is used to quantify the launch weight advantage of an ACES vehicle. Appendix A presents a more rigorous analysis of how ACES increases the Isp and reduces the mass ratio of a vehicle. As an introduction, fig. 2 (to be discussed more fully in Appendix A), illustrates how the apparent mass ratio of a stage is dramatically reduced from a LOX carry vehicle as the collected oxidizer purity can be enhanced by the LEA collection equipment. Figure 2 also shows the small velocity interval over which simple air collection can be better than carrying LOX from takeoff.

A. Oxidizer Collection Equipment

Air to be enriched to LEA enters through inlets that are either part of or separate from the propulsion system. Figure 3 shows the air to be bled from the main inlet of a ramjet type engine. This air is initially cooled by nitrogen waste in a heat exchanger (heat exchanger 1), then cooled by hydrogen (heat exchanger 2) to saturated vapor at approximately 10 -15 atm pressure. The separator generates LEA (90% O_2) and nitrogen vapor (about 98% N_2) at approximately 2-3 atm, 140°R. A distillation process (to be discussed later in Sec. IV) is used. The LEA is sent to storage tanks. The cold nitrogen vapor is

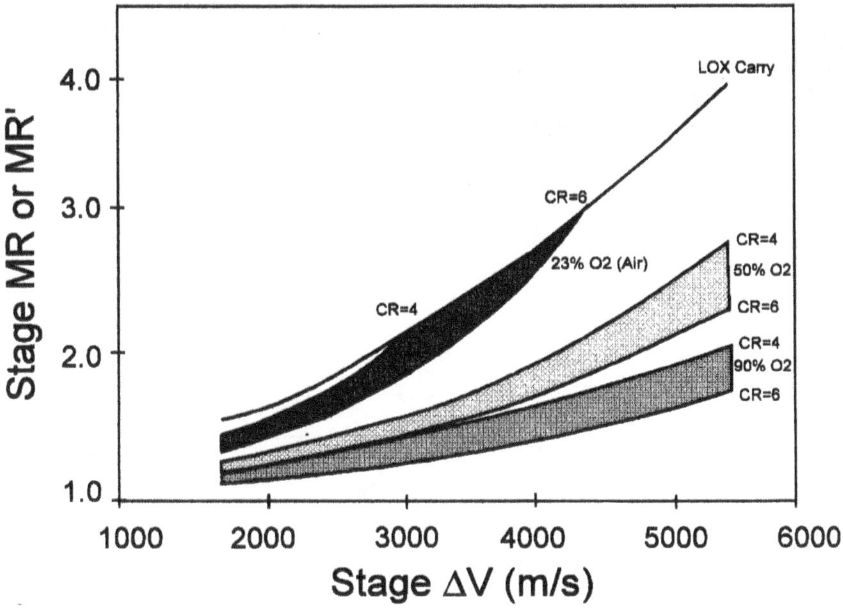

Fig. 2 Effect of collected oxidizer purity and collection ratio on apparent mass ratio.

compressed to propulsion system chamber pressure (approximately the inlet recovery pressure) and used to cool incoming air. Figure 3 suggests it is introduced directly to the propulsion cycle. Since the nitrogen is nearly at inlet recovery pressure and has been reheated to very near the inlet temperature, it can be discharged from the vehicle in a number of efficient ways, including a separate thrust nozzle, or mixed with the propulsion nozzle flow. I_{SP} values for the nitrogen waste are on the order of 140-160 s. Most of the drag incurred in capturing the nitrogen component of the collected air can be recovered as thrust. The net drag penalty incurred during LEA collection is mostly that of the collected LEA. During the LEA collection cruise phase, this drag imposes approximately a 25% penalty on the cruise specific impulse. For a ramjet at Mach 5, I_{SP} would be reduced from 3000 s to about 2300 s. Refrigeration required to collect the LEA comes from the cryogenic hydrogen fuel. The value of the collection ratio (*CR*) is set by the energy required to condense liquid oxygen. Typically, *CR* values range from 4 to near 6, the smaller value being typical of a system with few hydrogen expanders and using reduced effectiveness heat exchangers. The higher *CR* value typifies a system with extensive turbomachinery (including a cold air compressor), very high effectiveness heat exchangers, slush hydrogen, and para-ortho conversion of the hydrogen.

The weight of the LEA collection equipment tends to be on the order of 3 - 5% of vehicle takeoff weight. Specific weight of the equipment is usually in the range of 10 - 15 lbs equipment weight per pound per second of air flow. Thus, for the usual airflow rates of 1000 - 2000 lbs/s, equipment weight would range from 10,000 to 30,000 lb. The air separator (to be discussed fully in section IV). has a specific weight in the range of 3 - 5 lb/lb of air flow per second. The smallest specific weights of the LEA collection equipment are compatible with *CR* values on the order of 4. The highest weights are compatible with *CR* values of about 5.

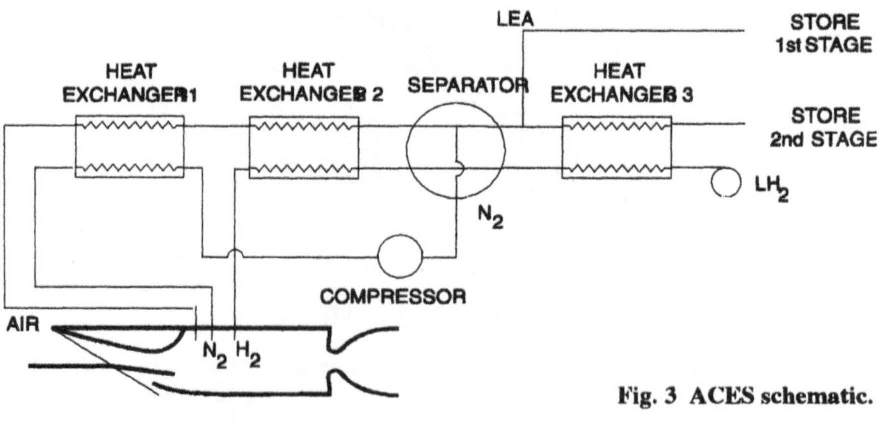

Fig. 3 ACES schematic.

B. Operating Regimes for Oxidizer Collection

The process of LEA collection is usually accomplished within the interval of Mach 1.5 at 25,000 ft to Mach 5.5 at 75,000 ft. LEA may be collected during a protracted acceleration within this speed interval (the usual and more efficient approach) or may be accomplished during cruise at a fixed speed. Whichever method is chosen, the LEA collection equipment is sized to permit operation at or near the design *CR* value at all times. The vehicle drag establishes the engine fuel flow rate requirement during LEA collection, and so the LEA collection rate should be *CR* times the fuel flow rate. For the sizes of vehicles presented in this chapter (400,000 - 1,000,000 lbs), the total LEA quantity ranges from 200,000 to 500,000 lb. and is collected at a rate of 250 - 450 lbs per sec. The LEA collection times range from 10 to 20 min. The air flow from which LEA is extracted ranges from 1000 to 2000 lb/s.

The speed range over which LEA collection occurs affects the component requirements of the LEA collection system. For example, initiating collection at Mach 1.5 requires a cryogenic air compressor to achieve the 10-atm pressure required by the air separator. Also, the process may not achieve a power balance because of work limits on hydrogen fuel turboexpanders used in the process. For operation at the high-speed end, near Mach 5.5, heat exchanger material temperature limits are approached for conventional materials and reasonable weights. Fuel heat sink becomes a problem and forces a undesired reduction of *CR*.

In a design study of an ACES vehicle, the second stage LEA quantity is fixed within fairly narrow limits once the second stage starting conditions are defined. During first-stage flight, fuel flow is governed by vehicle size/weight and choice of flight condition. The achievable *CR* value then defines the amount of LEA collect fuel required and the time needed to collect the LEA. Achieving a minimum launch weight for the vehicle involves a search for a flight condition or acceleration interval that minimizes the sum of LEA collection fuel and LEA collection equipment weight. The relation among *CR*, LEA equipment specific weight, and LEA collection flight condition will define an optimum LEA collection rate and flight condition.

III. Early Vehicle Systems Studies: The Aerospaceplane

Initial studies of an ACES vehicle, conducted as part of the USAF Aerospaceplane program, considered a single-stage-to-orbit vehicle.[2] The payload goal was 9072 kg (20,000 lb). By 1962, it was concluded that projected vehicle materials technology and propulsion weight requirements made a single-stage concept all but impossible for technology expected to be available by the 1975 - 1980 period. Study emphasis then turned to two-stage vehicles. These were studied in the 1962 - 1964 period and were assessed to be considerably less risky. The final vehicle design, designated the 1964 Point Design Vehicle, is shown in Fig. 4. Gross take-off weight was 317,518 kg

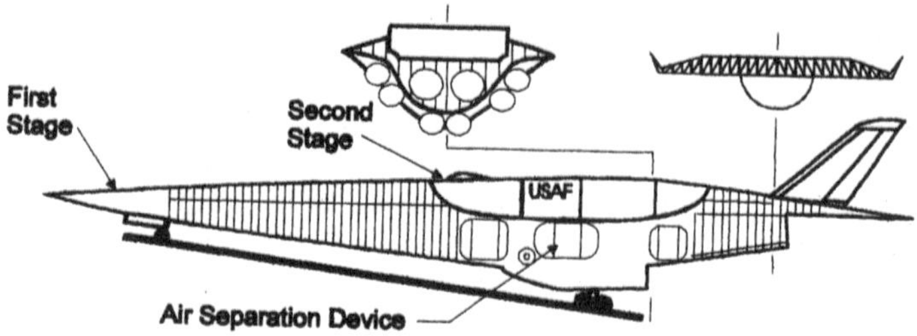

Fig. 4 General Dynamics 1964 Point Design Vehicle, final configuration from Aerospaceplane program.

(700,000 lb). It delivered a payload of 10,433 kg (23,000 lb) to a 555.6 km (300 nautical mile) polar orbit. Hydrogen fuel turboramjet propulsion was used to accelerate the vehicle to the LEA collection point near Mach 5. After the 20 - 30 min. LEA collection phase, the vehicle accelerated to the staging point using rocket and ramjet thrust. At staging, the second stage ascended to orbit on rocket propulsion using a ballistic path. The weight history and propulsion phases are shown in Fig. 5. Staging was accomplished at Mach 8.6 at 53,645-m (176,000-ft) altitude.

Because the 1964 Point Design Vehicle staged hypersonically, optimum solutions to important configuration and materials problems could be obtained separately for a hypersonic (Mach 8) vehicle and for a rocket-powered ascent/re-entry vehicle. As examples, the first stage was exclusively a Mach 8 aircraft, and as such it could employ superalloy construction on the underside and titanium on the upper surface. The second-stage configuration

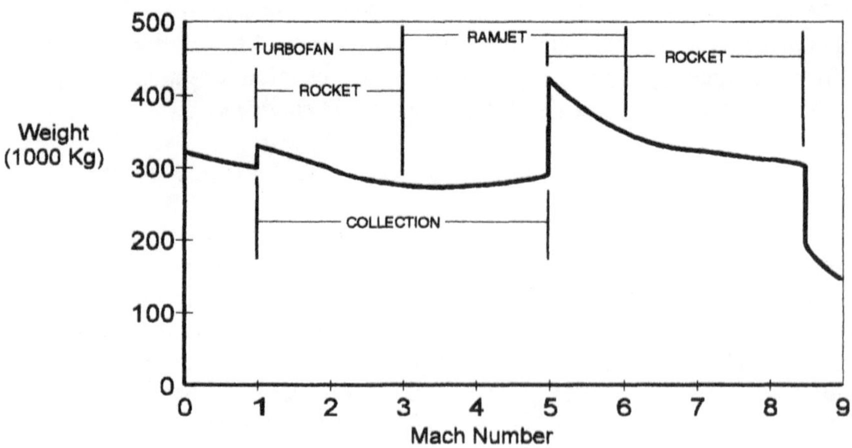

Fig. 5 1964 Point Design Vehicle propulsion modes.

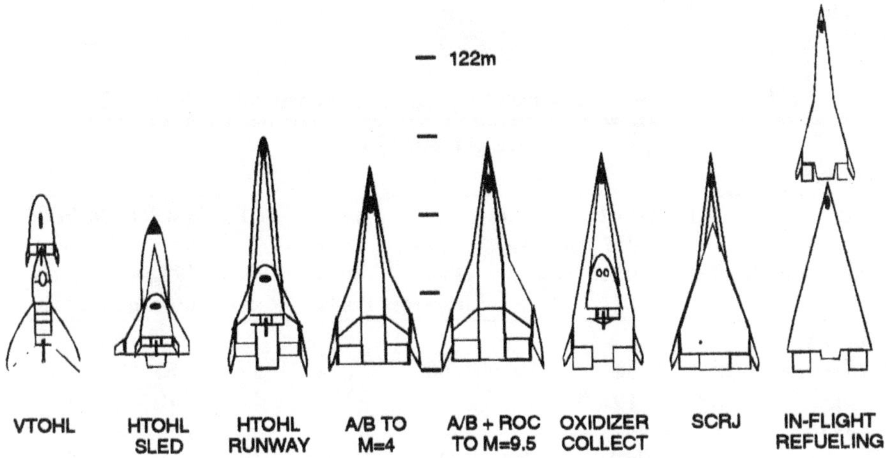

Fig. 6 Size comparison of 18,144 kg payload vehicles. (Ref. 2).

requirements were dominated by re-entry heating, thus its blunted shape. Coated refractory metals were used on the underside and superalloy and titanium on the upper surface. The relatively heavy LEA collection equipment only had to be accelerated to the staging velocity. The advantages of a two-stage vehicle are accentuated when the propulsion system or other fixed components are unusually heavy. The 1964 Point Design Vehicle represented the highest state of refinement reached in the vehicle studies. The work is documented in Ref. 2.

A later effort, conducted in 1967, examined propulsion alternatives for two-stage vehicles delivering 18,143.9 kg (40,000-lb) payloads to a low Earth orbit. This study[1] is particularly significant because it considers eight different concepts, including ACES, rocket-only concepts, and scramjets, all using common study ground rules. The all-rocket systems have the smallest inert weight but have the heaviest launch weight (due to stored LOX). The ACES vehicle and the scramjet have the largest inert weight except for the in-flight refueling concept. Scramjet and ACES have the smallest takeoff weight. Inert weight strongly dominates the cost picture. The ACES and scramjet vehicles have comparable costs that are the highest of all the concepts except for in-flight refueling. Figure 6 shows the vehicles studied and Table 1 shows weight statements.

I V. Air Collection and Enrichment Systems Component Developments

A. Air Separator

During the 1959 - 1961 period, various air separation concepts were studied for the ACES application. These concepts included cocurrent spray contactors,

Table 1 Weight Comparison for Vehicles Discussed in Reference 2

	Weight 1000 Kg							
Vehicle[a]	2	3	4	5	6	7	8	9
Item	VTOHL	HTOHL Sled	HTOHL Runway	Air Br	Air Br + Rocket	ACES	SSCRJ	In-Fl Refuel
Propulsion	Rocket	Rocket	Rocket	TFRJ[b] + Rocket	TFRJ Rocket	TFRJ+ Rocket	TRJ[c] + Scramjet	TFRJ+ Rocket
Stage 1								
Propulsion	32	38	41	68	81	89	65	73
Equipment & Misc	4	10	12	11	11	14	10	20
Total Inert Weight	95	113	138	181	201	191	173	225
Fly Back Fuel	20	24	27	20	20	6	17	7
Weight at Staging	114	138	165	201	221	196	191	231
Boost Propellent	473	436	518	105	208	146	111	7
Total Stage	588	573	683	306	429	343	301	721
Stage 2								
Structure	20	20	20	20	17	20	47	71
Propulsion	10	10	10	10	9	10	18	45
Equipment & Misc	7	7	7	7	6	7	9	10
Total Inert Weight	37	37	37	37	32	37	74	126
Payload	18	18	18	18	18	18	18	18
Weight in Orbit	55	55	55	55	49	55	93	144
Propellant	160	160	160	160	125	174[d]	87	128
Total Stage	215	215	215	215	175	229	179	272
Total Inert Weight	132	150	175	219	232	228	247	351
Gross Take-Off Weight	803	788	898	521	604	420	480	993

[a] Vehicle No.1 is the 2nd stage.
[b] TFRJ - Turbofan Ramjet, [c] TRJ - Turboramjet, [d]Includes collected LEA

nitrogen freezeout, centrifugation, chelation, molecular sieves, membranes, fractional distillation, vortex tubes, and chemical reactions. See Appendix B for description of these concepts. For ACES to be advantageous relative to LOX-carrying orbital vehicle approaches, its equipment weight had to be very small relative to the required mass of stored LEA. By 1961, it was concluded that the chemical reaction and fractional distillation concepts had the best prospects for meeting weight and volume goals for the ACES air separation system.[3]

References 4 - 6 discuss the chemical-reaction air separation concept. System studies indicated this separation concept to have potential for higher payload performance than the rotary distillation concept.[7] Because of the extremely high-temperature environment created by the oxygen extraction chemical reaction, however, it became evident that the chemical air separator would be a much more complex mechanical device than originally anticipated. Thus, the bulk of the air separator development effort was concentrated on developing the fractional distillation air separation concept.

The optimum air separation scheme for the ACES application proved to be a variation of the commercially used double column cryogenic fractional

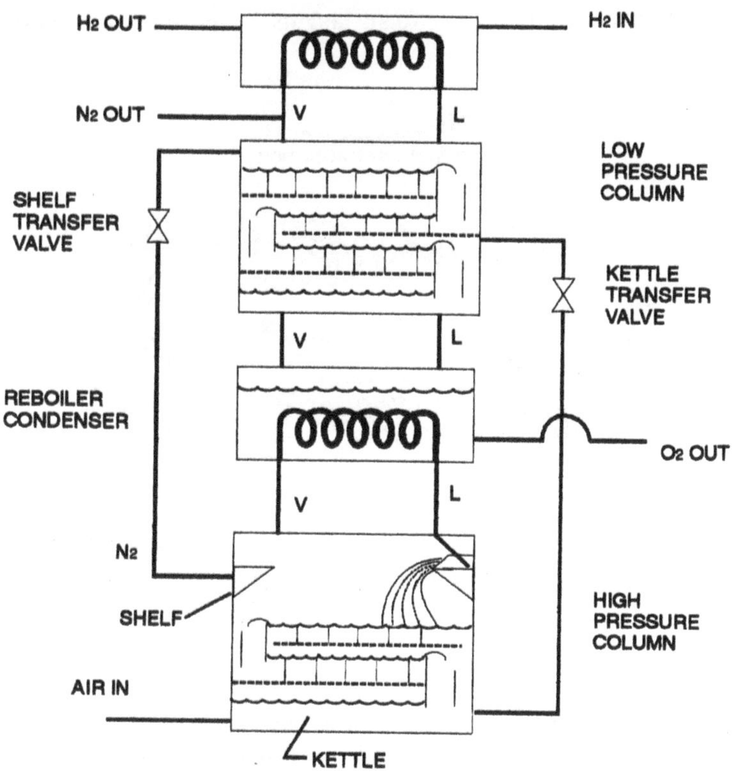

Fig. 7 Double column air separator.

distillation air separation process. Although the principles behind fractional distillation of air are simple, the actual process is somewhat complex (Fig. 7). The functional elements of this process include a high-pressure distillation column (a stack of perforated plates, called trays) to effect a preliminary separation, a low-pressure distillation column to provide product streams of the final desired purity, a reboiler-condenser (heat exchanger) in which the nitrogen-enriched vapor stream from the high-pressure column is condensed against the oxygen-enriched liquid product from the low-pressure column and subsequently used as liquid feed for both the high- and low-pressure columns, and a reflux condenser in which hydrogen is used to provide the required refrigerative capacity for the entire process. More detailed discussions of fractional distillation are found in the literature.[8-10]

Although the commercial cryogenic fractional air distillation process offers excellent oxygen recovery from the air stream and would make good use of the available refrigerative capacity of hydrogen, weight and volume requirements would be prohibitive for air vehicle applications. An invention by Nau and Campbell[11] solved this dilemna. The flat plates used in conventional 1-g distillation become cylinders rotating about their own axes. The liquid travels in a crossflow pattern from downcomer to downcomer (a drain tube) to the outside of the column, while vapor travels radially inward. As a result of the increased gravitational field, the foam heights in the trays are substantially

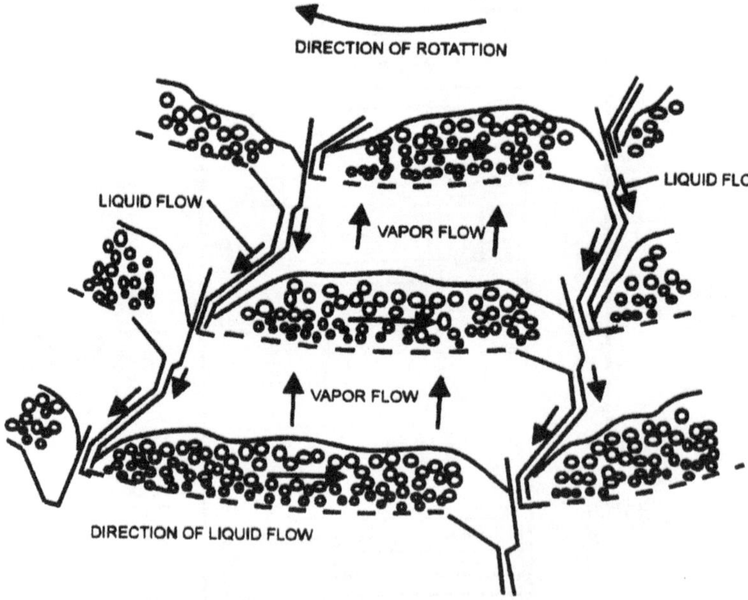

Fig. 8 Rotary high-g air separator trays for ACES. Vapor flows radially inward through the porous trays(plates). Liquid is impelled by the centrifugal field to flow outward and across the trays.

reduced, and vapor velocities can be increased. Thus, the weight and volume occupied by the columns are reduced, and the capacity of the process is dramatically increased (see Fig. 8). The reboiler-condenser also would be an extremely heavy, bulky component in 1-g use. To obtain desired weight and volume reductions in the reboiler-condenser, it is necessary to minimize resistance to heat transfer on both the boiling and condensing sides. In the rotary air separator, this was accomplished by applying a porous sintered metal surface to the boiling and condensing sides. Surface tension drew the condensate into the porous layer, and the gravitational forces then drained the condensed liquid from the porous matrix. The weight and volume reductions realized in the reboiler-condenser were a major accomplishment of the 1960s rotary air separator development program. The design of the reboiler-condenser is illustrated in Fig. 9.

The operating requirements for a rotary air separator for the 1964 Point Design Vehicle, along with weight estimates for a flight weight separator, are given in Table 2. Design studies of a full-scale rotary air separator established separator specific weight to be approximately 5 kg/kg/s of airflow. The diameter of the separator, which is dictated by the desired product purity, was approximately 3.05 m (10 ft); the length of the unit, which is determined by

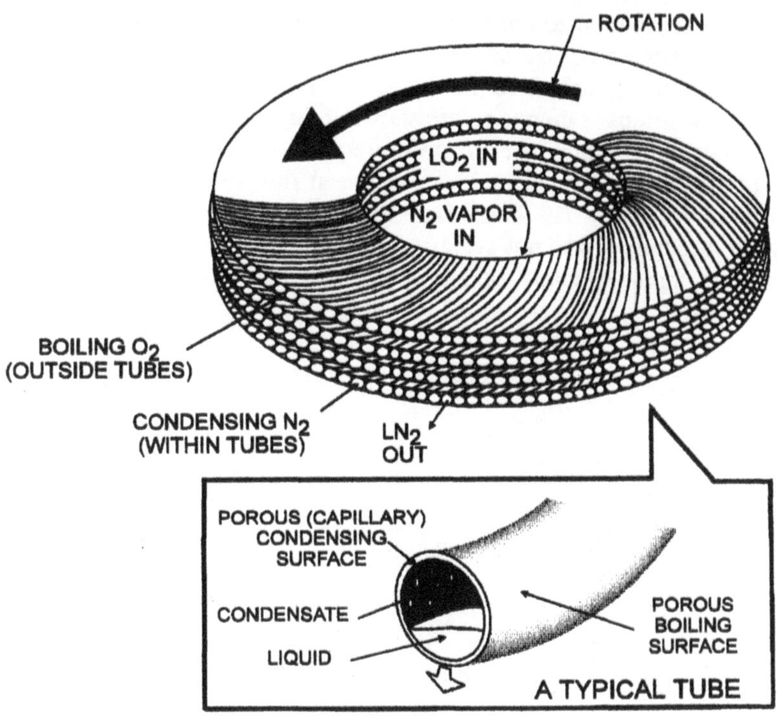

Fig. 9 Reboiler-condenser for rotary high-g air separator.

Table 2 ACES Rotary Air Separator Performance and Weight

Performance		
Throughput	945 kg/sec	(2083 lb/sec)
Inlet Pressure	15.1 atm	(222 psia)
Outlet Pressure	3.8 atm	(56 psia)
Waste Purity, wt. $\%N_2$	98	
Product Purity, wt. $\%O_2$	90	
Component Weights		
Columns	1225 kg	(2700 lbs)
Reboiler Condenser	1225 kg	(2700 lbs)
Other	1860 kg	(4100 lbs)
Total	4310 kg	(9500 lbs)

desired throughput capacity, was approximately 6.1 m (20 ft). To demonstrate the functional feasibility and confirm the weight and volume estimates of the cryogenic rotary air separator, several programs were conducted culminating in the design, fabrication, and test of a 45.36 kg/s (100 lb/s) boilerplate rotary air separator. The arrangement of the various components and the radial dimensions were similar to those of a full-scale unit. The internal components were made of aluminum, but stainless steel was used for all other components in order to simplify operation of the unit. Photographs of the boilerplate separator and various components are shown in Figs. 10 -16.

The predicted performance of the boilerplate separator, along with actual results from seven test runs conducted between July 1965 and April 1966, are presented in Table 3. The boilerplate tests were very successful. Although maximum throughput was not achieved due to limited power availability (600 hp available, approximately 900 hp required), the stable operation of the

Fig. 10 Scale model of flight weight rotary air separator, table-top model cutaway showing the internal details of the separator.

Fig. 11 Boilerplate rotary air separator test unit.

Fig. 12 Boilerplate rotary air separator test unit.

Fig. 13 Boilerplate rotary air separator, low pressure column.

Fig. 14 Boilerplate rotary air separator, closeup of tray assembly.

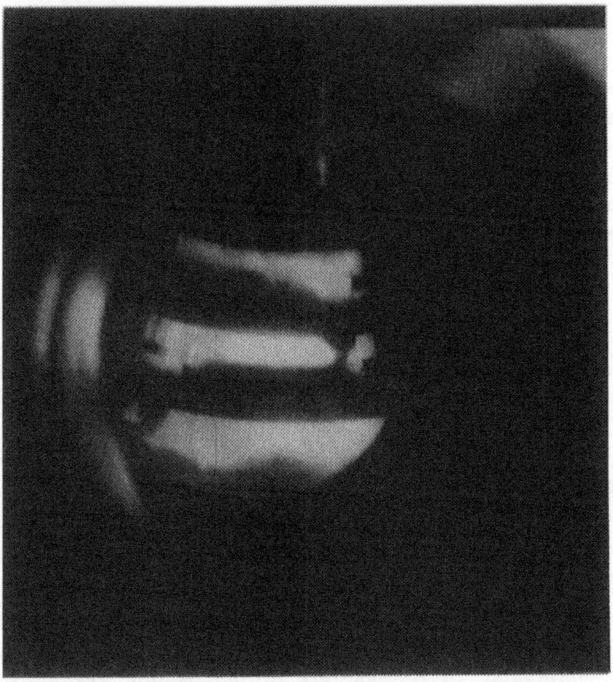

Fig. 15 Closeup of rotary distillation tray test unit showing liquid flow through downcomers and across trays. Tray spacing is approximately 5 cm.

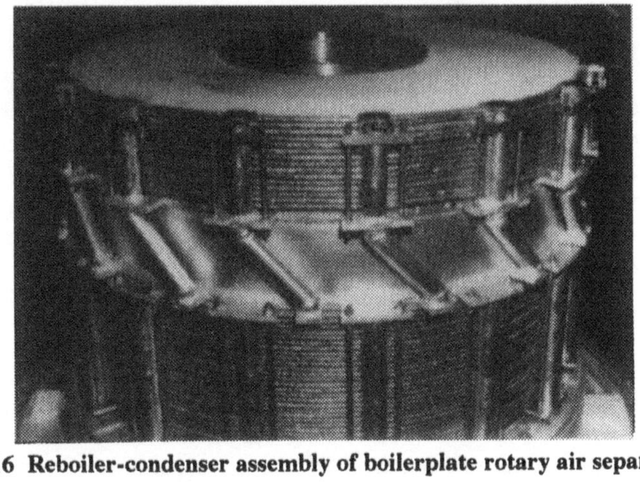

Fig. 16 Reboiler-condenser assembly of boilerplate rotary air separator.

Table 3 Boilerplate Rotary Air Separator Test Results

	Predicted Value	Actual Performance
Air Flow, kg/sec	45	23-35
Inlet Pressure, atm	16.3	11.2-14.6
Waste Pressure, atm	3.1	3.1-4.1
Product Purity, wt% O_2	90	80-94
Product Purity, wt% N_2	98	96-99
Design Speed, RPM	570	280-380
HP Column Tray Efficiency, %	80	63
LP Column Tray Efficiency, %	56	57
RBC Overall Heat Transfer Coefficient, W/m^2-K	370	476

system demonstrated that the operation of a double-column rotary air separator presents no major difficulty. Air separation efficiencies in the high-pressure column were somewhat lower than predicted, but it was believed this could be easily corrected through minor tray design modifications, and the overall heat transfer coefficient of the reboiler-condenser was even greater than predicted. The degree to which predicted performance of the boilerplate separator was achieved is indicative of the validity of the design correlations needed to design the various components of the separator. Thus, this very successful test program validated the weight and volume estimates for a flight-weight rotary air separator. Although hardware used in this program is no longer available, results from the rotary separator development program can be confidently used today as the basis for continuing development of the concept. The work is documented in Refs. 12 and 13.

A full-scale air separator, processing approximately 2000 lb/s of air, is about 10 ft in diameter and about 20 ft in length. It weighs about 9500 lb and requires about 18,000 hp. At first glance, these numbers suggest an impossibly large device, but comparison to a large turbofan engine is in order. The Pratt & Whitney JT9-D engine (used on the McDonnell-Douglas DC-10 aircraft) has an airflow of about 1500 lbs/sec, has a diameter about 10 ft, length of 20 ft, and weighs about 9,000 lb. This comparison illustrates that the air separator is about as compact and light as a familiar modern turbofan engine.

B. Heat Exchangers

The ACES vehicles studied in the 1960s required a complex heat exchanger system to cool air from the incoming stagnation conditions down to the saturated vapor condition required by the air separator. Even though heat exchanger use is common in conventional aircraft, the ACES heat exchanger system presented a unique requirement. Large quantities of air had to be

Table 4 ACES Heat Exchanger Materials

Maximum Metal Temperature, K	Material	Tube Wall Thickness, mm.
166	Niobium-1% zirconium coated alloy	0.152
1256	Hastelloy-X	0.102
811	Aluminum Alloys	0.102

cooled from temperatures as high as 1666.7K (3000°R) to temperatures as low as 111.1K (200°R). Unlike conventional aircraft heat exchangers, the ACES heat exchanger system was a major contributor to the overall weight of the propulsion system. Thus, developing very compact, lightweight, and high-performance heat exchangers was a primary goal of the propulsion system component development efforts undertaken in the 1960s.

The design of the ACES heat exchanger system was thoroughly studied by several researchers. The generally favored approach for ACES heat exchangers was a cross-counterflow tube and shell design using very thin-gauge materials. The materials selected for the various temperature regions along with achieved tube wall thicknesses for these materials are shown in Table 4. The cooling fluid, either hydrogen or waste nitrogen from the separation process, flowed inside the tubes, while the air to be cooled flowed outside. Figure 17 shows the generic design used. A total heat exchanger system specific weight of 10 kg/kg/s of air flow per second was estimated for the 1964 Point Design Vehicle. To confirm these weight

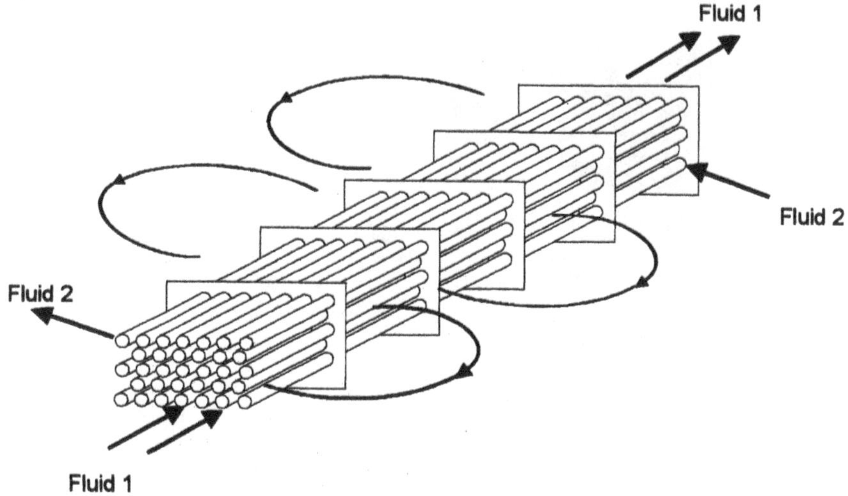

Fig. 17 General arrangement of cross-counterflow tube and shell heat exchanger.

Fig. 18 Cryogenic heat exchanger, 3.175-mm stainless steel tubes, 0.0508 mm wall.

estimates, several exploratory development programs were undertaken between 1962 and 1968. The objectives of these programs were to prove the validity of the various heat exchanger design and fabrication concepts and to demonstrate the overall performance of the heat exchangers. Because of funding limitations, full-scale systems were never demonstrated. Rather, modules sized for roughly 1/50 of the total airflow were generally evaluated. Figure 18 depicts a 444-tube core, 45.72-cm- (18-in) long module using 3.175-mm (1/8-in.) diameter, 0.0508-mm (2 mil) wall stainless steel alloy for low temperature application. A stainless steel heat exchanger module for use with air temperatures as high as 1388.9°K (2500°R) was also built. The core of this

Fig. 19 High temperature heat exchanger, L605 material, 2666.7 K inlet air temperature.

module had 3.048-m- (10-ft) long stainless-steel tubes, 3.175-mm (1/8-in) diameter and 0.0508-mm (2-mil) wall thickness. A heat exchanger module designed to handle air temperatures up to 2666.7K (4800°R) can be seen in Fig. 19. The core of this module was made up of 585, 2.54 -mm (1/10 in) diameter, 0.2032-mm- (8- mil) thick L-605 alloy tubes. Plans to evaluate a full heat exchanger system were made but never carried out.

The various tests were generally successful, thus proving the validity of the heat exchanger designs and fabrication techniques and the reliability of the elements of the heat exchanger system. A new specific weight estimate of 17.2 kg/kg/s was derived based on the experimental results. This represents a substantial increase from the original estimate. However, given the many advances in heat exchanger technology and fabrication techniques realized since the 1960s, substantial reductions to the total weight of the heat exchanger system are very likely, and the extensive design data base available from the 1960s programs will serve as the foundation for future programs. The heat exchanger work is documented in Refs. 14 - 16.

V. Recent Vehicle Systems Studies

A. Hydrogen Fueled Vehicles

Initially (about 1986), two-stage vehicles capable of delivering payloads of 29,483 kg (65,000 lb) (Vehicle B) and 68,039.6 kg (150,000 lb) (Vehicle C) were designed. These payloads are representative of a Shuttle class and a heavy-lift vehicle, respectively. The vehicles are fully recoverable and use horizontal runway launch and landing. The study ground rules specified that the sizes and vehicle weights would be determined that satisfied the payload goals.

Vehicle size and weights were estimated using the Preliminary Design and Weight Analysis Program (PDWAP). The vehicle sizing part of this code was developed by Carreiro.[17] The weight estimation part of the code was the NASA Weights Analysis of Advanced Transportation Systems Program (WAATS), which was enhanced and updated for use on a microcomputer. Reference 18 describes the WAATS code. To insure that the program accurately modeled a two-stage ACES vehicle, the weight adjustment coefficients were calibrated using the 1964 Point Design Vehicle weight information.Once the size and weights of the vehicles were established, they were flown using a two-degree-of-freedom trajectory code, ETOC1.[19] The inputs for the trajectory code consist of weight, aerodynamics, engine performance, and flight-path command data. Representative aerodynamic coefficients were obtained for both vehicles. The engine performance data were based on a turboramjet and obtained from Ref. 20. The flight- path command data controlled the vehicles acceleration to Mach 5. At a Mach 5 cruise, the ACES system was used to collect a predetermined amount of LEA. The vehicle then accelerated under rocket and ramjet power to Mach 8 at which point staging occurred. The

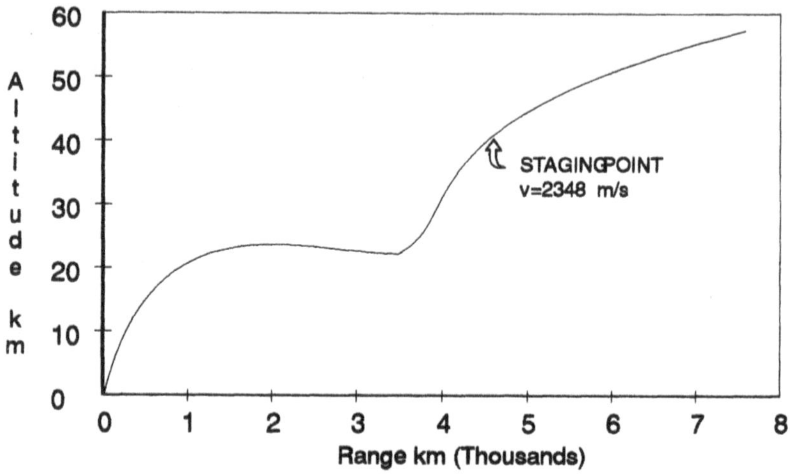

Fig. 20 Trajectory for vehicle C.

Table 5 Weight Statement for Shuttle-Class and Heavy Lift Vehicles

Weight, kg	Vehicle A		Vehicle C	
	Stage 1	Stage 2	Stage 1	Stage 2
Aero/Body Structure	60,395	16,198	122,300	29,824
Thermal Protection Sys.	20,842	8,695	43,560	15,106
Launch/Landing Gear	9,956	3,480	19,147	6,243
Propulsion	47,121	6,878	111,641	13,560
ACES	14,969		31,502	
Elect/Hydr/Avionics	12,131	3,764	20,227	5,760
Crew	356	356	356	356
Payload	29,484		68,040	
LH_2	138,562	28,662	364,329	59,653
Vehicle Dry Weight	165,414	39,015	348,377	70,493
GTOW	401,849		911,604	
Total LEA Collected	326,222		816,991	
Collect Air Flow (kg/sec)	1,103		2,100	
Collect Time (min)	23		27	
Maximum Vehicle Weight	626,875		1,427,134	
Orbital Mass	71,087		141,885	

second stage then accelerated to orbit using rocket propulsion. A typical two-stage trajectory is shown in Fig. 20.

The study process involved a series of iterations between PDWAP and ETOC1, proceeding until vehicles large enough to deliver the specified payloads were achieved. The resulting vehicles are summarized in Table 5 and Fig. 21. Both vehicles delivered a payload fraction of approximately 7.4%.

These studies along with the air separator development and heat exchanger development of the preceding sections were first discussed in Ref. 21.

Later (about 1988), interest turned to vehicles having a 4356-kg (10,000-lb) payload delivered to a 185.2-km (100 NM) polar orbit. A LOX carrying and an ACES vehicle were each designed to meet this payload.

The vehicles were chosen to illustrate the mathematical formulation discussed in Appendix A. Both vehicles were two stage, hydrogen fueled, and employed horizontal takeoff and landing. Airbreathing propulsion (turboramjets) was used to a staging point of Mach 5. Oxidizer collection took place at a constant Mach 5 cruise condition.

Vehicle size and weights were estimated from the methods of Refs. 17 and 18 and vehicle performance derived from Ref. 19. As before, engine performance was derived from Ref. 20. The resulting vehicles are summarized in Table 6 and Fig. 22.

The ACES vehicle has only about 61% of the takeoff weight of the LOX carrying vehicle and about 4.3% less dry weight. From Eq. (A19) a K value of 0.457 is obtained. Equation (A14) thus states that a LOX carrying vehicle will require 1.457 times the first stage mass of an ACES vehicle to reach staging. Figure 23 is a bar chart that arranges the weight statement items of Table 6 according to the mathematical development of Eqs. (A1 - A19. The figure illustrates the ACES first stage dry weight and total mass as discussed above

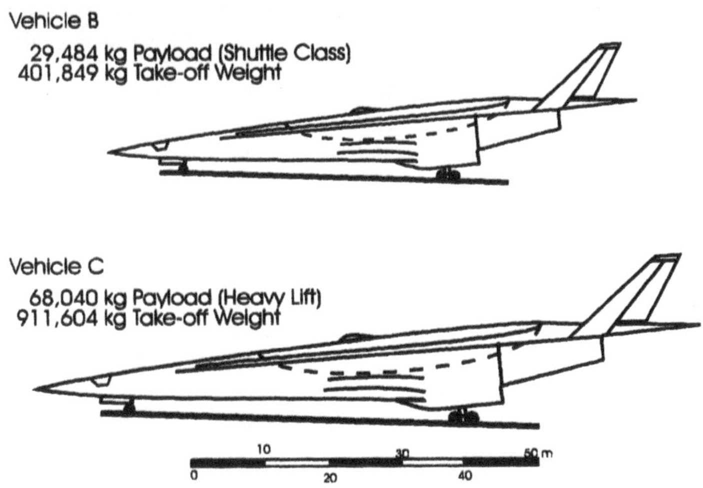

Fig. 21 Current study vehicles.

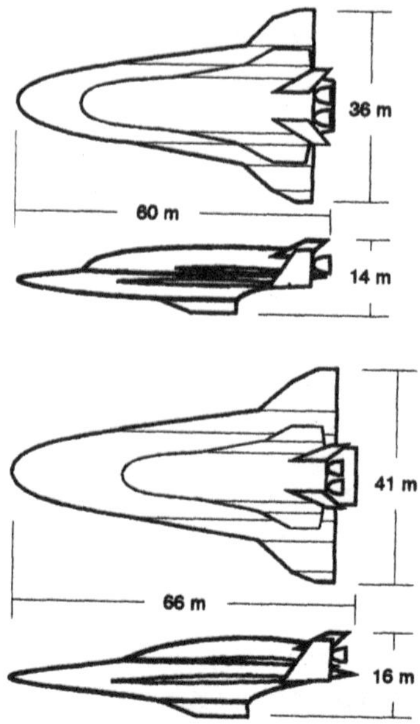

Fig. 22 Current study vehicles, 4535-kg payloads.

Table 6 Weight Statement for Study Vehicles (Weight in Kilograms).

	ACES Vehicle		LOX Carrying Vehicle	
	Stage 1	Stage 2	Stage 1	Stage 2
Aero/Body Structure	22,074	11,017	29,626	10,980
Thermal Protection System	6,637	3,645	8,301	3,638
Launch/Landing Gear	8,502	1,585	15,551	1,573
Propulsion	73,178	5,432	75,749	5,434
ACES	11,364			
Elect/Hydr/Avionics	7,341	4,126	7,035	4,097
Crew	543	512	616	511
Payload		4,535		4,535
LH_2	51,818	14,796	73,977	14,898
LOX				119,185
Vehicle Dry Weight	129,500	25,805	136,262	25,722
GTOW	228,086		375,706	
Total LEA Collected	119,856			

ACES Vehicle Collects LEA during Cruise at Mach 5.

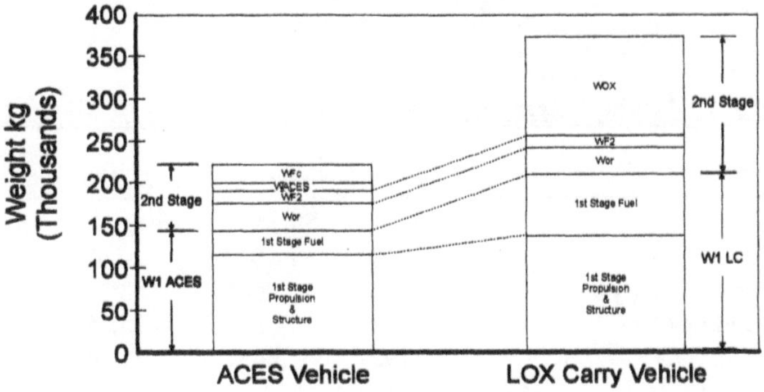

Fig. 23 Weight statement of Table 1 arranged to illustrate mathematical analysis of eqs. (A1 - A19). ACES vehicle collects LEA during cruise at Mach 5.

and illustrates how the starting weight of the second stage ACES vehicle is only one-half the weight of the LOX carrying vehicle. For this chapter the ACES vehicle was designed for constant speed oxidizer collection to illustrate the mathematical development leading to Eq.(A19).

The second stages of each vehicle have nearly the same effective I_{SP}, 393 s for the ACES vehicle, and 396 s for the LOX carrying vehicle. From Eq. (A11) the ACES second stage has an apparent mass ratio of 2.577, giving an apparent effective I_{SP} of 690 s. Reference 22 first reported this study.

In Ref. 23, an ACES vehicle designed to the same requirements as just given, collected LEA during acceleration from Mach 1.5 to 5.0, a more efficient approach that requires 22% less first-stage fuel and leads to a vehicle that has 8.6% less takeoff weight.

These vehicles are summarized in Table 7 and Fig. 24. The ACES vehicle has only about 56% of the takeoff weight of the LOX carrying vehicle and has about 9.3% less dry weight. From the study results, a K value of 0.65 is inferred from Eq. (A19). By Eq. (A14) this means that the LOX carrying vehicle will require 1.65 times the first-stage mass of an ACES vehicle to reach staging. The second stages of each vehicle are identical to those already discussed.

1. Two-Stage Scramjet and ACES Vehicle Comparison

Although recent ACES vehicle studies have concluded that ACES space boosters are competitive with scramjet powered vehicles, this observation has been primarily qualitative. Nau[1] showed quantitatively that an all-airbreathing (turboramjet to Mach 8 plus scramjet to Mach 20) vehicle could be launch-weight competitive with an ACES vehicle. In fact, he showed it to be the closest competitor to an ACES vehicle. In the years since this reference was published in 1967, much more data on scramjet performance is available to support a new examination of the competitiveness of scramjets with ACES.

TABLE 7 Weight Statement for Study Vehicles (Weight in Kilograms).

	ACES Vehicle		LOX Carrying Vehicle	
	Stage 1	Stage 2	Stage 1	Stage 2
Aero/Body Structure	18,705	11,017	29,626	10,980
Thermal Protection System	5,530	3,645	8,301	3,638
Launch/Landing Gear	8,173	1,585	15,551	1,573
Propulsion	72,990	5,432	75,749	5,434
ACES	11,338			
Elect/Hydr/Avionics	5,222	4,126	7,035	4,097
Crew	533	512	616	511
Payload		4,535		4,535
LH_2	40,225	14,796	73,977	14,898
LOX				119,185
Vehicle Dry Weight	121,958	25,805	136,262	25,722
GTOW	208,364		375,706	
Total LEA Collected	116,647			

ACES Vehicle Collects LEA during acceleration from Mach 1.5 to 5.

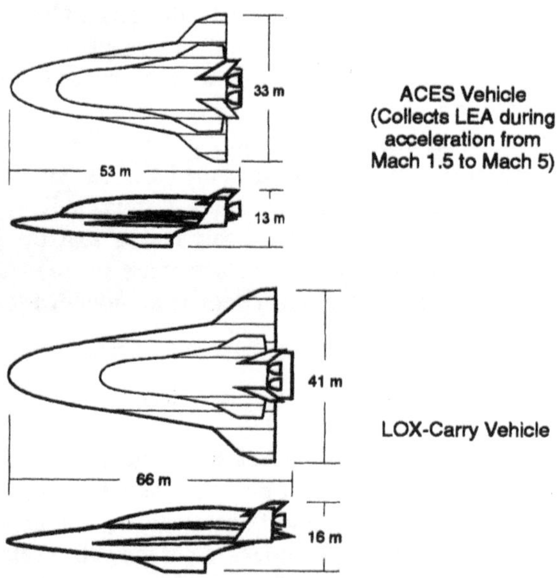

Fig. 24 Current study vehicles. 4535-kg payloads.

A two-stage-to-orbit (2STO) ACES and a 2STO scramjet were designed. Mach 8 staging was selected. Figure 25 presents a summary of these vehicles (labeled vehicles d and e) along with a comparison with those vehicles designed in Refs. 2 and 3 which staged at Mach 5 (vehicles a, b, and c). In those cases, Mach 5 staging was selected both for comparison with other studies and to demonstrate that vehicles with very attractive launch weights could be achieved within the technology of Mach 5 class turboramjets. In this comparison of ACES and scramjet vehicles, however, the authors recognized that Mach 8 staging might give a smaller launch weight for both systems if first-stage propulsion, vehicle structure, and thermal protection systems did not grow unreasonably. This assumption was proven valid by a noted 1.8% decrease in the launch weight of the Mach 8 staging ACES vehicle over that of the Mach 5 staging vehicle.

Summaries of the weight statements of these two-stage vehicles are shown in Table 8 and vehicle planforms are shown in Fig. 26. The Mach 8 staging ACES vehicle (vehicle d of Fig. 25) collects LEA during acceleration from Mach 1.5 to 5. Turboramjet performance assumptions were those given in Ref. 20. Turboramjet weights were consistent with near-term technology, i.e., thrust/weight levels of approximately 4 (inlet, engine, and nozzle included in weight). The second stage used hydrogen/LEA rocket propulsion. The scramjet vehicle (vehicle e) used identical turboramjet performance assumptions. Scramjet propulsion was used in the second stage from Mach 8 to Mach 20, followed by rocket boost to orbit. Scramjet weights and performance were varied to determine what levels were needed to achieve competitiveness with ACES. For scramjets, reasonable performance (I_{SP})

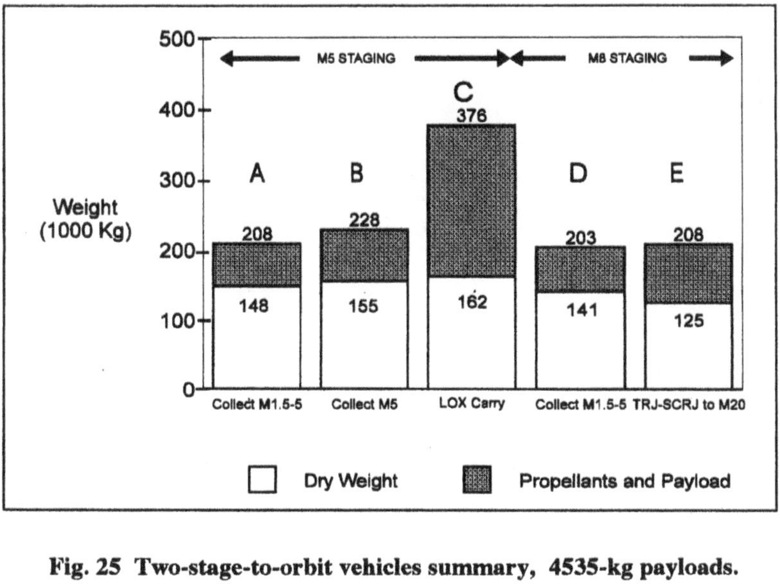

Fig. 25 Two-stage-to-orbit vehicles summary, 4535-kg payloads.

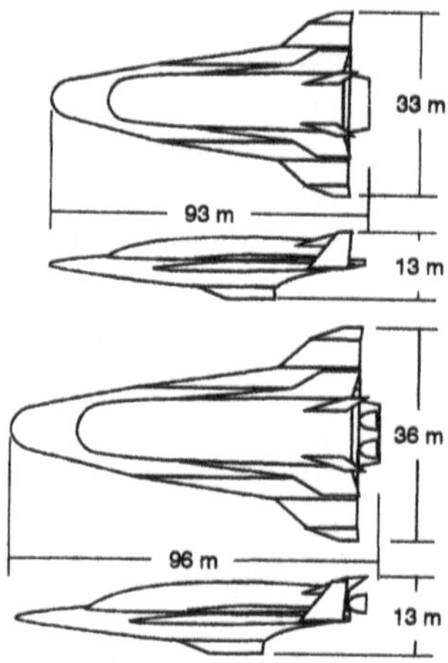

Fig. 26 Planforms of Two-stage-to-orbit current study vehicles. Vehicles stage at Mach 8.

Table 8 Weight Statement For Study Vehicles (Weight In Kilograms).

	ACES Vehicle		Scramjet Vehicle	
	Stage 1	Stage 2	Stage 1	Stage 2
Aero/Body Structure	22,880	10,957	14,082	16,974
Thermal Protection System	4,560	2,160	2,873	3,557
Launch/Landing Gear	4,908	1,675	4,966	1,676
Propulsion	66,725	11,844	63,596	7,570
ACES	5,443			
Elect/Hydr/Avionics	6,562	3,858	6,469	3,442
Crew	530	494	532	626
Payload		4,535		4,535
LH_2	46,245	9,712	23,122	30,521
LOX				23,123
Vehicle Dry Weight	111,078	30,494	91,986	33,219
GTOW	203,088		207,664	
Total LEA Collected	84,878			

Staging at Mach 8. ACES Vehicle collects LEA during acceleration from Mach 1.5 to 5.

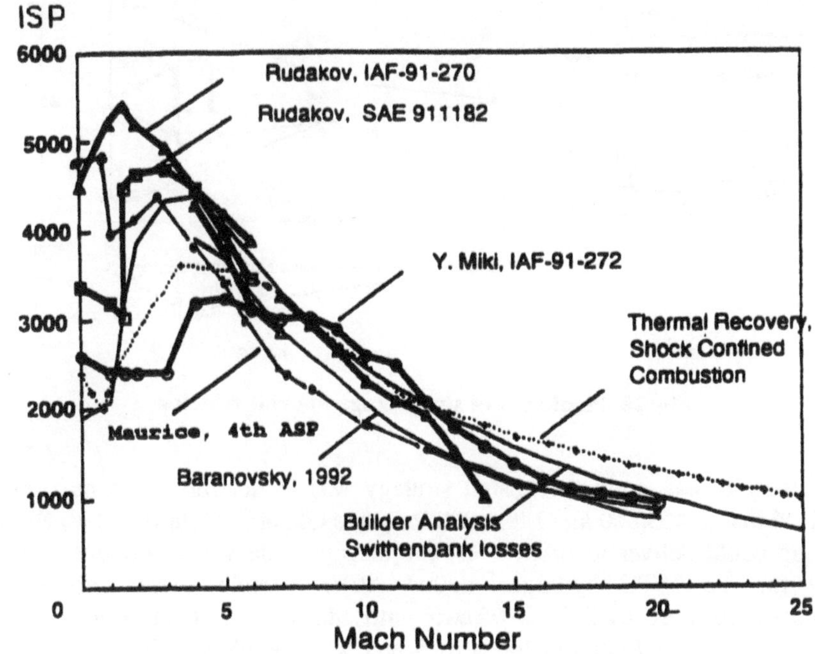

Fig. 27 Airbreathing specific impulse schedule.

projections to Mach 12 are available. However, scramjet performance above Mach 12 remains very speculative. Figure 27 shows the airbreathing I_{SP} assumptions for vehicles d and e. These assumptions are in line with those of Czysz et al.[24] and Vandenkerckhove[25].

The launch weight of the scramjet vehicle given the just stated propulsion ground rules was found to be equivalent to the launch weight of the ACES vehicle. However, this required airbreathing propulsion function to Mach 20. For the near term, ACES is still very competitive with scramjet for 2STO applications and all of its propulsion components can be developed in ground test facilities. This work was first reported in Ref. 26.

2. *Single-Stage Scramjet and ACES Vehicle Comparison*

Since it is recognized that single-stage-to-orbit (SSTO) capabilities may be achievable with very high-speed scramjet technology, the authors sought to examine the prospects that ACES could also enable a SSTO capability. A scramjet SSTO vehicle was designed and compared to an ACES SSTO vehicle previously designed. A summary weight statement for these two vehicles is shown in Table 9. Planforms are shown in Fig. 28.

The ACES SSTO vehicle is that previously reported in Ref. 27. The payload is slightly higher than that of the scramjet vehicle (5351 kg vs 4536

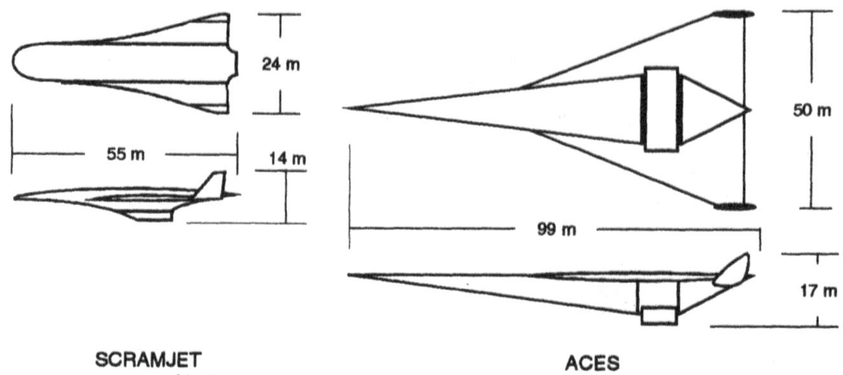

Fig. 28 Planforms of single-stage-to-orbit vehicles.

kg) since in that study the design strategy was to determine the maximum payload that a 453,600 kg (1,000,000 lb) gross takeoff weight (GTOW) SSTO vehicle could deliver to orbit. The propulsion mode was analogous to that used for the 2STO hydrogen fueled ACES vehicle already discussed. Turboramjets were used from takeoff until Mach 8. Collection took place while accelerating from Mach 1.5 to 5 and was completed while cruising at Mach 5. Rockets were turned on immediately at the end of collect, assisted the turboramjets to Mach 8, and were used to achieve orbit. The aerodynamic performance was also similar to that of the first stage of the 2STO ACES

Table 9 Weight Statement For SSTO Study Vehicles (Weight In Kilograms).

	ACES Vehicle	Scramjet Vehicle
Aero/Body Structure	76,882	31,824
Thermal Protection System	20,761	6,239
Launch/Landing Gear	8,280	4,954
Propulsion	51,382	43,053
ACES	22,125	
Elect/Hydr/Avionics	12,787	6,715
Crew	630	630
Payload	5,351	4,536
LH_2	254,895	71,679
LOX		37,273
Vehicle Dry Weight	192,217	92,785
GTOW	453,093	206,903
Total LEA Collected	671328	

vehicle. The scramjet vehicle uses turboramjet propulsion to Mach 8, scramjet propulsion to Mach 20, followed by rockets to orbit. The propulsion performance assumptions are the same as for the 2STO scramjet vehicle already discussed, however, more advanced turboramjet materials and structures were assumed, giving thrust/weight levels near 6 (inlet, engine, and nozzle included in the weight).

The GTOW of the scramjet SSTO vehicle is roughly 50% that of the SSTO ACES vehicle. If scramjets could be successfully operated to Mach 20 at the very optimistic structural weights and I_{SP} performance levels assumed for this study, they could be the lightest SSTO candidate. It must be recognized however, that ACES requires a much lower level of propulsion technology which can be developed totally in ground-test facilities. As such, ACES should not be discounted as a potential propulsion scheme for future airbreathing space boosters in favor of the more speculative scramjet. This work was reported in Ref. 26.

In the two preceding sections, comparison between 2STO and SSTO ACES and scramjet vehicles has been presented. In a 2STO mode, the ACES concept compares favorably with scramjet-powered options in terms of payload fraction and GTOW. An ACES space booster is about one-half the launch weight of a rocket-powered system. ACES shares with scramjet the prospects

Table 10 Weight Statement For Dual-Fuel (JP/LH_2) Vehicles (Weights In Lbs.

	Authors' study vehicles		Rockwell vehicle	
	Stage 1	Stage 2	Stage 1	Stage 2
Aero/body structure	56,082	24,882		
Thermal protection sys.	13,250	6,612		
Launch/landing gear	15,926	3,637		
Propulsion	114,158	10,541		
ACES	30,000			
Elect./hydra./avionic	18,024	9,931		
Crew	535	535		
Payload		10,000		10,000
JP-4	218,019	95,582	218,000	
LH_2	96,000		83,000	51,000
Vehicle dry weight	247,420	55,603	230,000	83,000
GTOW	723,694		682,000	
Total LOX collected	481,871		410,000	
Collect airflow, lb/s	2,128			
Collect time, min	17			
Collect Mach number	4-5.2		3-5.0	
Maximum vehicle weight	976,744		791,000	
Orbital mass		73,827		
Ascent time, min		33		

of operating flexibility that come with horizontal launch and runway-like operation, but it avoids the need for expensive flight development of the engine. In a SSTO mode, scramjet vehicles were found to be 50% lighter than their ACES counterparts. However, this weight savings is obtained only from very advanced technology assumptions for the scramjet vehicle and the need for flight development of the engine.

B. Dual-Fuel (LH_2/Hydrocarbon) Vehicle Systems Studies

An examination of the fuel usage of previously designed LH_2-fueled LEA-collect vehicles revealed that only 7 - 10% of the vehicle GTOW was the LH_2 needed to collect LEA. The remaining LH_2, about 40% of the GTOW, was used for acceleration to the LEA-collect speed and second-stage climb to orbit. It seemed that some large reductions in vehicle volume and dry weight might be achieved if a storable hydrocarbon fuel (JP) were used in the precollect airbreather acceleration, and in the postcollect and second-stage rocket phases. In addition to the volume reduction, minimizing the amount of LH_2 carried onboard to that required for in-flight LEA collection would greatly improve the ready alert capability. If the required quantity of LH_2 proved small enough, then separate LH_2 tank modules might be quickly "plugged in" to a vehicle already on alert and fully fueled with a storable hydrocarbon fuel. Significant advantages were seen for launch preparation times, LH_2 handling/fueling, and launch/maintenance equipment and crews. The dual-fuel ACES vehicles examined in this study only used LH_2 fuel while collecting LEA in a ramjet powered mode from Mach 4 to about Mach 5. A hydrocarbon fuel was used in both the precollect turboramjet and the postcollect rocket modes. Initial studies of the dual fuel approach first were discussed in Ref. 27 and later in Refs. 28 and 29.

It was recognized that the upper speed capability of hydrocarbon fueled turboramjets is limited compared to hydrogen fueled turboramjets. For the current study vehicle, takeoff and acceleration to Mach 4 was accomplished with hydrocarbon fueled turboramjets. At this point, the turboramjet engines switched to hydrogen fuel, and LEA collection was initiated and completed by Mach 5.2. Postcollect acceleration to the staging point, Mach 7, was then accomplished with JP/LEA rockets. No airbreathing propulsion was used beyond Mach 5.2. The JP/LEA rocket used for second-stage ascent to orbit was assumed to have a vacuum I_{SP} of 340s and a sea level I_{SP} of 312s. Compared to the all LH_2 vehicle of Ref. 23, the dual fuel vehicle is 62% heavier at launch but has about 3% less dry weight. The total amount of LH_2, 96,000 lb, is only about 13% of the launch weight of 723,694 lb.

In a related contract study by Yi et al.[30] at Rockwell, simultaneous use of JP and LH_2 during LEA collect led to an optimum LEA-collect flow rate that minimized the weight of LEA-collect equipment and JP fuel, thus leading to a minimum weight vehicle. Their study retained a LH_2 rocket for the second stage, giving a lighter vehicle but requiring more LH_2. Rockwell has

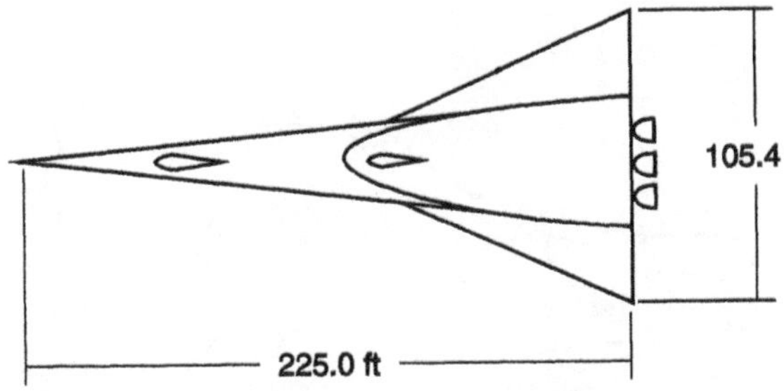

Fig. 29 In-house dual fuel LOX-collect vehicle, 10,000-lb payload.

suggested, by independent studies, that the total amount of LH_2 might be minimized enough so that the onboard tankage could have near-commercial Dewar insulation and thus have nearly unvented hold times on the order of days to a week.

In the Ref. 30 study, a dual-fuel vehicle of 682,000 lb GTOW was about 25% heavier at takeoff than an all-LH_2 vehicle and had about 11% less dry weight. They obtained a reduction of about 25% of the LH_2. The Rockwell vehicle used turboramjets to Mach 5, and collected LEA from Mach 3 to Mach 5. Staging was accomplished at Mach 5. This study found the minimum launch weight to be obtained with simultaneous use of JP fuel and LH_2 at a proportion of 25% JP and 75% LH_2. The minimum dry weight occurred when approximately equal flow rates of JP and LH_2 were used.

The dual-fuel vehicles are seen to reduce the amount of LH_2 required, thus easing the task of cryogenic servicing but, more importantly, permitting long-range subsonic flights (for self-ferry, orbiter transfer, etc) that are typical of aircraft and only require JP fuel and conventional airbase servicing. Unrefueled ranges are nearly U.S. transcontinental. Aerial refueling for greater ferry range would be practical because of the use of JP fuel.

A weight statement for these vehicles is shown in Table 10. The configurations are shown in Figs. 29 and 30.

The dual-fuel vehicles are seen to be in the weight class of large (Boeing 747-class) aircraft, can be operated subsonically as a conventional aircraft, and could stand on alert indefinitely with all but the LH_2 fuel loaded.

C. All-Hydrocarbon Vehicle Systems Studies

Noting that the dual-fuel LEA-collect vehicle did not exceed reasonable weight bounds, the next step, removing all LH_2, and thus all dependency on cryogenics fueling, was anticipated to yield a vehicle which could still be considered runway operable. The process for generating LEA in-flight was derived from commercial tonnage LOX production processes that employ stages of air compression and cooling followed by cryogenic distillation. Water carried onboard at takeoff and the heat sink of the fuel provide the

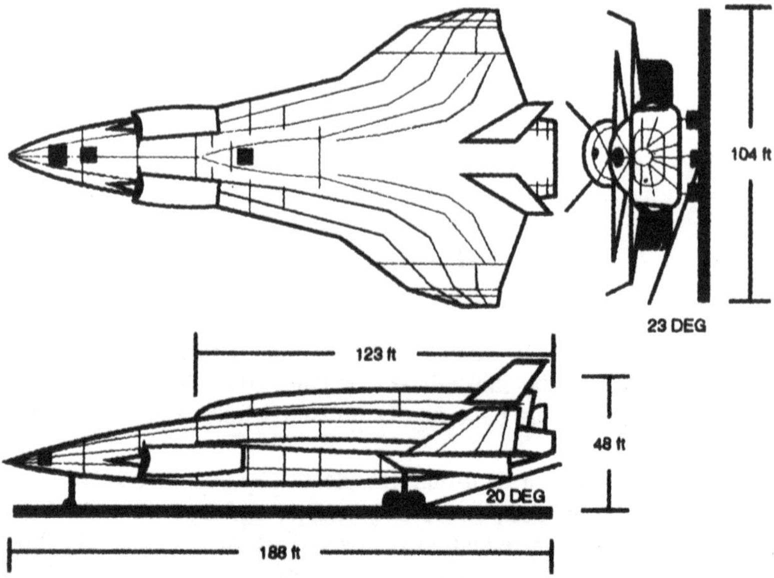

Fig. 30 Rockwell dual-fuel LOX-collect vehicle, 10,000-lb payload.

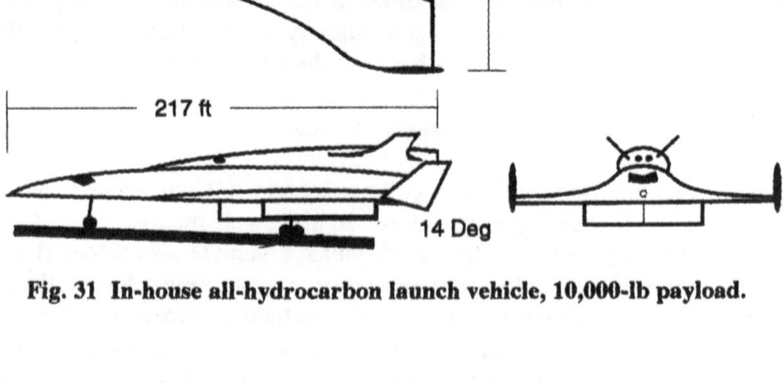

Fig. 31 In-house all-hydrocarbon launch vehicle, 10,000-lb payload.

required cooling capacity. Compression and expansion equipment along with heat exchangers and the air separator make up the LEA collection equipment. No cryogenics servicing or cryogenic storage facilities are required to support a flight.

A two-stage vehicle (Fig. 31) was designed. Its weight statement is given in Table 11. LEA collection is initiated at Mach 1.5 at about 25,000-ft altitude and proceeds as the vehicle accelerates and climbs along a constant dynamic pressure (1500 lb/ft^2) path to 60,000 ft. Acceleration is limited to slightly less than 0.1 g (3 ft/s^2) to provide the required 10.8 min of LEA collection time. For the baseline vehicle, 318,750 lb of LEA are collected. Water coolant in the amount of 159,375 lb are used and discharged overboard as high temperature steam. The total amount of fuel used during the LEA collection phase is approximately 150,000 lb. Its heat sink is partly employed for LEA collection. Acceleration from takeoff to initiation of LEA collection requires approximately 70,000 lb of fuel. The LEA-collect airflow is approximately 2000 lb/s giving a LEA production rate of approximately 450 lb/s. At the end of the LEA-collect phase the vehicle weight is approximately 940,000 lb, about 60,000 lb less than the takeoff weight. All of the coolant water and an approximately equal amount of fuel have been expended; in a sense they are

Table 11 Weight Statement For Author's Study All-Hydrocarbon Fuel Vehicle- Baseline Configuration, (Weights In Lbs.)

	Stage 1	Stage 2
Aero/Body Structure	23,815	18,843
Thermal Protection System	5,091	4,733
Launch/Landing Gear	20,378	3,245
Propulsion System	222,374	5,374
ACES	100,000	
Elect./Hydra./Avionics/Controls	20,897	10,225
Crew and Provisions	1,534	1,289
Fuel	253,000	106,250
Water	159,375	N/A
Vehicle Dry Weight	393,872	43,185
GTOW	977,980	N/A
Total Lox Collected		318,750
Collect Air Flow, Lb/S	2,000	
Collect Time, Min	10.8	
Collect Mach Number	1.5 To 3.5	
Max. Vehicle Weight	977,980	
Weight in Orbit		61,494
Ascent Time, Min.		24.0
Second Stage Ignition Weight		487,007

"exchanged" for a similar amount of LEA stored in the second stage. The ratio of LEA collected to water coolant expended is termed the collection ratio, *CR*. For the baseline vehicle, a *CR* value 2.0 was used. By comparison, LEA collection systems using LH_2 have *CR* values in the range of 4 - 6.

At the end of LEA collection the vehicle accelerates at maximum ramjet thrust and executes a pullup to reach Mach 4 staging at about 66,000 ft. The initial flight-path angle of the second stage is sufficient to allow a nearly ballistic ascent to orbit under rocket power. Vacuum I_{SP} of the rocket is 340 s. Second stage ignition weight is 487,000 lb. Initial acceleration of the second stage is slightly less than 1 g. Above 16,000 ft/s, rocket thrust is reduced to limit acceleration to 2 g's. This vehicle has substantial subsonic unrefueled range, 2200 NM at Mach 0.7/23,000-ft altitude. Aerial refueling could be used to achieve much longer ferry ranges, to maintain an aerial alert, and to allow an orbital mission to be initiated far from the takeoff location.

Figure 32 is a sketch of the process for generating LEA, first discussed in Refs. 31 and 32. This figure and the cycle process were derived from preliminary information provided by the Linde Co. Water and fuel serve as coolants. Compression equipment and turbines are part of the cycle. Basically, air is cooled from the inlet ram condition (about 1350°R maximum)

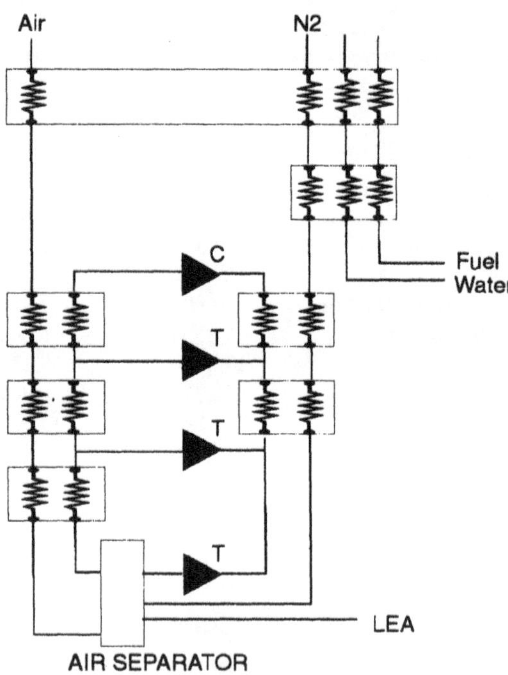

Fig. 32 Water and fuel coolant LOX generation process.

to the cryogenic condition where distillation separation can be carried out. The heat sink (water and fuel) is only available at ambient temperature. For this reason a series of heat pump cycles are required to pass the net refrigeration load (cooling and condensation of LEA) to a temperature above the water coolant and fuel temperature for rejection. Figure 32 is arranged so that in reading from top to bottom one sees temperatures generally decreasing from inlet air temperature levels down to air distillation temperatures. In the center part of the figure are shown three staged heat pump circuits which circulate a 98% N_2 stream. These circuits constitute a closed-loop liquefier. On the left-hand side of the figure it is seen that air is cooled in a succession of heat exchangers to a temperature low enough to enter the distillation unit. The closed- loop liquefier supplies the cold liquid N_2 reflux for the distillation unit. The N_2 flow rate in this liquefier loop can be several times the net outflow of N_2 waste. On the right-hand side of the figure it is seen that the liquefier loop raises the temperature of the waste N_2 stream to a level where heat can be rejected to boiling water. The water is further heated to near the incoming air temperature (top right of figure). The fuel in the amount demanded by the propulsion system for thrust is raised from ambient to a temperature near the incoming air. This probably would require use of an endothermic fuel to achieve enough fuel heat sink. As a safety measure, a design should pass fuel only through heat exchangers that are carrying nitrogen or water/steam. This would reduce any hazard due to fuel leakage. LEA collection ratios (*CR*) were varied from 1.5 to 2.5 and LEA collection equipment specific weights were

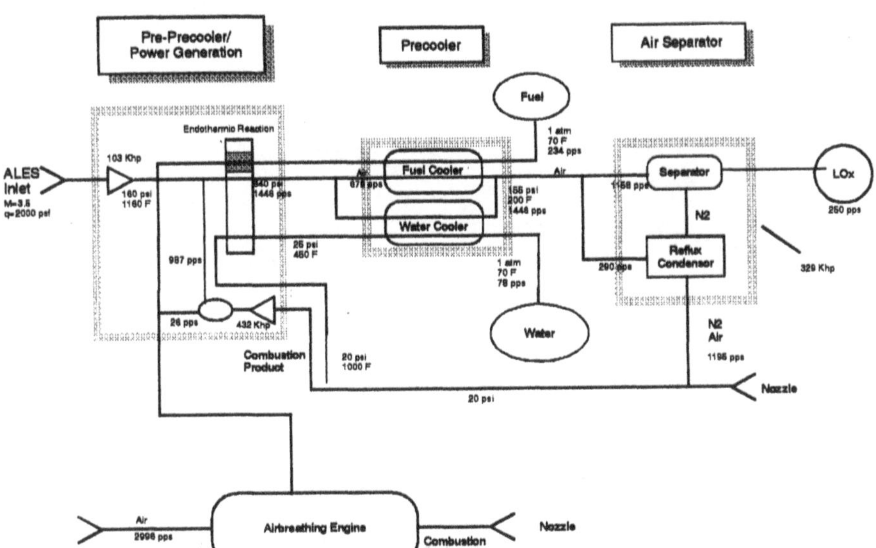

Fig. 33 Water and fuel coolant LOX generation process, Rockwell approach.

varied from 37.5 to 75 separately and in combination to test the influence of these factors on vehicle launch weight. The baseline specific weight of LEA collect equipment was 50 lb equipment per lb/s airflow. The vehicle launch weight was 977,980 lb using the baseline *CR* value and baseline LEA collection equipment weight. Using the worst case values the vehicle weight was 1,103,000 lb, still within the range of runway operable vehicles. It is noted that the Russian AN225 aircraft is rated at 1,300,000 lb takeoff weight. Refs. 28 and 29 discuss the all-hydrocarbon system. Reference 33 shows numerous examples of commercial LOX production processes that apply the principles just described.

Work conducted by Yi et al.[34] at Rockwell based on the final process cycle provided by Drenevich and Nowobilski[35] of Linde has shown that virtually no fuel is required for cooling, but a small amount is needed to generate power. Figure 33 shows the Rockwell process. This meant that the LEA collection rate could be optimized to yield a minimum weight vehicle that minimized the LEA collection equipment weight plus fuel burned during LEA-collect. Their analysis led to an optimum LEA-collect time of 20 min. A maximum *CR* value of 1.7 was obtained and a specific weight of 90 was estimated for the equipment. These values show lower performance and larger weight than the author's study above, but represent a great deal more design effort and are considered very accurate estimates. These results had not been available for use at the time of the Ref. 32 and 32 studies. The resulting vehicle, at 1,221,000 lb, is still a runway operable vehicle. The first-stage configuration is quite similar to the dual-fuel vehicle of Fig. 31, but the second- stage orbiter (Fig. 34) uses a more structurally efficient circular shape to accommodate the denser JP and LEA. A vehicle weight statement is shown in Table 12.

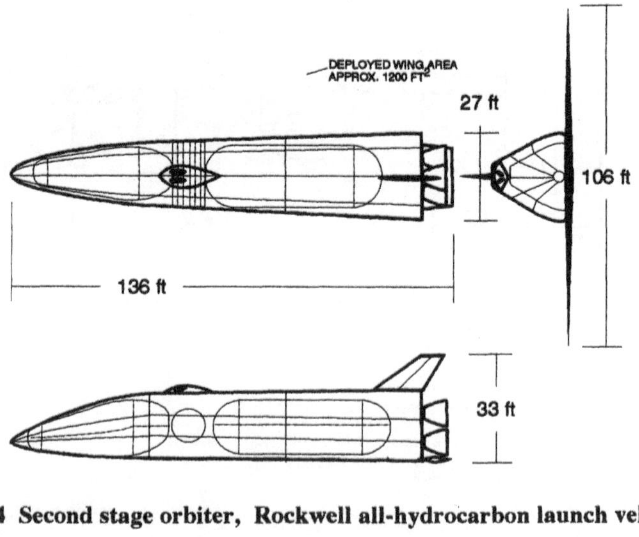

Fig. 34 Second stage orbiter, Rockwell all-hydrocarbon launch vehicle.

Rockwell selected Mach 5/65,000-ft altitude for staging, finding a significant weight advantage compared to Mach 4 staging.

D. Summary Of Dual Fuel and All-Hydrocarbon Studies

The foregoing section has described LEA-collect vehicle concepts that employ JP fuel for part or all of the LEA-collect process as a means to add airplane-like benefits to space launch by reducing dependency on cryogenic fuel in addition to the already desirable attribute of not requiring LOX onboard at takeoff. The previous studies of all-LH_2 two-stage LEA collect vehicles demonstrated that they are launch weight competitive (450,000-lb class) with two-stage turboramjet/scramjet vehicles at low-propulsion technical risk (no airbreathing function required above Mach 5 to 8). In fact, it was found that scramjet function at very high performance was required to Mach 20 to avoid weights greater than that of a LEA-collect vehicle. These same studies showed such significant weight advantage against LH_2 fueled LOX-carry vehicles that it was felt that the weight advantage could be traded for the operational advantages of substituting JP fuel for some of the LH_2. The dual-fuel LEA-collect vehicles achieved this goal. Their launch weights in the 700,000-lb class did not exceed LH_2 fueled LOX carry-vehicles (e.g., the German Sanger). Indeed the LOX carry vehicles, in attempting to substitute JP fuel, reached launch weights on the order of 1,700,000 lb (Ref 36). Because the dual-fuel LEA-collect vehicles had not reached the weight of the largest current aircraft, there was considerable margin for eliminating LH_2 as the

Table 12 Weight Statement For Rockwell All-Hydrocarbon Fuel Vehicle

Mated Vehicle, Lb X 1000		
	GTOW	1221
	Dry Weight	491
	Fuel Weight	619
	Water Weight	111
Carrier, Lb X 1000		
	Dry Weight	427
	Fuel Weight	506
	Water Weight	111
Orbiter, Lb X 1000		
	Staging Weight	503
	Dry Weight	54
	Fuel Weight	112
	Oxidizer Weight[a]	336
	Payload	10

[a]Oxidizer collected in flight

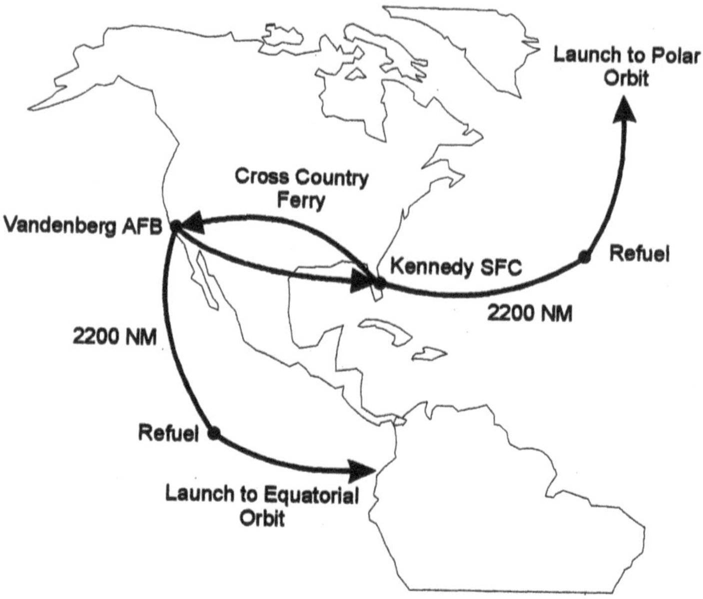

Fig.35 Launch options for vehicle that can employ subsonic aerial refueling of JP-type fuel.

LEA-collect refrigerant. The resulting vehicles (weighing up to 1,200,000 lb) have no cryogenics on board at takeoff.

These studies[27-35] have shown that dual-fuel or all-hydrocarbon vehicles are possible within the weights of the heaviest current aircraft. The use of hydrocarbon fuels enables such airplane-like operations as self-ferry and in-flight subsonic refueling to extend ferry range or to conduct a launch quite remote from the takeoff point. Hydrocarbon fuels permit a wider choice of operating sites, and make quick reaction flight to an alternate site more practical. The dual fuel concept minimizes the amount of LH_2 to be loaded to ready a launch. Indeed, as Yi et al.[30] suggest, Dewar-class LH_2 tankage may permit fully fueled launch holds of days.

The all-hydrocarbon LEA-collect vehicle concept achieves the most airplane-like operational characteristics by eliminating all cryogenics servicing and fluids storage requirements. Cryogen-free manned reusable space launch capability at reasonable payload fractions appears achievable because it employs significant airbreathing propulsion and does not carry LOX at takeoff. Rocket concepts that attempt to use ambient-storable propellants are forced to several stages, highly advanced materials and structures, and limited reusability. The available liquid propellant choices may have hazard and toxicity problems for manned flights. Hangar-like launch preparation facilities are practical for an all-hydrocarbon fueled vehicle, enabling all-weather preparations and launch, keeping preparations secure and unobservable, thus

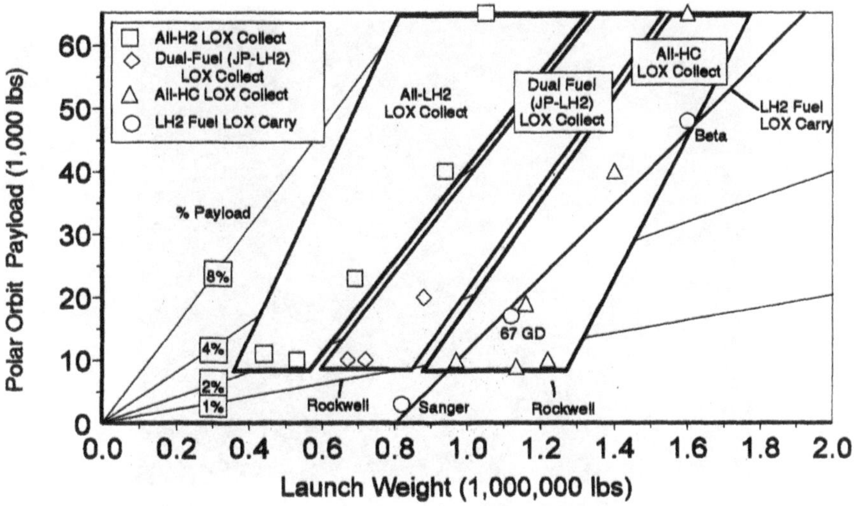

Fig. 36 Summary of LOX-collect vehicle performance.

enhancing covert or surprise launches. With sustained subsonic aerial refuelling, an all-hydrocarbon vehicle could sustain an extended ready-to-launch mode or could cope with unexpected delays in a launch countdown.

As an example of the flexibility that is available by use of hydrocarbon fuels, either the dual-fuel or the all-hydrocarbon fueled vehicle could conduct the launches suggested in Fig. 35. With subsonic aerial refueling of JP-type fuels, sufficient offset is provided to enable polar orbits from Cape Kennedy or to enable equatorial orbits from Vandenberg. Other orbit planes are available, depending on the number of successive refuelings. In addition to the nearly U.S. transcontinental ferry range suggested by Fig. 35, self-deployment to launch sites elsewhere in the world could be accomplished using aerial refueling support.

Parrington[37] shows that launch vehicles that can traverse as little as 700-N.M. range during first-stage ascent have the required offset to enable any orbit rendevouz within 2 h. He recommends a two-stage vehicle which achieves the required offset by subsonic cruise. He recommends staging at Mach 3. The military advantages for rapid rendevouz are easily reasoned.

Figure 36 shows the payload vs launch weight performance of the three classes of LEA-collect vehicles studied to date. Clearly, the weight penalty for avoiding LH_2 is great, but payload fraction advantages are significant compared to rocket systems and to LOX-carrying airbreathing systems. On-demand capability has been achieved by reusable vehicles that can operate from conventional runways.

Fig. 37 Photograph of Linde small scale structured packing test rig.

VII. Other Recent Related U.S. Efforts

Weimer[38] examined the feasibility of the vortex separator. The study results suggested that this device may have about the same weight and volume of the rotary separator already discussed.

Kennedy[39] conducted heat exchanger design studies, and examined candidate materials and manufacturing methods.

Acharya et al.[40] conducted design studies and small-scale experiments of a rotary air separator using a structured packing concept as an improvement to the original distillation tray approach. The structured packing concept consists of closely stacked corrugated and perforated sheets. The device behaves as a wetted film contactor, see figure 37 showing the small-scale test unit. The testing, with air and water as fluids, showed that large reductions in pressure drop were possible, leading to a wider operating range. Their design studies indicated that the volume of a rotary separator could be reduced by one-half and the weight reduced by one-third compared to the original rotary separator concept.

VIII. Conclusions

The broad-based ACES propulsion technology database that was generated in the 1960s has been reviewed. This technology offers the potential for relatively lightweight, fully recoverable launch vehicles capable of delivering substantial payloads to orbit. The ACES vehicle concept compares favorably with scramjet-powered options in payload fraction and is about one-half the launch weight of rocket-powered systems. ACES shares with scramjet the prospects of operating flexibility that come with horizontal launch and runway-like operation. ACES avoids the need for carrying LOX at takeoff that virtually mandates that rocket systems be vertically launched. ACES exhibits the effective performance of scramjets in the hypersonic regime without having to rely on scramjet function.

Appendix A - Mathematical Formulation

The performance advantages of utilizing ACES in two-stage space boost vehicles can be mathematically demonstrated. The first step is to examine the performance of the second stage rocket-propelled vehicle. The performance of this stage is given by

$$\Delta V = g_C I_{SP} \ln MR \tag{A1}$$

or its alternative form

$$MR = \exp\left[\Delta V/(g_C I_{SP})\right] \tag{A2}$$

The second stage vehicle mass ratio is expressed as

$$MR = (W_{OR} + W_{F2} + W_{OX})/W_{OR} \tag{A3}$$

For a rocket oxidizer to fuel ratio of O/F, the oxidizer and fuel quantities are

$$W_{OX} = W_{OR}[\exp(\Delta V/(g_C I_{SP})) - 1]/[1 + 1/(O/F)] \tag{A4}$$

$$W_{F2} = W_{OR}[\exp(\Delta V/(g_C I_{SP})) - 1]/[1 + O/F] \tag{A5}$$

During the first stage oxidizer collect phase, fuel in the amount W_{FC} is expended to collect and store oxidizer in the amount W_{OX} by the following relation

$$W_{OX} = CR \times W_{FC} \tag{A6}$$

The term CR is called the oxidizer collection ratio, that is the kilograms of oxidizer collected per kilogram of fuel used to collect oxidizer. Combining Eqs. (A4) and (A6) gives an expression defining the amount of fuel required to collect oxidizer

$$W_{FC} = W_{OR}[\exp(\Delta V/(g_C I_{SP})) - 1]/[CR(1 + 1/(O/F))] \tag{A7}$$

Thus at takeoff the total propellant required for the second stage is $W_{F2} + W_{FC}$. An "apparent" mass ratio, MR', can be formed

$$MR' = (W_{OR} + W_{F2} + W_{FC} + W_{ACES})/W_{OR} \tag{A8}$$

which can be combined with Eq. (A1) to define an "apparent" specific impulse, I'_{SP}, for the second stage

$$I'_{SP} = \Delta V/(g_C \ln MR') \tag{A9}$$

Note that the ACES equipment weight is charged to apparent mass ratio requirements in this accounting. Equations (A5), (A7), and (A8) can be combined to define MR' from the ΔV requirement and the actual rocket specific impulse

$$MR' = 1 + \frac{W_{ACES}}{W_{OR}} + [\exp(\Delta V/(g_C I_{SP})) - 1]\left[\frac{1+(O/F)/CR}{1+O/F}\right] \tag{A10}$$

or in terms of the actual mass ratio:

$$MR' = 1 + \frac{W_{ACES}}{W_{OR}} + [MR_{ACT} - 1]\left[\frac{1+(O/F)/CR}{1+O/F}\right] \tag{A 11}$$

Thus, the collection equipment weight, expressed as a fraction of the orbit weight, is seen to additively affect the apparent mass ratio.

At this point, it is instructive to examine the effects of CR, oxidizer purity (i.e., O/F ratio), and ΔV on MR'. CR values from 4 to 6, ΔV values up to 5486 m/s (18,000 ft/sec), and oxidizer purities of 23% O_2 (air), 50% O_2, and 90% O_2 were considered. Using stoichiometric O/F ratios, the following ideal (vacuum) rocket performance is assumed:

Propellant	O/F	Isp
23% O_2/H_2	34.25	240
50% O_2/H_2	16	345
90% O_2/H_2	8.89	450
100% O_2/H_2	8	470

The effective I_{SP} was taken as 85% of the values just given. Figure 2 illustrates the use of Eqs. (A10) and (A11) for the assumption of zero weight collection equipment. The uppermost curve is the required mass ratio of a LOX/H_2 rocket stage which carries LOX from takeoff. The other three groups of curves show how oxidizer collection can reduce the apparent mass ratio requirement to achieve a given ΔV. The effects of oxidizer purity and collection effectiveness are displayed. The curves for 23% O_2, i.e., simple air collection without enrichment, indicate that there can be a small benefit for simple air collection, if the desired ΔV does not exceed 2348 - 3048 m/s

(8,000 - 10,000 ft/s.) Enriching the collected air to 50% O_2 produces nearly a twofold savings in weight, and enrichment to 90% O_2, an even greater weight reduction. The effect of oxidizer collection equipment weight W_{ACES} can be readily inferred from the figure by adding a known value of W_{ACES}/W_{OR} to any point on the curves. For example, for $CR = 4$, 90% purity O_2, at a $\Delta V = 5486$ m/s (18,000 ft/s), the value of MR' is 2.078. Adding a typical value of $W_{ACES}/W_{OR} = 0.20$ increases MR' to 2.278. From Eq. (A9), I'_{SP} is 680 s.

The preceding treatment of the second stage performance is relatively accurate because the drag and gravity losses deducted from the ideal rocket specific impulse are small and are known with reasonable certainty for the usual high-speed, high-altitude staging conditions. The second step is to confirm that if such a large advantage to the second stage is seen for ACES, there is no offsetting penalty imposed on the first stage. Simple mass ratio analysis of the first stage is complicated by the fact that the drag and gravity losses are large and very much driven by the ascent path details. Unlike a conventional launch vehicle that applies all of its propellant directly to increasing velocity, an ACES vehicle cruises or accelerates slowly during part of the ascent to use some fuel (collect fuel) for increasing mass ratio instead of velocity.

The following comparison of launch weights of an ACES and a LOX carrying vehicle establishes the circumstances in which the ACES vehicle will have a smaller launch weight.

The gross takeoff weight of an ACES vehicle is:

$$GW_{ACES} = \lfloor W_{1ACES} \rfloor + \lfloor W_{ACES} + W_{OR} + W_{FC} + W_{F2} \rfloor \tag{A12}$$

The first bracketed term is those weight items charged to the first stage. Included in the term W_{1ACES} is all of the first-stage vehicle structure, all propulsion, and all fuel except that used to collect LEA. The second bracket groups the weight items considered to be the second stage. These consist of the oxidizer collection equipment weight W_{ACES}, the weight in orbit W_{OR}, the second-stage fuel W_{F2}, and the weight of fuel used to generate LEA for the second stage W_{FC}. This fuel is actually carried in the first stage. Note that W_{ACES} is actually in the first stage.

The gross take-off weight of a LOX-carrying vehicle is

$$GW_{LC} = [W_{1LC}] + [W_{OR} + W_{OX} + W_{F2}] \tag{A13}$$

The first bracketed term is the total of all first stage vehicle structure, all propulsion, and all fuel plus any first stage oxidizer. The second bracket, which is the weight of the second stage, includes the total weight in orbit W_{OR} and all of the second-stage propellants, W_{OX} and W_{F2}.

The first stage weight of a LOX-carrying vehicle will differ from the first stage weight of an ACES vehicle. A factor K may be introduced to relate these first-stage vehicle weights by the following relation

$$W_{1LC} = (1 + K)W_{1ACES} \tag{A14}$$

This states that the first-stage weight of a LOX-carrying non-ACES vehicle will differ by the multiplier $(1 + K)$ from an ACES vehicle. From Eq. (A14), K is defined by

$$K = (W_{1LC}/W_{1ACES}) - 1 \tag{A15}$$

Equation (A14) can be substituted into Eq. (A13) to give the following

$$GW_{LC} = [(1 + K)W_{1ACES}] + [W_{OR} + W_{OX} + W_{F2}] \qquad \text{A16)}$$

Using Eq. (A6) and the definition of O/F as W_{OX}/W_{F2}, Eq. (A12) becomes

$$GW_{ACES} = [W_{1ACES}] + \left[W_{ACES} + W_{OR} + W_{OX}\left(\frac{1}{CR} + \frac{1}{O/F}\right)\right] \tag{A17}$$

and Eq. (A16) becomes

$$GW_{LC} = [(1 + K)W_{1ACES}] + \left[W_{OR} + W_{OX}\left(1 + \frac{1}{O/F}\right)\right] \tag{A18}$$

Subtracting Eq. (A17) from Eq. (A18) to define the gross weight difference

$$\Delta GW_{LC} = KW_{1ACES} - W_{ACES} + W_{OX}\left(1 - \frac{1}{CR}\right) \tag{A19}$$

To make this subtraction, an assumption is necessary that W_{OR}, W_{OX}, and W_{F2} are equal for each type vehicle. Because the I_{SP} of a LEA/H_2 rocket is slightly lower than the I_{SP} of a LOX/H_2 rocket, this assumption has a small error. Another assumption is that each type vehicle has the same second stage starting velocity. Also, Eq. (A19) implies that all of the oxidizer (LEA or LOX) is used only for second stage acceleration.

Data from Ref. 1, partly tabulated in Table A1, are useful in considering the implications of Eq. (A19). Vehicles numbered 5 - 8 of Ref. 1 use airbreathing propulsion for their first stages. Table 2 is excerpted from that paper[1] and provides the vehicle data for the following discussion. Unfortunately, these vehicles do not conform, for purposes of Eq. (A19), strictly to the requirements of equal second-stage starting velocity or the requirement for use of LEA or LOX only in the second stage.

Vehicles 6 and 7 each have the same maximum airbreathing velocity, 1829 m/s (6000 ft/s). Vehicle 6 carries LOX for all rocket operating modes. It uses turbofan ramjets to 1829 m/s (6000 ft/s) followed by LOX/H_2 rockets to

Table A1 Comparison of 1st Stage Vehicles

Vehicle No. (per ref 1)	5	6	7	8
Propulsion	Air Br	Air Br + Rocket	ACES	SSCRJ
1st Stage Air Breathing Velocity, m/s	2,439	1,829	1,829	2,439
Staging Velocity, m/sec	2,439	2,896	2,439	2,439
Gross Takeoff Weight, 1000 Kg	522.7	604.5	420.5	481.8
Fuel/propellant used for 1st stage acceleration, 1000 kg	105	208.2	102.7[a]	111.4
Mass Ratio to staging	1.25	1.53	1.32	1.3
1st Stage I_{sp}, secs.	1,110	700	888	946
W_{ILC}, 1000 kg	306	429	-	301
W_1 ACES, 1000 kg	-	-	285.9[b]	-
K, from eq 14	0.07	0.5	0	0.05

[a]Excludes LEA collect fuel (W_{FC} = 43.6 K kg)
[b]Excludes LEA collect fuel (W_{FC} = 43.6 K kg) and LEA collection equipment (W_{ACES} = 13.6 K kg)

staging at 2896 m/s (9500 ft/s). Second stage propulsion is LOX/H_2 rockets. The first stage mass ratio is 1.525, giving a delivered I_{SP} of 700s. Vehicle 7 collects LEA for all rocket operation modes (an ACES vehicle). It accelerates using turbofan ramjets from takeoff to 1829 m/s. During part of this acceleration, LEA is collected and stored for rocket operation in the first stage and the second stage. Acceleration from 1829 m/s to the staging velocity of 2439 m/s (8000 ft/s) is accomplished with LEA/H_2 rockets. Second-stage propulsion uses LEA/H_2 rockets. The first stage has a mass ratio of 1.323, giving a delivered I_{SP} of 888s. For vehicle 7, the first stage mass ratio and specific impulse were inferred from the Ref. 1 data in the following manner. The collected LEA was estimated to be 217,727 kg (479,000 lb), requiring 43,636 kg (96,000 lb) of hydrogen coolant, i.e., a *CR* value of 5. The first stage holds 65,000 kg (143,00 lb) of LEA, and the second stage holds 152,727 kg (336,000 lb) of LEA. The fuel used directly for first-stage acceleration is 102,727 kg (226,000 lb). This fuel quantity is used to define the first stage mass ratio and specific impulse. Analysis of the data in Table 2 suggests the second stage fuel is 21,818 kg (48,000 lb), implying a rocket *O/F* of 7 and an effective I_{SP} of 389s. Applying this *O/F* ratio and I_{SP} to the first-stage rocket phase implies that the first stage begins rocket operation at 500,909 kg (1,102,000 lb) and ends (reaches staging, 2896 m/s) at 426,818 kg (939,000 lb). This analysis shows that the second stage weighs 77,273 kg (170,000 lb) at vehicle takeoff. The ACES equipment was assumed to weigh 13,636 kg (30,000 lb).

Vehicles 5 and 8 each have the same staging velocity of 2439 m/s (8000 ft/s) and use airbreathing propulsion to staging. Vehicle 5 uses turbofan ramjets in the first stage. It carries LOX to be used in second-stage LOX/H_2 rockets. It has a first-stage mass ratio of 1.251, giving an average I_{SP} of 1110s. Vehicle 8 uses turboramjets up to staging. The second stage uses scramjets to near orbital velocity (Mach 20). This vehicle has a first- stage mass ratio of 1.30, giving an average I_{SP} of 946s. Because vehicles 5 and 8 employ airbreathing to much higher speeds than vehicles 6 and 7, they have smaller first-stage mass ratios and are closer overall competitors to the ACES vehicle 7.

Applying the definition of K from Eq. (A14) to vehicle 6, a K value of +0.55 is observed relative to vehicle 7. By using rockets with stored oxidizer for all flight above 1829 m/s (6000 ft/s), vehicle 6 is seen to suffer a large first-stage weight penalty for carrying LOX.

Equation (A14) was applied to vehicle 5 which carries LOX in the second stage and uses high-performance airbreathing propulsion up to its staging velocity of 2439 m/s (8000 ft/s). Its relatively small K value of +0.070 implies that the very high-airbreathing performance used to a higher velocity than ACES vehicle 7 leads to first-stage weight reductions that offset the weight penalty imposed by carrying LOX in the second stage.

The scramjet vehicle 8, stages at 2439 m/s (8000 ft/s) and carries only a small amount of LOX, using airbreathing propulsion in each stage. Its small K value of +0.053 implies that its heavier second stage (179,000 kg vs the ACES vehicle 7 value of 77,273 kg at takeoff) imposes a first-stage weight penalty that is mostly offset by the weight reductions afforded by high performance first stage airbreathing propulsion used to a higher speed than vehicle 7.

As already stated, the vehicles of Ref. 1 do not compare one-to-one in a way that illustrates the implications of Eq. (A19) unambiguously. Even with these limitations, the following observations seem valid. For vehicles as related as 6 and 7, i.e., having the same first-stage airbreathing velocity, there is no circumstance by which a LOX-carrying vehicle could weigh as little as an ACES vehicle. Furthermore, the LOX-carrying vehicle pays a large first-stage weight penalty for carrying LOX. In comparing an ACES vehicle to vehicles having significantly greater airbreather velocity and performance such as 5 and 8, the ACES advantage is diminished but still remains. It is noted that if the ACES vehicle 7 had the same high-speed, high-performance first-stage airbreathing propulsion as vehicle 5, it would have had a smaller takeoff weight.

In Sec. V comparisons of ACES with LOX-carry and scramjet vehicles were made that adhered to the assumptions of the preceding mathematical development.

Appendix B - Description of Air Separation Concepts

This Appendix briefly describes the operating principles of the various air separation concepts examined for the original Aerospaceplane.

Fractional Distillation: This is the process by which the more volatile nitrogen is separated from oxygen. A vapor phase bubbles through successive pools (stages) of liquid, the two phases moving in an overall counterflow manner. The vapor is enhanced in nitrogen in successive stages, and the liquid phase is enhanced in oxygen in successive stages. Virtually any degree of oxygen-nitrogen separation may be achieved.

Vortex Tubes: In this concept, a liquid spray flows counter to a strongly swirling vapor flow. For cryogenic air, the phase contact enriches the vapor stream in nitrogen and enriches the liquid stream in oxygen. The swirling flowfield insures rapid separation of liquid and vapor phases. The process may be staged to achieve a high degree of nitrogen-oxygen separation.

Chemical Reactions: This concept uses the peroxide reaction of barium oxide or cobalt oxide to extract oxygen from air at high temperature. A rotating porous bed of oxide material would absorb oxygen from air by the peroxide reaction at high temperature and pressure and then rotate into a region of lower pressure where the peroxide could give up oxygen and revert to the oxide.

Co-current Spray Contactors: This is a cryogenic distillation device in which a liquid phase is sprayed into a vapor stream and flows in the same direction as the vapor. The vapor phase becomes richer in nitrogen and the liquid phase becomes richer in oxygen. At intervals (stages) the liquid phase is extracted from the flow and reintroduced upstream to provide a net overall counterflow of liquid and vapor. With sufficient stages any degree of oxygen-nitrogen separation may be achieved.

Nitrogen Freeze-Out: This is a cryogenic separation method which uses successive stages of partial freezing of liquid air to extract nitrogen as a solid precipitate from a liquid air stream. With staging and overall counter-current operation, significant separation of oxygen and nitrogen is possible.

Centrifugation: As studied for this application, air at nearly saturated vapor conditions would be subjected to ultracentrifuge acceleration. The rate of oxygen-nitrogen separation was small due to the nearly similar molecular weights.

Chelation: Chelates (organic-metallic compounds) were postulated to absorb and then release oxygen during applied pressure and temperature cycles of the chelate.

Molecular Sieves and Membranes: Air, in passing through a material of controlled pore dimensions, would separate into oxygen and nitrogen by the greater mobility (lower molecular weight) of the nitrogen molecule.

Acknowledgments

The authors thank the NASA Lewis Research Center for permission to use material generated in several recent contracts jointly funded by NASA and the US Air Force.

References

[1]Nau, R.A., "A Comparison of Fixed Wing Reusable Booster Concepts," Proceedings of the Society of Automotive Engineers Space Technology Conf., SAE Paper 670384, SAE Pub. P-16, Palo Alto, CA, May 9-12, 1967.

[2]Nau, R.A., "Aerospaceplane Propulsion System/Vehicle Integration Study," General Dynamics/Astronautics, GDA-63-1069, San Diego, CA, Vols. 1-16, 1963.

[3]Nau, R.A., "Study of a Propulsion Fluid System for an Aerospaceplane," General Dynamics/Astronautics, ASD-TR-61-699, Pt. I, San Diego, CA, Dec. 1961.

[4]Turner, J.R., "High Temperature Air Enrichment Program," Dynatech Corp., ASD-TR-61-699, Part I, Cambridge, MA, ASD-TR-63-859 (AD #342933), Sept. 23, 1963.

[5]Reti, A.R., and Turner, J.R., "Study of Oxides for Chemical Separation of Air into O_2-N_2 Components," Dynatech Corp., APL-TDR-64-1, Cambridge, MA, Vols. 1-3, Mar. 1965-to June 1966.

[6]Turner,J.R.,"High Temperature Air Enrichment Program, "DynatechCorp., APL-TDR-64-90 (AD #353898), Cambridge, MA, Sept. 15, 1964.

[7]Nau, R.A., "Aerospaceplane Propulsion System/Vehicle Integration Study - Evaluation of the Use of a Chemical Separation System for Air Enrichment," General Dynamics/ Astronautics, GDA-63-1069, San Diego,CA, Vol. 16, Oct. 1963.

[8]Robinson, C.S., and Gilliland, E.R., Elements of Fractional Distillation, McGraw-Hill, New York, 1950.

[9]McCabe, W.L., and Smith J.C., Unit Operations of Chemical Engineering, 3rd ed., McGraw-Hill, New York, 1973.

[10]Perry, J.H. (ed.), Chemical Engineer's Handbook, 5th ed., McGraw-Hill, New York, 1973.Sec. 13.

[11]Nau, R. A., and Campbell, S.A., "Rotary Separator", United States Patent 3,779,452, Dec. 1973.

[12]Gottzman, C.F., Notaro, J., and Olszewski, W. J., "Feasibility Study of a High Capacity Distillation Separator for an Air Enrichment System," Union Carbide Corp., ASD-TDR-63-665, Pt. 2, Tonawanda, NY, Feb. 1964.

[13]Gottzman, C.F., Notaro, J., and Olszewski, W. J., "Air Separator Test Program", AFAPL-TR-66-92, Tonawanda, NY , Oct. 1966.

[14]Buchmann, O.A., "Study of a Propulsion Fluid System for an Aerospaceplane Air Heat Exchange System, "Garrett Corp., AiResearch Manufacturing Co., ASD TR-61-699, Torrance, CA, Pt. 2, Vol. 5, Dec., 1961.

[15]Buchmann, O.A., "Exploratory Development of High Temperature Heat Exchangers," Garrett Corp., AiResearch Manufacturing Co., AFAPL-TR-67-144, Torrance, CA, March, 1968.

[16]Buchmann, O.A., "LACE-ACES Heat Exchanger Design Studies," Garrett Corp., AiResearch Manufacturing Co., APL-TDR-64-79, Torrance, CA, June, 1964.

[17]Carreiro, L.R., "PDWAP - Preliminary Design and Weight Analysis Program," Wright Research and Development Center, WRDC-TR-90-2005,Wright-Patterson AFB, OH, Jan. 1989.

[18]Glatt, C. R., "WAATS: A Computer Program for Weight Analysis of Advanced Transportation Systems," Aerophysics Research Corp., NASA CR-2420, Hampton, VA,Sept. 1974.

[19]Leingang, J.L., Donaldson, W.A., Watson, K.A. and Carreiro, L.R., "ETO - A Trajectory Program for Aerospace Vehicles", Wright Research and Development Center, WRDC-TR-89-2023,Wright-Patterson AFB, OH, June, 1989.

[20]Powell, T., "High Speed Propulsion Assessment - Task II Interim Report," United Technologies, Pratt & Whitney Div., FR-19753,West Palm Beach, FL, Feb. 1987.

[21]Leingang, J.L., Donaldson, W.A., Maurice, L.Q., and Carreiro, L.R., "Airbreathing Space Boosters Using In-Flight Oxidizer Collection", 1987 JANNAF Propulsion Meeting, San Diego CA.

[22]Leingang, J.L., Maurice, L.Q., and Carreiro, L.R., "Space Launch Systems Using Oxidizer Collection and Storage", 43rd Congress of the International Astronautical Federation, IAF-92-0664, Washington, D.C, Aug. 28-Sept., 1992.

[23]Maurice, L.Q., Leingang, J.L., and Carreiro, L.R., "Airbreathing Space Boosters Using In-Flight Oxidizer Collection", AIAA 28th Joint Propulsion Conf., AIAA Paper 92-3499, Nashville, TN, July 6-8, 1992.

[24]Czysz, P.A., "Space Transportation Systems Requirements Derived from the Propulsion Performance Reported in the *Hypersonic and Combine Cycle Propulsion Session* at the 1991 IAF Congress", 43rd Congress of the International Astronautical Federation, IAF-92-0858,Washington, DC, Aug.28 - Sept. 5, 1992 .

[25]Vandenkerckhove, J.A., "A Peep Beyond S.S.T.O. Mass Marginality", 43rd Congress of the International Astronautical Federation, IAF-92-0656, Washington, DC, Aug.28-Sept. 5,1992.

[26]Maurice, L.Q., Leingang, J.L., and Carreiro, L.R., "The Benefits of In-Flight LOX Collection for Airbreathing Space Boosters", 4th International Aerospace Planes Conference, AIAA Paper 92-5059, Orlando, FL 1-4 December 1992.

[27]Maurice, L.Q., Carreiro, L.R., Donaldson, W.A., and Leingang, J.L., "Airbreathing Space Boosters Using Advanced Air Collection Systems", 1989 JANNAF Propulsion Meeting, Cleveland OH., May 23-25, 1989.

[28]Leingang, J.L., Carreiro, L.R., and Maurice, L.Q., "On-Demand Reusable Space Launch Systems that Use In-Flight Oxidizer Collection", 1993 SAE Aerospace Atlantic Conf., SAE 93-1451, Dayton, OH April 20-23, 1993.

[29]Leingang, J.L., Carreiro, L.R., and Maurice, L.Q., "Eliminating LH_2 in LOX-Collect Space Launchers: Key to On-Demand Capability," AIAA 29th Propulsion Conference, AIAA Paper 93-1835, Monterey, CA, June 28-30, 1993.

[30]Yi, A.C., "JP-LH_2 Dual Fuel ALES Two-Stage-to-Orbit Concept, Rockwell Corp., Vol. IV, Contract NAS3-25559, Downey, CA, Dec., 1992.

[31]Leingang, J.L., and Carreiro, L.R., "All-Hydrocarbon Orbital Launch Vehicle", 1990 JANNAF Propulsion Meeting, Anaheim CA, Oct., 1990.

[32]Carreiro, L.R. and Leingang, J.L., "Advances in All-Hydrocarbon Launch Vehicle Concept", 1992 JANNAF Propulsion Meeting, Indianapolis IN, Feb. 24-27, 1992.

[33]Perry, J.H. (ed), "Chemical Engineers' Handbook," 3rd Ed., McGraw-Hill, New York, 1950, Section 25.

[34]Yi, A.C., "Hydrocarbon-Water ALES Two-Stage-to-Orbit Concept", Rockwell Corp, Vol V, Contract NAS3-23339, Downey, CA, Dec. 1992.

[35]Drenevich, R.F. and Nowobilski J.J., "Airborne Rotary Air Separator Study", Praxair-Linde Div., Rpt CR 191045, Tonawanda, NY, Sept. 1992.

[36]Weldon, V. and Fink, L., "Near-Term Two-Stage-to-Orbit, Fully Reusable, Horizontal Takeoff/Landing Launch Vehicle", 42nd IAFMeeting, BoeingCo., IAF-91-19A, Montreal, Canada, Oct. 1991.

[37]Parrington, A.J., "Toward a Rational Space-Transportation Architecture", Airpower Journal, Winter 1991, Maxwell AFB, AL.

[38]Weimer, R. F., "Vortex Tube Air Separator", Air Products & Chemicals, Inc., AFWAL TR 88-2070, Allentown, PA, Oct. 1988.

[39]Kennedy, R., "High Temperature Air Liquefaction Heat Exchanger Technology", Allied Signal Aerospace-AiResearch LAD, NASA CR-189182, Torrance, CA, Dec. 1991.

[40]Acharya, A., Gottzmann, C.F., and Nowobilski, J.J., "Airborne Rotary Air Separator Study", Union Carbide Corporation- Linde Div., NASA CR-189099, Tonawanda, NY, Dec, 1990.

Air Collection Systems

Vladimir V. Balepin*
Central Institute for Aviation Motors, Moscow, Russia

Nomenclature

C = mass concentration, specific heat capacity
C_N^N = nitrogen concentration in oxygen-depleted air
C_N^O = nitrogen concentration in oxidizer
C_O^A = oxygen concentration in air
C_O^N = oxygen concentration in oxygen-depleted air
C_O^O = oxygen concentration in oxidizer
d = humidity
d_H = hydraulic diameter of the channel
D = diffusion coefficient
G = mass flow rate
$\bar{h}$ = frost layer growth rate
H = altitude
I = specific impulse
K_0 = stoichiometric ratio of oxygen and hydrogen
K_{OX} = oxygen collection ratio
$\bar{m}$ = specific mass, steam mass flow to surface unit
M = Mach number, mass
q = dynamic pressure
r = heat of liquefaction
R = thrust

Copyright © 1994 by the American Institute of Aeronautics and Astronautics, Inc. All rights reserved.

* Senior Scientist, Hypersonic Propulsion Department. Currently, Visiting Researcher, Kakuda Research Center, NAL, Kakuda, Japan.

Re = Reynolds number
Sc = Schmidt number
T = temperature
V = speed
W = exhaust velocity
X = mass content of liquid phase
α_D = mass transfer coefficient
δ = payload fraction
Δ = difference
η = oxygen extraction coefficient

Subscripts
A = air
CR = cruise
F = flight
$FLOX$ = flight liquid oxygen
H = hydrogen
N = nitrogen, oxygen-depleted air
O, OX = oxygen, oxidizer
RJ = ramjet
TJ = turbojet

Superscripts
min = minimum
S = saturation line
$*$ = stagnation

Introduction

Proposals on atmospheric oxygen collection for oxidizer supply during rocket propulsion of an aerospace plane (ASP) trajectory have been under discussion since the early 1960s. A few British, American studies on this subject are known (see, e.g., Refs. 1 and 2). During the 1960s, Russia also provided a conceptual study. The United States also conducted successful experimental study of air separation technology that was considered suitable for flight conditions.[3]

A new impulse for investigation was given at the end of the 1980s. An optimistic prediction of Ref. 4 on payload fraction value prompted reseachers to study the problem more carefully using the experience accumulated during the ASP studies.

In-flight oxygen collection may be considered as the basis of a self-sufficient ASP concept, for example, as proposed in Refs. 3 and 5, as well as reserve of further development of the known concepts.

In ASP concepts under consideration, the power during the main part of the acceleration flight to orbital speed (meaning the kinetic energy increase) is, in general, provided by a hydrogen-oxygen rocket engine. Therefore, the oxygen mass fraction of the vehicle is significant — from 16% (in the two-stage German concept Sänger) up to 55% (in the single-stage British concept HOTOL) of initial mass.

The lack of requirement for onboard oxygen supply at start conditions permits a reduction in the initial mass for given payload mass, the amount of hydrogen needed for low-speed acceleration, wings, landing gear, and, in some cases, engines and tanks. For a given ASP initial mass, oxygen collection also permits an increase in payload fraction.

The present study considers conventional ASP flight scenarios — with a hydrogen-fueled turbojet (TJ), a ramjet (RJ), and a liquid rocket engine (LRE) — operating in series for a single-stage-to-orbit (SSTO) vehicle, and with TJ and RJ for the first stage and LRE for the second stage for a two-stage-to-orbit (TSTO) vehicle.

Currently, the development problem may be said to be in the phase of choosing an optimum scheme of powerplant and air separation system integration, and definition and development of key technologies.

The author concurs that "such schemes need a thorough analysis based on energy availability considerations in order to avoid expensive developments on a basically ineffective system," [6] and hopes that this study is a step in clarification for such availability analyses.

Separation of Oxygen from Air

The air making up the Earth's atmosphere is a mixture of many gases in relatively constant concentration. Only the carbon dioxide concentration and moisture content are variable. Nitrogen and oxygen make up 98.67% of the mass of dry air. Therefore, it is acceptable to consider the air as a binary mixture when oxygen collection and usage in rockets are discussed.

The most well known and widespread method of air separation is low-temperature fractional distillation, which consists of compressed air cooling, liquefaction, and separation in a distillation column.

Even single-stage distillation gives almost pure oxygen. However, in this case, only about 65% of the oxygen in the air may be extracted. A double-column distillation allows one to obtain "pure" oxygen and "pure" nitrogen; air is considered as a binary mixture of those two gases.

Neither double-column nor single-column installations can be used in flight conditions for rocket engine oxidizer production because of their mass and volume. According to Ref. 1, a commercially developed separator of this form weighs approximately 25 tons and produces about 1.3 lb/s of liquid oxygen. Despite progress in the technique, distillation columns remain very heavy. One of the reasons for the large mass and size is the use of low velocities of flow liquid and vapor streams, generated by gravitation at action. Low velocities demand large flow cross sections.

In the rotary boilerplate separator invented by Nau and Campbell of General Dynamics and described in Ref. 3, the flows occur in the increased "gravitational" field; thus, velocities may be higher. The capacity of the process is increased dramatically. A boilerplate separator for separation of 45 kg/s of airflow was tested successfuly in the early 1960s. At 90% oxygen purity, the separator specific mass was 5 kg /(kg/sec) of airflow.

Increased velocities occur in the vortex tube (VT) separator in which an inlet nozzle speed of sound can be reached. According to estimation based on experimental data, inlet air velocity in the vortex tube is about 40-60 m/s.

Therefore, a vortex separator may be lighter than a rotary separator. An additional reason for mass reduction is the lack of moving parts and overloads.

Prior to a brief description of the VT, it is necessary to introduce a few parameters of the air separation process.

The quality of oxygen extraction from the air is characterized by two independent parameters: concentration of oxygen obtained, C_O, and the oxygen extraction coefficient, η, the ratio of oxygen in the oxidizer obtained to the total oxygen amount in air undergoing separation.

$$\eta = \frac{G_O C_O^O}{G_A C_O^A} \tag{1}$$

For binary mixture elements, concentrations in both flows are determined by values of C_O^O and η as

$$C_N^O = 1 - C_O^O \tag{2a}$$

$$C_O^N = \frac{C_O^O C_O^A (1-\eta)}{C_O^O - \eta C_O^A} \tag{2b}$$

$$C_N^N = 1 - \frac{C_O^O C_O^A (1-\eta)}{C_O^O - \eta C_O^A} \tag{2c}$$

Note that at a given oxygen flow rate, the oxygen extraction coefficient determines the flow rate of separated air and, consequently, the mass and size of the air cooling and separation system.

In the experimental study of air separation in the VT that began in the mid-1970's,[7] an oxygen concentration value of C_O = 98% was reached at an extraction coefficient of $\eta = 0.42$. This means that 58% of the oxygen that was contained in the air and entered the system was lost with the oxygen-depleted air. Later, a scheme of a two-stage vortex separator was proposed. According to a preliminary estimation, a value of $\eta = 0.85$ may be reached in this scheme.

The present study considers a flight liquid oxygen (FLOX) plant based on VTs, with base values for calculation as C_O = 98% and η = 0.85. A brief, simplified description of air separation in the VT follows.

Air that is cooled to the saturation line and partially condensed passes through nozzle 2 of vortex tube 1 (Fig. 1). The nozzle is used for air injection and vortex generation.

In the VT chamber, the energy separation of the vapor-liquid mixture is accomplished. The liquid is thrown to the wall and flows to cone 3. Nitrogen is boiled off from the near-wall film and moves into the back flow region near the axis. From the back flow, oxygen is condensed into liquid film. The liquid layer enriched with oxygen goes to diffuser 4. The near-axis vapor flow moves from cone 3 to diaphragm 5. The flow now enriched by nitrogen discharges through outlet 6.

The air pressure ahead of inlet nozzle 2 is P_A = 3-6 atm, and the liquid phase fraction by mass, X_A is equal to 20-40%. The ratio of poor air and enriched airflows is controlled by a change of hydraulic resistance of outlet flows.

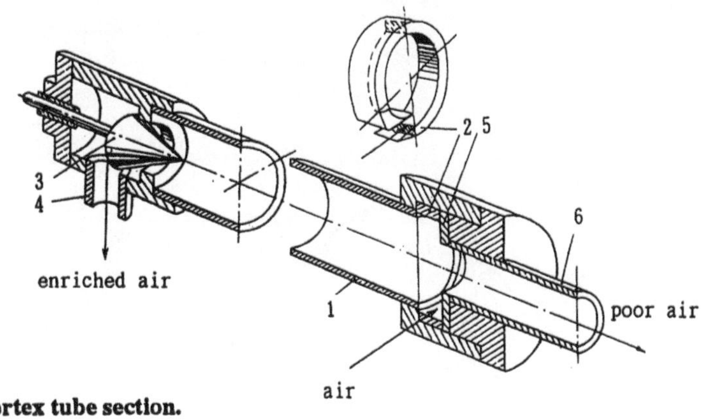

Fig.1 Vortex tube section.

The behavior of the parameters of air and separation products is shown as a T-s diagram (fig. 2).

Air from the initial state 0 is cooled along the isobar to saturation state 1 and is partially condensed to state 2. In this state, air passes through the VT nozzle. For better vortex generation, it is necessary to have the critical pressure ratio in the nozzle. Some decrease in pressure of separation products occurs in the VT chamber and flow outlets. It is proposed that the liquid fraction in oxygen flow accounts for about 80% (point 3Ox) and in poor airflow about 20% (point 3N). In experimental study, it is more convenient to come to point 2 by another way, which is shown in Fig. 2 by a dotted line. In this case, the air pressure in the initial state would be supercritical.

Just listing the features of the process in the VT shows that theoretical analysis of the whole phenomenon is very difficult. The main features are as follows:

1) Transonic flow of the two phase air through the inlet nozzle.

2) Interaction of two counterflows: direct two-phase flow and mainly gaseous backflow. Each flow has concentration and liquid content changing along the axis and radius.

3) Coexistence of the processes of condensation and evaporation under the conditions of radial and axial pressure gradients.

4) Discharge of the two two-phase flows.

Even for qualitative analysis, some data on temperature and pressure distribution inside the VT are necessary.

Full-scale experimental study of the VT rectifier should include "external" and "internal " testing. The former should give the effect of operational parameters and configuration of the tube on the final results characterized by concentration and recovery of the products. The latter should be aimed toward studying parameter distributions (pressure, temperature, and velocities) inside the tube. The results of the more complicated internal testing will be the basis of theoretical study.

Currently, a few aerospace and chemical companies are conducting experimental study of air vortex separation. Figure 3 shows a sample of the VT worked out at the Central Institute of Aviation Motors (CIAM).

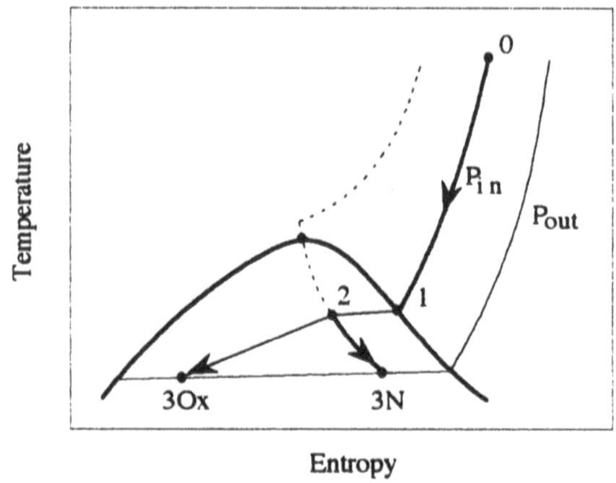

Fig. 2 T-s diagram of air separation process.

The advantages of the air rectifier based on vortex tubes are as follows: 1) high velocity (about 40-60 m/s in the VT inlet nozzle) of airflow that reduces the size and mass of the device; 2) lack of any moving parts; and 3) lack of significant pressure and temperature gradients, and small size of a single tube, allowing the possibility of thin-walled VT fabrication with light plastics compatible with low temperature. With this assumption, VT separator specific mass was estimated as 1.0-1.5 kg/(kg/s) of airflow and specific volume as 0.01-0.015 m^3/(kg/s) of airflow.

Figure 4 shows a draft of the vortex separator.

Air Cooling and Condensation

As noted, air is in a two-phase state before separation in a VT, with 20-40% of liquid by mass, at pressure P_A =3-6 atm.

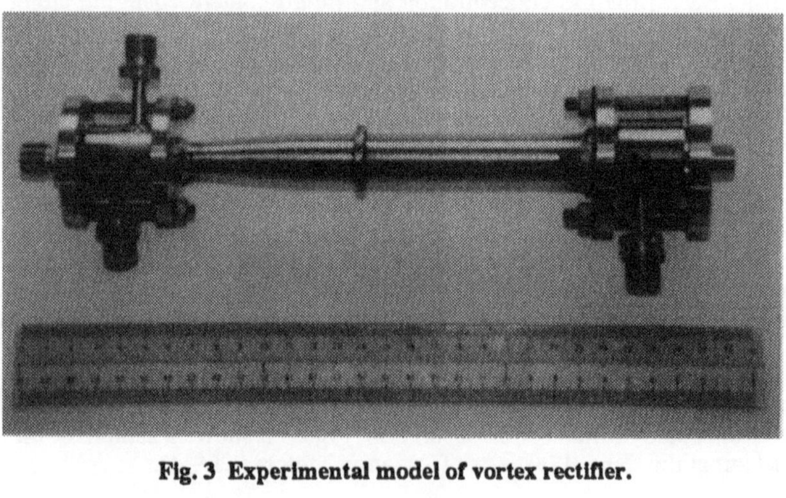

Fig. 3 Experimental model of vortex rectifier.

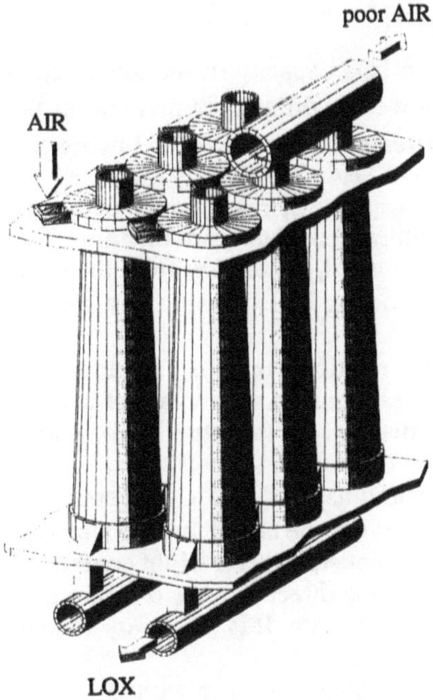

Fig. 4 Slice of vortex rectifier.

In flight conditions such pressure may be reached in ram air in the air intake at the flight Mach number range of 4-6 and altitude range of $H = 17...30$ km , and, of course, at all speed and altitude values when a compressor is made available.

The FLOX plant heat exchangers may have hydrogen and wasted flow of oxygen-depleted air as coolants. It is convenient to characterize the cooling system capacity by the ratio of cooled air and hydrogen flow rate, $\overline{G}_A$, by analogy with a liquid air cycle engine (LACE). There is no restriction on the $\overline{G}_A$ value from the point of view of heat balance in the heat exchangers.

The main coolant in the air condenser is liquid hydrogen from the tank. Hence, the heat balance of the condenser determines, as a rule, the upper limit of the $\overline{G}_A$ value. Oxygen in the rectifier outlet contains about 20% in the gas phase, which has to be condensed in a condenser for the remaining gas phase.

Thus, the collection ratio, i.e., the ratio of collected oxygen and total hydrogen flow rates, is determined from the heat balances of the air condenser and condenser of remaining gas phase in the oxygen:

$$K_{OX} = \frac{C_{PH}\left(T_A^S - T_H - \Delta T\right) + \Delta}{\dfrac{r_A X_A}{\eta C_O^A} + r_{OX}\left(1 - X_{OX}\right)} \quad (3)$$

and the air to hydrogen flow rate ratio is

$$\overline{G}_A = \frac{K_{OX}}{\eta C_O^A} \quad (4)$$

where C_{PH} is the average hydrogen specific heat capacity in the temperature range T_H to $(T_A^S - \Delta T)$. Here T_H is the lowest hydrogen temperature in the condenser, ΔT the minimal temperature difference between air and hydrogen, and Δ the contribution of additional methods of heat sink capacity increase.

Among the additional methods of heat sink capacity increase, which define the value of Δ, the following should be noted: utilization of the endothermic effect of H_2 para-ortho conversion, organization of slush H_2 circulation from the tank through the condenser, and intermediate H_2 cooling by expansion in a special turbine.

The use of hydrogen conversion for collection ratio increase is very attractive. The effect of the para-to-equilibrium transition is equal to an increase of average hydrogen specific heat capacity in the condenser by more than 40%, with a proportional increase of collection ratio.

An interesting discussion on thermodynamic aspects of H_2 conversion in application to a LACE cycle is given in Ref. 8. It may be added that the currently available conversion technology by catalysts is not suitable for flight conditions. By our estimation, the mass of the catalyst for the direct process of ortho-para conversion is about 1 ton/(kg/s) of hydrogen flow rate. It is necessary to seek new technologies of hydrogen conversion.

Hydrogen slush circulation and use of intermediate hydrogen expansion lead to significant complication and mass increase for a FLOX plant and may be reasonable only in special cases.

In current study the value of $K_{OX} = 4.19$ was taken as the base maximum value at $\Delta T = 0$, $\Delta = 0$, and $\eta = 0.85$. If a compressor is not included in the FLOX cycle, this corresponds to flight conditions with Mach number 5.0 at dynamic pressure $q = 75$ kPa and hydrogen temperature before the condenser $T_H = 14$ K.

FLOX Plant and Powerplant Integration

Some data show that oxygen collection during acceleration flight is more effective than during cruise flight at constant speed. In this case, the "exchange" of onboard hydrogen for atmospheric oxygen is the only result of the flight collection phase. However, it seems that these data were obtained without taking into account the FLOX plant mass, which increases significantly for varying regimes.

In the author's opinion, it is necessary to realize the utmost integration of the FLOX plant and powerplant. For instance, if there is a need to increase air pressure before separation, it is advisable to employ the TJ compressor, which fulfills its main function to this moment. It is advisable also to take air into the FLOX through the TJ air intake, to discharge depleted air through the TJ nozzle, and so on. Conversely, the FLOX plant heat exchangers may be used, in principle, for air cooling before the TJ compressor to improve TJ performance during the onerous regimes, such as the transonic flight regime.

Oxygen collection during acceleration requires a special compressor, or an increase in size of the TJ as well as air intake. A FLOX mass increase and complication of the control system are unavoidable in this case. An additional disadvantage is connected with an increased possibility of heat exchanger icing, because of high air humidity at low altitudes.

Of course, oxygen collection during acceleration has its own advantages, which need to be studied.

The current study considers oxygen collection in cruise flight at constant speed and altitude.

Characteristics of the integrated FLOX plant and powerplant and the FLOX mass and size are defined by five independent parameters: 1) total required thrust of the integrated powerplant and FLOX plant in cruise flight, R_{CR}; 2) total required amount of oxygen, M_{OX}; 3) specific impulse of the integrated system, I_{CR}; 4) oxygen collection ratio, K_{OX}; and 5) oxygen extraction coefficient, η.

The main parameters to be determined are FLOX plant mass, M_{FLOX}, and duration of collection in flight (or cruise flight range, L_{CR}). The values of thrust R_{CR} and specific impulse I_{CR} determine the value of minimal hydrogen flow rate in cruise flight G_H^{min}. The minimal FLOX plant mass and maximal duration of cruise are to be determined from values of M_{OX}, K_{OX}, η, and G_H^{min}.

If cruise duration corresponding to G_H^{min} is unsuitable for any reason, then the hydrogen flow rate may be increased without a thrust increase. Oxygen flow rate and FLOX mass increase proportionally. The oxygen collection time decreases in this case, but the FLOX mass growth is unfavorable especially for a single-stage ASP. Thus, specific impulse of the integrated system does not determine the amount of hydrogen spent in cruise flight but indicates FLOX mass and duration of oxygen collection.

The effectiveness of hydrogen usage during oxygen collection depends mainly on the oxygen collection ratio K_{OX}. In the final analysis, this coefficient sets the vehicle dimensions.

The combined system of the powerplant and the FLOX plant, which provides the thrust during ASP acceleration and cruise flight, and oxygen collection, consists of three units: turbojet, ramjet, and FLOX plant.

Depending on the manner in which each of these units interacts and the poor air utilization, three main schemes of powerplant and FLOX plant integration may be suggested.

Scheme 1 (Fig. 5a)

The turbojet does not operate during the air collection process.

The FLOX plant has a separate air intake (or a channel in a common air intake) and a separate nozzle (or a channel in a common nozzle), consisting of heat exchangers and rectifier, and may have its own turbocompressor. FLOX gives its contribution to total thrust. Depleted airflow is indicated in Fig.5 as N_2.

The ramjet provides the main part of the thrust during cruise flight. It is connected to FLOX only by a hydrogen line.

Such a scheme is the basis of the integrated system of the India-proposed ASP HYPERPLANE.[4]

A variant of this scheme with poor air reaction with a small amount of hydrogen before the FLOX nozzle seems very attractive. Such a scheme will be called below as Scheme 1a.

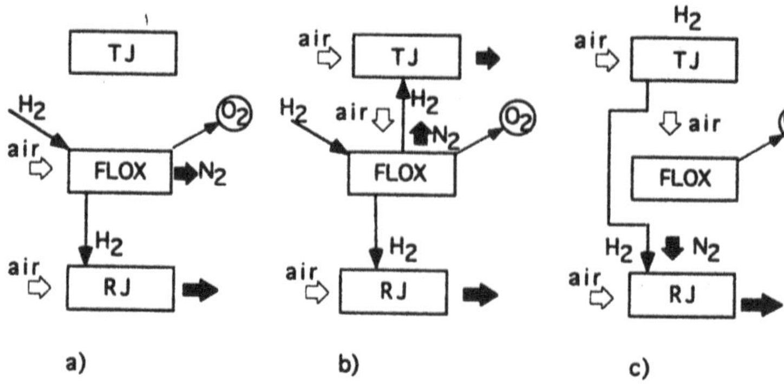

Fig. 5 Schemes of powerplant and FLOX plant integration.

Scheme 2 (Fig. 5b)

The turbojet is employed during the oxygen collection regime to obtain a pressure increase in the air separation cycle; poor air after oxygen extraction reacts with a small amount of hydrogen in theTJ combustion chamber to drive the turbine and provide additional thrust.

The FLOX plant has neither its own air intake nor nozzle, it consists of the heat exchangers, rectifier, and additional turbocompressor to compress poor air after the rectifier.

The ramjet operates the same as described in Scheme 1. Such a scheme was examined in Ref. 9 for application to the two-stage Sänger-type ASP.

Scheme 3 (Fig. 5c)

The turbojet is used during the oxygen collection regime only for a pressure increase in the air separation cycle. The compressor is driven by a hydrogen expansion turbine. In this case, the TJ is of the air-turbo-rocket (ATREX) type.

The FLOX plant separates air and increases the pressure of depleted air that follows into the ramjet combustion chamber.

The ramjet is the only thrust generator used during the oxygen collection regime. The ramjet working gases in this regime are air from the ramjet's own intake, depleted air from the FLOX plant, and hydrogen from the FLOX cooling system.

The main feature of Scheme 3 is poor air afterburning in the RJ combustion chamber. That is why poor air has to be compressed up to higher pressure than in Schemes 1 and 2.

Excluding some details, similar schemes of powerplant and FLOX plant integration were examined in Refs. 3 and 10.

It should be stressed that the main principle of the proposed classification is the manner of poor air utilization; other details of the known schemes may not be consistent with the schemes shown in Fig. 5.

As mentioned earlier, minimal value of hydrogen consumption providing cruise thrust is a very important parameter that defines collection time and FLOX plant mass. It is the reciprocal of specific impulse that does not have independent significance for the oxygen collection phase of flight. That is why G_H^{min} is used below to characterize the schemes' efficiency. Equations for three schemes of poor air utilization under discussion are given next. They are adjusted in comparison with equations presented in Ref. 11. Scheme 1:

$$G_H^{min} = \frac{R_{CR}}{I_{RJ} + \frac{K_{OX}}{\eta C_O^A}(W_N - V_F) - K_{OX} W_N} \tag{5}$$

Values of RJ specific impulse, I_{RJ} have to be taken according to known ramjet characteristics, and the value of poor air exhaust velocity W_N may be determined when pressure and temperature before the nozzle are given. Pressure depends on the existence and characteristics of the compressor for poor air and pressure drop in the heat exchanger. For a rough estimation, the value of poor air temperature $T_N = 0.6 T_A^*$ may be used, where T_A^* is air stagnation temperature after air intake. Scheme 2:

$$G_H^{min} = \frac{R_{CR}}{\left(1 - \frac{K_{OX}}{\eta C_O^A} \frac{1 - \eta C_O^A}{\alpha_N L_N}\right) I_{RJ} + \frac{K_{OX}}{\eta C_O^A}\left[\left(1 - \eta C_O^A\right)\left(1 + \frac{1}{\alpha_N L_N}\right) W_N - V_F\right]} \tag{6}$$

where L_N is the stoichiometric ratio of poor air and hydrogen flow rates; and α_N is the oxidizer excess coefficient in the mixture of depleted air and hydrogen ($\alpha_N \geq 1$ corresponds to either stoichiometric reaction between H_2 and O_2 contained in depleted air or oxygen excess).

The stoichiometric ratio is determined as follows:

$$L_N = K_0 / C_O^N \tag{7}$$

Data on oxygen concentration in poor air and stoichiometric ratios at different values of the extraction coefficient are shown in Table 1 (oxygen concentration in obtained oxidizer $C_O = 98\%$).

Table 1 Parameters of depleted air

η	0	0.4	0.6	0.8	0.9	1.0
C_O^N	0.2315	0.1534	0.1079	0.0571	0.0294	0
L_N	34.3	51.85	73.78	139.8	272.9	∞

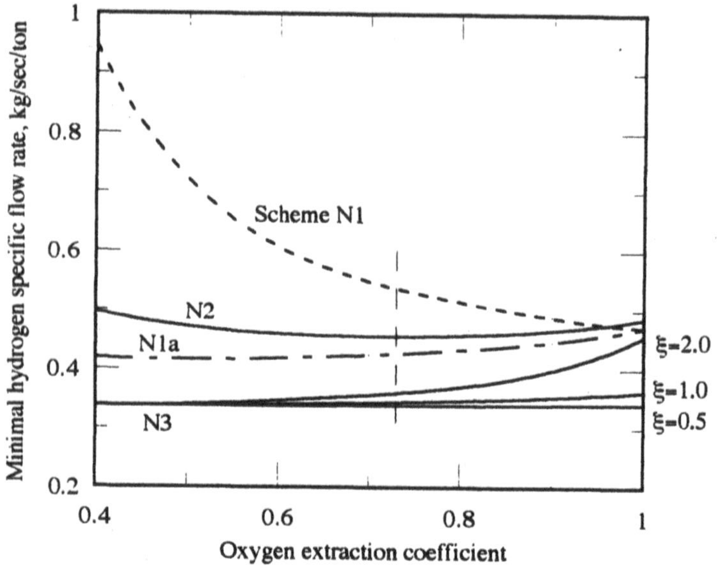

Fig. 7 Relation between oxygen extraction coefficient and minimal hydrogen specific flow rate for four schemes of poor air utilization.

Note that the curves for Schemes 1a and 2 connected with depleted air reaction are minimum at extraction coefficient values of η=0.5-0.7. It is close to the value for the single-stage commercial air rectifier. However, these values are outside the working area of the condenser. If a limit value of K_{OX} will be increased in any way, for example, by using para-ortho-conversion without significant condenser mass increase, low values of the oxygen extraction coefficient will be acceptable.

Table 3 shows the values of minimal hydrogen flow rates for each of the discussed schemes of integration with $\eta = 0.85$.

The difference between the best Scheme 3 and the far more simplistic Scheme 1a is only 22%. At the same scheme of the FLOX, it would mean the difference between FLOX mass values. However, the difference in the configuration of these schemes is significant. That is why other properties may be of more value.

Thus, in Scheme 3, it is the need for placing the bulky system of poor air distribution and mixing with mainstream air in the ramjet duct.

The advantages of Scheme 2 are apparent at relatively low speed of cruise flight M_{CR} or at high altitude H_{CR}. At M_{CR} = 5.0 the pressure level needed for

Table 3 Minimal hydrogen flow rates for different schemes

Scheme	1	1a	2	3
G_H^{min} (kg/s)/ton	0.500	0.438	0.460	0.350

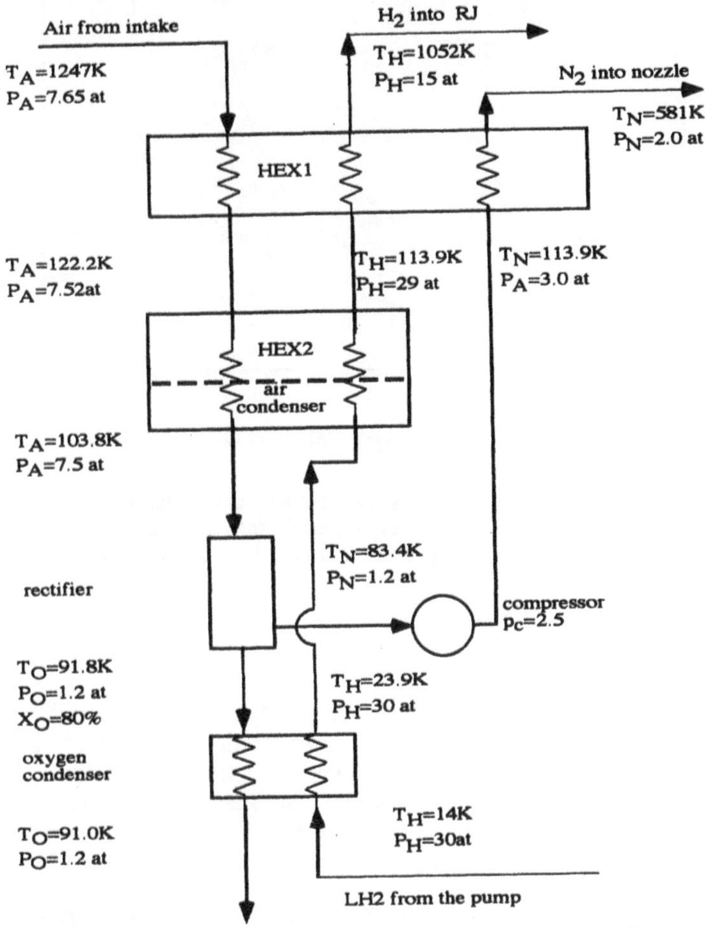

Fig. 8 Baseline FLOX plant scheme. Pointed parameters correspond to conditions of $M_f = 5.0$; $K_{OX} = 3.5$; $\eta = 0.85$.

air separation is attainable without the compressor at flight altitude of $H_{CR} < 24$ km.

Scheme 1 with poor air exhaust through the separate channel appears the list complex and is also lightweight. Its more economic modification, Scheme 1a, seems very attractive, too. The results discussed below are concerned mainly with Scheme 1.

FLOX Plant Mass Estimation

The FLOX system considered consists of the following units: three heat exchangers (two "H_2-air" and one "depleted air-air"); condenser; rectifier (bank of vortex tubes); and low-temperature turbocompressor to increase poor air pressure after rectification ($\pi_c = 2.5$-2.7).

Baseline scheme of the FLOX plant with pointed air and coolant parameters in the main stations is shown in Fig. 8.

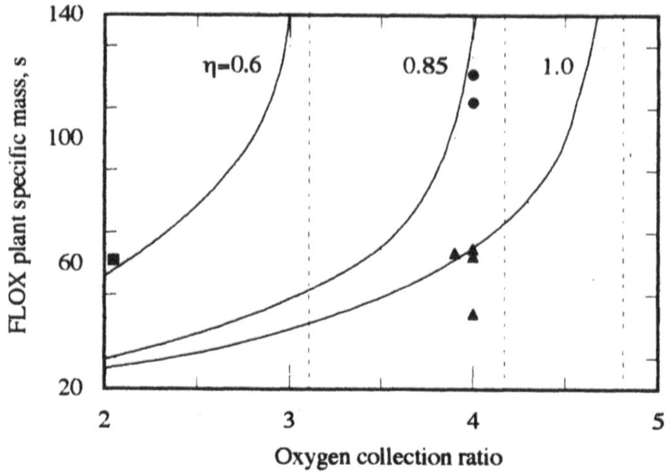

Fig. 9 Effect of oxygen collection ratio and oxygen extraction coefficient on FLOX plant specific mass (FLOX plant mass per 1 kg/sec of oxygen flow rate). n , according to Ref. 9; l , according to items 2 and 3 of Table 6; and s, according to Refs. 4, 10, and 15.

The mass of the heat exchangers and condenser was determined by the computer code of heat exchanger mass minimization, and mass values of the rectifier and turbocompressor were estimated by individual units calculation. As a result, generalized dependence of FLOX plant mass on the oxygen collection ratio and the oxygen extraction coefficient was obtained. This dependence for specific mass, i.e., FLOX plant mass divided by collected oxygen flow rate, is shown in Fig. 9.

FLOX mass goes to infinity when the collection ratio draws near the limit which is determined by condenser and low-temperature hydrogen heat exchanger heat balances. The value of this limit is proportional to the oxygen extraction coefficient. The limit value of K_{OX} for three examined values of η is shown in Fig. 9 by the vertical dotted line.

Figure 9 also shows the values of the FLOX plant specific mass as obtained by different authors. Despite the significant difference in schemes, operation conditions, and levels of FLOX mass estimation, good agreement of different data with results of proposed generalization may be noted. Such agreement allows one to consider the obtained relation as the best current summary of available data on FLOX mass and to recommend it for use in further studies of ASP with oxygen collection.

FLOX plant units mass for three values of collection ratio at $\eta = 0.85$ and values of specific frontal area (normal to flow) of inlet heat exchangers are listed in Table 4. Mass and frontal area values are divided by the oxygen flow rate.

Based on the data of Figs. 7 and 9, one may estimate the dependence of FLOX plant mass divided by the value of required cruise flight thrust. Such dependence for Scheme 1a is shown in Fig. 10.

Table 4 Mass of the FLOX plant elements

Collection ratio	2.2	3.5	3.8
Heat exchanger specific mass, s	18.5	52.2	72.8
Condenser specific mass, s	2.1	3.4	3.7
Rectifier specific mass, s	5.1	5.1	5.1
Turbocompressor specific mass, s	3.5	4.2	4.3
Total FLOX plant specific mass, s	29.2	64.9	85.9
Inlet frontal specific area, $m^2/(kg/s)$	0.235	0.381	0.427

FLOX Plant Configuration

A FLOX plant may be designed as a module structure. An example of such a module for separation of 50 kg/s of airflow rate is shown in Fig.11. This structure is designed according to size determined by calculation.

To decrease the module frontal area, the heat exchanger sections are located at an angle to the inlet airflow.

The flow turning to the required angle to the bank of tubes and flow smoothing out after the heat exchangers is provided by deflecting profiles.

Cooled air flows into the condenser where about 30% of the mass flow is converted to liquid. In fact, the flow of two-phase air in the gravitational field, a suitable condenser configuration, and the system of two-phase air distribution on the bank of vortex tubes need special study.

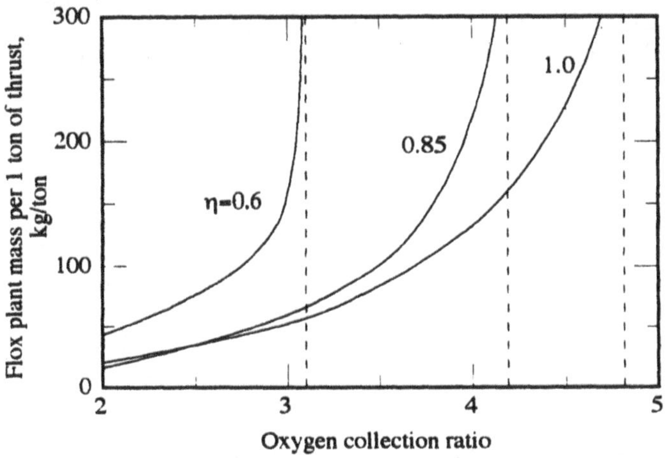

Fig. 10 Effect of oxygen collection ratio on the FLOX plant mass divided by the value of thrust on cruise flight regime.

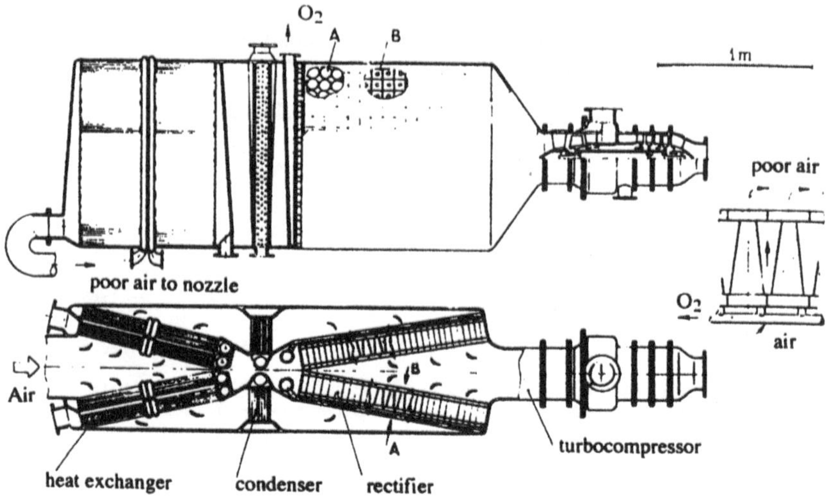

Fig. 11 FLOX plant structure, module for separation of 50 kg/s of airflow.

Two-phase air is distributed in the bank of vortex tubes, where air is separated on oxygen and poor air. Oxygen through the collector comes into the condenser of the remaining gas phase (not shown in Fig. 11) and passes into the LOX tank for usie on the rocket part of the trajectory.

Poor air with temperature of about 100 K comes into the turbocompressor where its pressure increases to provide its ability to work as coolant in the heat exchanger sections.

From the heat exchangers, poor air having temperature of about $0.6T_A^*$ goes to the ASP afterbody and wastes through the nozzle providing additional thrust or through the slits forming the near-wall layer in the nozzle of the main engine to decrease the drops and to provide thermal protection.

Heat exchanger sections and the bank of vortex tubes are installed on the central body, which contains control devices at constant low temperature and part of the collectors.

Efficiency of the Oxygen-Collecting ASP

In-flight atmospheric oxygen collection may be employed for SSTO as well as TSTO vehicles. In the latter case, a heavy FLOX plant is not lifted to the orbit but returns with the first stage. Using the principle of oxygen collection has some advantages for vehicles with a long cruise flight phase. The cruise regime allows one to consider deep integration of FLOX plant and powerplant cycles and to optimize all the structure elements for one regime. The estimations for FLOX plant application to TSTO and SSTO vehicles are given subsequently.

FLOX Plant Application to TSTO Vehicle of Sänger Type

Among the ASP projects under consideration, durable cruise flight is proposed only for the German two-stage vehicle, Sänger. The possibility of FLOX plant

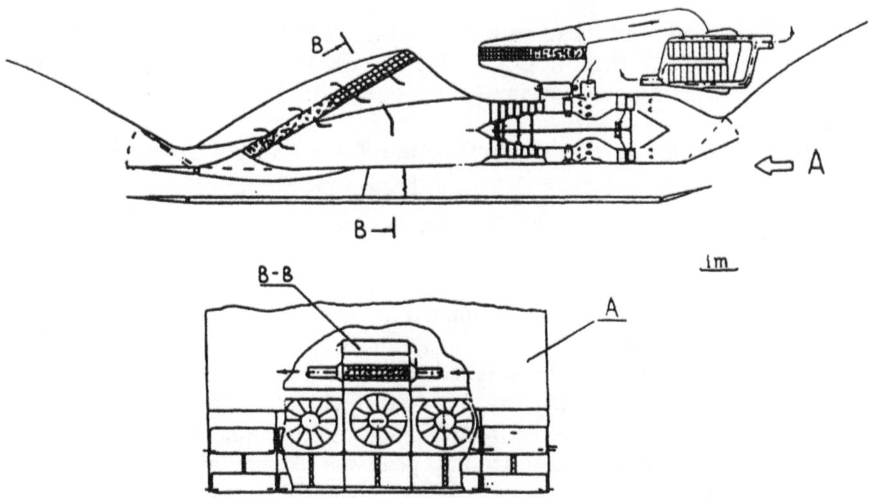

Fig. 12 Integrated FLOX plant configuration for the Sänger-type vehicle.

employment for this concept was examined in Ref. 9 with data from Ref. 12. More recent status reports on the Sänger program contain some modifications.

There are two directions to realize the advantage of FLOX plant using on first stage of the vehicle: 1) to increase the second stage mass with corresponding payload mass growth at the same initial mass as for LOX-carrying vehicle; 2) to reduce the initial vehicle mass by oxygen mass saving, hydrogen mass for initial acceleration saving, and some gains on structure element mass (taking into account the FLOX plant mass addition) at the same second-stage size and dry mass as for LOX-carrying vehicle.

Both directions have to lead to close results meaning the criterion of payload fraction value and choice should be done by criterion of cost of payload unit launching. In Ref. 9 a second direction was chosen for estimation.

An example is discussed where it is necessary to collect 56 tons of LOX within 2000 s of cruise flight at $M_{CR} = 4.5$, $H_{CR} = 26$ km. These data as well as power plant scheme correspond to Ref. 12. One of the five TJs serves as pressure generator for the FLOX plant at cruise regime and provides some thrust along with RJ. Turbojet thrust on this regime may be positive as well as negative. Thus, the FLOX plant and powerplant integrated scheme corresponds to Scheme 2 (Fig. 5b).

The variant of the FLOX plant and powerplant configuration is shown in Fig. 12. According to this scheme, only one TJ is employed in the FLOX cycle.

As mentioned earlier, the experimentally reached value of oxygen extraction coefficient is η=0.42 at oxygen concentration C_O = 98%. In the experimental study in progress, it is expected to reach a value of η = 0.6 with minimal complication to the rectifier structure. This value was assumed for FLOX plant mass estimation.

Depleted air (without 60% of extracted oxygen) and hydrogen stoichiometric reaction provide good working gas for the TJ turbine having a temperature of about 1900 K, taking into account components preheating in FLOX heat exchangers.

At $\eta = 0.6$ and the given oxygen flow rate $G_O = 28$ kg/s, the airflow rate through the FLOX plant is $G_A = 202$ kg/s. These flow rate values lead to a very low oxygen collection ratio required, $K_{OX} < 2.0$, and, consequently, to low FLOX plant mass.

The total mass of heat exchangers made of stainless steel (hot zone) and aluminum alloy (cold zone) was estimated as 1500 kg. Mass of the rectifier assembled of vortex tubes fabricated with plastics was estimated as 200 kg.

About 16.5% of initial vehicle mass reduction in comparison with the baseline concept is connected with empty oxidizer tank at start conditions. Total H_2 fuel mass of the first stage was reduced by 16%, i.e., 4.5% of initial ASP mass. Mass of engines, wings, and first-stage landing gear was reduced by 3.5% of initial ASP mass. Total initial mass reduction was about 24%, taking into account FLOX plant mass. Vehicle initial mass decreased from 340 to 260 tons. This is equal to the payload fraction increase by 1.3 times.

The relatively small amount of oxygen to be collected during the long cruise flight range reduces the benefit of oxygen collection but allows the following: avoid extra amount of hydrogen; propose a lightweight FLOX plant because the low value of K_{OX} is acceptable; and reduce the requirement on the air separation process because a low oxygen extraction coefficient value is acceptable.

Initial thrust of the vehicle may be provided by five identical TJs, the thrust of each being 230 kN (instead of 300 kN[12]). One of them should be connected with the FLOX plant. Structure of this particular TJ differs from that of others only in the air bypass system (Fig. 12).

Deep integration of the FLOX plant and powerplant cycles leads to the following advantages: TJ engine on the cruise flight regime turns from "passenger" to an active element of the FLOX plant; special FLOX plant air intake is not required; and if the turbojet generates any positive thrust, it is possible to reduce RJ mass and size by displacing design Mach number.

Much higher efficiency of TSTO collecting LOX during acceleration was obtained in Refs. 3 and 15.

FLOX Plant Application to SSTO

The efficiency of an SSTO oxygen-collecting vehicle was examined in Ref. 11. The task was set to compare two vehicles of 350 tons initial mass: LOX-collecting vehicle and LOX-carrying vehicle. The baseline configuration was taken from Ref. 13. The final dimensions of both vehicles are shown in Fig. 13.

Launching to the orbit with an altitude of 200 km and an angle of inclination of 47 deg was considered.

The task was set so that the thrust-to-weight ratio and thrust per unit of inlet area on maximal engine regime during takeoff were given as basic data. Inlet area and wing specific loading were determined by solving equations of mass balance and by matching available and required vehicle volumes.[14]

The combined powerplant consists of an ATREX for the initial part of the flight ($0 \leq M \leq 5$), a dual-mode scramjet for additional acceleration ($5 < M \leq 14$)

and a hydrogen-oxygen rocket engine for final acceleration and vehicle placement into orbit. Oxygen collection occurs during ramjet operation.

The main purpose of the study[11] was to determine the advantage of an oxygen-collecting ASP over the LOX-carrying ASP by using the payload fraction, i.e., to determine the difference between payload fraction values.

The results of plane and powerplant parameter optimization are given in Table 5.

The same calculation was made for similar vehicles of 260 tons initial mass. The trend of payload fraction increase when the absolute value of initial mass increase was confirmed for the LOX-carrying ASP as well as for the LOX-collecting ASP. The difference between payload fraction of both vehicles at initial mass of 260 tons was 4.1% of the initial ASP mass.

Reference 11 was the first attempt to take into account the real effect of oxygen collection ratio K_{OX} on oxygen-collecting ASP efficiency. The K_{OX} increase is favorable to reduce the hydrogen amount wasted during cruise flight, but as shown in Fig. 9 the K_{OX} limit value is not very high, at $\eta = 0.85$ it is about $K_{OX} = 4.2$. When the collection ratio goes to this limit, the FLOX plant mass increases very rapidly and restricts the payload fraction.

Figure 14 shows the effect of K_{OX} on the difference between payload fractions of LOX-collecting ASP and LOX-carrying ASP without taking into account FLOX plant mass, $\Delta\delta^*$; FLOX plant mass divided by ASP initial mass, $\overline{m}_{FLOX}$ and the difference between payload fraction of two examined vehicles taking into account FLOX plant mass, $\Delta\delta$.

Comparison of the results showed that the value of payload fraction of the oxygen-collecting ASP exceeds one for the LOX-carrying vehicle by 6.2% of the initial ASP mass.

The maximal value of ASP with oxygen collection corresponds to $K_{OX} = 3.5$ at slight decrease when K_{OX} decreases.

Table 5 Parameters of the oxygen-collecting and LOX-carrying ASPs

		Oxygen-collecting ASP	LOX-carrying ASP
Mach number of FLOX plant operation, M_{CR}		4.5-5.5	
Mach number of rocket engine start, M_R		11.5-12.5	13.0-14.0
Oxygen collection ratio, K_{OX}		3.3-3.7	
Onboard components total consumption, % of initial mass	ATREX	8	7
	RJ during collection	17	
	Dual-mode scramjet	19	17
	LRE	9	46
	Total	56	65
Initial vehicle density, kg/m^3		96.2	160
Cruise and collection time, min		27	
Total flight time, min		53	19
Cruise range, km		2700	
Total range, km		6750	3100

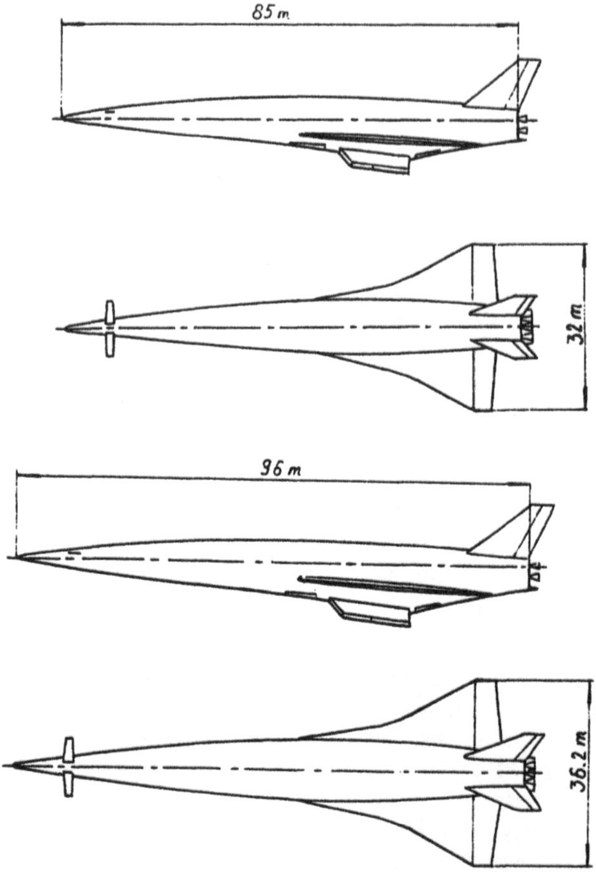

Fig. 13 Comparison of LOX-carrying (upper) and LOX-collecting (lower) vehicles.

To conclude the efficiency analysis, it should be noted that a twofold increase of ascent-acceleration range may be used with the FLOX plant for launching the vehicle into orbit with a smaller inclination angle than was indicated. This possibility can be regarded as a reserved means for payload increase.

Comparison of Known Data

Despite the long-term discussion on oxygen-collecting ASP, it should be certain that the question is far from resolved. That is why it is very interesting to compare available data. Although there is a significant difference in approaches, in chosen operation conditions, in ASP's initial mass, this comparison forms a summarizing matrix of parameters that may be very useful in future studies.

Table 6 shows the data of available studies published during last 5 years. Most of the data confidently show the increase of payload fraction in comparison with LOX-carrying vehicles and the trend of relative payload fraction increase when absolute initial ASP mass increases.

Table 6 Comparison of the data on oxygen-collecting ASPs

	Single-stage ASPs							Two-stage ASPs		
	1	2	3	4	5	6	7	8	9	10
	Ref. 4, revised concept	Estimation of the data of Ref. 4: CIAM	Estimation of the data of Ref. 4: Tupolev D.B.	Ref. 10	Ref. 11		Ref. 15	Ref. 15	Ref. 9	Ref. 3
ASP initial mass, tons	100	100	100	300	260	350	401.8	911.6	260	208.4
Payload fraction, %	15			8.0	+4.1*	+6.2*	7.34	7.46	1.28	2.18
Collection ratio	4.0	4.0	4.0	4.0	3.5	3.5	4.0	4.0	1.88	
Extraction coefficient	1.0	0.85	0.85	1.0	0.85	0.85	0.93	0.93	0.6	
Assigned value of oxygen concentration, %		95	95	90	98	98	90	90	98	
Average flow rate of collected oxygen, kg/s	90.6	100	100	183	96.6	134	237.5	486	28.0	
Ratio of collected oxygen mass to initial ASP mass, %	86.0			73.0	60.2	61.9	81.2	89.6	21.5	56.0
Fraction of hydrogen in initial ASP mass, %	45.4			56.7	54.2	53.0	41.6	46.5	34.1	26.4
Fraction of FLOX plant mass in initial ASP mass, %	4.0	12.1	11.2	3.87	2.75†	2.8†	3.73	3.46	0.65	5.44
Ratio of collected oxygen mass to FLOX plant mass	21.5	8.5	9.0	18.9	20.4	20.6	21.8	25.9	33.1	10.3
Ratio of FLOX plant mass to oxygen flow rate, s	44.2	121.0	112.0	63.6	67.0	67.0	63.0	64.8	60.7	
Initial ASP density, kg/m^3 ‡	77.9			87.6	84.5	96.2				
Cruise Mach number	3.5-8.0			5.0	5.0	5.0	5.0	5.0	4.5	1.5-5.0
Oxygen collection time, s	1000			1200	1620	1620	1380	1620	2000	

* Difference between oxygen-collecting and LOX-carrying ASP payload fractions.

† Determined at maximal value of oxygen flow rate during collection.

‡ For items 4,5,6, fullness coefficient for ASP volume was assigned as $\varphi = 0.8$.

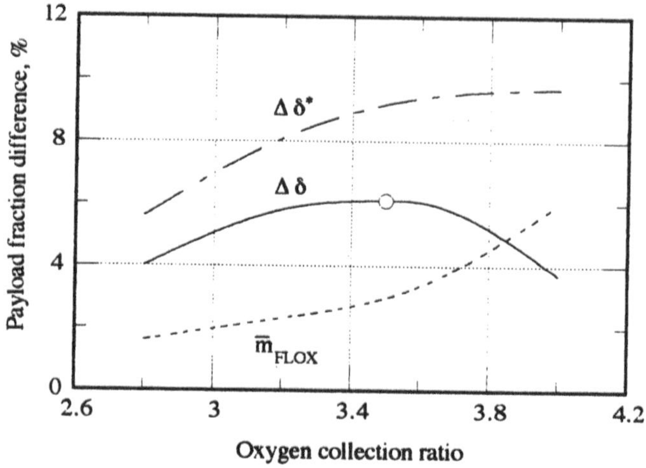

Fig. 14 Effect of oxygen collection ratio on ASP effectiveness.

Influence of Main Parameters of Flight on an Oxygen-Collecting Vehicle Payload Fraction

This section is based on the calculation made by a well-developed approach in the course of study[11] and on more recent approximate estimation.

The estimation was made for an SSTO vehicle of 260 tons of initial mass mentioned previously with baseline values of flight scenario parameters as follows: Mach number of the transition on rocket mode, $M_R = 12.0$; purity of collecting oxygen, $C_O = 98\%$; altitude of cruise flight, $H_{CR} = 24.0$ km; Mach number of cruise flight, $M_{CR} = 5.0$; oxygen collection ratio, $K_{OX} = 3.5$; and rocket engine mixture ratio, $K = 6.0$.

The listed parameters were changed over some range to determine their influence on the payload fraction. Figure 15 shows the effect of variation of these parameters on relative payload fraction. The value of the payload fraction at baseline parameter values is assumed to be 1.0 and is marked on the charts.

The existence of the maximum of the curve of payload fraction vs Mach number of the transition on the rocket mode for the ASP with oxygen collection may be explained by the same reasons as for the LOX carrier, however optimal value of M_R is higher for the last one (see Table 5). At M_R lower than optimal, the amount of onboard and collecting components required for LRE increases; at M_R higher than optimal scramjet and vehicle thermal protection penalty increases. (See Fig. 15a.)

It is clear that maximal values of the specific impulse and payload fraction may be reached at pure oxygen. Average decrease of the payload fraction in the considered range of C_O is about 1.5% per 1% of oxygen admixtures. Note that for TSTO vehicles of Sänger type whose LRE is on duty from $M_R = 6.8$, payload fraction decrement is about 3.2% per 1% of admixtures.[9] (See Fig. 15b.)

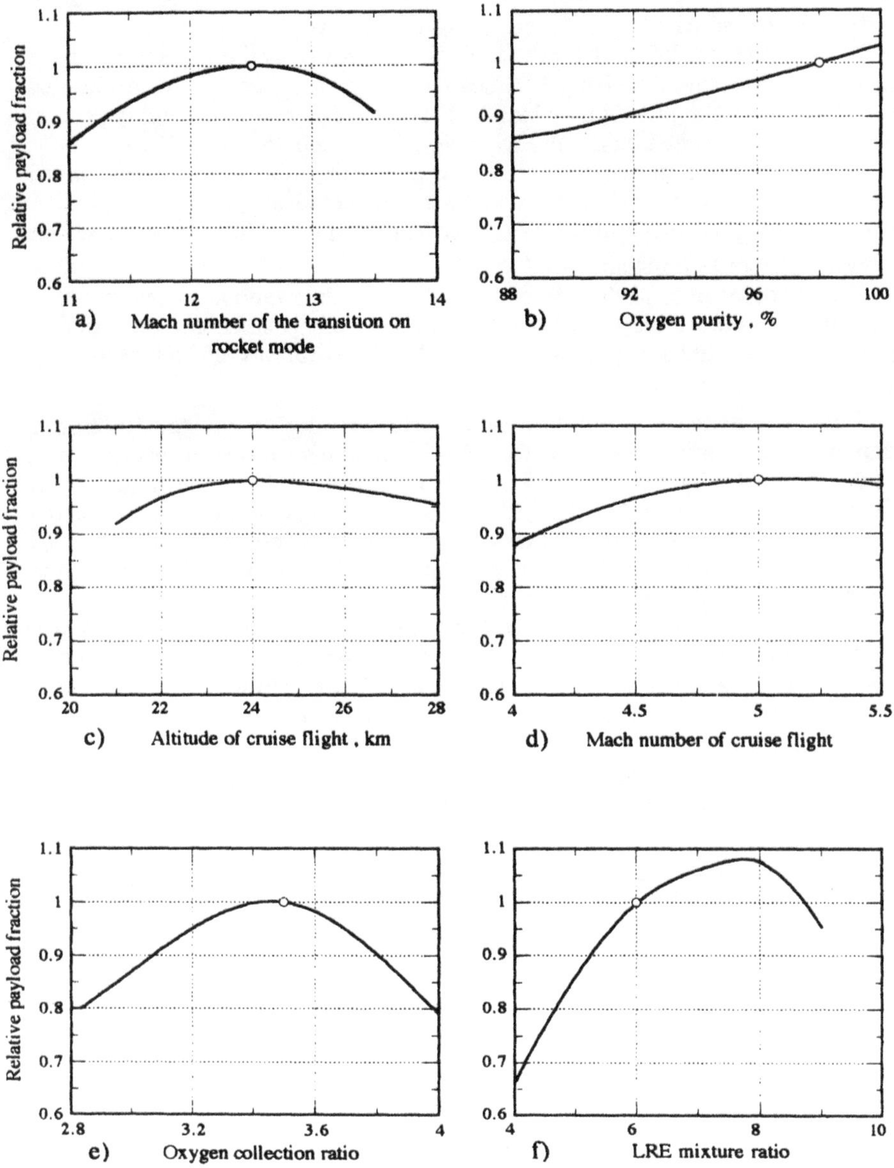

Fig. 15 Effect of flight parameters on relative payload fraction of oxygen-collecting ASP.

The influence of the altitude of cruise flight is moderate. For altitudes higher than optimal, the necessity to increase air intakes, engines, and the FLOX plant has an effect. Payload fraction reduction at lower than optimal H_{CR} is influenced by vehicle drag and required thrust increase.

The gentle slope of the curve at high altitude removes the possibility of the negative influence of cruising vehicles on the ozone layer. If it were the real problem, the flight altitude may be established as high as 28 km. Eighty percent of atmospheric ozone is concentrated under this altitude. (See Fig. 15c.)

A payload fraction decrease while oxygen collection occurs at low-speed cruise flight is connected with expenses on collected oxygen acceleration. A decrease at higher than optimal M_{CR} may be explained by increase of losses on inlet impulse of refining air. (See Fig. 15d.)

The dependence of payload fraction on the collection ratio has a maximum in the range considered only when real FLOX plant mass is taken into account (see Fig. 14). Note that the payload fraction decreases rapidly at increased K_{OX} values. (See Fig. 15e.)

First, the increase of mixture ratio corresponding to payload fraction at maximum in comparison with the value corresponding to the maximum of specific impulse by more than 1.5 should be noted. In Ref. 16, the value $K = 7.6$ is indicated as optimal for the LRE of the Sänger system because below this value the resulting structure and tank mass penalty are more influential than the increase of specific impulse, Whereas the impure oxygen application optimal muxture ratio increases too. (See Fig. 15f.)

For the oxygen-collecting vehicle, this tendency is confirmed by the existence of the minimum in the dependence of the total amount of hydrogen needed for the aerospace plane on the LRE mixture ratio (Fig. 16). In this area, the vehicle has minimal size hydrogen tank and engines, and, consequently maximal payload fraction.

Note that the dependence of the total amount of hydrogen on the collection ratio has no such minimum (Fig. 17) and the K_{OX} increase is restricted by

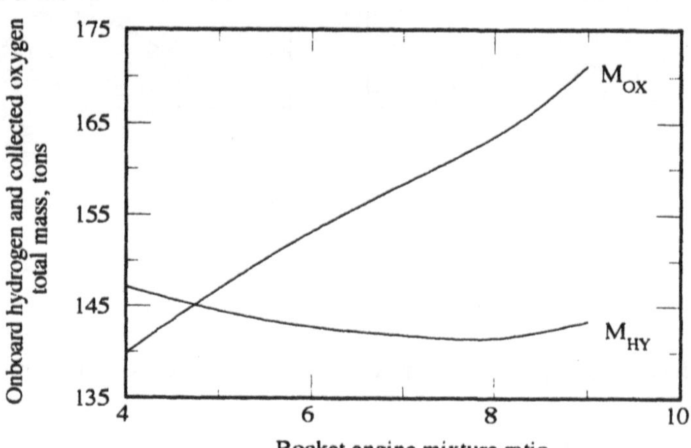

Fig. 16 Influence of LRE mixture ratio on total required amount of onboard hydrogen and collected oxygen for ASP of initial mass M_0 = 260 tons.

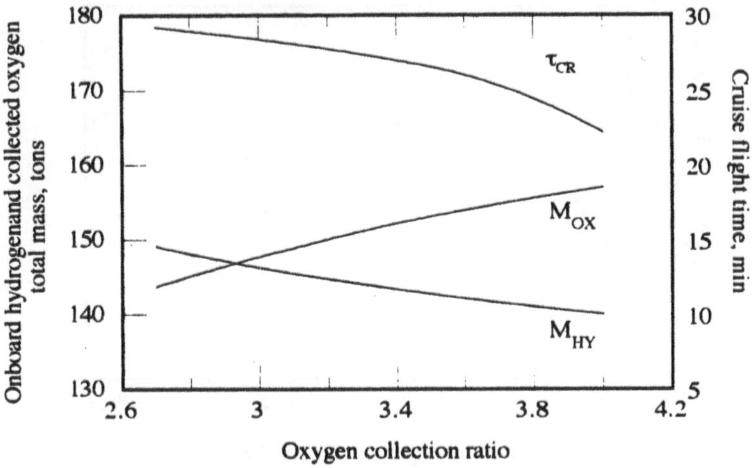

Fig. 17 Influence of oxygen collection ratio on total required amount of onboard hydrogen, collected oxygen, and collection time for ASP of initial mass M_0 = 260 tons.

FLOX plant mass (Figs. 14 and 15d). According to Fig. 17 while the K_{OX} increase total amount of oxygen required increase due to vehicle dry mass growth, however, the collection time decrease because of hydrogen flow rate on collection regime and FLOX capacity increase.

The influence of the average specific impulse of powerplants in the range of M=0-5.0 on the difference of the payload fractions of the oxygen-collecting ASP and LOX-carrying ASP was estimated (Fig. 18).

The linear connection between specific impulse and engine specific mass was assumed. The weak influence of specific impulse of low-speed engines on payload fraction difference may be noted. It should be emphasized that it is not a question of low-speed engine efficiency, it is only comparison of the vehicles with and without oxygen collection having the same low-speed engines.

According to Fig. 18, the engines with middle values of specific impulse and specific mass may be considered as candidates for oxygen-collecting ASP. Among these engines are the precooling air cycle engine and LACE cycles.

The FLOX plant seems to be well integrated with cooling systems of these cycles. The most attractive in this meaning are the engines with deep air cooling and liquefaction. Among the best studied are (Fig. 19) the widely known liquid air cycle engine;[17] the air-turbo-rocket with deep air precooling,[18] providing air precooling to the saturation line, well integrated with the ramjet cycle; and the air-turbo-rocket with air liquefaction.[19]

After main duty of these engines as thrust generators up to M = 5-6, their cooling systems may be employed for an oxygen collection cycle. In such cases, the heat exchangers, while being the heaviest parts of the structure, would be of dual use.

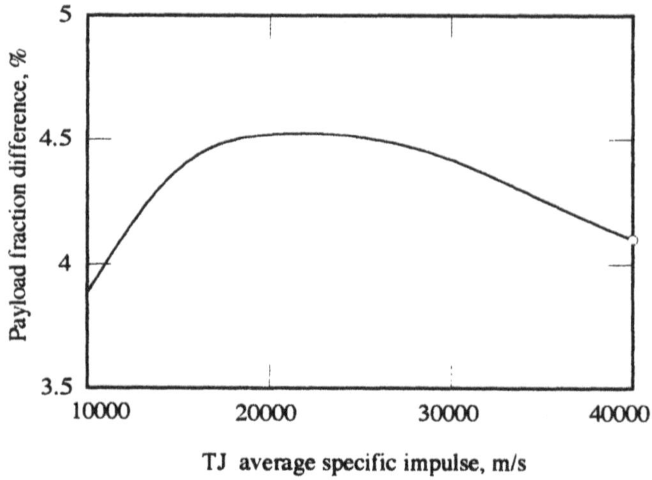

Fig. 18 Influence of specific impulse of low-speed engine on the difference of payload fractions of oxygen-collecting ASP and LOX-carrying ASP, $M_0 = 260$ tons.

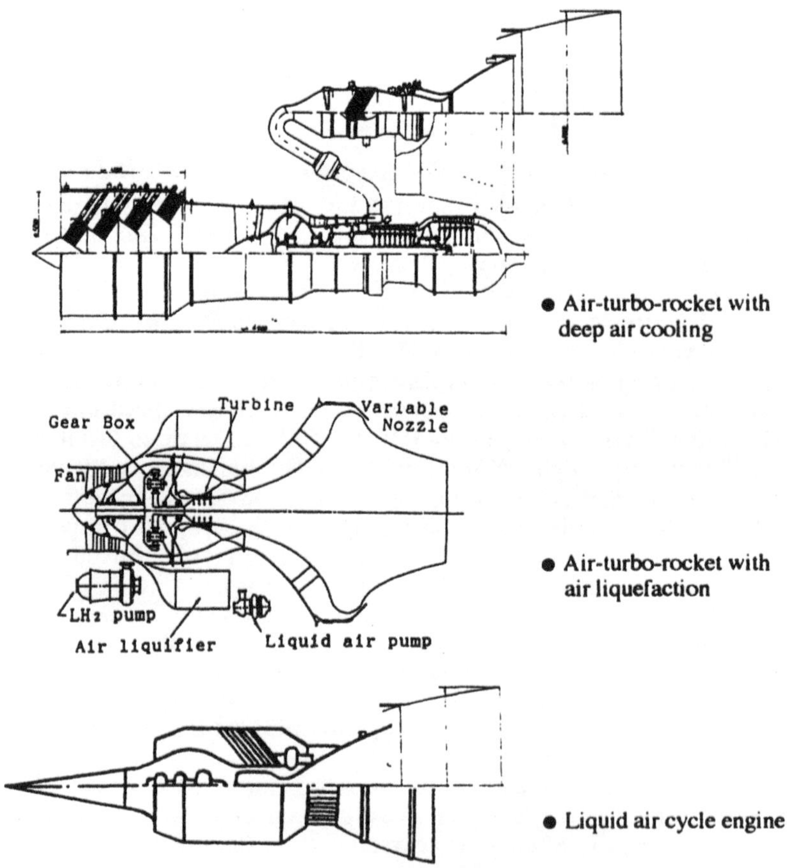

Fig. 19 Schemes of PACE and LACE engines.

According to Ref. 17, the air liquefaction system of LACE is intended for full liquefaction of approximately 16.5 kg/s of air per 1 kg/s of hydrogen at $M = 5.0$. The total hydrogen flow rate in this regime is about 100 kg/s.

If such system to connect with air rectifier and to include cruise part in flight scenario, the oxygen collection ratio at $\eta = 0.85$ will be $K_{OX} = 3.25$ and the required amount of oxygen will be collected during 400-500 s.

In this case, the total mass of the powerplant, including LACE, scramjet, and the FLOX plant, increases insignificantly in comparison with the powerplant of the LOX-carrying vehicle because the mass of heat exchangers used for LACE makes up about 80% of FLOX plant mass.

Estimation of FLOX Heat Exchanger Icing

One of the important questions restricting the enthusiasm for precooled and LACE cycles is the more than high probability of heat exchanger icing. Atmospheric moisture content is maximal at sea level and rapidly decreases when altitude increases. The dependence of annual average humidity for middle latitudes may be approximated by two curves:

$$\begin{aligned} d &= 7\times 10^{-3} \times 10^{-H/4.35} && \text{at } 0 \le H \le 10\,\text{km} \\ d &= 2.07\times 10^{-4} \times 10^{-H/12.95} && \text{at } 10 < H \le 30\,\text{km} \end{aligned} \tag{11}$$

According to this expression, air humidity at sea level is more than that in the altitude of 24 km by 2400 times. However, the total time of the oxygen-collecting vehicle cruise flight is about 2000 s and calls for icing estimation.

Such estimation may be made based on experimental data obtained by the author together with E. A. Folomeev and S. M. Galkin in CIAM.[20] Two of the experimental curves reflecting frost layer growth are shown in Fig. 20.

Study has shown the possibility of heat-mass transfer analogy used for the channels of different shapes in the wall temperature range T_W = 110-273 K and

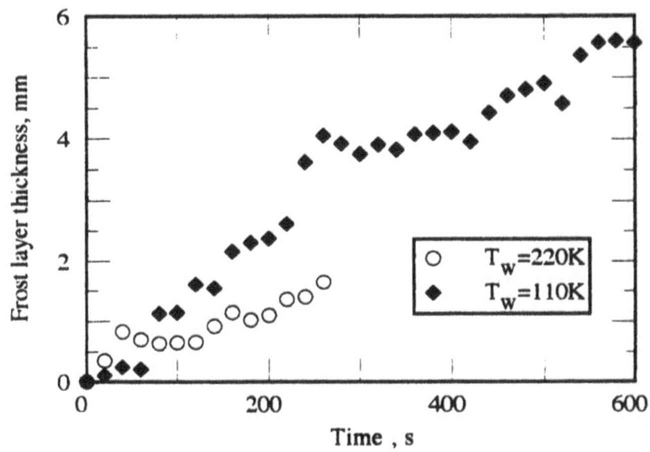

Fig. 20 Frost layer growth rate at different wall temperatures: d = 10 g/kg; Re = 10^5.

Reynolds numbers of $Re = 5\times10^3 - 3\times10^5$. The expression for diffusion Nusselt number is

$$Nu_D = 0.045\,Re^{0.8}\,Sc^{0.43} \tag{12}$$

As distinct from known approaches, that consider the Nusselt number as

$$Nu_D = \frac{\alpha_D d_H}{D} = \frac{\overline{m} d_H}{(\rho_{V\infty} - \rho_{VW})D} \tag{13}$$

the possibilities to ignore the value of near-wall steam density ρ_{VW} and to consider the specific mass flow to the surface as $\overline{m} = \overline{h}\rho_{FR}$ was substantiated.

In this case, the Nusselt number may be expressed as

$$Nu_D \approx \frac{\overline{h}\rho_{FR} d_H}{\rho_{V\infty} D} \tag{14}$$

where $\rho_{V\infty}$ is the steam density in air flow, ρ_{VW} the near-wall steam density, and ρ_{FR} the frost layer density.

Setting equal the right-hand sides of Eqs. (12) and (14) and assuming $Sc = 1$ the equation for the frost layer growth rate becomes

$$\overline{h} = \frac{0.045 Re^{0.8}\,\rho_{V\infty} D}{\rho_{FR} d_H} \tag{15}$$

This equation satisfactorily describes the frost growth process until the outside surface temperature reaches the frost melting point value, as occurred in the example with a higher wall temperature (Fig. 20). Maximal layer thickness is connected with initial wall temperature. According to a proposed approach it is not necessary to determine near-wall vapor density, and the value of frost layer density $\rho_{FR} = 100$ kg/m^3 assumed for obtaining criterial equation is applicable for similar conditions.

The cold zone of the "air-poor air" shell and tube heat exchanger was chosen for estimation of the frost growth rate under conditions of cruise flight at an altitude of 24 km. The behavior of the air, coolant, and average wall temperature vs tube row numbers is given in Fig. 21.

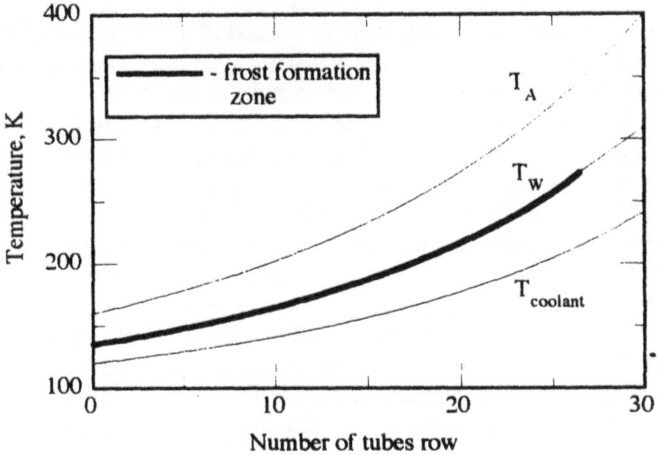

Fig. 21 Behavior of air, coolant, and average wall temparature vs tube row numbers in cold zone of "air-depleted air" heat exchanger.

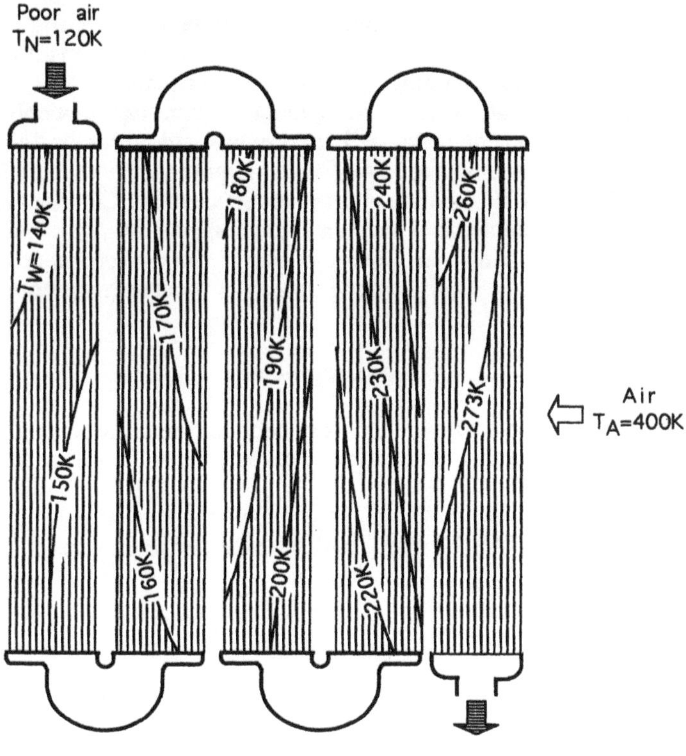

Fig. 22 Heat exchanger matrix with wall temperature isotherms

Such types of heat exchangers usually consist of a few zones of tubes with coolant parallel flow in each zone. Coolant after one zone flows into the collector and then turns to another zone. Figure 22 shows the nonscaled matrix of the heat exchanger operates under the same conditions as shown in Fig. 21.

Two-dimensional calculation was conducted to determine the wall temperature distribution. Such calculations allow one to predict icing zones more exactly and to analyze tube thermal expansion (shrink).

Estimation by Eq. (15) showed that the frost layer growth rate is in the range of $\bar{h} = (1.5\text{-}6.0) \cdot 10^{-5}$ mm/s, increasing from the colder zone to the warmer. In other words, the anticipated thickness of frost layer formation during a typical cruise period of time is about 0.1 mm. It seems that such a frost layer cannot be an obstacle for proper system operation. Moreover, in some experiments, an increase of local heat transfer coefficient up to seven times was obtained due to a thin frost layer.[20] It is very attractive to employ this effect for heat exchanger size reduction.

Concluding Remarks

Calculation studies show the possibility of a significant increase of payload fraction for an aerospace plane employing collection of atmospheric oxygen for use on the rocket part of the flight. Air separation technologies suitable for flight application have been identified and various typical conditions have been examined. Well-substantiated estimations of the mass units making up the air cooling and separation systems were carried out. Several possible variants of the FLOX plant and the powerplant integration have been examined.

Experimental confirmation and development of identified technologies are required. The main current tasks of experimental study are as follows: 1) development of an air rectifier based on the technique of vortex tubes; 2) search for methods of heat transfer increase in order to reduce the FLOX plant mass and size; and 3) search for methods of oxygen collection ratio increase that leads to hydrogen flow rate reduction. Application of para-ortho conversion provided by methods suitable for use under flight conditions is of fundamental importance.

Acknowledgments

The author gratefully acknowledges the cooperation with N. P. Dulepov's group of CIAM in trajectory estimations and with E. V. Tjurikov (CIAM) in FLOX structure analysis.

Thanks are also due to NAL of Japan Kakuda Research Center staff, especially K. Kamijo and M. Yoshida for discussion and assistance.

Addendum

The initial phase of experimental study of air separation with a vortex tube (VT) was completed in the Kakuda Research Center in

Japan.[21] Almost pure oxygen (C^O_O) = 98.7%) was obtained in several stable regimes with very good reproducibility of test results. However, oxygen recovery was rather low (h ≈ 5%) in this series of tests. Some other test regimes yielded an oxygen concentration C^O_O= 80–85% with a recovery factor h = 30–36%. A very important result for in-flight oxygen collection feasibility is that the highest VT efficiency value was obtained at a pressure ratio close to 2 rather than the anticipated value of 5–6. The pressure ratio in the VT is the ratio of the air pressure in front of the VT to the oxygen-depleted air pressure.

Some structural changes were planned as a result of the initial test phase that are aimed at obtaining an oxygen concentration $C^O_O > 90\%$ at a recovery factor h = 45–55% in a further study of the single stage vortex tube.

References 21 and 22 introduced the concept of a SSTO rocket combined propulsion scheme based on a LRE integration with an oxygen collection system. A rocket, having much higher takeoff thrust-to-weight ratio than a horizontally starting vehicle, is very sensitive to engine mass. In the case of an oxygen collecting rocket, the engine mass includes the FLOX plant mass. Therefore the collection ratio is restricted significantly to a lower limit than considered earlier. The optimal value was estimated as K_{OX} = 1.5–2.0. In fact, the amount of oxygen from the very start, and its mixture ratio is higher than the collection ratio. However, oxygen collection means a smaller amount of onboard oxygen is required. The main feature of this concept is that an air separation system provides significant positive thrust during the initial acceleration (up to 40% of total thrust at starting conditions) and is shut off shortly after this thrust becomes negative. This thrust is generated by a special device that may be called a depleted air augmentor (DAA). The DAA resembles a conventional turbojet, however the depleted air temperature in front of the compressor is near saturation conditions. Such utilization of the depleted air provides the advantage of some freedom in the selection of the recovery factor. At a recovery as low as h = 50–60% the propulsion system still has an attractive performance. With a lower recovery factor, weight of the air separation system at a given collection ratio becomes unacceptable. According to a preliminary estimation a combined propulsion system, including an oxygen collection system, provides a payload fraction benefit for a single-stage rocket of up to 2.5% of the vehicle initial mass compared to a rocket with a simple LRE propulsion. Other relevant studies in this connection are Refs. 23 and 24.

The concept of a SSTO rocket propulsion system consisting of a thermally integrated liquid rocket engine and a deep-cooled turbojet has been proposed and analyzed.[25] The DCTJ/LRE propulsion provides an increase of the payload fraction of a SSTO rocket to the LEO of up to 2.5% of the vehicle initial mass compared to the simple rocket propulsion system. This concept is considered as near-term technology because it does not require the sophisticated technologies related to a LACE cycle. It is appropriate to mention this concept here because it is formally derived from the combined cycle FLOX/LRE. In the current case the oxygen recovery factor h = 0 and air liquefaction is not needed. Thus the air precooler and the compressor are relatively light, and this ultimately leads to the benefit mentioned.

In Ref. 25, a simple idea was introduced concerning icing of a precooler, originally formulated in the CIAM. It was shown that icing is unlikely when the air temperature is lower than the water dew point, or the vapor pressure is lower than the vapor pressure at the triple point. The latter condition provided an icing-free regime during several tests, each lasting about 20 minutes, conducted in the CIAM, with dense ice appearing immediately after the vapor (partial) pressure rose above the triple point pressure. Based on a physical model of ice formation, which was verified by basic experiments, the icing limits in a flight speed–altitude map were estimated. Accordingly, icing is unlikely to be a problem for oxygen collection systems operating in a high altitude cruise-flight. References 25 and 26 provide speed–altitude icing limits for certain ambient conditions.

The Institute for Space and Astronautical Science of Japan, in cooperation with four major Japanese heavy industry companies, has developed the advanced ATREX cycle engine that is the first in the world to be precooled in practice. Insofar as a precooler is the key unit of an oxygen collection cycle, a LACE cycle, a deep-cooled cycle, the unique practical experience obtained during the ATREX development study on precooler integration, configuration, fabrication, and operation can be expected to become useful for all of those cycles as well. Currently a precooler is fabricated to chill down air to 160 K at sea level static conditions and when the engine ground test including the precooler is underway.[26]

References

[1]Jeffs, R. A. and Beeton, A. B. P., "Liquid Air Cycle Engines for High-Speed Aircraft," *Journal of the British Interplanetary Society*, Vol. 19, 1963-64, pp. 484-490.

[2]Nau, R. A., "A Comparison of Fixed Wing Reusable Booster Concepts," SAE Paper 670384, SAE Publ. P16, 1967.

[3]Maurice, L. Q., Leingang, J. L., and Carreiro, L. R., "Airbreathing Space Booster Using In-Flight Oxidizer Collection," AIAA Paper 92-3499, 1992.

[4]Gopalaswami, R., Gollakota, S., Venugopalan P., Narathinam M., and Sivathanu Pillai, A., "Concept Definition and Design of a Single Stage to Orbit Launch Vehicle-HYPERPLANE," Paper IAF-88-194, 1988.

[5]Leingang, J. L., Carreiro, L. R., and Maurice, L. Q.,"On-Demand Reusable Space Launch Systems That Use In-Flight Oxidizer Collection," SAE Paper 931453, 1993.

[6]Czysz, P. and Murthy, S. N. B., "Energy Analysis of High-Speed Flight Systems," *High-Speed Flight Propulsion Systems*, edited by S. N. B. Murthy and E. T. Curran, Vol. 137, Progress in Astronautics and Aeronautics, AIAA, Washington, DC, 1991, pp. 143-235.

[7]Suslov, A. D., Chizhikov, Yu. V., and Ivanov, S. V., "Development and Investigation of a New Type of Air-Separating Apparatus - A Vortex Rectifier," *Chemical and Petroleum Engineering*, Vol. 16, Sept.-Oct. 1980, pp. 507-511.

[8]Billig, F., S., "Propulsion Systems from Takeoff to High-Speed Flight," *High-Speed Flight Propulsion Systems*, edited by S. N. B. Murthy and E. T. Curran, Vol. 137, Progress in Astronautics and Aeronautics, AIAA, Washington, DC, 1991, pp. 21-100.

[9]Balepin, V. V. and Tjurikov, E. V., "Integrated Air Separation and Propulsion System for Aerospace Plane with Atmospheric Oxygen Collection," SAE Paper 920974, 1992.

[10]Roudakov, A. S.,Gatin, R. Yu., Dulepov, N. P., Koralnik, B. N., Harchevnikova, G. D., and Yugov, O. K., "Analysis of Efficiency of Systems with Oxidizer Liquefaction and

Accumulation for Improvement of Aerospaceplane Performance," Paper IAF-91-270, 1991.

[11]Balepin, V. V., Dulepov, N. P., Folomeev, E. A., Harchevnikova, G. D., Tjurikov, E. V., and Avramenko, A. Yu., "Flight Oxygen Plant for Aerospace Plane: Thermodynamic and Integration Aspects," SAE Paper 931452, 1993.

[12]Kuczera, H. and Koelle, D. E., " Propulsion Systems Optimization for a Two-Stage Winged Launcher System,", Paper IAF-88-246, 1988.

[13]Yamanaka, T., "Space Plane Research Activities in Japan," AIAA Paper 89-5008, 1989.

[14]Yugov, O. K., Dulepov, N. P., and Harchevnikova, G. D., "The Analysis of Efficiency of Hypersonic and Combined Engines in the Propulsion System of the SSTO Vehicles," Paper IAF-90-258, 1990.

[15]Maurice, L. Q., Leingang, J. L., and Carreiro, L. R., "Airbreathing Space Boosters Using In-Flight Oxidizer Collection", *AIAA Journal of Propulsion and Power*, July 1991.

[16]Koelle , D. E. and Kuczera, H., "SÄNGER Space Transportation System - Progress Report 1990," Paper IAF-90-175, 1990.

[17]Miki, Y., Togawa, M., Tokunaga, T., Eguchi, K., and Yamanaka, T., "Advanced SCRAM-LACE System Concept for SSTO Space Plane," Paper IAF-91-272, 1991.

[18]Rudakov, A. S. and Balepin, V. V., "Air Precooling for Aerospaceplane Engine," *Aerospace Engineering*, Aug. 1991, pp.29-30.

[19]Sakata, K., Minoda, M., Yanagi, R., and Nouse, H., "Hypersonic Turbomachinery-Based Air-Breathing Engines for the Earth-to-Orbit Vehicle," *AIAA Journal of Propulsion and Power*, Vol. 7, Jan.-Feb. 1991, pp. 108-114.

[20]Balepin, V. V., Folomeev, E. A., Galkin, S. M., and Svetlakov A. L., "Cryogenic Heat Exchangers - Key Technology for Precooled Turbojet Engines," SAE Paper 911183, 1991.

[21]Balepin, V. V., Hashimoto, T., Watanabe, Y., and Kamijo, K., "Air Separation Technology for Space Vehicle Application," International Symposium on Space Technology and Science, Yokohama, Paper ISTS 94-a-03, 1994.

[22]Balepin, V. V., Yoshida, M., and Kamijo, K., "Rocket Based Combined Cycles for Single Stage Rocket," SAE Paper 941166, 1994.

[23]Vandenkerchkhove, J., and Czysz, P., "SSTO Performance Assessment with In-Flight LOX Collection, Paper IAF-94-X.4.426, 1994.

[24]Balepin, V. V., Czysz, P., Maita, M., and Vandenkerchkhove, J., "Assessment of SSTO Performance with In-Flight LOX Collection," AIAA Paper 95-6047, 1995.

[25]Balepin, V. V., Folomeev, E. A., Galkin, S. M., Tjurikov, E. V., "Rocket Based Combined Cycles for Vertical Take-Off Space Vehicles," AIAA Paper 95-6078, 1995.

[26]Tanatsugu, N., Sato, T., Balepin, V., Naruo, Y., Mizutani, T., Kashiwagi, T., Hamabe, K., Tomike, J., Minami, R., "Development Study on ATREX Engine," Paper IAF-95-S.5.01, 1995.

Pulse Detonation Engine Theory And Concepts

Thomas Bussing * and George Pappas †
ASI (Adroit Systems, Inc.), Bellevue, Washington 98004

I. Introduction

Airbreathing engines can be classified according to the type of combustion process employed in the device. The combustion process can be characterized as either steady (quasisteady) or unsteady. Propulsion systems may be further classified according to whether a deflagrative or detonative mode of combustion is utilized. The following is a brief description of several important concepts relevant to the study of pulse detonation engines (PDEs).

A. Deflagration

Deflagration is the most common type of combustion. It is characterized as a subsonic combustion process relative to the reactants. A deflagration reaction propagates at relatively low flame speeds. The propagation of a deflagration flame in a pre-mixed mixture is governed by the laminar or turbulent diffusion of unburned gases ahead of a flame and burned gases behind the flame. Deflagration flame speeds are on the order of one or more meters per second for most typical hydrocarbon/air mixtures. The most readily observed deflagrations produce small decreases in pressure and can be modeled as nearly isobaric, or constant pressure, processes. For example, combustion in a gas turbine can be modeled as a constant pressure process. A more complete description of deflagrative combustion is provided in section II.

B. Detonation

A detonation is a supersonic combustion wave that typically propagates at a few thousand meters per second relative to an unburned fuel/air mixture. Detonation is a much more energetic phenomenon than deflagration and produces large overpressures. A detonation wave compresses a fluid, increasing

---

Copyright © 1995 by the American Institute of Aeronautics and Astronautics, Inc. All rights reserved.

*Vice President, Manager Aero. & Astro. Sciences Group. Member AIAA.

†Member AIAA.

its pressure and density, in addition to increasing its temperature. Detonations can be approximated as supersonic shock waves that initiate and are closely coupled to a thin flame front or combustion region. As a result of the high-speed nature of a detonation wave, detonation closely approximates a constant volume combustion process. A more complete description of detonation combustion is provided in section II.

C. Steady-State Engines

The most widely used class of engines are steady (or quasi-steady) state combustion propulsion systems. These engines are characterized by a continuous, constant pressure, deflagration combustion process. Turbojets and ramjets are examples of this class of engine.

D. Unsteady Engines

The second class of engines is based on unsteady combustion operation. In these relatively simple devices, combustion occurs in an intermittent manner and is governed by the fuel/air flame speeds and combustion chamber characteristics. Unsteady propulsion systems can be based on either deflagration combustion (pulse jets) or detonation combustion (pulse detonation engines).

E. Pulse Jets

Pulse jets were one of the first intermittent combustion jet engines ever studied. The German V-1 "Buzz Bomb" used in World War II was one of the earliest and most notable pulse-jet-powered devices. In the 1950s, the U.S. Project Squid research effort examined the feasibility of pulse jet engines.[1,2] Since then, pulse jets have received diminished levels of attention.

The performance of pulse jets is defined by the properties of deflagration waves. Low deflagration flame speeds limit the maximum combustor chamber pressures and engine operating frequencies that can be achieved.

F. Pulse Detonation Engines

PDEs have received considerable attention over the past decade. These devices use detonation waves that propagate through a premixed fuel/air mixture and produce large chamber pressures and, thereby, thrust.

PDEs offer the potential for high performance because the rapid detonation process results in constant volume combustion with very high operating frequencies and combustor chamber pressures. PDEs are predicted to be very efficient and offer good thrust characteristics from low subsonic to high supersonic flight regimes. PDEs also have the potential to operate at very high energy densities, allowing the use of simple and compact combustor designs.

II. Detonation Physics

Detonation is a much more energetic and violent phenomenon than deflagration and results in large pressure and temperature increases, because the combustion region is closely coupled to a supersonic shock wave.

A detonation wave can be approximated as a shock wave that is followed closely by a deflagration. The Chapman-Jouguet (C-J) and Zeldovich-von Neumann-Doring (ZND) models to be discussed in this section represent detonations as simple planar or spherical waves. However, detonation waves are actually complex, oscillatory phenomena with three-dimensional cellular structures.

Basic Detonation Wave Physics

Detonation waves can be modeled as discontinuities in the flow of an ideal gas at which heat addition occurs[3]. This treatment of detonation waves is a reasonable approximation for studying engine performance since the significant changes in the fluid properties occur over very short distances. The modeling describes the fundamental physics of simple one-dimensional detonations in a constant area duct (Fig. 1). The analysis is based on the conservation equations for mass, momentum, and energy across the detonation, and on the equation of state.

Mass continuity:

$$\rho_1 u_1 = \rho_2 u_2 \tag{4}$$

Momentum conservation:

$$P_1 + \rho_1 u_1^2 = P_2 + \rho_2 u_2^2 \tag{5}$$

Energy equation:

$$h_1 + \frac{u_1^2}{2} + q = h_2 + \frac{u_2^2}{2} \tag{6}$$

where

1 = state of unburned gas upstream of detonation wave
2 = state of burned gas immediately downstream of detonation wave
P = pressure
ρ = density
u = one-dimensional velocity
h = enthalpy of the gas (per unit mass)
q = heat addition (per unit mass)

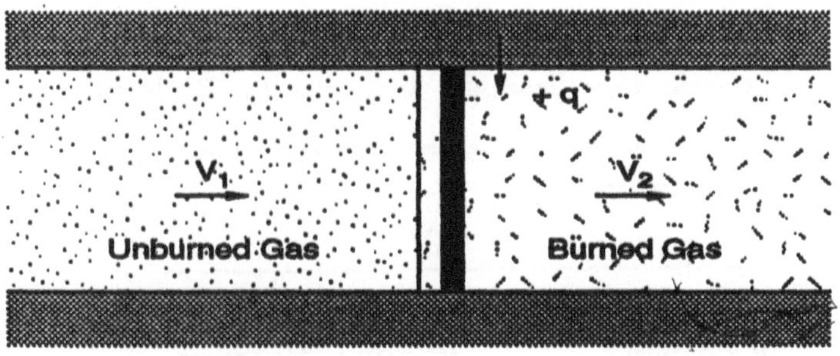

Fig. 1 Detonation wave in constant area duct.

Combining the continuity and momentum equations (4) and(5) yields

$$(\rho_1 u_1)^2 = (\rho_2 u_2)^2 = m^2 = -\frac{P_2 - P_1}{\nu_2 - \nu_1} \tag{7}$$

$$\frac{\left[\frac{P_2}{P_1} - 1\right]}{\left[\frac{\nu_2}{\nu_1} - 1\right]} = -\frac{m^2}{P_1 \rho_1} = -\mu \tag{8}$$

where

ν = specific volume, $= 1/\rho$
μ = wave speed

Since $(\rho_1 u_1)^2$ and $(\rho_2 u_2)^2$ are always positive, the terms $(P_2 - P_1)$ and $(\nu_2 - \nu_1)$ must be of opposite sign. Two combustion processes, deflagration and detonation, are indicated (see Table 1). Equations 7 and 8 define a Rayleigh line with a constant negative slope $(-\mu)$ in the pressure-specific volume plane.

A deflagration results in a net decrease in both pressure and density as a result of the combustion process. The most commonly observed types of deflagrations are classified as being weak because they are nearly isobaric and result in only small decreases in pressure (e.g., constant pressure gas turbine combustion).

In contrast, a detonation produces both an increase in pressure and an increase in fluid density. This is because the strong shock that precedes the flame region of the detonation wave both compresses and pressurizes the unreacted gas prior to the heat addition process.

The continuity, momentum, and energy equations (4-6) can be combined and rearranged to yield the Hugoniot expressions for deflagrations and detonations:

$$h_1 - h_2 + q = \frac{1}{2}(P_1 - P_2)(\nu_2 + \nu_1) \tag{9}$$

Table 1 Combustion processes indicated by Rayleigh line.

Combustion process	ΔP	$\Delta \nu$
Deflagration	Negative	Positive
Detonation	Positive	Negative

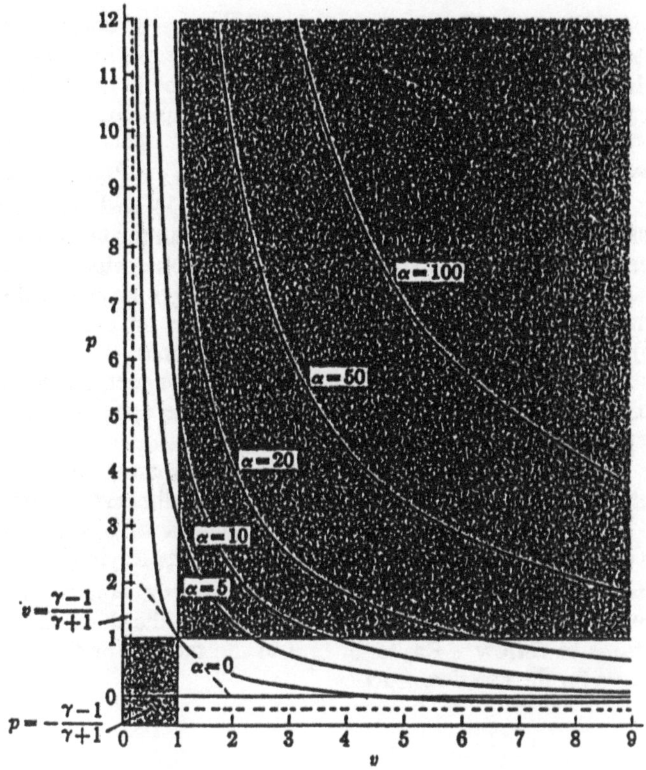

Fig. 2 Hugoniot diagram (y = 1.4) (Ref. 15).

$$e_1 - e_2 + q = \frac{1}{2}(P_1 - P_2)(v_2 - v_1) \qquad (10)$$

where

e = internal energy of gas per unit volume
h = enthalpy, $= e + p/\rho = e + pv$

Hugoniot equations (9) and (10) relate the thermodynamic properties upstream and downstream of a combustion region. Figure 2 is the Hugoniot diagram for a system with a ratio of specific heats of 1.4. The parameter a is a dimensionless heat of reaction ($\alpha = q\rho_1/p_1$). The possible final states of the system are defined by the intersection of the Rayleigh line with the Hugoniot curve. Since the Rayleigh line is a straight line with a negative slope that passes through point (1,1), the end states that lie in the shaded regions are physically meaningless. Therefore, the acceptable end states are divided into two distinct branches of the Hugoniot curve: the upper/detonative branch and the lower/deflagrative branch.

Approximately 20 years after the discovery of constant supersonic velocity detonation waves in gaseous explosives, Chapman in 1899 and Jouguet in 1905 independently determined that the tangency point of the Rayleigh line and the Hugoniot curve for the detonation reaction products represented the stable end state for self-sustaining detonation waves.[4] This Chapman-Jouguet condition for detonation is indicated by point B (P_+, v_+) in Fig. 3. It can be seen graphically that a Rayleigh line through point (1,1) will not intersect the Hugoniot curve if the magnitude of its slope is less than the tangent line's slope. This implies a minimum wave speed, μ, for detonations, which corresponds to the C-J condition. At higher wave speeds (i.e., larger Rayleigh line slopes), there are two possible detonation end states: one on curve A-B (strong detonations) and another on curve B-C (weak detonations).

Table 2 gives the properties of Chapman-Jouguet, strong, and weak detonations. Note the subscripts + and - refer to the upper (detonation) and lower (deflagration) C-J conditions, respectively.

In the early 1940s, Zeldovich, von Neumann, and Doring each independently formulated similar models for the one-dimensional structure of a detonation wave. The basic ZND model consists of a shock wave that compresses a gas from its initial state (P_1, ρ_1) along the Rayleigh line to a high-pressure state on its Hugoniot curve called the von Neumann spike. A homogeneous chemical reaction zone closely-follows the shock and causes the flow conditions to follow a Rayleigh line to the tangent or C-J point on the Hugoniot curve for the reaction products [4].

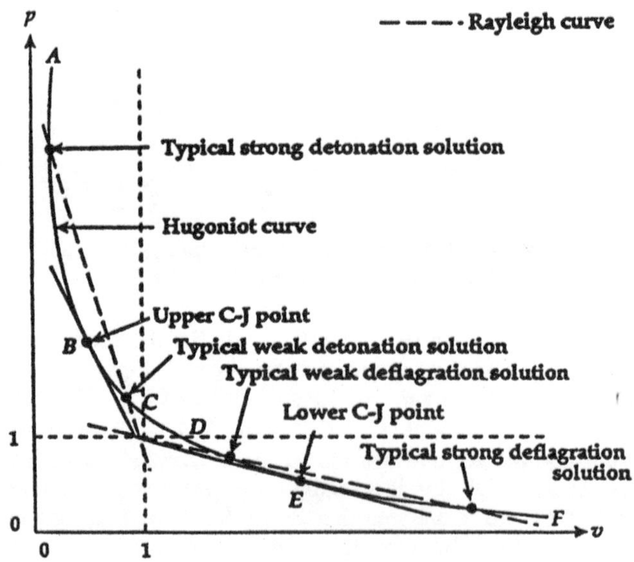

Fig. 3 Rayleigh line-hugoniot curve flow solutions (Ref. 15).

Table 2 Properties of Hugoniot curves (Ref. 15).

	Section in Fig. 2–5	Pressure ratio $p = (p_\infty/p_0)$	Velocity and density ratios $v = (v_\infty/v_0) = (\rho_0/\rho_\infty)$	Propagation Mach number $M_0 = (v_0/a_{f,0})$	Downstream Mach number $M_\infty = (v_\infty/a_{e,\infty})$	Remarks
Strong detonations	Line *A-B*	$p_+ < p < \infty$	$v_{min} < v < v_+$ ($v_{min} > 0$)	$M_{0+} < M_0 < \infty$	$M_\infty < 1$	Seldom observed; requires special experimental arrangement.
Upper Chapman-Jouguet point	Point *B*	$p = p_+$ ($p_+ > 1$)	$v = v_+$ ($v_+ < 1$)	$M_0 = M_{0+}$ ($M_{0+} > 1$)	$M_\infty = 1$	Usually observed for waves propagating in tubes.
Weak detonations	Line *B-C*	$p_1 < p < p_+$ ($p_1 > 1$)	$v_+ < v < 1$	$M_{0+} < M_0 < \infty$	$M_\infty > 1$	Seldom observed; requires very special gas mixtures.
Weak deflagrations	Line *D-E*	$p_- < p < 1$	$v_1 < v < v_-$ ($v_1 > 1$)	$0 < M_0 < M_{0-}$	$M_\infty < 1$	Often observed; $p \approx 1$ in most experiments.
Lower Chapman-Jouguet point	Point *E*	$p = p_-$ ($p_- < 1$)	$v = v_-$ ($v_- > 1$)	$M_0 = M_{0-}$ ($M_{0-} < 1$)	$M_\infty = 1$	Not observed.
Strong deflagrations	Line *E-F*	$0 < p < p_-$	$v_- < v < v_{max}$ ($v_{max} < \infty$)	$M_{0min} < M_0 < M_{0-}$ ($M_{0min} > 0$)	$M_\infty > 1$	Not observed; forbidden by considerations of wave structure.

1. Detonation Wave Profile

The pressure distribution within a detonation chamber with one closed and one open end has a characteristic shape. Figure 4 depicts a detonation that was initiated at the closed end of a tube and is propagating toward the open end. The valved end is assumed to be closed during detonation. Other tube detonation initiation options are possible but are beyond the scope of this paper. These options include the following: 1) initiating detonation at the open end of a tube with one closed end, and 2) detonating in a tube with two open ends (i.e., valveless designs). The detonation wave can be modeled as a strong shock which triggers combustion, and a thin flame front in which heat addition occurs just behind the shock. The shock front moves at the detonation velocity, VD, relative to the gas and increases the pressure and temperature of the gas from its previous values of P_1 and T_1.

The region of unburned gas just behind the shock is a stable high-pressure region known as the von Neumann spike. This region represents the ignition delay; its thickness (i.e., the detonation induction length) is dictated by the chemical kinetics. The duration of the spike with respect to a fixed point is very short (e.g., on the order of 1 μs).

Once the chemical reaction kinetics are initiated and heat is added to the flow, as a result, the temperature increases and the pressure decreases. The thickness of the heat addition region is determined by the time required to complete the combustion reactions. At this point the burned gas is at state 2. In accordance with a detonative process, the temperature, pressure, and density at state 2 are significantly greater than at state 1.

The pressure and density in the stable detonation wave (P_2, ρ_2) are significantly lower than in the von Neumann region between the shock front and chemical reaction zone. However, the detonation wave temperature (T_2) just after the flame region is significantly higher than in the von Neumann spike.

In closed-tube detonations, an expansion region exists behind the heat addition region (see Sec. III. B). Rarefaction waves emanate from the closed end to ensure that the normal velocity at the wall is zero. As a result of the expansion, most of the burned gas in the detonation tube is at a pressure, P_3, which is significantly lower than the pressure just behind the detonation wave (P_2).

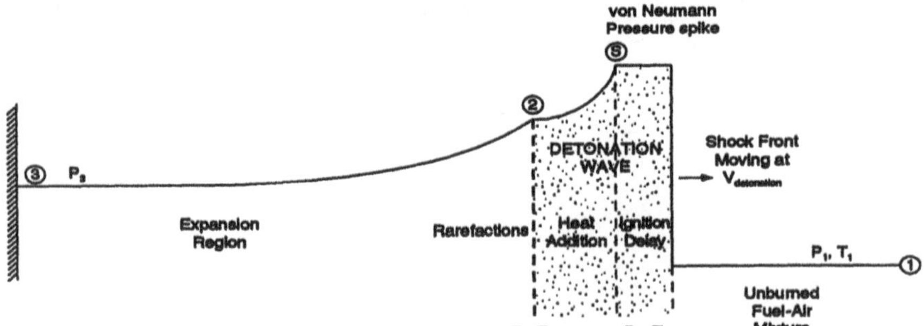

Fig. 4 Detonation chamber pressure profile.

2. Properties Across the Detonation Wave

The transitory von Neumann pressure increase across the leading shock of the detonation wave can be expressed from the continuity, momentum, and energy equations (4-6) and the perfect gas relations as

$$\frac{P_s}{P_1} = 1 + \frac{2\gamma_1}{\gamma_1 + 1}(M_1^2 - 1) \tag{11}$$

where

γ_1 = ratio of specific heats in unburned gas
M_1 = Mach number of detonation wave relative to unburned gas

The following relations can be written for the pressure and temperature ratios across a C-J detonation wave in an ideal gas:

$$\frac{P_2}{P_1} = \frac{1 + \gamma_1 M_1^2}{1 + \gamma_2} \tag{12}$$

$$\frac{T_2}{T_1} = \frac{m_2}{m_1}\frac{\gamma_2}{\gamma_1}\frac{1}{M_1^2}\left[\frac{1 + \gamma_1 M_1^2}{1 + \gamma_2}\right]^2 \tag{13}$$

where

γ = ratio of specific heats
m = molecular weight of gaseous mixtures

An important parameter required in the calculations of the fluid properties within the detonation tube is the velocity (V_{2C}) or Mach number (M_{2C}) of the burned gas relative to the walls of the detonation chamber. For C-J detonations, the Mach number of the detonation wave relative to the burned gas is unity (i.e., the burned gas moves away from the detonation at the speed of sound at state 2). Relating the speed of the burned gases just behind the detonation to the speed of the wave and accounting for the initial velocity component resulting from fuel/air injection yields:

$$M_{2C} = (M_0 + M_1)\left[\frac{m_2}{m_1}\frac{\gamma_1}{\gamma_2}\frac{T_1}{T_2}\right]^{\frac{1}{2}} - 1 \tag{14}$$

where

M_{2C} = Mach number of burned gas w.r.t. the detonation chamber wall
M_0 = Mach number of unburned gas w.r.t. the detonation chamber wall
M_1 = Mach number of detonation wave w.r.t. the unburned gas

The value of M_{2C} is in the high subsonic range for detonations initiated within tubes that contain a fuel/air mixture with no initial velocity. If the initial velocity of the unburned gas is not zero, then it is possible for M_{2C} to be greater than 1.

The pressure at the closed end is reduced to P_3 by an expansion zone that forms to ensure the normal velocity is zero at the closed end. The pressure just behind detonation wave P_2 is reduced to the plateau pressure P_3 at the wall. P_3 can be deduced from the following pressure ratio:

$$\frac{P_3}{P_1} = \frac{P_2}{P_1}\left[1 - \frac{\gamma_2 - 1}{2} M_{2c}\right]^{\frac{2\gamma_2}{\gamma_2 - 1}} \tag{15}$$

Figure 5 depicts three pressure ratios associated with a detonation wave that propagates toward the exit of a detonation tube. P_S is the von Neumann spike pressure, P_2 the pressure just after the detonation wave heat addition, and P_3 the static pressure that acts at the closed end of the tube. This value has been decreased substantially from P_2 as a result of the rarefaction region behind the detonation wave.

For an airbreathing hydrogen PDE system, the detonation Mach velocity is approximately 1950 m/s, which yields a Mach number of roughly 4.8. This results in a peak shock pressure ratio (P_S/P_1) of roughly 25, a peak detonation pressure ratio (P_2/P_1) of 15, and a chamber pressure ratio (P_3/P_1) of approximately 6.

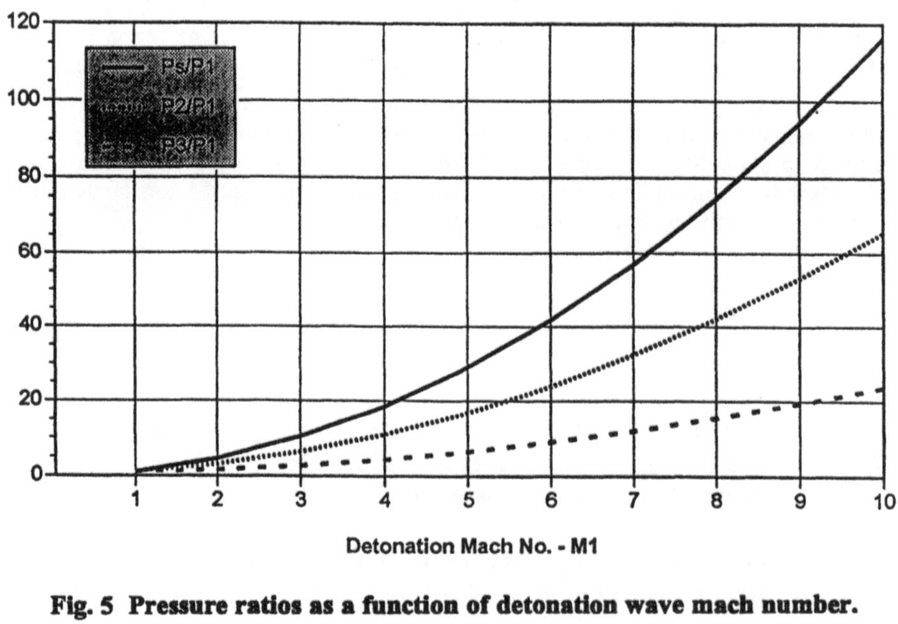

Fig. 5 Pressure ratios as a function of detonation wave mach number.

III. PDE Cycle Operation

The engine cycle of a pulse detonation engine can be described simply according to the fundamental processes that occur during that cycle. The idealized PDE cycle comprises several distinct events or processes:

1). A detonation combustion chamber is filled with a detonable mixture.

2). Detonation is initiated at the closed end of the combustor.

3). The detonation wave propagates through the combustor and exits at the open end.

4). The burned gases in the combustor exhaust through a blowdown process.

When the conditions within the combustor reach a specified state, the combustor is recharged with a fresh detonable mixture, and the cycle is repeated.

In the following sections, a basic description of a generic airbreathing PDE cycle operation is given. The detonable mixture used in the device is assumed to be a gaseous mixture of fuel and air. The discussion also applies to rocket operation and to systems that use other detonable mixtures (e.g., fuel/oxidizer).

A. Detonable Mixture Fills Combustor

A valve that seals one end of the detonation chamber will open to start the detonation engine cycle. A fuel/air mixture is admitted into the chamber at pressure P_1 and temperature T_1 determined by the flight conditions and the air inlet design. (See Fig. 6)

The fuel/air injection process can impact the net thrust generated by the engine. High combustor inlet Mach numbers decrease thrust performance because of the strong dependence of combustor performance on the Mach

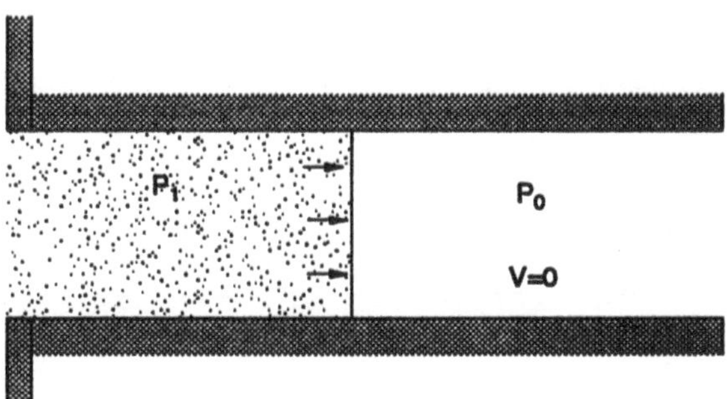

Fig. 6 Detonable mixture fills combustor.

number of the unburned gases before the onset of detonation. In addition, the detonation initiation power requirements tend to increase with higher fuel/air flow speed. Alternatively, imposing very low chamber Mach numbers requires capturing and decelerating the external air stream, which can result in prohibitive performance losses caused by increased drag effects. Low injection Mach numbers will also degrade performance by increasing the time required to fill the chamber and reducing the potential engine cycle rates.

B. Detonation Is Initiated at Closed End

After the fuel/air mixture enters the chamber, a valve at the end of the combustion chamber is closed in preparation for detonation initiation. The valve seals the combustor head when the downstream fuel/air mixture is still at some finite distance from the open end of the chamber. The valve timing ensures that the fuel/air mixture and the detonation wave reach the combustor exit simultaneously, and thus prevents any of the unburned mixture from escaping. As illustrated in Fig. 7, the degree to which the detonation tube is filled with fuel/air is calculated from the overall length of the tube and the relative velocities of the injected fuel/air mixture and detonation wave.

After the valve is closed, a detonation wave is initiated immediately in the fuel/air mixture region near the closed end of the chamber (Fig. 8). An expansion zone is created between the closed end and the detonation wave since the velocity must be zero at the closed end wall. Rarefaction waves are generated at the closed end of the detonation chamber and proceed toward the exit. The rarefaction's originate at the closed end and satisfy the constraint of zero axial fluid velocity normal to the wall. The strength of the expansion region is a function of the axial velocity of the burned gases behind the detonation wave, which must be decelerated to satisfy the closed-end boundary condition.

Two factors contribute to the axial velocity of the burned gas: 1) the initial velocity of the injected fuel/air mixture, and 2) the velocity imparted to the burned fuel/air mixture by the detonation wave. Both components of axial velocity are directed toward the open end of the detonation chamber. The detonation wave moves away from the burned gases at the local speed of sound (C-J condition). However, both the detonation wave and the burned gases move in the direction of the exit.

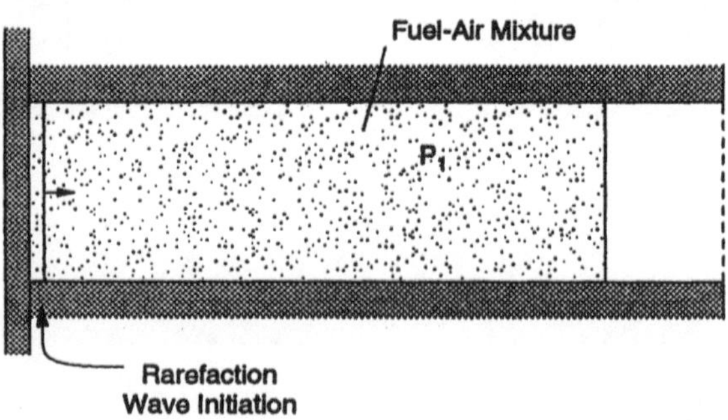

Fig. 7 Detonation tube is closed at one end.

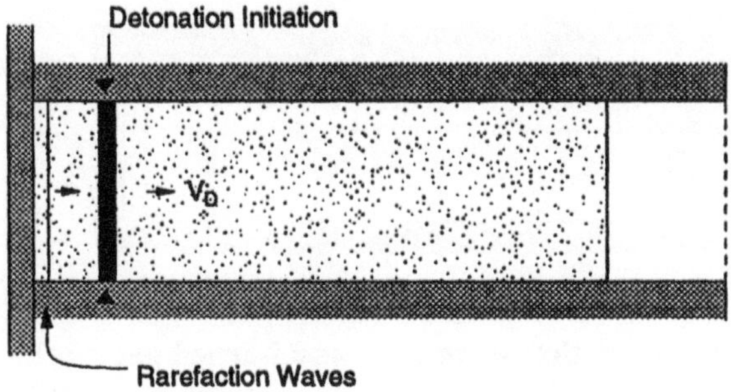

Fig. 8 Detonation is initiated at the closed end.

C. Detonation Wave Propagation

Ideally, a detonation wave will proceed toward the open end of the chamber at the C-J detonation velocity, V_D, of the mixture. The region ahead of the detonation will contain the unburned gas at state 1. Just behind the detonation wave, the burned gas will be at a significantly higher temperature and pressure (state 2). The burned gas near the closed end of the detonation tube will be at state 3, which has a lower temperature and pressure than the burned gas immediately behind the detonation (state 2). This is caused by the expansion region generated behind the detonation wave, see figure 4. As shown in Fig. 9, the remainder of the burned gas within the expansion region between the detonation wave and the closed end will be at some intermediate condition between states 2 and 3.

D. Burned Combustor Gases Are Expelled

When the detonation wave exits the combustor, the chamber contains burned gas (i.e., combustion products) at elevated temperatures and pressures. Conditions along the length of the chamber range from state 3 at the closed end to state 2 at the open end. Pressure increases from P_3 at the closed end to P_2 at the open end. The axial velocity distribution varies accordingly from zero at the closed end to potentially supersonic values near the exit.

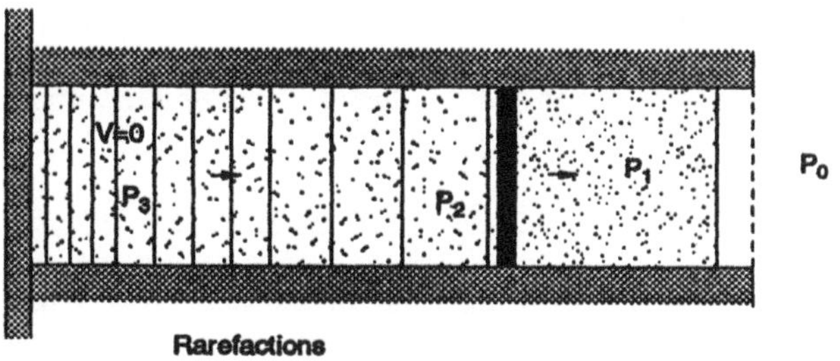

Fig. 9 Detonation wave propagation.

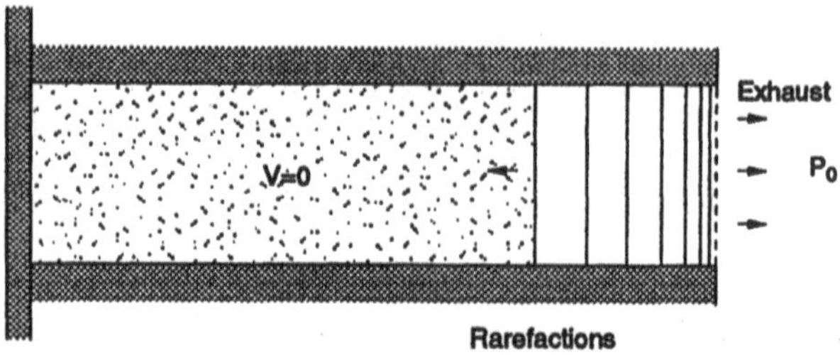

Fig. 10 Detonation wave exits and burned gas discharges.

As the detonation wave exits the tube, a pressure differential exists at the open end. This pressure difference creates a series of rarefaction waves that propagate into the tube and expel the burned gas (Fig. 10). The rarefaction waves travel into the tube at the speed of sound of the combustor gas mixture.

Initially, as the detonation wave exits the detonation tube, the gas within most of the tube is at a pressure near P_3 and the velocity is near zero. An expansion process, associated with the backward running rarefaction wave, then accelerates the fluid toward the exit.

As the gas in the detonation chamber expands, the driving pressure differential along the chamber diminishes. The unsteady blowdown process is characterized by a series of compression and rarefaction waves that are alternately created and reflected. The blowdown process is self-aspirating, and the flowfield at the combustor exit alternates between outflow and inflow (i.e., reverse flow). The pressure and temperature eventually decay to ambient levels and the exhaust velocity decays to zero.

E. Simple Airbreathing PDE Cycle

The detonation cycle can be repeated when the pressure within the chamber falls below a specified injection pressure. At this point, the chamber can be filled with a fresh fuel/air mixture and the previously described detonation cycle restarted. Figure 11 shows the sequence of events and phenomena that occur within the detonation tube over an entire detonation cycle.

F. Basic Engine Operation

The cycle frequency for a pulse detonation engine is the inverse of the time required to complete a full detonation cycle. Cycle time can be defined as the time required to fill the detonation tube with a fresh fuel/air mixture, initiate detonation, allow the detonation to traverse the length of the tube, and then allow the burned gases to expand and reduce the chamber pressure to the refill level:

$$T_{CYCLE} = T_{DETONATION} + T_{EXPANSION} + T_{FILL} \tag{16}$$

Certain assumptions are inherent in the expressions for detonation, expansion, and fill times. It is assumed that detonation is initiated at the in tant the detonation chamber is closed, and also that a negligible amount of time is required to initiate detonation. An estimation of the detonation, rarefaction, and

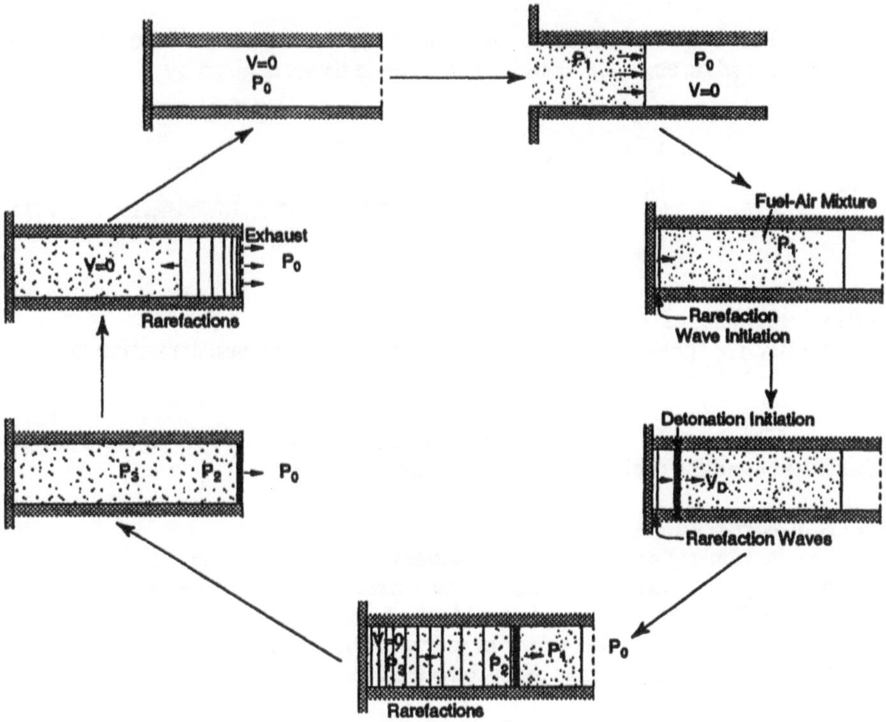

Fig. 11 Summary of detonation cycle.

fill times can be made. The time required for the detonation wave to traverse the device is computed readily from the detonation tube length and the velocity of the detonation wave with respect to the combustor walls. The time required to fill the combustor can also be computed readily from the dimensions of the device and the fuel/air injection velocity. Obtaining an estimate for the expansion or blowdown time is not quite as straightforward because of its unsteady nature. However, the timescale for the gas discharge process is determined by the speed of sound of the burned gas at state 3. A rough estimate of the expansion time can be made by computing the time required for sound waves to traverse the length of the device over several iterations. A more precise calculation of the cycle time can be obtained from unsteady CFD codes (see Sec. V).

The instantaneous or peak thrust of the PDE device can be approximated by examining the pressure differential acting on the interior surfaces of the PDE detonation chambers. The thrust that acts during the thrusting part of the cycle can be estimated for a detonation chamber of constant cross-sectional area A:

$$F = A(P_3 - P_1) - \text{drag forces} \tag{17}$$

The drag forces include all inviscid and viscous forces associated with flow about and through the vehicle. The average thrust is given by

$$F_{AVG} = \frac{\int_0^{T_{CYCLE}} F(t)\,dt}{T_{CYCLE}} \approx F\left[\frac{T_{DETONATION} + T_{EXPANSION}}{T_{CYCLE}}\right] \qquad (18)$$

where

$F(t)$ = instantaneous or unsteady thrust

F = effective "peak" thrust that acts over the detonation and expansion time

The specific impulse for a PDE can be computed from the average thrust level and the fuel consumption or mass flow rate.

G. Non-Airbreathing PDE Cycle

A non-airbreathing pulse detonation engine is very similar to an airbreathing cycle except that an onboard oxidizer, rather than external air, is mixed with the fuel to create a detonable mixture.

The processes that occur in a detonation chamber over a single cycle are fundamentally the same as those described for the airbreathing system.

The cycle is composed of several distinct events or processes.

1) The detonation chamber is charged with a fuel/oxidizer mixture.
2) Detonation is initiated at the closed end of the combustor by a detonation initiator.
3) The detonation wave progresses through the combustor at thousands of meters per second (i.e., at several times the speed of sound).
4) The detonation wave exits the combustor and initiates the generation of a series of rarefaction waves at the open end, which travel into the chamber and cause the burned gases to exhaust.
5) When the pressure within the combustor drops to the appropriate levels, the chamber is recharged with fresh fuel and oxidizer.

VI. Engine Efficiency

A pulse detonation engine relies on the periodic detonation of a fuel/air mixture to produce thrust. The PDE concept differs from typical airbreathing and rocket engines that rely on steady-state, isobaric deflagration combustion.

Detonation waves propagate at supersonic speeds of several thousand meters per second. This flame speed allows detonation combustion to occur as a constant volume process, whereas, traditional deflagration combustion is a constant pressure process. The constant volume detonation process can be modeled as part of a Humphrey cycle, whereas the constant pressure deflagration process can be described as part of a Brayton cycle.

Figures 12 and 13 compare the temperature, pressure, and volume characteristics of the Brayton and Humphrey cycles. The Brayton cycle (0-1-4-

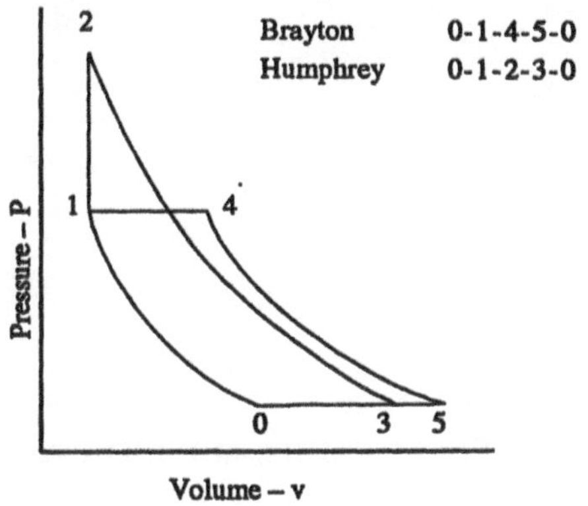

Fig. 12 Pressure-volume Cycle.

5-0) consists of two constant pressure processes (1-4 and 5-0) and two isentropic processes (0-1 and 4-5). The Humphrey cycle is similar, except that the constant pressure combustion process (1-4) is replaced by a constant volume heat addition process (1-2).

The efficiencies of the constant pressure Brayton and constant volume Humphrey cycles can be computed from the pressure-volume and temperature-entropy diagrams in Figs. 12 and 13. The efficiency of a cycle is defined as the useful work output divided by the total heat energy input.

The efficiency of the Brayton cycle depends only on the temperature change during either of the two isentropic compression or expansion processes (i.e., $T_0/T_1 = T_4/T_5$):

$$\eta_{BRAYTON} = 1 - \frac{T_0}{T_1} \tag{1}$$

The efficiency of the Humphrey cycle is given as:

$$\eta_{HUMPHREY} = 1 - \gamma \frac{T_0}{T_1} \left[\frac{\left(\frac{T_2}{T_1}\right)^{\frac{1}{\gamma}} - 1}{\frac{T_2}{T_1} - 1} \right] \tag{2}$$

The efficiency of the Humphrey cycle depends not only on the isentropic compression temperature ratio, T_0/T_1, but also on the ratio of specific heats, g, and the temperature change caused by the constant volume combustion (i.e., the detonation temperature ratio T_2/T_1).

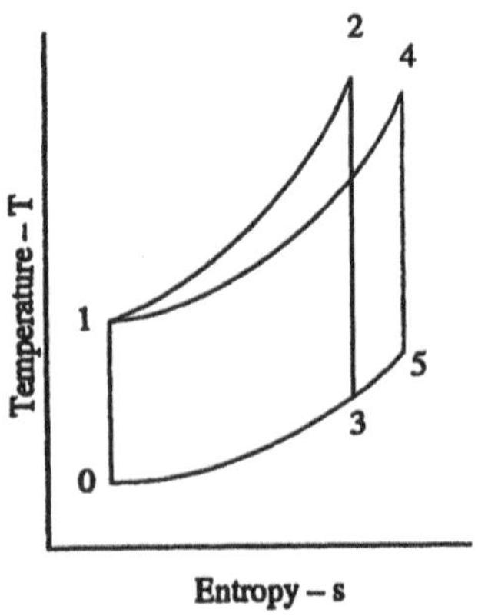

Fig. 13 Temperature-entropy cycle

The difference between the Brayton and Humphrey cycle efficiencies is the following T_0/T_1 multiplier:

$$\gamma\left[\frac{\left(\frac{T_2}{T_1}\right)^{\frac{1}{\gamma}}-1}{\frac{T_2}{T_1}-1}\right] \tag{3}$$

The value of this expression is always less than one for typical detonation combustion. As a result, the typical efficiency of a detonation cycle is greater than the efficiency of the Brayton cycle.

A comparison of Brayton and Humphrey cycle efficiencies can be made using a representative detonation combustion process and approximating the ratio of specific heats (γ) as a constant over specific parts of the cycle. An equilibrium chemistry calculation for stoichiometric hydrogen/air yields a detonation temperature ratio T_2/T_1 of 10.2 and specific heat ratios of 1.4 in the unburned gas and 1.16 in the burned gas. The Humphrey detonation cycle efficiency was calculated for average cycle γs of 1.4 and 1.16, which represent the upper and lower limits of the varying specific heat ratio.

Figure 14 shows the calculation of cycle thermal efficiency as a function of the compression ratio (P_1/P_0). The actual detonation cycle efficiency would lie somewhere between the two limiting detonation curves ($\gamma = 1.4$ and 1.16). The constant volume detonation process clearly offers a significant improvement in efficiency as compared with the isobaric combustion cycle. Wortman et al.[5] have shown a similar trend.

For example, at a compression ratio of 6, for example, the constant volume process offers a 30-50% improvement in thermal efficiency over the constant pressure cycle. Replacing the constant pressure deflagration combustion process in an engine with a detonation reaction would yield a 30-50% improvement in fuel consumption, assuming that the same propulsive efficiency could be maintained.

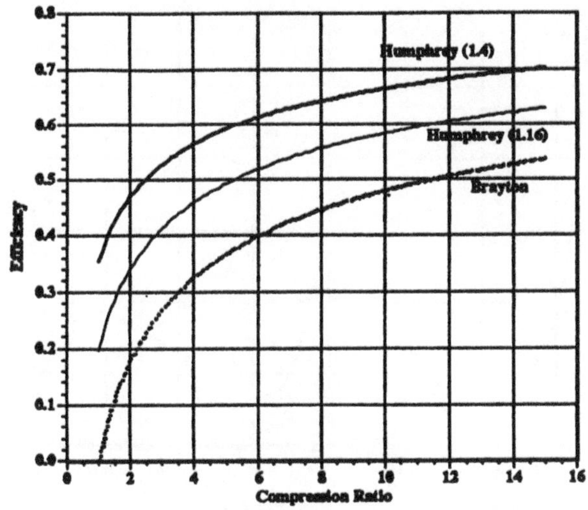

Fig. 14 Thermal efficiency of Brayton isobaric and Humphrey detonation cycles.

V. Computational Assessment Of PDE Unsteady Characteristics

The results of some preliminary performance calculations are presented in this section. The unsteady results for a simple blowdown test case are presented as well as hydrogen/air and hydrogen/oxygen detonation calculations. A simple cylindrical geometry (2 cm in diameter and 20 cm in length) was assumed for the detonation combustion chamber.

A one-dimensional CFD code was used to study the processes that occur in the detonation combustor. The conditions in the detonation tube (e.g., pressure, temperature, Mach number, density, etc.) were estimated as a function of time and location within the tube. The concentration and distribution of the various gaseous species present in the combustion chamber were also computed.

The unsteady CFD computation was also essential for providing a realistic approximation of the unsteady phenomena associated with detonation engine operation. For example, the behavior of the expansion processes within the combustor could be characterized and visualized.

CFD Model

A modified version of the Mozart quasi-one-dimensional CFD code, developed by Cambier[6] at NASA-Ames, was used in the unsteady calculations. Mozart computes the one-dimensional inviscid flowfield (with area variation) using the Euler equations. The code uses an implicit, first-order Total Variation Diminishing (TVD) algorithm for multiple species.

Real gas dynamics and nonequilibrium kinetics are computed, and thermal equilibrium is assumed. The momentum density, stagnation energy density, and seven species (H, O, H_2, O_2, OH, H_2O, and N_2) are convected.[6] A

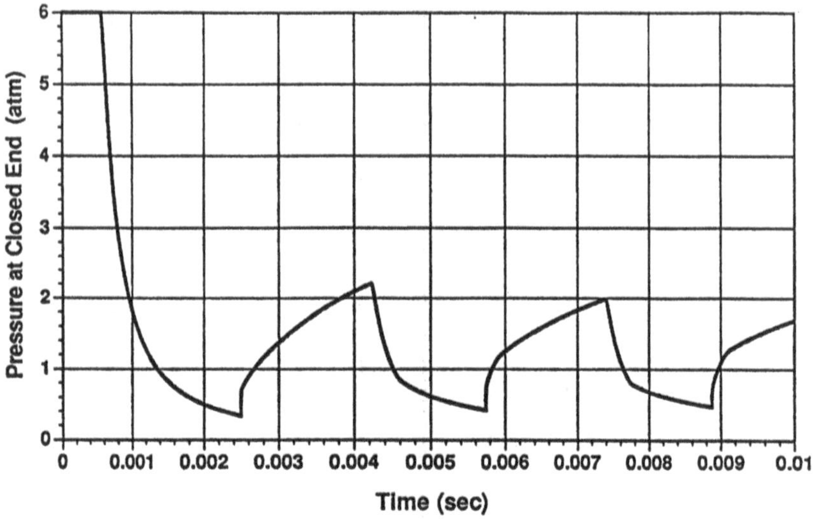

Fig. 15 Unsteady nonreacting blowdown.

seven-species, 18-reaction hydrogen chemistry model was used in the calculations.

1. Non-Reacting Blowdown Test Case (Nonreacting Air)

The goal of the simulation was to examine the unsteady behavior of the blowdown process, independent of any combustion or chemistry effects. A detonation wave can be conceptualized as a constant volume process that increases the temperature and pressure of gases in the combustion chamber, after which the burned combustion products are expelled through an unsteady blowdown process. To study the unsteady behavior of the blowdown process, a simple non-reacting blowdown test case was computed using the one-dimensional CFD tool.

A cylindrical detonation chamber (2 x 20 cm) was filled with air at a pressure of 6 atm and a temperature of 300 K. The pressurized air discharged through the open end of the cylinder to ambient conditions (1 atm, 300 K).

Figure 15 shows the static pressure at the closed end of the tube as a function of time. The closed-end wall pressure remains at 6 atm until the rarefaction's generated by the expansion at the exit reach the closed end at approximately 600 μs. The closed-end pressure drops to less than 1 atm and then oscillates around the ambient level of 1 atm. The self-aspirating nature of the blowdown process is illustrated in the figure. The amplitude of the pressure oscillations in the chamber would decay because of viscous damping, and the ambient pressure level would be achieved eventually. Note: Any decay in the magnitude of the pressure oscillations evident in the calculation results from numerical viscosity and smoothing since the algorithm is inviscid.

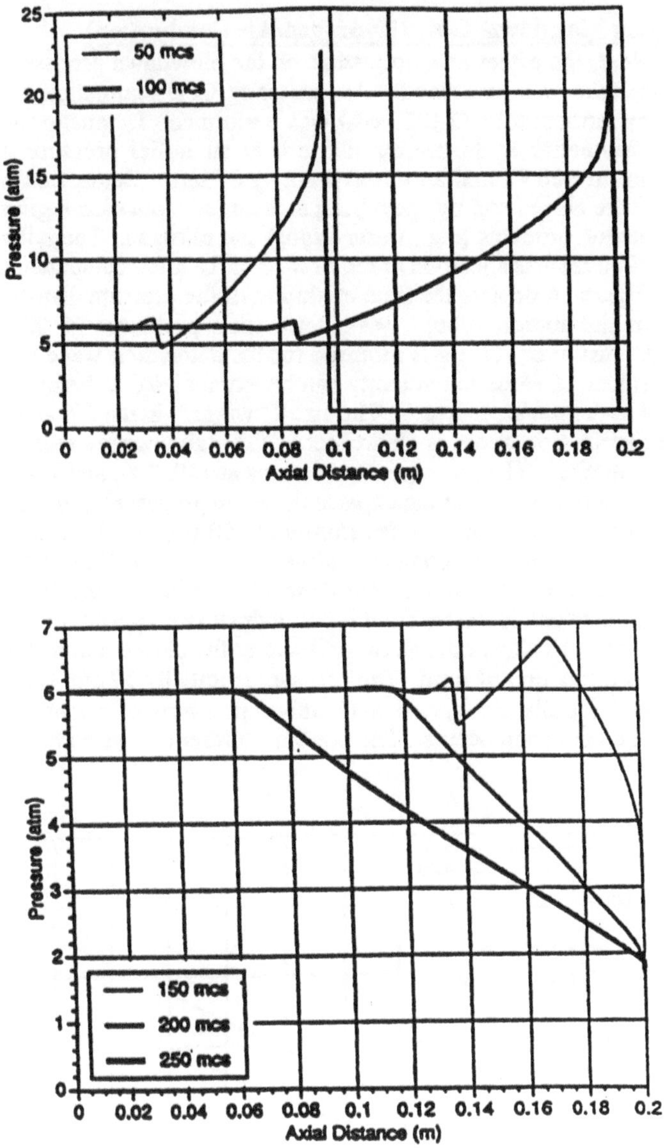

Fig. 16 Detonation tube pressure distribution: a) 50 and 100 μs; and b) 150, 200, and 250 μs.

2. Reacting Blowdown Test (Hydrogen/Air Combustion)

Next, the effect of combustion on the blowdown processes within the detonation tube was examined. An unsteady CFD simulation of detonation within a cylindrical tube (2 x 20 cm) was performed. Detonation was initiated in a stoichiometric hydrogen/air mixture at an initial pressure of 1 atm, a temperature of 300 K, and an initial velocity of zero. Detonation was directly initiated at the closed end by specifying an energetic initiation region containing hot combustion products (e.g., water vapor) and nitrogen. The initiation region (30 atm, 3000 K) was specified at the first 10 of the 1000 computational nodes.

Figure 16 depicts the time evolution of the pressure distribution within the 20-cm detonation tube. Peak detonation pressures of 22-23 atm are predicted. Just over 100 μs is required for the detonation wave to traverse the entire chamber. Detonation velocity can be extrapolated as being roughly 1950 m/s in the hydrogen/air mixture. The small "wiggle" in the 50- and 100-μs mcs pressure distribution traces is an artifact of the high-pressure region required to initiate detonation. The pressure distributions at 150, 200, and 250 μs show the detonation wave exiting and the expansion region progressing into the tube. The velocity distribution at later times (500 and 750 μs, and 1, 2, and 3 ms) also provides insight into the unsteady blowdown process (Fig. 17). The self-aspirating nature of the device is represented by the largely negative velocities at 1, 2, and 3 ms which indicates fluid is being drawn into the combustor.

Figure 18 shows the static pressure at the closed end of the detonation chamber as a function of time. The pressure is initially 30 atm (in the initiation region) and rapidly decays to a chamber pressure of about 6 atm as the detonation combustion occurs. The pressure further decays and oscillates about

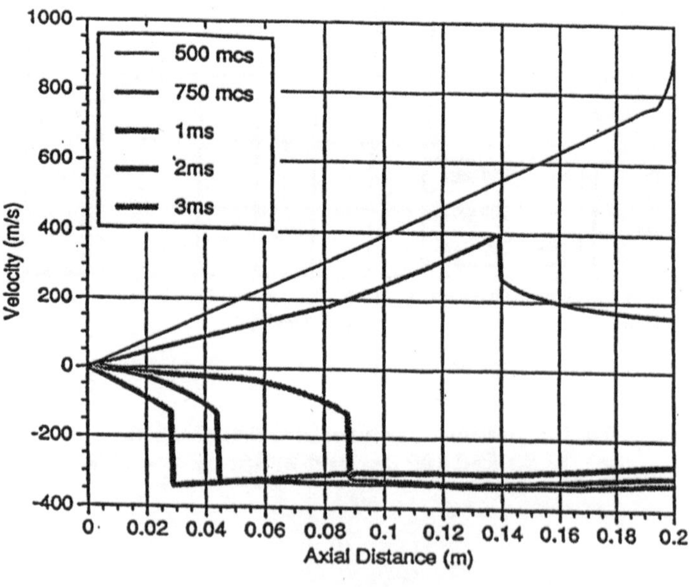

Fig. 17 Detonation Tube Velocity Distribution at 0.5, 0.75, 1, 2, and 3 ms.

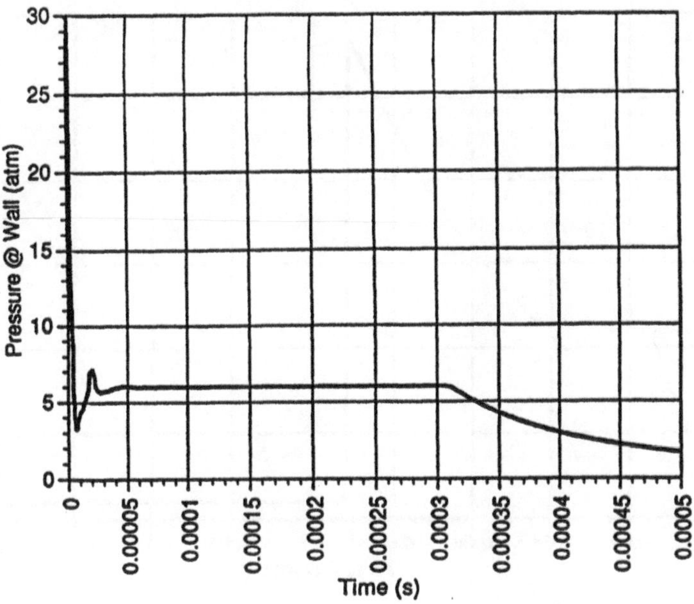

Fig. 18 Pressure history at closed-end.

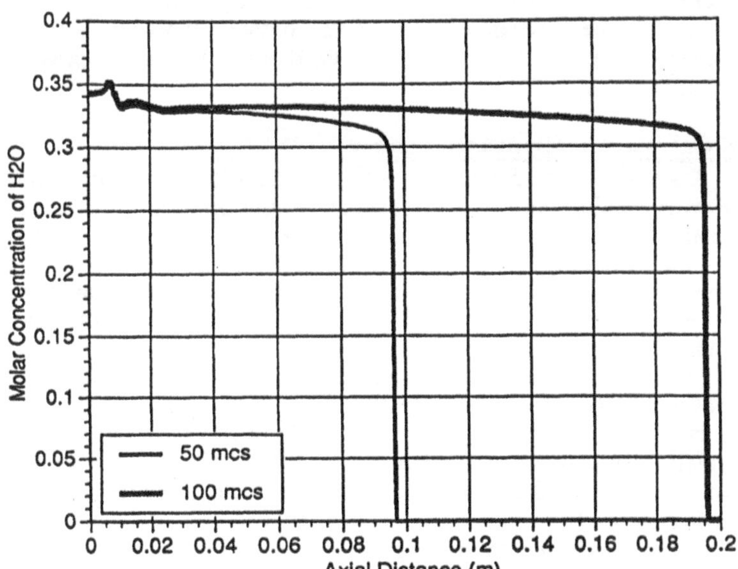

Fig. 19 Water vapor distribution at 50 / 100 μs.

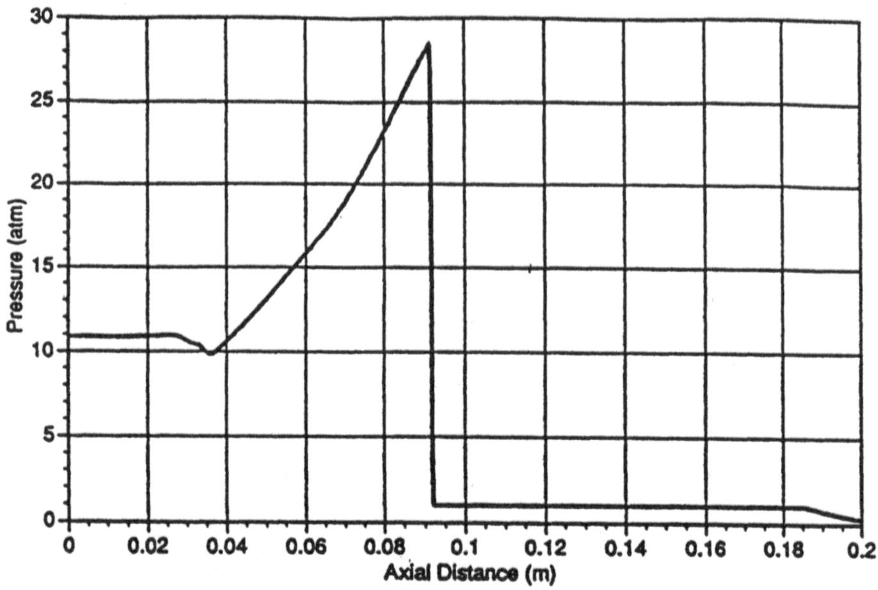

Fig. 20 Detonation tube pressure at 32 μs (rocket mode / closed-end initiation).

the 1-atm ambient pressure as the detonation wave exits and the blowdown process is initiated.

Figure 19 is an example of the chemistry and species calculations that are typical for these detonation calculations. The detonation combustion process converts most of the molecular hydrogen and oxygen into water vapor. Small molar concentrations of other species (e.g., O, H, and OH) are also present in the combustion products.

3. PDE Calculations (Hydrogen/Oxygen)

A simulation of detonation within a stoichiometric hydrogen/oxygen mixture was performed as a preliminary examination of rocket mode performance. The calculation assumed that the vehicle was operating at an altitude of 70,000 ft with ambient conditions of 0.04 atm and 218 K. A stoichiometric hydrogen/oxygen mixture at 1 atm and 200 K was assumed to be at rest in the detonation chamber. Detonation was initiated at the closed end ($P = 30$ atm, $T = 3000$ K).

Figure 20 indicates the detonation wave has traversed almost half the length of the tube (i.e., 9 cm) in only 32 μs. As expected, the detonation wave velocity in the hydrogen/oxygen mixture ($V = 2850$ m/s) is greater than the hydrogen/air detonation velocity of approximately 1950 m/s. The expansion of the 1-atm mixture into the 0.04-atm environment causes the linear decrease in pressure near the exit.

Rocket mode operation results in higher post detonation pressures than airbreathing operation (Fig. 21). The chamber pressure is approximately 11 atm (i.e., $P_3/P_1 = 11$). It should be noted that a performance benefit is derived by

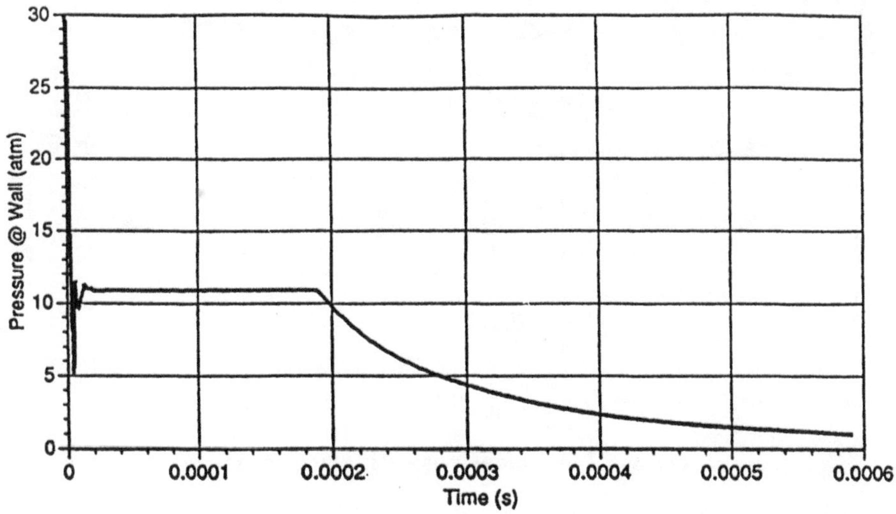

Fig. 21 Closed-end pressure history (rocket mode / closed-end initiation).

using a low mixture initial temperature of 200 K. The chamber pressure ratio P_3/P_1 would have been roughly 7 if an initial temperature of 300 K had been assumed.

VI. PDE Design Issues

A PDE is composed of several unique components. One possible grouping of components is shown in Fig. 22. For the purposes of discussion, the PDE will be broken into 10 components. These include the following:

1). Inlet
2). Fuel, air, and onboard oxidizer distribution manifolds
3). Inlet/detonation chamber interface
4). Fuel source
5). Oxidizer source
6). Fuel/air/oxidizer mixing section
7). Detonation initiation section
8). Detonation chamber(s)
9). Detonation chamber/nozzle interface
10). Nozzle

Several of these components have certain attributes that can be very different from conventional propulsion systems. For example, the ability to initiate a detonation quickly in a very short distance is one of the primary engineering considerations. The following section will briefly outline several of the considerations required to properly design a functioning PDE. The objective is to educate the reader on the many issues critical to implementing detonation

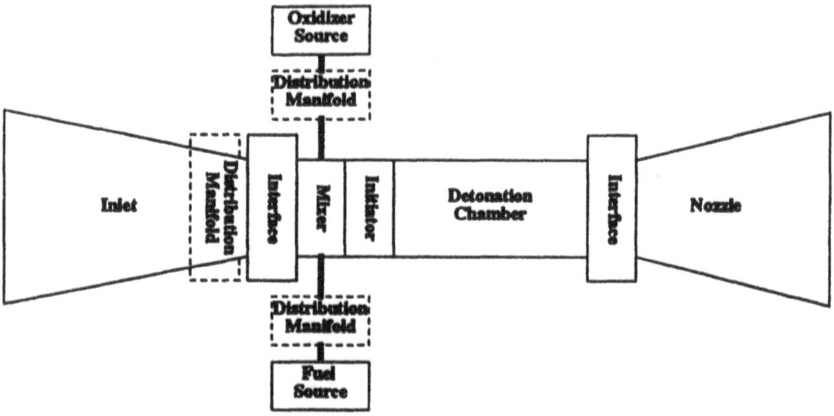

Fig. 22 Schematic of PDE components.

combustion in practical propulsion systems. As of the writing of this paper, many of the PDE design details are either proprietary or covered by other restrictions. The solutions to many of the issues presented here are described in several pending patents. Therefore, the following description is limited to information in the open literature and is intended only to highlight important issues.

A. Inlet

The inlet design for a PDE is highly dependent on whether the inlet is operating in an unsteady or steady manner. Unsteady inlets have seen limited use notable in the V-1 pulse jet engine. More recently several researchers, including Lynch,[7] Loth,[8] and Eidelman et al.,[9] have evaluated the characteristics of unsteady inlets. Lynch[7] and Eidelman et al.,[9] performed detailed numerical evaluation of PDE devices where the blast wave from the detonation chamber is expelled from an open inlet. As expected, an inlet operating in this mode would incur significant external flowfield losses. An alternative approach described by Loth[8] considers valveless designs where the inlet/combustor geometry is optimized to minimize unsteady perturbations within the inlet. These geometries include aerodynamic one-way valves or rectifiers. This approach offers promise but tends to leak and has poor off-design performance. A third approach is to design an inlet manifold large enough to dampen the pressure transients so that the flow at the inlet lip is in a steady state. This approach tends to increase the total volume of the engine and may not be practical for volume-limited applications. Steady-state inlets have seen application on essentially every airbreathing propulsion system. They are relatively simple to analyze and their performance characteristics are well understood. There are at least two approaches that can be used to operate a PDE inlet in a steady-state mode. The first involves bleeding excess air between detonation cycles through a well-designed inlet/detonation chamber manifold. Excess air would be dumped overboard in such a manner as to mitigate any inlet or performance losses. The second approach involves using multiple detonation combustors and decoupling the predominantly steady inlet flowfield from the unsteady flow within the combustors. In this case, detonation chambers would be filled and fired in an alternating manner, with one chamber firing as another is being fueled.

Mechanical valves may also be used in many of these inlet configurations.

B. Fuel, Air and Onboard Oxidizer Distribution Manifolds

The fuel, air, and oxidizer manifolds combine the various fuel and oxidizer fluids to produce a uniform, detonable mixture. The manifolds must be designed to ensure that a high degree of macroscopic mixing occurs. Detonation processes are very sensitive to stoichiometry, particle/droplet size, and the local degree of mixing. The design of a given manifold can be difficult.

C. Inlet/Detonation Chamber Interface

The inlet/detonation chamber interface serves to separate the inlet flowfield from the detonation processes. The separation can occur through either "aerodynamic" means, as described earlier, or by mechanical valves. Mechanical means include "flapper" valves used on the V-1 pulse jet, spring-loaded cone valves used in internal combustion engines, and other valving arrangements considered outside the scope of this paper.

D. Fuel Source

The ignition characteristics of the fuel are critical to the detonation process. The fuel can be stored in several different forms, but must be delivered to the fuel manifold as a gas, liquid, and/or solid of sufficiently small droplet/particle size to permit the formation of a stable detonation. For example, solid particles or liquid droplets must be no larger than several tens of microns to ensure that the particle breakup time is compatible with the detonation time scale.

E. Oxidizer Source

A separate oxidizer source may be required to reduce the time and physical distance required for detonation initiation. Several concepts use a predetonation region containing fuel and a pure oxidizer to trigger the detonation of a less detonable mixture such as fuel and air.

F. Fuel/Air/Oxidizer Mixing Section

The fuel/air/oxidizer mixer section incorporates a device or process to mix the fuel, air, and/or oxidizer mixture constituents to the microscopic length scales required for detonation. The device may contain mechanical and/or fluid dynamic mixers to achieve the desired level of mixing. Note the various combustible mixture constituents could also have been pre-mixed and then admitted into the detonation chamber. Many possibilities exist, but are considered outside the scope of this paper.

G. Detonation Initiation Section

The role of the detonation initiation section is to create and establish a stable self-sustained detonation. The resulting detonation wave would travel into the detonation chamber to trigger detonation in the primary fuel/air mixture. There are several methods that can be used to initiate a detonation. Detonation can be initiated indirectly by creating a deflagration that undergoes a rapid deflagration-to-detonation transition process. Direct detonation initiation can be accomplished using concepts such as predetonators, laser initiators, and plasma torches. The predetonation concept has been investigated by Helman et al.[10] and Bussing and Pappas.[11,12] The device offers mechanical simplicity but requires a separate oxidizer source. The laser and plasma torch concepts each require power supplies that can be complex and heavy. The interested reader is referred to Bussing and Pappas,[11,12] for a more detailed description of detonation initiation methods. Note that there are many possible detonation initiation/detonation chamber configurations.

H. Detonation Chamber

The detonation chamber is a simple device designed to sustain and confine the detonation process. Several issues are essential to ensure proper operation of the detonation chamber. These include a structure capable of sustaining the detonation blast wave pressures and temperatures. The design must also include a mechanism for transferring the forces generated to the vehicle. In addition, the refueling and purging strategies must be understood and properly implemented. It is also necessary to prevent preignition of the fuel/air mixture before the detonation chamber is fully fueled. Considerable progress has been made in this area, but it is considered outside the scope of this paper.

I. Detonation Chamber/Nozzle Interface

The primary consideration in designing the detonation chamber/nozzle interface is the need to ensure that the flow remains attached.

J. Nozzle

Designing a nozzle for implementation into a PDE is a nontrivial process. However, certain characteristics of the detonation chamber exhaust process may be utilized to simplify the design process. The flowfield at the detonation chamber exit remains approximately constant for a period of time after the detonation wave has exited the combustor. This behavior allows the nozzle shape to be designed for quasisteady operation. As with steadystate devices, the PDE nozzle is designed to expand the combustor gases to ambient pressure at the design point.

1. Design Issue: Location of Detonation Initiation

One design issue that has received considerable attention in the literature has been the question of where to initiate detonation in a given combustion chamber. At least two extreme options for initiation location exist: closed- and open-end detonation initiation.

The PDE closed-end detonation initiation cycle has been described previously. The most obvious difference between the open- and closed-end initiation cycles is the location of detonation initiation and the propagation of the detonation wave through the combustor (Fig. 23). The refueling processes in

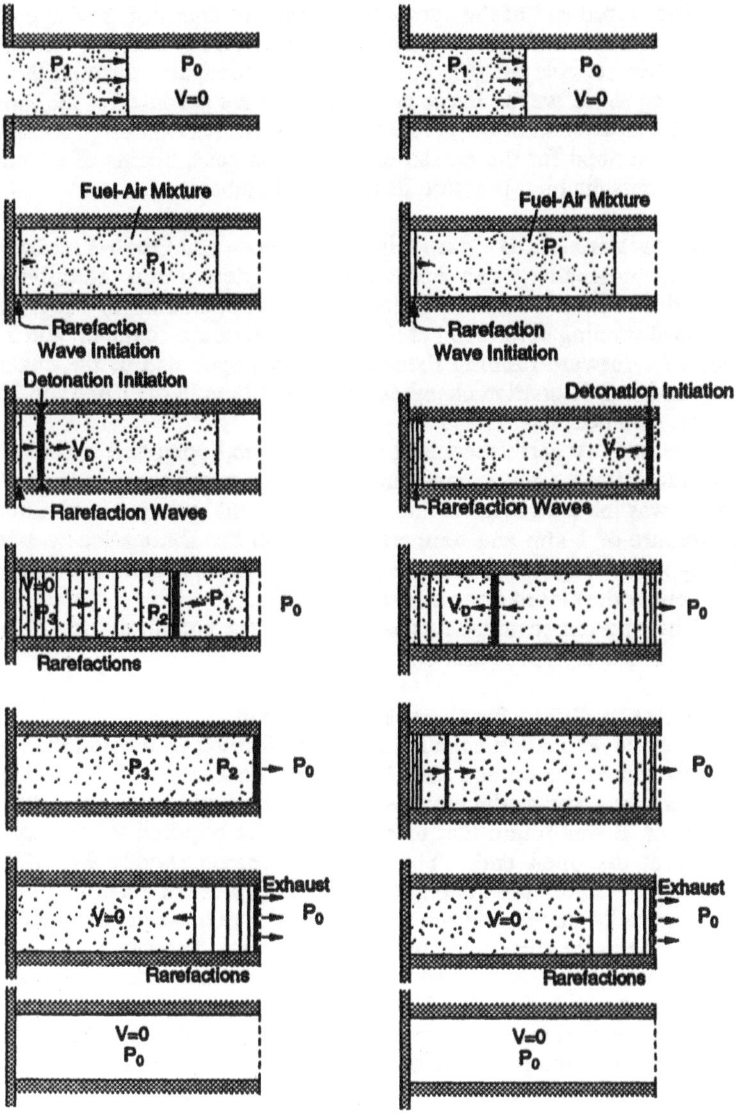

Fig. 23 Comparison of closed- and open-end initiation detonation cycles.

both cycles are essentially the same. In the open-end initiation cycle, however, detonation is initiated near the exit of the combustor. The detonation then propagates toward the closed end at the C-J detonation velocity. The burned gases behind the detonation also travel toward the closed end at a slower speed.

The closed end of the combustion chamber does not "see" the effect of the detonation wave until the supersonic process actually reaches the wall. The detonation then reflects off the closed end and propagates through the burned gas as a strong shock wave. Large peak pressures are produced at the thrust wall located at the closed end of the detonation chamber. An exhaust process, similar to the one described for the closed-end initiation case, begins after the strong shock wave exits the high-pressure detonation chamber.

A. One-dimensional CFD Comparison of Closed- and Open-End Initiation

The unsteady behavior within the detonation chambers of a hydrogen/air PDE was evaluated for two modes of operation: 1) direct initiation of a rearward-running detonation at the closed end of the chamber, and 2) direct initiation of a forward-running detonation at the open end of the chamber. A simple cylindrical detonation chamber geometry (2 cm in diameter and 20 in cm length) was assumed.

Preliminary performance of the closed- and open-end initiation cycles was compared for a simple one-dimensional, constant-area, 1000-node grid. Detonation was initiated in a stoichiometric quiescent hydrogen/air mixture at an initial pressure of 1 atm and temperature of 300 K. Detonation was initiated directly by specifying an initiation region with a high pressure and temperature region at either the closed or open end of the tube.

The one-dimensional unsteady combustor conditions for closed-end detonation initiation were presented in section 2.

B. Chamber Conditions for Open-End Initiation

A static detonation calculation was performed with detonation initiated at the open end of the combustor. The initial conditions of the hydrogen/air mixture were the same as the closed-end initiation problem (1 atm, 300 K, $U = 0$). However, it was found that more energy was required to directly initiate detonation at the open end. The initiation region (nodes 990-1000) was specified to have a pressure of 60 atm and a temperature of 3000 K.

Figure 24 shows the progression of the detonation wave toward the closed-end. As in the closed-end initiation case, peak detonation wave pressures are approximately 22.5 atm. The velocity distribution within the tube for the first 250 μs is shown in Fig. 25. Notice the velocity of the burned gases trailing behind the detonation wave is negative with respect to the chamber wall, which indicates the detonation products are traveling toward the closed end. After 100 μs, when the detonation impacts the closed-end wall, the velocities within the tube are positive, indicating the fluid is moving toward the open end. The combustor fluid velocity remains positive until the blowdown process begins and self-aspiration occurs.

The pressure at the closed end of the tube remains at 1 atm until the detonation reaches it at approximately 100 μs or 0.1 ms (Fig. 26). Closed-end static pressure peaks at a value of almost 44 atm, which is significantly higher than the 30-atm peak pressure that occurred during closed-end initiation.

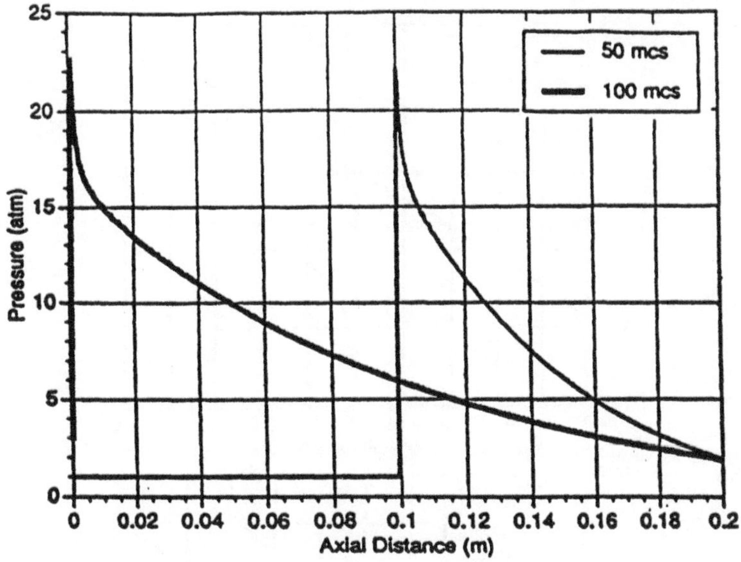

Fig. 24 Detonation tube pressure distribution at 50 and 100 μs (static conditions / open-end initiation).

C. Direct Comparison of Closed- and Open-End Initiation Cycles

A direct comparison of the performance of the closed- and open-end detonation initiation was made using static chamber conditions. Stoichiometric hydrogen/air at 1 atm, 300 K, and zero velocity was specified initially in both cases. In addition, the initial conditions in both detonation initiation regions were set to 60 atm and 3000 K to provide a basis for direct comparison.

Figure 27 compares the pressure history at the closed end of the detonation tube for both open- and closed-end detonation initiation. As observed earlier, the peak pressure is significantly greater in the open-end initiation case (43 vs 30 atm). After jumping to 43 atm at 100 μs, the pressure decays steadily to 1 atm for the open-end initiation case. The closed-end initiation simulation reaches a closed-end pressure of 6 atm in less than 50 μs and remains at that value until past 300 μs. Both closed-end pressure traces reach 1 atm in approximately 600 μs.

The closed-end pressure was integrated over time to compute the internal thrust and specific impulse over one detonation cycle. Time integration of the pressure revealed equivalent thrust production and fuel efficiency for both cases since the same amount of hydrogen was consumed in both cases. Neither initiation scheme offered a clear performance advantage that could be discerned from this numerical simulation.

In practice, however, there are at least two drawbacks in attempting to initiate detonation directly at the open end. First, the structural dynamics and force transfer issues are exacerbated because, the pressure peaks and fluctuations are significantly greater in the open initiation scheme. In addition, a larger critical energy is required to initiate detonation directly at the open end,

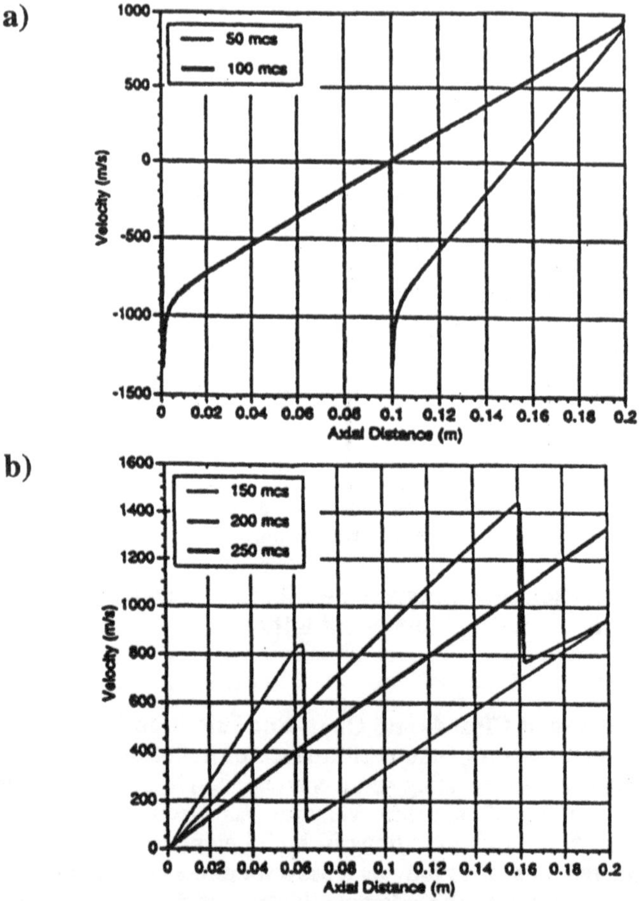

Fig. 25 Detonation tube velocity distribution (static conditions / open-end initiation): a) 50 and 100 μs; and b) 150, 200, and 250 μs.

because the closed-end wall and corners are not present to confine the gaseous mixture and thus augment the detonation initiation process.

The degree to which a detonable mixture is confined plays an important role in defining direct and indirect detonation initiation requirements. Detonation chamber geometry and initiation source energy deposition characteristics and location are also important factors in determining initiation energy requirements.

VII. PDE Experimental Testing Issues

Several issues critical to the development of an operational PDE must be understood if a viable pulse detonation engine is to be built and tested. These issues include a variety of detonation initiation and propagation characteristics,

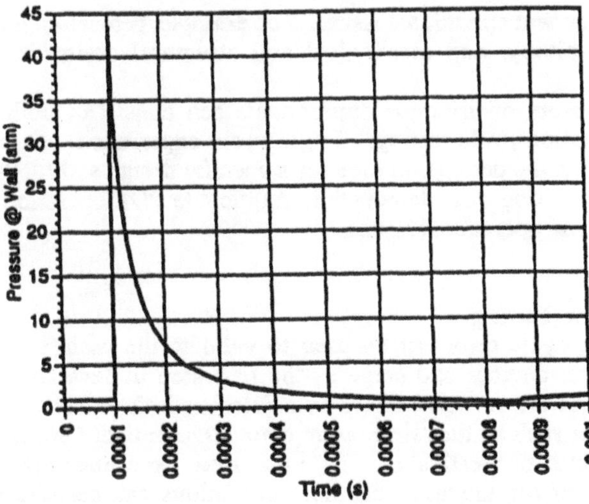

Fig. 26 Pressure history at closed end of detonation tube (static / open-end initiation).

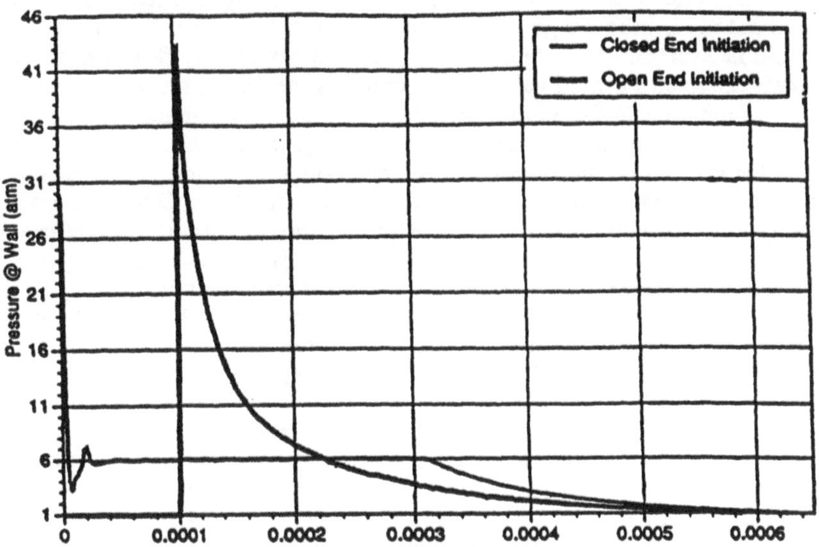

Fig. 27 Comparison of closed-end pressure history (closed- vs. open-end initiation).

as well as design and operational issues. For example, detonation initiation, fuel and oxidizer mixing, and mechanical and structural design issues can be evaluated.

PDE proof-of-principle experiments can consist of both single and multiple detonation cycle testing. Single-cycle experiments can measure the conditions within the detonation tube for a specific design and fuel/air mixture. Multiple-cycle testing can demonstrate multicycle PDE operation and assess critical design and operational issues.

A. Single-Cycle Testing

Single-cycle tests can be used to validate the basic performance of pulse detonation devices and serve as the first step in developing a detailed understanding of the complex issues associated with the operation of the PDE concept. Issues such as fuel/air mixture detonation initiation requirements and detonation wave properties can be measured experimentally. Detailed measurements of the unsteady chamber conditions can be used to estimate preliminary engine performance and to validate analytical and computational predictions.

Single-cycle testing can use a simple experimental setup consisting essentially of an instrumented shock tube. Figure 28 is a representative schematic for an experimental detonation shock tube. This particular tube assembly is composed of sections that are 4 ft in length and contain ports for fueling lines and pressure transducers.

The experimental apparatus can consist of fuel tanks and monitoring equipment, predetonation initiator module, primary detonation tube, sensors and instrumentation, tube purge and evacuation systems, and a high-speed data acquisition system. Fuel tanks, high-pressure regulators, flow meters, and remotely operated solenoid valves can provide the desired gas mixtures.

Single-cycle detonation tests can be used to determine the detonation initiation energy and power requirements for a given detonable mixture. One end of the detonation tube can contain a detonation initiation device such as a high-energy spark ignitor or a predetonation chamber. Detonation wave velocity and detonation tube pressures are important parameters that can be measured in single fire detonation tests. High-frequency response pressure transducers can be used to estimate the unsteady static pressure history within the detonation tube. Multiple pressure transducers distributed along the length of the primary detonation chamber can record the progress of the detonation wave as it traverses the tube. Figure 29 is a sample wall pressure trace from detonation testing.

Detonation wave speeds can also be computed from pressure transducer time-of-arrival data. A high-speed multichannel data acquisition system can sample and store the transducer information. The presence of detonation combustion can be verified by placing smoke-coated metal foils along the inner walls of the shock tube. Smoke foils can be imprinted with the characteristic diamond-shaped cellular structure if detonation is produced during a test. Measured peak pressures can also be used to confirm the presence of detonation.

The single-cycle testing can measure and verify fuel detonability and the unsteady conditions within a single detonation cycle. Specific impulse can

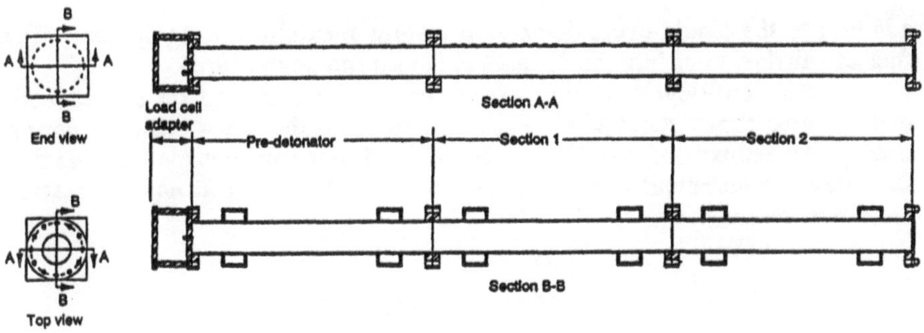

Fig. 28 Representative detonation tube schematic.

be calculated from pressure and fuel consumption measurements. In addition, multicycle performance parameters, such as average thrust and mass flow rate, can be estimated based on the single detonation data.

B. Multicycle Testing

Repetitive, multiple detonation testing can be used to study the continuous multicycle operation of a pulse detonation device. Critical operational and mechanical design issues can be evaluated.

Multicycle testing involves the continuous fueling and firing of a detonation tube device. Fuel and oxidizer manifolds and mixers are required in

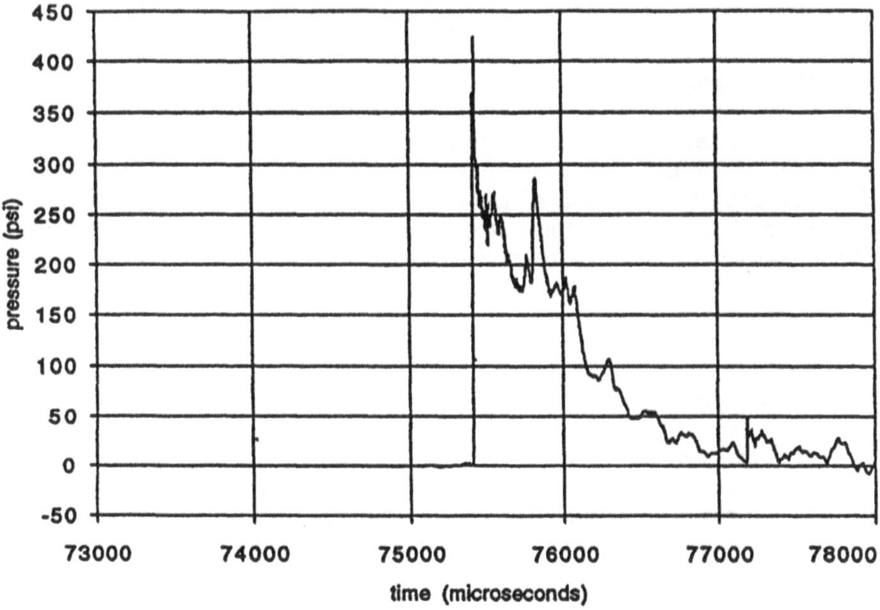

Fig. 29 Representative pressure transducer trace from detonation tube testing.

addition to the single-cycle detonation testing apparatus and equipment. The fuel and airflow rates into the detonation device can be measured.

The multicycle testing assembly can be mounted on a load cell-instrumented thrust stand as a means of measuring the thrust achieved during testing. An estimate of specific impulse of the device can be made from the fuel flow rate measurement and the thrust measured by load cell and/or pressure transducer readings.

VIII. Summary

Pulse detonation engines (PDEs) are mechanically simple and offer significant performance benefits. These devices operate in a constant volume combustion mode that is fundamentally more efficient than conventional constant pressure combustion engines. A PDE operates in a cyclical manner: a detonation wave increases the temperature and pressure of a gas in a combustor which then discharges to produce thrust. The cycle can be repeated after the combustor is recharged with a detonable fuel/air or fuel/oxidizer mixture. Experimental and computational results to date support these claims. However, more detailed testing is required to validate fully the feasibility of the concept. The design of an operational PDE is a complex process that involves the integration of components that may operate in steady and unsteady modes.

Appendix

1. PDE Research Review

A review of the available propulsion-related pulse detonation research is useful in obtaining an understanding of the state-of-the-art in PDE development. Examination of the research indicates several trends in the experimental and computational study of detonation as applied to propulsion. These trends can be summarized as follows.

A. Experimentation

Many researchers have encountered difficulties in attempting to study and characterize the performance of pulse detonation devices. These difficulties were caused primarily by inadequate detonation initiation energy and poor fuel/air mixing.

Some investigators have attempted to initiate detonation in mixtures with insufficient amounts of energy and have, therefore, unwittingly created deflagrative burning devices. Typically, detonation was not achieved at all, or was achieved only after a lengthy transition from a deflagration. Nicholls et al.,[13,14] Krzycki,[15] and Wortman et at.[5] each attempted to detonate fuel mixtures which require relatively high initiation energies requirements (e.g., propane/air) using low-energy spark plug ignitors. Helman et al.[10] may have achieved detonation in an ethylene/air mixture by first initiating detonation within a smaller tube containing a readily detonable ethylene/oxygen mixture. The following is a summary of the key references.

Key PDE Experimental Tests

1). Heinrich, Hoffmann[16] (1940): The first work on intermittent detonation devices for propulsion began in Germany in the late 1930s. Acetylene/oxygen/water mixtures were detonated in November 1939.

Hoffmann developed a single-reaction propulsion device based on intermittent detonative combustion. This device did not have any valves or regulating mechanisms. One difficulty in achieving operation of the device was in controlling suitable precompression of the mixture and attaining automatic intermittent operation.

The procedure for operating the device was as follows: filling, ignition to detonation, and exhaust. Holes were bored in the long detonation tubes to allow gas jets to escape. The progress of the combustion wave and the onset of detonation were indicated by the height of the gas jets escaping through the holes.

Experiments with liquid hydrocarbon fuel (i.e., benzene) and oxygen were performed. Issues such as fuel atomization, evaporation, and mixing were found to be critically important. The optimum length of the detonation tube and the most advantageous arrangement of ignition points were determined.

2). J. A. Nicholls et al.[13] (1957): The feasibility of a reaction device operating on intermittent gaseous detonation waves was examined. A simplified theoretical analysis of a pulse detonation engine was performed. Preliminary experiments using a single detonation wave were conducted. Impulse tests were performed by suspending the detonation tube as a pendulum. Cyclic detonation tests were also performed using hydrogen/air mixtures. Thrust, fuel flow, airflow, and temperature measurements were made over a range of operating conditions.

Specific impulses of over 2100s were obtained experimentally and agreed well with Nicholls' predicted values. However, Nicholls' theoretical model was extremely conservative, and agreement with his experimental data may have resulted because direct detonation may not have been achieved. The device was probably operating in both a deflagration and detonation mode.

3). L. J. Krzycki[15] (1962): An analytical and experimental investigation was conducted to determine the operating characteristics of an intermittent detonative device. Gaseous propane/air mixtures were ignited by an automotive spark plug at frequencies from 0 to 60 Hz. Idealized relationships were derived analytically and experiments were performed to find the effect of fuel/air ratio, airflow rate, and spark frequency on device performance. The author concluded, "The device did not appear promising as a thrust producing mechanism because of its low operating frequencies."

The experimental device consisted of three assemblies: air manifold, fuel injector, and detonation tube. The detonation tube was made from either steel or aluminum tubing, 6 ft in length with an internal diameter of 1 in. A propane/air mixture was injected at the closed end of the tube via a reverse flow injector to provide adequate mixing and turbulence. The spark electrode was located 15 in. downstream from the fuel injection plane. A reaction balance was used to measure the average thrust of the system; however, pressure measurements were not made.

Several observations point to the fact that the majority of the combustion process that occurred in the device was deflagrative and not detonative. First, the power output of the spark plug was less than 0.065 J, enough to initiate deflagration but orders of magnitude less than required to initiate detonation directly. Second, the aluminum detonation tube failed structurally after 2 min. of operation, and the steel tube began bending due to its own weight as a result of excessive heat transfer, at a point 8 in. from the exhaust end. It is believed the onset of detonation or transition from a deflagration occurred at this point.

In addition, it was noted that the detonation tube fired at a lower frequency than the spark generator. It was concluded from this that the relationship between spark electrode discharge and detonation wave formation was a function of the wave interaction occurring within the tube. Two thrust peaks were observed as a function of frequency, which indicates the device was operating in a resonant mode. Both of these observations also strongly support the fact that the device was operating in a primarily deflagrative pulse jet mode.

4). J. A. Nicholls et al.[14] (1966): The feasibility of a rocket motor that utilizes a detonation wave propagating in an annular combustor chamber was examined. The intent was to introduce propellants continuously through injectors and to expel the exhaust through an annular nozzle. It was concluded from the analyses that the rotating detonation wave engine concept was feasible and worth consideration in view of the possible performance advantages.

Experiments utilizing gaseous hydrogen/oxygen and gaseous methane/oxygen mixtures were conducted in a small-scale annular rocket motor. Methane was chosen because of its lower deflagration velocities, higher detonation pressure ratio, and better photographic characteristics. A breakable diaphragm extending across the annular chamber and a baffle plate extending outside the chamber were required to achieve a single detonation wave propagating in one tangential direction.

The fuel and oxidizer were kept separate in the manifold and injected through small discrete orifices, impinging in the combustor, to provide a total propellant mass flow of approximately 1 lb/s. A detonation wave was created by a transition from a deflagration, which was initiated by a flush-mounted spark plug with a spark energy of less than 30 mJ (0.03 J) obtained by discharging a 1-μF capacitor at 250 V.

A unidirectional detonation wave was obtained by using a frangible diaphragm closure across the combustion chamber adjacent to the spark plug. The diaphragm was strong enough to contain the initial deflagration, but weak enough to burst when struck by the initial detonation wave after one revolution.

Examination of the data also revealed that in all cases the pressure pulse transmitted through the diaphragm did not generate a detonation wave on the second rotation, but instead decayed. Much of the data was anomalous and seemed to indicate a steady-state Chapman-Jouguet detonation was not established. The desired continuously cycling detonation wave was not achieved.

5). L. H. Back et al.[17] (1982): A variation of the pulse detonation engine concept was originated at the Jet Propulsion Laboratory (JPL) in the early 1970s. The JPL concept involved intermittent detonation of solid high explosives, instead of gaseous mixtures, at the apex of a nozzle. When the nozzle is filled with a high-density gas, the detonation shock wave accelerates the ambient gas and causes it to discharge, along with the explosive gas, from the nozzle at high velocity.

The nozzle detonation device was conceived as a means of propulsion within the high-pressure atmospheres of solar system planets such as Jupiter. Conventional engine designs cannot achieve high performance in these planetary atmospheres as ambient pressures can be greater than 50 atm.

Experimental data were obtained for various nozzle geometries: long cone, short cone, straight, and firing plug. A small quantity of Delta sheet "C" explosive was detonated by a microdetonator in a firing-plug cavity within a nozzle expansion cone. Specific impulse data were calculated for a variety of ambient gas mixtures and pressures. Hydrogen/helium mixtures (used to simulate Jupiter's atmosphere), and pure helium, nitrogen, and carbon dioxide gases were tested at pressures ranging from about 1 to 70-80 bars.

6). D. Helman et al.[10] (1986): An exploratory experimental study of an airbreathing propulsion device based on detonative combustion of a gaseous mixture was conducted. Detonation within a chamber containing an ethylene/air mixture was initiated by a detonation tube containing an ethylene/oxygen mixture. Intermittent detonation was achieved within the naturally aspirated chamber at a frequency of 25 Hz.

Ethylene was used as the fuel, and air was the main oxidizer in the main detonation chamber. Detonation was initiated in this main chamber by a primary detonation tube that contained a limited volume of an ethylene/oxygen mixture. Two different primary detonation tubes with volumes of 2 and 4% of the total reaction chamber volume were tested. Primary detonation was initiated in these 10-and 20-cm-long tubes by an automobile spark ignition system. The spark plug was installed at the front end of the detonation tube, and the other end of the detonation tube was connected to the front end of the detonation chamber. High-response piezoelectric pressure transducers were used for pressure measurements inside the detonation tubes and chamber.

Ethylene and oxygen were injected intermittently, and the spark plug was fired as soon as the solenoid gas valves were closed. The maximum system operating frequency was dictated by the 25-Hz maximum cycling frequency of the solenoid valves. The fuel/oxygen ratio was varied between 25 and 35%, and the injection pressure, which controlled the flow rate, was varied from 9 to 30 psig.

Specific impulse was calculated from the pressure time history and the amount of fuel consumed. The maximum effective (i.e., time-averaged) pressure in the detonation chamber was 100 psig, which yielded a specific impulse between 1000 and 1400s. These performance figures are rather poor and may indicate that direct detonation initiation was not achieved.

7). A. Wortman et al.[5] (1992): The feasibility of generating transverse detonation waves in a high-speed channel flow was experimentally studied. A

4-ft-long detonation duct with a 1 x 12-in. cross section was operated in a blowdown mode using compressed air reservoirs. Liquid or gaseous propane was injected through injectors in the duct walls. A row of spark plugs in the duct wall was used to ignite the propane/air mixtures.

Simple experiments were performed indicating that pressure ratios of 10 to 12 were attainable. Difficulties with fuel injection and mixing required the tests to be operated with restricted airflow (Mach 0.5) and duct pressures barely above ambient. This resulted in pressure ratios of 5 to 6, and operating frequencies below 10 Hz. The most successful configuration was that in which liquid propane was injected through a nozzle transversely to the air stream, while an air jet was directed across it and in the direction of the spark plugs.

A high scatter in the data was caused by problems with the fuel injection system and the extreme difficulties with the mixing of propane and air. Injection of propane vapor proved impractical because, even with a relatively large valve, it was not possible to transfer sufficient quantities of fuel to the air stream during the injection period. Liquid injection proved extremely difficult because the propane vapor expanding from the nozzle reached Mach numbers of 2 and propelled the liquid droplets over very large distances from the injection point. In addition, the spark energies used were orders of magnitude less than required for direct initiation of detonation (approximately 200-300 kJ for propane/air).

Detonation was not achieved because the propane fuel stoichiometry, mixing, and initiation energy and power requirements were not satisfied.

B. Computational Modeling

Cambier and Adelman[6] and Eidelman et al.[18] have computationally analyzed the performance of PDEs. Several simplifying assumptions were made to study a complete pulse detonation engine concept. Phenomena such as inlet losses, detonation initiation, and fuel/air mixing were not modeled. These preliminary numerical studies did not address the issues critical to the design of an operational PDE device.

The application and utilization of detonation in propulsion systems incorporates many complex phenomena that have not been well understood. In addition, validation of the pulse detonation engine as a viable and efficient mode of propulsion has not been well established. The following is a summary of the key references.

Key PDE Numerical Simulations

1). J. L. Cambier and H. G. Adelman[6] (1988): This numerical study examined the cyclic charging and combustion rates of the pulsed detonation wave engine (PDWE). Ideal engine performance for a PDWE was predicted. The analytical modeling of a PDWE combustor used a computational fluid dynamic (CFD) flow solver with finite-rate hydrogen/air chemistry.

A one-dimensional analytical model of the combustor was used, and several combustor/nozzle shapes were analyzed. A tube 6 cm in diameter and 50 cm in length was used to model the combustor, and a 43-cm-long diverging nozzle was used to increase the exit diameter to 23 cm. The diverging design allowed the detonation/shock wave to be expelled from the nozzle, which created a reflected rarefaction wave that traveled back to the inlet area and assisted in recharging the combustor with fresh air and fuel. When the pressure at the inlet fell to 3.5 atm, the combustor was recharged rapidly with a combustible mixture. The entire engine cycle required 1.5 ms and resulted in a predicted cyclic firing rate of 667 Hz.

A stoichiometric mixture of hydrogen and air at a pressure of 3 atm filled the first 40 cm of the tube. A detonation wave was created by raising the temperature of the first 2 cm of the combustible mixture to 1500 K. Rapid combustion was initiated in this region, raising the gas temperature to 3000 K and the pressure to a value greater than 20 atm. Calculations showed the fresh mixture moving at speeds up to 350 m/s, while the burned gases reached a speed of 1800 m/s and a Mach number of 2.5 at the nozzle exit.

The numerical results were scaled and compared with the 1957 Nicholls experiment, which produced a cyclic combustion rate of 25 Hz, a thrust of 3.2 lb, and a specific impulse of 2100 s. Scaling the area and cyclic rate still showed the calculation produced 1.87 times more thrust than the experiment. Cambier et al. attributed this improved performance to the fact that the simulation represented the maximum ideal performance and did not realistically model losses caused by the inlet, fuel/air injection, and mixing. It may also have resulted because Nicholls did not achieve direct detonation in the 1957 experiment.

Overall performance was calculated by integrating the instantaneous thrust and the fuel flow rates. This simplified simulation predicted that a PDWE could produce very high specific impulses on the order of 6500 s. The specific impulse of the PDWE was predicted to be 800 s for operation in a rocket mode with an onboard oxidizer.

2). S. Eidelman et al.[18] (1990): Unsteady simulations of a specific airbreathing pulsed detonation engine concept were made. The PDE concept utilized a cylindrical detonation chamber with a constant 8-cm diameter. Two engine lengths of 8 and 16 cm were studied to ascertain the impact on thrust production.

An air inlet at the beginning of the combustor remained open during the detonation cycle. As a result, the efficiency of the PDE design was dependent on the interaction of the surrounding flow with the internal flow dynamics.

A stoichiometric ethylene/air mixture was used. The assumption of inviscid flow was also made to minimize the computational effort required to simulate the PDE flow phenomena. The computational tool solved the Euler equations for an unstructured grid and allowed dynamic grid refinement.

Detonation initiation was not modeled; rather, it was assumed that a triggering device initiated detonation at the start of each cycle. Steady subsonic flow conditions were established in the detonation chamber, and planar detonation was initiated at the rear, or exit, of the chamber. The burned mixture was ejected from the open air inlets in the combustor as well as from the exit at the rear of the combustor. Air recharged the detonation chamber cause by inlet flow and self-aspiration (i.e., reverse flow) from the aft end of the chamber resulting from strong over expansion.

Thrust was predicted to scale linearly with detonation chamber volume and operating frequency. Specific fuel consumption was computed from the simulations to be approximately 1 lb/(lb-h) as compared with 1.6-1.8 lb/(lb-h) for present-day small turbojets. The predicted operating range of this PDE concept was Mach 0.2 to 3.

3). T. R. A. Bussing and G. Pappas[19] (1994): This paper was written by the authors as an introduction to the pulse detonation engine concept. The basic detonation physics and intermittent cyclical operation of a simple PDE are described. Unsteady one-dimensional CFD analyses were performed for stoichiometric hydrogen/air and hydrogen/oxygen mixtures. A high-pressure temperature region was used to initiate detonation directly at the closed end of a cylindrical chamber, 2 cm in diameter and 20 cm in length.

4). E. D. Lynch et al.[7] (1994): CFD analyses of the pulse detonation engine concept and of detonation initiation were performed. The baseline geometry for the PDWE analyses was similar to that analyzed by Eidelman et al.[9]: an axisymmetric, 8-cm-long, 8-cm-diam detonation tube with a simple, straight annular inlet. The inlet was varied in length from 1 to 12 cm, and cases with freestream Mach numbers of 0.8 and 2 were analyzed. The Rockwell USA code solved the Reynolds-averaged Navier-Stokes equations in the axisymmetric PDE calculations. A seven-step, six-reaction hydrogen/air chemistry model was used.

Detonation was initiated at the aft, or open, end of the tube, which contained a stoichiometric hydrogen/air mixture. A forward-propagating detonation wave was generated and reflected off the thrust surface at the front end of the chamber. Incident detonation shock pressures of 23-30 atm, which are higher than predicted by the Chapman-Jouguet model, were created. These higher pressures were attributed to the use of chemical kinetics, which allowed the precursor shock component of the detonation wave to be captured. An extremely high, but short-lived, pressure of over 100 atm was created as the detonation wave reflected off the thrust surface.

Because of the open inlet design, the detonations were expelled from the inlet entrance and the detonation tube exit, reducing the pressure in the tube and allowing aspiration of fresh air. The detonation reflection and expansion processes were found to be relatively independent of the particular PDWE design. However, the time required to scavenge combustion products was noticeably affected by the inlet length and design.

In addition to the PDWE calculations, a series of analyses were made of detonation wave initiation in a closed-end tube. The data was compared with the experimental results of Carrier et al.[20] for detonation initiation of acetylene/oxygen mixtures using an Ar F laser. One-dimensional, inviscid analyses with mechanistic finite-rate chemistry were completed for stoichiometric C_2H_2-O_2 mixtures at initial conditions approximately matched to the 400 and 800 Torr pressures used in the experiments. The peak pressure ratios for the computation and experiment were nearly identical in the 1-atm initial pressure case whereas the computed pressure rise was 8-12% higher than experimentally measured for the 0.5-atm case. An ignition delay time of 2 μs was computed and was shorter than the 4 μs seen in the experiments.

2. Pulse Jet Research Review

During the late 1940s and early 1950s, extensive research was conducted in the area of pulse jet propulsion. Several hundred technical papers were written from the time of the German V-1 Buzz Bomb era to the U.S. Project Squid program.

Pulse jets are intermittent combustion devices, as are PDEs; however, pulse jet combustion is deflagrative, not detonative, in nature. Pulse jets may be considered the precursor to the concept of pulse detonation engines. The reader is referred to the many research papers listed in the Reference and Bibliography sections for an accounting of pulse jet accomplishments.

IX. Addendum

Considerable progress has been made toward the development of a practical Pulse Detonation Engine (PDE). These results are summarized in Refs. 21–24.

Reference 21 describes a novel Rotary Valve Multiple Combustor Pulse Detonation Engine (RVMPDE) concept that can operate at higher cycle frequencies and higher flight Mach numbers, and will have lower inlet losses than previous PDE concepts. It consists of several PDE combustors, or detonation chambers, coupled to an air inlet and fuel source via a rotary valve. The valve serves to isolate the steady operation of the air inlets and fuel systems from the unsteady nature of the detonation process. A rotor valve is preferable over reed valves due to its lower pressure loss and higher mechanical load capability. The rotary valve allows the fueling of some of the PDE combustion chambers while detonation occurs in other PDE combustor chambers. In this way, the fueling system can operate in a steady state mode in conjunction with an unsteady combustion process.

Reference 22 provides an overview of a PDE performance model. The model uses a consistent force accounting methodology to determine the implications of air-breathing PDE operation and design strategies. The model includes inlet, mixer, combustor, and nozzle contributions to PDE performance attributes. Techniques for modeling individual engine components are presented, along with an analysis of the implications of utilizing multiple- or single-chamber designs for various on-design flight conditions.

Reference 23 describes a set of shock tube experiments performed to support the design of a prototype hydrogen-fueled PDE. Quiescent mixtures of both hydrogen/oxygen and hydrogen/air were detonated at ambient initial pressures and temperatures in specially designed shock tubes of approximately 2- and 6-in. internal diameter. Measured parameters include wall pressure histories at various points along the tubes, thrust force histories, detonation wave velocities, deflagration to detonation transition (DDT) distances, and detonation transmission (DT) distances. In addition, the effect of mixture stoichiometry and DDT augmentation devices are studied. Quasi-one-dimensional, unsteady, chemically reacting numerical simulations of selected experiments were performed to aid in the interpretation and validation of the experimental results. Specific data derived from these experiments have enabled the design of a "boiler plate" prototype PDE.

Finally, Ref. 24 describes a series of tests designed to characterize the detonation characteristics of fuels for the development of a hydrocarbon PDE. Mixtures of propane/air and propane/oxygen were detonated in 2- and 6-in. diameter tubes. Mixtures of propane/oxygen were ignited using a spark ignitor. Mixtures of propane/air were ignited using the detonation wave created in a propane/oxygen detonation. Shchelkin spirals were inserted in the tubes to increase detonability. Parameters measured include velocity, pressure, and the distance required for a deflagration to DDT. The parameters were measured as a function of mixture stoichiometry and tube dianeter. The velocities and pressures are compared with analytical results where possible.

References

[1]Foa, J. V., "The R.P.I. Wave Engine," Project Squid Conference on Wave Engines and Pulse Jets, Rept. AD-A952624, Princeton University, Princeton, NJ, Nov. 1954.

[2]Foa, J. V., Elements of Flight Propulsion, Wiley, New York, 1959.

[3]Williams, F. A., Combustion Theory: The Fundamental Theory of Chemically Reacting Flow Systems, Addison-Wesley, Reading, MA, 1965.

[4]Tarver, C. M., "Chemical Energy Release in One-Dimensional Detonation Waves in Gaseous Explosives," Combustion and Flame, Vol. 46, 1982, pp. 111-133.

[5]Wortman, A., Othmer, P., and Rostafinski, W., "Detonation Duct Gas Generator Demonstration Program," AIAA Paper 92 3174, July 1992.

[6]Cambier, J. L., and Adelman, H .G., "Preliminary Numerical Simulations of a Pulsed Detonation Wave Engine," AIAA Paper 88-2960, July 1988.

[7]Lynch, E. D., Edelman, R., and Palaniswamy, S. "Computational Fluid Dynamic Analysis of the Pulse Detonation Engine Concept," AIAA Paper 94-0264, Jan. 1994.

[8]Loth, E., Loth, J., and Loth, F., "High Efficiency Detonation Internal Combustion Engine," AIAA Paper 92-3171, July 1992.

[9] Eidelman, S., Grossmann, W., and Lottati, I., "A Review of Propulsion Applications of the Pulsed Detonation Engine Concept," AIAA Paper 89-2446, July 1989.

[10]Helman, D., Shreeve, R. P., and Eidelman, S., "Detonation Pulse Engine," AIAA Paper 86-1683, June 1986.

[11]Bussing, T. R. A., and Pappas, G., "A Novel Very-Low-Cost Airbreathing Propulsion System," SBIR Phase I Final Report, USAF Contract F-33615-92-C-2270, Dec. 1992.

[12]Bussing, T. R. A., and Pappas, G., "A Novel Hydrogen Fueled Propulsion System," SBIR Phase I Final Report, NASA Contract NAS1-19875, Aug. 1993.

[13]Nicholls, J. A., Wilkinson, H. R., and Morrison, R. B., "Intermittent Detonation as a Thrust-Producing Mechanism," Jet Propulsion, May 1957, pp. 534-541.

[14]Nicholls, J. A., Cullen, R. E., and Ragland, K. W., "Feasibility Studies of a Rotating Detonation Wave Rocket Motor," Journal of Spacecraft, Vol. 3, June 1966.

[15]Krzycki, L. J., "Performance Characteristics on an Intermittent-Detonation Device," NAVWEPS Rep. 7655, AD-284312, U.S. Naval Ordnance Test Station, China Lake, CA, June 1962.

[16]Hoffmann, H., "Reaction-Propulsion Produced by Intermittent Detonative Combustion," German Research Institute for Gliding, Rept. ATI-52365, Aug. 1940.

[17]Back, L. H., Dowler, W. L., and Varsi, G., "Detonation Propulsion Experiments and Theory," AIAA Paper-82-1115, June 1982.

[18]Eidelman, S., Grossmann, W., and Lottati, I., "Air-Breathing Pulsed Detonation Concept: A Numerical Study," AIAA Paper 90-2420, July 1990.

[19]Bussing, T. R. A., and Pappas, G., "Pulse Detonation Engine Performance," AIAA Paper 94-0263, Jan. 1994.

[20]Carrier, G. F., Fendell, F. E., and Chou, M. S., "Laser-Initiated Conical Detonation Wave for Supersonic Combustion. III," AIAA Paper 92-3247, July 1992.

[21] Bussing, T. R. A., "A Rotary Valve Multiple Pulse Detonation Engine (RVMPDE)," AIAA Paper 95-2577, July 1995.

[22]Hinkey, J. B., Bussing, T. R. A., and Kaye, L., "Shock Tube Experiments for the Development of a Hydrogen-Fueled Pulse Detonation Engine," AIAA Paper 95-2578, July 1995.

[23]Bratkovich, T. E., and Bussing, T. R. A., "A Pulse Detonation Engine Performance Model, AIAA Paper 95-3154, July 1995.

[24]Ting, J. M., Bussing, T. R. A., and Hinkey, J. B., "Experimental Characteristics of the Detonation Properties of Hydrocarbon Fuels for the Development of a Pulse Detonation Engine," AIAA Paper 95-3155, July 1995.

Bibliography

Abouseif, G. E., and Toong, T. Y., "On Direct Initiation of Gaseous Detonations," Combustion and Flame, Vol. 45, Jan. 1982, pp. 39-46.

Adamson, T. C., Jr., and Olsson, G. R., "Performance Analysis of a Rotating Detonation Wave Rocket Engine," Acta Astronautica, Vol. 13, 1967, pp. 405-415.

Bach, G. G., Knystautas, R., and Lee, J. H., "Direct Initiation of Spherical Detonations in Gaseous Explosives," 12th International Symposium on Combustion, July 1968.

Back, L. H., "Application of Blast Wave Theory to Explosive Propulsion," Acta Astronautica, Vol. 2, 1975, pp. 391-407.

Baklanov, D. I., Shimevin, D. G., Kiselev, Yu. N., Mironov, E. A., and Popov, V. A., "Certain Technical Aspects of Using the Detonation Combustion Mode," Fizika Goreniya i Vzryva (Combustion, Explosion and Shock Waves), Vol. 12, Jan.-Feb. 1976, pp. 47-52.

Beck, E. J., Jr., "Heat Transfer at High Pulsating Frequencies in a Pulse-Jet Engine," Naval Civil Engineering Laboratories, Rept. AD 835 638, June 1968.

Beeson, H. D., et. al., "Detonability of Hydrocarbon Fuels in Air," Dynamics of Detonations and Explosions: Detonations, edited by A. L. Kuhl, J.-C. Leyer, A. A. Borisov, and W. A. Sirignano, Vol. 133, Progress in Astronautics and Aeronautics, AIAA, Washington, DC, 1989.

Benedick, W. B., Guirao, C. M., Knystautas, R., and Lee, J. H., "Critical Charge for the Direct Initiation of Detonation in Gaseous Fuel-Air Mixtures," Dynamics of Explosions, edited by J. R. Bowen, J.-C. Leyer, and R. I. Soloukhin, Vol. 106, Progress in Astronautics and Aeronautics, AIAA, Washington, DC, 1985.

Berchtold, M., and Gardiner, F. J., "Introduction and Status Review of the Aerodynamic Wave Machine," Project Squid Conference on Wave Engines and Pulse Jets, Rept. AD-A952624, Princeton University, Princeton, NJ, Nov. 1954.

Billig, F. S., "Propulsion Systems from Takeoff to High-Speed Flight," High Speed Flight Propulsion Systems, edited by S. N. B. Murthy and E. T. Curran, Vol. 137, Progress in Astronautics and Aeronautics, AIAA, Washington, DC, 1991.

Bitondo, D. et al., "Preliminary Performance Studies of the Multi-Jet," Aerophysics Development Corp., Rep. 2000-1-R1, Jan. 1953.

Bitondo, D., and Perper, D., "Non-Stationary Combustion Studies," AFOSR TR-58-144, AD-204734, Curtiss-Wright Corp., Aug. 1958.

Bjerketvedt, D., Sonju, O. K., and Moen, I. O., "The Influence of Experimental Condition on the Reinitiation of Detonation Across an Inert Region," Dynamics Of Explosions, edited by J. R. Bowen, J.-C. Leyer, and R. I. Soloukhin, Vol. 106, Progress in Astronautics And Aeronautics, AIAA, Washington, DC, 1986, pp. 3-17.

Bollay, W., and Bitondo, D., "Preliminary Progress on the Experimental Performance of the Multi-Jet Engine," Project Squid Conference on Wave Engines and Pulse Jets, Rept. AD-A952624, Princeton University, Princeton, NJ, Nov. 1954.

Boni, A. A., Wilson, C. W., Chapman, M., and Cook, J. L., "A Study of Detonation in Methane/Air Clouds," Acta Astronautica, Vol. 5, 1978, pp. 1153-1169.

Borisov, A. A., Saneev, E. L., Khomik, S. V., Fomin, I. B., Kasainov, B. A., and Veyssiere, B., "On Detonation of Aluminum Suspensions in Air and Oxygen," 23rd International Symposium on Combustion, July 1990.

Briggs, A. T., "Multi-Reed Valve Combustion Chamber Analysis of Test Data of Report 1097," Lycoming (The Aviation Corporation), Rept. AD 38670, May 1947.

Bull, D. C., Elsworth, J. E., and Hooper, G., "Initiation of Spherical Detonation in Hydrocarbon/Air Mixtures," Acta Astronautica, Vol. 5, 1978, pp. 997-1008.

Burgner, G., "Feasibility of Pulsejets and Intermittent Combustion Devices as Modern Propulsion Power Plants," Naval Weapons Center, Rept. AD 845 634, Oct. 1968.

Bussing, T. R. A., Pappas, G., and Kaye, L., "Pulse Detonation Engine Preliminary Design Considerations," AIAA Paper 94-3220, June 1994.

Bussing, T. R. A., "An Analysis of a Generic Air Breathing Hypersonic Vehicle," Ninth International Symposium on Air Breathing Engines, Athens, Greece, Sept. 1989.

Bussing, T. R. A., and Lidstone, G. L., "An Improved Computational Model for a Scramjet Propulsion System," AIAA Paper 87-2078, June 1987.

Bussing, T. R. A., "A Mathematical Model of a Solid Fueled Hybrid Gas Generator Low-Volume Ramjet," ERNO Raumfahrttechnik GmbH, Sept. 1979.

Bussing, T. R. A., and Eberhardt, S., "Chemistry Associated with Hypersonic Vehicles," AIAA Paper 87-1292, June 1987.

Bussing, T. R. A., and Murman, E. M., "A Finite Volume Method for the Calculation of Compressible Chemically Reacting Flows," AIAA Journal, Vol. 26, Sept. 1988, P. 1070-1079.

Bykovskii, F. A., and Mitrofanov, V. V., "Detonation Combustion of a Gas Mixture in a Cylindrical Chamber," Fizika Goreniya i Vzryva (Combustion, Explosion and Shock Waves), Vol. 16, Sept.-Oct. 1980, pp. 107-117.

Cambier, J. L., Adelman, H. G., and Menees, G. P., "Numerical Simulations of a Pulsed Detonation Wave Augmentation Device," AIAA Paper 93-1985, June 1993.

Chan, C.K., Lau, D., Thibault, P. A., and Penrose, J. D., "Ignition and Detonation Initiation by Shock Focusing," American Institute of Physics, 1990.

Crowe, R. K., "Pulsating Combustion Device Miniaturization," M.S. Thesis, Naval Postgraduate School, Monterey, CA, Rept. AD-A-035 707, Dec. 1976.

Dahab, O., Kapuscinski, M., and Wolanski, P., "Influence of Dust Parameters on Detonation Velocity, Structure, and Limits," Dynamics of Detonations and Explosion: Detonations, edited by A. L. Kuhl, J.-C. Leyer, A. A. Borisov, and W. A. Sirignano, Vol. 133, Progress in Astronautics and Aeronautics, AIAA, Washington, DC, 1991.

Dynin, E. A., "One Dimensional Stability of Stationary Compression and Combustion Waves," Fizika Goreniya i Vzryva (Combustion, Explosion and Shock Waves), Vol. 17, No. 4, July-Aug. 1981, pp. 85-91.

Eick, W., "Shutterless Pulso-Ramjet Engine," Foreign Technology Division, Rept. AD-763 173, June 1973 (German translation).

Eidelman, S., and Grossmann, W., "Pulsed Detonation Engine Experimental and Theoretical Review," AIAA Paper ,92-3168, July 1992.

Eidelman, S., Grossmann, W., and Lottati, I., "Computational Analysis of Pulsed Detonation Engines and Applications," AIAA Paper 90-0460, Jan. 1990.

Fendell, F., Mitchell, J., McGregor, R., Magiawala, K., and Sheffield, M. "Laser-Initiated Conical Detonation Wave for Supersonic Combustion. II," AIAA Paper 92-0088, Jan. 1992.

Glassman, I., Williams, F. A., and Antaki, P., "A Physical and Chemical Interpretation of Boron Particle Combustion," Proceedings of Twentieth International Symposium on Combustion, 1984, pp. 2057-2064.

Goldman, Y., and Timnat, Y. M., "A Critical Review of Pulsating Combustion," Technion Israel Institute of Technology, Rept. AD-B 062 657, Nov. 1981.

Good, F. A., "Design and Testing Development of the Weber Valveless Pulse Jet Engine Series," Project Squid Conference on Wave Engines and Pulse Jets, Rept. AD-A952624, Princeton University, Princeton, NJ, Nov. 1954.

Grimm, B. F., "Performance Study of Giannini Power Plant Package on a Pursuit Aircraft," Rept. ATI-56735, March 1948.

Gudkovich, V. N., Granovski, E. A., and Shtessel, E. A., "Ignition of Rich Propane-Oxygen Mixtures," Fizika Goreniya i Vzryva (Combustion, Explosion and Shock Waves), Vol. 16, Nov.-Dec. 1980, pp. 93-97.

Hertzberg, M., and Cashdollar, K. L., "Introduction to Dust Explosions," Industrial Dust Explosions, American Society for Testing and Materials STP 958, Philadelphia, 1987, pp. 5-32.

Hertzberg, M., Zlochower, I. A., and Cashdollar, K. L., "Metal Dust Combustion: Explosion Limits, Pressures, and Temperatures," Proceeding of Twenty-fourth International Symposium on Combustion, Sydney, Australia, July 1992.

Hertzberg, M., Zlochower, I. A., and Cashdollar, K. L., "Explosibility of Metal Dusts," Combustion Science and Technology, Vol. 75, 1991, pp. 161-165.

Hertzberg, M., Conti, R. S., and Cashdollar, K. L., "Spark Ignition Energies for Dust-Air Mixtures: Temperature and Concentration Dependencies," Proceeding of Twentieth International Symposium on Combustion, 1984, pp. 1681-1690.

Hertzberg, A., and Russo, R., "The Investigation of a Resonant Wave Engine," Project Squid Conference on Wave Engines and Pulse Jets, Rept. AD-A952624, Princeton University, Princeton, NJ, Nov. 1954.

Jaeger, D. H., "Monthly Progress Report for November 1948 on Ducted Pulse Engine," Marquardt Aircraft Co., Rept. ATI-50175, Nov. 1948.

Jost, W., Explosion and Combustion Processes in Gases, McGraw-Hill, New York, 1946.

Kailasanath, K., and Oran, E. S., "Power-Energy Relations for the Direct Initiation of Gaseous Detonations," Dynamics of Shock Waves, Explosions, and Detonations, edited by J. R. Bowen, N. Manson, A. K. Oppenheim, and R. I. Soloukhin, Vol. 94, Progress in Astronautics and Aeronautics, AIAA, New York, 1984.

Kamm, W. I. E., "Proposal for Initial Construction of an Experimental High Pressure Pulse Jet," Rept. ATI-71627, Continental Aviation and Engineering Corp., Detroit, MI, Nov. 1946 (German translation).

Karo, A. M., Hardy, J. R., and Walker, F. E., "Theoretical Studies of Shock-Initiated Detonations," Acta Astronautica, Vol. 5, 1978, pp. 1041-1050.

Kauffman, C. W., Srinath, S. R., Tezok, F. I., Nicholls, J. A., and Sichel, M., "Turbulent and Accelerating Dust Flames," Proceedings of Twentieth International Symposium on Combustion, 1984, pp. 1701-1708.

King, M. K., "Ignition and Combustion of Boron Particles and Clouds," AIAA Paper 82-4174, March 1982.

King, M. K., "Predictions of Laminar Flame Speeds in Boron-Oxygen-Nitrogen Dust Clouds," Proceedings of Fifteenth International Symposium on Combustion, 1974, pp. 467-477.

Klose, A., and Whitman, N. D., "Experimental Jet Engine," Rept. ATI-11634, G.M. Giannini and Co., July 1945.

Knystautas, R., Guirao, C., Lee, J. H., and Sulmistras, A., "Measurements of Cell Size in Hydrocarbon-Air Mixtures and Predictions of Critical Tube Diameter, Critical Initiation Energy, and Detonability Limits," Dynamics of Shock Waves, Explosions, and Detonations, edited by J. R. Bowen, N. Manson, A. K. Oppenheim, and R. I. Soloukhin, Vol. 94, Progress in Astronautics and Aeronautics, AIAA, New York, 1984.

Korovin, L. N., Losev, S. A., and Smekhov, G. D., "Combustion of Natural Gas in a Commercial Detonation Reactor," Fizika Goreniya i Vzryva (Combustion, Explosion and Shock Waves), Vol. 17, May-June 1981, pp. 68-71.

Laderman, A. J., Urtiew, P. A., and Oppenheim, A. K., "Effect of Ignition Geometry on Initial Flame Acceleration in a Spark Ignited Explosive Gas," Combustion and Flame, Vol. 6, Dec. 1962, pp. 325-335.

Laderman, A. J., "Detonability of Hydrogen-Oxygen Mixtures in Large Vessels at Low Initial Pressures," AIAA Journal, Vol. 4, Oct. 1966, pp. 1784-1790.

Laderman, A. J., Urtiew, P. A., and Oppenheim, A. K., "Effect of Ignition Geometry on Initial Flame Acceleration in a Spark Ignited Explosive Gas," Combustion And Flame, Vol. 6, Dec. 1962, pp. 325-335.

Lancaster, O. E., ed., Jet Propulsion Engines, Princeton University Press, Princeton, NJ, 1959.

Lawrence, H. R., and Weatherston, R., "The Capabilities of the Shrouded Wave Engines," Project Squid Conference on Wave Engines and Pulse Jets, Rept. AD-A952624, Princeton University, Princeton, NJ, Nov. 1954.

Lee, J. H., "On the Transition from Deflagration to Detonation," Dynamics of Explosions, edited by J. R. Bowen, J.-C. Leyer, and R. I. Soloukhin, Vol. 106, Progress in Astronautics and Aeronautics, AIAA, New York, 1986, pp. 3-17.

Lee, J. H. S., "Initiation of Gaseous Detonations," Annual Review of Physical Chemistry, Vol. 28, 1977, pp. 75-104.

Lee, J. H., Knystautas, R., and Yoshikawa, N., "Photochemical Initiation of Gaseous Detonations," Acta Astronautica, Vol. 5, 1978, pp. 971-982.

Lee, J. H., and Knystautas, R., "Laser Spark Ignition of Chemically Reactive Gases," AIAA Journal, Vol. 7, Feb. 1968, pp. 312-317; also AIAA Paper-68-146, 1968.

Leingang, J. L., "Ideal Pulsejet with Constant Volume Combustion," AFAPLM, Feb. 1976.

Libis, N., and Goldman, Y., "Study of a Pulse Ramjet Based on Twin Valveless Combustors Coupled to Operate in Antiphase," ISABE Rept. 93-7038, 1993.

Lidstone, G. L., and Bussing, T. R. A., "Fuel Injection Analysis of Scramjet Combustors", NASP CR 1115, March 1991.

Lobanov, D. P., Fonbershtein, E. G., Ekomasov, S. P., and Tregub, A. M., "Detonation of Gasoline-Air Mixtures," Fizika Goreniya i Vzryva (Combustion, Explosion and Shock Waves), Vol. 12, May-June 1976, pp. 446-450.

Lockwood, R. M., Sargent, E. R., and Beckett, J. E., "Pulse Reactor Thrust Augmented Intermittent Jet Lift-Propulsion System, Final Report," Advanced Research (Hiller Aircraft Corporation), Rept. AD 825 218, Feb. 1960.

Lockwood, R. M., "Ducted Pulse Jets for High Subsonic Speed, High Altitude Propulsion," Project Squid Conference on Wave Engines and Pulse Jets, Rept. AD-A952624, Nov. 1954.

Logan, J. G., Jr., "Valveless Intermittent Engines for Subsonic Propulsion Characteristics," Project Squid Conference on Wave Engines and Pulse Jets, Rept. AD-A952624, Nov. 1954.

Logan, J. G., "Summary Report on Valveless Pulsejet Investigation," Project Squid Technical Memorandum, A-952 972, Cornell Aeronautical Laboratory, Oct. 1951.

Logan, J .G., "Suggested Forms of Air Duct Motors Utilizing Intermittent Combustion-Part III," Rept. ATI-61716, March 1948.

Loth, J.L., Montgomery, L., and Chandran, R., "Pulse Jet One-Way Valve Performance," AIAA Paper 92-3169, July 1992.

Manildi, J. F., "Experimental Jet Engine," Rept. ATI-11643, July 1945.

McJones, R. W., and Lockwood, R. M., "Experimental Investigation of Intermittent Jet Propulsion Cycles Using Single-Explosion Techniques," Project Squid Conference on Wave Engines and Pulse Jets, Rept. AD-A952624, Princeton University, Princeton, NJ, Nov. 1954.

Milak, G. C., "Performance of the Four-Inch Valve Pulsejet Engine Under Velocity," Project Squid Conference on Wave Engines and Pulse Jets, Rept. AD-A952624, Princeton University, Princeton, NJ, Nov. 1954.

Morrison, R. B., "A Shock Tube Investigation of Detonative Combustion," Engineering Research Institute, University of Michigan, Ann Arbor, 1955.

Murray, S. B., and Lee, J. H., "The Influence of Physical Boundaries on Gaseous Detonation Waves," Dynamics of Explosions, edited by J. R. Bowen, J.-C. Leyer, and R. I. Soloukhin, Vol. 106, Progress in Astronautics and Aeronautics, AIAA, New York, 1986, pp. 3-17.

Natan, B., and Gany, A., "Ignition and Combustion Characteristics of Individual Boron Particles in the Flowfield of a Solid Fuel Ramjet," AIAA Paper 87-2034, July 1987.

Nikolaiev, Yu. A., "Theory of Detonation in Wide Tubes," Fizika Goreniya i Vzryva (Combustion, Explosion and Shock Waves), Vol. 15, May-June 1979, pp. 142-149.

O'Brien, J. G., "The Pulsejet Engine-A Review of Its Development Potential," Rept. AD-787 439, Naval Postgraduate School, Monterey, CA, June 1974.

Ostrander, M. J., Hyde, J. C., Young, M. F., Kissinger, R. D., and Pratt, D. T., "Standing Oblique Detonation Wave Engine Performance," AIAA Paper 87-2002, June 1987.

Pangburn, D. F., "A Comparison of Combustion Modes In Intermittent Jet Engines," Repts. TR-AE5504, and AD-79389, Rensselear Polytechnic Institute, Dec. 1955.

Persechino, M. A., "Preliminary Experimental Investigation of a Vertical Lift Valveless Pulsejet Engine Device," Project Squid Conference on Wave Engines and Pulse Jets, Rept. AD-A952624, Princeton University, Princeton, NJ, Nov. 1954.

Plaster, M., McClenagan, R. D., Benz, F. J., Shepherd, J. E., and Lee, J. H. S., "Detonation of Cryogenic Gaseous Hydrogen-Oxygen Mixtures," Dynamics of Detonations and Explosions: Detonations, edited by A. L. Kuhl, J.-C. Leyer, A. A. Borisov, and W. A. Sirignano, Vol. 133, Progress in Astronautics and Aeronautics, AIAA, Washington, DC, 1991.

Polevichek, Ye. P., "Simplified Theoretical Model of the Process of Combustion in the Combustion Chamber of a Pulse-Jet Engine," Air Force Foreign Technology Division, Rept. AD-B-046 572, March 1980 (translation of Russian 1977 Paper).

Polevichek, Ye. P., "Mathematical Model of the Process of the Filling of the Combustion Chamber of a Pulse Jet," Air Force Foreign Technology Division, Rept. AD-B-047141, Oct. 1979 (translation of Russian 1977 Paper).

Porter, C. D., "Cornell Valveless Pulsejet Engine Tests and Further Developments," Naval Research Laboratory, Rept. AD 17602, Aug. 1953.

Porter, C. D., "Development of a Valveless Pulsejet with High CT and Consideration of the Multi-Jet Engine," Project Squid Conference on Wave Engines and Pulse Jets, Rept. AD-A952624, Princeton University, Princeton, NJ, Nov. 1954.

Pratt, D. T., Humphrey, J. W., and Glenn, D. E., "Morphology of a Standing Oblique Detonation Wave," AIAA Paper 87-1785, June 1987.

Pusch, W., and Wagner, H. G., "Investigation of the Dependence of the Limits of Detonability on the Tube Diameter," Combustion and Flame, Vol. 6, Sept. 1962, pp. 157-162.

Rudinger, G., "On the Performance Analysis of the Ducted Pulsejet," Project Squid Technical Memorandum, ATI-122 295, Cornell Aeronautical Laboratory, Oct. 1951.

Sanders, J. C., "A Preliminary Evaluation of the Explosion Jet Propulsion Engine, NACA Memorandum to Navy, ATI-20974, Aug. 1944.

Science Applications International Corp., "Propulsion Applications of the Pulsed Detonation Engine," Final Report # , Sept. 1990.

Shchelkin, K. I., and Troshin, Ya. K., Gasdynamics of Combustion, Translation of "Gazodinamika goreniya," NASA TT F-231, Moscow, 1963.

Shepherd, J. E., "Chemical Kinetics of Hydrogen-Air-Diluent Detonations," Dynamics of Explosions, edited by J. R. Bowen, J.-C. Leyer, and R. I. Soloukhin, Vol. 106, Progress in Astronautics and Aeronautics, AIAA, New York, 1986, pp. 3-17.

Staiger, K., and Kamm, W. I. E., "Development of Intermittent Jets for Aircraft," USAF Air Materiel Command, ATI-6916, Nov. 1947.

Torda, P., "Approximate Theory of Compressible Air Inflow Through Reed Valves for Pulse Jet Engines," Project Squid Technical Report ATI-59330, Sept. 1948.

Totten, J. K., "Final Summary Technical Report on the Calendar Year 1963 Ramjet Technology Program," Rept. AD 351 398, Marquardt Corp., June 1964.

Tulis, A. J., and Selman, J. R., "Unconfined Aluminum Particle Two-Phase Detonation in Air," Dynamics of Shock Waves, Explosions, and Detonations, edited by J. R. Bowen, N. Manson, A. K. Oppenheim, and R. I. Soloukhin, Vol. 94, Progress in Astronautics and Aeronautics, AIAA, New York, 1984.

Ul'yanitskii, V. Yu., "Galloping Mode in Gas Detonations," Fizika Goreniya i Vzryva (Combustion, Explosion and Shock Waves), Vol. 17, Jan.-Feb. 1981, pp. 118-124.

Varsi, G., Back, L. H., and Kim, K., "Blast Wave in a Nozzle for Propulsive Applications," Acta Astronautica, Vol. 13, 1967, pp. 405-415.

Vasilev, A. A., Nikolaev, A., and Ul'yanitskii, V. Yu., "Effect of Initial Temperature on Gas Detonation Parameters," Fizika Goreniya i Vzryva (Combustion, Explosion and Shock Waves), Vol. 15, Nov.-Dec. 1979, pp. 149-152.

Vasilev, A. A., "Analysis of Cell Parameters of Multifront Gas Detonation," Fizika Goreniya i Vzryva (Combustion, Explosion and Shock Waves), Vol. 13, May-June 1977, pp. 404-408.

Vasiliev, A. A., and Nikolaev, Yu., "Closed Theoretical Model of a Detonation Cell," Acta Astronautica, Vol. 5, 1978, pp. 983-996.

Velazquez, J. L., "Design Compatibility of Tip-Mounted Pulse-Jet Engines and Helicopter Rotors," Project Squid Conference on Wave Engines and Pulse Jets, Rept. AD-A952624, Princeton University, Princeton, NJ, Nov. 1954.

Veyssiere, B., "Structure of the Detonations in Gaseous Mixtures Containing Aluminum Particles in Suspension," Dynamics of Explosions, edited by J. R. Bowen, J.-C. Leyer, and R. I. Soloukhin, Vol. 106, Progress in Astronautics and Aeronautics, AIAA, New York, 1986.

Wilson, C. W., "Pulsed Laser Initiated Chemical Explosions," AIAA Paper 80-1382, July 1980.

Wortman, A., "Detonation Wave Augmentation of Gas Turbines," AIAA Paper 84-1266, June 1984.

Yang, C. H., "Theory of Ignition and Auto-Ignition," Combustion and Flame, Vol. 6, Dec. 1962, pp. 215- 225.

Zel'dovich, I. B., Kogarko, S. M., and Simonov, M. N., "An Experimental Investigation of Spherical Detonation of Gases," Soviet Physics - Technical Physics, Vol. 1, Aug. 1957, pp. 1689-1713.

Analysis of the Pulse Detonation Wave Engine

E. D. Lynch* and R. B. Edelman†
Rockwell International, Canoga Park, California 91309-7922

I. Introduction

The pulse detonation wave engine (PDWE) is an intermittent energy conversion device that takes advantage of the higher availability of energy produced by constant volume combustion relative to systems based on the conventional constant pressure combustion process. The basis for this advantage derives from the reduction of energy absorbed by the working fluid in the form of pressure–volume energy. Thus, detonative combustion wherein the volume decreases is actually more efficient thermodynamically than both constant volume combustion and "pulse," or deflagration, combustion based systems. The operation of detonative combustion in a practical engine involves a repetitive cycle of fuel and oxidizer injection, mixing, detonation initiation and propagation, and combustion product scavenging.

Using the constant volume, Humphrey cycle as the ideal reference cycle, a 30–50% increase in thermodynamic efficiency relative to the steady flow, Brayton cycle is indicated, depending on precompression levels, cf. Refs. 1 and 2. For airbreathing propulsion systems this translates into I_{sp} levels of over 3600 s corresponding to a specific fuel consumption (SFC) of 1 lbm/lbf-h or less. Application of the detonative mode of combustion in rocket systems yields an I_{sp} improvement on the order of $\sqrt{\gamma}$ times the conventional constant pressure combustion based I_{sp}. This corresponds to at least a 5–10% increase in I_{sp} for a hydrogen–oxygen rocket. Realizing the performance potential of the PDWE, however, requires achieving rapidly the key flow processes, such as fuel–oxidizer mixing and initiation of detonation over the cycle.

II. History of Intermittent Combustion Propulsion Devices

Intermittent combustion for propulsion applications is not new, becoming famous with the introduction of the German V-1 missile powered by a subsonic combustion pulse jet, Fig. 1a. The following steps are involved in the basic operation of this engine.

Copyright © 1995 by the American Institute of Aeronautics and Astronautics, Inc. All rights reserved.

*Senior Engineering Specialist, Computational Fluid Dynamics, Rocketdyne Division. Senior Member AIAA.

†Chief Scientist, Combustion, Rocketdyne Division, Member AIAA.

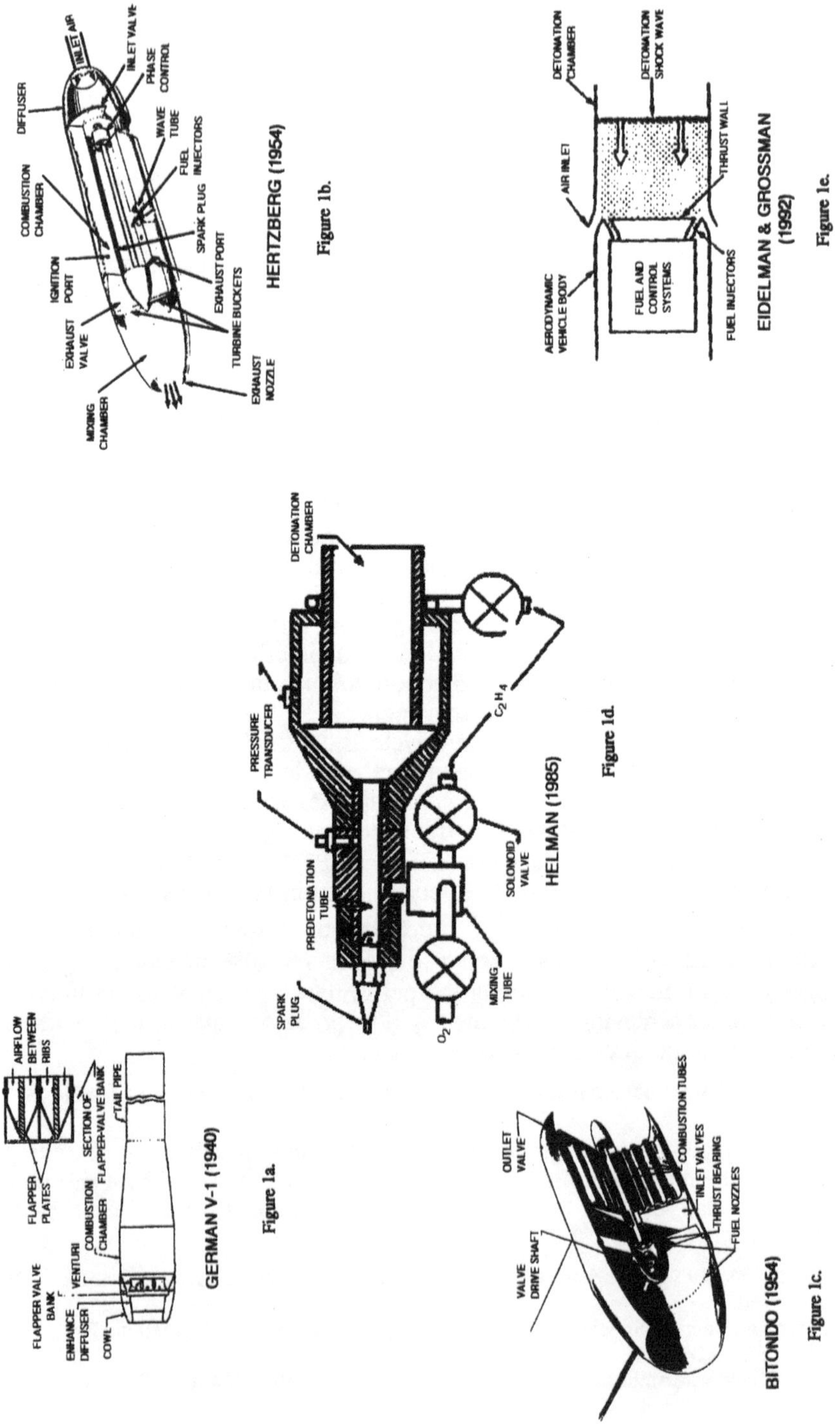

Fig 1 Various concepts highlight wave engine features.

1) Air enters the combustion chamber through a reed type inlet valve and is mixed with fuel, which is injected continuously, to provide a suitable air/fuel ratio. The air within the chamber consists of two components: a fuel-enriched charge near the valve and a column of air or exhaust gas in the exhaust pipe.

2) Upon ignition, rapid combustion of the charge occurs near the valve, but expansion of the gases is resisted by the column of air in the exhaust pipe. The increased pressure closes the valve and forcibly ejects the column of gases out of the tail pipe.

3) As the ejected gases gain momentum, the pressure in the chamber falls below the atmospheric pressure, thereby inducing a fresh charge of air to enter through the inlet valve. During its passage into the chamber, the air is enriched with fuel which vaporizes on mixing with the hot gases in the chamber.

4) The exhaust gases flowing through the tail pipe, now reduced in momentum due to the atmospheric back pressure, reverse their direction of flow and return to the chamber, producing a slight precompression, thereby closing the inlet valve.

5) The combination of precompression and mean mixture temperature of the returned, relatively hot exhaust gases permits spontaneous ignition to occur, and the whole combustion cycle recommences.

Most known pulse jets have a combustion frequency of 45–200 Hz depending on the following factors. 1) The acoustic properties of the combustion chamber and exhaust pipe: an elongation of the tail pipe causes a reduced frequency so that a long engine normally works slower than a short one. 2) The amount of fuel supplied during each combustion cycle: an increase in the fuel supply typically results in an instantaneous decrease in frequency since more combustion product gas has to be expanded out through the exhaust pipe where the gas speed is already sonic. 3) The flow and temperature conditions in the combustion chamber: ignition will not take place before the air–fuel charge has been brought in contact with a heat source of sufficient capacity.

In the early-to-mid 1950's, two efforts were underway in the United States to explore the development of wave engines for missile[3] and helicopter[4] applications. These engines employed rotating valves at the inlet and exit of their respective multitube combustors as shown in Figs. 1b and 1c, respectively. Furthermore, both engines used common inlets and exhaust collector/nozzle components. These features were intended to simplify the construction and to minimize length and weight while performing essentially in steady flow since sequential banks of tubes would always be exposed to the inlet and exhaust nozzle at any instant. Although these two engines differ in detail, their operation is similar. At the beginning of the Hertzberg and Russo[3] cycle, air is flowing through the engine, and fuel is added to form a combustible mixture. The exhaust valve is closed, stopping the flow and causing a hammer shock to move upstream and compress the combustible mixture. When all of the combustible charge is compressed, the mixture is ignited and burning occurs at a constant rate. After undergoing complete combustion, the exhaust valve is opened, allowing the hot gases to expand in a propulsive jet. Meanwhile, the hammer shock strengthened by combustion continues to propagate upstream and is followed by the interface separating the hot and cold gases. When the shock reaches the engine inlet, the inlet valve is closed, causing the shock to be fully reflected into the duct. This process generates thrust, and the reflected wave aids

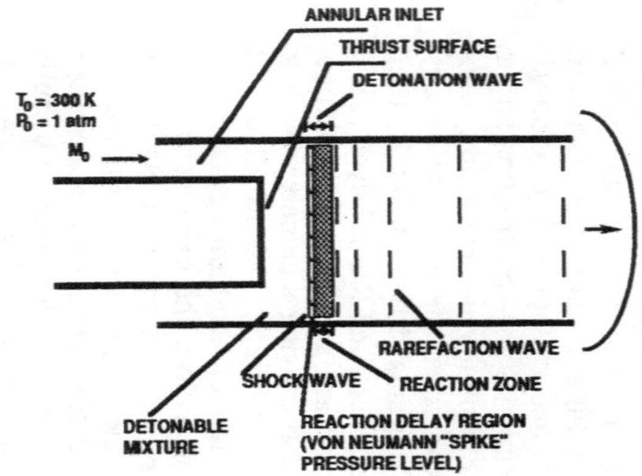

Fig. 2a Internal flow characteristics shortly after rear end initiation.

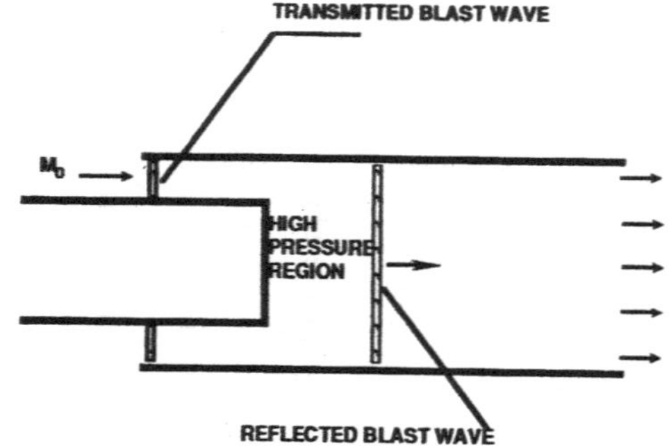

Fig . 2b High-pressure region at thrust surface due to detonation wave reflection; note that since heat release is completed, reflected and transmitted waves are blast waves.

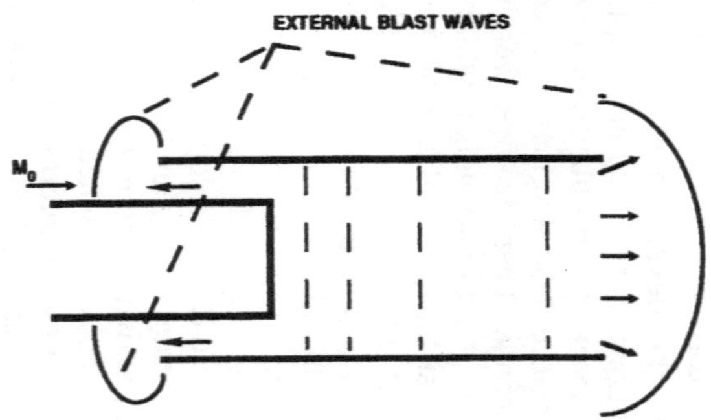

Fig. 2c Expansion of transmitted and reflected blast waves.

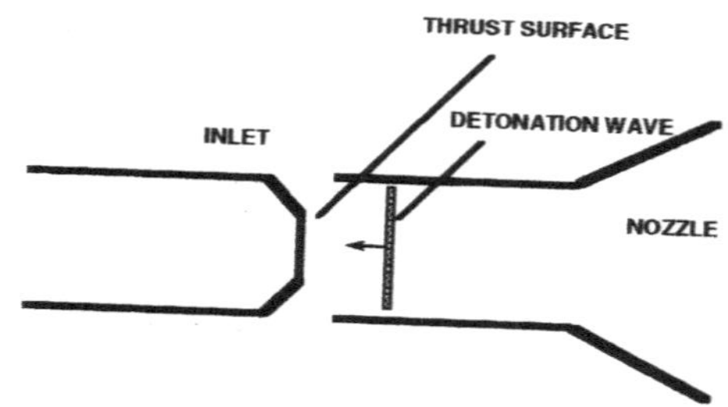

Fig. 2d Inlet and nozzle options for added performance evaluation.

Fig. 2 PDWE design for computational analyses.

in scavenging the engine. The upstream propagating expansion wave resulting from the discharge of hot gases drops the engine pressure and, thus, augments the introduction of the fresh charge which commences upon opening the inlet valve. The cycle is then repeated.

The major problem encountered in developing these particular subsonic combustion, pulse jet-like engines was low-combustion efficiency. Although it is evident that mixing and kinetic limitations were encountered, the efforts were terminated before these issues could be addressed. The major difference between the Hertzberg and Russo[3] and the Bitondo et al.[4] engines was the use, by Bitondo et al., of insulated and, hence, heated tube walls to provide autoignition along the entire tube length in order to achieve constant volume combustion.

The recent attention given to the detonation mode of combustion is based on the order of magnitude higher heat release rates and higher cycle peak pressures attainable with detonative heat release relative to the deflagrative, or pulse-jet mode of operation. Furthermore, the frequency and, hence, thrust is not controlled by the acoustics of the system or by altering the mixture ratio as is the case in pulse jets; the firing frequency in the PDWE can be directly controlled by the rate of detonation initiation at optimum equivalence ratio. These characteristics provide a fundamental improvement in cycle efficiency, system performance, and throttleability. Because the combustion process takes place supersonically, as a thermodynamically efficient, quasiconstant volume process, the pulse detonation engine can achieve better specific fuel consumption relative to more conventional steady flow engines. Moreover, the elimination of flowpath turbomachinery and the consolidation of the compression, combustion, and thrust generation processes within a single component without the need for high levels of precompression (with consequent high losses) to peak operating pressures provides the basis for a propulsion system with a high thrust-to-weight ratio (about 20:1), high thrust-to-cross-sectional area, and the potential for relative simplicity with a high degree of reliability.

Figures 1d and 1e show the test article operated by Helman et al.[5] and a concept put forth by Eidelman and Grossmann,[6] that was recently tested in Sweden.[7] Reference 5 explored the advantage of stratifying the mixture and implemented a predetonation chamber at the front end of the test article that used oxygen as the oxidizer. The main combustion chamber used ambient air, aspirated into the chamber from the rear of the unit. The predetonation chamber and the main chamber were fueled with relatively reactive ethylene. The repetitive operation of the unit involved simultaneously filling the predetonation chamber with fuel and oxygen and the main chamber with fuel. On the completion of filling, the supply valves were closed and a detonation wave was initiated in the predetonation chamber using a spark to initiate the process. The resulting blast wave exiting this chamber was intended to provide direct initiation of the ethylene–air mixture in the main chamber. Since primary thrust is developed at the front end of the unit, a nozzle is not employed nor needed to backpressure the system. The benefit of the added thrust using a short divergent nozzle was not explored, although it would affect the overexpansion and rear air aspiration process used in this concept. Analysis of high-frequency pressure measurements showed rapid detonation formation in the predetonation chamber while only partially complete, i.e., sub-Chapman–Jouguet (C–J) waves, were produced in the main chamber. Nevertheless, frequencies up to 25 Hz were obtained, limited by the solenoid valves used in these tests. Limited tests in which oxygen enrichment was used at the interface of the predetonation and main chambers showed substantial improvement in the quality of the detonations and, hence, higher pressures produced in the main chamber, indicating the critical importance of achieving rapid C–J detonation formation in a practical device. Specific impulse values of only 1000–1400 s were estimated based on the cycle averaged pressure data obtained from the nonoxygen enriched test series. The more recent work of Eidelman and Grossmann[6] showed that I_{sp} levels of at least 3600 s are possible based on detailed two-dimensional, complete combustion, transient computational fluid dynamics (CFD) analyses of the valveless concept shown in Fig. 1e. Starting with the chamber filled with a detonable mixture, the operation of this concept can be described as follows with the aid of Fig. 2.

1) Energy is deposited in the mixture (viz., through a spark or an array of sparks, a predetonation tube, or a laser system) at a sufficiently fast rate to induce rapid transition to detonation. The detonation wave shown in Fig. 2a consists of a nonreacting shock followed by an ignition delay region at essentially constant pressure and temperature corresponding to the adiabatic shock conditions, i.e., the Von Neumann "spike" state. This region is followed by rapid subsonic combustion (deflagration) relative to the shock front, a process that accelerates the flow through a pressure drop. The flow then reaches sonic conditions due, in this case, to the rarefaction wave following the reaction zone created to satisfy the exit boundary condition. This weakens the wave and helps to establish the Chapman–Jouguet detonation wave speed which is the most stable and most often observed in practice. Thus, although excessive initiation energy will overdrive the detonation wave initially, the trailing rarefaction wave will accelerate the flow behind the detonation wave until it reaches sonic speed, relative to the front, i.e., the Chapman–Jouguet condition. To maintain an overdriven condition, external energy input from behind would be required, a process that is not of practical interest.

2) The detonation wave interacts with the front end of the chamber creating a local pressure rise two and a half to three times the pressure rise across the incident wave. Figure 2b shows that a major portion of the wave reflects off the thrust surface, and the remainder is transmitted forward into the inlet. The reflected and transmitted waves are adiabatic shock or blast waves, since heat addition is essentially completed when the incident wave reaches the thrust surface. The transmitted wave momentarily unstarts the inlet flow until it is sufficiently weakened by the rarefaction occurring behind it or is expelled from the inlet as shown in Fig. 2c, and the inflow is re-established. This process depends on the external stream dynamic pressure, the inlet geometry, and the rate of expansion of the internal flow as the reflected shock (blast wave) propagates downstream toward the exit of the engine.

3) The flow trailing the reflected shock partially scavenges the combustion products when it is expelled at the exhaust end as illustrated in Fig. 2c. The flow following the expelled wave typically encounters a lower pressure and quickly induces an expansion wave to propagate upstream into the chamber, evacuating additional combustion products. Depending on the inlet and detonation chamber geometries, the internal flow becomes overexpanded and vitiated air is aspirated back into the chamber. This back end air induction

process was demonstrated experimentally by Helman et al.[5] in their demonstration of the valveless PDWE concept shown in Fig. 1d.

4) The cycle is completed through the filling process during which time the chamber is recharged with fuel and air or with oxidizer in the case of a rocket application. Fuel injection can be accomplished in concert with air induction; in this valveless concept fuel–air mixing and control would benefit from both front- and rear-end fuel injection.

Whereas rear detonation initiation offers more than a factor of two increase in pressure rise due to reflection off the front thrust wall, the practical benefits in terms of cycle averaged thrust and the implementation of rear initiation represent areas of continued evaluation. However, the concept has been tested in a freejet-like configuration in Sweden using "practical" hydrocarbon fuels.[7] Few details were given although a frequency on the order of 4 Hz was reported.

An apparent limitation experienced in the intermittent deflagrative and detonative combustion concepts described here is the completeness of reaction in general, and the rapid attainment of fully developed C–J detonations in PDWE devices. This area needs focused attention and is a major subject of this chapter.

The performance potential and the capability to produce thrust from static to supersonic flight conditions is the basis for renewed interest in exploiting the PDWE for aircraft and missile applications.[2,5–12] However, as previously cited, this performance potential has yet to be fully realized as is apparent from the reviews provided by Eidelman and Grossmann[6] and Eidelman et al.[11] Previous tests of the PDWE concept reported cycle frequencies (up to 35 Hz) insufficient to produce thrust levels sufficient for aircraft applications which are possible at high cycling rates (100–200 Hz). In particular, the speed of the fuel and oxidizer valves and the inability to achieve rapidly Chapman–Jouguet quality detonations were the most obvious limitations in previous PDWE tests. Moreover, the time needed to mix the fuel and oxidizer and to scavenge the combustion products represents additional time constraints in the PDWE cycle. Several reasons are apparent for why such limitations in previous tests were present and the performance potential of the PDWE has not been realized. The lack of a focused multidisciplinary approach to solving the issues in fuel injection, mixing, rapid detonation initiation, fast acting valves, and controls has limited the PDWE cycle time, where all flow processes must be completed rapidly, e.g., on the order of 0.01 s.

Furthermore, the previous description given of the valveless concept highlights the complexity of the operation of an airbreathing pulse detonation wave engine. Quantifying the operation in sufficient detail to derive realistic performance estimates and practical engine designs requires an in-depth understanding of the relationships between geometry and the highly transient, wave dominated, nonuniform reacting flow. In addition, minimizing and controlling internal drag and heat loads including local hot spots are also important to successful engine design and operation. Because these processes are highly interactive and the timing of injection, detonation initiation, and inlet and exhaust valve operation, when used, are critical, CFD is ideally suited to support successful PDWE test and development. As computer technology and computational techniques have improved, more in-depth analyses of the PDWE have appeared (e.g., Bussing and Pappas,[2] Lynch, et al.,[8] Eidelman et al.[9,10]

and Cambier and Adelman[12]) to understand the flow path physics involved in the PDWE operation, the PDWE design and operational features generating optimum performance, and the cycle frequencies and I_{sp} characteristic of the PDWE.

A description of these analyses is provided in Appendix A, but in synopsis these analyses indicate that the potential cycle frequencies and performance of the PDWE are significantly above those previously achieved in testing (e.g., cycle frequencies were above 100 Hz). However, predominantly these analyses have employed simplifying assumptions of one-dimensional flow with mechanistic kinetics or multidimensional inviscid flow with complete combustion; these assumptions do not allow simultaneous description of all of the PDWE flow path physics including the Von Neumann spike in the detonation wave, recirculation regions which trap combustion products, or the initial detonation initiation process. In this chapter the approach is taken to employ the USA code, a multidimensional, nondissipative shock capturing, computational fluid dynamics code with mechanistic kinetics, together with grids with similar number of gridpoints as relatively fine CFD grids for engineering problems, to provide the insights necessary to develop a rapid design tool for the design and evaluation of critical features needed for successful pulse detonation wave engine development.

Analytical general purpose, general geometry, design tools such as the USA code with a range of turbulence and combustion models are important to PDWE development in several areas. First, because instrumentation in rapid firing PDWE experiments is difficult and limited, interpretation of the measurements and of the test operation relies heavily on analysis. Second, analysis is required to supply performance predictions for application studies. Finally, detailed flowfield analysis can provide a description of the PDWE flow processes which must be fully understood to develop a design system relating geometry, fuel and oxidizer injection for optimum mixing, detonation initiation and timing requirements. This chapter concentrates on two flow processes crucial to PDWE operation, scavenging time and detonation initiation, but also examines such issues as transient mixing and structural design criteria.

III. Computational Methodology

The computational methodology is designed to be adaptable to multidimensional CFD problems of significant complexity but also to be applied for simple problems in a time-constrained, workstation-based, design environment. Simulating all of the critical flow processes in the PDWE prescribes several challenges for the computational methodology.

1) The wide variation of time scales between detonation initiation and propagation and combustion product scavenging and fuel–air mixing suggests application of as large Courant–Friedrichs–Lewy (CFL) numbers and time steps as possible while maintaining time accuracy.

2) Simulating the detonation initiation process and the kinetic processes inherent in the Von Neumann spike calls for application of detailed mechanistic kinetic models.

3) In the scavenging process, recirculation regions are created which can trap combustion products, which may serve as a source for predetonation. Modeling these separation regions in this transient, nonequilibrium turbulent flow is a substantial challenge, embracing a range of time and turbulent length scales.

4) Capturing the motion of reacting shock (i.e., detonation) waves while minimizing the number of gridpoints employed is a problem of current research interest.

The Unified Solution Algorithm (USA) CFD code, described in detail in Chakravarthy and Osher[13] and Palaniswamy et al.[14] allows a range of options for time advancement including point-implicit Runge–Kutta, block and scalar approximate factorization, and a number of relaxation methods. In the multidimensional computations described here, a block approximate factorization procedure was employed; once a detonation wave was formed, the CFL number which could be achieved without inducing instabilities was 0.5–10, based on the smallest cells in the grid, depending on the particular flow transients. Two or three global time iterations per time step were used whenever significant differences from the result with one global time iteration emerged. For one-dimensional computations with extended kinetic mechanisms, the inversion of the large block matrices required by approximate factorization proved to have too large a cost per time step; hence, a point-implicit, two-stage Runge–Kutta procedure was adopted instead. Until the characteristic Zeldovich–Von Neumann–Doring (ZND) detonation structure was achieved in these one-dimensional computations, the CFL number required for accuracy was less than 0.1; after the ZND structure appeared, the CFL number could be raised as high as 0.5, at which point slight numerical oscillations could appear in the small cells near the wall where the detonation was initiated.

Over the past 15 years, methods have appeared combining approximate solutions to the Riemann problem with limiters applied when obtaining the convective fluxes through higher order interpolation. Total variation diminishing (TVD) methods allow capturing and tracking nonreacting shock waves without inducing unphysical oscillations or adding sufficient dissipation to degrade the solution quality.[15] In these computations, several types of convective term TVD limiters having different properties in how shocks are "compressed" or smeared out were employed. In the multidimensional computations a variant of the min–mod limiter with a moderate compression coefficient (cf., Ref. 13) was applied in tracking the detonation waves; experiments with different limiters in one-dimensional computations suggest this was a good choice since a variant of the compressive "superbee" (cf., Ref. 16) TVD scheme seemed to enhance artificially physical compressions and the min–mod scheme with the lowest possible compression coefficient proved too diffusive to capture the Von Neumann spike completely. This limiter treatment was combined with the Lax–Friedrichs Riemann solver, which appeared to be more robust than the more standard frozen-γ-based Roe Riemann solver in areas such as the reaction zone where γ changed very rapidly.

The grids employed in the computations were reasonably fine but of the same order of grid density as those typically employed in multidimensional CFD analyses. Algebraic means were employed for grid generation with clustering at no-slip wall boundaries.

To prevent smearing of the shock wave from artificially igniting reactions in front of the shock for nonadaptive grids relatively coarse with respect to chemical kinetic processes, as described by LeVeque and Yee,[17] the chemical source term was evaluated in the neighborhood of shock waves from the minimum flow conditions in front and behind the shock. This procedure is designed to reduce the computational stability restrictions imposed on the grid fineness and time step by the stiff source terms.[18] For the grid densities employed in the one-dimensional analyses presented later, this method yielded similar results in terms of wave speeds and flowfield profiles to the more conventional method of source term evaluation but captured a larger component of the near nonreacting Von Neumann spike and seemed more robust numerically, especially with coarser grids.

The approach to describing the chemical kinetics of the fuels H_2, CH_4, C_2H_2, and C_2H_4 was to employ the simplest level of mechanistic kinetics appropriate in multidimensions but to use initially extended mechanisms for one-dimensional computations, using these results to examine where simplifications such as quasiglobal modeling (e.g., Edelman et al.[19]) might be employed for multidimensional work on practical, heavy hydrocarbons. In particular, the seven-reaction, 6-reactive-species simplification of the H_2–air model of Drummond et al.[20] with the Oldenborg et al.[21] third body enhancement coefficients was applied in the multidimensional computations; in one-dimensional work, this model was compared to the results of the 8-reactive species, 18-step model of Drummond et al.[20] and a 1-step global model. The C_2H_2–O_2 and CH_4–O_2 computations were completed with the 100-reaction and 25-reaction mechanisms of Miller et al.[22] and Chelliah et al.[23] with the pressure dependence of some of the rate constants and enhanced third body recombination neglected initially. The C_2H_4–O_2 computations were carried out with the 93-step mechanism of Westbrook et al.[24] A more detailed description of the levels of kinetic modeling available for CFD analysis is provided in Appendix B.

Viscous turbulent transport of mass, momentum, and energy was described by the Goldberg[25] 0-equation variant of the Baldwin–Lomax model. This model was not meant to reproduce all facets of the transient, nonequilibrium turbulent flow in the PDWE cycle but instead to give a first-order description of such processes as mixing layer development and enhanced species transport through turbulence. The application of variants of the two-equation turbulence model represents ongoing work to describe better the transient mixing processes in the PDWE cycle.

IV. Baseline Pulse Detonation Wave Engine Analyses with Straight Inlet

Described in Fig. 2 is a generic PDWE design, consisting of an axisymmetric detonation chamber with a relatively simple, angled or straight, annular inlet, and possibly a short nozzle. The CFD analyses were conducted to understand the dominant flow features of the PDWE cycle in variants of this design including the elements of detonation initiation and reflection, blowdown of the detonation chamber pressure to a low level, scavenging of the combustion products, and creation of an environment suitable for mixing of fresh fuel and air to form a detonable mixture for the subsequent cycle.

The baseline PDWE analyses were completed on variations from a configuration similar to that analyzed by Eidelman et al.[10]: an axisymmetric, 8-cm-long detonation tube with a simple, straight, annular inlet (see Fig. 2a). The inlet was varied in length from 1 to 12 cm, the detonation tube diameter was varied from 8 to 16 cm, and cases at freestream Mach numbers of 0.8 and 2 were analyzed. In the initial analysis for an 8-cm-radius detonation tube, approximately 9000 gridpoints concentrated in the detonation tube, where a grid of 117 × 52 was used, were employed to analyze the inlet, the detonation tube, and a region aft of the tube exit designed to yield a reasonable approximation to the nearfield of the aft end aspiration process. This grid resolution was sufficient to define the fluid mechanics processes and to provide a first-order description of the finite rate chemistry in the detonation region. Full Navier–Stokes computations with viscous no-slip wall boundary conditions and freestream characteristic boundary conditions imposed at the inlet entrance were conducted.

Subsequent computations, designed to give a better description of the external flow and of the aspiration process at longer physical times, were also completed with approximately 9000 gridpoints, again concentrated in the detonation tube, where a grid of 192 × 30 was employed, but with the computational boundaries moved far from the PDWE (approximately 10 engine radii from the centerline and 10 engine lengths downstream) to minimize any possible influence from the imposition of the characteristic boundary conditions at a finite distance. At this distance the blast waves were able to leave the computational domain without inducing significant unphysical transients. A more uniform grid with inviscid slip conditions was applied to prevent excessively small time steps and to allow faster advancement in physical time at the same computational cost; viscous effects were allowed to be present only in free shear mixing layers, not boundary layers. (This approach serves to make easier computations with the widely disparate time scales of detonation propagation and scavenging but must be revisited to analyze skin friction losses and heat loads, in particular.) The chemical source term was modeled through the modified 7-step model of Drummond et al.[20] described earlier, and the turbulent viscous fluxes were modeled algebraically.

A forward-propagating detonation wave was initiated from the aft end of the tube into a stoichiometric H_2–air mixture at time $t = 0$. Various means were tried to initiate this detonation wave within a short distance. Turbulent transport, an overpopulation of radicals, and closing off the aft end of the detonation tube to prevent the induced expansion from overtaking the detonation wave were all found to enhance detonation transition. Eventually, near direct transition was achieved within a short distance (of order 1/10 the length of the detonation chamber) by raising the initial temperature and pressure of a sufficient volume of the unburnt stoichiometric mixture. Preliminary indications are that the level of energy deposition depends on the means taken to enhance detonation transition. This result illustrates the critical importance of the initiation mechanism and the various ways in which it can be controlled.

As the wave propagated forward at an average speed of 1800 m/s climbing to near the Chapman–Jouguet velocity as the wave approached the thrust surface, it remained basically one-dimensional, as is illustrated in Fig. 3 by the conversion of the detonable mixture to H_2O as evidenced by the H_2O contours in the top-half the detonation shock, created during the initiation transients, and in the wall boundary layer, were apparent in the analysis with no-slip boundaries. In Fig. 3, the detonation wave is propagating to the left, the annular inlet is at left, the thrust surface is located at $x = 0$, and at the tube exit where the wave was initiated ($x = 3.14$ in.), the exit has been opened to the atmosphere and a blast wave is beginning to transfer combustion products into the external air flow.

A wave structure similar to that predicted by the ZND theory was created, as is seen in Fig. 3 where the H_2O contours show the partially reacting shock, followed by a region of near completion of reaction, and then an expansion required to satisfy the boundary condition at the tube exit. Because of the Von Neumann spike generated by the inclusion of kinetics, the incident shock pressure of approximately 25 atm was higher than the Chapman–Jouguet pressure of approximately 16 atm. Accounting for the differences in the chemical properties of H_2 and C_2H_4 and for the appropriate thrust surface area, the thrust results reported by Eidelman, et al.[10] are consistent with the Chapman–Jouguet complete combustion pressure rise.

As the detonation wave reflected off the thrust surface slightly more than 0.04 ms after initiation at the aft end, an extremely high but short-lived pressure was created. This pressure, a consequence of the Von Neumann spike reflecting into the partially burnt gas, is apparent in Fig. 4, which compares the thrust surface pressure for the baseline straight inlet analysis to the scaled Eidelman et al., complete combustion results for a similar, but not identical, geometry. The

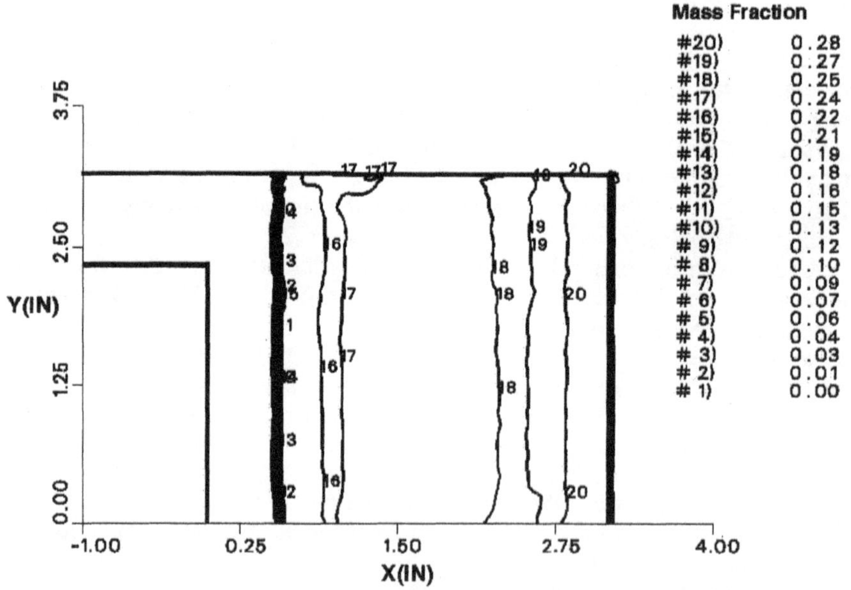

Fig. 3 H_2O contours for baseline analysis 0.036 ms after detonation initiation.

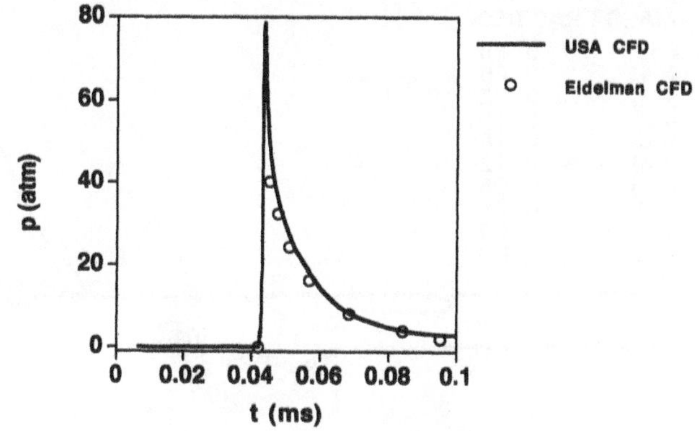

Fig. 4 Average pressure on thrust surface for computational analyses at M = 0.8.

high-pressure spike, which appeared in all of these analyses with finite rate combustion, is not created under the assumption of complete combustion. The reflection of the Von Neumann spike and associated combustion region may be recognized from the change in slope between the combustion region and the following expansion.

The static pressures at $t = 0.066$ ms in Fig. 5 show that after interaction with the thrust surface, a reflected shock wave and a transmitted shock wave are created, with the transmitted wave continuing up the inlet, tending to unstart it momentarily. The change in area on entering the inlet from the combustor side creates a mechanism for the production of transverse waves that make the pressure on the thrust surface increasingly nonuniform. At the exit the underexpanded initial wave is expanding, carrying with it combustion products.

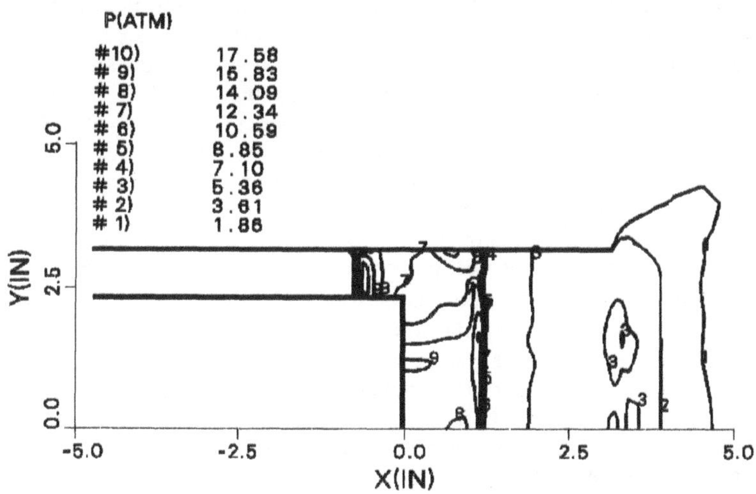

Fig. 5 Static pressure for baseline analysis at t = 0.066 ms showing nonuniform pressure on thrust surface.

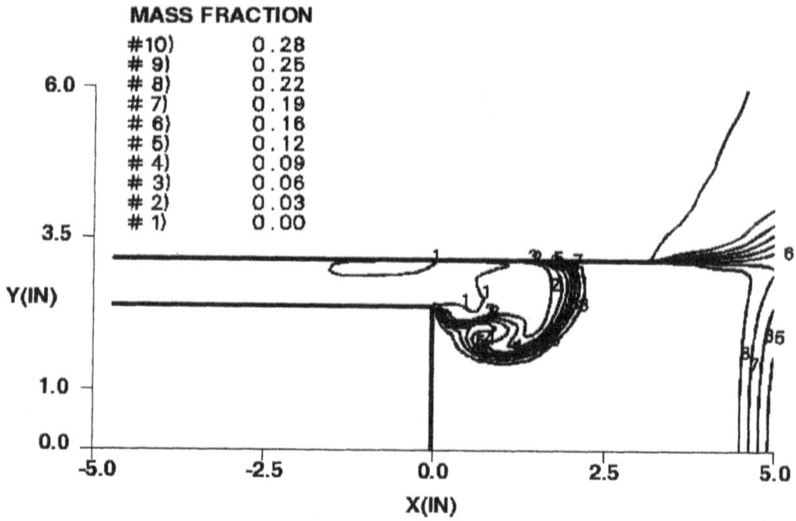

Fig. 6 H_2O contours showing aspiration of fresh air in baseline analysis.

Eventually, the blast waves are expelled from both the inlet entrance and the detonation tube exit, reducing the pressures in the detonation tube and allowing aspiration of fresh air (illustrated in Figs. 6 and 7 by the reduction of the concentration of H_2O in the detonation tube). Figure 6 shows that fresh air is initially swept in through the inlet; a recirculation region has begun to form behind the thrust surface at $x = 0$ through an unsteady mixing layer which mixes the fresh air with the combustion products left in the detonation chamber. Figure 7 shows at a later time that the shear layer has continued out the exit, allowing fresh air also to enter through the aft end of the device. The time at which fresh air aspiration occurs is dependent on the design and freestream conditions. In the aft end, the expansion following the blast wave tends to aspirate not only fresh air but also combustion products previously expelled from the detonation tube. After the blast waves are expelled from the PDWE, the thrust surface pressures for the different analyses begin to diverge depending upon the freestream conditions and particular inlet length.

This particular PDWE geometry also offers a natural mechanism for enhanced mixing in the time-dependent vortices seen in the H_2O contours (Fig. 6) in the unsteady shear layer near the combustor base. Figure 8, showing the velocity vectors during the front end aspiration process, illustrates that while most of the flow is directed through the engine, a vortex is forming near the entrance to the combustor. Whereas these vortices, created by the sudden inflow of air resulting from the relatively low internal pressure, could be used effectively by injection of fuel near this unsteady mixing layer, this region could also serve as a source for premature combustion initiation. The design of injector systems that through proper timing and placement of fuel introduction achieve adequate distribution and micromixing without premature ignition represents an important application of

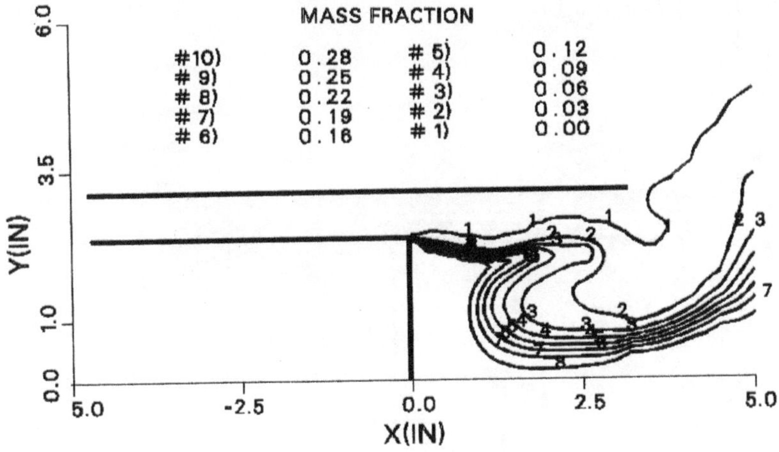

Fig. 7 H_2O contours for baseline analysis after beginning of aft end aspiration.

CFD currently being carried out in support of proper design. Clearly, fuel reactivity must be considered in concert with geometric features.

The time to scavenge combustion products was noticeably affected by the inlet design and, in particular, the inlet length. Of particular concern was the substantial time (up to several milliseconds depending on freestream conditions) needed to bring fresh air into the low-pressure recirculation region containing residual products of combustion, existing behind the thrust surface (see Fig. 9 which shows the velocity vectors at long time); this recirculation region was, as expected, present in all of the analyses, being even more significant under the lower, $M = 0.8$ freestream conditions.

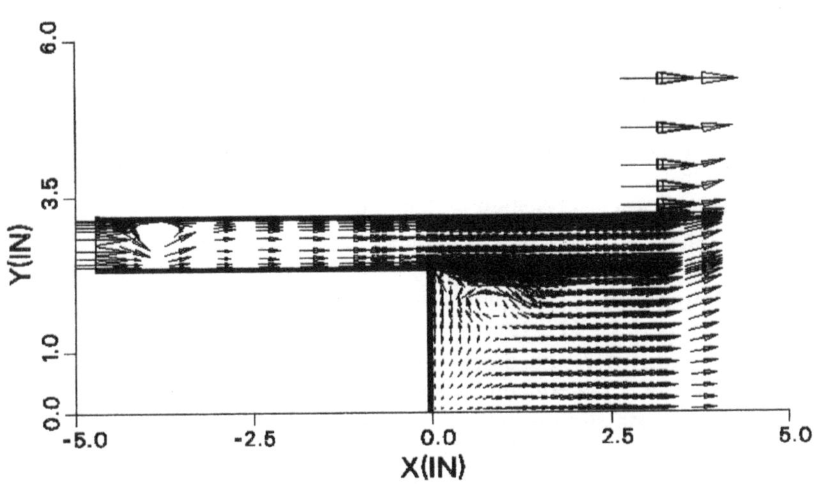

Fig. 8 Velocity vector field during forward end aspiration showing unsteady vortices.

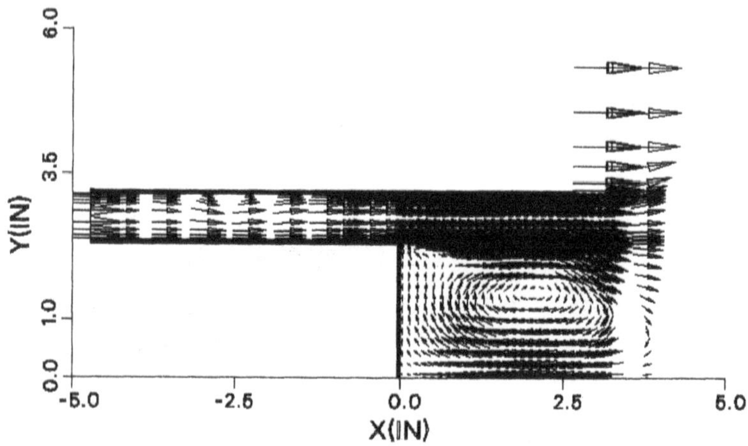

Fig. 9 Recirculation region created behind thrust surface for straight inlet.

V. Computations for Scoop Inlet Configuration

In the straight inlet analyses, a large recirculation region was created behind the thrust surface upon the inflow of fresh air. Significant time was required to entrain fresh air and to purge the reaction products from this region. As already mentioned, allowing such a region of hot combustion products to remain could serve as a source of preignition in the PDWE cycle. This is consistent with the average recirculation residence times being several times longer than the convective time of a particle flowing over the recirculation zone. Hence, as described in Lynch and Edelman,[26] an alternative design (similar to Fig. 2d without the nozzle attachment) has been considered where the inlet is shaped to flush this region of combustion products and also to reduce the loss associated with the eddy formation process.

To allow direct comparison to the straight inlet results at $M = 0.8$, the geometry and freestream conditions were made similar: an axisymmetric detonation tube initiated at the back-end. The previous straight inlet was, however, replaced by an open "scoop type" inlet angled at 45 deg.

In the computations of this geometry with the USA code, approximately 8370 gridpoints were used overall with a more concentrated 120×41 grid in the detonation tube. The full Navier–Stokes equations were employed with the turbulent viscous fluxes modeled algebraically and inviscid wall boundary conditions imposed in these initial calculations to allow faster advancement in physical time as described earlier. The chemical source term was modeled through the modified 7-step model of Drummond et al.[20] and in the external flow characteristic boundary conditions were applied far from the PDWE. These approximations were similar to those imposed in the straight inlet computations to allow a direct comparison to those results.

At time $t = 0$ the last few cells in the detonation tube, filled with a stoichiometric H_2–air mixture, were ignited by raising the pressure at

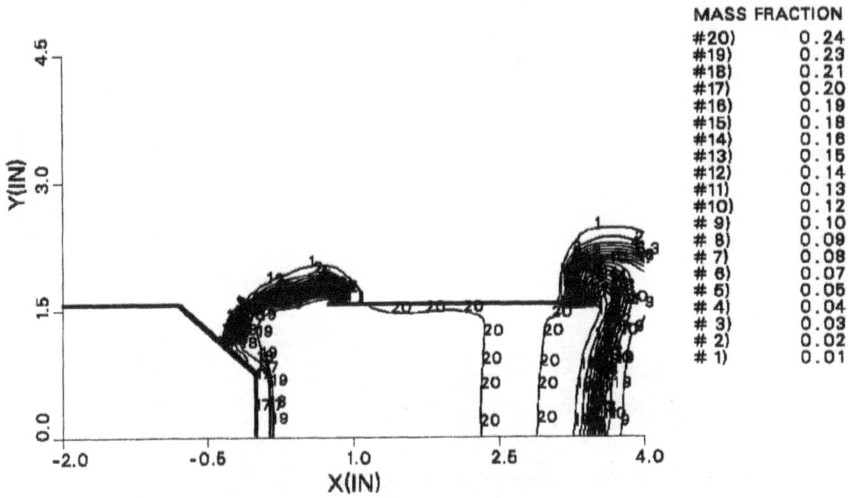

Fig. 10 H_2O mass fraction contours, 45-deg angled, scoop inlet, t = 0.045 ms.

quasiconstant volume to 30 atm, inducing rapid reaction and creating a detonation wave propagating up the combustor and a blast wave propagating into the external flow. Until the reflection of the detonation wave off of the front surface, the flow is characteristic of a nonreacting shock, followed by a region of

As is seen in Fig. 10, showing the H_2O contours in the top-half of the detonation tube at $t = 0.045$ ms, a point where the detonation wave is reflecting off the front surface, near complete reaction is obtained behind the detonation wave. (In Fig. 10, the freestream flow is from left to right, the tip of the detonation wave is at approximately x = -0.25 in., propagating from right to left, and a blast wave is propagating out the aft end into the external airstream. The thrust surface is at x = 0, the scoop inlet is located between x = -0.75 and 0.75 in., and the detonation tube exit is at 3.14 in.) After reflection of the detonation wave off the front surface, a complex multidimensional shock structure is generated by reflections off the centerline and the chamber walls; moreover, the shape of the thrust surface makes the shock reflect over time and, thus, broadens the peak in the average thrust surface pressure. As the internal blast waves exit the domain and become external blast waves, multidimensional rarefactions are created behind the blast waves, carrying combustion products into the external flow and creating vortex structures which trap combustion products .

Figure 11 shows this multidimensional shock structure through the pressure contours at $t = 0.096$ ms, a point where the reflected detonation wave is exiting the detonation chamber and the initial wave has exited the inlet and become a spherical blast wave, its front at x = -2 in. The tip of the blast wave created in the early initiation process is now far aft of the engine but, being spherical, is also interacting with the blast wave which has exited the inlet. Expansions created behind the blast waves are beginning to bring the pressure behind the thrust surface below atmospheric. In Fig. 12, showing the H_2O contours at $t = 0.41$ ms, fresh air is being brought into the engine through the inlet, and

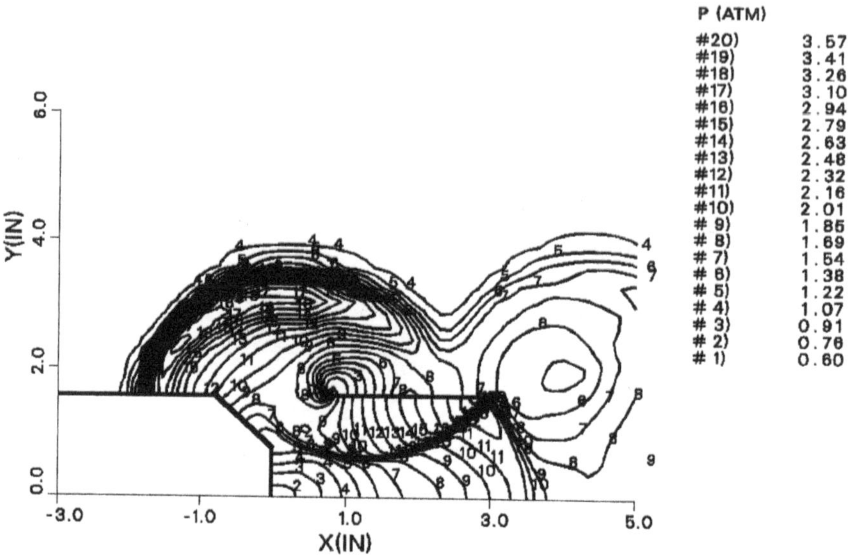

Fig. 11 **Static pressure contours, 45-deg angled inlet, t = 0.096 ms.**

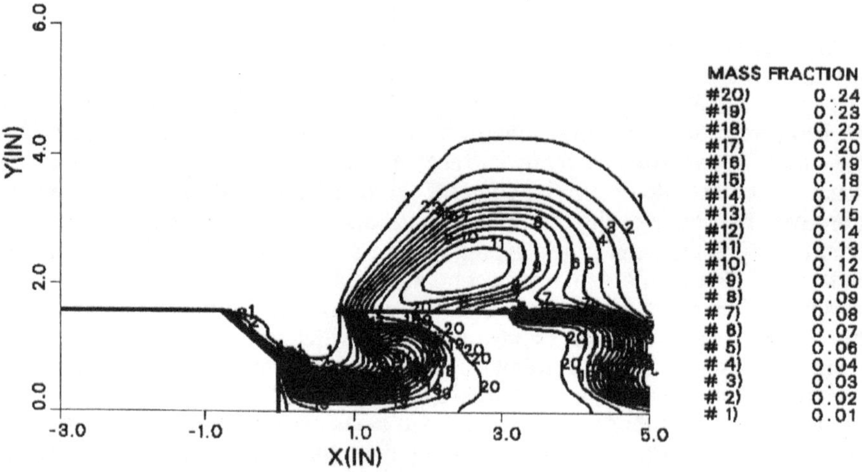

Fig. 12 **H_2O mass fraction contours, 45-deg angled inlet, t = 0.41 ms.**

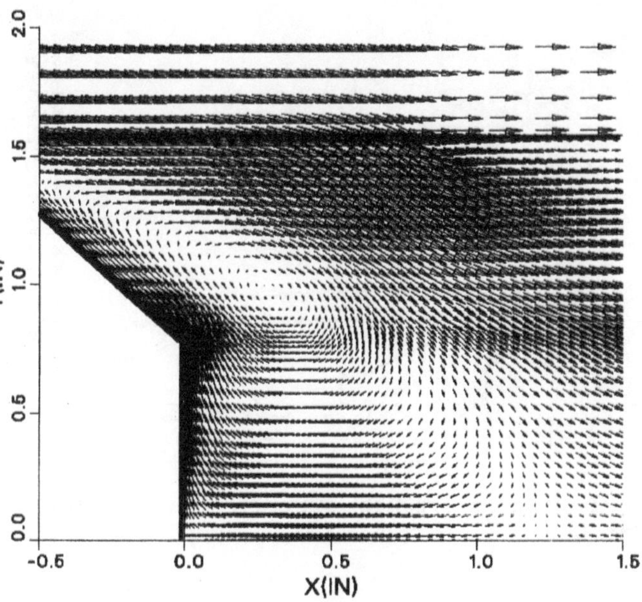

Fig. 13 Velocity vectors, 45-deg angled inlet, t = 0.61 ms.

the scavenged combustion products are being swept downstream. A stagnation point is created approximately 1 in. behind the thrust surface due to the axisymmetric character of the flow, and a recirculation region is beginning to form behind the thrust surface, trapping some hot combustion products. However, as is seen in Fig. 13, showing the velocity vectors behind the thrust surface at t = 0.61 ms, this recirculation region is not nearly as extended as in any of the straight inlet cases. As a consequence of this shortening of the recirculation zone, as seen in Fig. 14 presenting the H_2O contours, by t = 1.03 ms nearly all combustion products have been scavenged from the combustor (which ends at 3.14 in.). Moreover, additional shaping of the inlet and a more physical treatment of viscous entrainment effects are expected to reduce the scavenging time further. Indeed, the optimum inlet design will require that careful attention be paid to sizing, matching, and, in particular, shaping to minimize losses and the trapping of combustion products. Furthermore, aerodynamic rectification to achieve the thermodynamic advantages of valves within a valveless concept represents a potentially fruitful avenue of research for advanced lightweight, fixed geometry, PDWE designs for airbreathing missile applications where simplicity and cost are major drivers.

VI. Detonation Initiation In Hydrocarbon Systems

Whereas hydrogen is a primary candidate for high-speed aircraft, volume and weight are important to missile applications where size, range, and acceleration impose special requirements on the propulsion system. Hydrocarbon-based fuels represent practical candidates that can meet these requirements, including

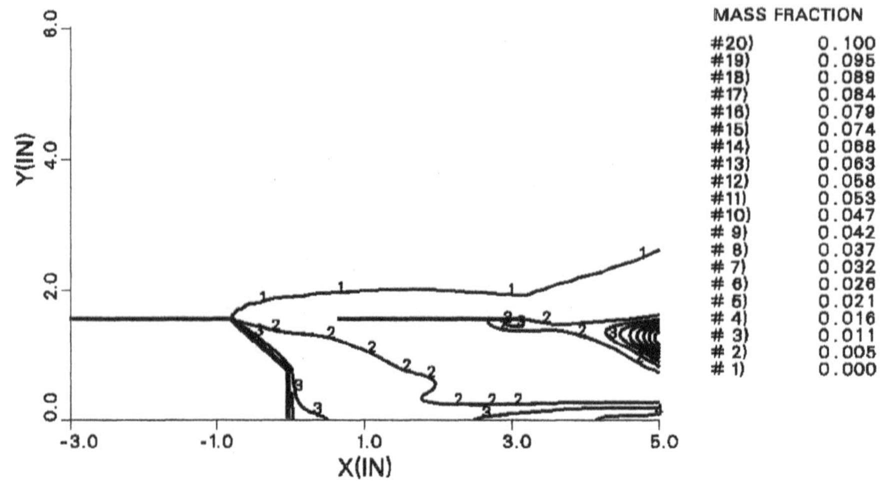

Fig. 14 H_2O **mass fraction contours, 45-deg angled inlet,** t **= 1.03 ms.**

storability. Many storable hydrocarbon fuels can be more difficult to detonate than hydrogen, requiring longer lengths to achieve transition from ignition to detonation. The time needed to attain Chapman–Jouguet detonation conditions lengthens the PDWE cycle time and, hence, limits the duty cycle; increasing the cycle frequency is crucial to optimizing PDWE performance because thrust scales with frequency. Also, reducing the quantity of energy to be deposited in the flowfield to achieve detonation is important to limit the size and weight of the ignition system.

To understand the relative energy input requirements to achieve rapid ignition-to-detonation, a series of one-dimensional transient fluid dynamics analyses were undertaken with the USA code, employing the methodology described in Section III. Four fuel–oxidizer systems are considered in Secs. VII–IX: C_2H_2–O_2 and H_2–O_2, readily detonable mixtures; C_2H_4–O_2, where ethylene is a fuel characteristic of reactive hydrocarbons; and CH_4–O_2, where methane is an extremely stable fuel and is typically difficult to detonate. Although detonation in fuel–air systems is of ultimate interest, because of the more readily available fuel–oxygen database, this was the focus of these initial studies. However, the fuel–oxygen results are of direct relevance to both pulse detonation rocket systems and the predetonation tube of Helman et al.[5] The relevance of one-dimensional analysis of the complex three-dimensional cellular structure found in detonation waves is discussed in Sec. X.

VII. Detonation Initiation in C_2H_2-O_2 Mixtures

Initially, a 30-cm section of the planar detonation experiments of Carrier, et al.[27] for stoichiometric C_2H_2–O_2 mixtures under conditions of large initial energy input were simulated with a grid clustered at the closed end. The Ar F excimer laser employed in the experiments to photo-dissociate the C_2H_2 was

Table 1 Peak pressure: stoichiometric C_2H_2-O_2 mixtures

Initial tube pressure, torr	Peak pressure, atm			
	Experiment	Coarse grid min-mod moderate compression	Fine grid compressive TVD	Fine grid diffusive min-mod scheme
200	10		≈13	
400	21	≈23	≈26	≈22
800	49	≈47	51	≈42

simulated by initializing the problem with a portion of the C_2H_2 at the closed tube end dissociated to C_2H and H, whereas the energy input of the excimer laser was simulated by raising the pressure and temperature of the fuel–oxidizer mixture. This fuel conditioning procedure was later tested for the less reactive hydrocarbons, where it reduced the level of energy deposition required for detonation. This reduction in the level of energy deposition for the "conditioned" fuels varied noticeably depending on the kinetic dissociation rate of a particular fuel, being no more than 20% for ethylene but being extremely large for methane. If, however, sufficient energy were deposited in the flowfield over a sufficiently large volume, little difference in behavior was observed between the conditioned and unconditioned fuel–oxidizer mixtures indicating that an optimum sensitization level is likely to exist.

In these analyses a ZND structure reproducing the classical similarity solution was generated in the computations with the computed wave speeds of 2545 m/s at 800 torr and 2450 m/s at 400 torr and peak pressures (see Table 1) at high initial tube pressures very close to the experimentally measured values of 2492 m/s at 800 torr and 2446 m/s at 400 torr but with ignition delay times slightly too short (i.e., of order 2 μs compared with the measured values of 3.8 μs at 800 torr, 4.6 μs at 400 torr, and 8.5 μs at 200 torr) and with too slow a decline in peak detonation pressure with initial tube pressure. Hence, the grid was refined to consist of 410 clustered gridpoints over 5 cm to represent more accurately the initiation process. As can be seen in Fig. 15 describing the development of a stoichiometric C_2H_2–O_2 detonation at 800 torr through the pressure profile at various instants in physical time, an initial input of approximately 20 atm pressure and 3000 K over an extremely short distance of approximately 0.0001 m caused a vigorous reaction, which in turn created a blast wave. This shock wave enhanced reaction behind it, creating an overdriven detonation wave (indicated by the pressure spike at $t = 0.6$ μs), which oscillated between overdriven and underdriven conditions (illustrated by the oscillations in the pressure) before settling into an essentially stable condition with only slight

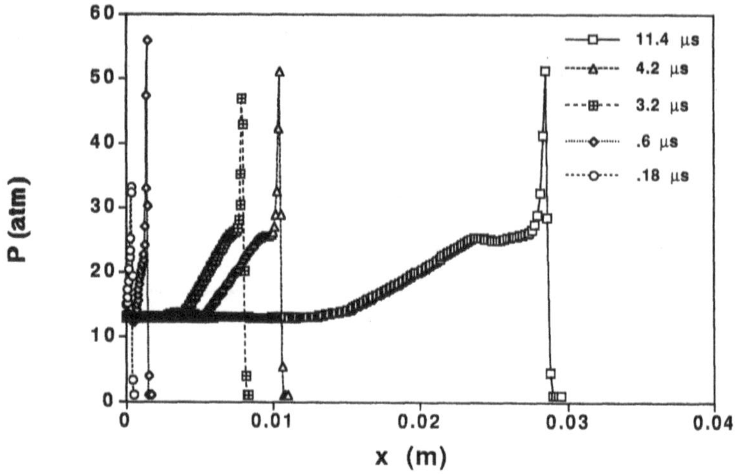

Fig. 15 Detonation development seen through pressure profiles in stoichiometric C_2H_2–O_2 mixture, 800 torr initial pressure.

oscillations in the peak pressure at approximately 4 μs at 800 torr and 6 μs at 400 and 200 torr; this suggests a longer ignition delay than the previous results where a Chapman–Jouguet wave was created approximately 2 μs after initiation at both 400 and 800 torr and is more in line with the experimental results.

The final ZND peak pressure in the computations was approximately 51 atm with the most compressive TVD scheme as opposed to 49 atm in the experiments and 47 atm in the previous coarser grid analyses. However, the peak pressure was found to depend on the limiter employed in the TVD scheme; for the most diffusive min–mod limiter, a peak pressure of approximately 42 atm was obtained (cf., Fig. 16 which shows the difference in the pressure profiles for the most compressive and most diffusive TVD schemes after they have achieved a ZND structure). These results suggest for this problem with a nonlinear source term, the most compressive scheme may be artificially compressive and the most diffusive min–mod scheme is too diffusive. Thus, a moderately compressive TVD scheme appears best, keeping shocks sharp, but not artificially compressing them. The change in peak pressure with initial tube pressure is shown in Table 1. Little diminution in the peak pressure is noted with initial pressure, suggesting that this result may be due to the Lindemann-form pressure dependent kinetics, which were not treated in these initial analyses.

Of some interest is the formation of a secondary shock wave behind the initial shock as the detonation wave travels downstream. As noted by Eidelman and Sichel,[28] this behavior results from increasing energy release in the later portion of the reaction zone. In hydrocarbons this type of energy release pattern can appear because of the initial conversion of fuel to CO, which is subsequently converted to CO_2. In Fig. 17, which shows the major species distributions at a time of 0.6 μs, where the ZND structure is still developing, the shock induces a reaction zone which rapidly converts C_2H_2 and O_2 to CO; this is followed by oxidation to CO_2 and a region of recombination where further CO_2 is produced.

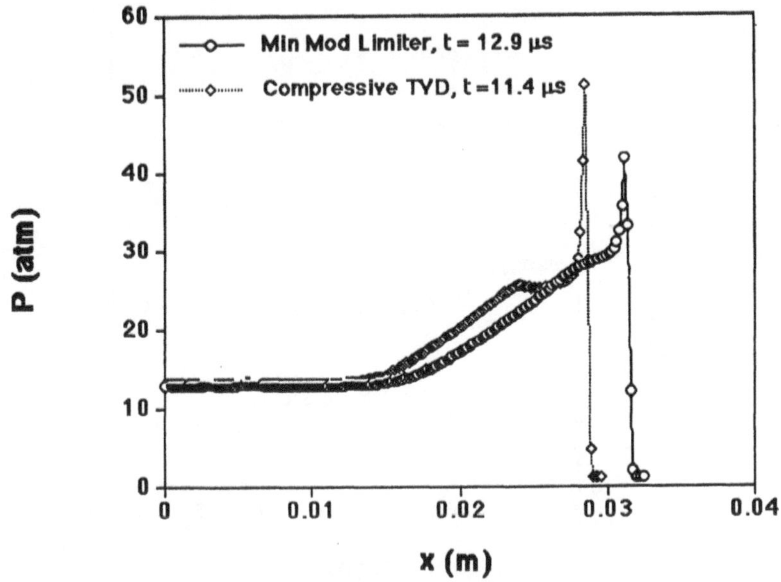

Fig. 16 Effect of treatment of convective term on detonation pressure profile in stoichiometric C_2H_2–O_2 mixture.

Figure 16 shows that the compressive TVD scheme enhances development of these secondary shock waves and, although capturing these physical phenomena, may be artificially compressing these trailing shock waves under these circumstances. This may be a manifestation of the one-dimensional treatment of the inherent three-dimensional structure described in Sec. X and could account for the predicted overcompression.

VIII. Detonation Initiation in H_2-O_2 Mixtures

Analyses were also conducted of H_2–O_2 and C_2H_4–O_2 mixtures to understand the type of behavior to be expected in tests of other fuels with the necessary reactivity for PDWE operation. Figure 18 illustrates the development of the ZND detonation structure through the transient pressure profiles of a stoichiometric H_2–O_2 mixture under initial conditions of 300 K and 800 torr with the 18-step modified Drummond et al.[20] H_2–air kinetics model. Figure 19 showing the development of the peak detonation pressure with time indicates that the steady ZND structure is similar in the results using the 18-step and 7-step models of Drummond et al.,[20] with a slightly higher peak pressure of 28–29 atm in the 7-step model (which does not include peroxide reactions) as opposed to 26–27 atm in the 18-step model, justifying use of the 7-step model in the multidimensional computations. Larger initial energy input reduces slightly the time required to achieve the pseudosteady detonation structure but does not affect the structure achieved. A steady postburning pressure of 20 atm and postexpansion pressure of approximately 8 atm were observed. For the H_2–

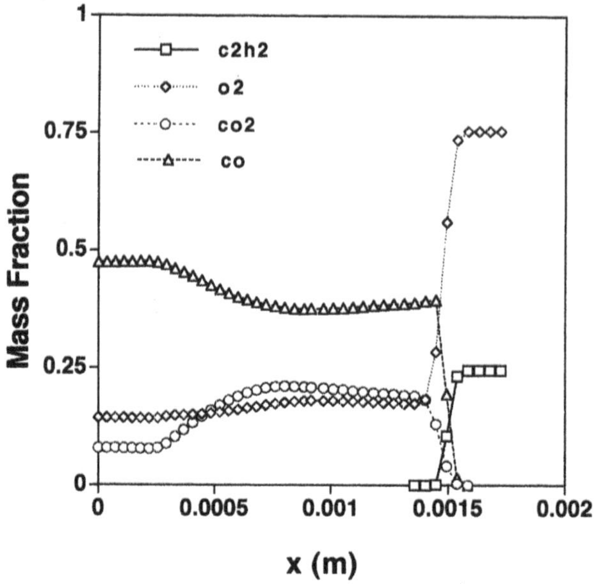

Fig. 17 Species composition in stoichiometric C_2H_2–O_2 detonation, t = 0.6 μs.

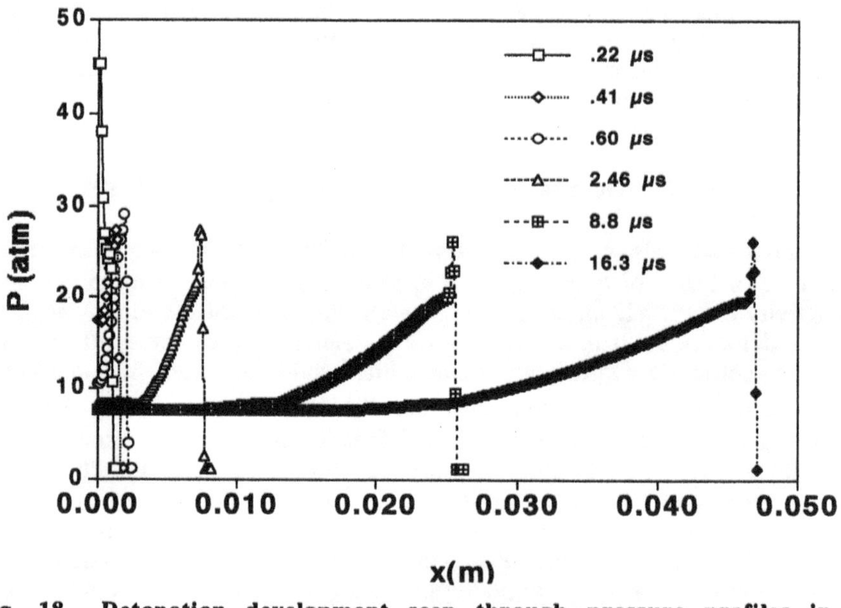

Fig. 18 Detonation development seen through pressure profiles in stoichiometric H_2–O_2 mixture, 18-step Drummond et al.[20] kinetics.

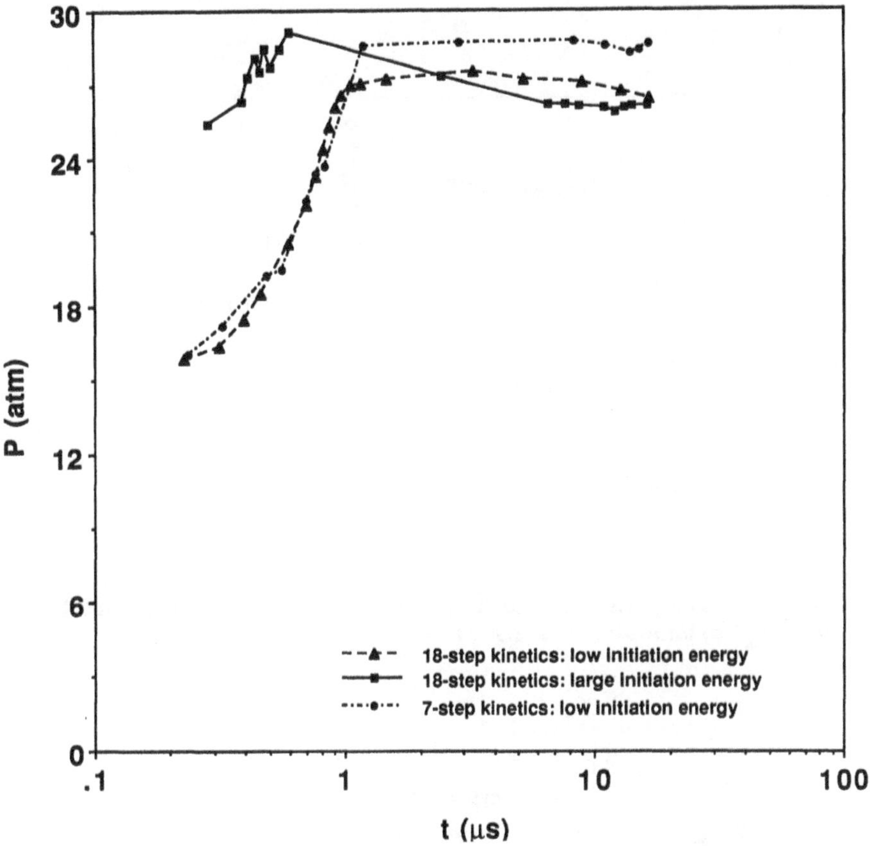

Fig. 19 Comparison of peak detonation pressure for 7-step and 18–step Drummond et al.[20] H_2–air kinetics.

O_2 mixture, conditions of 20 atm and 3000 K would detonate the system, if the energy input were spread over a large enough volume. As for C_2H_2, this level of energy input was of the same order of magnitude as that seen in experiments to achieve direct transition to detonation.

In the steady detonation structure the temperature peak (Fig. 20, showing the temperature and pressure at $t = 2.9$ μs in the 7-step model computation) and reaction zone lie slightly behind the shock. This is followed by a region of recombination and expansion where the temperature is nearly constant. The gas velocity profile (Fig. 21) at $t = 2.9$ μs where the ZND structure has still not completely stabilized shows the end of the rarefaction waves almost exactly one-half of the distance from the shock front to the closed end wall. This result indicates that the classical similarity solution is a useful approximation to the propagating flow structure; however, the similarity solution does not show the Von Neumann spike. Additional deflagrative transients created in the formation of the detonation are still visible in the region of near zero velocity, but these waves were found to decay as the ZND structure stabilized.

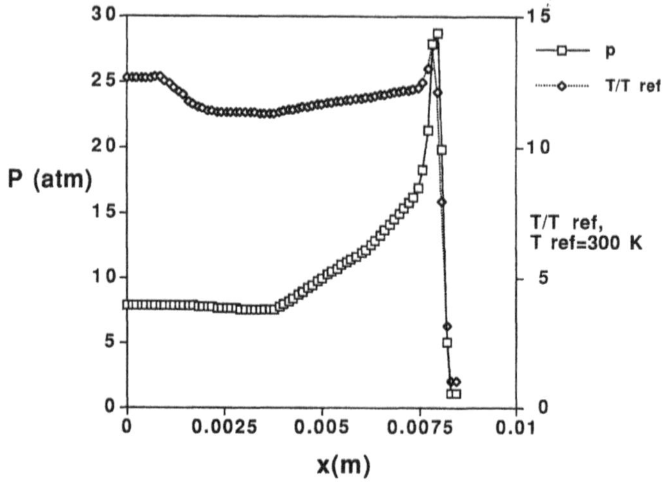

Fig. 20 Temperature and pressure profiles for stoichiometric H_2-O_2 detonation, t = 2.9 μs.

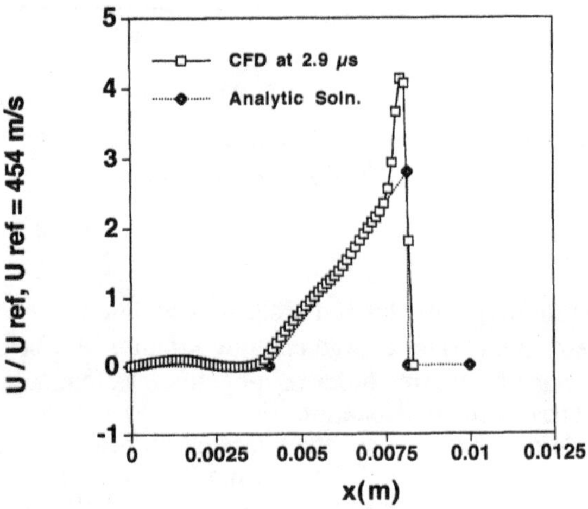

Fig. 21 Gas velocity profile for stoichiometric H_2-O_2 detonation, t = 2.9 μs.

IX. Detonation Initiation in C_2H_4 and CH_4 Mixtures

Ethylene is a relatively reactive, nonstorable hydrocarbon fuel which could be used as a reference fuel in stepping up to high molecular weight hydrocarbon fuels of practical interest. Figure 22 shows the computed pressure profiles for the initial stages of detonation transition after deposition of energy in a quiescent C_2H_4–O_2 mixture. As expected, the C_2H_4–O_2 system was more difficult to detonate than H_2–O_2 and C_2H_2–O_2, requiring initial pressures greater than 30 atm together with elevated temperatures in a small volume. This energy input was sufficient to induce rapid burning in this domain, raising the pressure high enough to create a strong shock wave (with a pressure of about 70 atm), which decays but then induces reaction behind the wave (see the description in Lee[29]). If the reaction is sufficiently strong, the shock then becomes an overdriven detonation wave, which decays to a ZND structure with Chapman–Jouguet wave speed (see Fig. 23 showing how the pressure profile becomes self-similar at long times). These results suggest that optimum energy addition has not been achieved and that reducing the characteristic overdriven initiation would be beneficial to efficient ignitor design. Nevertheless, the transients seen in Fig. 23 show some similarities to the physical processes described in the computations of Clarke, et al.[30] and Sileem, et al.,[31] who completed computations for a related problem with finer grid resolution in the reaction zone but a simpler, global-type kinetics model.

As observed earlier with C_2H_2–O_2 detonation, the heat release profile and convective term treatment promotes the development of secondary shock waves, which were found by Eidelman and Sichel[28] to enhance detonation transition.

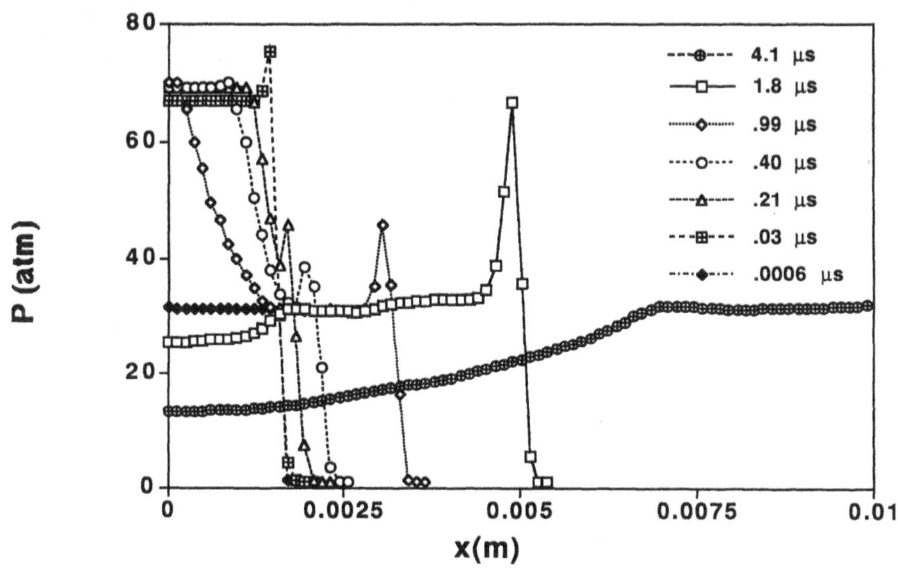

Fig. 22 Detonation initiation seen through pressure profiles in stoichiometric C_2H_4–O_2 mixture, 800 torr initial pressure.

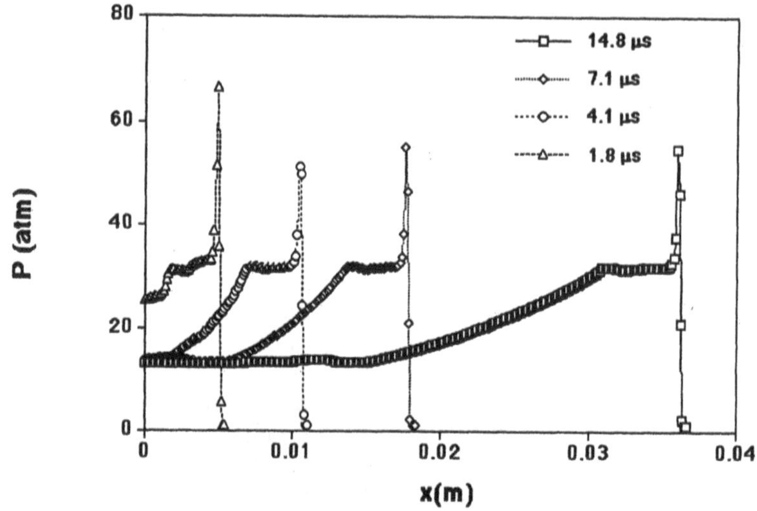

Fig. 23 Detonation propagation seen through pressure profiles in stoichiometric $C_2H_4-O_2$ mixture, 800 torr initial pressure.

Possibly because of the formation of these secondary waves, the $C_2H_4-O_2$ mixture still has not achieved a completely steady structure by the end of the computational domain, a factor to be considered in the initiation of C–J detonation waves with practical, heavy hydrocarbon fuels.

The stability of methane is based on the high-activation energy in the formation of the methyl radical rendering methane mixtures difficult to detonate. This is demonstrated in the transient pressure profiles in Fig. 24, showing that a mixture of CH_4 and O_2 (without additional radicals) could not be detonated computationally, no matter what energy input was applied. Because of the delay

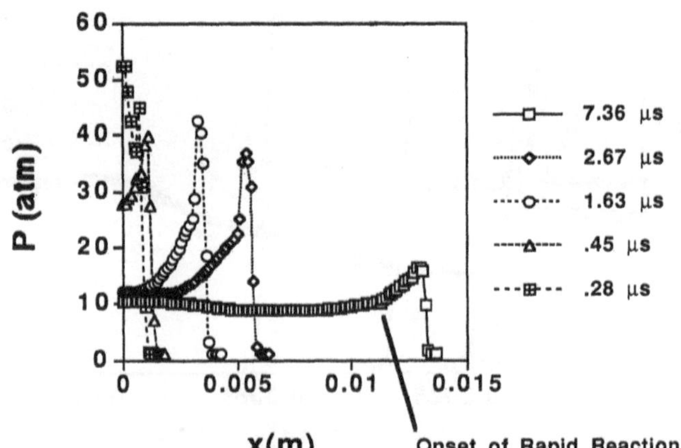

Fig. 24 Pressure profiles for detonation initiation in unsensitized stoichiometric CH_4-O_2 mixture, 800 torr initial pressure.

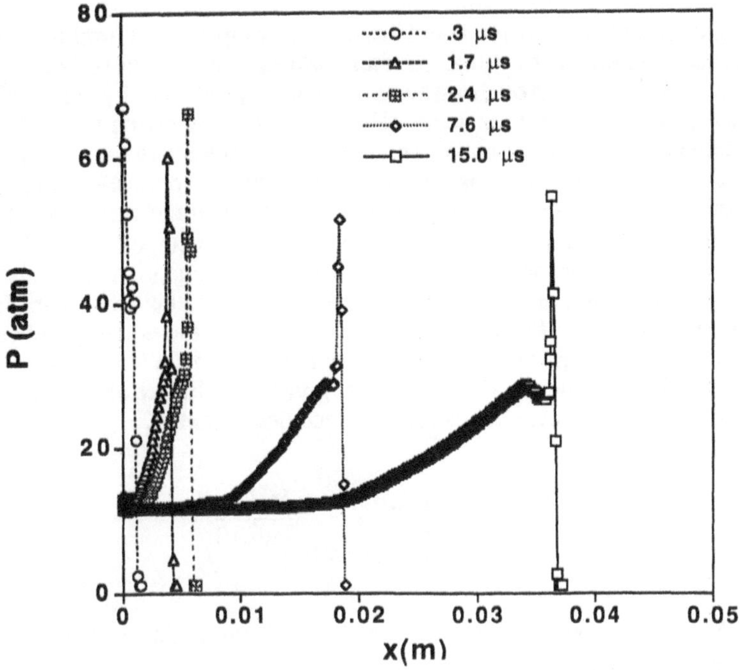

Fig. 25 Pressure profiles for detonation initiation in sensitized stoichiometric CH_4–O_2 mixture, 800 torr initial pressure.

in forming CH_3, burning was not rapid enough to convert the blast wave into a stable detonation wave. The blast wave eventually decayed to an acoustic wave followed far behind by a region of rapid reaction.

When the same input of energy was deposited in a small volume and the region ahead of this volume was "sensitized" with a substantial fraction of methyl radicals, however, an overdriven detonation was achieved, demonstrating the efficacy of fuel conditioning (Fig. 25). Further analysis could help optimize the level of radical addition needed to produce direct initiation with a reasonable energy deposition level. Although methane is of most interest to nonpropulsion applications, heavy hydrocarbons are of more interest to propulsion. Nevertheless, this example, using methane whose kinetics are well enough known for this purpose, has shown the importance of sensitization for practical fuels as well.

X. Direct Detonation Initiation Theory

Although it is noteworthy that planar one-dimensional kinetics calculations are useful in evaluating chemical sensitivity effects on detonation initiation and propagation, they do not provide new understanding of the structure of detonation waves. The dynamics underlying the global characteristics of detonation waves are highly three dimensional and transient, comprised of multiple wavelet interactions producing a cellular structure whose origin is not fully understood. Thus, although the aerothermodynamic states produced by an established detonation wave can be quite accurately predicted by invoking the ZND model, there exists no complete fundamental theory to predict the requirements for detonation initiation. This has led to a semiempirical approach to the problem of defining the critical energy requirement to initiate and to sustain a detonation in a particular fuel–oxidizer environment. Lee[29] developed a correlation for direct initiation based on the conservation of energy equation that depends on ignition delay data, autoignition temperature, and the Chapman–Jouguet conditions. To close the equation Lee also assumed the equality between the initiation source energy and the chemical energy due to the heat liberated by the mixture. The result of this analysis gives the initiation energy and a relationship for the critical radius for sustained detonation as,

$$E_o = \frac{\rho_o k_j I}{2} \frac{\left(C_o M_s^*\right)^{j+3} \tau\left(M_s^*\right)^{j+1}}{\left\{1-\left[1/2\left(M_s^*/M_{CJ}\right)^2\right]^{1/(j+1)}\right\}^{j+1}} \tag{1}$$

where k_j = 1, 2π, and 4π for j = 0, 1, and 2 for planar, cylindrical, and spherical geometries, respectively. Also,

$$I = \frac{(Q + e_o)}{(j+1)\, C_o^2 M_{CJ}^2}$$

and M_s^* is the shock Mach number corresponding to the autoignition temperature and τ is the ignition delay time. The critical radius given by this energy based analysis is

$$R_s^* = \frac{\rho_o k_j I}{2} \frac{M_s^* C_o \tau\left(M_s^*\right)}{1-\left[(j+1)/2\, C_o^2/Q \left(M_s^{*2} I - e_o/(j+1) C_o^2\right)\right]} \tag{2}$$

where C_o, ρ_o, and e_o are the ambient mixture sound speed, density, and specific internal energy, respectively, and Q is the heat of combustion.

Application of Eq. (1) was limited to mainly fuel–O_2 mixtures as shown in Fig. 26. Results of the present one-dimensional kinetics calculations are also given in Fig. 26 for comparison with the data and Eq. (1). In all cases, the trends shown by Eq. (1) are in agreement with the data as are the limited results of the present one-dimensional analysis, which were obtained from computing the initial energy input from the volume of cells with energy deposition, the prescribed initial pressure and temperature in those cells, and an estimate of the mixture specific heat. In refining the comparison between CFD and theory, a wider range of initial temperatures and pressures, as well as a greater range of energy deposition levels (through different computational grids and initial conditions) must be considered.

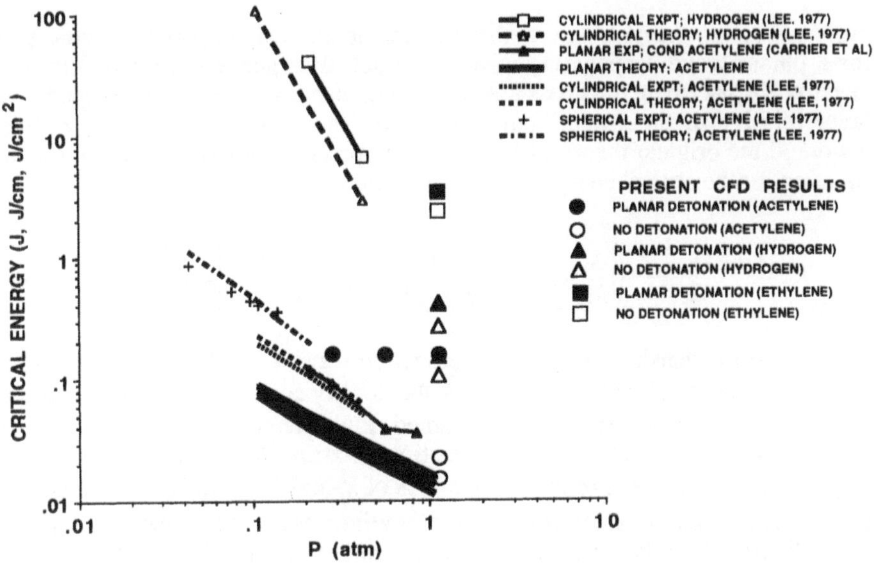

Fig. 26 Critical detonation energy vs pressure: stoichiometric fuel–oxygen mixtures.

However, as pointed out by Thibault[32] more recent work by Murray[33] concluded that certain of the assumptions are overly restrictive, imposing a high degree of uncertainty in application particularly outside of the highly sensitive fuel–oxygen systems considered in assessing Eq. (1). Of particular concern is the "static" assumption involved in equating the source energy to the chemical energy of the mixture. This assumption does not capture the dynamic character of the detonation wave, a feature of particular importance in less sensitive mixtures where air instead of oxygen is the oxidizer. Murray's work included an extensive review and assessment of the literature and other data, including his own. He concluded that flow divergence on the passage of the mixture through the wave front is a more fundamental and, hence, more general basis for the determination of the initiation energy requirement. In particular, Murray found that by combining the fractional increase in flow area relative to the wave front with the chemical reaction length equal to the cell length L_C, a nearly universal value was determined for the critical fractional increase in flow area over the cell length, viz.,

$$\xi_{L_c} = \Delta A/A \cong 0.2 \tag{3}$$

where A is the initial flow area and ΔA is the change in flow area over the length of a cell. Values of ξ greater than 0.2 result in a nonpropagating detonation in most cases. Murray's discovery that the cell length is a reasonable reaction length was demonstrated for spherical and cylindrical waves having a natural divergence as they propagate outward, as well as for planar waves propagating in nonrigid, deformable confinements. It is important to note that for a planar wave propagating in a rigid duct, divergence is provided by the negative displacement thickness associated with the development of the boundary layer relative to the wave front. Based on this indirect evidence, Murray reworked the original theory of Lee and coworkers and arrived at the following equation for the critical energy for direct initiation:

$$E_c = \frac{\rho_o k_j R_s^{j+1} Q}{2} \left\{ \left[\frac{M_s}{M_{CJ}}\right]^2 + \frac{\rho_s}{\rho_o}\left[\frac{j+1}{q+j+1}\right]\left[1 - \left(1 - \frac{L_c}{R}\right)^{q+j+1}\right] - 1 \right\} \tag{4}$$

where ρ_0 is the density ahead of the detonation wave, Q the heat of combustion, M_S the critical shock Mach number, ρ_S the density corresponding to the critical Mach number, q the exponent in an assumed single parameter exponential spatial variation of the postshock, precombustion mixture density and L_C the cell length. The geometry is reflected in values of k_j and j, where $k_j = 1, 2\pi$ and 4π for $j = 0, 1,$ and 2 spanning planar, cylindrical, and spherical waves, respectively. Clearly, in using Eq. (4) a number of empirical parameters are required beyond that apparent in Lee's result given by Eq. (1). In Eq. (1), the critical Mach number M_S is defined by the autoignition temperature whereas in Eq. (4) it is defined by direct measurement for which Murray deduced

$$M_S \cong 0.7\, M_{CJ} \tag{5}$$

Furthermore, in Eq. (1) the ignition delay time appears explicitly whereas in Murray's relationship the cell length L_C is required, a result of replacing the equality of source and chemical energies by the empirically deduced critical expansion rate over the cell length, viz.,

$$\frac{R_s}{L_c} = \frac{\rho_{CJ}/\rho_o - 1}{(1.2)^{1/j} - 1} \tag{6}$$

which, it should be noted is valid only for cylindrical and spherical detonations. For planar waves in rigid tubes, Eq. (6) must be replaced by a relationship involving the divergence due to the negative boundary-layer displacement growth relative to the wave front. Finally, a relationship is required for the critical radius which, through Eq. (6), is obtained from the cell length L_C. From analysis of smoked foil data on cell size and the critical tube diameter, D_C, for the transmission of detonation waves from planar to spherical, it was found that

$$\lambda \cong 0.7\, L_c \cong D_c/13 \tag{7}$$

where λ is the cell width. The critical tube diameter, D_C, was correlated in terms of chemical kinetic parameters via the standard global relationship for ignition delay time in the form (see Appendix B),

$$D_c = k\, U_s [O_2]^a [\text{fuel}]^b RT \exp(E_{act}/RT) \tag{8}$$

where U_s is the critical velocity and a and b were taken from literature values obtained from kinetics studies whereas k and E were obtained from fits to detonation wave data, primarily by Moen et al.[34] (as described in Murray[33]). Results for several hydrocarbon fuel–air mixtures are shown in Fig. 27, for spherical detonations. It was found that direct application of Eq. (4) consistently gives conservative results and, as shown in Fig. 27, adjustments of k values were required. Nevertheless, the trends are very well predicted and provided that care is used in applying Eq. (4), it is a good starting point for the design of ignition systems for PDWEs. Of particular note is the importance of geometry with planar initiation requiring the minimum energy for initiation as implied by Eq. (6) where $R_s/L_c \sim 4\, L_c$ for cylindrical waves and $R_s/L_c \sim 8.4\, L_c$ for spherical waves given the typical value of the Chapman–Jouguet density ratio being on the order of 1.8. Furthermore, the design of the initiator also involves consideration of the energy density and rate of energy deposition to ensure that the blast wave produced by the initiator does not decay prior to achieving the critical radius for sustained propagation. Multiple arrayed sparks, predetonation blast tubes, and chemically sensitized fuels are important considerations in the design of appropriate ignition systems. Finally, as shown by Shchelkin,[35] augmenting the ignition-to-detonation process through turbulence generation is also critical to operation at the rapid firing rates needed to achieve the high-performance potential of pulse detonation engines.

XI. Summary

A major goal of this chapter is to illustrate the importance of analysis in providing insights on how to overcome potential limitations in achieving the remarkable performance made possible by harnessing the efficacy of detonative combustion.

The capability of CFD using the USA code to model the flow processes critical to PDWE operation such as combustion product scavenging and detonation propagation has been shown. A treatment of the source term which promotes capturing of the Von Neumann spike, even on grids which show some smearing of the shock wave, has been demonstrated and is important to structural considerations. A range of nondissipative treatments of the convective term have been tried, with probably the best results with the Lax–Friedrichs Riemann solver with the min–mod limiter with a moderate value of the compression parameter. Computations with extended kinetic mechanisms for H_2 and hydrocarbon fuels show the correct detonation behavior, and the energy input requirements correspond to measured relative detonability, $C_2H_2 < H_2 < C_2H_4 << CH_4$, and are in rough accord with quasitheoretical models of detonation initiation requirements. This was achieved without resolving the structural details, i.e., the cellular structure of the detonation waves. Whether or not this holds up is a matter of our continuing studies. However, the pressure dependence of the reaction rates may need to be included to predict the correct ZND peak pressure for low initial pressures. Simplification of the H_2–O_2 mechanism to eliminate peroxides was found to have only a small effect on the predicted detonation initiation and propagation characteristics.

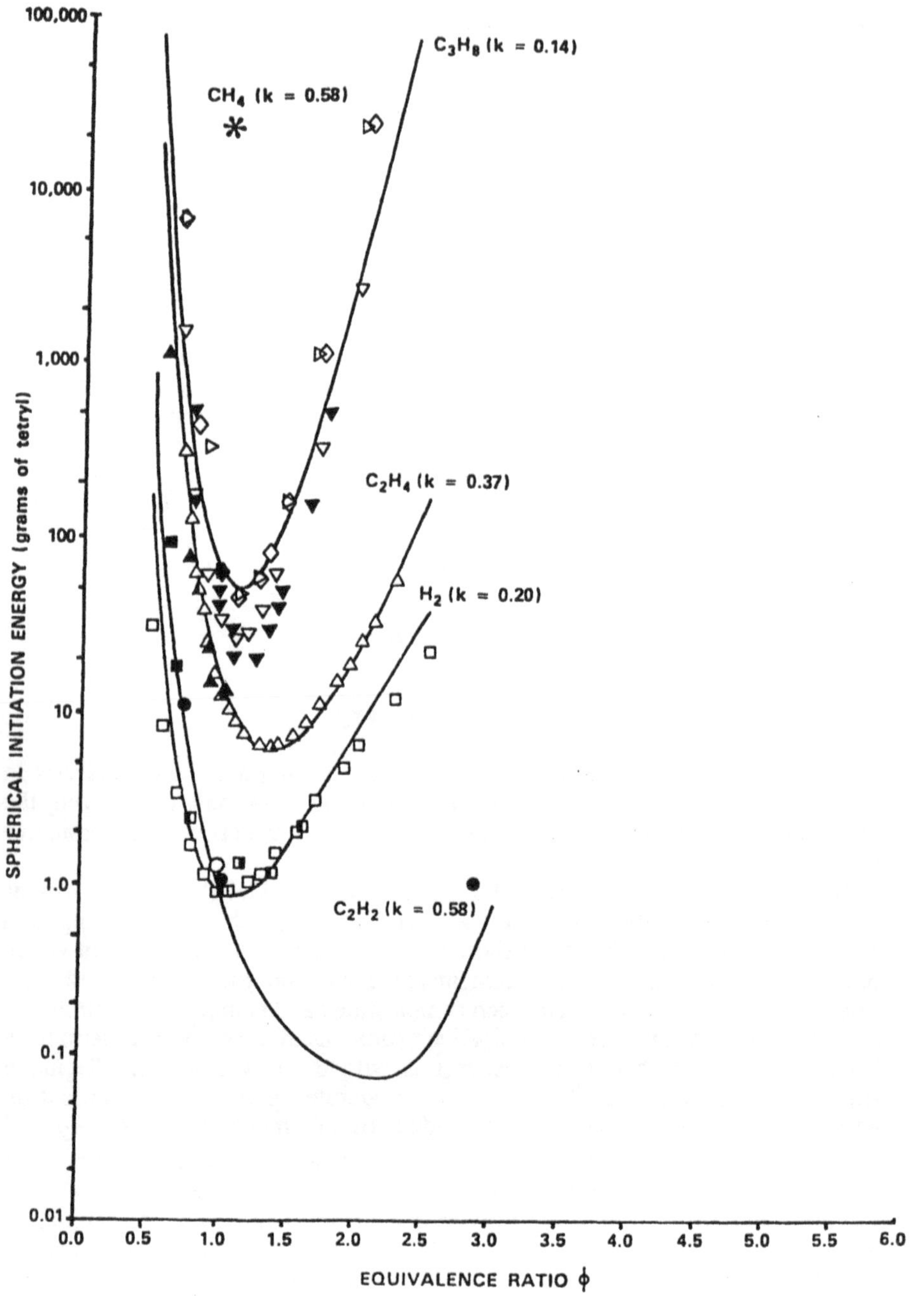

Fig. 27 Spherical initiation energy (grams of Tetryl) vs stoichiometry for various fuel–air mixtures, data as reported by Murray,[33] used with permission.

The angled scoop-type inlet has been identified as a means of reducing substantially the time needed to scavenge the combustion products. It was also shown that upon inflow of fresh air, vortices are created which enhance fuel–air mixing and which must be tailored through careful geometric design to prevent preignition. Finally, the ability to achieve rapid transition to detonation can be considerably enhanced by fuel conditioning including sensitization with appropriate additives.

Although analyses of the type described here are particularly important to the understanding and design of pulse detonation engines, they do not replace experiments to evaluate and refine the observations and conclusions derived from this work. Achieving the high-performance potential at high-firing rates is the overall goal that must be demonstrated experimentally with measurements based on adequate time and spatial resolution of key parameters. Pressure and direct thrust measurements are most important although temperature and species determination should also be given considerable attention. Of fundamental and practical importance is the development of the turbulent mixing field under these highly transient conditions. Thus, fuel injection and mixing in the highly transient environment of the PDWE is another critical process that must be harnessed to achieve adequate micromixing within the time dictated by the operating frequency. Injector design and placement criteria that provide sufficient premixing prior to ignition are required to ensure that direct, repeatable, high-quality (C–J) detonations occur under high-frequency, repetitive operation. Parametric testing in direct-connect hardware offers a sound approach to the development of designs to meet these requirements.

Finally, for sufficiently large engines multitube concepts will be necessary to avoid excessive lengths and diameters that would have a negative impact on operability and performance. In this case airbreathing engines would require valves similar to those used by Hertzberg and Russo[3] and Bitondo et al.[4] (Fig. 1) in their subsonic combustion engines. Although this adds some weight and complexity, it offers positive control on the flow process. Furthermore, flow rectification using fluidic control technology offers the potential for valve reduction, or elimination, and for thrust recovery through appropriate inlet design. Although much remains to be done, many apparent design options exist which reduce the risk in developing PDWE engines capable of high performance relative to competing systems.

Appendix A: Computational Analyses of the Pulse Detonation Engine

Most previous analyses of the PDWE have employed one of two computational methods. In a series of papers, Eidelman and Grossmann,[6] Eidelman et. al.,[7, 9–11] Lottati et al.,[36] and Lottati and Eidelman,[37] put forth a computational methodology in which an adaptive unstructured grid system was combined with a second-order Godunov scheme to solve the multidimensional Euler equations. The dynamic adaption scheme combined with the convective term treatment allowed capturing the detonation waves propagating into the external flow with little shock smearing. Also, the unstructured grid was useful in modeling relatively complicated geometries. The method suffers from the three problems typically encountered with unstructured grids: the relative computational inefficiency of unstructured grids relative to structured grids, the difficulty in implementing implicit time advancement schemes, and the difficulty in viscous term treatments, especially in gridding near the boundaries. Since these are the subjects of current research, however, an unstructured grid methodology may be the method of choice in future PDWE simulations. Avoiding the stiffness due to finite rate chemistry, combustion was simulated in the computations by replacing the ethylene fuel and oxidizer behind the detonation wave by the combustion products CO_2 and H_2O, i.e., complete combustion was assumed in their analysis. This precludes observing the Von Neumann spike and kinetic limitations and can not be used to examine detonation wave cellular structure regardless of other numerical refinements.

The baseline PDWE design geometry was a concept without valves at either the aft or forward ends of the detonation chamber. Designs with forward and aft end initiation were considered with aft end initiation appearing to generate larger average thrust. Parametrics of the inlet design were conducted to determine how to minimize wave drag, and a grill at the front end of the detonation chamber was found to aid in achieving a uniform predetonative mixture in the detonation chamber. Also, analyses were performed of chamber plus aft body configurations, of the PDWE installed in a missile, and of the test geometry of Eidelman et al.[7] which consisted of a valveless, aft end initiated PDWE with a smoothly shaped, scoop inlet. Operation from static $M = 0$ to supersonic external flow conditions was considered; a small decrease was observed in performance with increasing Mach number because of increased wave drag.

Significant conclusions from Refs. 6, 7, 9–11, 36 and 37 were that fresh air refill could be achieved within 6–7 ms (150 Hz operating frequency) even at static conditions and that performance in terms of SFC was of order 1 lb/h-lb, competitive with small gas turbines. Thrust was observed to scale with cycle frequency and detonation chamber volume. Another important observation was that propagation of the detonation waves into the external flow kept the pressure on the thrust surface elevated longer but hindered the reduction of the internal pressure below atmospheric and thus lengthened the time needed for scavenging the combustion products and for refilling with fresh air. This result has important implications in one-dimensional analyses of the flow.

The other type of analyses, one-dimensional, finite rate combustion calculations, were first undertaken by Cambier and Adelman[12] using an operator-splitting method treating the convective terms with a second-order TVD scheme, the diffusive fluxes with central differencing, and the chemical source term point implicitly. An implicit time advancement scheme was allowed in later code implementations;[38] advancing the solution with the large steps possible with implicit methods is especially important with multidimensional, highly stretched grids. Cambier and Adelman analyzed detonation chambers of various area schedules filled with a stoichiometric H_2–air mixture. Possibly because of the simplifications made in their analyses, the PDWE was found to be capable of an extremely high I_{sp} (about 6500 s) and a cycle frequency of up to 667 Hz. A conventional converging–diverging nozzle generated performance inhibiting shock waves, whereas a simple expanding nozzle promoted rapid scavenging and fresh air refilling. Using the MOZART code developed by Cambier and

Adelman,[12] Bussing and Pappas[2] and Bussing, et al.[39] conducted similar analyses of the PDWE cycle for valved airbreathing and rocket PDWE applications; their calculations showed little performance advantage to the forward-propagating, back-end-initiation PDWE design. Furthermore, no impediments can be seen from the results of their computations to achieving the cycle frequencies of 100–200 Hz important in aircraft applications.

Computations have also been performed for some variants from PDWE concepts previously analyzed and tested. Eidelman et al.[40] improved the efficiency of the unstructured, adaptive grid methodology described earlier by combining it with a structured grid for the external far-field flow. With this methodology they analyzed a PDWE fit inside a wing: their computations indicated the configuration would generate significant lift and thrust while operating at a 200 Hz frequency, although no direct performance comparisons were made to other propulsion systems.

Cambier et al.[38] applied a multidimensional version of their methodology described earlier to analyze a use of pulse detonation to enhance scramjet performance. In this design, the pulse detonation wave augmentor (PDWA), mixing is enhanced through interaction of shock waves with mixing layers, thereby generating azimuthal vortices. Thus, the PDWA is an alternative high-speed engine design to the conventional scramjet and oblique detonation wave engines.

Appendix B: Chemical Kinetics Models for Propulsion Oriented Computational Fluid Dynamics Codes

In fluid dynamic computations of the pulse detonation wave engine, one of the key modeling issues is how to describe the combustion kinetics. Over the past 25 years, a hierarchy of modeling concepts have been developed for propulsion applications with H_2 and a range of hydrocarbon fuels.[41,42] A partial list of models appearing in the literature is presented in Table B1.

At the simplest level of modeling are the equilibrium combustion and global kinetic modeling concepts. In the equilibrium combustion model, all kinetic rates are assumed infinitely fast, and so the combustion reactants are in equilibrium with the products with the composition at each point being controlled by the local temperature, pressure, and degree of mixing. The degree of mixing can be determined from solving transport equations for hydrogen, oxygen, nitrogen, and carbon element conservation — which have no source terms because of the requirement elements can not be destroyed. Eliminating the chemical source term enables more rapid transient CFD analysis through moderating any potential chemical stiffness from fast reactions. In many situations involving propulsion with hydrocarbon fuels, however, because of ignition delays, partial oxidation, or incomplete recombination, the steady state is not well represented by the equilibrium combustion model. Within the context of the PDWE, assuming equilibrium combustion would prevent observing the Von Neumann spike and the cellular structure, key features of the detonation wave structure caused by the finite rate kinetic time delay, and would not allow treating the detonation initiation process.

Table B1 Partial list of combustion kinetic models

	Global	Quasiglobal	Reduced Mechanistic	Extended Mechanistic
H_2	Rogers and Chinitz[45]			Oldenborg et al.[21] Drummond et al.[20]
CH_4	Westbrook and Dryer[42]	Westbrook and Dryer[42]	Bilger et al.[47] Rogg[48]	Skinner et al.[49] Chelliah et al.[23]
C_2H_4	Westbrook and Dryer[42]	Singh and Jachimowski[46] Edelman et al.[19]		Westbrook and Dryer[42]
	Hautman et al.[43]	Westbrook and Dryer[42] Edelman and Fortune[44]		Wilk et al.[50]
C_2H_2	Westbrook and Dryer[42]	Westbrook and Dryer[42]		Miller et al.[22] Warnatz et al.[51] Kern et al.[52]
C_nH_m		Edelman and Fortune[44]		

Another relatively simple modeling concept is the global model, in which the entire reaction system is assumed to behave as a process consisting of a very few irreversible or reversible reaction steps. For example, a one-step reaction model of a generic fuel would be represented as fuel + oxidizer $\rightarrow$ products, proceeding at the finite rate

$$W_o = AT^n \exp(-E_{act}/RT)\,[\text{Fuel}]^a\,[\text{Oxidizer}]^b$$

where a and b may or may not be the stoichiometric coefficients. The parameters a, b, E_{act}, A, and n are set to reproduce flame properties, such as the ignition delay, over a range of conditions of temperature T, pressure p, and fuel equivalence ratio ϕ.

As described by Westbrook and Dryer,[41,42] in the one-step global model for hydrocarbon combustion,

$$C_n H_m + \left(n + \frac{m}{4}\right) O_2 \rightarrow n\, CO_2 + \frac{m}{2} H_2O$$

$$W_o = A \exp\left[-E_{act}/RT\right]\,[C_nH_m]^a\,[O_2]^b$$

the parameters A, E_{act}, a, and b are fit to reproduce flame speeds for laminar diffusion flames at different ϕ and T. With two-step global hydrocarbon models,

$$C_n H_m + \left(\frac{n}{2} + \frac{m}{4}\right) O_2 \rightarrow n\, CO + \frac{m}{2} H_2O$$

$$CO + \frac{1}{2} O_2 = CO_2$$

the $[CO]/[CO_2]$ ratio, assumed zero in the one-step model, can also be reproduced.

The ultimate complexity in global models is the four-step global model of Hautman et al.[43] designed to reproduce the flame speed, the $[CO]/[CO_2]$ ratio, and the $[H_2]/[H_2O]$ ratio.

As with the equilibrium combustion model, global models require solving few additional transport equations, making them ideal for three-dimensional CFD analysis; however, they should be applied with caution especially when analyzing problems where ignition or recombination are major issues. Essentially, they are designed to correlate flame properties and may fail when used outside the range for which they were designed. Moreover, since important intermediates are not included in the species composition, global models tend to smear out the details of the reaction zone.

Because of the increase in length in the reaction induction zone, PDWE computations with a H_2–O_2, one-step global model predicted the same qualitative behavior as computations with extended mechanistic models but showed a slightly larger ZND spike. Also, some differences between the global and mechanistic models appeared in the steady composition after the passing of the shock and expansion waves; this was due to neglecting such species as OH under high-temperature conditions, and to the "tuning" of the kinetics of the forward and backward reactions under these conditions.

Extended mechanistic models, which attempt to model the actual intermediate chemical steps in the fuel oxidation process, represent the highest level of chemistry modeling. Because the actual physics is modeled, mechanistic models are well suited for problems, such as ignition, where a predictive rather than a correlative capability is required. For heavy hydrocarbon fuels, however, extended mechanistic models may require several hundred kinetic steps and 30–100 chemical species. Hence, two methods have been employed to retain the predictive capability of mechanistic models within a more computationally efficient methodology. Reduced mechanistic models simplify the mechanism through elimination of reactions unimportant under typical engine conditions and through application of the steady state and partial equilibrium approximations. For high-temperature conditions, H_2–O_2 kinetic models without peroxides (but with appropriate third-body recombination efficiencies, e.g., the seven-step, 6-reactive-species model described in the text), have been found to give results relatively similar to more detailed models. Development of reduced models may be fairly time-consuming, however, and the resulting model may not be a simple mass action law expression. Thus, reduced models have only been developed and applied for a few fuels.

A more popular method for completing combustion analyses of heavy hydrocarbon fuels is the quasiglobal model[19,42,44] where a few global steps to pyrolyze and to partially oxidize the fuel are combined with a detailed mechanistic model for a "lighter" system, lower in the chain, such as the wet CO–H_2 system.

Thus, quasiglobal models work well where thermal cracking of the fuel is fast compared with equilibration between the lighter hydrocarbon components. Because all of the important species are included, quasiglobal models have the potential to represent accurately the steady state compositions but may show inaccuracies in the transient composition of trace intermediates.

Because of their computational efficiency and their ability to capture most of the important physics, the quasiglobal and reduced mechanistic models currently are the best approach for most PDWE analyses with extended mechanistic models reserved for specialized problems such as ignition-to-detonation studies and global models employed in three-dimensional, transient, full Navier–Stokes analyses with several hundred thousand grid points.

Acknowledgments

The NAS system has provided computer time in furtherance of this work, which has benefited these investigations. Also, suggestions on numerical strategies have been provided by S. Palaniswamy of the Rockwell Science Center. The authors are also indebted to Paul Thibault of Combustion Dynamics LTD for pointing out important literature on the subject of direct detonation initiation, particularly the research of Murray.

References

[1]Wortman, A., Othmer, P., and Rostafinski, W., "Detonation Duct Gas Generator Demonstration Program," AIAA Paper-92-3174, 1992.

[2]Bussing, T. R. A., and Pappas, G., "An Introduction to Pulse Detonation Engines," AIAA Paper-94-0263, 1994.

[3]Hertzberg, A., and Russo, R., "The Investigation of a Resonant Wave Engine," Project Squid Conference on Wave Engines and Pulse Jets, Univ., AD-A952624, Prince ton, NJ, Nov. 1954.

[4]Bitondo, D., Kumm, E. L., and Bollay, W., "Preliminary Performance Studies of the Multi-Jet," Aerophysics Development Corp., Rept. 2000-1-R1, Pacific Palisades, CA, 1953.

[5]Helman, D., Shreeve, R. P., and Eidelman, S., "Detonation Pulse Engine," AIAA Paper-86-1683, 1986.

[6]Eidelman, S., and Grossmann, W., "Pulsed Detonation Engine Experimental and Theoretical Review," AIAA Paper-92-3168, 1992.

[7]Eidelman, S., Grossmann, W., Gunners, N. E., and Lottati, I., "Progress in Pulsed Detonation Engine Development," AIAA Paper-94-2721, 1994.

[8]Lynch, E. D., Edelman, R. B., and Palaniswamy, S., "Computational Fluid Dynamic Analysis of the Pulse Detonation Engine Concept," AIAA Paper-92-0264, 1994.

[9]Eidelman, S., Grossmann, W., and Lottati, I., "Computational Analysis of Pulsed Detonation Engines and Applications," AIAA Paper-90-0460, 1990.

[10]Eidelman, S., Grossmann, W., and Lottati, I., "Air-Breathing Pulsed Detonation Engine Concept; A Numerical Study," AIAA Paper-90-2420, 1990.

[11]Eidelman, S., Grossmann, W., and Lottati, I., "Review of Propulsion Applications and Numerical Simulations of the Pulsed Detonation Engine Concept," Journal of Propulsion, Vol. 7, No. 6, 1991, pp. 857–865; also AIAA Paper 89-2446, 1989.

[12]Cambier, J. L., and Adelman, H. G., "Preliminary Numerical Simulations of a Pulsed Detonation Wave Engine," AIAA Paper-88-2960, 1988.

[13]Chakravarthy, S., and Osher, S., "A New Class of High Accuracy TVD Schemes for Hyperbolic Conservation Laws," AIAA Paper-85-0363, 1985.

[14]Palaniswamy, S., Ota, D. K., and Chakravarthy, S. R., "Some Reacting Flow Validation Results from USA-Series Codes," AIAA Paper-91-0583, 1991.

[15]Chakravarthy, S. R., "High Resolution Upwind Formulations for the Navier-Stokes Equations," Lecture Series 1988-05, Computational Fluid Dynamics, Von Karman Institute of Fluid Dynamics, March 7–11, 1988.

[16]Roe, P. L., "Some Contributions to the Modeling of Discontinuous Flows", Large Scale Computations in Fluid Mechanics, Vol. 2, Lectures in Applied Mathematics 22, edited by B. Engquist, S. Osher, and R. C. J. Sommerville, American Mathematical Society, Providence, RI, 1985, pp. 163–193, 1985.

[17]LeVeque, R. J., and Yee, H. C., "A Study of Numerical Methods for Hyperbolic Conservation Laws with Stiff Source Terms," Journal of Computational Physics, Vol. 86, No. 1, 1990, pp. 187–210.

[18]Engquist, B., and Osher, S., University of California, Los Angeles, private communication, Sept. 1992.

[19]Edelman, R. B., Farmer, R. C., and Wang, T.-S., "Combustion and Emissions of Synthetic Fuel Components: Analysis and Modeling," American Chemical Society, Symposium Series 217, edited by W. Bartok, American Chemical Society, Washington, D.C., 1983.

[20]Drummond, J. P., Rogers, R. C., and Hussaini, M. Y., "A Numerical Method for Supersonic Reacting Mixing Layers," Computer Methods in Applied Mechanics and Engineering Vol. 64, 1987, pp. 39–60.

[21]Oldenborg, R., Chinitz, W., Friedman, M., Jaffe, R., Jachimowski, C., Rabinowitz, M., and Schott, G., "Hypersonic Combustion Kinetics," NASP TM-1107, NASA Langley Research Center, May 1990.

[22]Miller, J. A., Mitchell, R. E., Smooke, M. D., and Kee, R. J., "Toward a Comprehensive Chemical Kinetic Mechanism for the Oxidation of Acetylene: Comparison of Model Predictions with Results from Flame and Shock Tube Experiments," 19th International Symposium on Combustion, Combustion Inst., Pittsburgh, PA, 1982, pp. 181–196.

[23]Chelliah, H. K., Law, C. K., Ueda, T., Smooke, M. D., and Williams, F. A, "An Experimental and Theoretical Investigation of the Dilution, Pressure and Flow-Field Effects on the Extinction Condition of Methane-Air-Nitrogen Diffusion Flames," 23rd International Symposium on Combustion, Combustion Inst., Pittsburgh, PA, pp. 503–511, 1990.

[24]Westbrook, C. K., Dryer, F. L., and Schug, K. P., "A Comprehensive Mechanism for the Pyrolysis and Oxidation of Ethylene," 19th International Symposium on Combustion, Combustion Inst., Pittsburgh, PA, 1982, pp. 153–166.

[25]Goldberg, U. C., "Separated Flow Treatment with a New Turbulence Model," AIAA Journal, Vol. 24, No. 10, 1986, pp. 1711–1713.

[26]Lynch, E. D., and Edelman, R. B., "Analysis of Flow Processes in the Pulse Detonation Wave Engine," AIAA Paper-94-3222, 1994.

[27]Carrier, G. F., Fendell, F. E., and Chou, M.-S., "Laser-Initiated Conical Detonation Wave for Supersonic Combustion III," AIAA Paper-92-3247, 1992.

[28]Eidelman, S., and Sichel, M., "The Transitional Structure of Detonation Waves in Multi-Phase Reactive Media," Combustion Science and Technology, Vol. 26, No. 5, 1981, pp. 215–224.

[29]Lee, J. H. S., "Initiation of Gaseous Detonation," Annual Review of Physical Chemistry, Vol. 28, 1977, pp. 75–104.

[30]Clarke, J. F., Kassoy, D. R., and Riley, N., "On the Direct Initiation of a Plane Detonation Wave," Proceedings of the Royal Society of London, A, Vol. 408, 1986, pp. 129–148.

[31]Sileem, A. A., Kassoy, D. R., and Hayashi, A. K., "Thermally Initiated Detonation Through Deflagration to Detonation Transition," Proceedings of the Royal Society of London, A, Vol. 435, 1991, pp. 459–482.

[32]Thibault, P. A., Combustion Dynamics LTD., Alberta, Canada, private communication, July, 1994.

[33]Murray, S. B., "The Influence of Initial and Boundary Conditions on Gaseous Detonation Waves," Ph.D. Thesis, Dept. of Mechanical Engineering, McGill Univ., Montreal, Canada, Dec. 1984.

[34]Moen, I. O., Bjerketvedt, D., Jenssen, A., and Thibault, P. A., "Transition to Detonation in a Large Fuel-Air Cloud," Combustion and Flame, Vol. 61, No. 3, 1985, pp. 285–291.

[35]Shchelkin, K. I., "Detonation of Gases in Rough Tubes," Zhural Tekhnizheskoi Fiziki, 1947, Vol. 613.

[36]Lottati, J., Eidelman, S., and Drobot, A., "A Fast Unstructured Grid Second Order Godunov Solver (FUGGS)," AIAA Paper-90-0649, 1990.

[37]Lottati, I., and Eidelman, S., "A Second-Order Godunov Scheme on a Spatially Adapted Triangular Grid," Applied Numerical Mathematics, Vol. 14, 1994, pp. 353-365.

[38]Cambier, J.-L., Adelman, H. G., and Menees, G. P., "Numerical Simulations of a Pulsed Detonation Wave Augmentation Device," AIAA Paper-93-1985, 1993.

[39]Bussing, T. R. A., Hinkey, J. B., and Kaye, L., "Pulse Detonation Engine Preliminary Design Considerations," AIAA Paper-94-3220, 1994.

[40]Eidelman, S., Lottati, I., and Grossmann, W., "A Parametric Study of the Air-Breathing Pulsed Detonation Engine," AIAA Paper-92-0392, 1992.

[41]Westbrook, C. K., and Dryer, F. L., "Simplified Reaction Mechanisms for the Oxidation of Hydrocarbon Fuels in Flames," Combustion Science and Technology, Vol. 27, No. 1, 1981, pp. 31–43.

[42]Westbrook, C.K., and Dryer, F.L., "Chemical Kinetic Modeling of Hydrocarbon Combustion, " Progress in Energy and Combustion Science, Vol. 10, No. 1, 1984, pp. 1–57.

[43]Hautman, D. J., Dryer, F. L., Schug, K. P., and Glassman, I., "A Multiple-Step Overall Kinetic Mechanism for the Oxidation of Hydrocarbons," Combustion Science and Technology, Vol. 25, No. 5, 1981, pp. 219–225.

[44]Edelman, R. B., and Fortune, O. F., "A Quasi-Global Chemical Kinetic Model for the Finite Rate Combustion of Hydrocarbon Fuels with Application to Turbulent Burning and Mixing in Hypersonic Engines and Nozzles." AIAA Paper-69-86, 1969.

[45]Rogers, R. C., and Chinitz, W., "Using a Global Hydrogen-Air Combustion Model in Turbulent Reacting Flow Calculations," AIAA Journal, Vol. 21, Nov. 4, 1983, pp. 586–592.

[46]Singh, D. J., and Jachimowski, C. J., "A Quasi-Global Reaction Model for Ethylene Combustion," NASA Langley Research Center, private communication, March 1993.

[47]Bilger, R. W., Esler, M. B., and Starner, S. H., "On Reduced Mechanisms for Methane-Air Combustion," Reduced Kinetic Mechanisms and Asymptotic Approximations for Methane-Air Flames, edited by M. D. Smooke, Springer-Verlag, New York, 1991, Chapter 5, pp. 86–110.

[48]Rogg, B., "Sensitivity Analysis of Laminar Premixed CH_4-Air Flames Using Full and Reduced Kinetic Mechanisms," Reduced Kinetic Mechanisms and Asymptotic Approximations for Methane-Air Flames, edited by M. D. Smooke, Springer-Verlag, New York, 1991, Chapter 8, pp. 159–192.

[49]Skinner, G. B., Lifshitz, A., Scheller, K., and Burcat, A., "Kinetics of Methane Oxidation," Journal of Chemical Physics, Vol. 36, No. 8, 1973, pp. 3853-3861.

[50]Wilk, R. D., Pitz, W. J., Westbrook, C. K., and Cernansky, N. P., "Chemical Kinetic Modeling of Ethene Oxidation at Low and Intermediate Temperatures," 23rd International Symposium on Combustion, Combustion Inst., Pittsburgh, PA., 1990, pp 203–210.

[51]Warnatz, J., Bockhorn, J., Moser, A., and Wenz, H. W., "Experimental Investigations and Computational Simulation of Acetylene-Oxygen Flames from Near Stoichiometric to Sooting Conditions," 19th Symposium on Combustion, Combustion Inst., Pittsburgh, PA., 1982, pp. 197–209.

[52]Kern, R. D., Xie, K., Chen, H., and Kiefer, J. H., "High Temperature Pyrolyses of Acetylene and Diacetylene Behind Reflected Shock Waves," 23rd Symposium on Combustion, Combustion Inst., Pittsburgh, PA., 1990, pp. 69–75.

High-Speed Flight Thermal Management

V. J. Van Griethuysen*
U.S. Air Force Wright Laboratory
Wright-Patterson Air Force Base, Ohio 45433-7251

M. R. Glickstein, Ph.D.†
Pratt & Whitney
West Palm Beach, Florida 33410-9600

D. H. Petley‡
NASA Langley Research Center
Hampton, Virginia 23681-0001

H. J. Gladden§
NASA Lewis Research Center
Cleveland, Ohio 44135-3191

and

D. L. Kubik¶
McDonnell Douglas Aerospace
St. Louis, Missouri 63166-0516

I. Introduction

As flight speeds of aircraft increase beyond supersonic speeds into the hypersonic regime, technical difficulties associated with aircraft design and integration grow exponentially. Several design aspects associated with high-speed flight contribute to these difficulties. These aspects include aerodynamic design of the aircraft, design and integration of the propulsion system, and thermal management of the

---

This paper is declared a work of the U.S. Government and is not subject to copyright protection in the United States.

*Mechanical Engineer, Propulsion Systems Branch, Advanced Propulsion Division

†Senior Project Engineer, Advanced Engineering Programs

‡Aerospace Engineer, System Analysis, Hypersonic Vehicle Office, Member AIAA

§Aerospace Engineer, Heat Transfer Branch, Internal Fluid Mechanics Division, Member AIAA

¶Unit Manager - Engineering, New Aircraft and Missile Products integrated flight system. As flight speed increases, the severity of the thermal environment increases rapidly until, at speeds above approximately Mach 4, the ambient air temperature relative to the aircraft is too hot to utilize ram air as an effective coolant. As the heat sink capability of the air disappears, alternate cooling resources must be employed to provide thermal control of the airframe and propulsion systems. The aircraft fuel becomes the primary cooling resource at speeds above Mach 4.0, providing both coolant capability for local absorption and removal of heat and heat sink capacity for storing the absorbed heat. Various fuel types are considered for application in high-speed aircraft, as the cooling capability varies significantly among the differing types. The ability to provide adequate thermal management becomes a critical requirement as speeds increase into the hypersonic flight regime. As a result, the operational speed of the aircraft may be limited by the choice of fuel, alternate coolant, or overall cooling schemes.

Thermal management, what is it? The term has different meanings to different people. To some, it is a series of heat exchangers. To others, thermal management is the ablative material covering the leading-edge surfaces of missiles, or it is the environmental control system of an aircraft. To still others, it is the radiator on a satellite or the open bay doors of the Shuttle during orbit. These are just subsets and localized regions of what constitutes a thermal management system. Other components or factors that are included in a thermal management system (TMS) are the coolant flow network, insulation, active cooling devices, thermal protection systems, and all heat sources and sinks. Thermal management systems vary, depending on the system and applications they support. To obtain a complete picture of thermal management, one needs to look on a global scale of a system or vehicle. Thermal management is an integrated global approach to the task of satisfying the total cooling and thermal energy requirements of a system, within the limits of the resources to that system.

The continual increase in flight vehicle performance requirements, such as increased speed, combined with improved thermal efficiency and lower gross weight, puts a great demand on the efficient use of resources onboard the vehicle. When vehicle speeds reach hypersonic levels, the high-aerothermal load environment and the need to minimize weight drive the designer to seek ways of maximizing the use of all available resources. The concept of thermal management has been an important part of flight propulsion systems since the inception of the jet engine. It is only with the recent interest in hypersonic flight, however, with particular focus on single-stage-to-orbit (SSTO) propulsion systems, that thermal management has been recognized as a critical element, tying together a multitude of disciplines.

Broadly stated, thermal management is the control of thermal energy throughout a system, while the system is in operational use. Thermal energy or heat is derived from a number of sources, such as aerodynamic heating during high-speed flight within the atmosphere, from internally generated heat sources such as the combustor in the propulsion subsystem, or from the sun on the vehicle surface. In some respects, thermal management (TM) can be considered as an accounting method of thermal energy, ensuring that all heat gains and losses are known and accounted for. It is necessary to assure that material and structure temperature limits are not exceeded, and desired component and subsystem operational temperature ranges are sustained. Not only is cooling included to accomplish this but also insulating, and at times, heating. This sounds simple in theory, yet in practice may be complex, depending on the overall system, size, level of thermal loads, material temperature limits, and applications.

For high-speed flight, vehicle TM is the control of thermal energy, resulting in acceptable structural and component temperatures throughout the entire flight, including pre- and postflight. Coolant distribution systems, or active cooling systems, thermal protection systems (TPS) (i.e., insulation, ablatives), and thermal/power transfer devices (i.e., heat exchangers, pumps) for all relevant airframe and engine components, subsystems, and structures are included in an overall TMS. A strong interface is required with structures, materials, propulsion, aerothermodynamics, subsystems, vehicle controls, and environmental control and life support system (ECLSS) disciplines.

With the advent of high-speed flight, particularly hypersonic speeds, TM issues come to the forefront of concern during the design stage. There are several steps the designer must take in order to develop a complete TMS that has been integrated on a total vehicle basis. Within these steps, iterations are often necessary, due to the complexities and interdependencies between the various disciplines.

II. Historical Perspective

An interesting review of the early history of flight propulsion is given by Edwards in Ref. 1. Edwards begins with the reciprocating engines used in the 1903–1914 time period by the pioneers in heavier-than-air flight and continues through the introduction and development of the jet engine. A particularly noteworthy chart (Fig. 1) shows the tremendous increase in speed with the advancement of the gas turbine engine during the 1950s and 1960s. It is in this context that TM began to assume some significance in the design process, even though the concept was not specifically identified or fully developed.

Vehicle performance enhancement goals have been met since the introduction of the jet engine by increasing the thrust-to-weight ratio

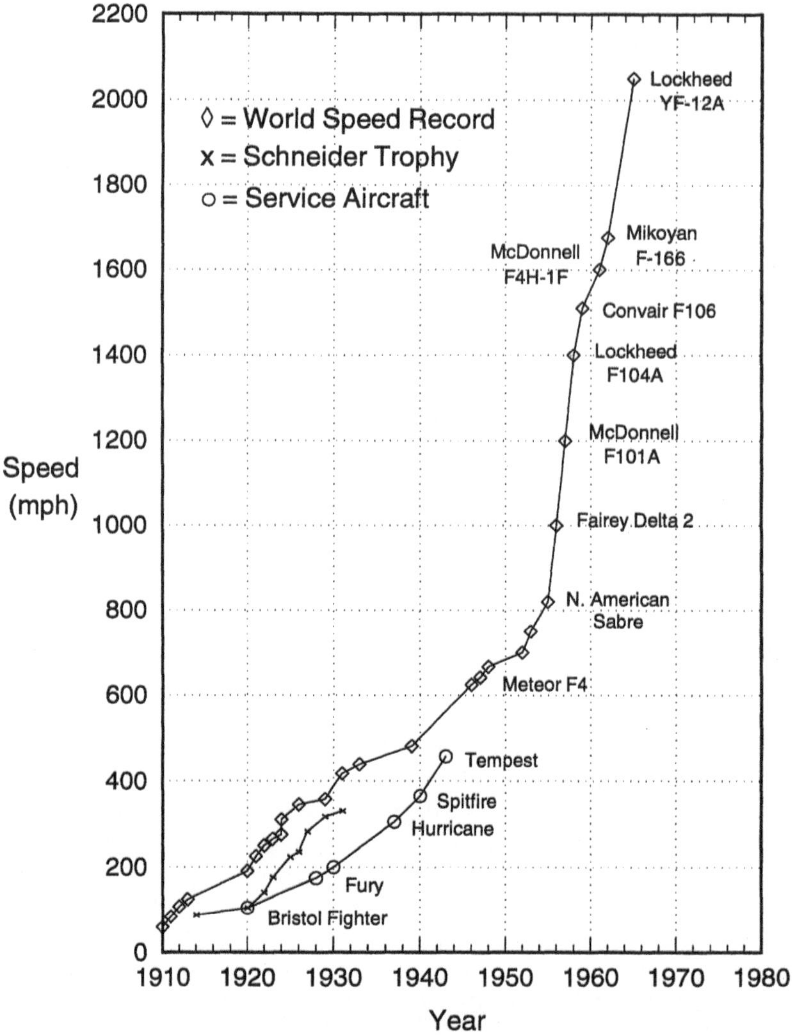

Fig. 1 Aircraft speeds, 1910-1970

and lowering specific fuel consumption of the propulsion system. This has driven industry and government research to find lightweight, high-temperature, high-strength materials and to increase cycle operating temperatures and pressures. As a result, cooling of hot-section components and the use of high-temperature, corrosion-resistant materials became a high priority.

Over the past several decades, propulsive cycles have progressed from turbojet to high bypass ratio turbofans while core cycle pressures and temperatures have been constantly increasing. The engine performance benefits that can be obtained by increased cycle gas

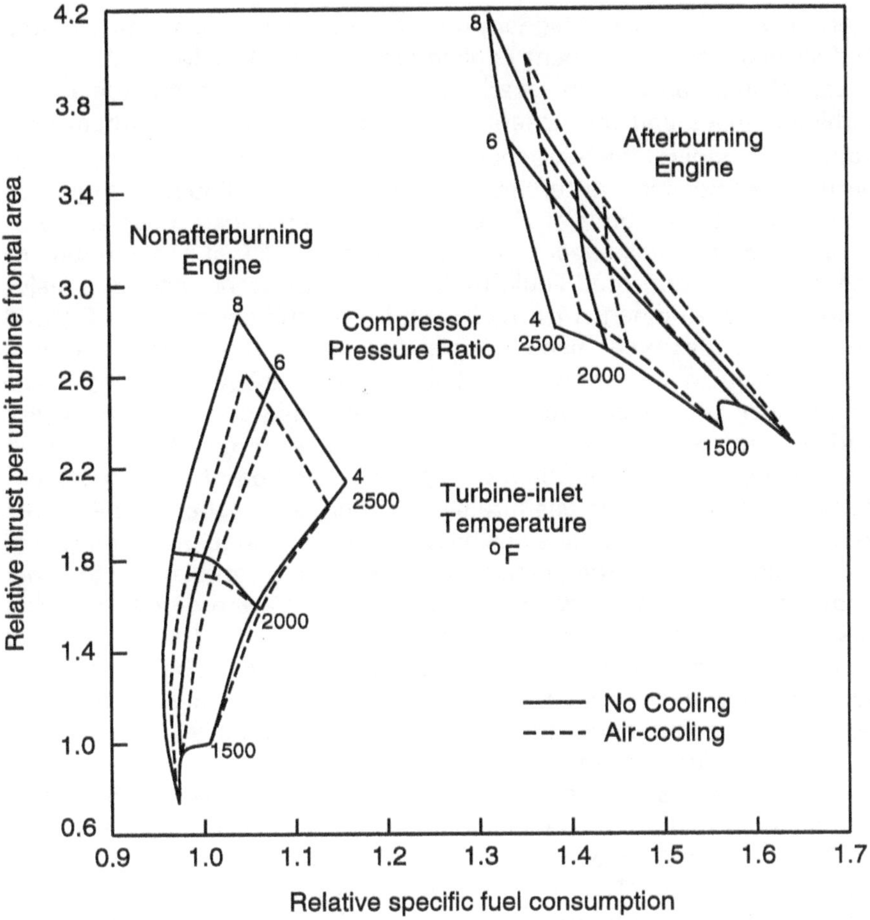

Fig. 2 Effect of air-cooling on turbojet engine performance for various turbine inlet temperature and compressor pressure ratios at flight Mach number of 2 in the stratosphere

temperature and pressure, and the adverse effects of bleeding air from the compressor to cool the engine hot sections are discussed in Refs. 2–5. Figure 2, reproduced from Ref. 5, depicts the effect of cooling air bleed on turbojet engine thrust and specific fuel consumption. This figure shows a decrease in net relative thrust and an increase in relative specific fuel consumption when air cooling is considered. Compressor bleed air has been the coolant, however, because of its limited impact on cycle efficiency and minimal weight impact. Other types of heat exchangers have been investigated; however, they do not compare favorably with the bleed air approach for most applications investigated to date because of their increased weight. The tradeoffs on the thermodynamic cycle of higher pressure ratio, higher gas temperature, and air bleed for component cooling are discussed by Moore.[6]

Esgar[7] summarizes two decades of research into hydrogen fueled gas turbine engines and the benefits to be gained from the fuel heat sink capacity in cooling the hot-section components. When the compressor air used for cooling the turbine is refrigerated with liquid hydrogen as a heat sink, the amount of air required can be reduced by up to 50%. This can substantially decrease the thrust specific fuel consumption of the engine. Flight-rated liquid hydrogen pumps were seen as a major hurdle to development of a hydrogen-fueled turbine engine. Ground support facilities were also recognized as potential problems but were not included in the study. Heat exchanger design technology was found to be at an acceptable level; however, their safety and reliability had not been demonstrated. Many other studies (not referenced) have been made to determine the most effective method of cooling system components so that the gas turbine can operate at higher temperatures and pressures. The conclusions have generally been to use compressor bleed air, despite the resulting lower cycle efficiency to cool components. These conclusions, however, have prompted extensive research into higher temperature materials and efficient cooling schemes in order to minimize bleed air requirements. The use of other types of intermediate heat exchangers and coolant fluids to maintain acceptable component temperatures has generally imposed an excessive weight penalty on the propulsion system.

The Shuttle, also called the Space Transportation System (STS), has components that are reusable and expendable. This required a variety of TMSs, which have developed and evolved over years of flight experience. The STS employs a reusable TMS, whereas the large expendable external hydrogen tank has a TMS that is designed for one flight. The solid rocket boosters (SRB) do not need a TMS, as they are separated at low dynamic pressure when the aerodynamic heating is low.

The STS has both passive and active thermal management components. The TPS used over most of the exterior surface, is made up of ceramic tiles and blankets. Tile sizes vary over the vehicle surface, with the maximum size being approximately 6 in. by 6 in. The gaps between the tiles were designed into the system to allow for the difference in thermal contraction of the tiles and the aluminum structure during orbital cool down. Tile densities of 9 and 22 lb/ft^3 were used in the original TPS design, although tiles with lower densities have been used as the design evolved. Tiles are mounted with a silicone rubber adhesive and a strain isolation pad made of high-temperature nylon. Nylon strips with high-temperature capability are also mounted underneath the tile gaps.[8] The tiles reach their design temperature limit on reentry at high angle of attack, with the tiles on the bottom being exposed to much higher heat flux than those on the top. A black, high-emissivity coating is applied to the outer surface and sides of the tile for locations on the vehicle where large amounts of radiative cooling is needed. Black-coated tiles are used for most of the bottom surface of the STSs and are rarely needed on the top surface.

Gaps between the tiles and steps (as a result of different tile heights) on the surface change during flight. Angles and arrangements of gaps were carefully controlled to minimize gap heating. Many of the gaps, however, had to have fillers. Analysis of flight data has shown that the most severe gap heating occurs at locations of forward facing steps (relative to the front of the vehicle) and occurs late in reentry, while the surface flow transitions from laminar to turbulent, long after peak heating to the tile surface.[8] The effects of steps and gaps on Shuttle tile heating have been studied extensively, using data from wind tunnels and flight data.[8,9]

STS leading edges and the nose cap are made of carbon/carbon to withstand the high thermal and mechanical loads, where peak temperatures are above 1800°F. Also, carbon/carbon has ablative characteristics that provide desirable fail-safe capability. The only significant active cooling used on the STS is in the Space Shuttle Main Engine (SSME) where hydrogen is used to cool the nozzles.

The external hydrogen tank has a low-temperature expendable insulation system. The main purpose of the insulation is to minimize the boil off of hydrogen while on the launch pad and in flight. Lightweight polyurethane is used because of its low density and very low-thermal conductivity. This low-strength insulation is adequate because it must survive only one low-dynamic pressure ascent.

A SSTO type vehicle must function in an environment that ranges from a stationary position on the runway through a flight speed of Mach 25 and then function as a spacecraft. The TM of the vehicle and its propulsion system over such a wide range of conditions challenges the limits of every discipline, including thermal control. The thermal control requirements at Mach 16 are quite different from the on-orbit requirements of space. Several volumes of AIAA Progress in Astronautics and Aeronautics have been written on spacecraft thermal control.[10-14]

Currently, both commercial and military vehicle development programs are expanding the scope of thermal management to be more inclusive of the thermal energy and power balances throughout the vehicle. Hudson and Levin[15] discuss the need to optimize the use of fuel as a heat sink in both the engine and the airframe of military aircraft. The authors show that a proper integration of the control systems improves reliability, permits higher engine inlet temperatures, and reduces the need for ram cooling air. Even this, however, falls short of the definition of thermal management given in the introduction.

Rued, et al.,[16] and Petley and Jones[17] discuss TMS requirements for high-supersonic/low-hypersonic propulsion systems such as the air-turbo ramjet. Rued, et al.,[16] for instance, conclude that these types of engines must be actively cooled using a combination of air and fuel as heat sinks. The heat sink capacity of the fuel is sufficient to meet the overall cooling demands. Advances in high-temperature, lightweight materials, however, are necessary in order to reduce the complexity, weight, and power requirements of the system.

The intent of this discussion was to show that TM concepts have been considered for several decades on a piecemeal and fragmented basis. Even though this is not a comprehensive review of the history of airplane and flight development nor of TM applications, it does give the reader a sense of the evolution of this subject to the present.

III. System Perspective

A. Heat Loads

The design of high-speed aircraft and their propulsion systems is dependent on the aeroheating environment. Aeroheating results from the high-speed flow of air around the aircraft. Compression of the air at or near the stagnation regions and friction between the fluid streaming along the airframe surface convert kinetic energy of motion into heat within a thin layer of air blanketing the body surface. The temperature of this layer increases with the square of the speed. Since this high-temperature layer is concentrated in the air at the surface of the aircraft, heat will flow readily from this boundary layer to the aircraft. The rate this heat transfer occurs also increases with aircraft speed.

As a result of the increased aeroheat loads, thermal related issues and design problems assume a growing consideration in overall high-speed aircraft development. As a consequence of this heating, thermal protection and control become more complex and impact the design requirements of most high-speed aircraft components and subsystems. The crew, equipment subsystems, propellant, and structure must be cooled and/or protected to some degree. The onboard propellant will become involved in TM schemes to mitigate the design impacts of the external aeroheating. These factors all impact the weight, volume, performance, and cost of the high-speed aircraft design.

The high-speed aircraft flight trajectory envelope, body size, contour, and surface characteristics determine the aeroheating (or cooling) and is the basis for selection of surface materials and structural concept, TPS design, and subsystem heat load definition. The severity of the heating depends on many factors, of which the most important is the state of the boundary-layer flow. The boundary layer determines the magnitude of the local heating rates that is dependent on the adiabatic wall enthalpy, as well as the forced convection heat transfer coefficient.

An effective heat sink temperature can vary, depending on the circumstance under consideration. For example, the effective sky heat sink temperature can vary from about 410°R (surface looking at the sky from sea level) to about 200°R (surface looking at the sky from above at an altitude of 100,000 ft.). An aircraft surface looking at Earth will see sink temperatures ranging from 450 to 560°R depending on ground characteristics and season. In the case where the particular aircraft surface is adjacent and "sees" a hot surface, the radiation equilibrium temperature calculation should be modified to account for the radiation geometry and emission exchange factors.

Most high-speed aircraft that operate in the high-heating, high-speed flight regime may have insulation, lightweight structural, or passive TPS components. Thus, the surfaces of the aircraft can be assumed near adiabatic, and equilibrium temperature is a good first-order indication of surface temperature. Heat loads from other sources, such as turbomachinery and subsystems, within the vehicle must be identified. Areas that require temperature control and/or cooling also must be identified. These may include the fuel tank, avionics, and ECLSS.

B. Structural Thermal Response

The structural thermal response of high-speed vehicles is determined by the magnitude of heating (external and internal), thermal properties of the structure and the TMS design. The design of the TMS depends mainly on the magnitude of the surface heat flux, time required for the mission, and thermal properties of the structural material. Several structural materials are compared in Table 1.

Materials with high-strength and high-temperature capability tend to be more dense so that it may be better to use a low-temperature material with insulation or cooling. In addition to averting structural material temperature limits, insulation or active cooling are used at heated surfaces to prevent overheating components or payloads and excessive thermal gradients or distortions of the structure. If the mission time is short enough and/or the heat flux is low enough, a simple TPS design is used, and the thermal response of the structure can be calculated by transient conduction analysis. For longer flight times or higher heat fluxes, active cooling must be used and a coolant distribution system is needed. As shown in Fig. 3, cooling systems use fuel as the heat sink, either directly or indirectly through a secondary cooling loop. Often the cooling system can be analyzed with the assumption of quasi-steady flow so that a series of steady-state analyses at fixed points in time are adequate. A transient thermal fluid

Table 1 Structural Materials

	Al 2024	Titanium 6242	Inconel 617	ACC
Max. use temp., °F	350	1000	1600	3000
Operating temp., °F	250	800	1300	2700
Strength F_y, psi	33,000	96,000	40,000	40,000
Density ρ, lb_m/in^3	0.1	0.164	0.302	0.071
Specific strength (F_y/ρ)	330,000	585,000	132,000	563,000

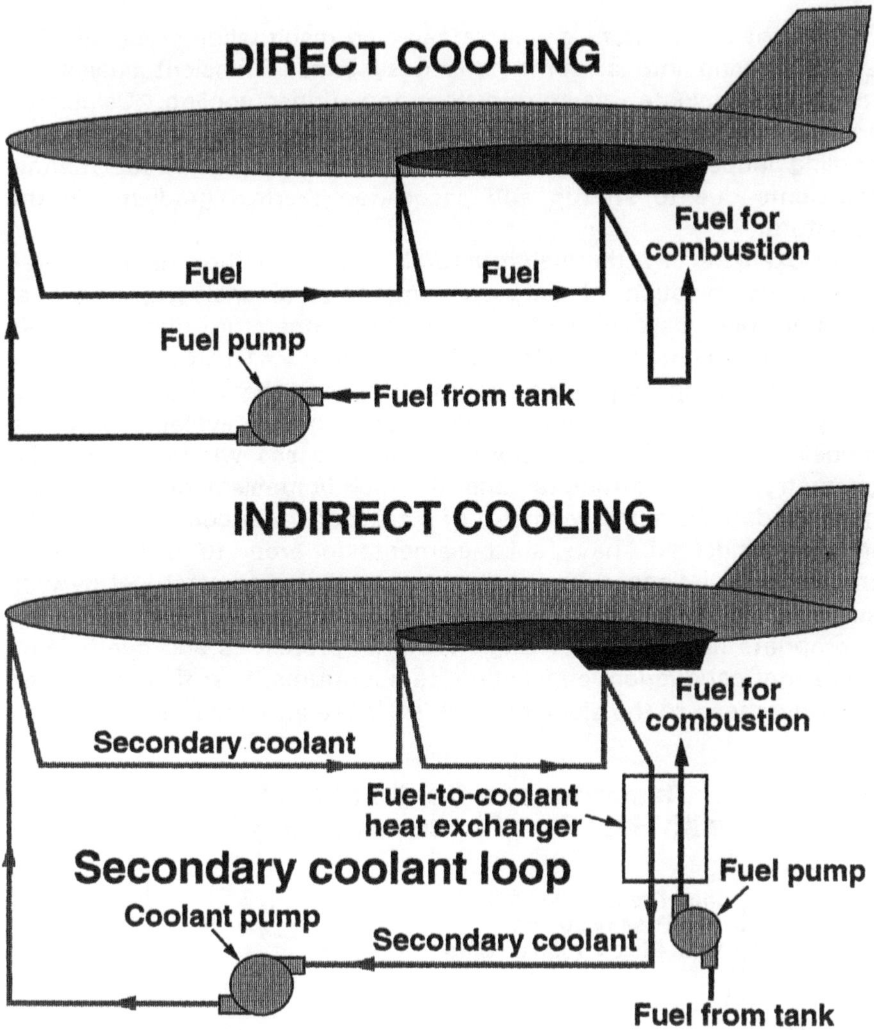

Fig. 3 Direct and indirect cooling systems

analysis is necessary for rapidly changing coolant flow rates or heating rates.

Temperature gradients in the structure may be a problem at locations on a vehicle where the heat flux to the structure is high and/or no insulation or active cooling is used. Temperature gradients are often significant at leading edges and all surfaces that are in the engine flow path or exposed to the external plume of heated gases. Studies of a proposed SSTO vehicle showed that either TPS or active cooling concepts were needed on most of the windward side, leading edges, all surfaces in the engine, and the afterbody external nozzle.[18]

Significant temperature gradients can also result when cryogenic fuel starts flowing into a warm cooling system. Transient analysis is needed to evaluate this startup flow and sudden cooling. Calculated transient response of a cooling panel is shown in Fig. 4. Secondary cooling loops, as shown in Fig. 5, can be used to reduce thermal transients due to startup and associated thermal gradients in the structure.

An example of a thermal/structural analysis method demonstrates how a combination of finite-element, thermal analyses, and heat transfer codes can be used to determine temperature distributions in large structural models.[19] Here, with the aid of PATRAN (developed by PDA Engineering), a pre- and postprocessor to thermal and structural analyses software programs, an engineer is able to develop appropriate models to determine temperatures and stresses within a particular geometry. Both structural and thermal finite-element models are generated of the same geometry of a component or components. The thermal model will have solid elements in order to facilitate heat transfer calculations, whereas the structural elements can get by with surface elements. Engineering judgment is used to determine appropriate element grid sizing. Material properties and coefficients (radiation, conduction, convection, temperatures, heat flux) are added and referenced to the elements. When these and any other requested

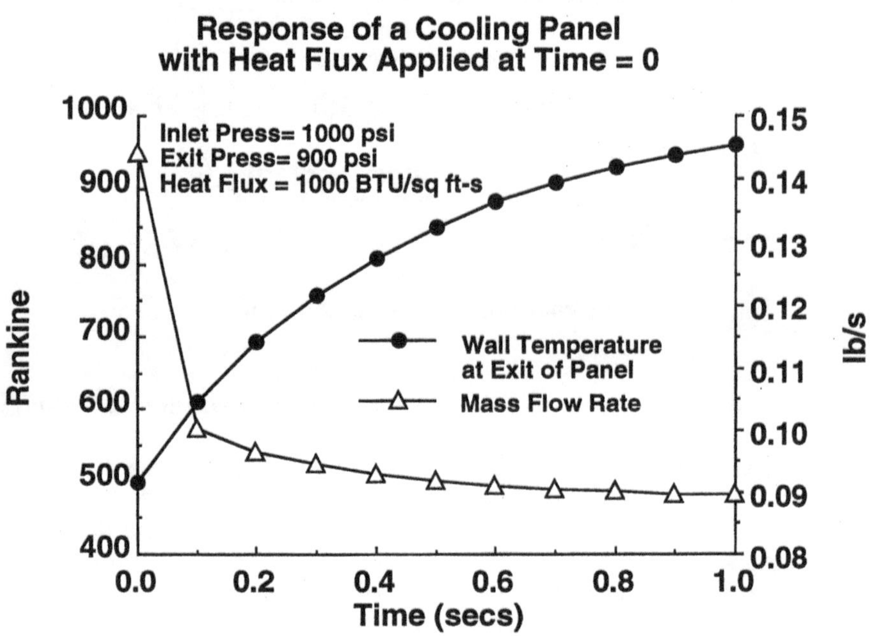

Fig. 4 Transient response of a cooling panel

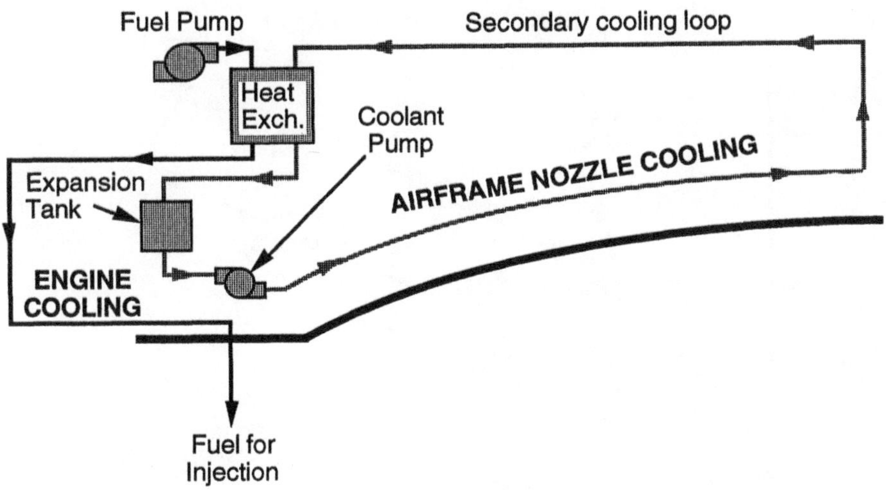

Fig. 5 Secondary cooling loop for an external nozzle

data or parameters have been input, the thermal analyzer software is run. The resulting temperature distributions can be plotted and checked for reasonableness. Once the thermal engineer is satisfied with the results, the data can be applied to the structural analyzer to check for stresses.

C. Propulsion System

Since Hypersonic vehicles must operate over a very broad flight envelope, from sea level takeoff to high-speed (Mach 8 and above) cruise at extreme altitudes, the design of the vehicle and propulsion system requires striking a balance in the selection of performance optimization conditions and operational characteristics (Fig. 6). The propulsion system must be capable of satisfying the thrust requirements of the vehicle over the entire flight envelope. Because of the operational limitation of gas turbine engines to speeds generally below Mach 4, imposed by current material limitations, and the inability of ramjets and scramjets to operate at low flight speeds, propulsion systems for high-speed flight must operate under different modes by incorporating several propulsion concepts. As an example, a vehicle designed for Mach 5.0 flight might use an integrated turbojet-ramjet or turboramjet, operating in turbojet mode during acceleration from takeoff to Mach 2–3, with the ramjet mode starting operation at speeds above Mach 1.5–2.

The propulsion TMS includes the design of a thermally and power balanced fuel system, providing protection to the internal surfaces of

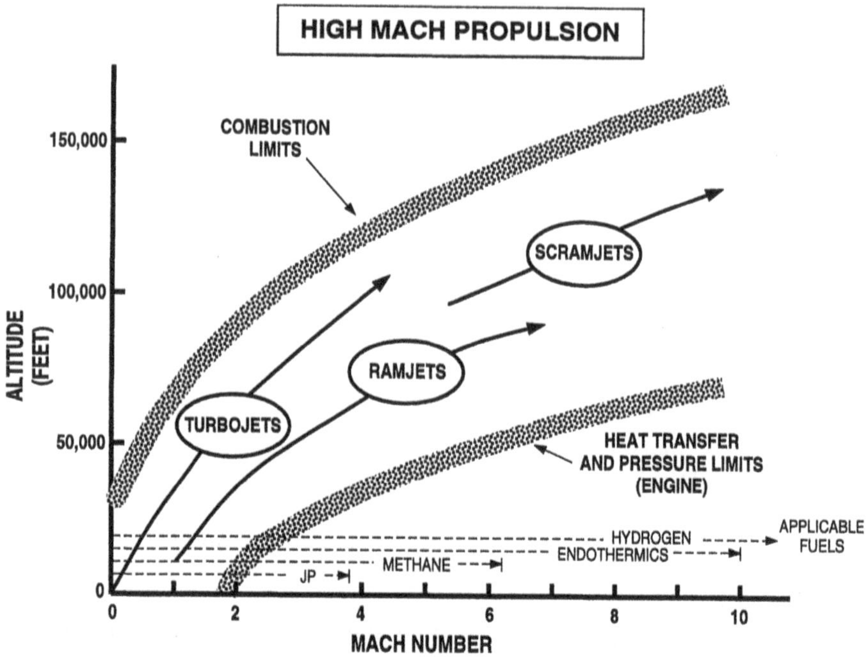

Fig. 6 Flight envelope of high-speed airbreathing engines

the engine flow path, all related components and the engine external surfaces (unless formally agreed to be covered by the airframers). Fuel is used as a coolant for the entire aircraft before it is supplied to the engine. Two types of fuel, cryogenic and hydrocarbon endothermic fuels, have been used to provide the cooling needed for airbreathing hypersonic flight. The required fuel flow to the propulsion system influences the design of the entire vehicle cooling system. Heat loads inside the engine, particularly in the combustion region, are generally so large that they have a strong impact on the engine design. Conservative estimates indicate that above Mach 20, the fuel will have cooling requirements that exceed the requirements for combustion by as much as four times.[20] Steps and gaps (where structural panels join) must be minimized inside the hypersonic airbreathing engine because of the high dynamic pressures and temperatures. Forward facing steps in particular, must be avoided whenever possible because of the amplified heating that these steps produce.[8] Large amounts of engine surface area exposed to heating may mean that the fuel flow will be higher than the ideal amount needed for combustion in order to satisfy the cooling requirements. As might be expected, the cooling system for the engine must be optimized for minimum weight and minimum fuel flow for cooling.

The heat loads inside the engine are often high enough to justify the use of film or transpiration cooling. However, additional fuel or other coolant must be carried on the vehicle to be used for these cooling approaches. The weight and volume penalty associated with the additional fuel or coolant comes at a significant price, which must be compared with the advantages of the added cooling and/or performance provided, to justify their use. These types of cooling are typically considered for areas where the heat flux is approximately 1000 Btu/ft^2-s or higher. Film cooling may be used to achieve better propulsion performance, such as greater thrust, regardless of the level of heating at the surface. Leading edges and combustor internal surfaces are the most likely locations for these cooling methods.

Film cooling is accomplished by injecting fuel parallel to the wall, such that part of the fuel cools the wall and reduces the incident heat flux until it is mixed and burned. Film cooling also reduces skin friction, while providing an increase in engine thrust. Transpiration cooling is similar to film cooling, except that the fuel flows through a porous wall, entering the propulsion flow path with relatively low velocity. Transpiration cooling could be used in place of film cooling where it shows some advantage in weight, required coolant flow, or engine performance. Gap seals and seals between moving parts can also be effectively cooled with transpiration.

The heat loads associated with cowl and strut leading edges are a concern, as these heat fluxes can be greatly intensified by shock impingement, particularly when a shock-on-shock interaction (multiple shocks converging and impinging on a surface at the same location) is present. A single shock impingement can result in heat loads that are more than double the undisturbed level. Shock-on-shock occurs on the cowl leading edge when an oblique bow shock comes off the vehicle's inlet compression ramp or nose and impinges on the bow shock wave of the leading edge.[21] The phenomena of shocks converging on a leading edge was investigated by Edney during the late 1960s as described in Ref. 22 and subsequently investigated by others.[23-26] With the reinterest of hypersonic flight in the 1980s and 1990s, the phenomena of shock impingements came into focus again during the National Aero-Space Plane (NASP) program. This was due to concern over the projected heat flux amplification rates and their impact on cooling mechanisms to handle the resulting temperatures. During experimental tests, heat transfer rates as a result of extreme shock impingement conditions to the leading-edge surfaces were amplified over nonshock conditions by various factors. In 1988, Wieting[27] and Holden et al.,[28] saw amplification factors from 6 to 30 or more. Investigations were made to determine methods of reducing the peak amplification factors. One approach by Glass et al.,[29] was to look at the effects of the cowl sweep angle. Experimental tests revealed a decrease of peak amplification factors by approximately 40% for a sweep angle of 45 deg. Wieting[30] summarizes this further, indicating that changing the flow angle can further reduce the peak an additional 25%, resulting in a peak heat transfer rate of 24,500 Btu/ft^2-s.

For engine flow path performance, it is desired that the cowl leading edge be as thin and razor sharp as possible. From a thermal view point, it is desired to have the leading edge be as blunt as possible to facilitate heat dissipation and cooling. For the thin leading edge, experiencing a multiple shock interaction places a particularly heavy burden on the structure, at the very location where the wall thickness is small. This also makes it difficult to identify potential cooling concepts not only to handle the peak heat fluxes but also to be small enough to fit within the confined space. Ultimately, there must be a compromise between the engine flow path requirements, the design of the cooling concept, the material selection, and the manufacturing techniques.

D. Airframe

The airframe is an important part of the integrated approach to total vehicle TM. The airframe employs TPS concepts against the severe external environments and extreme temperature gradients between hot external surfaces and colder internal fuel heat sinks. These gradients influence structural design and can be most severe when cryogenic fuel and oxidizers are carried onboard. The vehicle TMS design and control features integrate the propellent storage and handling, and equipment and crew cabin thermal control and structural thermal control systems with the propulsion system.

Propellant system design support is accomplished through local and vehicle level TM analysis, establishing the fuel tank wall temperature response with the engine operating and nonoperating through all mission phases. The design process will size the fuel tank and subsystem component insulation requirements, taking into account total fuel heat load and local heat source or heat sink impacts. This analysis will be accomplished for all fuel and/or oxidizer types or molecular species that need to be part of a total integrated assessment. Fuel tank thermodynamics are an integral part of the tank external structure TMS analysis process as the tank provides a heat sink for the structure. The fuel TMS concept selection analysis may be impacted by the following requirements or integration considerations: 1) fuel tank geometries/structural concept/materials/structural attachment, 2) mission profiles/fuel depletion rates, 3) transient fuel state/stratification/location in tank, 4) ullage stratification/pressurant gas characteristics/ullage pressure management (ullage spray), 5) fuel slosh/slosh control baffles, 6) acceleration forces/fuel and pump(s) location, 7) tank insulation characteristics/external heat loads/energy management limits, and icing, 8) ullage gas-liquid interface and wall interface convective heat transfer coefficients, 9) tank pressure management/pump feed requirements, 10) fuel heat sink/recirculation, 11) fuel heat storage/phase change characteristics, 12) fuel boil off and venting design impacts, 13) residuals left in empty tank during flight, 14) empty tank thermal/pressure control, 15) ground support/ fueling/ ground hold, 16) mission and takeoff abort safing, 17) post landing heat soak and safing, and 18) access/inspection/instrumentation/ handling/mission cycle life.

Structure and TMS integration is accomplished with concept selection and acreage distribution as part of a flight trajectory thermal boundary tradeoff design process. Design and structural integration of selected TPS concepts are accomplished along with detailed airframe design. Thermal protection may be passive, active, or phase-change in nature, and may be located external to or designed as part of the external primary structure. The passive systems generally include insulation and air gap layers to limit heat transfer. The active systems use fluid or vapor flowing through structural heat exchangers or heat pipes, transferring and removing heat. The phase-change systems use a combination of the two approaches. Typically, passive systems and insulation are considered over active systems because of their lower weight, increased reliability, easier production, better inspectability, and overall lower risk. Phase-change systems offer weight advantage over passive active systems in specific applications but may require development. Whether passive, active, or phase-change systems are selected, trade and sensitivity studies need to be conducted to optimize the final design. The TPS design process includes determining maximum surface temperatures, selecting materials and concepts, optimizing the TPS designs, and conducting development tests. If thermal protection is required, the type and amount of coverage is a function of airframe aerodynamic and exhaust heating analysis results. These data provide radiation equilibrium surface temperatures for the vehicle or component's external hot surfaces so that maximum surface temperatures can be identified. Radiation equilibrium temperature is defined as the temperature of an external surface when the convective heat input is in balance (or equilibrium) with the radiation heat loss to a given sink temperature. Maximum temperatures are then used to select TPS materials and concepts and to support mission/ configuration/trajectory trades.

Rocket propelled vehicles typically fly at low dynamic pressure on ascent and have insulation systems that are sized by reentry heat loads. Airbreathing propulsion vehicles, such as ramjet or scramjet powered vehicles, must fly ascent trajectories that have higher dynamic pressures to achieve efficient engine performance, with their insulation and active cooling systems sized by both ascent and reentry. For an airbreather, the ascent trajectory must be chosen so that the dynamic pressure and required weight of thermal protection are optimized. Active cooling is often required on leading edges and on the surface of external nozzles where the hot combustion gases impinge.

The weight per square foot of active cooling systems must be minimized, particularly for a large external nozzle, such as would be used on a SSTO vehicle. Indirect cooling with a secondary coolant has been studied for weight savings on nozzles and inlet ramps of hypersonic aircraft. The secondary coolant fluid may be chosen to be compatible with lightweight materials yet may not be compatible with the fuel. This approach has been applied to Mach 5 cruise aircraft and to an SSTO vehicle, as described in Refs.[31-33]. An example of an actively cooled inlet structure design for a Mach 5 cruise vehicle is shown in Fig. 7. An integral cooling panel design with indirect cooling of silicone oil was used to minimize weight. Cooling panels for external nozzles that are constructed with insulating ceramic materials have also been considered to minimize weight or coolant flow requirements. Metallic and ceramic cooling panels are compared in Fig. 8.

In the conceptual design phase, simple, quick, and reasonably accurate methods are used for airframe fuselage, inlet, and exhaust

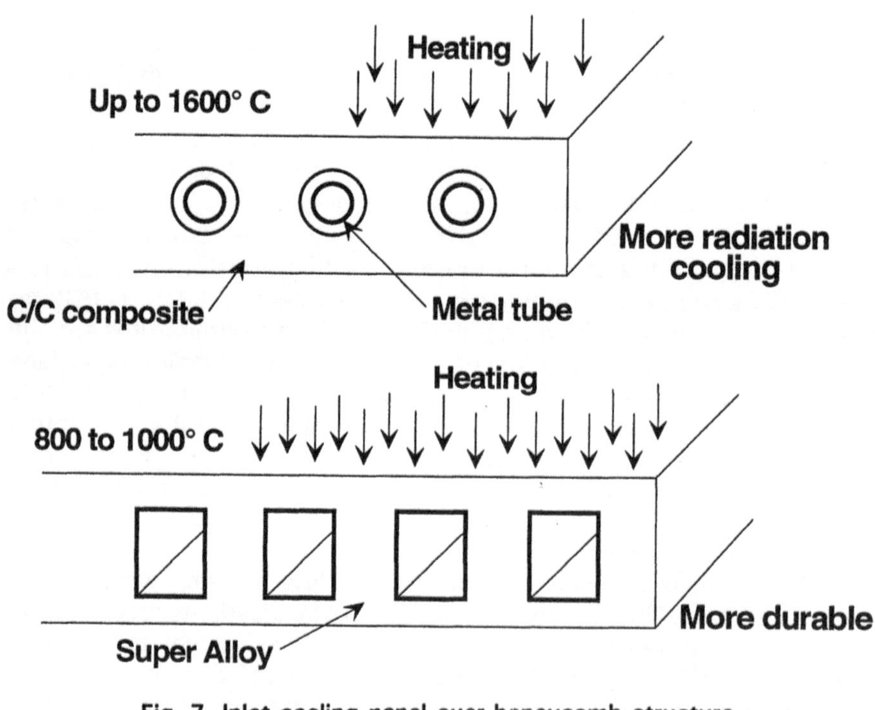

Fig. 7 Inlet cooling panel over honeycomb structure

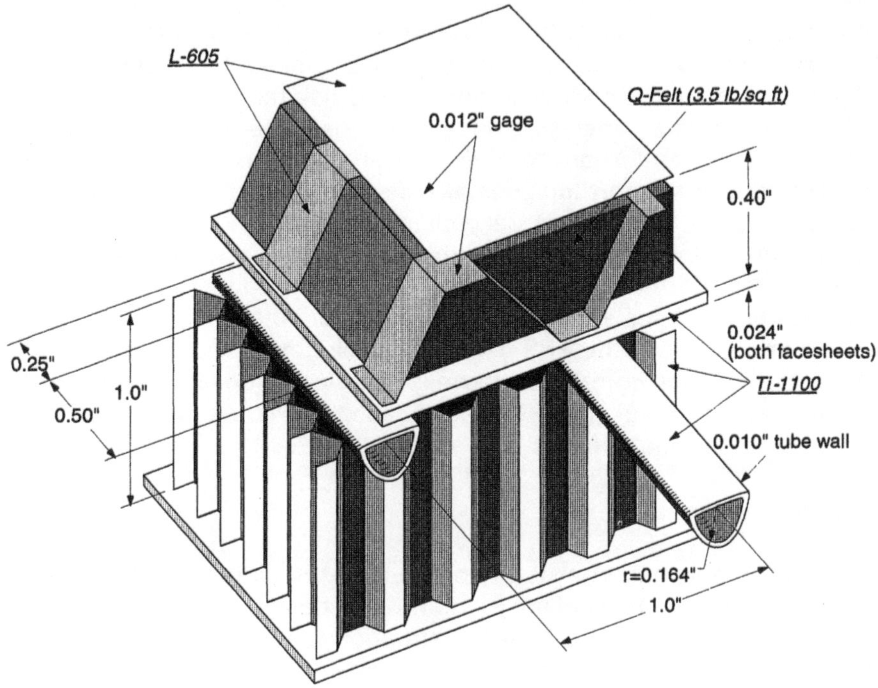

Fig. 8 Cooling panel designs

heating analyses. For preliminary design, the conceptual design methods are again applied in combination with more accurate, yet time consuming, approaches to improve accuracy of the design database. Computational fluid dynamic (CFD) analyses and wind tunnel tests of the configuration are also introduced to access three-dimensional flowfields, shock impingement, flow separation, and real gas effects. Detailed design includes preliminary design tools and database, combined with local phenomena testing and analyses and flight test data.

After maximum temperatures are determined for the vehicle or component, an analysis is conducted to determine whether thermal protection is required and if so, which type should be used and how it should be integrated into the overall design. Studies are conducted to assess the advantages and disadvantages of protected vs hot structures, and the use of active vs passive cooled or phase-change TPS, for the particular application. Candidate concepts and material systems are selected by the design team. Simplified thermal mathematical models for the candidate concepts are developed for evaluation under transient flight environmental design conditions. For complete analysis, ground hold and turn-around times are also included. The candidate systems are assessed and ranked for weight, cost, reliability, producability, inspectability, and risk factors.

Detailed thermal analysis models are used to accomplish more extensive optimization of the active cooling and passive TPS designs. Detailed design support includes assessment of such features as TPS fasteners, gaps, and structural splines and joints. More extensive verification tests are likewise conducted to assure a viable design and to validate performance of particular features. Coupon, component, and sub- and full-scale development, and ground and flight tests may be required to verify the TPS design. Development tests may include either separate or combined thermal, physical property, acoustic or dynamic environment measurements based on requirement definition and simulation requirements.

E. Performance

Thermal management system analyses that track the entire vehicle's energy can be a mechanism for improved design and performance. Total mass flow energy assessment includes ingredients such as internal energy, system entropy, work, gravity, and kinetic, chemical and radiation emission energy, all related to some reference state.

Evaluating performance of various TMS concepts must ensure that design requirements are satisfied by each alternative. Requirements include vehicle structural temperature and temperature gradients for specific load conditions, propellant tank thermal protection and energy management, engine and flow path thermal protection and coolant flow management, subsystem components and leading-edge thermal control. Vehicle operational considerations, such as ground operation, takeoff and ascent, cruise, and descent and landing requirements also impact the TMS concept selection process. The vehicle geometry and engine integration can significantly influence local and global heating and TMS design requirements.

The TMS controls the routing of the fuel (and oxidizer) through the network of airframe and engine heat exchangers. The flow is managed to prevent any of the cooling panels, local structures, and components from exceeding maximum allowable material temperatures. To maximize efficiency of the available coolant flow, the TMS should control the flow distribution to the individual panels. The heating distribution on the airframe and engine is a function of trajectory, vehicle stability, and configuration, with peak heating occurring at various locations at different Mach numbers. Local shocks and shock interactions impact local heating levels, driving the cooling requirements for that area. These localized heating spikes and discontinuities must be closely managed. The resulting peak heating is further compounded with increased flight speed, where the coolant flow rate required can exceed the flow required for stoichiometric combustion. Thus, for efficient use of the fuel as a coolant, the flow can be minimized by maintaining heat exchanger temperatures close to their material temperature limits and by using a cooling network to serially allow for higher fuel temperatures throughout the system.

Once all of the vehicle's heating and cooling requirements and the temperature limits have been identified, priorities are set to establish the cooling loop, or network routing, and flow rates. If the fuel is to be used as the coolant, then combustor temperature requirements must also be considered during the design of the cooling network. Plumbing and pumping requirements are established, including valves and piping size.

IV. Airframe/Propulsion Thermal Integration

A. Integrated Cooling Network and Coolants

Design studies, conducted to evaluate the feasibility of aircraft systems for hypersonic flight, have defined several interesting features related to the use of fuel for flight propulsion and cooling. First, the choice of fuel for a selected application must consider the defined mission and its operational requirements. Because thermal capability of fuels vary over such a wide range, the fuel must be chosen with due consideration to the desired flight speed and attendant environmental impact. Furthermore, the aircraft design is strongly impacted by the fuel choice, since the significant properties (i.e., density, heat of combustion, heat sink, boiling point, etc.) vary widely for the range of available fuels. The choice of fuel may impact the vehicle size and configuration, as well as the type of fuel storage and handling systems both in the aircraft and in ground supply systems. The most significant feature that has been illuminated by studies to date is the requirement for a very high level of integration between the fuel system and the airframe/propulsion thermal management systems.

The design of an integrated cooling network for a hypersonic vehicle begins with the selection of potential fuels and/or coolants to be used. Several classes of liquid fuels are currently available for use in aircraft, including conventional blended hydrocarbon fuels, cryogenic fuels, and pure or special blends of hydrocarbons. Each of these fuel classes has characteristics that lend it to certain applications. The fuel to be selected has the highest possible combustion energy per pound with adequate cooling capacity to accommodate the highest heating conditions of the flight. For flight Mach numbers of approximately 8 or less, endothermic hydrocarbon fuels, such as methylcyclohexane (MCH), have sufficient cooling capacity. Secondary coolant loops may have significant advantages in the lower Mach number range. They may be compatible with lightweight structural materials or designs, and they may provide additional safety or flexibility of operation.[32] For Mach numbers above 6–8, cryogenic fuels are needed[34] and the heat fluxes are high enough to require that the fuel be pumped directly through the cooling panels to cool the internal surfaces of the engine and other hot spots of the vehicle.

Conventional blended hydrocarbon fuels in current aircraft include Jet-A (in commercial use), JP-4, JP-5, JP-7, and JP-8 (in military use). Utilization of these fuels in high-speed aircraft applications results in minimum impact on logistic (distribution, storage, and availability) support requirements but does not provide very high levels of thermal sink capacity, due to limitations imposed by the thermal stability characteristics of the fuels. This imposes limitations to the use of such fuels, if their application is based on current conventional technology, to flight systems operating below approximately Mach 4.0

Cryogenic fuels, such as liquid hydrogen or liquid methane, are fuels that have very low boiling points, requiring special techniques for storing and handling. These fuels offer very large heat sink potentials, compared to conventional aircraft fuels, thus providing the capability for application into the hypersonic flight regime. Hydrogen, in particular, exhibits sufficient thermal control capability for flight applications up to orbital velocities, whereas methane or liquified natural gas (LNG) can be applied at speeds up to the Mach 5–6 range. These benefits, however, are achieved at the price of severe logistical difficulties, as compared to current aircraft operations with conventional fuels. Cryogenic fuels require refrigerated storage to avoid fuel loss due to boiling as a result of heat leakage from the surrounding environment. Also, fuel handling systems require high levels of thermal insulation/isolation to avoid the same problem. The low density of cryogenic fuels, coupled with their storage and handling requirements, results in significant impacts on aircraft design and operational characteristics.

Although liquid hydrogen has the lowest density of the cryogenic fuels, it has the highest energy per pound and cooling capacity. It is the liquid fuel of choice for the highest Mach number flight, and the only one that has been seriously considered for SSTO vehicles. A better fuel for an SSTO vehicle would be a mixture of liquid and slush hydrogen because it is more dense and has higher heat sink capability than liquid hydrogen. More research is needed on containment tanks and efficient production methods before slush can be used on an operational vehicle.[18] Thermal properties of some cryogenic, hydrocarbons, and endothermic fuels are compared in Table 2.

As already mentioned, another class of fuels that appears to offer a large potential for use in the low hypersonic regime at flight speeds up

Table 2 Properties of Potential High-Speed Fuels

	LH2	LNG	JP-7	JP-8	MCH
Boiling point, °F	-423	-259	360-484	284-572	213
Freezing point, °F	-435	-296	-47	-53	-196
Net heating value					
Gravimetric, Btu/lb_m	49,951	21,000	18,875	18,400	18,649
Volumetric, Btu/gal	29,180	82,059	123,500	124,378	119,746
Heat Sink					
Chemical, Btu/lb_m	0	0	0	0	894
Physical, Btu/lb_m	6,057	1,427	508	450	1,031
Specific gravity at 60°F	0.075	0.45	0.79	0.81	0.77
Critical pressure, psia	188	673	306	340	504
Critical temp., °F	-400	-166	746	772	570
Composition					
Aromatics, vol %	0	0	4	20	0
Napthenes, vol %	0	0	10		100
Paraffins, vol %	0	98+	80	78	0
Olefins, vol %	0	<2		2	0
Sulfur, ppmw	0	<1	60	500	<5
Type of Fuel	Cryogenic		Hydrocarbon		Endothermic

to approximately Mach 8–10 are endothermic fuels. These are hydrocarbon fuels that absorb heat during controlled chemical reactions, resulting in very large heat sink potentials, relative to conventional aircraft fuels. This capability has been demonstrated with several different types of hydrocarbons and, typically, involves some form of decomposition or chemical reforming of the original fuel molecular structure. An example of this process is the dehydrogenation of MCH to toluene, with a resulting chemical heat sink of 894 Btu/lb_m. Although the logistic requirements of such fuels are similar to conventional JP fuels, the thermal system requirements are different. The controlled reforming of hydrocarbons is performed in the presence of appropriate catalysts, chosen for their ability to selectively drive the reforming reactions in the direction to produce desired reaction products. The desired reactions, in turn, are selected as those that produce large heat sink potential and yield reaction products with good combustion characteristics. Systems designed to use endothermic fuels therefore require special catalytic components in which the required heat transfer and reactions take place.

The heat sink capacity of a fuel (per unit mass or volume) depends on several factors but is basically defined by the difference in energy levels between the original storage state and the energy at the maximum allowable temperature. The maximum allowable temperature of the fuel in a cooling system is limited by (the lower of) either the thermal stability of the fuel or by the material capability of the system structure. The energy level of a fuel at a specified temperature depends on the chemical composition of the fuel at that temperature. Therefore, the difference in energy between the two different temperature levels depends on whether the fuel retains the same chemistry (sensible heating) or undergoes a change in chemistry, as in endothermic reforming. The total heat absorption capacity that is available for thermal control of an aircraft system also depends on the rate of fuel flow, defined by the propulsion energy requirement and the heating value of the fuel. Therefore, the potential of a selected fuel to provide the necessary heat sink for a specified application depends both on the fuel properties and on the methods used to capitalize the fuel potential.

An interesting option to direct fuel cooling for high Mach number flight is to use water for cooling. Water cooling systems have been considered for a tank-volume-limited hypersonic test vehicle, which would use hydrogen for fuel and reach a Mach number of 12–15 with a test time of 10 s. Both open- and closed-loop water cooling systems were considered. The systems were designed to maintain water pressures high enough to prevent boiling in the cooling system. The open system used an ambient temperature helium tank with an initial pressure of 15,000 psi to maintain pressure in a 3600 psi water tank for supply of high-pressure water to the cooling system. By the end of the test, the helium tank pressure would decrease to 3600 psi. The closed-loop water system was designed to meet the requirement of ideal flow of hydrogen to the engine for propulsion with minimum venting of cooling fluid. The water in the closed-loop system would be at 2000 psi initially. Water would be pumped around the loop with the water pressure increasing until a pressure of 6000 psi is reached and venting of steam begins. The system was sized so that very little venting of steam was needed during the 10-s test time. Comparison between water cooling and hydrogen cooling showed that water cooling allowed significantly more test time for the same total tank volume and vehicle size, although the vehicle weighed more when water cooling was used. Cost and reliability advantages were identified for components, such as turbopumps, which could be designed for water and/or stream rather than hydrogen. It was also evident that the water cooling systems could be started more rapidly and with less complexity than the hydrogen cooling system that depended on starting an aerothermal-heat driven thermodynamic cycle for the turbopump that pumps the hydrogen.[35]

B. Integration, Synergism and Optimization

The elements of TM have been practiced for many years, but the requirement for a highly integrated approach to the design process has come into sharp focus with the recent interest in high-speed flight. The design and integration process involves simultaneous consideration of a range of technical disciplines, including 1) heat transfer in solids and fluid processes; 2) thermodynamics, including simultaneous chemical processes; 3) fluid dynamics and aerodynamics; 4) materials science and structural analysis; and 5) fabrication and manufacturing methods.

The basic requirement of thermal management that must be considered over the entire flight envelope of an aircraft system is the satisfaction of the local cooling requirements of all components in the aircraft and propulsion system. This includes being within the total heat absorption resources available and within the design constraints of the integrated airframe/propulsion system. This also means that two resource capabilities must be considered in the implementation of a thermal system design, the ability to act as a local coolant and the ability to act as a heat sink.

Heat sink resources for use in flight systems may be air, fuel, or radiation to the external environment or may be expendable, such as ablative materials or dedicated heat sinks. Whereas all of these find use in different applications, air and fuel are of primary importance in manned systems designed for hypersonic flight, with fuel becoming the primary sink at high-speed conditions. Radiation also plays a major role in such applications by providing autonomous thermal control of the external structure of the aircraft, thereby controlling surface temperatures without requiring active cooling. Expendable heat sinks, in contrast, are used in expendable flight systems, such as missiles and solid rocket boosters. Coolants that act to locally absorb heat from a heat rejecting surface may be a primary heat sink (direct cooling), or a secondary coolant, such as air or a recirculated fluid (indirect cooling).

The cooling technology defines the method by which heat may be removed from a surface or fluid stream and ultimately transferred into the primary heat sink. The technology selected for use in a particular application must be compatible not only with the thermal resources but also with the surrounding materials and environmental conditions, while effectively providing the required degree of thermal control. As already mentioned, two general approaches are available for cooling with a fluid heat sink, namely, direct and indirect cooling. Direct cooling employs the primary heat sink to absorb heat from the heat source. The cooling mechanism used may be convection with air or with fuel, and the fuel may exhibit either frozen (fixed) or reacting chemistry. In addition, mass transfer cooling may be used, with the primary heat sink being injected on or through the surface being cooled, ultimately mixing with the fluid heat sink (transpiration or film cooling).

In the use of indirect cooling, a secondary fluid absorbs heat from the heat source and ultimately rejects the heat to the primary heat sink. This method may employ air as an expendable secondary coolant, with the air initially cooled by the heat sink, subsequently used to cool the hot structure, and finally dumped into the heat source stream. Alternately, a reusable secondary coolant can be used in a pumped recirculation system, convectively absorbing heat from the heat source and rejecting it to the heat sink. Another alternative is with the use of a heat pipe to transport heat from one location to an appropriate heat sink location. The selection of the particular method to be employed depends on several factors, including the available sink, the condition of potential secondary coolants available, and the thermal environment to be controlled. An example of the procedures involved in configuring a thermal management system for a defined application, and the rationale for selection of the design parameters, will be illustrated in the following section.

C. Design Integration of a High-Speed Flight System

Integration of a flight system for high-speed flight entails defining various requirements, including 1) the aircraft mission, 2) the resulting flight envelope or flight profile, 3) the selected fuel (based on logistic, operational, and thermal requirements), 4) the propulsion system and its performance characteristics, 5) the thermal management system, and 6) the vehicle controls system.

Attainment of an integrated flight system design, satisfying the various defined requirements, is a complex iterative process involving many design considerations. The desired goal of this process is the design of an optimized flight system capable of performing the defined aircraft mission. This design is optimized if the defined optimization parameter (i.e., aircraft takeoff gross weight) is minimized for the integrated design.

The design process, necessary to achieve the desired goal, can best be performed in a structured iterative approach. In an overview of the total process, the mission requirements are set, the vehicle and propulsion systems are integrated, and a mission optimization analysis is performed to define the vehicle and propulsion system sizes, as well as the desired flight profile. These results, in turn, allow definition of the thermal requirements of the integrated flight system. An iterative design process is performed to configure and design a TMS management system capable of satisfying the thermal requirements of the flight system over the entire operating range of its flight profile. If this is not found to be achievable within the limits of the selected design conditions (selected fuels, materials, technology, etc.), then the conditions of the optimized flight design (i.e., vehicle size, flight profile, cost, etc.) may need to be relaxed or modified to allow satisfying the mission requirements. It should be noted, however, that it is not necessarily a given that a desired mission can always be satisfied within the constraints of a defined state of available technology. Relaxation of design requirements away from an optimized design may allow a closer approach to satisfying a mission, but the end result of the process may be the definition of critical technologies (i.e., inadequate to meet the requirements of the defined goals). This result would simply show that further advances must be achieved in the critical technologies (such as cooling techniques, materials, structures, etc.) before the desired mission capable vehicle can be designed.

Upon definition of the flight system and its thermal requirements, a general approach for the design and integration of the thermal management system involves a number of discrete steps, which may be described in an outline format to provide a guide to the overall procedure. The design process may be described by the following procedural steps

1) Identify the thermal requirements of the flight system.
2) Identify thermal resources available.
3) Define an integrated TMS concept.
4) Design the defined TMS at a selected most critical design point.

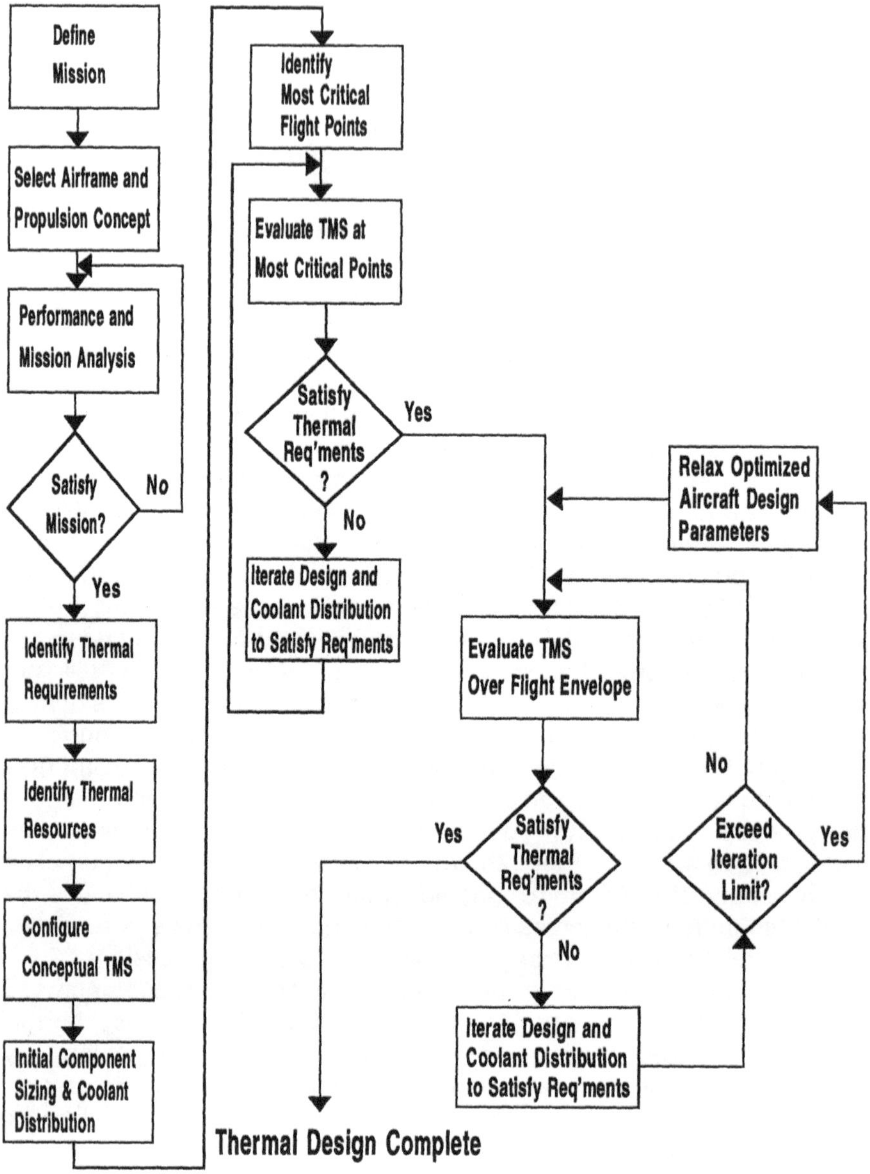

Fig. 9 Design process for Thermal Management System

5) Evaluate a model of the selected thermal system over the flight envelope.
6) Modify the system's components as required to satisfy thermal control requirements.

The total procedure for the design and integration of a flight system to satisfy a defined mission is illustrated by the process flow chart shown in Fig. 9. This illustration includes the vehicle and propulsion integration, as well as the thermal system integration. Also shown is the fallback logic to allow relaxation of the aircraft design and flight parameters, in the event that a satisfactory optimized design is not achieved or obtainable.

Other considerations for optimizing the entire TMS of a high-speed vehicle is the integration of the propulsion TMS with the airframe TMS. This is necessary to ensure that the required coolant flow rates and temperatures into the engine inlet are met. Having a TMS that integrates both the engine and airframe cooling requirements will also serve to keep the vehicle weight down by reducing the amount of duplication. This also results in an efficient and optimized cooling network, taking into consideration the varying temperatures and cooling requirements, if accomplished successfully. Tradeoff analyses will be necessary to ensure that the coolant routing not only satisfies the cooling requirements but also satisfies weight, reliability, and maintainability requirements. To efficiently conduct the tradeoff studies, engineering computer models are employed. A brief discussion of computer modeling is covered later in this chapter.

Optimization of a TMS for a hypersonic airbreathing vehicle is highly dependent on the types of propulsion systems used, the mission of the vehicle, and design parameters, such as cost and reusability/expendability. The fuel flow rates available for cooling are determined by the types of propulsion systems used. The complete flight trajectory must be specified, whether it be dominated by acceleration or a cruise segment. For a short flight, transient effects are important, and most of the thermal protection problems can be solved with the use of insulation or the addition of small amounts of thermal mass or ablatives (in locations where the ablation does not interfere with the engine flow path or boundary layer conditions). In severe cases, active cooling may be required. The TPS is weight optimized to maintain all materials within their allowable temperature limits, although minimization of structural temperature gradients may also be important (this type of optimization was done for the STS). For longer flights of high Mach numbers, active cooling must be used.

The optimization of a cooling system must consider all of the requirements of the fuel system and the integrated structural design, which will include materials with differing thermal requirements. A

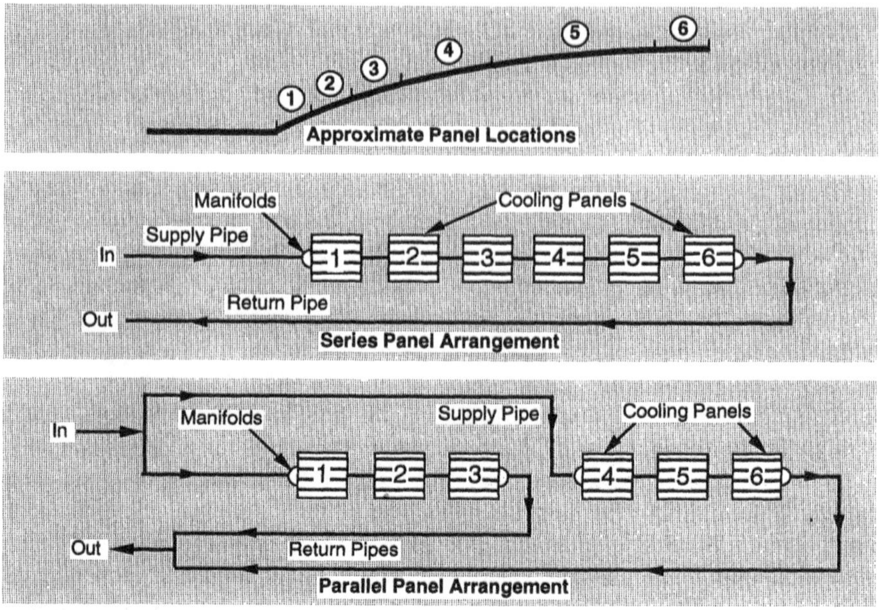

Fig. 10 Two alternative coolant routing designs for an external nozzle

coolant routing design (plumbing arrangement) must be formulated, consistent with the temperature and thermal stress limits of the materials selected for the structure and cooling panels. Thermal management models can aid in this process. The best routing is determined using engineering judgment, as well as analyzing and modeling the most promising options. Two options that were considered for coolant routing to be used on the airframe external nozzle of a SSTO vehicle are shown in Fig. 10. The cooling system weight must be minimized, as well as the coolant flow rate, if the supply of coolant is limited.

V. Thermal Management System Design and Integration Example

A sample procedure will be performed in this section, as an example of the process of designing a TMS for integrating the thermal control requirements of a selected aircraft. The aircraft and its defined mission are selected for purposes of illustration only and do not represent any specific aircraft or mission requirement. The operational and aircraft system parameters, however, are consistent with the requirements of the assumed illustrative mission. As the thermal requirements are strongly dependent on other system design features, this example will briefly illustrate the entire design process associated with integrated

Table 3 Aircraft Performance Parameters

Range	Exceed 2000 n mile
Flight Trajectory	Minimum acceleration to cruise, descent included in range requirement
Speed (cruise)	Mach 5.0
Altitude	For best range
Payload (include crew of two)	10,000 lb_m
Fuel load	As required for range +5% reserve

propulsion system for high-speed flight. The illustrated design example, however, will refer only to the ramjet portion of an integrated turboramjet system.

A. Identify Mission Requirements

The selected aircraft mission is high-speed penetration for military reconnaissance over unfriendly territory. The required aircraft performance parameters are defined as Table 3.

The optimum aircraft flight profile would normally be established by a mission analysis of the integrated flight vehicle defining the best flight trajectory and vehicle size required to satisfy the mission requirements. For illustrative purposes of this example, the appropriate flight profiles are assumed and are shown in Figs. 11–13. The vehicle is assumed to accelerate at a rate for minimum fuel burn up to a 1200-psf dynamic pressure flight path, then gradually decrease the dynamic loading at higher speeds to limit the internal pressures in the ramjet (to 140 psia). The cruise condition is attained at Mach 5.0/85,000 ft altitude, producing a dynamic pressure of 800 psf. As shown in Fig. 12, the altitude gradually increases during a 30-min cruise leg, a result of vehicle weight loss due to fuel burn, with constant thrust. The flight profiles shown would normally require definition of the vehicle and the propulsion system as part of an iterative evaluation but are assumed for this example.

The selected airframe concept is a lifting-body configuration, with fuselage integrated propulsion systems. Two-dimensional mixed compression inlets are integrated below the vehicle, utilizing precompression on the vehicle lower surface. The propulsion systems consist of four engine pods, with individual inlet supplies. Inlet performance is assumed to match the Military Standard (Mil-E-5007D)[36] inlet recovery schedule,

$$P_{t2}/P_{t1} = 1.00 \text{ from } M = 0 \text{ to } 1.0$$
$$P_{t2}/P_{t1} = 1.00 - 0.076(M-1)^{1.35} \text{ from } M > 1.0 \text{ to } 5.0$$
$$P_{t2}/P_{t1} = 800 / (M^4 + 935) \text{ for } M > 5.0$$

where P_{t2} is the total pressure at the compressor inlet and P_{t1} is the free stream total pressure.

The airframe aerodynamic performance is assumed to yield a lift/drag *(L/D)* ratio of 5.5 at cruise. The airframe structure provides a fuel fraction of 0.50, defining the fuel load available at takeoff.

The selected propulsion system concept includes the following features: 1) dual stream over/under turboramjet (turbojet low-speed accelerator and ramjet for high speed), 2) common integrated single expansion ramp nozzle (SERN) for each engine, 3) nozzle expansion includes the vehicle base area, 4) nozzle design for expansion to cruise ambient pressure condition, and 5) isolation of the turbojet accelerator engine when operating in ramjet-only mode.

The next step in the vehicle integration process is performance of a mission analysis study to establish the optimum flight profile and to determine the size of the vehicle, the propulsion, and the fuel load. The trajectory optimization is performed as part of this study to maximize mission performance, minimize the total vehicle size/weight, and minimize the structural and thermal demands.

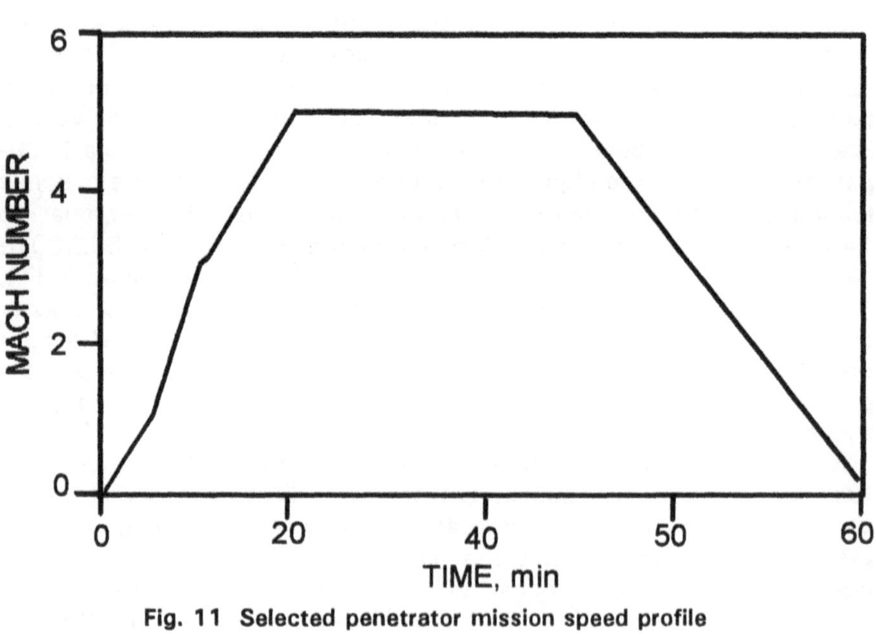

Fig. 11 Selected penetrator mission speed profile

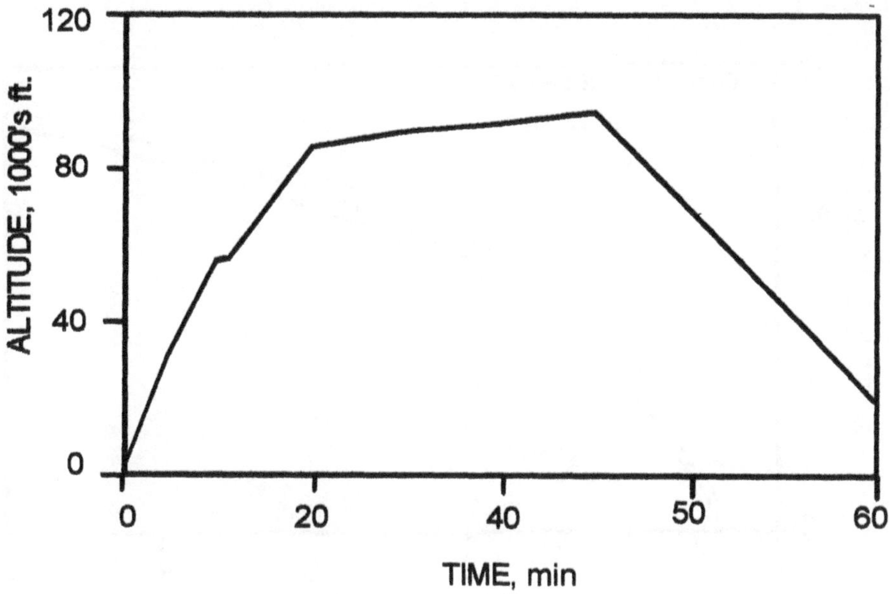

Fig. 12 Altitude profile for selected mission

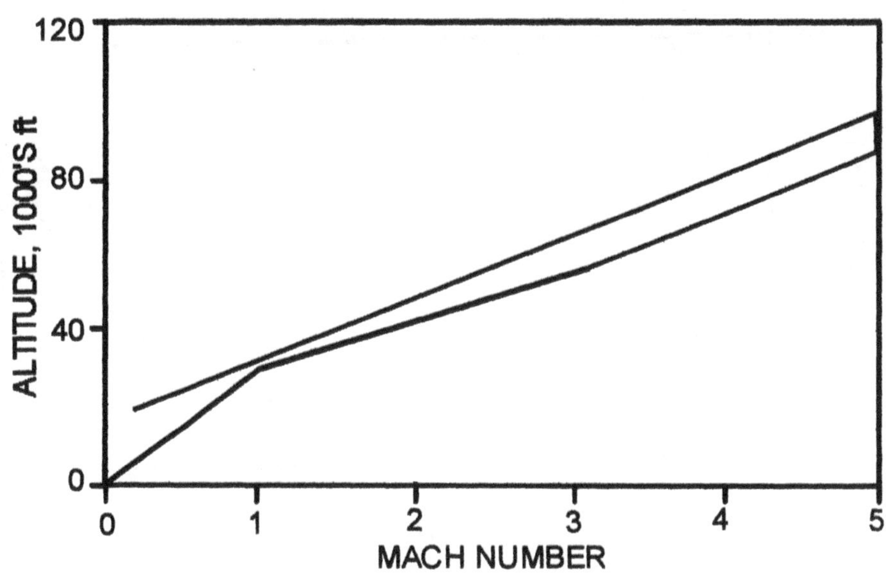

Fig. 13 Flight profile for selected mission

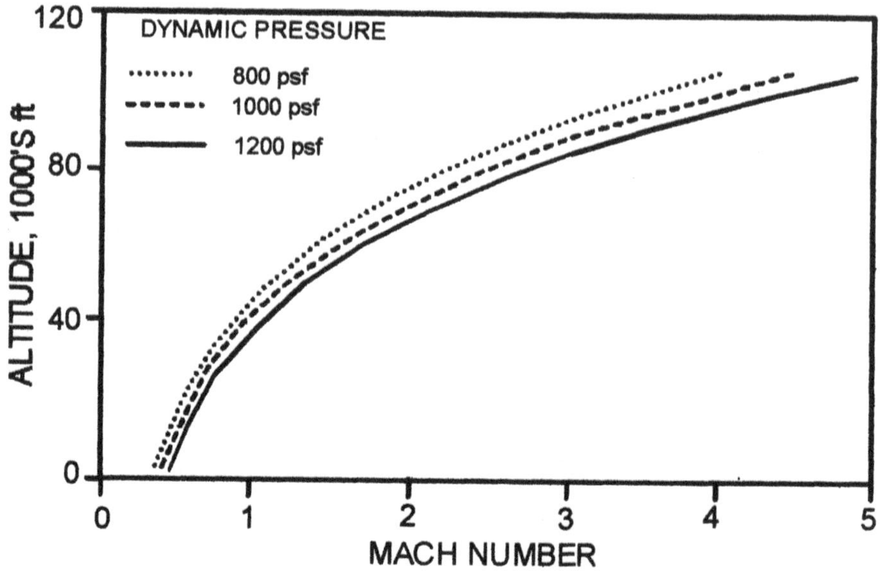

Fig. 14 Flight profiles for constant dynamic pressure

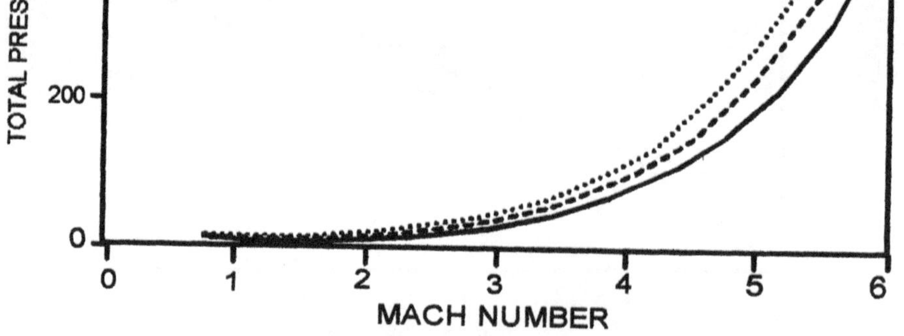

Fig. 15 Free stream total pressure profiles for constant dynamic pressure

As a result of the mission analysis, the propulsion performance is established that is necessary to satisfy the mission requirements of the airframe. The results of these (assumed) studies to satisfy the defined mission yield the following definition of the airframe and propulsion requirements: 1) takeoff gross-weight (TOGW) = 250,000 lb_m, 2) vehicle empty weight = 120,000 lb_m, 3) initial fuel weight = 120,000 lb_m, 4) payload = 10,000 lb_m, 5) assume acceleration as defined by selected flight profile, 6) assumed fuel burned during acceleration = 33% of total fuel, 7) weight at beginning of cruise = 210,000 lb_m, 8) thrust = drag = 210,000/5.5 = 38,182 lb_f total, and 9) desire 10% thrust margin at beginning of the cruise leg.

The flight trajectory determined by the mission optimization study provides detailed definition of the optimum flight path for achievement of the required mission with a minimum vehicle size. To understand the impact of the selected flight on the thermal and structural system requirements, it is necessary to consider the effect of aerodynamic loading on these systems. Fig. 14 illustrates the flight paths defined by constant dynamic pressure trajectories. The dynamic pressure q (in pounds per square foot) is the kinetic energy of the air relative to the aircraft, based on the ambient local air density and aircraft velocity. Acceleration at constant dynamic pressure approximates an optimum flight path, and the flight path illustrated in Fig. 13 could be well represented at speeds above Mach 1.5 by a constant q = 1000 psf path. Flight at higher dynamic pressures results in higher total air pressures and higher internal engine pressures, as shown in Fig. 15,

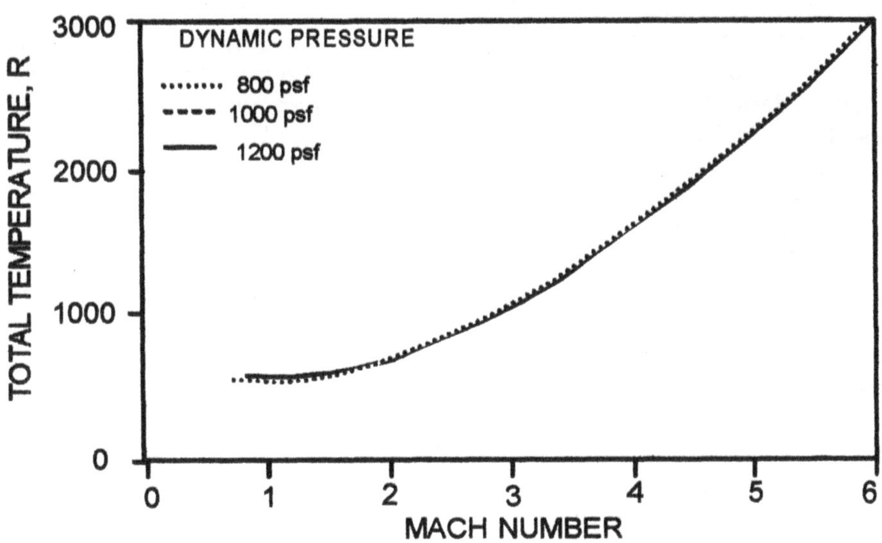

Fig. 16 Total temperature profiles for constant dynamic pressure

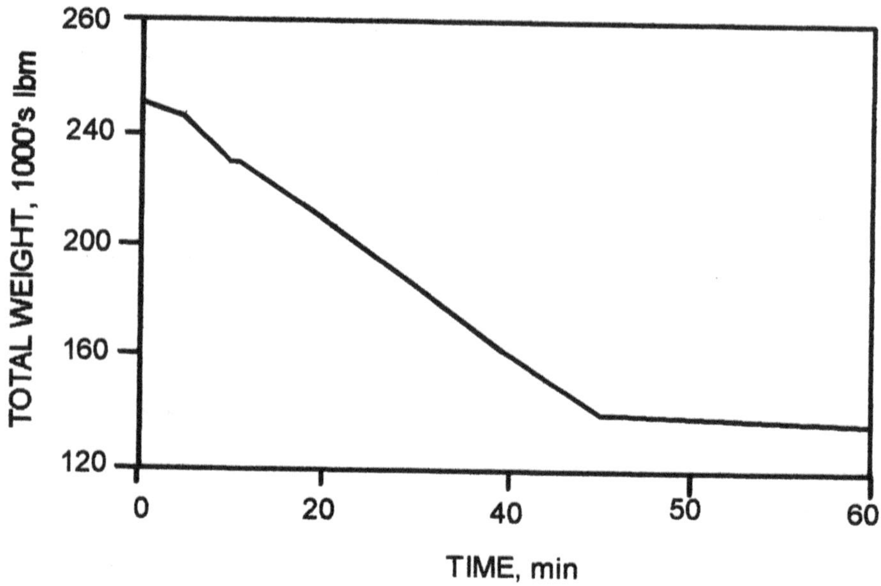

Fig. 17 Vehicle weight change during flight due to fuel burned

which in turn yields higher thrust and better installed propulsion performance. On the other hand, higher internal engine pressure results in higher structural and thermal loading, requiring heavier structures and better cooling. Thus, a compromise is required in the selection of the flight path (altitude and speed). In this example, the aircraft reaches 1200 psf during climb, with the initial cruise at Mach 5.0 at 85,000 ft (800 psf). The variation of total temperature is nearly independent of dynamic pressure, depending primarily on flight speed, as shown in Fig. 16, reaching nearly 2300°R at the Mach 5.0 cruise condition.

As a result of the mission analysis, the gross vehicle weight (including fuel) is defined over the flight path. As seen in Fig. 17, the total weight continuously decreases as fuel is burned, resulting in a corresponding decrease in vehicle drag (and required thrust) if the flight is maintained at constant altitude. The variation in vehicle weight can also be accommodated by allowing the vehicle to climb at constant power and speed, ultimately reaching an altitude at which thrust and lift balance the aircraft weight and drag. The variation in thrust required to maintain the aircraft at a constant speed and altitude is illustrated in Fig. 18. As shown in this figure, there is a significant reduction in required thrust over the 30-min duration of the cruise leg, beginning 20 min into the flight. Since the captured inlet airflow is constant when flying at constant air speed and altitude (if the aircraft

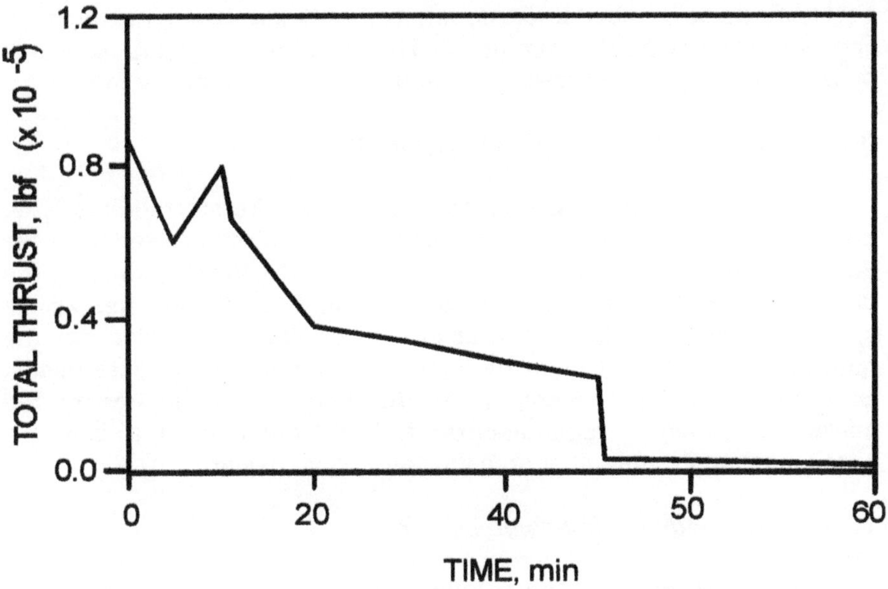

Fig. 18 Variation in thrust requirement over flight profile

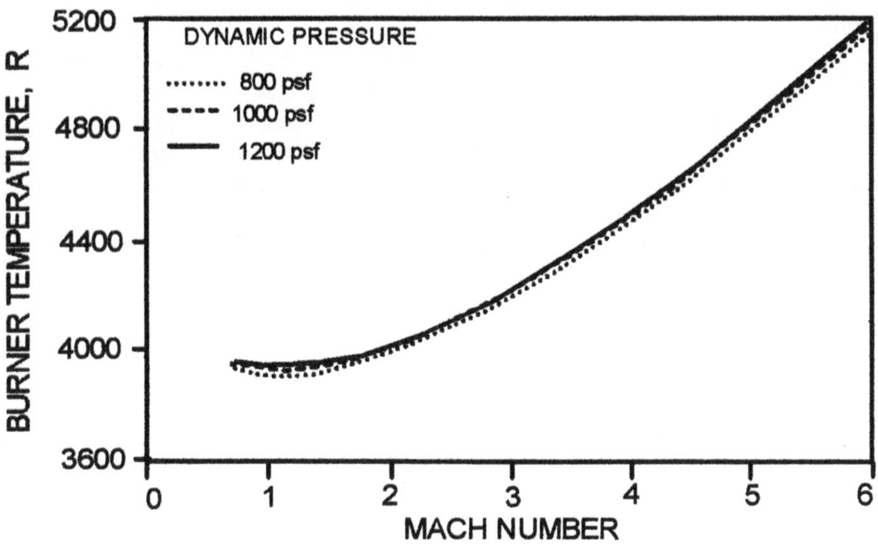

Fig. 19 Variation of burner pressure with flight speed

angle-of-attack is maintained constant), variation in engine thrust is achieved by varying the fuel flow. Therefore, the available heat sink capacity of the fuel decreases continuously over the cruise leg.

B. Propulsion Performance and Engine Design

The design of the engines, to satisfy the aircraft requirements, consists of several steps. The process begins with the selection of the design point and its corresponding requirements. In this example, the design point will be the beginning of the cruise, and the engines will be sized to provide sufficient thrust at that condition. Because of the nature of the selected flight path, the cruise point is the determining condition for sizing the engine and its inlet. Determination of the aerothermodynamic requirements and selection of the physical configuration are the initial steps in the engine design.

1. Aerothermodynamic Engine Design

Based on the results of the mission analysis, the required total engine thrust at the beginning of cruise is 38,182 lb_f at the selected flight conditions (Mach 5.0 and 85,000 ft). The cruise conditions are selected for this example to limit the internal burner pressure at an acceptable level with regard to structural requirements. As shown in Fig. 19, the assumed inlet recovery characteristics (Military Standards Mil-E-5007D)[36] yield a burner pressure at a cruise of 96 psia.

With the selection of four engine modules, the net initial cruise thrust is 9545.5 lb_f/engine. It is desirable to size the engines with excess thrust at the beginning of cruise to provide margin for acceleration and maneuvering capability. The maximum thrust for this design is selected as 10,606 lb_f/engine, providing 10% margin at the beginning of cruise. With this thrust capacity, the vehicle accelerates to the cruising speed and altitude at maximum power, and the engines are then throttled back to 90% power for cruise. At the selected cruise, the ambient conditions are T_{am} = 410°R and P_{am} = 0.3167 psia.

Ideal maximum thrust of an engine is achieved if the propulsive stream is expanded isentropically to ambient pressure. At the cruise condition, the nozzle pressure ratio (ratio of burner pressure to ambient pressure) requires an expansion ratio of 24 to achieve ideal performance. Assuming that the fuel is a generic endothermic hydrocarbon, with the same heating value as current aircraft fuels, the required engine size and performance can now be determined. Basing the engine design on the high-speed conditions at the end of acceleration (Mach 5.0 and 85,000 ft), with the burner operating at maximum power, the thrust is dependent on the burner temperature and nozzle expansion ratio. The burner temperatures across the

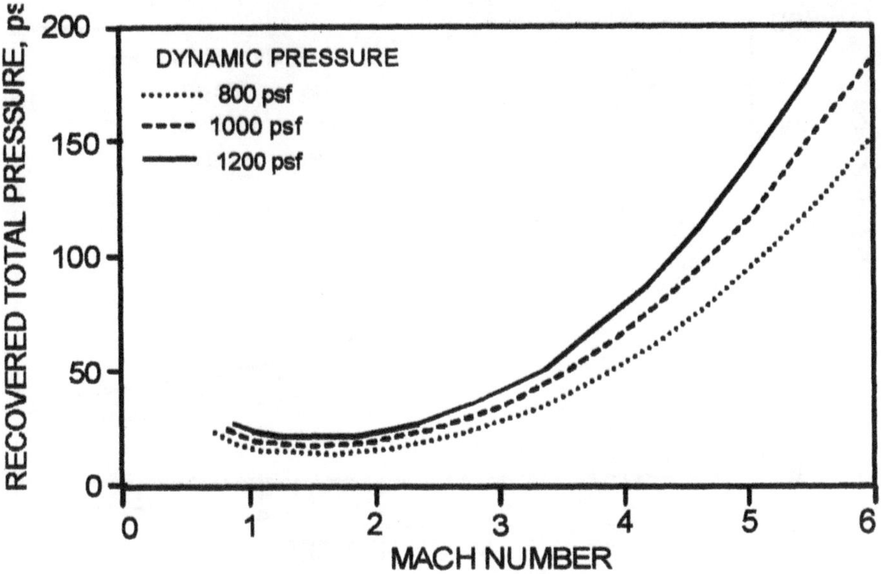

Fig. 20 Variation of burner temperature with flight speed

acceleration profile are shown in Fig. 20, and the design point temperature is seen to be 4800°R. Evaluating the (chemical) equilibrium expansion of stoichiometric combustion products (air/hydrocarbon fuel) from 96 psia to ambient pressure, the gross specific thrust (ideal nozzle thrust) is

$$F_g/W_{tot} = \text{exit velocity}/g_c = 244.4 \text{ lb}_f\text{-s/lb}_m \text{ of products}$$

where W is mass flow rate.

The net installed specific thrust (nozzle thrust minus ram drag, including losses due to nozzle inefficiencies) is

$$F_n/W_{air} = (F_g C_v/W_{air} - V_0/g_c) = 109.8 \text{ lb}_f\text{-s/lb}_m \text{ of air}$$

where C_v is the velocity coefficient for the nozzle and V_0 is the undisturbed free-stream velocity.

The net installed specific thrust (thrust/fuel flow rate) at this operating condition is found to be 1666 $\text{lb}_f\text{-s/lb}_m$, based on stoichiometric fuel flow (f/a = 0.06591). The air and fuel flows needed to produce the required thrust are

$$\begin{aligned}\text{Required air flow} &= \text{engine thrust/net specific thrust}\\ &= 10606/109.8 = 96.59 \text{ lb}_m/\text{s}\end{aligned}$$

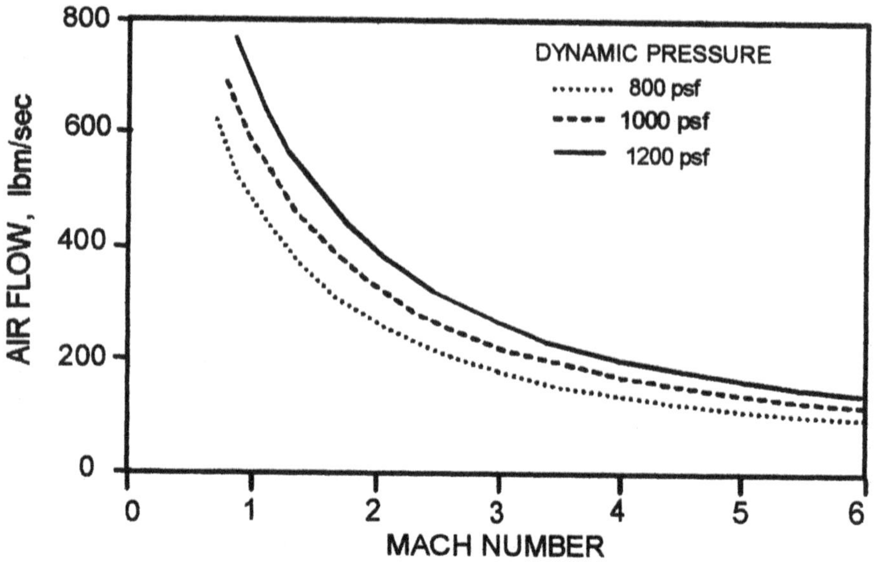

Fig. 21 Variation on constant Q paths for 1440 in^2 inlet area

In hypersonic flight, the airflow captured by an inlet varies with speed and dynamic pressure, as shown in Fig. 21 (for a 10-ft^2 inlet). From the data shown, the inlet capture area required to provide the necessary airflow at the design flight condition is 9.42 ft^2/engine. With the inlet capture area fixed at the design condition, the captured inlet airflow is defined over the remainder of the supersonic flight envelope. This is the airflow available to produce thrust, unless provision is made to bypass some of the captured flow. Since the airflow will remain constant over the selected cruise path, thrust variations will be achieved by varying the fuel flow rate supplied to the burner. At the end of acceleration, the required fuel flow is 6.37-lb_m/s/engine (maximum power), reduced to 5.73 lb_m/s for cruise power. By the end of cruise, with a reduced required thrust due to reduced vehicle weight, the fuel flow is approximately 3.82-lb_m/s/engine.

2. Design of Engine Geometry

Although the engine design point is selected at the Mach 5.0/85,000-ft cruise condition, selection of the physical configuration must consider operational requirements across the entire flight envelope. The variable inlet and the nozzle geometry are required to operate across the wide range of flight conditions. The inlet is assumed to have a fixed capture area but variable internal geometry to facilitate starting and other requirements. The selected configuration of the exhaust nozzle is a two-dimensional SERN, with variable throat area. The turbojet accelerator engine, used for takeoff and low-speed flight, is mounted in a bay above the ramjet, sharing the same inlet and nozzle. The following features are selected as design ground rules

1) The turbojet operates from takeoff to Mach 3.0.
2) The ramjet operates from Mach 1.5 – 5.0.
3) The turbojet is shut down and protectively cocooned at speeds above Mach 3.0.

The following design details arise as a result of the above ground rules, and the selected inlet size. At Mach 3.0, with only the ramjet operating, the ramjet must pass all of the captured airflow. Therefore, Mach 3.0/55,000 ft is the condition with the highest airflow and requires the largest nozzle throat area. At this point airflow = 145.5 lb_m/s for each engine and throat area (A^*) = 4.488 ft^2.

For combustion stability, the burner flow area must be sized to provide an acceptably low speed for stable combustion. This results in selecting a maximum velocity of Mach 0.56 in the burner area. At higher flight speeds, the throat is closed down and the burner velocity is lower. This selection results in a burner area of 800 in^2. The chosen cross section is 40 in. wide by 20 in. high. Thus, the throat height is 16 in. at Mach 3.0.

At the Mach 5.0 cruise condition, the ramjet configuration is as shown in Table 4. The selected ramburner configuration at cruise is illustrated in Fig. 22. The curved expansion surface is integrated with the base of the vehicle, thus providing the large required nozzle ratio, with minimum additional structure.

C. Thermal Management System Design

The process of designing a thermal management system for a given application begins with identification of the requirements of the system, and the resources available for management of the

Table 4 Configuration of Ramburner at Cruise Design Point

LOCATION	AREA in^2	WIDTH in^2	HEIGHT in
Burner	800	40	20
Throat	136	40	3.41
Nozzle Exit	3220	40	82

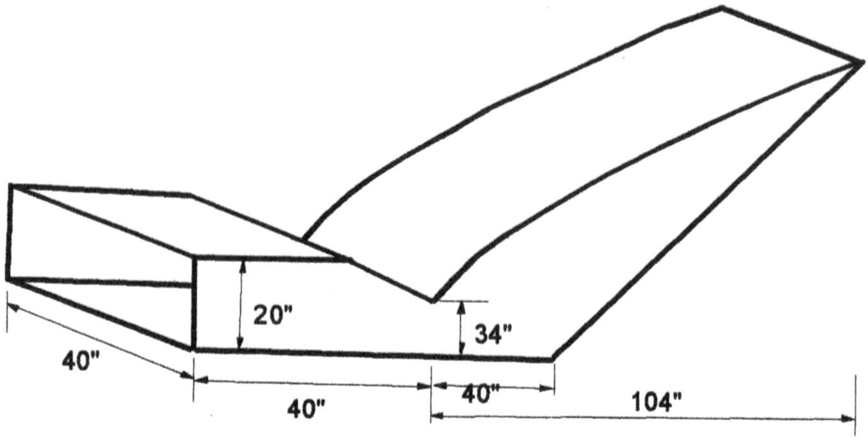

Fig. 22 Two-dimensional SERN nozzle sized for single engine

requirements. A conceptual approach is then defined, modeled for analysis, and designed at selected critical design conditions. When satisfying the design conditions, the model is evaluated over the broader flight envelope to test the general satisfaction of the system.

1. Identify Thermal Requirements

Thermal management requirements of high-speed aircraft generally are imposed by three types of loads, internally generated aircraft loads, aerothermal heating, and propulsion induced loads. These can be further broken down into the following detailed requirements

1) Internally generated aircraft demands are composed of ECLSS, aircraft avionics, aircraft flight controls, and auxiliary power systems.
2) Aerothermal loads are composed of aerodynamic heating of airframe structures and external engine surfaces and cowl, heat leakage into fuel tanks and fuel system, and heating of inactive propulsion system components.
3) Propulsion generated loads are composed of engine combustors, nozzles and exhaust systems, and related avionics and controls.

These various thermal requirements must be quantified, defining the loads and any restraints on allowable temperature levels, prior to the design of a thermal management system. This process requires definition and evaluation of the thermal characteristics of the various vehicle subsystems and their behavior at representative flight conditions over the operating envelope.

For the purpose of this exercise, the magnitude of the various thermal requirements contributed by the vehicle and by the low-speed propulsion components are specified as assumed values, and the loads of the high-speed propulsion system are evaluated. The components of the assumed vehicle internal loads in British thermal units per second are 50 for ECLSS, 120 for avionics, 95 for actuation systems for a total load of 265 BTU/s. The thermal loads (in British thermal units per second per engine) on the low-temperature propulsion subsystems are composed of inlet cooling of 720 and inert turbojet cooling of 350 for a total load of 1070 BTU/s/engine.

In configuring the thermal management system to accommodate the various requirements, the vehicle internal loads generally have the lowest required temperature control levels and, as a result, must take first priority with regard to access to the cooling resources. The lower temperature propulsion components, such as the inlet and the low-speed accelerator engine, take second access to the heat sink resources. The high-temperature propulsion components, such as the ramjet cowl leading edge, the combustor, and the nozzle, must operate at much higher temperatures than the other subsystem components and, therefore, typically have last access to the cooling resources.

The first step in defining the thermal control requirements of the propulsion system is the definition of the aerodynamic environment throughout the burner and thrust nozzle. Because the nozzle must have variable geometry to operate efficiently over the broad flight

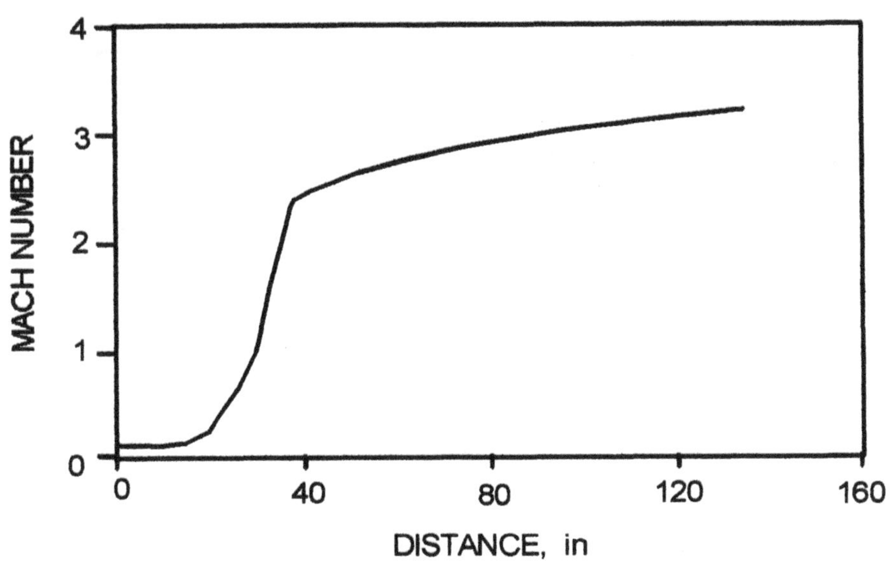

Fig. 23 Mach number variation along upper surface of nozzle

envelope, a particular flight condition must be selected as the design point, and the nozzle geometry corresponding to that flight condition is used to define the internal aerodynamics and the resulting thermal environment. The design point for the TMS should be the condition at which the most critical demands are placed on the system. There are, however, two types of critical thermal demands, making the selection somewhat fuzzy. Critical requirements occur at the condition placing the highest local heat fluxes on the system components and also may occur at conditions of reducing engine thrust, when the available fuel flow for use as the cooling sink is reduced. Thus, the design process must take into consideration local cooling requirements, as well as total available heat sink. As a first estimate of the critical condition, the beginning of cruise is typically a good starting point, with the aircraft at its maximum speed and the engine thrust reduced to a power level from the maximum power acceleration, as required to maintain flight speed. This results in a high level of thermal load, with a reduced heat sink capacity.

The aerodynamic environment throughout the nozzle can be evaluated for the selected SERN type by using one-dimensional analysis in the subsonic section and method of characteristics in the supersonic section of the nozzle. This approach, when applied to the upper expansion surface of the nozzle shown in Fig. 22, yields the Mach number profile shown in Fig. 23. Application of the simplified boundary analysis method of Spaulding and Chi[37] yields the heat flux shown in Fig. 24 (for an assumed constant temperature of 1000°F).

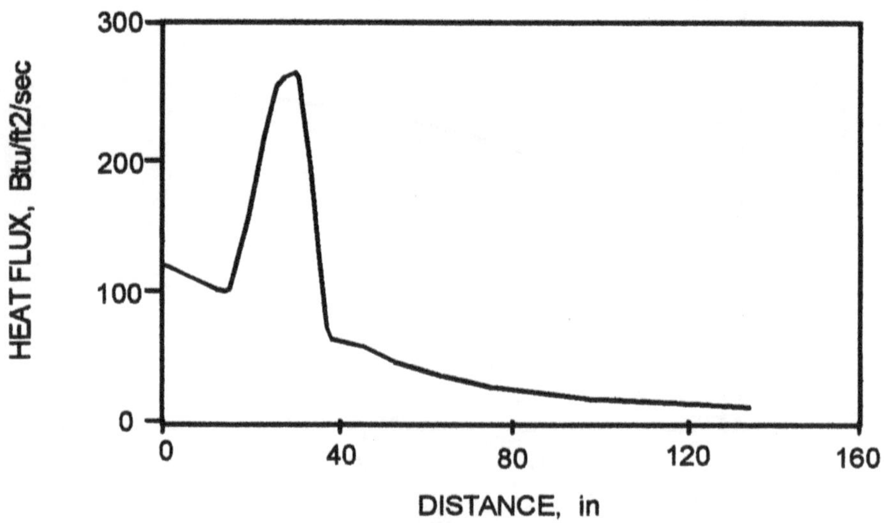

Fig. 24 Heat flux variation along upper surface of nozzle

2. Define Thermal Management System

With several components of the thermal requirements defined, the next step in the design process is the definition of an integrated system concept that can satisfy the requirements within the limitations of the available cooling resources. The cooling resources that are available for thermal control across the flight profile include the following: 1) various air sources at low flight speeds, 2) fuel and other expendable stores at high flight speeds, and 3) radiation from the external surfaces to space.

The central objective of the thermal system design procedure is matching the thermal demands and available resources in a manner that most efficiently capitalizes on the potential of the resources. Two points must be considered throughout the design process: 1) providing a local heat sink with sufficient capability to absorb the heat load demand, and 2) insuring that temperature constraints of flight system components are observed. Therefore, configuration of a conceptual integration demands that the temperature sensitivities of the various components be considered while selecting the order in which cooling resources will be utilized. Table 5 illustrates the major components that require thermal control in an aircraft system, with an approximate range of limiting temperatures associated with each of the various components or subsystems, and provides a guideline for the ordered configuration of the thermal management system integration. Based on the sequential order suggested by Table 5, a conceptual thermal management system is configured to integrate the required

Table 5 Arrangement of system for best use of resources

Component application	Temperature range of component application °F	Fuel temperature range
Burner fuel nozzle	1200–1600	Max allowable fuel temperature
Hot exhaust sections	600–1400	⇑
Conditioning of cooling air	800–1200	‖
Engine ECS (accelerator)	<600	‖
Hydraulic and lube system	<500	⇓
Life support and avionics ECS	<200	Tank exit temperature

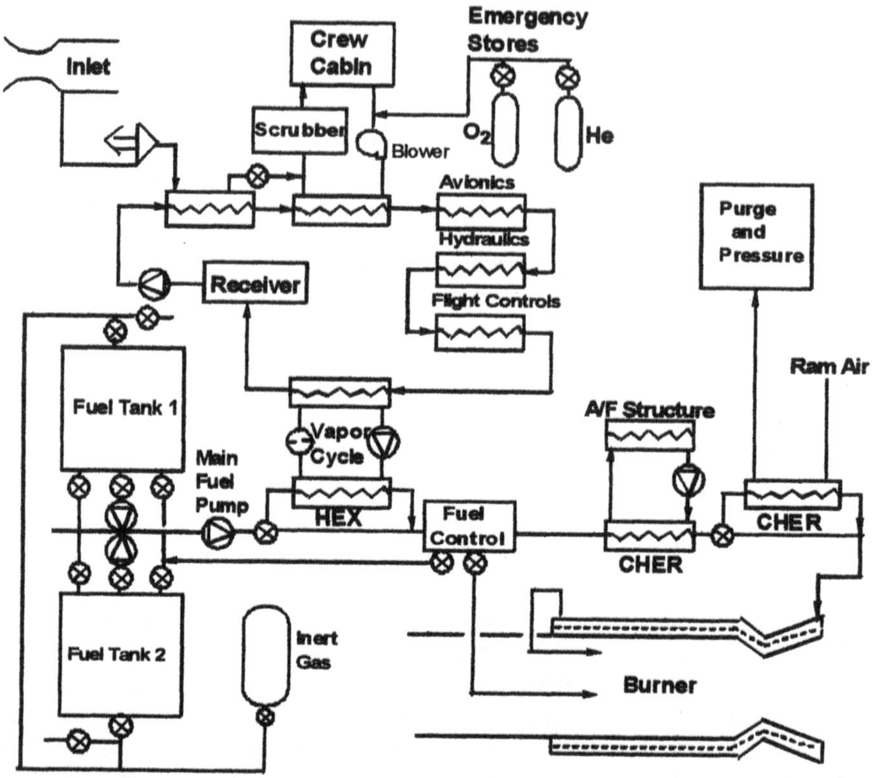

Fig. 25 TMS with direct fuel cooling of ramburner and nozzle

components with the fuel and other subsystems, as illustrated in Fig. 25.

In the system arrangement shown in Fig. 25, fuel is used in a series of applications, sequentially providing cooling for the aircraft environmental control system (ECS), for critical aircraft structures (such as the internal surfaces of the propulsion inlets and ducts), and finally for the hot structure of the burner and thrust nozzle. As illustrated, a variety of thermal subsystems interface with the fuel system and ultimately reject heat into the fuel. These subsystems provide the capability of maintaining critical components, such as aircraft avionics, at the low temperatures required for acceptable operation and life of sensitive electronics. In contrast, hot-burner structures can operate at sufficiently high temperatures to allow direct cooling with heated fuel.

3. *Thermal Management System Model*

Whereas the illustrated thermal management system shown in Fig. 25 represents a conceptual physical arrangement, it does not in itself

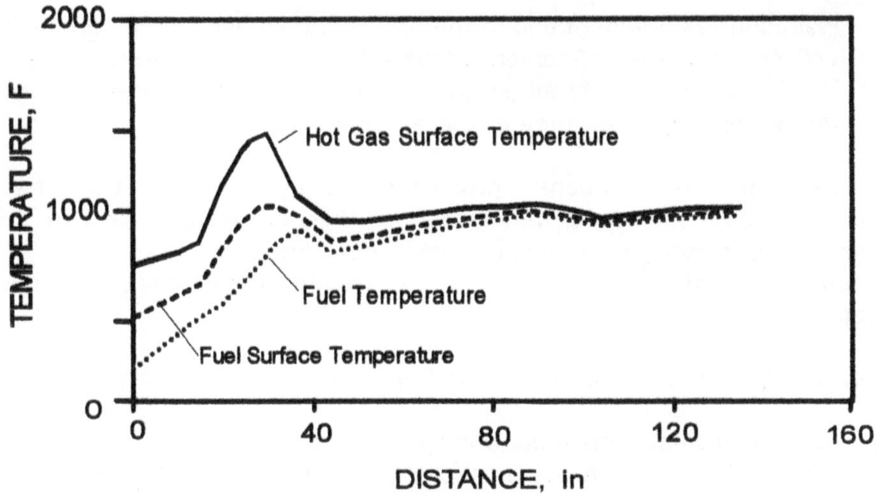

Fig. 26 Temperature profiles on upper surface of nozzle

define the ability of the system to satisfy the operational requirements of the aircraft and propulsion system throughout the flight envelope. Evaluation of the thermal system performance requires that the physical model be simulated by a sufficiently accurate computational model, with capability to represent the physical processes occurring within the various system components. The computational model, is in turn, used to develop the detailed design of the integrated TMS and then used to simulate the system throughout its required range of operation.

The sequence of the design process consists of three steps, in the following order

1. Identify the most thermally critical flight condition, based on either maximum heat flux, shortage of available heat sink, or a combination of the two conditions.
2. Configure a computational model of the selected system concept and perform an iterative design synthesis of the subsystem components, starting at the initial condition of the cooling resource (i.e., fuel in the fuel tanks) and preceding along the coolant path as it flows through the system. Adjust the geometry of the components as necessary to satisfy their local performance requirements. This procedure ultimately results in a design that satisfies the initial critical design conditions.
3. Exercise the resulting system design over the flight conditions defined by the aircraft flight envelope. If additional thermal

inadequacies are identified in the process, then each such condition will require further refinement of the system, until the thermal system is adequate to satisfy all aircraft and propulsion requirements over the entire operating range of the flight envelope.

If the described sequential process results in a design that satisfies all local and operational requirements, then design closure can be assumed, although the design may not necessarily represent an optimized configuration (i.e., minimum weight, lowest cost, long life, etc.).

4. Design Evaluation at the Most Critical Thermal Condition

For this example, the most critical design point is assumed to occur at the beginning of cruise, as the engine thrust is reduced from the maximum power level (for acceleration) to a reduced level (sufficient to match aircraft drag). The selected conditions are 90% power at Mach 5.0/85,000-ft altitude. The aircraft weight and drag at this condition determine the propulsion requirements and the corresponding fuel flow.

With the propulsion cooling system concept selected, as shown Fig. 25 (direct fuel cooling of burner and nozzle structure), the fuel flow is distributed within these structures in accordance with surface areas and heat fluxes. The fuel passage sizing and internal configuration within each structural element is initially selected, and thermal effectiveness of the system is evaluated. Areas of inadequate cooling are identified, the relevant cooling geometry is modified, and the thermal response to the change is evaluated. This process continues, with possible redistribution of the fuel flow, until the design satisfies all thermal requirements within the structure.

In the first design iteration, the thermal requirements of the ramjet burner and nozzle are considered to vastly outweigh the requirements of the aircraft ECS, and all design effort is thus focused on the ramburner area. The fuel conditions available as coolant supply to the burner are estimated and provide a starting condition for the design. As a starting design, the fuel flow is split into four parallel streams, flowing the same direction as the hot-combustion gas. The results of the first design iteration are illustrated in Figs. 26–29. Figures 26 and 27 show the predicted temperature profiles along the length of the upper and lower surfaces in the ramburner. These figures show the temperature on the hot-gas surface and on the fuel wetted surface, and the temperature of the heated fuel. The distance scales begin at the location of the fuel injectors, and the maximum wall temperature coincides with the maximum heat flux location. The temperature of the fuel is noted to exhibit anomalies downstream of the throat, with

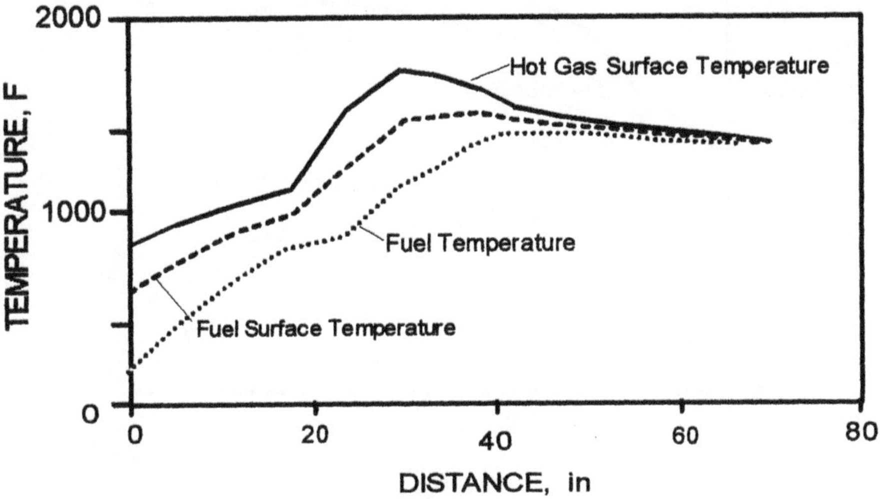

Fig. 27 Temperature profiles on lower surface of nozzle

a reduction in temperature occurring as the fuel is heated. This is a result of endothermic reaction of the fuel, with the heat absorption of the fuel exceeding the heat input of the hot gas. This is further illustrated in Figs. 28 and 29, in which the fuel temperature and enthalpy profiles are described along the fuel flow paths. As expected, the fuel enthalpy continuously increases as the fuel is heated, whereas the fuel temperature depends on the relative levels of heat absorption and endothermic heat sink capability.

Examination of the evaluation results shown in Figs. 26 and 27 indicate that the temperatures of the structures are not balanced on the upper and lower surfaces. Determination of the adequacy of the cooling design requires structural evaluation of the burner walls at those temperatures and under the structural loading imposed by gas and fuel pressures and by the thermal strains. The largest stresses occurring in the cooled structure are due to the thermal gradients within the cooled walls. Detailed evaluation of the temperature profiles across the cooled walls indicate that the unheated surface of the fuel passage (adjacent to the outside of the burner) remains approximately at the temperature of the fuel, whereas the average temperature of the heated wall is the average of the temperatures shown in the figures. The resulting thermal stresses in the region of maximum heat flux is nearly 100 ksi for conventional burner type nickel alloys and, therefore, exceeds the allowable stresses at the predicted temperatures for this particular design. The implication of this observation is the need to simultaneously consider both the temperature and structural requirements while developing the design of the thermal management

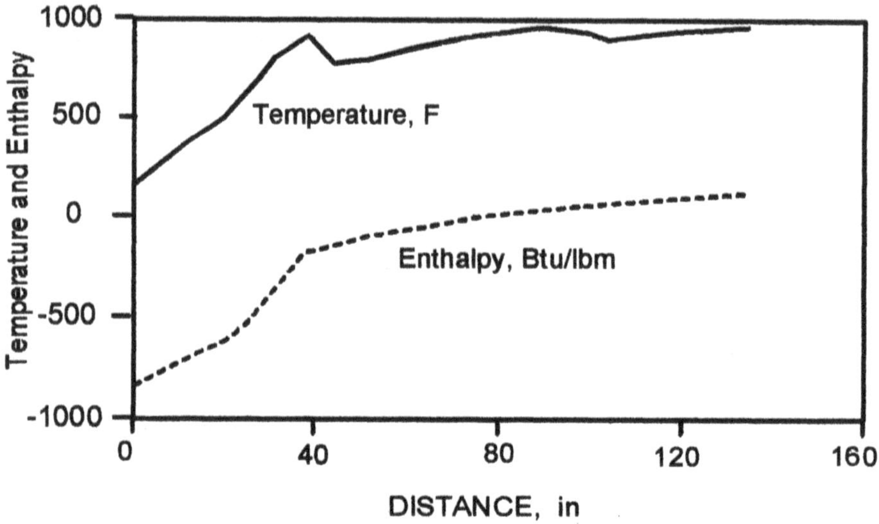

Fig. 28 Fuel temperature and enthalpy variation over upper surface of nozzle

system. As applied to this example, the conclusion at this point is that the design closure has not been attained, and another pass through the design iteration process will be required.

5. *Integration Process to Attain Design Closure*

Development of a satisfactory design for an integrated aircraft thermal management system is achieved through an iterative process, in which the satisfaction of the following design goals are used as a guide to defining design concepts and attaining design closure.

1) All airframe and propulsion component and subsystem level requirements must be simultaneously satisfied.
2) Requirements must include both thermal and structural considerations.
3) The TMS must satisfy all operational requirements over the entire flight envelope.
4) Operational capability of the integrated system must allow meeting the specified aircraft mission goals.

6. *Thermal Management Modeling*

The TMS must be analytically modeled in sufficient detail to insure that the hypersonic vehicle can withstand the severe aerodynamic heating in the atmosphere, while providing other essential functions. Depending on the type of vehicle, the heating conditions to be considered may be for ascent, cruise, or descent. The TMS may employ a combination of insulation and passive and active cooling systems to survive the heating conditions. Typically, power for the vehicle is obtained by a thermodynamic power generation device, such as a turbine, in the cooling system. Analysis of the TMS is best accomplished using automated computer codes formulated for the task. The analysis requires interfaces with other disciplines that have a strong influence on the overall design. The TM design and analysis discipline require customized computer codes that include, as a minimum, the airframe TPS and the active cooling system using fuel as a heat sink for both the engine and airframe.

The airframe TPS must be sized to withstand the aerodynamic heating occurring during the trajectory or mission. An iterative transient analysis is required unless the insulation is applied over a cooled panel; in this case, a steady-state analysis may be adequate. Surface heat flux distribution as a function of time is a primary input to this analysis. Typically, the heat flux distributions are calculated for three or more surface temperatures, so that the local heat flux at any point on the surface of the vehicle may be obtained by interpolating the heat flux based on actual surface temperatures. For expediency, a radiation equilibrium temperature boundary condition may be used, in place of interpolating heat flux with reasonable results for insulation sizing. When this approach is used for an uninsulated surface on the vehicle, however, the surface temperature and absorbed heat flux will be overpredicted. The range of types of insulation that may be used are selected based on thermal and mechanical properties and any other important design considerations, such as weight. The insulation is then sized, using the heat flux and insulation properties. The insulation thermal properties that must be known are thermal conductivity, specific heat, and surface radiation emissivity. The insulation is sized by iterative transient analysis, starting with the minimum practical thickness and increasing until the structure maximum temperatures are within allowable limits or the insulation is found to be inadequate. Insulation is deemed inadequate when the temperature limit or maximum thickness limit is reached and more insulation thickness is needed to protect the structure. In this case, active cooling is required. General purpose heat transfer programs have been used in an automated logic loop to complete this analysis process. In the future, it is anticipated that this analysis will be done with a user friendly computer program that is, for example, highly integrated with aerodynamics, trajectory, and structural analysis programs.

Active cooling is used where TPS is insufficient. Active cooling is generally needed for airframe and engine leading edges, internal

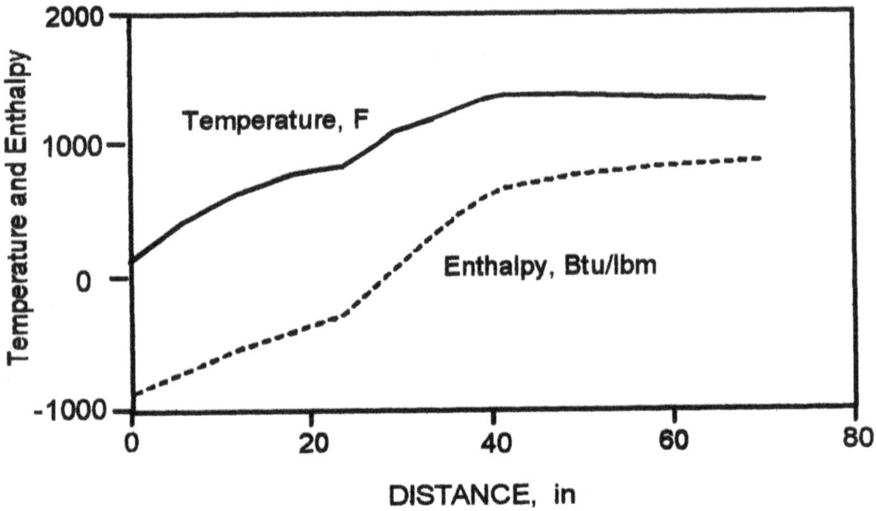

Fig. 29 Fuel temperature and enthalpy variation over lower surface of nozzle

surfaces of the hypersonic airbreathing engine, and the external nozzle if it is used in the vehicle design. Heat flux distributions and coolant flow conditions are the primary inputs to analyze the active cooling system. The types of materials used for cooling are selected based on thermal, mechanical, and weight considerations. The complete properties of the fuel entering the cooling system, the fuel pressure at the engine injectors, and the maximum fuel temperature at the fuel injectors are set based on vehicle design. A baseline cooling system is designed, and a steady-state thermal/fluid flow analysis of the system is completed for maximum heat loads and coolant flow. Coolant routing and cooling panel designs are iterated, and the steady-state analyses repeated until acceptable results are obtained for the cooling system. The design is acceptable when the temperatures and stresses are acceptable throughout the system. The design may be further iterated using an optimization scheme to obtain an acceptable cooling system that also offers minimum weight and/or uses minimum fuel flow for cooling. The need to minimize the coolant flow occurs primarily when the fuel flow for cooling the vehicle exceeds the ideal fuel flow to the engine for combustion. In an actual vehicle design, it is usually necessary to repeat this process many times, as overall vehicle design changes are typically needed throughout the process of design, analysis, and optimization of a hypersonic vehicle.

In the past, modeling of thermal management has been primarily limited to laborious, one-dimensional, steady-state analysis of individual components or subsystems of a TMS, using finite-element or finite difference analyzers. The resulting data would then be passed on for analysis of the next component or subsystem and so on. Cooling system analyses were often simplified in order to keep data input and exchange at manageable levels, and minimize computational run times. Detailed analyses would then be accomplished at areas that had been identified as having potentially high temperatures and/or stresses. A reason for this approach was in large part due to the limitations of the available computers, storage memory, and supporting software. As a result, the codes available to the TM designer did not lend themselves to timely iterations, or easy interactions with the other disciplines, or quick concept comparison studies.

There are currently several methods that are being used by industry and the government to conduct thermal and flow analyses. Some of the more well known codes are Systems Improved Numerical Differencing Analyzer/Fluid Integrator (SINDA/FLUINT), Thermal Radiation Analysis System (TRASYS), P/THERMAL and PATRAN. SINDA is a finite difference analyzer, and the related FLUINT is an internal flow system analyzer. Also related to SINDA is P3/THERMAL, which is a combined finite-element and difference thermal analysis code that offers radiation view factors, thermal hydraulic networks, a convection correlation library, and nonlinear problem solving capabilities. TRASYS analyzes thermal radiation diffuse and grey surfaces and determines view factors, overall radiation interchange, and environmental heating for orbiting spacecraft. PATRAN, a CAE software code, integrates geometry generation, finite-element modeling, analysis applications, materials databases and output graphics. Like most codes, these take time to learn, take practice to achieve accuracy in analysis capability, and require frequent use to stay proficient. Typically, these codes must be used in conjunction with other codes, in order to analyze a complete TMS. This makes the iterative process arduous and time consuming. In addition to these commercially available codes, there are codes that have been developed by companies that are for internal use only and not available to the general public.

Methods presently used for TM modeling have greatly improved, although are still lacking in many areas. Custom computer programs and general purpose heat transfer programs have been used for the active cooling system analysis. When the general purpose SINDA program was used, however, extensive supplemental programs were needed to accommodate the fuel flow part of the analysis. Analysis of cooling with the fuel (or other secondary coolant) requires the capability to model a system of series and/or parallel connected panels and other fluid components, such as valves to the level of detail and accuracy needed. Compressible flow with heat addition, area change,

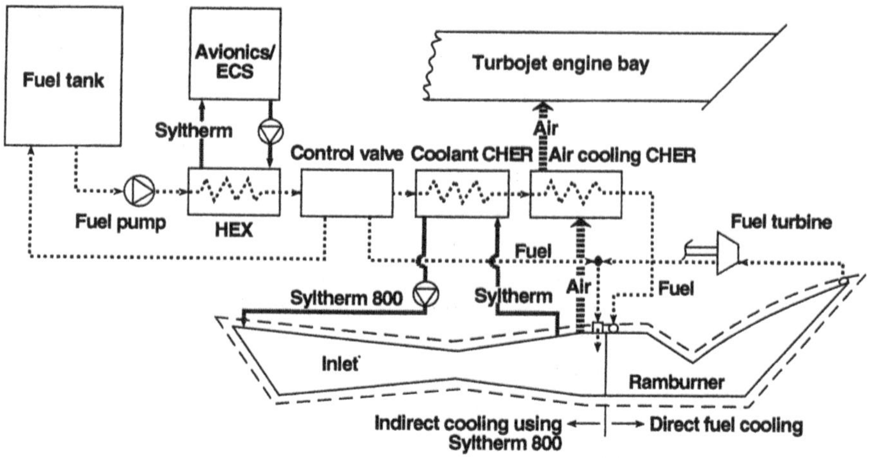

Fig. 30 TMS using endothermic fuel

and friction must be modeled. The future for cooling system analysis methods is expected to be improved with graphics, faster responses, greater detail, and improved accuracy. An advanced computer language capability that allows a subroutine to call itself (recursion) is currently in use in a custom cooling system analysis code, under development by NASA. This custom code will completely replace the SINDA dependent code in the near future.

Programs, such as the former NASP program, served as a stimulus for advancing the development of TM analysis codes by the Air Force, NASA, and industry. Better heat load calculation capabilities have been added, for example, to the Aerodynamic Preliminary Analysis System (APAS) and Aeroheat codes. Computer programs with the capability for detailed transient and steady-state analysis of cooling systems were developed at organizations such as NASA Langley Research Center, Pratt and Whitney, Rockwell International, Science Applications International Corporation and others.[39] An engineering code, funded by the Air Force, known as the Vehicle Integrated Thermal Management Analysis Code (VITMAC), is under development to specifically model high-speed vehicle TMS.

Present codes have been applied to hypersonic cruise vehicles using endothermic fuel and to orbital access vehicles with one or two stages. Cooling systems for Mach 5 cruise aircraft that use endothermic fuel were designed and analyzed. The cooling system for a Mach 5 cruise aircraft that uses endothermic fuel is shown in Fig. 30. Both direct and indirect cooling were used, utilizing individual advantages. Endothermic fuel must be exposed to a catalyst while it is heated in order to promote the required endothermic reaction of the fuel. This was

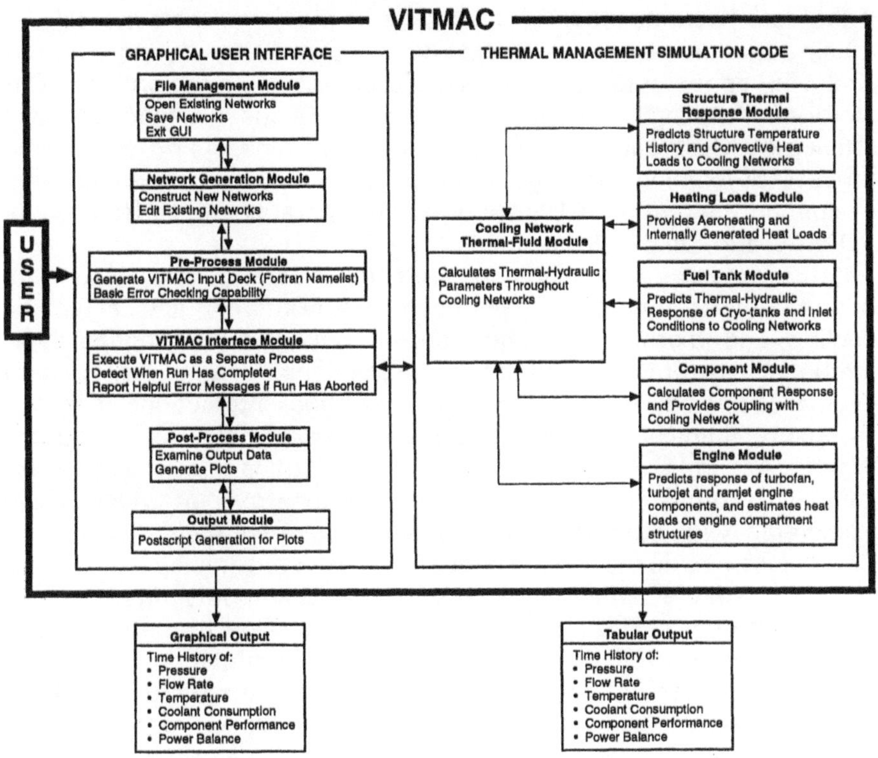

Fig. 31 Overview of VITMAC's TM modules and GUI

accomplished either with a catalytic heat exchanger reactor (CHER) or the catalyst was incorporated inside the cooling panels.[33] Concepts for orbital access vehicles included at least one stage and used cryogenic fuel for propulsion and cooling. When there is cryogenic fuel, not only does the fuel tank need to be modeled but also the fuel as it flows through the network, since it is often used for other things, such as cooling inlet air. When slush hydrogen is used, it is important to consider how the slush will be used to best advantage as a heat sink as it flows from the tank to the fuel injectors in the engine. Melting of the slush in the tank by recirculation of liquid hydrogen back to the tank must be considered.

A complaint frequently heard about many of the thermal analyzer codes is that it takes a long time to learn how to use the codes and to become proficient in their use. Software interface methods, known as graphical user interfaces (GUI), are being used to aid in this problem. These GUIs assist the user in learning a program quickly with the aid of menus, icons, push buttons, and scrollbars. The GUI also makes it easier to step through an application and visually see it on the screen.

If the user has a question on the usage of the code, there typically is a help function. GUIs can also be an aid to the user during post-run analyses, as the data can be presented in such a way to enhance and pictorially present the data. When developed effectively and efficiently, GUIs can save valuable manhour time during the overall process of learning the code, input of data, making corrections, and evaluating the output. GUIs are being incorporated into codes such as VITMAC and SINDA, for example.

Improvements particularly needed in modeling, are heat flux boundary-layer condition predictions (both internal and external), cooling and thermal protection system analysis, and optimization methods. The precision of heat flux predictions, for example, at sharp engine leading edges experiencing multiple shock interactions, is crucial to ensure there is sufficient cooling at critical moments during the flight. Accurate methods for prediction of transition location from laminar to turbulent flow are needed to reduce over or under design associated with boundary-layer condition uncertainties. Analysis of TPSs must include the effects of stagnation heating in steps, gaps, and seals in airframe panels to ensure structural integrity and survival. Cooling system analysis must include the capability to model several different types of active cooling and thermal protection methods. Thermal management analysis must be coupled to structural analysis to provide better design life predictions. Optimization methods must be more automated, include as many design variables as possible, and facilitate iterative trades with the other disciplines (e.g., propulsion, aerodynamics, structures, and subsystems). In addition, optimization methods must also include the ability to look at the impacts within the TMS at various locations of the vehicle when changes happen at other locations or when off-design conditions occur.

There have been major changes in the computer and software industry during the last few years, enabling greater capability and flexibility to the TM modeler. With the increased memory, computational speeds, and supporting GUIs, what once required a mainframe computer can now be accomplished on workstations and high-end personal computers. The TMS design process covered earlier in this section lends itself to computational modeling. VITMAC has been designed to take advantage of the progress in the computer industry, in addition to its fundamental approach in modeling thermal management systems. This particular code has the ability to couple the effects of a vehicle's aerodynamic and internally generated heat loads with the structure thermal responses and the coolant network steady-state and transient thermal-hydraulic responses for the propulsion system and airframe.

For example, VITMAC is organized in a modular fashion, with two primary macro-modules: a TM simulation code and a GUI. Fig. 31

shows an overview of the VITMAC architecture, including the outputs. As described in Ref. 40, the TM macro-module is comprised of six modules, providing all of the thermal-fluid parameters necessary to evaluate a TM network. This includes fuel/coolant flow rate, temperature and pressure distributions, as well as airframe and engine structure temperatures. With the use of embedded icons for the various components, the user is literally able to put a network together on the computer screen with the aid of a keyboard, mouse, icons, and pull-down menus. Inputting data is a matter of entering specified data on the icon's or component's data menu. Data, such as thermophysical properties for a number of structural and thermal protection materials and friction factors and loss coefficients for a wide selection of plumbing fixtures (i.e., valves, entrances, exits), are included in a database within the code, allowing the user to select a value or to input a value. This code has the flexibility to have multiple loops (serially and/or in parallel), coolant sources, heat sinks, branch merge points, connections, mass addition, and coolants. Heat fluxes and loads can easily be added. Structures, comprised of single, multiple, or composite materials, coatings, insulation, and/or gaps, can be easily modeled. If a material, component, or coefficient is different from the values currently embedded in the code, the user can easily add them.

When a component, system, or a complete TMS needs to be evaluated, the analyst or engineer puts together a model of the item through a step by step process, entering data as needed. Once the model is completed, the code will preprocess the data and run. Assuming a successful run, the code then gives the user the option to look at various outputs. The user can save the results and go on to look at another configuration or make changes in the current configuration to see the impacts of the change. The code lends itself to parametric or tradeoff studies, eventually leading to an optimized system. The capability to address issues related to weight impacts, fuel flow rate, and vehicle range are not currently in the program, but there are plans to add this capability.

Modeling of TMSs in the future will be easier and less time consuming than today. This will be attributed largely to the advances being made in software and graphic user interfaces, in addition to computer hardware development. Engineering, finite difference/element, and CFD codes being developed and used today will be further refined and expanded to handle the systems of tomorrow. As more ground and flight tests are conducted, the codes will become more thoroughly validated and achieve greater accuracy as more application and verification runs are accomplished. Input of data will be easier with GUIs, including the capability to input data, such as a cooling network, with the aid of voice commands.

With the recent downsizing of engineering staffs in many aerospace industry and government offices, there will be a greater need for less labor intensive design tools, while maintaining and improving computational capability and accuracy. This will lead to tools that require less time to learn, apply, and run, while meeting the requirements of the thermal engineer and analyst. Computational run times will be decreasing through more efficient coding, improved computer hardware, or a combination of both.

VI. Issues, Problems, Concerns

When breaking ground for new technology, there are always obstacles to overcome. The development of high-speed flight systems and auxiliary systems is definitely no exception. In addition, computational and prediction methods are undergoing development, enhancements, and verification. The degree of uncertainty of some of the prediction methods are unknown. Until flight tests are conducted that more closely match the vehicle configuration and flight conditions, full validation of the predictions will not be obtained. Since there are no ground test facilities that completely simulate flight conditions, run at the high Mach number regimes and produce flight-like heat loads, verification of heat load predictions and systems operation is and will be difficult. This makes performance testing of high-temperature TMS components under realistic operating conditions a challenge.

The TM design and analysis process has been applied to SSTO vehicles and hypersonic cruise vehicles with reasonable success; however, many problems remain to be solved. The first basic problem is management and communication of the TM tasks within the organization. For some hypersonic vehicles, TM should be a top level trade discipline to be considered at the conceptual stage along with such things as aerodynamics, propulsion, and trajectory. TM has been recognized as a discipline in the design of hypersonic vehicles; however, the definition of the thermal management working group in the organization has often been difficult. Management reorganizations have resulted in thermal management groups being formed, dissolved, and reformed along with a changing definition of the thermal management task. The thermal management group is desperately needed for top level trades and to design and integrate active cooling systems into a vehicle. Without a well-defined group, TM design may be delayed until after the vehicle design is frozen so that little optimization is possible without costly redesign.

There are difficulties in obtaining accurate heat loads for the design. Aeroheating loads, for example, must be determined as accurately as possible, due to the domino effect they have on the overall design process. This in itself is not a trivial task, particularly for complex vehicle shapes. If the predicted aeroheating loads are too low, the structure could be in danger of exceeding material temperature limits, potentially resulting in a catastrophic conclusion to the vehicle. If at the other extreme, the predicted heat loads are too high, the vehicle will be overdesigned and suffer large takeoff gross weight penalties. Factors to be reconciled for accurate predictions include boundary-layer and flow conditions, (i.e., laminar, turbulent, or transitional flow), compressible or incompressible gas, and shock interactions. Since the vehicle will be in flight, these factors must be dealt with dynamically in terms of time, velocity, altitude, trajectory, and physical location on the vehicle. Some other aerothermal points related to high-speed vehicle structural design are covered in Ref. 41. The location of transition from laminar to turbulent flow can only be approximated, and this has a strong effect on heat load. Airframe leading-edge heat loads are not adequately known. The engine cowl leading-edge heating due to a shock-on-shock condition is very high (on the order of 50,000 $B/ft^2/s$ for sharp leading edges) and is particularly difficult to predict. For all leading edges, better analytical methods are needed to account for the chemical kinetics of dissociation, catalytic effects of the wall, and shock interactions as they effect heat loads. Improved ways to calculate the effect of shocks on heating of airframe and internal engine surfaces are also needed. More work is also needed to quantify the heat load reductions that can be obtained from film and transpiration cooling concepts.

There are difficulties in predicting convective cooling for high-heat fluxes that occur at leading edges and shock impingement locations. Currently, the most promising cooling concepts for high-heat flux are curved channel cooling and impingement cooling, although straight channel cooling must be used where hot spots occur on flat surfaces. Convective heat transfer in curved and straight channels was studied using existing experimental correlations and CFD in Refs. 42 and 43 with limited success. The result of these studies was to emphasize the need for more experimental data.

VII. Concluding Remarks

There have been misconceptions in the past about what thermal management is and includes. In the past, the term was used on a localized basis. With high-speed flight, designers can no longer take this narrow focus but instead must look at thermal management on a system level. With that in mind, thermal management can no longer be left as an after thought in the design process but must be an active part from the early design stages throughout the vehicle design iterations. It is hoped that this chapter has helped clarify for the reader what thermal management covers. Although the discussion has been geared toward high-speed flight, the principles and concepts can be easily applied to other regimes.

From the discussions within this chapter, the reader should have a clearer understanding of thermal management. Also, the reader should come away with a greater perspective and should start viewing this technology on a broader scale as it pertains to a high-speed flight system or any other system where there are large thermal loads. From the historical discussion of thermal management as it pertains to flight, we have seen that this technology has been included on a piecemeal basis during earlier flight designs. Since the thermal loads were not a major factor, thermal management was not a major driver in the design process. As we entered the world of high-speed flight and began to experience the effects of aerodynamic heating, this changed. The development of a hypersonic vehicle is where one started seeing TM included during the early design stages.

The system perspective covered the impacts of heat or thermal load predictions on design and the domino effects it can have with prediction uncertainties. As flight speeds increase, the need for an integrated approach to thermal management also increases, due to the interdependencies with structures, materials, propulsion system, airframe, fuel tank, and subsystems. Following this, it has become clear that it is necessary to have the airframe and engine cooling networks integrated in order to optimize the overall TMS while reducing unnecessary redundancies. The complexity of designing an integrated cooling network and TMS requires computer codes that can aid the designer to conduct trade studies in order to develop an optimized TMS. These codes will have to be applied and validated with test data to assure accuracy in modeling and predictions.

From the preceding discussion, it has become obvious that thermal management is a complex technical multidiscipline. There are many technical issues that must be addressed in order to develop a successful TMS. As this is a relatively new field, TM will undergo a number of growing pains and various levels of acceptance by program managers. It is hoped that these discussions will help to clarify what thermal management is and some of the challenges that will be faced, and also stimulate interest in this area.

VII. Acknowledgements

The authors would like to thank Pratt and Whitney for contributing Figs. 9 and 11-29 and Science Applications International Corporation for Fig. 31. The authors would also like to thank Mr. David Buckwalter for his help in the preparation of some of the figures and Mr. W. Lee Bain, III for his encouragement and support during the preparation of this chapter.

References

[1]Edwards, J. L., "By the Application of Power," *Aeronautical Journal*, Vol. 79, 1975, pp. 59–84.

[2]Esgar, J. B., and Ziemer, R. R., "Review of Status Methods and Potentials of Gas-Turbine Air-Cooling," NACA RM E54123, 1956.

[3]Esgar, J. B., Livingood, J. N. B., and Hickel, R. O., "Research on Application of Cooling to Gas Turbines," American Society of Mechanical Engineers, ASME Paper 56-SA-54, Jun. 1956.

[4]Esgar, J. B., and Ziemer, R. R., "Effect of Turbine Cooling with Compressor Air Bleed on Gas-Turbine Engine Performance," NACA RM E54L20, 1955.

[5]Esgar, J. B., and Slone, H. O., "Gas-Turbine-Engine Performance When Heat from Liquid-Cooled Turbines is Rejected Ahead of, Within, or Behind Main Compressor," NACA RME56B09, 1956.

[6]Moore, A., "Gas Turbine Engine Internal Air Systems: A Review of the Requirements and the Problems," American Society of Mechanical Engineers Winter Annual Meeting, ASME Paper 75-WA/GT-1, Dec. 1975.

[7]Esgar, J. B., "State of Technology on Hydrogen Fueled Gas Turbine Engines," NASA TM-X-71561, May 1974.

[8]Petley, D. H., Smith, D. M., Edwards, C. L. W., and Carlson, A. B., "Analysis of Gap Heating Due to Stepped Tiles in the Shuttle Thermal Protection System," NASA TP 2209, Aug. 1983.

[9]Avery, D. E., "Experimental Aerodynamic Heating to Simulated Space Shuttle Tiles in Laminar and Turbulent Boundary Layers with Variable Flow Angles at a Nominal Mach Number of 7," NASA TP 2307, Aug. 1985.

[10]Bevens, J. E., (ed.), *Thermophysics: Applications to Thermal Design of Spacecraft,* Vol. 23, Progress in Astronautics and Aeronautics, AIAA, New York, 1971.

[11]Lucas, J. W. (ed.), *Heat Transfer and Spacecraft Thermal Control,* Vol. 24, Progress in Astronautics an Aeronautics, AIAA, New York, 1971.

[12]Lucas, J. W., (ed.), *Fundamentals of Spacecraft Thermal Design,* Vol. 29, Progress in Astronautics and Aeronautics, AIAA, New York, 1972.

[13]Hering, R. G., (ed.), *Thermophysics and Spacecraft Thermal Control,* Vol. 35, Progress in Astronautics and Aeronautics, AIAA, New York, 1974.

[14]Collicott, H. E., and Bauer, P. E., (eds.), *Spacecraft Thermal Control, Design, and Operation,* Vol. 86, Progress in Astronautics and Aeronautics, AIAA, New York, 1983.

[15]Hudson, W. A. and Levin, M. L., "Optimizing Aircraft Fuel Thermal Management," *Aerospace Engineering*, Feb. 1987.

[16]Rued, K., Mark, H., and Goetz, G., "Heat Management Concepts for Hypersonic Propulsion Systems," 27th JANNAF Propulsion Conf., AIAA Paper 91-2493, 1991.

[17]Petley, D. H. and Jones, S. C., "Thermal Management for a Mach 5 Cruise Aircraft Using Endothermic Fuels," Aircraft Design, Systems and Operation Conf., AIAA Paper 90-3284, 1990.

[18]Waldman, B., and Harsha, P., "The First Year of Teaming: A Progress Report," AIAA-91-5008, Dec. 1991.

[19]Wells, T. A., "Thermal/Structural Analysis and Aircraft Design," *Aerospace Engineering*, Vol. 12, Dec. 1992, pp. 9–12.

[20]Committee on Hypersonic Technology for Military Application, *Hypersonic Technology for Military Application*, Air Force Studies Board, National Academy Press, Washington, DC, 1989, p. 17.

[21]Sobel, D., "Cowl Leading Edge Heat Transfer in the Presence of Shock Impingement," AIAA 2nd International Aerospace Planes Conf., AIAA Paper 90-5256, Orlando FL, Oct. 29–31, 1990.

[22]Edney, B. E., "Anomalous Heat Transfer and Pressure Distribution on Blunt Bodies at Hypersonic Speeds in the Presence of an Impinging Shock," Aeronautical Research Institute of Sweden, FFA Rept. 115, Stockholm, Sweden, 1968.

[23]Keyes, J. W., and Hains, F. D., "Analytical and Experimental Studies of Shock Interference Heating in Hypersonic Flow," NASA TN D-7139, May 1973.

[24]Ryan, B. M., "Summary of Aerothermodynamic Interference Literature," Naval Weapons Center, Technical Note 4061-160, China Lake, CA, Apr. 1969.

[25]Craig, R. R., and Ortwerth, P. J., "Experimental Study of Shock Impingement on a Blunt Leading Edge with Application to Hypersonic Inlet Design," Air Force Aero Propulsion Lab., AFAPL-TR-71-10, Wright-Patterson AFB, Ohio, Apr. 1971.

[26]Korkegi, R. H., "Survey of Viscous Interactions Associated with High Mach Number Flight," *AIAA Journal*, Vol. 9, No. 5, 1971, pp. 771–784.

[27]Wieting, A. R., "Experimental Study of Shock Wave Interference Heating on a Cylindrical Leading Edge," *AIAA Journal*, Vol. 27, No. 11, 1989, pp. 1557–1565.

[28]Holden, M. Wieting, A. R., Moselle, J., and Glass, C., "Studies of Aerothermal Loads Generated in Regions of Shock/Shock Interaction in Hypersonic Flow," AIAA Paper 88-0477, Jan. 1988.

[29]Glass, C. E., Wieting, A. R., and Holden, M. S., "Effect of Leading Edge Sweep on Shock-Shock Interference Heating at Mach 8," AIAA 27th Aerospace Sciences Meeting, Reno, NV, Jan. 9–12, 1989.

[30]Wieting, A., "Shock Interferences Heating in Scramjet Engines," AIAA Second International Aerospace Planes Conf., AIAA Paper 90-5238, Orlando, FL, Oct. 29–31, 1990.

[31]Jones, S., and Petley, D., "A Comparison of Cooling Methods for the Airframe Nozzle of a Single-Stage-to-Orbit Aircraft," AIAA Paper 91-5036, Dec. 1991.

[32]Petley, D. H., and Jones, S. C., "Thermal Management for a Mach 5 Cruise Aircraft Using Endothermal Fuel," *Journal of Aircraft*, Vol. 29, No. 3, 1992, pp. 384–389.

[33]Pegg, R. J., Hunt, J. L., Petley, D., H., and Burkhardt, L., "Design of a Hypersonic Waverider-Derived Airplane," AIAA Paper 93-0401, Jan. 1993.

[34]Petley, D. H., and Hunt, J. L., "Hydrogen for Hypersonic Airbreathing Vehicles," Hydrogen Energy Progress IX, Proceedings of the 9th World Hydrogen Energy Conf., Manifestations et Communications Internationales, Paris, Jun. 1992, pp. 1651–1660.

[35]Petley, D. H., and Dziedic, W. M., "Water Cooling System for an Air-Breathing Hypersonic Test Vehicle," AIAA Paper 93-1984, Jun. 1993.

[36]Department of Defense, "Military Specification: General Specification for Engine Aircraft, Turbojet and Turbofan," Mil-E-5007D, Amendment 2, Washington, D.C., Oct. 1982, pp. 12–13.

[37]Spalding, D. B., and Chi, S. W., "The Drag of a Compressible Turbulent Boundary Layer on a Smooth Plate With and Without Heat Transfer," *Journal of Fluid Mechanics,* Vol. 18-Part 1, Jan. 1964, pp. 117–143.

[38]Anon., "Registration Information," Fifth Annual Thermal and Fluid Analysis Workshop, Aug. 16–20, 1993.

[39]Petley, D. H., Jones, S. C., and Dziedzic, W. M., "Integrated Numerical Methods for Hypersonic Aircraft Cooling Systems Analysis," AIAA Paper 92-0254, Jan. 1992.

[40]Wassel, A. T., Issacci, F., and Van Griethuysen, V. J., "An Integrated Modeling Approach for Hypersonic Aircraft Thermal Management," AIAA Paper 95-6022, Apr. 1995.

[41]Martellucci, A., and Harris, T. B., "Assessment of Key Aerothermal Issues for the Structural Design of High Speed Vehicles," *Thermal Structures and Materials for High-Speed Flight*, Vol. 140, Progress in Astronautics and Aeronautics, AIAA, Washington, DC, 1992, pp. 59–91.

[42]Petley, D. H., and Jones, S. C., "Comparison of Heat Pipe and Channel Flow Cooling of a Scramjet Cowl Leading Edge," NASP 1125, Aug. 1990.

[43]Dziedzic, W. M., Jones, S. C. Gould, D. C., and Petley, D. H., "Analytical Comparison of Convective Heat Transfer Correlations in Supercritical Hydrogen," *Journal of Thermophysics and Heat Transfer*, Vol. 7, No. 1, 1993, pp. 68–73.

Energy Management and Vehicle Synthesis

P. Czysz*
St. Louis University, Cahokia, IL 62206
and
S.N.B. Murthy [1]
Purdue University, West Lafayette, IN 47907

I. Introduction

The major drivers in the development of launch vehicles for the 21st century are reduction in cost of launch vehicles, payload in orbit, and operations, reusability on a nearly continuous basis, capability to recover the major parts of the vehicle and all of the payload when forced to abort the mission, and improvements in vehicle and launch readiness. Such requirements become variously qualified and constrained in each nation by its state and commerce policies, geography, and other considerations.

An approach to the conceptual design of such vehicles is still largely a matter of debate, although several design synthesis methods have been developed (Refs. 1-3). The difficulty is in rationalizing needs, capabilities, and opportunities. While it is fully recognized that airbreathing propulsion has a crucial role in meeting the stated goals for launch vehicles, and the vehicle in that

---

* Oliver L. Parks Endowed Chair, Parks College
[1] Senior Researcher, School of Mechanical Engineering
Copyright © 1995 by the American Institute of Aeronautics and Astronautics, Inc. All rights reserved.

case needs to be fully integrated in design, functions, and operation, one needs some reasonable estimates of available and required maturity in various technologies before a conceptual design can be rationalized. However, there are invariably ambiguities and controversies associated with estimates of available and required maturation levels, whether propulsion-propellant schemes, configuration geometries, materials and structures, flight management, or controls are considered, individually or collectively.

In the development of atmospheric flight vehicles at lower speeds, it has become accepted practice to adopt variants of a methodology, developed about a quarter of a century ago (Ref. 4), for conceptual design. The method is based on historical data on design, test results, and operational experience. In the case of hypersonic vehicles the total experience from flight vehicles designs, test data, analysis and predictions, and flight vehicle experience is quite small, and certainly cannot yield statistically weighted correlations for evolving an optimum concept. In addition, the number of critical conditions to be met at the higher speeds up to launch speeds is large, and, at the same time, there are over-riding demands in regard to certain technologies such as propulsion and materials and structures. It is precisely for these reasons that the launch vehicle technology has continued in its present form with all rocket schemes, including, through various developments, such improvements as partial recoverability, or reuse capability with recovery and unpredictable refurbishing.

It is clear that the approach to the conceptual design of a hypersonic vehicle needs to rest on (a) whatever data and projections can be established, including results from preliminary study capabilities that may become available (for example, Ref. 5), and (b) a recognition of the fact that the most significant gains must be realized from propulsion-propellant technology. The latter points to an overwhelming constraint as well as a means of overcoming deficiencies in other respects.

Based on such reasoning, a methodology has been developed that focuses on energy management while taking into account the structural-material industrial capability and the configuration concept geometrical parameters. The methodology is illustrated through the establishment of design spaces that become available for various parameter sets in the case of a SSTO vehicle, starting with a reference vehicle and various sets of data and correlations.

Energy analysis for-high speed flight systems has been discussed in Ref. 6, and that article provides a useful introduction here. Other background may be found in Refs. 7 - 10.

This paper is devoted to the discussion and illustration of a methodology for energy-centered evolution of the total launch vehicle. Section II deals with energy analysis of airbreathing propulsion, and shows how a spectrum of engines can be evolved. (Appendix A presents details regarding engine cycle analysis, and the method of establishing net energy addition to the working fluid by combustion of fuel, taking into account the internal losses.) Section II deals with some aspects of propulsion-structural material interface, and delineates the relation between the spectrum of airbreathing engines and the choice of materials. (Appendix B is devoted to examples of available data and projections on materials and structural loads.) In Sec. IV various aspects of matching propulsion performance and vehicle configuration are discussed. (Examples of available data and projections regarding these are provided in Appendix C.) Then, in Sec. V, the methodology for vehicle synthesis is described with emphasis on the concept and utilization of design spaces. Finally in Sec. VI, a discussion is provided on application of the methodology for vehicle synthesis based on the establishment of a series of design spaces.

II. Energy Analysis of Airbreathing Propulsion

Considering combined-cycle engines with an airbreather, one of the main parameters affecting the propulsion-propellant concept and its impact on vehicle design is the extent to which the airbreathing capability is utilized in different flight regimes along the trajectory. There are a number of factors that lead to various types of constraints on the inclusion of airbreathing: the local ambient conditions of air, including its entropy, the trajectory, the kinetic energy of intake air entering at the flight speed; the internal mixing, drag, and thermodynamic nonequilibrium losses during flow and heat addition by combustion of a fuel; and the efficiency of various processes and components of the engine. Finally, these factors yield an *energy utilization effectiveness* and the extent to which the *available energy* is utilized.

The energy effectiveness of an engine is the amount of energy availability input that is converted to useful output. The limit for the energy effectiveness is set by the equivalent Carnot cycle, or the Second law of thermodynamics. The energy effectiveness of an engine is related to the energy effectiveness of the components and the processes of the engine.

In the same fashion, one can set up relations for the change in the true thermodynamic efficiency of an engine in terms of the true thermodynamic efficiency of the components and the processes of the engine.

The energy effectiveness of a process or a component is dependent on the loss of energy availability in it for given input to it. The measure of such energy effectiveness is the extent to which the energy availability change in the process or the component affects the output of the engine, which should be considered in terms of the energy availability converted to useful work output in the engine, with all of the processes and components, except the one under consideration, operating in an ideal fashion without loss of energy availability. The energy effectiveness is high when the true thermodynamic efficiency is high.

Next, it is of interest to examine the manner in which an engine, or, more definitively, its components and processes, can be optimized to obtain the maximum possible conversion of input energy availability into output in the engine. It is common engineering practice to analyze the thermodynamic process of an engine based on its representation in the form of a thermodynamic cycle. The complexity of the engine determines the complexity of the thermodynamic cycle. The complexity of the engine depends on the number of parameters governing its structure, the flowpath of one or more working fluids utilized in the engine, and other related processes. However complex the engine, it can be represented by a thermodynamic cycle, each part of which is a simple process. One can then proceed to optimize the parameters

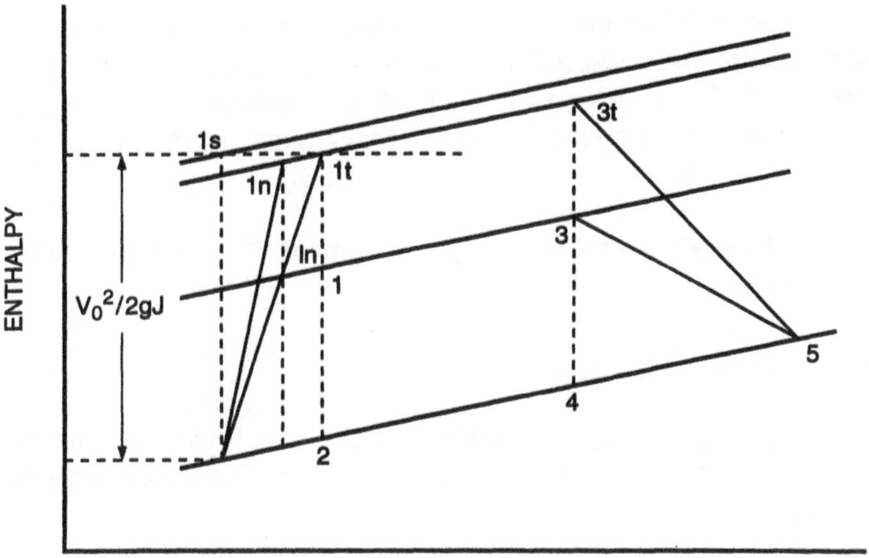

Fig. 1 Brayton cycle diagram.

of each process or part of the cycle for obtaining the highest engine output.

Considering the Brayton cycle engine illustrated in Fig. 1, one can ask if the cycle parameters can be optimized based on energy availability considerations so as to realize a desired maximum in the output, e.g., the largest thrust power output, under a given set of constraints. A solution, of course, is a combination of processes, each of which causes no change in energy availability - an impractical solution. However, noting that the total energy availability is made up of a number of components, it is possible to choose one of the components and regulate changes in energy availability relative to that component in several processes such that the overall change in energy availability is minimal. In general, this requires, a number of assumptions in the analysis and, correspondingly, in practice, in the choice of governing parameters for different processes. Such analysis may be conducted with respect to each of the components of energy availability, and the results compared to choose the best control scheme from a particular point of view.

In this section, various aspects of energy availability analysis are discussed before a spectrum of airbreathing engines is presented for inclusion as part of a combined-cycle engine. Appendix A presents some background and additional considerations.

A. Energy Availability Requirement for Airbreathing Propulsion

1. Some Constraints

1. Fuel combustion energy input. The amount of energy, Q, input to the air by fuel combustion in the combustor may be written as

$$Q = Q_c\,(f/a)_s\,\phi_c \tag{1}$$

where the calorific value of the fuel, the fuel-to-air ratio, and the fuel equivalence ratio are denoted by Q_c, f/a, and ϕ_c, respectively, and subscript s denotes the stoichiometric condition. Now, although Q_c, the heat released per unit mass of fuel, varies substantially for different fuels, the value of Q in Brayton cycle engines, based on energy release per unit mass of air, varies insubstantially. For example, the heat input to air may be assumed to be as follows per unit mass flow of air in the case of different fuels: most hydrocarbons: 1270-1370 Btu/lbm; hydrogen and acetylene: 1505-1570 Btu/lbm; and unconventional fuels: 2200-2400 Btu/lbm.

Normalizing Q with respect to the enthalpy of ambient air H_0, which varies only slightly from 100,000 to about 300,000 ft altitude over the range of 98 to 118 Btu/lbm, and thus may be assumed, as in Ref. 10, to have a mean value of 100 Btu/lbm in the relevant upper latitudes, one can write

$$\frac{Q}{H_0} \cdot \frac{H_0}{V_0^2/(2gJ)} = 75.19\left(\frac{V_0}{1000}\right)^{-2} \tag{2}$$

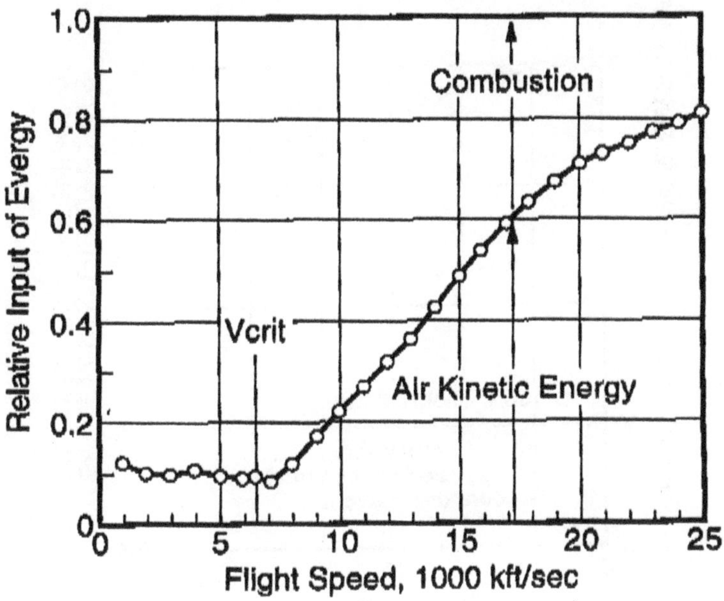

Fig. 2 Fraction of energy supplied by combustion energy and air kinetic energy as a function of flight speed.

It is assumed here that $2gJQ$ is equal to $75.19 \times 10^6 \; (ft / s)^2$ for the case of hydrogen fuel.

The amount of kinetic energy available in the air entering the engine is shown in Fig. 2 as a fraction of the total energy supplied to the engine by the air flow and fuel combustion for the case of hydrogen fuel. It can be observed that at about a flight (or engine entry) velocity of 6000 ft / s, the amount of energy supplied by the fuel diminishes continuously as the flight speed increases. Including Carnot losses and the internal mixing losses (discussed in Appendix A), one can estimate the ratio K / C, where $K = V_0^2 / (2gJH_0)$, and $C = Q / H_0$, as a function of flight velocity in different cases, as shown in Fig. 3. Thus, it is extremely

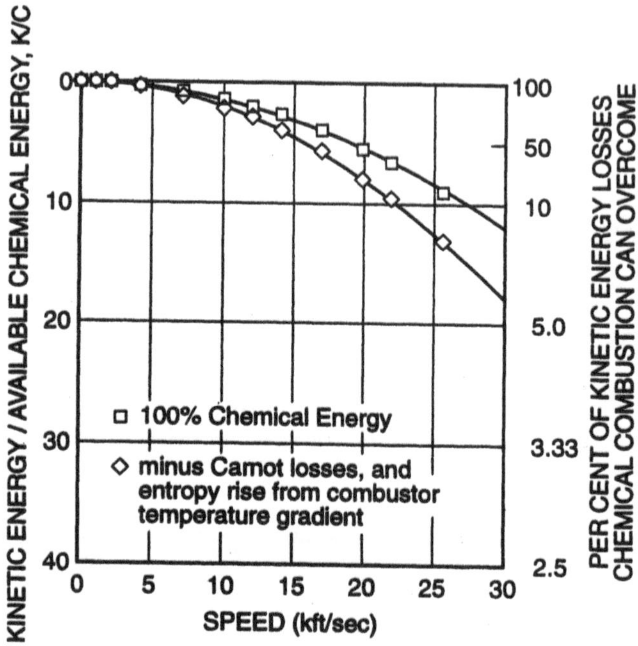

Fig. 3 Ratio of kinetic energy and chemical energy as a function of flight speed.

important to avoid loss of the kinetic energy of air and the combustion-generated energy input as the flight speed increases.

Now the velocity at entry to the combustor must be retained at supersonic values above the critical velocity given by Eq. (A.10) since it is necessary to limit the diffusion of air to the optimum value given by Eq. (A.9), and thus to realize the highest thrust from the engine for a given amount of fuel input and given values of inlet air compression and thrust nozzle expansion efficiencies. If the air is compressed over and above the value corresponding to ψ_{opt} then the maximum output from the engine cannot be obtained.

In general, there is a minimum value of pressure in the combustion chamber below which a given fuel does not burn in a stable fashion. In the case of hydrogen fuel, this value of pressure is known to be about 0.5 (sea-level atmospheric pressure).

In addition to the thrust output from an engine, insofar as flight is concerned, one should also consider the effect of the combustor entry pressure on the *propulsive efficiency*, which may be defined as the ratio of thrust work output to the amount of chemical energy that is converted to the net kinetic energy obtained, with respect to a stationary observer, by expansion of the products of combustion in the thrust nozzle.

The air compression in the inlet thus has to be based on all three of the foregoing factors, the maximum compression for obtaining the highest thrust, according to Builder's analysis; the minimum value of pressure required for stable combustion in the combustor; and the influence of the maximum pressure of the cycle on the propulsive efficiency that can be obtained. There is, however, another extremely significant consideration, namely, the presence of nonequilibrium in the engine flow along its flowpath as the flow velocity increases.[11] It is possible to carry out an optimization of the Brayton cycle processes based entirely on considerations of minimizing nonequilibrium effects.

2. Nonequilibrium Effects. It is well known that the entropy of atmospheric air increases as the altitude increases relative to the value at sea level or the absolute zero of temperature. Thus, there is a natural degradation in air quality. Above an altitude of about 180,000 ft, it appears that energy addition by combustion, and possibly other means, may not prove useful.

At higher altitudes, above about 100,000 ft in Brayton cycle engines operating with combustion, one finds appreciable chemical nonequilibrium effects as the air is compressed in the inlet and attains higher temperatures, energy is added by combustion in the combustor, and finally the products of combustion expand rapidly through the thrust nozzle. In the altitude range of 100,000 - 150,000 ft, it has been shown[12] that the nonequilibrium effects may be mitigated substantially by the addition of hydrogen. However, above that range, the addition of hydrogen is not effective.

In general, it is essential to carry out detailed calculations on the state changes of the working fluid and the associated nonequilibrium losses. However, the main findings on the applicability of airbreathing engines can be established based on near-chemical equilibrium assumptions, noting that additional first-order changes may become necessary on taking into account nonequilibrium effects.

Some related correlations are included in Appendix A.

3. Thrust-to-Drag Ratio. The thrust-to-drag ratio is an important parameter in the case of a flight vehicle operated with an airbreather, which deliberately flies through the atmosphere over the denser altitudes to be able to take in the required mass of air with an intake of small cross-sectional area. The ΔV lost as a result of vehicle drag for an airbreathing vehicle is large. The magnitude of this loss has no relation to that for a rocket that suffers no appreciable atmospheric drag effects. The specific impulse of a rocket motor is very nearly that of the intrinsic propellants used in it. In the case of an airbreathing vehicle designed for acceleration, the specific impulse is a function of flight and engine operating conditions; furthermore, an effective specific impulse must be considered that includes the thrust-to-drag ratio, viz.

$$I_{SPE} = I_{SP} \frac{(T/D) - 1}{T/D} \tag{3}$$

The thrust-to-drag ratio being greater than unity for an accelerator, the effective specific impulse is less than the specific impulse of the propellants utilized in an airbreathing engine.

It is now of interest to establish a means of comparing one airbreathing engine with another, and the performance of an engine at two different flight operating conditions.

Engine effectiveness discriminant. In the case of a vehicle with a cruise mission, the thrust generated must equal the drag. Then, considering the range ΔR obtained with a mass of fuel equal to ΔW_P, one can write

$$\frac{\Delta W_P}{\Delta R} = \frac{\dot{W}_P}{V} = \frac{D}{V I_{SP}} = \frac{\overline{W}}{V I_{SP}(L/D)} \tag{4}$$

where $\dot{W}_P / V$ is the increase of *specific range*, and $VI_{SP}(L/D)$ is the *range factor*. The weight $\overline{W}$ denotes the mean weight of the vehicle over the cruise range.

In the case of an accelerating vehicle, the thrust generated must be greater than the drag, and the acceleration is then proportional to the excess of thrust over the drag. A discriminant may then be postulated between the effectiveness of two engines in terms of a parameter as given by

$$\frac{\Delta W_P}{\Delta V} = \frac{\dot{W}_P}{a} = \frac{T - D}{I_{SPE}} \frac{\overline{W}}{g(T - D)} = \frac{\overline{W}}{gI_{SPE}(L/D)} \tag{5}$$

where a, $\dot{W}_p / a$, and gI_{SPE} are the acceleration, the inverse of specific ΔV (ft.lb-s / lbm of fuel), and the velocity increment factor (ft / s), respectively.

Considering a rocket motor vehicle with mean values of *T/D* equal to 10 and I_{SP} equal to 470 s, an airbreathing vehicle that matches the performance of the rocket in an acceleration mission then must meet the following requirements in the mean.

$(T/D)_{airbreather}$	$(I_{SP})_{airbreather}$, s
2.2	770
2.0	846
1.8	958
1.0	1128

Thus with T/D equal to about 2.0, the mean specific impulse of an airbreather has to be about twice that of a rocket motor to provide the same acceleration capability.

A somewhat similar result has been obtained by Builder in the form of a *thrust coefficient* θ for the engine; this is discussed in Appendix A.

Internal losses of the engine. The amount of energy or energy availability added to the working fluid in the combustion chamber by the combustion of the fuel becomes reduced because of a) the skin friction and heat transfer to the walls, and b) the mixing between the fuel and the air. The "a" losses depend on, in addition to other factors, the use of a cowl and the provisions for its motion (in relation to the shock at the nose). The "b" losses depend on the use of struts or other means for fuel injection. Some tentative estimates on these losses and their implications are discussed in Appendix A.

In summary, an airbreather must be considered in terms of three parameters, as follows:

1) I_{ASP} in relation to air mass flow, which should be as high as possible;
2) $I_{SP} = (\theta J Q_c) / V_0$, which should be about twice the value of I_{SP} for a rocket; and
3) $I_{SPE} = \frac{\theta J Q_c}{V_0}\left[\frac{(T/D) - 1}{T/D}\right]$, which should be equal to that of a rocket.

The challenge is to realize this in practice.

B. Spectrum of Airbreathing Engines

On the basis of the energy being available by the combustion of a fuel, and taking into account the kinetic energy available in air at different flow speeds, one can identify five regimes of operation, as shown in Table 1. The first column shows the ratio of heat added by combustion to air, Q_A, to the kinetic energy of airflow, both normalized with the enthalpy of ambient air, H_0. The third column shows the same quantity with Q_A corrected for an estimated loss of energy resulting from mixing, ΔE_{mix}. The fourth and fifth columns include some remarks about the need for fuel mass and energy, and the extent of the sum total of all engine-related drag losses.

Table 1. Possible regimes of operation of airbreathing engines

$\frac{Q_A}{KH_0}$	V_0 *ft/s*	$\frac{Q_A - \Delta E_{mix}}{KH_0}$	Fuel mass and fuel energy	All engine drag losses
1.00	7,143	0.9440	Negligible	Negligible
0.5000	10,101	0.4414	Important for acceleration	Can limit maximum thrust
0.2000	15,972	0.1445	Necessary to provide thrust	Can prevent acceleration
0.1429	18,898	0.0911	Essential for acceleration	Must be minimized
0.1000	22,588	0.0538	No longer can overcome losses (entropy rise)	May prevent practical airbreather

For flight speeds less than 7143 ft / s (2164 km / s), fuel mass and energy requirements are minimal, as well as the mixing losses. As shown in the third column, it can be estimated that about 94% of the energy available is used for overcoming internal drag and providing excess thrust for acceleration.

As the flight speed increases, Q_A becomes progressively a smaller fraction of the kinetic energy of air, and the effect of $\Delta E_{mix} / H_0$ becomes increasingly greater. In the regimes of the energy ratio decreasing from 0.20 to 0.10, the performance is so sensitive to energy losses that even a modest increase in losses can easily cause the thrust available for acceleration to become zero.

Figure 4 is a representation of the five flight regimes in altitude-velocity space. It includes a band for flight trajectory, and also an indication of the change in entropy of air with altitude. As the entropy of air increases, an increase in the fuel equivalence ratio becomes less effective in overcoming chemical nonequilibrium effects. The map shows the five regimes given in Table 1, and identifies ten *potential, successful* regimes for

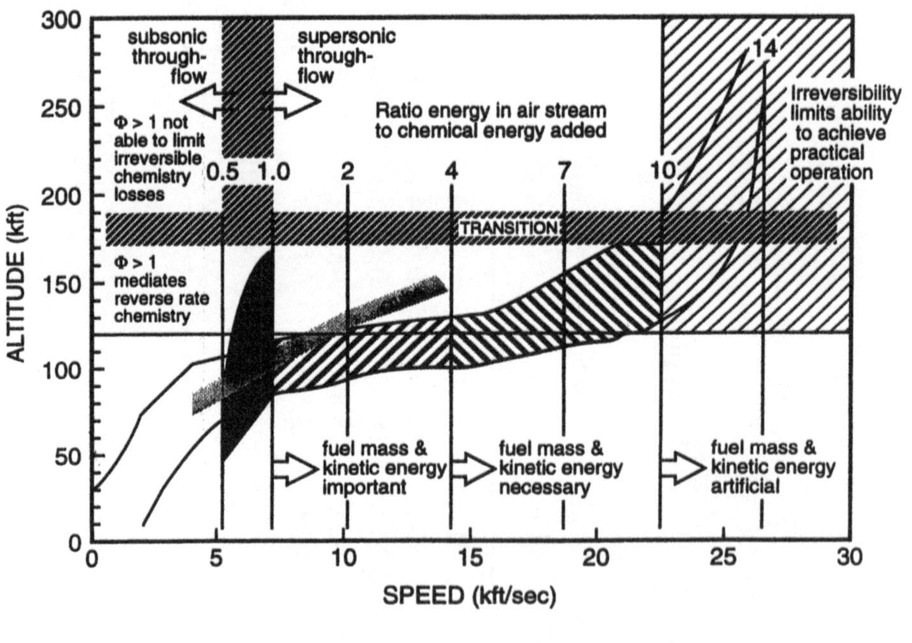

Fig. 4 Spectrum of flight regimes in altitude-flight speed space.

accelerating flight and three potential, successful regimes for cruise flight, taking into account operation with the fuel equivalence ratio being less than or equal to unity and also over the unity value.

Three of the low energy ratio regimes of Table 1 are identified in Fig. 4 as those where fuel mass and kinetic energy are important, necessary, and critical, in that order. In the *important* region, the engine derives about 14% of energy input from fuel mass, and kinetic and thermal energy while it still derives its energy input largely from fuel combustion. In the *necessary* region, fuel mass and energy may need to provide 30-40% of the energy input. Finally in the *critical* region, about 70-100% of the input energy must come from fuel mass and energy. It may also be observed in Fig. 4 that cruise flight at flight speeds above 8000 ft/s requires operation at a higher altitude than acceleration flight, and, therefore, requires a larger fuel equivalence ratio for successful operation. The region in Fig. 4 with the energy ratio between 0.5 and 1.0, or, equivalently, at flight speeds between 5000 and 7000 ft / s, is the region for optimal transition from subsonic to supersonic operation of the combustor, according to Builder's estimate.

It is of some interest to discuss the engine regimes in terms of the entropy of air as a function of altitude and the implications to chemical energy availability. This can be done by identifying various altitude-entropy regimes (see Table 2).

Airbreathing propulsion becomes progressively more difficult and complex as one considers increasingly higher flight speeds. Combined-cycle engines are effective up to a high flight speed; however, the airbreathing part of the engine becomes less effective as flight speed increases.

III. Propulsion-Structural Material Interface

In this section, an attempt is made to describe an approach to 1) relating structural material capability to propulsion possibilities in the altitude-speed space, and 2) determining the structural weight that is feasible as a function of design surface temperature with a given choice of materials and thermal management and protection systems. The discussion under item 1

Table 2. Altitude-entropy regimes

Altitude, *h*	entropy, *S/R*	Implications
$h < 100{,}000$ ft	$21 < S/R < 31.1$	Conventional chemical energy input engine without significant nonequilibrium.
$100{,}000 < h < 120{,}000$ ft	$31.1 < S/R < 33.1$	Some frozen state; excess hydrogen can reduce frozen equilibrium effects.
$120{,}000 < h < 170{,}000$ ft	$33.1 < S/R < 35.7$	Frozen state up to 10%. Significant amount of excess hydrogen may be necessary.
$h > 170{,}000$ ft	$S/R \geq 35.7$	Large proportion of gas in frozen state. Excess hydrogen is incapable of restoring equilibrium.

is directly related to industrial capability in materials and structures and that under item 2 to the selection of vehicle configuration. Both of those issues appear in the following sections.

The choice of materials is based on past history and limited projections. Although the historical evidence and projections can be modified with other data and reasoning, the approach to determining the inevitable interactions between the choice of propulsion and structural materials in the context of an integrated vehicle system is considered as basic.

The historical data and projections are discussed in Appendix B.

A. Preliminary Considerations

1. Material Strength. The *specific strength* of a material is defined as the ratio of the tensile strength (lbf / in^2) to density (lbf / in^3), and thus in units of inches.

Now, considering the fact that the angle of attack of a flight vehicle and the local heat transfer rate at the vehicle surface are functions of altitude and flight speed, the skin temperature can be related to a parameter defined by the ratio of lift force L to planform area S_p given by the following.

$$\frac{L}{S_p} = N_z \left(\frac{\text{weight}}{S_p}\right) - \left(\frac{\text{centrifugal force}}{S_p}\right) \tag{6}$$

Here N_z is the normal load factor, a dimensionless quantity, and the second term on the right is a correction for the curvature of the trajectory. A given value of the parameter corresponds to a specific constant skin temperature.

2. Vehicle Surface Heating. As a simplification, it is assumed that a turbulent boundary layer is present 5 ft downstream of the nose. This location is treated as the reference location for thermal balance on the vehicle. The thermal balance accounts for the heat generation at the surface caused by skin-friction heating,

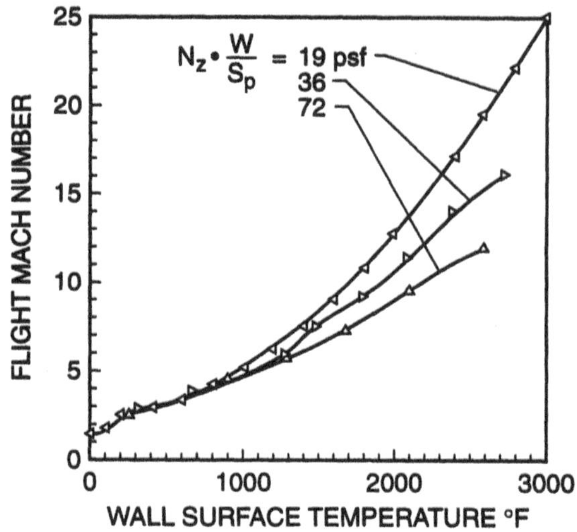

Fig. 5 Variation of wall surface temperature as a function of flight Mach number using wing loading as a parameter.

radiative transfer from the skin to the outer surface, and heat transferred internally from the surface.

The wall surface temperature as a function of flight speed is shown in Fig. 5 with $N_Z W / S_p$ as a parameter utilizing the data on materials given in Appendix B. Even when a modest turn is required for the vehicle in the atmosphere, it becomes necessary to reduce the flight speed as the normal load factor increases. The vehicle design must be based on the product of the normal load factor and the wing loading, or else a significant reduction in speed may become necessary, and there may be insufficient fuel to complete the mission. It is also important to note that a simulation of a launch vehicle must be carried out in an ascent mode rather than in a descent mode, since both dynamic pressure and surface temperature are high in an ascent mode.

B. Propulsion-Material Interface

It is now possible to reexamine Fig. 4 in relation to the structural materials and their strength. First, it is necessary to select a range of wing loading applicable to hypersonic cruise and

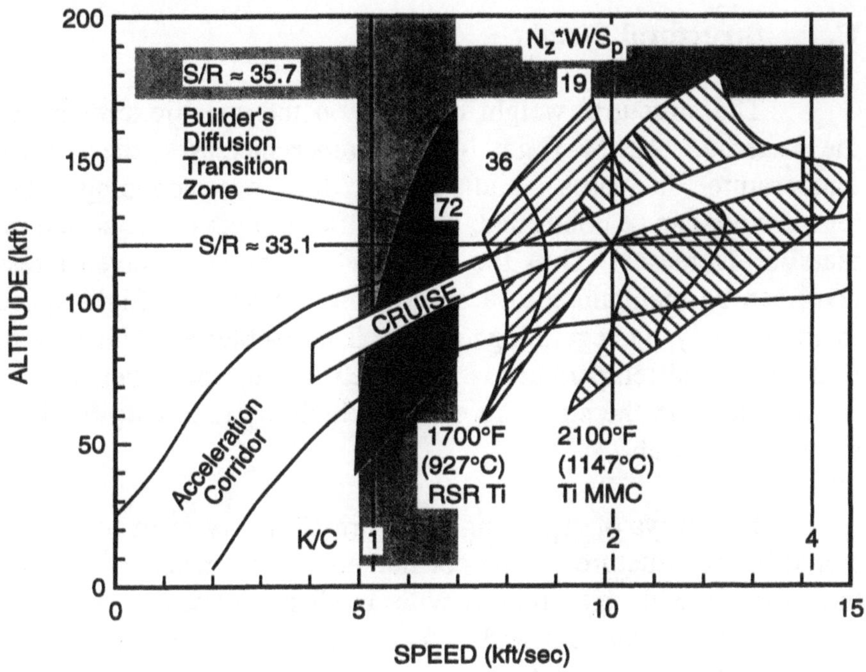

Fig. 6 Material capabilities and propulsion capabilities combined in altitude-flight speed space.

accelerating vehicles. It is then possible to establish a range for the parameter defined in Eq. (17). Finally, in terms of that parameter, it is possible to construct a diagram, as shown in Fig. 6, depicting the material capabilities in an altitude-flight speed diagram including propulsion capabilities.

Several observations can be made from Fig. 6. The constant temperature regimes shown are directly dependent on the assumed wing loading ranges. Practical design spaces seem to be feasible with titanium alloys for propulsion systems, except the engine module, in the energy ratio *(K / C)* range of 1.5-4, corresponding to a flight Mach number of up to 12 or 14. The altitude-flight speed space for the utilization of a material increases as the flight speed increases, although this depends on the wing loading chosen and affects the structural weight.

C. Structural Weight

The structural weight depends on the specific strength of the material. As the flight Mach number increases, the surface temperature increases, and unless the corresponding skin temperature can be controlled, the specific strength of the structural material decreases. Thus the structural weight for a given stress level can be maintained to a nearly constant level as flight speed increases only by regulating the metal temperature. The way to limit the metal temperature is to prevent the thermal energy from accumulating in the structure of the vehicle by removing the heat from the material.

Two ways of removing the thermal energy from becoming stored in the structure can be considered: 1) by means of cooling the structure with, e.g., fuel or with another intermediary between the structure and the fuel, and 2) by blocking the heat from entering the structure by an insulation-type nonstructural thermal shield that operates at radiation equilibrium temperature.[13-16] For example, with 95% of the incoming thermal energy being radiated away from the structure, it is possible to visualize about 2.8% being absorbed in the thermal shield while the remaining 2.2% of the heat enters the structure, and needs to be removed by internal cooling. In the use of an insulation-type thermal heat shield, one has a constraint in terms of the time allowable before the structure attains an equilibrium temperature. The duration of time over which thermal protection is effective is a function of insulation material properties, and the thickness employed, whereas the requirements vary for different types of missions and vehicles.

In summary, a spectrum of airbreathing engines has been evolved in the altitude-flight velocity space using the trajectory, $\Delta S / R$, and C / K as parameters. Although energy availability is the single, comprehensive, unambiguous parameter that should be used to describe the state of the working substance in an engine, the engine parameters may be optimized for best output based only on specific components of energy availability. The effectiveness of a propulsion system depends upon the local thrust-to-drag ratio of the engine as well as the total thrust output and the specific impulse. As the flight speed increases, airbreathing propulsion becomes considerably more complex than at lower speeds.

The choice of propulsion then depends on the materials and thermal management schemes available. The specific strength of the material must be examined in relation to both the product of the normal load factor and wing loading and also the fraction of heat permitted to enter the structure out of that generated by skin friction.

IV. Propulsion Performance-Vehicle Configuration Interface

In a complex system with several components, it is often considered, as in the case of a relatively low speed aircraft, that an optimal system can be derived on the basis of simply combining optimized elements. However, a complex system, by definition, involves many interactions among the elements, and thus a hypersonic flight system requires convergence of several components and features. The most direct way of establishing this is by using *sensitivity coefficients*, referred to as structural bond coefficients, external bond coefficients, and interaction bond coefficients.[6] Those coefficients represent the interactions between each component and the input, the total system, and other components. In evolving an optimal total system, it then may happen that all of the components are not necessarily operating at peak efficiency at various times.

In this section, the interest is in demonstrating the relation between propulsion selection and vehicle configuration, geometry and size selection.

A. Propulsion-Propellant Performance Index

One may define the operational empty weight, W_{OE}, of a flight vehicle as

$$W_{OE} = W_E + W_{PY} \tag{7}$$

where W_E, the empty weight of the vehicle, is given by

$$W_E = W_{ST} + W_{EN} + W_{SE} \tag{8}$$

Here W_{ST} is the structure weight, W_{EN} the engine weight, and W_{SE} the scaleable equipment weight, all of which scale with the size and geometry of the vehicle. In Eq. (7), W_{PY} denotes the payload weight. It is then possible to define the takeoff gross weight of the vehicle as follows.

$$W_G = W_{0E} + W_P \tag{9}$$

where the propellant weight is denoted by W_P. The weight ratio is then written as

$$R_W = W_G / W_{0E} \tag{10}$$

It then follows that

$$W_P = W_{0E}(R_W - 1) \tag{11}$$

and, expressing W_P in terms of the volume $\mathcal{V}_P$ and density ρ_P of the propellant, Eq. (11) may be rewritten as

$$W_{0E} = \frac{\rho_P}{R_W - 1} \frac{\mathcal{V}_P}{S_p} S_p \tag{12}$$

In Eq. (12) the first term on the right refers to the propulsion-propellant subsystem, the second to the configuration geometry and size, and the third is the planform area. Thus, the equation characterizes the overall vehicle system. A sizing program can then be developed to match, through iterative procedures, the volume required for the propellant and the volume available in the chosen vehicle configuration. Some other useful relations in this connection are the following, obtained from Eqs. (9), (12), and (8):

$$W_E = W_{0E} \Big/ \left[1 + (W_{PY} / W_E)\right] \tag{13}$$

$$W_E = \frac{\rho_P}{R_W - 1} \frac{v_P}{S_p} \frac{S_p}{1 + (W_{PY} / W_E)} \tag{14}$$

and

$$\frac{W_{ST}}{S_W} = \frac{\rho_P}{R_W - 1} \frac{v_P}{S_p} \frac{W_{ST} / W_E}{1 + (W_{PY} / W_E)} \frac{S_p}{S_W} \tag{15}$$

where the wetted area of the vehicle, S_W, is introduced on both sides of Eq. (15). The terms W_{ST} / W_E and W_{PY} / W_E have to be treated as parameters or obtained from historical data, e.g., of the 1960-1970 time period, and in the form of projections, as discussed in Appendix C. One then has in Eq. (15) a basis for relating the configuration, the propulsion performance, the propellant, the propellant requirement, and the aerodynamic performance in a vehicle sizing program.

Table 3 summarizes a series of relations that form the basis for a particular sizing program.[13] The weights are expressed in terms of three main parameter groups: 1) referred to as the *principal driver* for the configuration concept convergence; 2) related to *size and geometry;* and 3) referring to the *payload and structural weight fractions*. The principal driver is, in fact, related to propulsion energy availability. The sizing program of Ref. 13 is utilized in the various illustrations throughout the rest of this paper.

One can then obtain what may be referred to as the *propulsion performance index*. It can be observed in Eqs. (14) and (15) that the propulsion-propellant-related parameter and the size- and geometry-related parameter are given, respectively, by

$$\frac{\rho_P}{R_W - 1} v_P$$

and

$$\frac{\rho_P}{R_W - 1} \left(\frac{\rho_P \left(1 + \frac{0}{F}\right)}{1 + \frac{0}{F} \cdot \frac{\rho_P}{\rho_0}} \right) \frac{1}{W_P}$$

Here *(0 / F)* denotes the oxygen to fuel ratio. Now, based on the so-called rocket equation, one can write

$$\Delta W_F \ \alpha \ \exp[\Delta V / (g I_{SPE})] + 1$$

Hence, one can define the *propulsion performance index* or the *propulsion-propellant index* as follows:

$$I_{PP} = \frac{\rho_P}{R_W - 1} \ \frac{\rho_P\left(1 + \frac{0}{F}\right)}{1 + \left(\frac{0}{F}\frac{\rho_P}{\rho_0}\right)} \left[\exp\frac{\Delta V(T/D)}{\{g I_{SP}[(T/D) - 1]\}} - 1\right]^{-1} \tag{16}$$

propellant index	propellant ratio in engine cycle	energy required

The index is a representation of the effectiveness of the propulsion system. It embodies the propellant selected, the propellant (oxygen-to-fuel) ratio utilized in the engine cycle, and the aeropropulsion performance of the vehicle.

B. Propulsion-Propellant Concepts

The propellants consist of intrinsic (or carried onboard) oxidizer and fuels, and air taken from the atmosphere when an airbreathing engine is utilized. In general the onboard oxidizer is oxygen, and the fuel is either all hydrogen or a combination of hydrogen and hydrocarbon in parts.

The propulsion system may consist of some form of combined-cycle engine, with separate rocket and airbreathing engines of different types, or with a truly combined-cycle engine, operating variously along the flight trajectory from takeoff to orbital launch velocity and back to Earth for landing of the recovered vehicle.

A number of propulsion concepts can be examined during vehicle synthesis. A brief discussion on possibilities is included in

Appendix C. All of them are combined-cycle engines, except for the all-rocket engine and the all-airbreather engine. It is assumed that in all cases the fuel is 50% slush hydrogen, and the onboard oxidizer is triple point oxygen. The data given in Appendix C for the propulsion systems are subsequently utilized in Sec. VI in the construction and discussion of design spaces for launch vehicles.

C. Conceptual Configuration Geometric Parameters

In Eqs. (12), (14), and (15), one can notice that the volume of the fuel appears as $\mathcal{V}_P / S_p$ or volume per unit planform area. It is useful to express this parameter in terms of the total volume of the vehicle and the ratio S_p / S_W where S_W is the wetted area of the vehicle. Thus the propellant volume fraction may be expressed by

$$\upsilon_F = \frac{\mathcal{V}_P}{\tau} S_p^{-n_P} \tag{17}$$

where n_P includes the correction for the propellant volume.

Next, writing the *area ratio*

$$\omega = S_W / S_p \tag{18}$$

and the *structure factor* μ as

$$\mu = W_{ST} / S_W \tag{19}$$

one can define a structure loading factor by

$$K_{ST} = \mu / (S_p)^{n_{ST}} \tag{20}$$

where n_{ST} includes the correction for the structure.

Furthermore, one can relate the propellant volume fraction to the payload volume fraction, noting that

$$\upsilon_P = \upsilon_T - \upsilon_{PY} \tag{21}$$

where subscript T indicates the total vehicle.

Table 3. Size-weight relations

$$W_G = W_R \cdot \underbrace{\left[\rho_P \big/ (W_R - 1)\right]}_{(2)} \cdot \underbrace{\left(\mathscr{V}_P / S_p\right)}_{(3)} \cdot S_p$$

$$W_{0E} = \underbrace{\left[\rho_P \big/ (W_R - 1)\right]}_{(2)} \cdot \underbrace{\left(\mathscr{V}_P / S_p\right)}_{(3)} \cdot S_p$$

$$W_E = \underbrace{1 \big/ \left[1 + (W_{PY} / W_E)\right]}_{(1)} \cdot W_R \cdot \underbrace{\left[\rho_P \big/ (W_R - 1)\right]}_{(2)} \cdot \underbrace{\left(\mathscr{V}_P / S_p\right)}_{(3)} \cdot S_p$$

$$\frac{W_{ST}}{S_W} = \underbrace{(W_{ST} / W_E) \big/ \left[1 + (\dot{W}_{PY} / W_E)\right]}_{(1)} \cdot W_R \cdot \underbrace{\left[\rho_P \big/ (W_R - 1)\right]}_{(2)} \cdot \underbrace{\left(\mathscr{V}_P / S_p\right)}_{(3)} \cdot S_p$$

(1) Payload and geometry; (2) Principal driver of configuration convergence other than size; Geometry and propulsion energy analyses; (3) Size and geometry

Table 4. Size-structure factor-weight relations

$$\underset{(1)}{W_G / S_p} = \underset{(2)}{\tau v} \cdot \underset{(3)}{\left[\left(W_R \rho_P \right) / \left(/W_R - 1 \right) \right]} \cdot \underset{(4)}{S_p^{0.5717}}$$

$$\underset{(1)}{W_E} = \underset{(2)}{\tau v} \cdot \underset{(3)}{\left[\rho_P / \left(W_R - 1 \right) \right]} \cdot \underset{(4)}{S_p^{0.5717} / \left[1 + \left(W_{PY} / W_E \right) \right]}$$

$$\underset{(1)}{\mu} = \underset{(2)}{K_{ST} \tau v / \omega} \cdot \underset{(3)}{\left[\rho_P / \left(W_R - 1 \right) \right]} \cdot \underset{(4)}{S_p^{0.7097} / \left[1 + \left(W_{PY} / W_E \right) \right]}$$

(1) Weight, materials, structure, and industrial capability, (2) Configuration concept, and slenderness, (3) Total energy required, propulsion system concept, propellant, and aerodynamics, (4) Size, slenderness, and payload

It is then possible to express W_G, W_{0E}, W_E, and μ as shown in Table 4 by making use of the relations given in Table 3. The new relations given in Table 4 include the following three factors: 1) configuration concept and slenderness, 2) propellant- and propulsion-related quantities, and 3) payload, and vehicle size and slenderness. Thus, Table 4 provides the fundamental relations required in establishing a sizing program that accounts for vehicle, propulsion, payload, and structural weight.

D. Hypersonic Configuration Characteristics

Next it is necessary to include the vehicle concept in terms of shape and size. This can be done by identifying certain definitive concepts and determining their characteristics. If another vehicle concept needs to be examined, it should, hopefully, be possible to relate such a vehicle concept to one of the definitive, reference concepts. In the following, four vehicle concepts have been chosen as reference concepts, namely, 1) a right circular cone, 2) a blended body, 3) a winged body, and 4) a waverider.

In addition to the characteristics defined earlier, namely, τ, ω, S, υ, and K_{ST}, other characteristics of interest are as follows:

$$T = S_p / \upsilon_T^{2/3}, \tag{22}$$

$$G = \omega / \tau\upsilon, \tag{23}$$

βC_{D0}, and $(L/D)_{TRM}$, where the subscript *TRM* indicates the transition Mach number from airbreathing propulsion to rocket propulsion; thus, e.g., $(L/D)_{12}$ indicates (L/D) at the flight Mach number of 12 when the transition Mach number is 12. These characteristics are given in Appendix C.

E. Industrial Capability Index

Earlier in Sec. IV.A, a definition was provided for the so-called *propulsion performance index*, I_{PP}, Eq. (16). In a similar fashion one may refer to μ, the ratio of structural weight to wetted area, as the *structural index*. Next, based on μ, one may establish what may be referred to as the *industrial capability index*, I_{TC}.

Considering the weight of the structure in relation to the empty weight of the vehicle, one can combine Eqs. (14) and (15) to obtain what may be referred to as the *structure index*, μ, given by the following; the structure index is the same as the structure factor, Eq. (19).

$$\mu = W_{ST} / S_W \tag{24}$$

$$\mu = K_{ST} \frac{\tau \upsilon}{\omega} \frac{\rho_P}{R_W - 1} \frac{S_p^{0.7097}}{1 + \left(W_{PY} / W_E\right)} \tag{25}$$

Thus μ is defined by the product of three factors pertaining to 1) configuration concept and slenderness, 2) propulsion, propellant, and aerodynamics, and 3) payload and vehicle size.

One can then rewrite Eq. (25) to obtain an expression for an index that may be referred to as the *industrial technology capability index*, or the *industrial capability index*, I_{TC}, in the following forms:

$$I_{TC} = \left[\frac{\rho_P}{R_W - 1}\right] / \mu \tag{26}$$

$$= \frac{\omega}{\tau \upsilon} \frac{1}{K_{ST}} \left[\frac{1 + \left(W_{PY} / W_E\right)}{S_p^{0.7067}}\right] \tag{27}$$

$$= \left[\frac{K_W}{K_{ST}} \frac{1}{\tau \upsilon}\right] \left[\frac{1 + \left(W_{PY} / W_E\right)}{S_p^{0.7067}}\right] \tag{28}$$

Thus I_{TC} is a product of two factors, one related to the configuration concept and slenderness, and the other to the payload and vehicle size and slenderness. In addition, the index I_{TC} is, on

Table 5. Structure index - industrial technology capability index relations

$$I_{TC} = 10\frac{\rho_P}{(W_R - 1)\mu} = \frac{10}{K_{str}}\frac{K_W}{\tau K_V} \cdot \frac{1 + W_{PY}/W_E}{S_p^{0.7097}}$$

$$I_{TC} = 10\frac{\rho_P}{(W_R - 1)\mu} = 10\frac{K_W}{0.228\tau^{1.206} K_V} \cdot \frac{1 + W_{PY}/W_E}{S_p^{0.7097}}$$

• Energy	• Systems	• Geometry	• Size
• Propulsion	• Materials		
• Propellant	• Structures		
• Aerodynamics	• Manufacturing		
	• Integration		

Technical Capability | **Geometry Matched To Technical Capability**

one hand, related to all of the technologies of relevance to the design and functioning of the vehicle, and, on the other hand, to the concept geometry and size parameters, the latter including the payload mass ratio.

The relations for μ and I_{TC} are summarized in Table 5. In the case of I_{TC}, a factor of 10 has been introduced to obtain numbers greater than unity for easy comprehension; it is clear that I_{TC} is suitable only for comparative assessments. Unless I_{TC} is greater than or equal to the product of the concept geometry factor and size (including payload mass) factor, it is clear that there is no possibility of successfully realizing the design aimed at in the concept with the given industrial technology capability.

Some other broad conclusions from Table 5 are that the specific geometry and size determine the level of technology capabilities required to converge to a design of the size and geometry selected. Specific technology capabilities or goals may be looked upon as fixing the range of geometry and size possibilities available to obtain design convergence. The smaller the size of the vehicle, the greater the technology required. As payload fraction increases the technology requirement increases. An intriguing possibility from the relations in Table 5 is that W_{PY} / W_E can be zero or negative with certain combinations of assumed or available technology capability and desired concept configuration geometry. Finally, as G [see Eq. (23)] increases, the technology requirement also increases under three different possibilities: 1) ω increase with S_W increase, 2) τ decrease with slenderness increase, and 3) υ_P decrease with propellant volume fraction decrease.

Some numerical examples also may be useful. If $I_{PP} = 4.01$ and $\mu = 3.5\,lbm/ft^2$, then $I_{TC} = 11.45$. Then if I_{PP} has to reduce to 3.4 for some reason, the structural index for obtaining the same value of I_{TC} becomes 3.0 lbm / ft^2. This decrease implies that much stronger materials or much better structural design is required. However, if I_{PP} can be raised to 4.6, the structural index for the same I_{TC} would be 4.0 lbm / ft^2, which is a gain of about 14% over $\mu = 3.5\,lbm/ft^2$. Thus propulsion performance provides a substantial margin in structural design and permits the use of less sophisticated materials.

F. Illustrations

Historical data and projections for hypersonic vehicle concept configurations are given in Appendix C.

Based on the relations given in Eqs. (C.23a to C.23c), one can establish, as a first approximation, that any vehicle concept with a transition to rocket propulsion at less than 18,000 ft / s and is smaller than 2000 ft^2 in planform area will be a challenge. It would require a gross weight of about 187,000 lbm, with an empty weight of 63,500 lbm for a payload weight of 10,000 lbm. Similarly, an airbreathing vehicle that is designed for minimum size would, for zero payload, require a 2100-ft^2 planform area with a gross weight of 140,000 lbm and an empty weight of 46,000 lbm.

Now, Eq. (C.23a) can be mapped in $S_p - I_{PP}$ space, as in Fig. 7, using τ and payload as parameters over the ranges $0.20 \leq \tau \leq 0.63$, and 0 ton $\leq$ payload $\leq$ 15 metric tons. Thus, the figure provides a relation among planform area, τ, W_{PY}, and the propulsion-propellant index, I_{PP}.

In Fig. 8 an overlay of I_{PP} for different propulsion systems was obtained using the sizing routine of Ref. 13. Thus, one can obtain from Fig. 7 the propulsion-propellant index required if a certain type of propulsion system is to be incorporated into a vehicle of given planform area and τ values. This figure is based on data given in Table C.1, with a value of μ equal to 3.5 lbm / ft^2, and correcting the I_{PP} for the drag change caused by a change in τ. The figure shows the possible configurations for propulsion systems ranging from *AARR* to $AAAR_{22.2}$, the possible configurations for propulsion systems ranging from $AAAR_{15.4}$ to *AAAR*, and the all-rocket single-stage-to-orbit (SSTO) and the all-airbreather SSTO. The all-airbreather vehicle shows convergence for heavy payloads and stout configurations. In contrast, the all-rocket vehicle has a broader range of payload and is less stout. (A

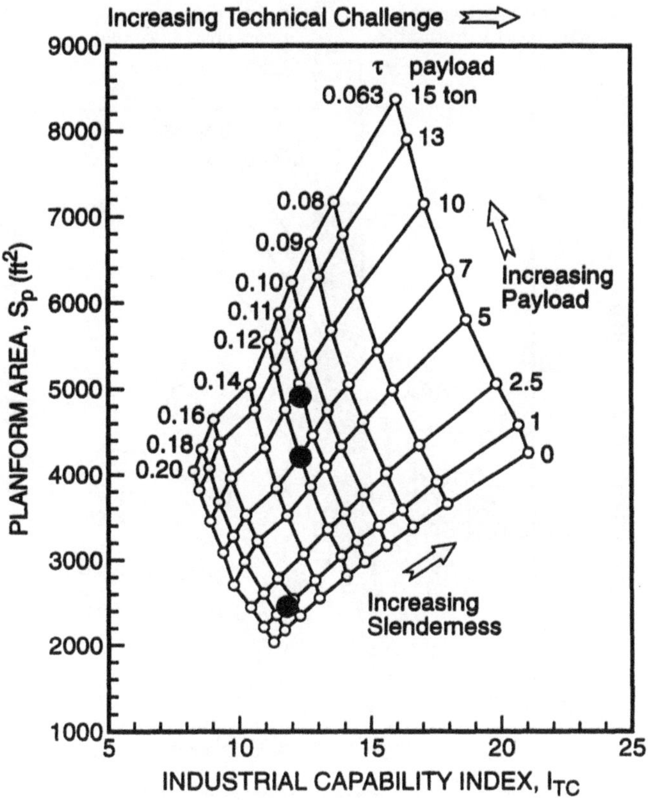

Fig. 7 Relation between τ and W_{PY} shown in planform area-industrial capability index space.

configuration is referred to as stout when its τ value is large, stoutness being the opposite of slenderness.) The greatest flexibility seems to exist for the rocket-based combined-cycle (RBCC) systems, and the successfully converged systems, indicated by solid symbols, are generally in the RBCC propulsion system range.

V. Vehicle Synthesis

Vehicle synthesis is the task of combining the vehicle components to obtain a converged vehicle design that meets the mission needs under given choices and constraints pertaining to the capability of the components as well as to the mission needs. In

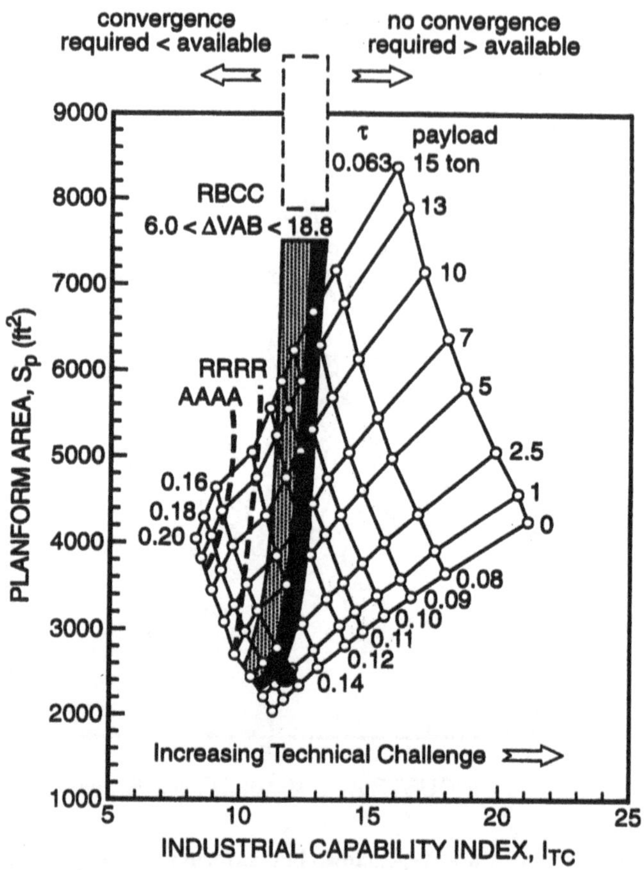

Fig. 8 Relation between τ and W_{PY} including propulsion feasibility in planform area-industrial capability index space.

this paper, two components have been identified for the vehicle, namely, the aerodynamic body of the vehicle and the propulsion-propellant system (subsystem, scheme, or element, as it may be referred to). The components are described by a number of variables pertaining to their features, and the processes associated with them, that ultimately affect vehicle performance. The choices and constraints related to the components include, among others, their performance capability, the structural-material strength limitations or the structural index, and the available industrial capability for manufacturing the components such that they yield specified performance. These may also be treated as variables in preliminary or conceptual design. Other choices and constraints pertaining to the mission may also be treated as variables in preliminary design. The task of synthesis then is obtaining a matching of all of the variables of the vehicle system such that the vehicle performs the desired mission.

Each of the variables obviously has a range of values. Thus, one chooses variables and their range. It is obviously necessary to base such a choice on rational grounds, in many cases including projections based on available data and predictions. However, there are many instances in which the objective may be to set up goals in performance or capability, and then the choice of allowable variables and their ranges may need to be flexible.

When the mission needs or specifications increase or become extensively defined and detailed, then, the problems of matching become complex, and the number of possibilities for matching or synthesizing becomes smaller. Furthermore, in many cases, there may be a specific, major requirement for the vehicle. If that requirement is made the principal objective of design for the vehicle, one may have to establish special combinations of variables that ultimately meet the requirement. However, it is entirely possible that there is no convergence of the design within given ranges of variables. A useful approach to adopt is to examine what extension of ranges do, in fact, lead to a converged design.

In connection with vehicle synthesis, one often comes across the expression “vehicle optimization”. In the case of aircraft, spacecraft, and missiles, which involve nearly the same number of variables as in hypersonic technologies and vehicles, there is a vast body of historical data and experience in optimizing chosen features of the vehicle to meet desired objectives. However even in the case of those vehicles, the optimization cannot be carried out strictly on mathematical grounds.

An example may be useful here. The internal shape of a thrust nozzle, for instance, can be optimized if one demanded maximum thrust for given inlet and outlet conditions under the requirement that the length or the surface area of the nozzle must be a minimum. Such a problem can be set up from first principles based on well-established mathematical methods. If one, in addition, would like to take into consideration factors such as installation mechanical-structural needs, associated gas leakage, and heat loads, several types of problems may arise: lack of experience in similar situations; lack of data on the complexities; inability to describe the processes and complexities in a suitable analytical form, and unavailability of computational tools for a reasonable level of predictions. In the case of hypersonic vehicles, all of those factors seem to arise in most aspects and stages of preliminary design. It seems prudent, therefore, to aim at convergence of vehicle design rather than optimization in a strict mathematical sense. It would not be incorrect to say that even in the case of other, more traditional, especially low speed vehicles, the arts and skills supporting a successful design are at least equal to the rational (mathematical) methods.

The task of synthesizing a hypersonic vehicle thus consists in evolving a methodology for obtaining a converged design, and for establishing possibilities of options with improvement in the attainable indices and capability for either the given mission goals or those that are superior to them in specific directions. This section is devoted to a description of the broad features of the methodology for vehicle synthesis or design convergence. In the next section, illustrations are provided for the application of the methodology.

A. Outline of Methodology

An aircraft synthesis method has been available for many years,[4] and it may be useful to become familiar with it.

The methodology for hypersonic vehicle convergence presented here may be illustrated as in Fig. 9. It is assumed that data sets have been generated, based on past experience and extensions as well as on predictions from sizing programs, on capabilities in propulsion, fuels, materials, and industrial manufacturing. A reference vehicle is postulated for the given mission with adequate definition of key parameters. Then, a series of design spaces are constructed with respect to various parameters. A vehicle design convergence is sought based on the influence of various parameters on the vehicle performance as depicted by the design spaces.

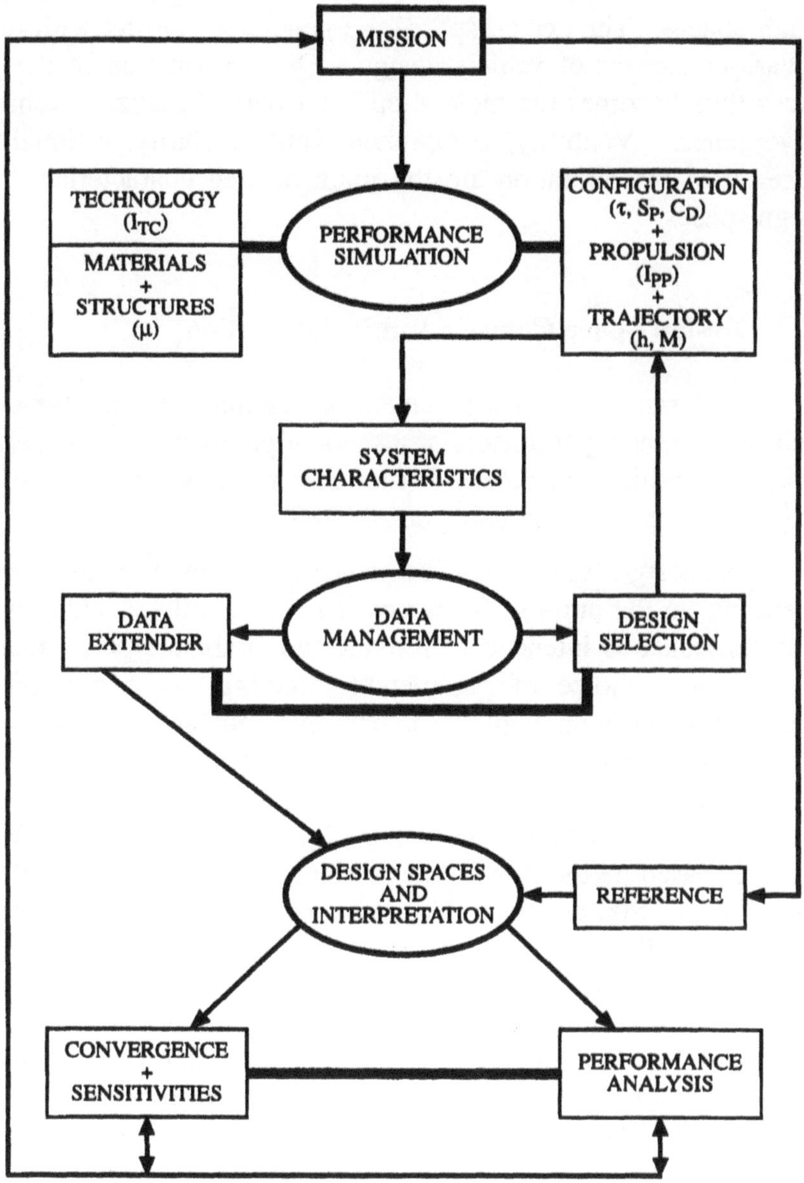

Fig 9. Methodology for flight vehicle synthesis.

It can be seen from Fig. 9 how the reference vehicle characteristics are matched at the level of interpretation of the design spaces. The design spaces are generated starting with data on various aspects of vehicle design. The construction of design spaces thus becomes the most significant part of realizing vehicle convergence. Visibility, comprehensiveness, clarity, rationality, and ease of interpretation are the main desired characteristics of design spaces.

B. Design Space Concept and Its Utilization

A *design space* is a parameter space that may involve two or more individual parameters or groups of parameters. In general, it may be multidimensional. It indicates the available choice in design for the parameters being considered.

A design space can be constructed in various ways, depending on the purpose for which it is to be utilized. One type of design space is intended to provide the attainable performance with a given choice of parameters affecting the performance. Thus, for a hypersonic inlet, for example, one may consider an enthalpy-freestream entropy space with the contraction (area) ratio and the number of shocks included as parameters to examine the design space available at a particular flight Mach number (see Fig. 10). It is based on the flowfield equations of mass and momentum conservation, the entropy gain equation, and the shock-wave relations. It illustrates the influence of the choice of contraction ratio and the number of shock waves on the losses, including the loss of energy availability in the inlet during the chosen mode of diffusion. Similar diagrams can be constructed for other flight conditions, and one can examine the results with respect to an altitude-flight Mach number space indicating a flight trajectory band. The results shown in Fig. 10 have been obtained under a number of assumptions, e.g., a calorically perfect, equilibrium gas, and constraints in configuration. The entire procedure adopted for the inlet also may be applied to other components of an engine, including a combined-cycle engine. Such a procedure is described in Refs. 14 and 15. A set of design spaces becomes established

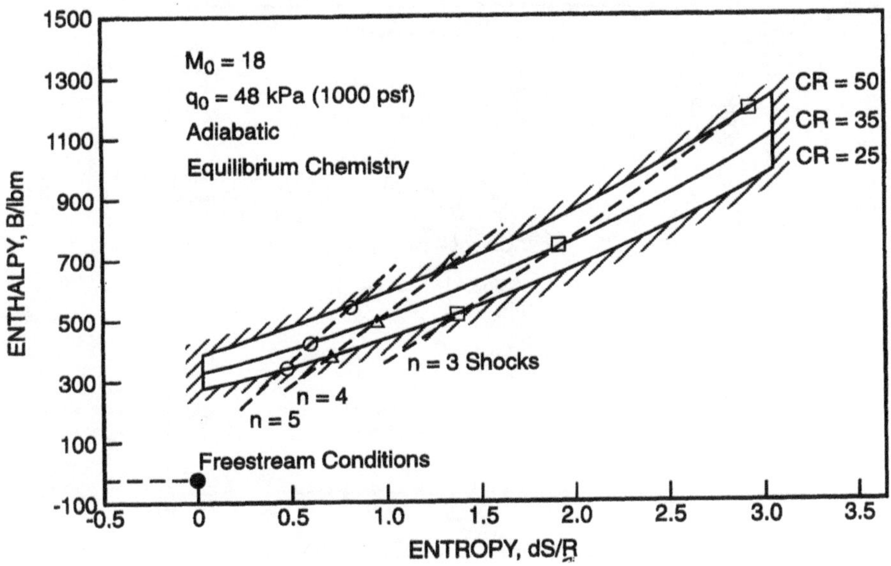

Fig. 10 Inlet design possibilities.

that show the attainable performance for a series of choices of parameters affecting the performance, some of which are utilized as variables and others as parameters in the construction of the design spaces.

In the current analysis, a design space is constructed in a slightly different fashion, and this proves particularly useful in assessing the possibility of convergence of a vehicle that can meet the desired goals with reference to the indices, capability, configuration concept and its details, and various weight factors.

For example, a vehicle may be specified in terms of the payload that it should deliver to a given orbit under various constraints, options, and limits. One can then choose a *reference vehicle* that, under available indices and capability, is able to perform the mission with a set of configuration and propulsion-propellant concepts. One can then construct a series of design spaces in terms of the parameters affecting the vehicle design such that each design space shows the effects of departing from the chosen reference conditions. In particular, it is of interest to establish the possible *margin* in the choice of various parameters

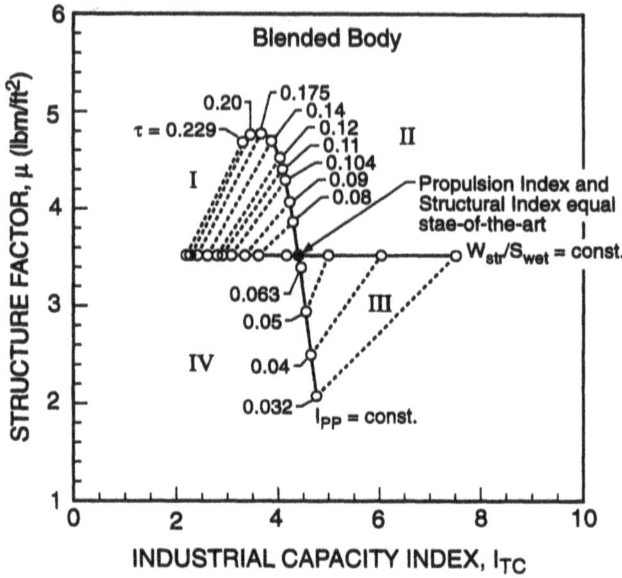

Fig 11. Example of a design space.

with reference to the assumed, available indices, capability, and concept and propulsion choices; also, simultaneously, one can establish from the design space the need for improvements in the indices and capability before certain design changes can be realized.

As an example, one can construct a design space with μ and I_{TC} as variables, using τ as a parameter, as illustrated in Fig. 11. The figure includes reference values for μ and I_{TC} at point *R*. As τ is increased, it is shown that μ increases in numerical value while I_{TC} decreases. By definition, an increase in μ permits the use of less sophisticated materials, and a decrease in I_{TC} allows lower industrial capability.

In Fig. 11 an arched line is drawn, the points over which represent a constant value of I_{PP}, e.g. a reference value. The region left of the arched line represents a requirement of propellant-propulsion performance that is less than the available or reference value. It is possible to identify four regions in the design space shown in Fig. 11: I indicating the available space for practical design application with reference to the chosen values of

μ, I_{PP}, and I_{TC}; and II, III, and IV where one requires improvements in μ, I_{PP}, or I_{TC} or both. Considering region *I*, margins in both μ and I_{TC} become available provided a higher value of τ is acceptable for the vehicle, i.e., if the vehicle shape can be made less slender or more stout. In fact, there is a value of τ for which the margin in μ is the highest, and another for which the margin in I_{TC} is the highest; the latter reduces the margin in μ to zero. In regions *II* and *III*, one clearly needs an improvement in both industrial capability and propellant-propulsion capability relative to the reference value. In region IV, in the same fashion one can see the need for an improvement in material-structure availability, although the I_{PP} requirement is within the reference value. Region III presents the most severe challenge.

Now, one can consider other variables or combinations of them, and establish a series of design spaces for the vehicle. In each case various margins and limits become evident. Hence, one can utilize the combination of such design spaces to establish the possibility of vehicle design convergence with respect to the entire set of parameters.

Before proceeding further, it may be useful to recall the elements of the vehicle synthesis procedure of the previous sections.

In Sec. II, an attempt has been made to evolve a spectrum of airbreathing engines in flight altitude-flight speed space using the ratio of fuel combustion-released energy to the kinetic energy of air entering the intake of the engine, and the entropy level of the air as parameters. The engines considered are restricted to the Brayton cycle variety. If other types of engines had been considered, the spectrum would have changed, at least in certain regimes of flight speed, along with the admissible variation in flight trajectory. In addition, other parameters would have become significant.

Based on the concept of engine effectiveness, it has been shown that an airbreathi g engine should, in relation to a rocket, provide a fuel specific impulse that is about twice that of a rocket, an effective specific impulse equal to that of a rocket, and as high a value as possible of specific thrust based on air mass flow. In addition, the engine effectiveness is related to energy effectiveness based on energy availability considerations.

The next major consideration is that of materials and their available structural strength. In Sec. III an attempt has been made to examine the material-propulsion interface. Taking into consideration the unavoidable limitations on structural strength, and the need for thermal management for the vehicle as the flight speed increases, and for the propulsion system as the thermal equivalent of the kinetic energy of the inlet air increases, one could identify in the spectrum of engines the range of application of materials of different structural strength. The structural material density and strength as well as the thermal management system add a number of new parameters for consideration.

In Sec. IV, the propulsion-vehicle configuration interface has been considered, and several measures of performance are introduced, namely, the structure index, μ, the propulsion-propellant index, I_{PP}, the mean propellant density, $\bar{\rho}_P$, and the industrial technology capability index, I_{TC}. In defining these in addition to the usual weight ratios, the size and geometry of the vehicle are identified by $\mathcal{V}_T$. $\mathcal{V}_P$, S_p, and, S_W directly and in the form of ratios such as υ_F equal to $\mathcal{V}_P / \mathcal{V}_T$, τ, and S_p / S_W, each of which has a distinct significance.

Regarding the flight vehicle, four reference concept configurations, and a number of propulsion concepts have been introduced; the latter include various possibilities for combining airbreathing engines and rocket motors, noting that combined-cycle engines include airbreathing engines. In the case of concept configurations, the drag coefficient with L / D equal to zero is a parameter. The engines are characterized by the parameters determining engine effectiveness.

It is now intended to proceed to developing *design spaces* through a combination of which one may arrive at a convergence of the vehicle system to meet a given set of mission demands under a given set of constraints. The design spaces involve the vehicleand the propulsion-propellant parameters, generally in a multiparameter space. Each design space developed helps to meet a particular design requirement by indicating the space in which to look for a possible solution relative to that requirement. The set of design spaces then should lead to vehicle system convergence.

A notable feature in the illustration provided in Sec. VI is that, while the methodology of vehicle synthesis is general enough, the type of vehicles considered is kept limited in that the number of concept configurations and propulsion-propellant configurations is restricted. This is not because of the inapplicability of the methodology in other cases, but because of the need to concentrate on a few configurations that could be xamined with the limitations in available data, the results from vehicle sizing routines, and the scope for projections.

VI. Parametric Design Spaces

In the current analysis, a hypersonic launch vehicle system has been considered only in terms of the flight vehicle and its propulsion-propellant subsystem. Even so, a large number of parameters enter into the description of the system and its performance. Up to this point, we have concentrated on obtaining a vehicle synthesis by evolving a methodology for combining a vehicle configuration with a propulsion-propellant configuration. The methodology has rested on the correlation of vehicle and propulsion-propellant parameters using available data and projections and estimates from sizing routines. The correlations have then been applied to variations in a selected reference vehicle. The result is a set of options for a class of vehicles represented by the reference vehicle.

A. Mission and Reference Vehicle

An SSTO flight vehicle system is considered as an example. The vehicle is expected to carry a payload of 20,947 lbm with a weight ratio W_R equal to 2.70. The vehicle configuration concepts are the four reference shapes considered earlier, blended body, winged body, waverider, and right circular cone. The propulsion-propellant concepts consist of various types of rocket-based combined-cycle engines, including the all-rocket and the all-airbreather engines; these are the same as discussed in Table C.1.

Table C.1. Series of ten propulsion systems

Propulsion configuration	ΔV_{AB} kft/s	*WR*	I_{SP}	I_{SPE}	*T/D*	ΔV losses ft/s	rho ppl lbm/ft3	*O/F*	Index*
RRRR	0.0	7.971	383	455	6.35	4,788	26.02	6.00	3.73
ARRR	3.0	6.645	420	521	5.17	6,137	20.28	3.94	3.59
AARR	6.4	5.285	478	621	4.34	7,674	17.674	3.04	4.06
RRAR	8.9	4.924	499	670	3.91	8,784	16.87	2.89	4.30
RAAR	12.4	3.604	621	872	3.47	10,361	12.35	1.66	4.74
AAAR15.4	15.4	3.297	667	972	3.19	11,710	10.32	1.16	4.49
AAAR18.8	18.8	2.940	738	1,119	2.93	13,238	8.35	0.70	4.30
AAAR22.2	22.2	2.735	791	1,247	2.73	14,766	6.87	0.37	3.96
RAAA	22.6	2.978	729	1,155	2.71	14,946	6.96	0.39	3.52
AAAA	25.6	2.704	800	1,309	2.57	16,294	5.84	0.15	3.43

Payload - 30,000 lbm; hydrogen = 5.13; oxygen = 81.0; *average value = 4.01, and standard deviation = 0.366, 9.13%

For examining design spaces, a set of reference conditions is chosen as follows: $\tau = 0.10$; $\mu = 3.5\ lb/ft^2$ based on the current and assumed value of I_{TC} relative to the use of materials and structures with a surface temperature of about $1000^{\circ}C$; $I_{TC} = 11.7$; $I_{PP} = 4.09$; $\mathcal{V}_P = 34{,}924\ ft^3$; $T/D = 3.2$ at $M_0 = 12$; $I_{SP} = 1164\ s$; and $I_{SPE} = 800\ s$.

In the following, five design spaces are considered.

1) The relation between the structure index μ and the propulsion-propellant index I_{PP}, noting that μ is included in I_{TC}, and I_{PP} includes the vehicle drag associated with τ. Hence, the design space utilizes μ and I_{PP} as variables with τ as a parameter.

2) The relation between the structure index and vehicle planform area, based on the characteristics of four reference concept geometries for constant values of $\mathcal{V}_{PY}/\mathcal{V}_T$ and $\mathcal{V}_P/\mathcal{V}_T$ for a range of values of τ.

3) The relation between W_{PY} and vehicle size or τ for various values of I_{TC}.

4) The relation between W_{PY} and ΔV of airbreathing propulsion for various concept configuration geometries with τ as a parameter.

5) The relation between μ and S_p for various values of T/D and τ which reflects the influence of I_{PP} and I_{TC} on W_{PY} and $\mathcal{V}_T$, as shown in items 1 and 3 above.

It will be observed that the foregoing five design spaces involve several indices of performance, size parameters of vehicle, and energy effectiveness of propulsion systems, in addition to the prescribed requirements for the mission. The variables and the parameters are chosen to illustrate and to discuss various aspects of vehicle design. In most of the following illustrative cases, the structural index μ is treated as a variable to bring out the implications of the choice of materials from various points of view.

Fig 12. Available structure factor— industrial capability design space for different configuration geometries.

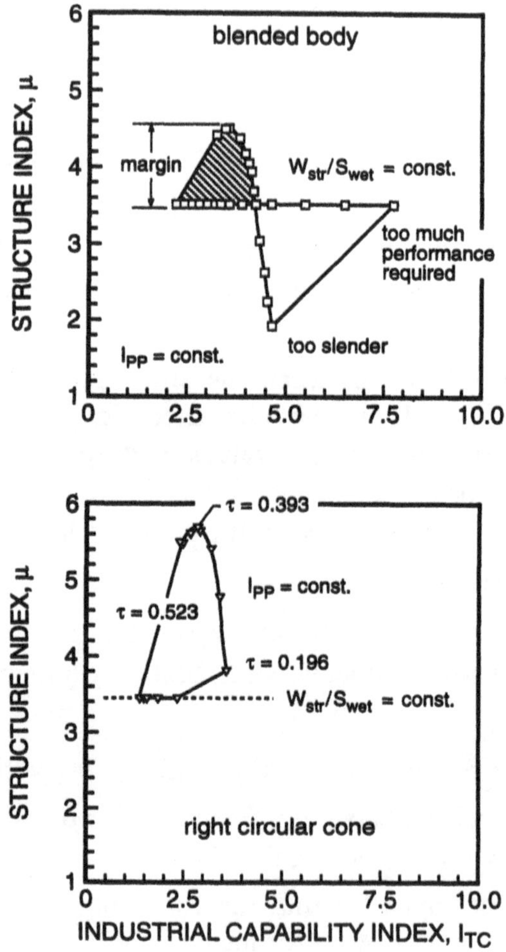

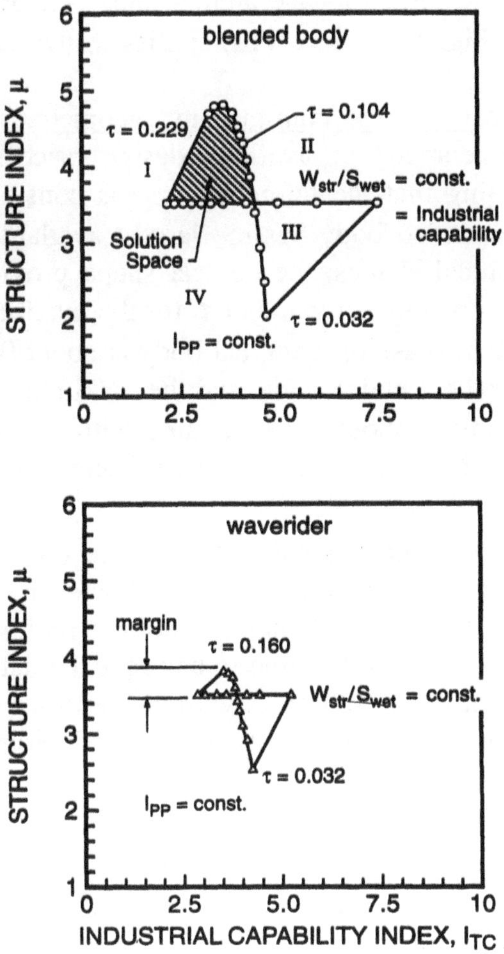
blended body
τ = 0.229
τ = 0.104
τ = 0.032
I
II
III
IV
W_str/S_wet = const.
= Industrial capability
Solution Space
I_PP = const.
waverider
margin
τ = 0.160
τ = 0.032
W_str/S_wet = const.
I_PP = const.
STRUCTURE INDEX, μ
INDUSTRIAL CAPABILITY INDEX, I_TC

B. Available $\mu - I_{PP}$ Design Space

Referring to Fig. 12, considering the line μ = *const*, the values of τ corresponding to I_{PP} are determined on either side of the reference point assumed. With constant values of τ, the maximum possible values of μ are determined corresponding to I_{PP} values and joined by the arched line. All of the discussion presented for Fig. 11 in Sec. V.B. applies in the current, identical case.

1. Influence of configuration geometry. The effect of configuration geometry on available design space is illustrated in Fig. 12 assuming that the reference I_{TC} is constant in all cases. Whereas the blended body has the largest available design space among nonconical shapes, the conical shape provides the largest design space. The highest value of τ for the maximum value of μ permissible in the case of a conical body is about 0.393 compared with 0.175 for the blended body, and the maximum value of μ for the conical body is about 5.8 compared with 4.9 for the blended body. Even in the case of a winged body there is a margin in μ of about 1.0.

The use of the conical body may be confined to vertical takeoff and vertical landing missions; however, the interesting feature is that the material and technology requirements are so small for this case that it becomes a natural candidate concept for a vehicle that must be reusable without refurbishment, a more stringent requirement than one with accepted refurbishment after each mission.

2. Other Associated Implications. Margin in μ and size. An interesting question is whether the vehicle size in terms of the planform area S_p changes appreciably if one tries to take advantage of the margin in μ at a chosen value of τ. In examining this, one notices that μ has an effect on W_E, the empty weight. Hence, a design space can be constructed for W_E and S_p, using τ as a parameter, for two different values of μ in the case of each of the four reference configuration geometries, as in Fig. 13. In the calculations, I_{PP} has been obtained at two values of μ equal to 3.5 and 4.09 lb / ft^2.

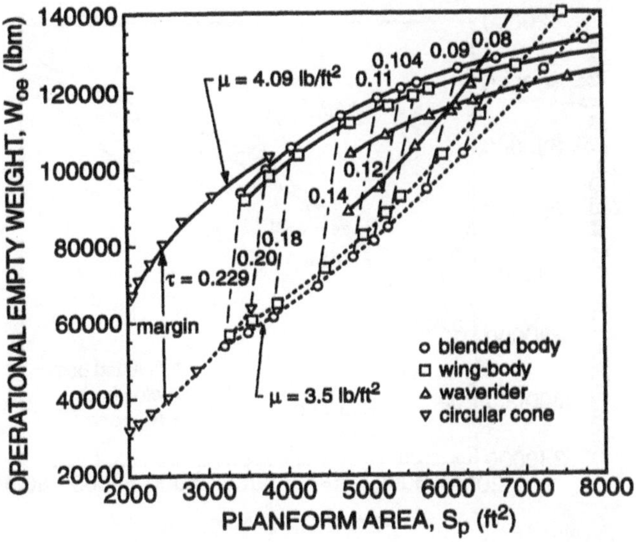

Fig. 13. Effect on empty weight shown in a design space with τ and μ as parameters.

It is immediately apparent that while S_p does not change much, there is an increase in W_E with μ for a constant value of τ. It appears that some compromise is necessary in the amount of margin in μ utilized in a design. At the same time, with any use of the margin, there is a gain in cost and manufacturing resulting from the feasibility of using less sophisticated materials and structures.

Margin in μ and gross weight. Next we to assess the change in W_G, the gross takeoff weight, resulting from utilizing the margin in μ. As in the case of W_E one can obtain the change in W_G resulting from increasing μ, using τ as a parameter. The result is illustrated in Fig. 14. The I_{PP} values required have been established at the two values of μ as in Fig. 13. The extent of the increase in the case of three of the configuration concepts are shown in the figure.

There is likely to be a significant decrease in cost even for adopting a part of the margin available in μ. As long as such cost savings can be realized, increases in W_E and W_G may be

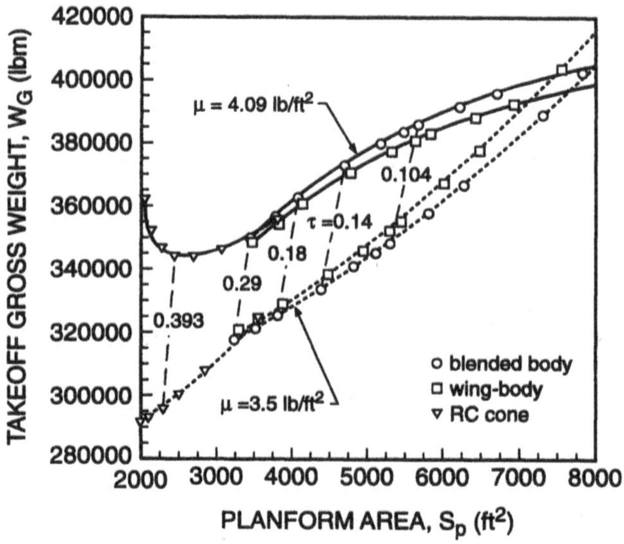

Fig. 14 Effect on gross weight of varying (μ, S_p, and τ).

acceptable in themselves. It will be noted that, in determining the margin in μ, both I_{PP} and I_{TC} have been taken into account.

I_{TC} and size. Based on the correlations given in Eq. (C.23a) in Appendix C, the variation of size, in terms of planform area, with I_{TC} was presented in Figs. 7 and 8, using τ and W_{PY} as parameters, for the case of a blended body. Such a figure can be extended to other configuration concepts, as shown in Fig. 15. Figure 15 also shows values for the size of vehicles with zero payload.

It can be seen that the blended body has the largest design space area among the noncircular cross-sectional shapes, whereas the waverider has the smallest, mainly because of the range of applicable τ values being quite small, although the L/D values are high. The right circular cone derives its advantage from the high values of μ that can be utilized. The variations of size are applicable for various propulsion options, from all-airbreather to all-rocket. One of the observations from the results presented is that aerodynamic efficiency in terms of drag reduction does not seem to be a major driver in itself in hypersonic flight vehicles. If the thrust required for vehicle acceleration is available with a high energy effectiveness of the propulsion-propellant system, compromises may be feasible both with respect to vehicle weight and shape.

C. Configuration Concept Geometry

In discussing configuration concepts, use is made of the following variable.

$$S = S_W / v_T^{2/3} = \omega / \tau^{2/3} \tag{29}$$

Both S, the vehicle slenderness, and ω, the ratio S_p / S_W, increase as τ decreases, i.e., as the planform area increases for a given total volume of the vehicle. Attention is drawn to the eight aerodynamic, structural, and size characteristics presented for the four reference configuration concept geometries in Appendix C, Figs. C.5 to C.7. Those characteristics are made use of in the following. The propulsion-propellant configuration is chosen to provide airbreathing propulsion up to the flight speed of 22,000 ft / s. Thus, an all-rocket engine is assumed to be used beyond that flight speed.

Based on the range of τ utilized for various configuration concepts in Fig. 15, one can then determine the variation of μ as a function of S_p using τ as a parameter, as shown in Fig. 16; μ and τ are related in I_{TC}. The payload is assumed to be the same in all cases. Figure 15 also shows the reference value of μ, and the points along the line of constant μ indicate, by definition, a series of values of (τ, S_p) for which the margin in μ is zero.

One can observe from the figure that the highest permissible μ and the lowest acceptable I_{TC} occur at τ equal to about 0.175 for the blended body and the winged body, at about 0.160 for the waverider, and at about 0.393 for the right circular cone. The larger τ values, associated with less slender bodies, seem to provide a larger margin in μ. Thus, adopting a margin of 15% above the reference value of μ, that is using $\mu = 4.03\ lbm/ft^2$, the waverider is entirely eliminated while the

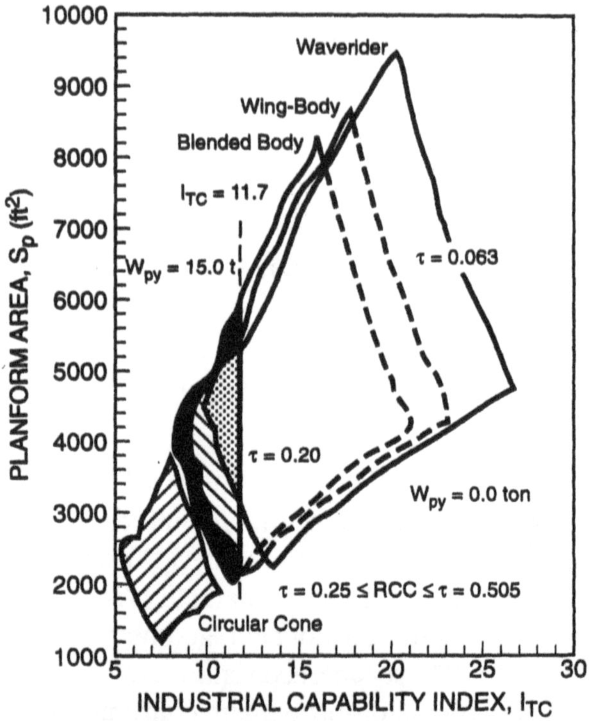

Fig. 15 Effect of configuration geometry on the relation between τ and W_{PY} planform area-industrial capability space.

wing body and the blended body show a very limited remaining margin; however, there is a clear advantage in the case of the right circular cone in the required I_{TC}. The right circular cone is, of course, restricted to vertical takeoff and vertical landing missions.

It is also of interest to consider the variation of μ with respect to W_G, as in Fig. 17. A significant finding here is that the highest value of μ arises in the case of the nonsymmetric body shapes. In addition, the highest values of μ arise at about the same value of W_G, whereas in the case of the right circular cone, a considerably lower value of W_G seems possible. The conclusion from this is that the choice of configuration concept geometry depends on the availablility of I_{TC}. In other words, given a value of I_{TC}, the configuration concept geometry and the value of τ or slenderness determine the realizable payload weight with a chosen propulsion-propellant concept.

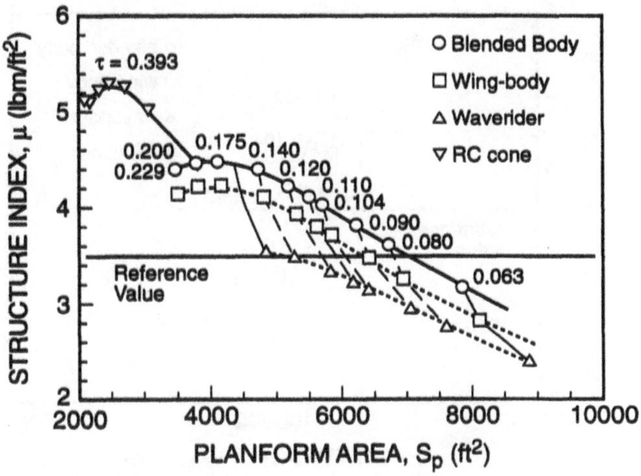

Fig. 16 Effect of configuration geometry on permissible structure index—using τ and S_p as variables.

Two implications to note in preliminary system design exercises are as follows: 1) One must not start with a preconceived configuration concept until one has assessed the available margin in μ and the associated value of τ in light of the available technology capability; and 2) when one obtains such numerical values as zero or negative payload, one must reexamine the configuration concept geometry, the slenderness (τ), μ and the I_{TC}, and the I_{PP} of the propulsion-propellant concept chosen. The zero-payload case need not, *in any rational approach*, end in despair.

Based on the results given in Fig. 17, when a reference configuration is sought, the choice should be based on τ being not less than 0.393 for a right circular cone and not less than 0.175 for a blended body. Those configurations will not be the lightest in terms of W_G, but they correspond to the I_{TC} available with the reference value $\mu = 3.5\ lbm/ft^2$, and the I_{PP} available with the propulsion-propellant concept under consideration. Every progress in I_{TC} and I_{PP} can be expected to lead to a lower W_G and a higher payload ratio, provided one utilizes an appropriate configuration concept. If the limits to improving I_{TC} appear

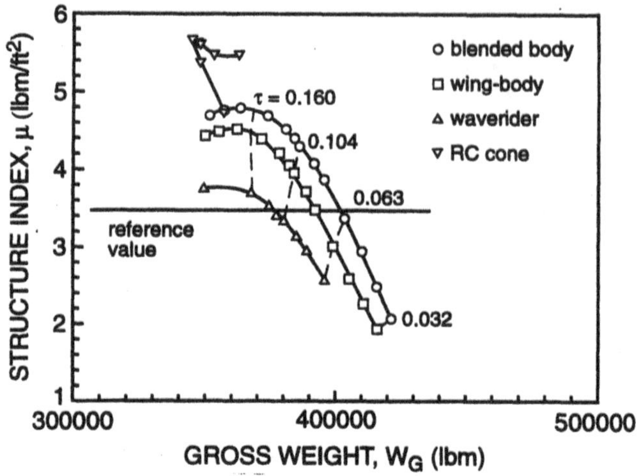

Fig. 17 Effect of configuration geometry on resulting gross weight using τ as a parameter.

insurmountable, one can only investigate possibilities for improvements in I_{PP}, and energy utilization effectiveness.

To examine the possibilities for other missions involving horizontal takeoff and horizontal landing, one may present Fig. 17 in a slightly different form as in Fig. 18, where μ is shown as a function of the gross weight planform loading or W_G / S_p. Utilizing a 15% margin in μ, one can see that the available design space for the nonsymmetric configuration concepts is, in fact, quite small when the takeoff wing loading is held in the range of 70 - 100 lbm / ft^2. (Some improvements may be possible with takeoff lift enhancing devices, while landing loads are generally small relative to takeoff loads.)

D. Payload Weight and Vehicle Size

It may be pertinent to recall here the following observations from the earlier sections:

1) The payload affects the I_{TC} required, as shown in Fig. 15, noting the assumption that the propellant mass is proportional to payload mass.

2) As the payload decreases, W_E does not decrease in direct proportion, as shown in Fig. C.8.

And, 3) for a given value of τ, a decrease in planform area demands an increase in available I_{TC}; see Eq. (C.23a).

In the following, all of the vehicles are assumed to have an SSTO mission. The range of payload considered is 0 - 45,000 lbm. The propulsion-propellant concept is assumed to consist of airbreathing propulsion up to a flight speed of 22,200 ft / s. Initially a blended-body configuration concept is considered, and, later, others are included.

The variation of μ with respect to S_p is shown in Fig. 19, using τ and the payload as parameters. It can be observed in the figure that a 7- ton payload is feasible with a value of $\mu = 3.5$ or an equivalent I_{TC}; however, even allowing τ to increase to 0.2 from about 0.08 does not allow a margin of 15% in μ at the peak value of μ. With a margin of 15%, the smallest vehicle size can allow a payload of 4 metric tons, and a *payload installed density*

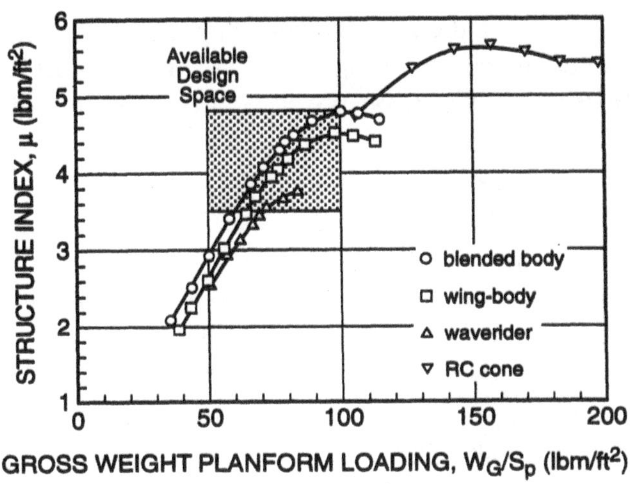

Fig. 18 Effect of mission requirements such as horizontal take-off and horizontal landing.

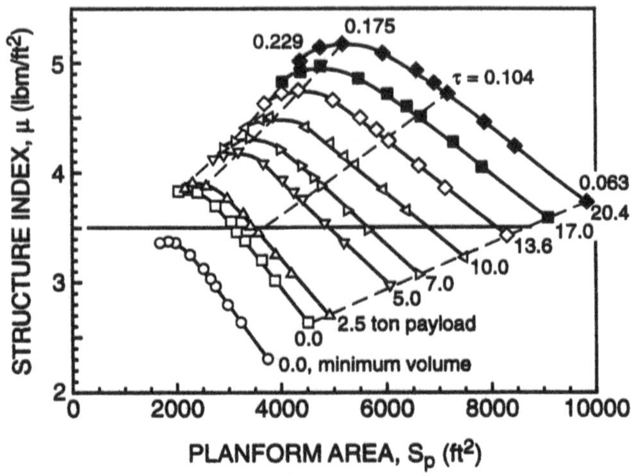

Fig. 19 Effect of choice of τ and W_{PY} in structure index-planform area space. Note the zero-payload values.

$(W_{PY} / \mathcal{V}_T)$ of 6.0 lbm / ft^3, noting that τ becomes about 0.22. If a smaller sized vehicle is attempted, then a reduction in μ arises and hence an increase in the value of I_{TC} that is needed, and τ becomes larger. A 20-ton payload vehicle can be attempted with the same margin in μ but in a rather slender vehicle, with practically no margin left to account for any uncertainties in I_{TC} and also I_{PP}.

Considering the zero payload case, some additional remarks are warranted in continuation of those made at the end of Sec. VI.C. In Fig. 19 it will be observed that two sets of values are denoted for the zero payload case: 1) the higher μ values, where $W_{PY} = 0$ but there is volume available in the vehicle for adding the payload, and 2) the lower μ values corresponding to the case where no volume is provided for any payload, *per se,* as in a demonstrator in which an instrument bay is provided that is also suitable for carrying a small, high density payload. It is clear that case 2 is not feasible with the reference value of μ and the corresponding I_{TC}. In case 1, there is very little margin in μ, and any attempt to make use of it tends to increase the value of τ. Thus, a zero-payload demonstrator may be as difficult to build as a modest payload vehicle, say with $W_{PY} = 2.0$ tons which may be equal to the mass of an instrument and data acquisition system.

In Fig. 20, the zero payload case is shown with a minimum volume for all of the four reference concept configurations. It is clear that the right circular cone does provide a margin in μ, whereas none of the other concept configurations is feasible at the reference μ value and require, therefore, considerably higher I_{TC} than the reference value. From the point of view of a demonstrator vehicle, if the objective is restricted to proving, e.g., the performance of the propulsion system -- since I_{PP} has been included in obtaining the results -- then a conical body would be satisfactory within the available I_{TC}. If the objective included assessment of aerodynamic performance or takeoff and landing, it seems, according to Fig. 20, that one needs to consider a vehicle with a payload of about 2 - 4 tons.

The thermal load and the thermal management required are also significant issues in the development of a small vehicle. Thus, there may be a need to reconsider, even as reference, the available I_{TC} and μ.

E. Propulsion System Concept

There are reasons to believe that a propulsion concept based on a rocket-based, combined-cycle engine that can be

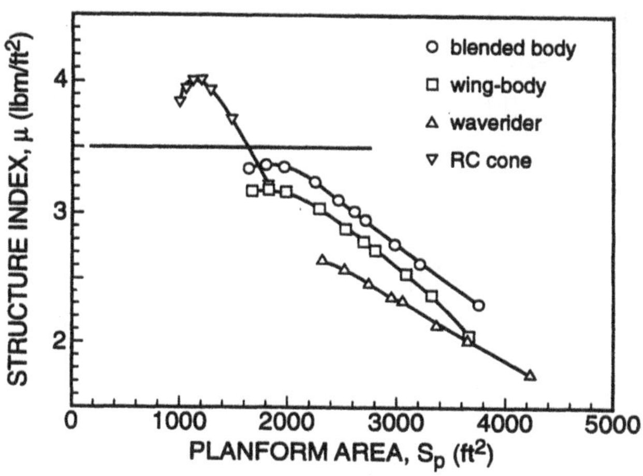

Fig. 20 Zero payload vehicle with various configuration geometry choices.

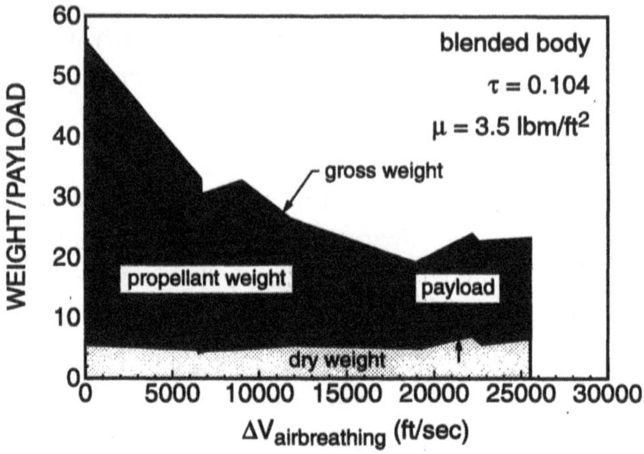

Fig. 21 Weights in terms of payload weight as a function of ΔV obtained with airbreathing propulsion for a specific vehicle.

operated in different modes over various parts of a flight trajectory may be superior to a combination of currently available separate engines[5]. For example, estimates are available for SSTO and two-stage-to-orbit (TSTO) vehicles for the weights as a function of ΔV of the airbreathing part of the propulsion system, as shown in Fig. 21. In this case, W_E is nearly constant, at about six times the payload, up to $\Delta V = 15{,}000\ ft/s$, and then increases to a value of seven to eight times the payload, noting that this applies to a blended-body configuration with $\tau = 0.104$ and $\mu = 3.5\ lbm/ft^2$, the reference values utilized throughout this section. The magnitude of W_G changes with ΔV on account of the change in W_P. It may be noted that, in these estimates, a cross-range requirement was included arbitrarily.

It is now possible to deduce the variation of μ as a function of S_p utilizing ΔV for airbreathing propulsion, and τ as parameters, as shown in Fig. 22. The lowest margin in μ occurs when the propulsion concept is entirely an airbreather. The best margin seems to occur at $\Delta V = 15{,}400\ ft/s$, the value increasing with τ, and then the all-rocket propulsion system is better than the all-airbreather.

For the payload reference value, it is also interesting to see in Fig. 23 that W_E varies in a narrow band with respect to S, and

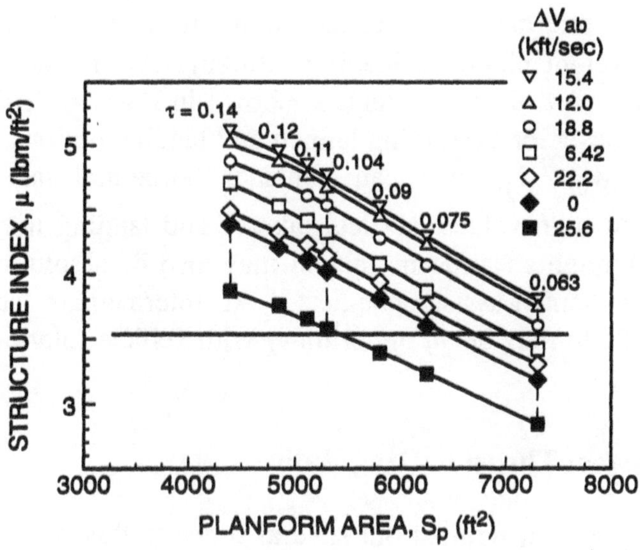

Fig. 22 Effect of ΔV obtained with airbreathing propulsion on margin in structure index.

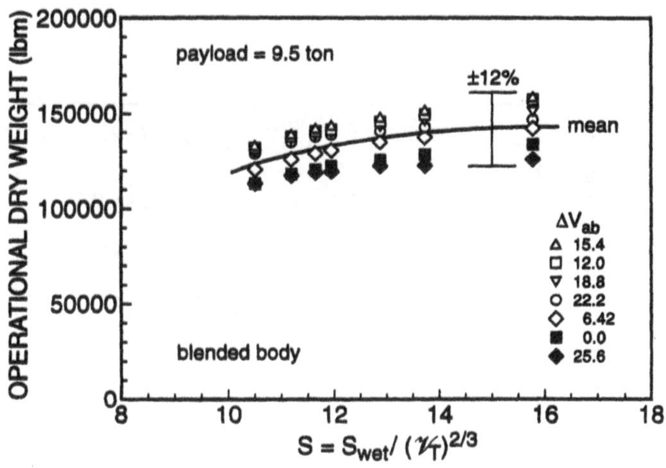

Fig. 23 Variation of operational dry weight as a function of geometry for a vehicle with a given payload for different ΔV obtained with airbreathing propulsion.

thus a tentative, average value estimated may prove adequate in initial analysis.

Next one may consider the variation of μ with respect to the gross weight planform loading, utilizing ΔV for airbreathing propulsion and τ as parameters, as shown in Fig. 24. This figure provides a basis for examining launch and landing options. At low values of W_G / S_p, one can consider horizontal takeoff and landing. At high values, vertical takeoff and landing may be the choice, although an airborne launch may also be a solution along with other launch-assist schemes. At intermediate values of W_G / S_p, there may be an opportunity with vehicle rotation during launch.

F. System Thrust to Drag Ratio

It has been pointed out several times in this paper that the availability of large thrust is a major requirement in any launch vehicle, especially when airbreathing propulsion is included; in fact, there may be no substitute for thrust from many different considerations.

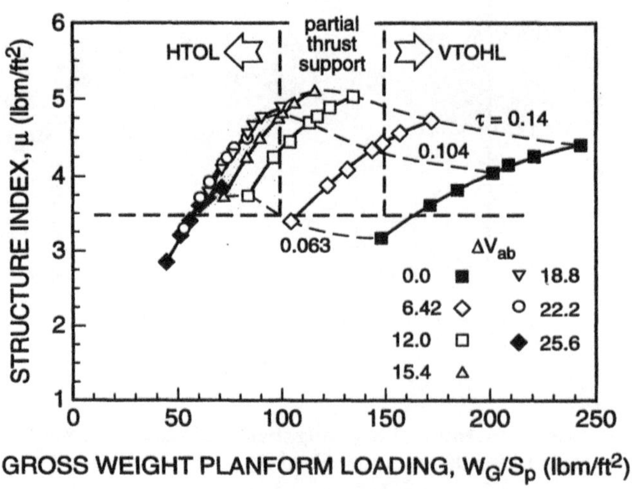

Fig. 24 Influence of launch and landing requirements on vehicle parameters for different ΔV obtained with airbreathing propulsion.

Considering a blended body, one can obtain the relation between μ and S_p, using T/D and τ as parameters, as shown in Fig. 25. The T/D value is a measure of the average over the trajectory. Three observations from the figure are as follows:

1) The margin available in μ is a function of T/D.

2) The slenderness of the vehicle, as represented by τ, is a major consideration in obtaining an increased margin with a given value of T/D.

3) There is an imperative need to assess any propulsion-propellant concept simultaneously with respect to available technology capability, the choice of vehicle size, and the flight trajectory.

In Secs. VI.B to VI.F, an attempt has been made to show how, in a set of parametric spaces, various aspects of the vehicle design can be examined within the context of chosen configuration concepts and propulsion-propellant concepts, starting from a set of reference parameters. There is no implication in that effort that a specific choice is better than others. A methodology for seeking

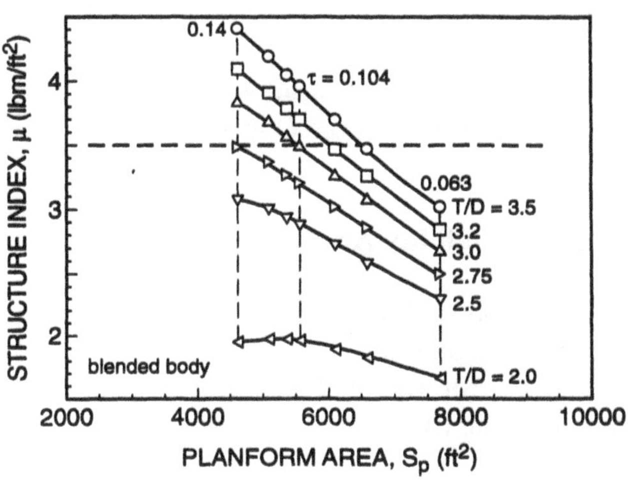

Fig. 25 Primary influence of thrust-to-drag ratio and τ on margin in structure index for a blended body configuration.

what design spaces are useful in answering useful questions in the evolution of the vehicle, and the type of guidance that one can gain by rational means from the design spaces have been the primary interests. The illustrations provided clearly demonstrate the importance of realizing the best and most effective performance of the propulsion system and gaining the most margin in regard to materials and structures or the associated industrial capability, whether one is attempting a demonstrator for a specific purpose or a total launch vehicle. Conventional approaches to configuration concept evolution in aeronautics may not be appropriate for accelerating launch vehicles. In general, a reduction in the size of a vehicle for a given mission must not become the means for meeting the need for cost reduction in accomplishing a mission. There is no substitute for the thrust work that must be realized utilizing an appropriate propellant-propulsion system.

Acknowledgement

One of us (SNBM) acknowledges valuable partial support under NASA Grant NAG 1-1601 (Mr. J.Wiedner, technical monitor) for part of the work on energy requirement analysis.

Appendix A: Thermodynamic Considerations

A. Energy Availability Analysis

The energy availability (ε) per unit mass of the working fluid in an engine at some location in the flowpath, x, relative to a given environment (for which the state and flow parameters are denoted with subscript 0) can be written as follows:

$$\varepsilon_{0x} = c_p(T_x - T_0) + \left(V_x^2 - V_0^2\right) / 2gJ - T_0 \Delta S_{0x} \quad \text{(A.1)}$$

Here *T, V*, and *S* denote the temperature, velocity, and entropy, respectively. The entropy change between *0* and *x* is denoted by ΔS_{0x}. The local energy content per unit mass of the working fluid relative to the same environment is given by the expression

$$E_{0x} = c_p(T_x - T_0) + \left(V_x^2 - V_0^2\right)^2 / 2gJ \qquad \text{(A.2)}$$

The difference between ε_{0x} and E_{0x} is the difference between the entropy at *0* and *x*, when the system is adiabatic between *0* and *x*. The availability flux change between two locations, 1 and 2, in the engine may then be written as follows, assuming no mass flux change between 1 and 2.

$$\dot{\varepsilon}_1 - \dot{\varepsilon}_2 = \frac{d}{dt}\left[c_p(T_2 - T_1) + \left(V_2^2 - V_1^2\right)/2gJ - \left(T_1 \Delta S_{12}\right)\right] \qquad \text{(A.3)}$$

Here $(\dot{\ })$ indicates rate per unit time. The gain in entropy between 1 and 2 is zero only if the process between 1 and 2 is isentropic.

As defined in Eqs. (A.1) and (A.2), the energy availability and energy consist of several components. Furthermore, the entropy change may be expressed in detail in terms of temperature and pressure changes; e.g., in the case of a perfect gas under the assumption of constant specific heats in Eq. (A.1), one obtains

$$\Delta S_{0x} = \frac{R}{T_0} \ln \left[\left(\frac{T_x}{T_0}\right)^{\gamma/\gamma-1}\left(\frac{p_0}{p_x}\right)\right] \qquad \text{(A.4)}$$

A change in the energy availability of the working fluid between two stations in an engine can be expressed under adiabatic conditions in terms of the work (or capability for performing work) added or extracted and the associated, irrecoverable loss of energy availability. This then leads to the concept of an efficiency for the process between the two stations, the *true thermodynamic efficiency*, accounting for the irrecoverable loss. In view of the comprehensive nature of the definition of energy availability and the unambiguous nature of the concept of true thermodynamic efficiency, or *Second law efficiency*, so named because of its origin in the Second law of thermodynamics, it is most appropriate to compare different thermodynamic processes on those bases. Thus, whereas the *energy-based efficiency* is given by

$$\eta_E = 1 - \frac{\text{loss}}{\text{input}} \tag{A.5}$$

the *true thermodynamic efficiency* is written as

$$\eta_\varepsilon = 1 - \frac{\text{loss} + \text{destruction}}{\text{input}} \tag{A.6}$$

where all of the quantities are expressed in terms of energy availability. The "destruction" component refers to the irreversible loss, or irreversibility, whereas the "loss" component can, in principle, be recovered either within the system under consideration or from the environment of the system depending on the nature of the loss.

1. Builder's Analysis of the Brayton Cycle

The Brayton cycle, illustrated in Fig. 1, applies to a gas turbine engine, as well as the ramjet and scramjet engines. A particular approach to the analysis of that cycle was presented by Builder.[16] The approach has been discussed in Ref. 6. Builder sets himself the task of optimizing the cycle parameters to obtain the maximum thrust output with given ambient conditions and input of fuel combustion energy. He introduces the following assumptions:

1) The working fluid everywhere in the cycle is a perfect gas in chemical equilibrium with constant specific heats.

2) The heat addition by combustion of fuel occurs under the conditions of constant static pressure, constant velocity of flow, or $V_1 = V_3$, and the heat added, Q, is equal to $(H_3 - H_1)$.

3) The compression in the inlet and the expansion in the nozzle are assumed to occur with the following efficiencies:

$$\eta_c = (H_1 - H_2) / (H_1 - H_0) \tag{A.7}$$

and

$$\eta_e = (H_3 - H_5) / (H_3 - H_4) \tag{A.8}$$

The nozzle expansion is assumed to be carried to the ambient pressure.

Then it can be shown that the thrust nozzle exit velocity, V_5, and, hence, the thrust, are a maximum when

$$\psi^2 = \left(H_1 / H_0\right)^2 = \frac{\eta_c \eta_e}{1 - (\eta_c \eta_e)} \frac{Q}{H_0} \tag{A.9}$$

There are important implications of the analysis with respect to energy availability considerations.

Implications of optimal compression. Builder's analysis provides valuable insight to the determination of the combustor entry velocity, V_1, required. Referring to Fig. 1 and Eq. (A.9), it is possible to consider a value of V_0 at which, under given conditions of efficiencies and heat addition in the burner, the optimum value of ψ is just equal to the maximum value of ψ possible. This value of V_0, denoted by V_{crit} and called the *critical flight velocity*, represents the demarcation between near maximum diffusion and, hence, a subsonic value of V_1 and optimal diffusion with a supersonic value of V_1. It is given by the following with $\psi_{opt} = \psi_{stag}$, the value of ψ with $V_1 = 0$.

$$V_{crit} = \sqrt{2gJ(\psi_{opt} - 1)} \tag{A.10}$$

Thus, below this value of flight velocity, the emphasis is on diffusion and recovery of momentum in the form of pressure, recognizing the need for a low enough value of velocity at entry to the burner, and the highest value possible for the peak pressure of the engine. At flight velocities above this value, the combustor entry velocity continually increases, noting that, at the same time, a minimum pressure is required for sustained combustion. Introducing $K = V_0^2 / (2gJH_0)$, Eq. (A.10) may be rewritten, in general, as

$$\frac{V_1}{V_0} = \left[1 - \frac{\psi_{opt} - 1}{K}\right]^{1/2} \tag{A.11}$$

It is then of interest to note that as V_0 increases, ultimately V_1 approaches, in principle, V_0 with a significantly small diffusion in the inlet. In practice, efficiencies η_c and η_e may change substantially as the air entry velocity changes, and, hence, the application of Eqs. (A.10) and (A.11) requires a new estimate of ψ_{opt} at each flight speed, taking also into account the change in combustion heat addition.

The existence of a critical value of flight velocity has important implications to dual-mode ramjets and scramjets, both in operation as well as in development testing of a dual-mode combustor.

B. Concepts of Thermodynamic Efficiency

In addition to the concepts of energy-based efficiency, and the true thermodynamic efficiency introduced in Eqs. (A.5) and (A.6), a number of other concepts have been utilized in engineering practice.

Referring to Fig. A.1, illustrating an air compression process in terms of enthalpy and entropy, a number of types of efficiency are often defined based on energy considerations. In the case of a diffusing inlet, the performance is expressed in terms of several types of efficiency, e.g., the so-called kinetic energy efficiency, process efficiency, static pressure efficiency, and diffusion or compression efficiency. These may be defined in the adiabatic case as follows with reference to Fig. A.1.

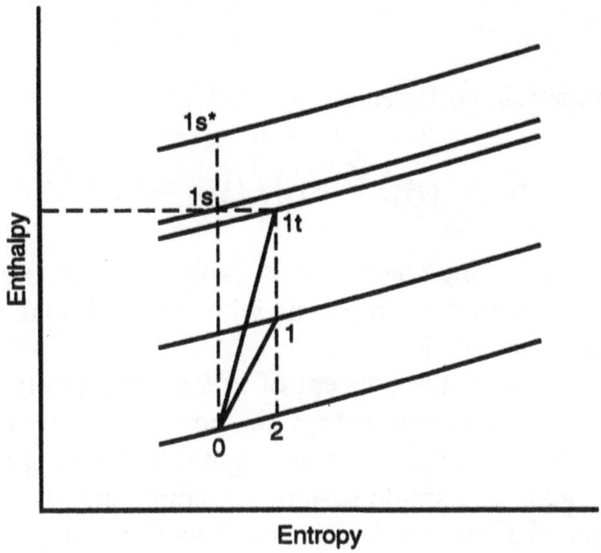

Fig. A.1. Thermodynamic diagram for the inlet process.

Kinetic energy efficiency:

$$\eta_k = V_{e1}^2 / V_0^2 \tag{A.12}$$

where V_{e1} is the velocity obtained by the isentropic expansion of air from state 1 to ambient pressure.

Process efficiency:

$$\eta_{kD} = (H_1 - H_2) / (H_1 - H_0) \tag{A.13}$$

Static pressure efficiency:

$$\eta_p = p_1 / p_{1s} \tag{A.14}$$

and

Compression efficiency:

$$\eta_c = (H_b - H_0)/(H_i - H_0) \qquad (A.15)$$

Each of those includes some level of ambiguity and noncomprehensiveness compared with the rational efficiency related to irrecoverable loss.

Among them, the concept of kinetic energy efficiency has led to the so-called kinetic energy method of analysis that can be applied to other components of an engine, including a combustor, and to an engine as a single system. A comprehensive discussion of the basis and application of the kinetic energy method of analysis can be found in Ref. 17. A brief discussion follows here on the relation between the true thermodynamic efficiency, η_ε, and the kinetic energy efficiency, η_k, for a diffusion process in an inlet.

It follows from Eq. (A.12) that

$$\eta_k = (\varepsilon_{11} - \varepsilon_2)/(\varepsilon_{1S} - \dot{\varepsilon}_0) \qquad (A.16)$$

$$= \left[1 - \frac{\varepsilon_2 - \varepsilon_0}{\varepsilon_{1S} - \varepsilon_0}\right]^{-1} \eta_\varepsilon \qquad (A.17)$$

where the true thermodynamic or rational efficiency is introduced based on Eq. (A.17). Thus the kinetic energy efficiency is different from the true thermodynamic efficiency by the factor involving $(\varepsilon_2 - \varepsilon_0)$. Whereas ε_2 is uniquely determined by (ε_1, p_1, h_1, and p_0), ε_2 has no independently unique relation to ε_0. Thus, the kinetic energy efficiency has no absolute significance, although the final state of diffusion processes, such as *1s*, *1t*, and *1n*, can be compared with one another on the basis of the kinetic energy corresponding to V_0, V_{1t}, and V_{1n} obtained in the three cases, respectively. Another way of recognizing this is to observe that, while *0*, *2*, and *2n* are states at the same pressure value, they differ arbitrarily in enthalpy and entropy values. The kinetic energy efficiency is thus a specific performance parameter that can become comprehensive and unique only with additional information. For example, the process efficiency, [Eq. (A.13)], may be given in addition, and this may lead to a unique measure of the performance of the inlet compression process, but then a comparison between two inlets or two stations in an inlet on the basis of two performance parameters becomes unwieldy in application, and explanation of performance changes.

Similar arguments can be generated in the case of the other efficiencies defined by Eqs. (A.14) and (A.15). Neither equation provides the same, comprehensive information as true thermodynamic efficiency, with equal clarity or unambiguousness. For example, it can be shown that, in the adiabatic case,

$$\eta_{kD} = \left(\eta_k V_0^2 - V_1^2\right) / \left(V_0^2 - V_1^2\right) \qquad \text{(A.18)}$$

implying that η_{kD} depends on η_k, or both efficiencies are needed to establish the states of the fluid across the process.

In the case of a combustor as well, the same type of arguments apply to similarly defined efficiencies.

However, in the case of a combustor, an attempt has been made[18] to compare the performance of two combustors, or of a single combustor at two stations along the flowpath by means of the so-called *thrust work potential* of the working substance. Since the thrust work potential is directly proportional to the thrust and the flight velocity, it is adequate to consider the local thrust potential. In the terminology given in Fig. A.2, the thrust potential of the working substance at a station x in the combustor is given by the following.

$$T = \left[\left(\rho V^2 + P\right)A\right]_x - \left[\left(\rho V^2 + P\right)A\right]_0 \qquad \text{(A.19)}$$

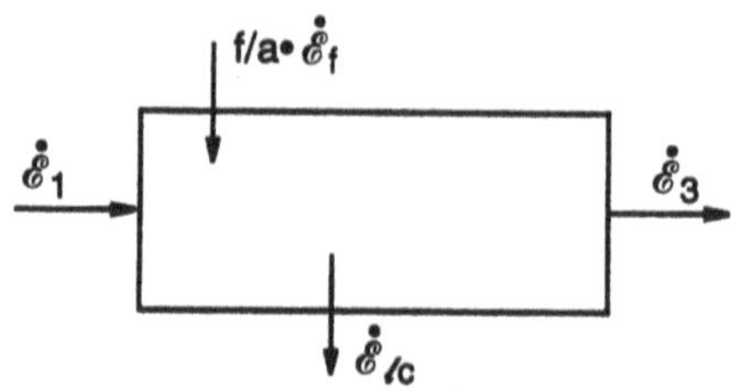

$\dot{\mathscr{E}}_1$ = energy availability of the ambient atmosphere
+ kinetic energy of inlet air
− energy availability loss in inlet

$\dot{\mathscr{E}}_f$ = energy availability of fuel in storage
+ kinetic energy added
+ thermal energy added
+ heat released by combustion
+ energy availability of products when brought to the state of the ambient atmosphere

$\dot{\mathscr{E}}_{lc}$ = energy availability loss

Fig. A.2. Schematic of energy availability flow in a combustor.

Based on the facts that the engine operates as an open cycle, and that the maximum possible expansion of the combustion products is to the external ambient pressure, it is clear that the thrust provides a measure of the energy in the combustion products that can be converted to work by an *expansion process*. However, it is not the measure of the energy availability in the combustion products relative to the local, external, ambient conditions. In addition, a specific value of the thrust or the thrust work can be obtained by numerous combinations of pressure, temperature, and ratio of specific heats, and it is not a unique indicator of the state of combustion products at location x. For example, with two different values of heat input and two conditions of nonequilibrium, there are values of pressure for which the thrust and the thrust potential are the same relative to a given environment. When rather small changes in heat release and pressure are extremely significant, as in the case of scramjet burners, for example, it is not adequate to consider apparently characteristic parameters of the state of the gas for comparison of the effectiveness of a combustor.

It may be worth noting here that the energy availability at x relative to the ambient conditions is absolutely unique. This is the reason that it is a useful parameter for both analysis of a system and comparison of one thermodynamic state with another. In addition, each part of the energy availability has a unique significance and can be utilized for comparative studies. If one attempted a similar comparison of individual contributions to thrust (for instance, that caused by momentum and that resulting from pressure difference) it evolves that the state of the fluid does not become fully and uniquely defined, unless the density or temperature of the fluid in the combustor is also specified. In other words, the momentum and pressure of the combustor fluid do not specify the internal energy of the combustor fluid or its density. If the thrust that can be generated is given along with the entropy difference between the combustor fluid and the ambient atmosphere, it still does not specify the internal energy of the combustor fluid.

Considering the static enthalpy and the kinetic energy as fluid state parameters, they may be nondimensionalized by the stagnation temperature, which provides a definite measure of a gas state in any adiabatic process for a calorically perfect gas. The nondimensional parameters then provide a unique space for examining many individual processes as well as a total engine, provided one also specifies the stagnation or the static pressure of the gas. In comparing the state of the working substance, e.g., at two locations of a combustor or in two combustors, the nondimensional parameters and the pressure together also provide the required information for energy availability determination. They may, therefore, be utilized together in comparative or specific analysis[19].

C. Energy Availability Changes in the Brayton Cycle and Builder's Analysis

The Brayton cycle is illustrated in Fig. 1. Now, the availability of the working fluid undergoes the following changes in the inlet, combustor, and nozzle:

$$\dot{\varepsilon}_1 - \dot{\varepsilon}_0 = c_p(T_1 - T_0) + \left(V_1^2 - V_0^2\right) / 2gJ - T_0 \Delta S_{01} \tag{A.20}$$

$$\dot{\varepsilon}_3 - \dot{\varepsilon}_1 = c_p(T_3 - T_1) + \left(V_3^2 - V_1^2\right) / 2gJ - T_1 \Delta S_{13} \tag{A.21}$$

and

$$\dot{\varepsilon}_3 - \dot{\varepsilon}_5 = c_p(T_3 - T_5) + \left(V_3^2 - V_5^2\right) / 2gJ - T_3 \Delta S_{35} \tag{A.22}$$

Equation (A.21) may be rewritten in the form

$$\dot{\varepsilon}_3 - \dot{\varepsilon}_1 = Q - T_1 \Delta S_{13} \tag{A.23}$$

where ΔS_{13} is only a function of T_3 / T_1 or, equivalently, Q. An increase in Q gives rise to an increase in T_3 and thus a rise in thrust output as well as in entropy gain. Under the assumptions of the analysis, the increase in T_3 or the gain in entropy does not affect the criterion for maximizing the thrust, Eq. (A.9). However, Eq. (A.9) does imply that the thrust is a maximum only when the loss of availability in the air compression inlet is related in a specific fashion to the loss of availability in the thrust nozzle. This can be seen by rewriting Eq. (A.9) as:

$$\eta_c = \frac{\psi^2}{\psi^2 + Q / H_0} \frac{1}{\eta_e} \tag{A.24}$$

Two observations from Eq. (A.24) are as follows. First, when either η_c or η_e is equal to unity, the other can be obtained from Eq. (A.24). Second, there is no meaning to the case $\eta_c = \eta_e = 1$, since then both the inlet and the nozzle operate under the isentropic condition, and the output is merely a function of T_3 and, hence, of Q. In this, case the gain of entropy in the combustor only reduces the availability in the engine exhaust. Incidentally, Eqs. (A.9) and (A.24) also ensure, under the given conditions, the least loss of enthalpy and, in this case, the least loss of energy availability in the engine exhaust relative to the external ambient conditions.

It is significant to note here that Builder's approach to maximizing thrust output with a given input in the cycle follows the general principles stated in Sec. II.A.1. A component of energy availability is chosen, and a relation is established for the changes in that component in two different parts of the engine cycle. The relation is unique and provides a basis for the design and control of the engine.

D. Energy Input by Fuel Combustion

Equation (2) provides the ratio of fuel combustion energy input, in the case of hydrogen fuel, in relation to the kinetic energy of air at entry to the engine, the atmospheric air entering the engine at the local flight velocity.

Including Carnot loss caused by the inability to expand to ambient temperature, one can write

$$\frac{Q_A}{H_0}\,\frac{H_0}{V_0^2/(2gJ)} = 51.02\left(\frac{V_0}{1000}\right)^{-2} \qquad \text{(A.25)}$$

where Q_A is the actual chemical energy supplied to air.

The energy supplied becomes further reduced because of various reasons, most notably internal drag, and mixing losses in bringing air and fuel into contact at the molecular level, as pointed out, for example, by Swithenbank[20]. An estimate of internal losses due to the use of struts for fuel injection, based on such experimental data, is presented in Table A.1.

Table A.1. Internal losses due to fuel injection struts

Fuel injection configuration	$\frac{\Delta E_{mix}}{KH_0}$	Relative drag
Five vertical struts	0.119	3.24
One horizontal strut	0.0647	1.76
No internal struts	0.0456	1.24
Shock confined (cowl retracted)	0.0361	1.00

The net chemical energy supplied to air accounting for the internal losses may be estimated as

$$\frac{Q_A - \Delta E_{internal}}{H_0 K} = \frac{82{,}440\left(\frac{V_0}{1000}\right)^2}{\frac{V_0}{1000}^{2.231} - 1615.8} \tag{A.26}$$

In Sec. II.C.1, the need for choosing the inlet compression of air and the addition of energy to air by fuel combustion with reference to the setting in of chemical nonequilibrium in the working fluid, and the importance of accounting for nonequilibrium effects in the analysis of airbreathing propulsion systems as the altitude of flight increases were discussed.

The working fluid is found in various states of nonequilibrium beginning with the air inlet, during the combustion process in the burner, and ending with combustion products discharging through the thrust nozzle. It is interesting to make an estimate of the effect of such nonequilibrium, e.g., in terms of the resulting changes in entropy and energy availability.

A method of correlating the entropy change caused by flow expansion in different nozzles, during which the gas is in a state of frozen equilibrium, consists in expressing the entropy change in terms of the entropy level, Σ, corresponding to the frozen equilibrium state and a correction based on some geometrical parameters of the nozzle, as follows[21].

$$\frac{S}{R} = \Sigma - 0.4 \ln\left[\tan\theta / \left(r^* / r_{ref}\right)\right] \tag{A.27}$$

where r^* is the nozzle throat radius, r_{ref} a reference radius of 1 in, and θ the nozzle expansion angle. For a representative ramjet configuration, the second term on the right-hand side is about −1.1. Thus, if Σ is equal to 32.0 and 34.6 for 3.0 and 10.0% frozen equilibrium, S/R becomes 33.1 and 35.7, respectively.

Now, the freestream entropy of air increases with altitude. This introduces a constraint on the maximum altitude for airbreathing propulsion. In addition, the entropy change of the fluid under practical conditions of operation of a Brayton cycle engine is nearly constant and varies only with the ambient air value. Thus, the value of entropy at entry to the expansion nozzle as well as the change in the nozzle because of frozen equilibrium predominantly are functions of altitude, except for the effects of engine geometry changes. Thus, one has a method of estimating a correction to energy availability resulting from chemical freezing during nozzle expansion.

E. Energy Effectiveness According to Builder

Effectiveness of an engine, one of its components, or a process in it is the measure of the impact of a change in the input to the engine, or the performance of a component or process included in the engine, on the engine output under given operating conditions.

Equation (5) provides a measure of the performance of an installed engine in flight in terms of propellant consumption per unit change in ΔV. Builder has defined a so-called *energy conversion factor* or *thrust coefficient*, θ, which serves the same purpose of comparing the performance of two engines, or of one under two different operating conditions.

Referring to Builder's analysis, one can write

$$\theta = \frac{TV_0}{JQ_c\dot{W}_P} = \frac{TV_0}{JQ\dot{W}_a} \tag{A.28}$$

and, equivalently,

$$\theta = \frac{I_{SP} V_0}{J Q_c} = \frac{I_{ASP} V_0}{J Q} \tag{A.29}$$

Here Q_c and Q represent the amount of energy supplied per unit mass of fuel and air, respectively. I_{ASP} denotes specific thrust with respect to air mass flow.

Considering the limits of ψ_{max} and ψ_{opt}, the energy conversion factor in the two cases, taking into account fuel mass addition to the working fluid, may be written as follows.

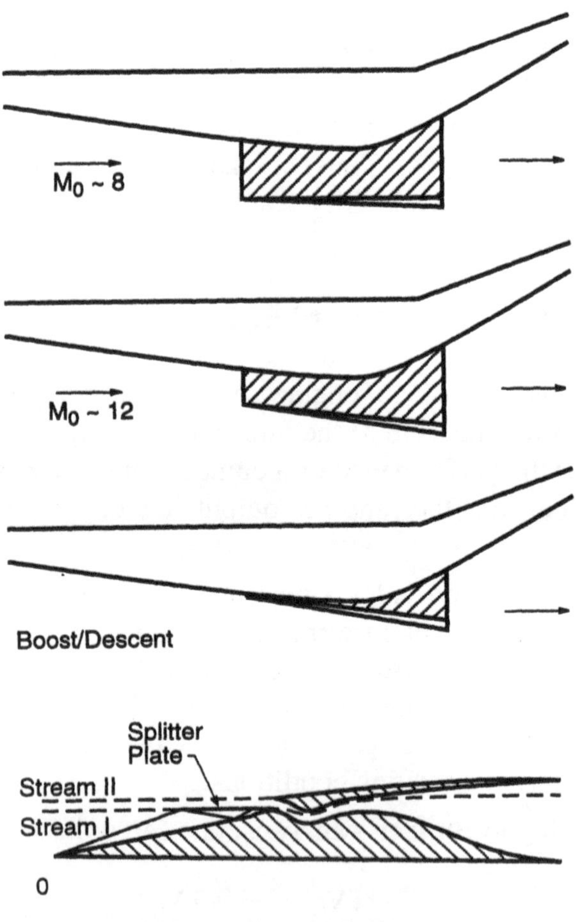

Fig. A.3. Schematic of conbustor entry geometry variability.

$$\theta\psi_{max} = \frac{2K}{C}\left[\left(1 + f_s\phi_c\right)\left\{\eta\left(1 + \frac{C}{K+1}\right)\right\}^{1/2} - 1\right] \quad (A.30)$$

and

$$\theta\psi_{opt} = \frac{2K}{C}\left[\left(1 + f_s\phi_c\right)\left\{1 + \eta\left(\frac{C}{K}\right)\frac{\psi_{opt} - 1}{\psi_{opt}}\right\}^{1/2} - 1\right] \quad (A.31)$$

where $\eta = \eta_c\eta_e$, and $f_s = f/a$, the fuel to air ratio, under stoichiometric conditions.

F. Energy Loss Caused by Internal Drag

The energy added to the working fluid in the combustor by the combustion of fuel is given in Eq. (A.25).

Referring to Fig. A.3, an estimate can be made of the internal losses, $\Delta_{internal}$, caused by the presence of a movable cowl and the fuel injection struts, as shown in Eq. (A.26) and Table A.1. Accounting for such internal losses, the resulting changes in

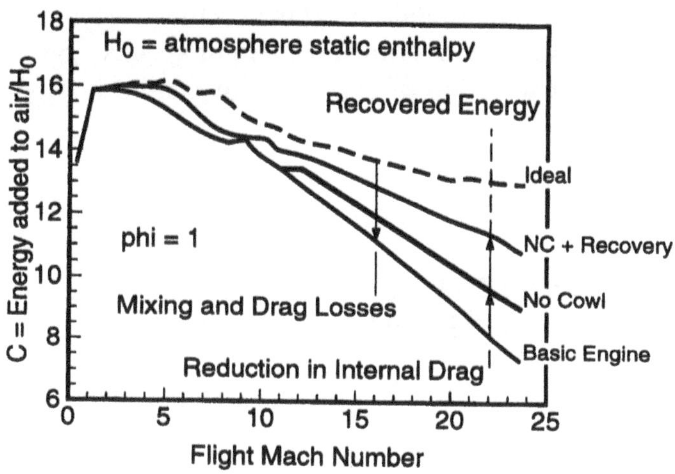

Fig. A.4 Gross energy added by combustion of fuel and losses as a function of flight Mach number.

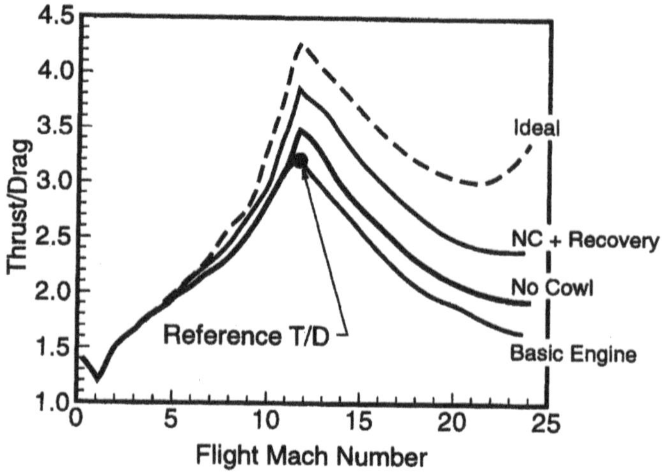

Fig. A.5. Ideal and realized thrust-to-drag ratio as a function of flight Mach number.

the energy added to air by combustion are shown as a function of flight Mach number in Fig. A.4. The resulting changes in thrust-to-drag ratio of an engine in flight are illustrated in Fig. A.5.

G. Propulsion Concepts for Vehicle Synthesis Study

Considering the flight trajectory from takeoff to orbital launch velocity, it is found, in general, that different types of engines may be suitable over different flight speeds. From various points of view, a selection of engines is discussed in Appendix C, Sec. A.

Appendix B:

Historical Data, and Projections on Materials and Structures

P. Czysz

The properties of materials available for designing structures set a number of opportunities as well as constraints in the determination of the feasibility of structures for vehicles and for propulsion systems. Whereas specific materials with known properties can be considered at the preliminary or conceptual design stage, an examination of the applicability of materials of different ranges of properties based on historical data and reasonable projections provides much greater credibility, and also renders the analysis more amenable to changes.

The data presented in the following have been based on Refs. 22 – 30.

Figure B.1 presents the specific strength of structural material as a function of material temperature along with specific data on materials for a number of past flight vehicles. The aircraft considered are unprotected hot structure vehicles. Thermodynamic equilibrium exists in those vehicles among the friction heat generated, the radiative heat transfer, and the input to the structure. In some cases, fuel is used as a heat sink to maintain the skin

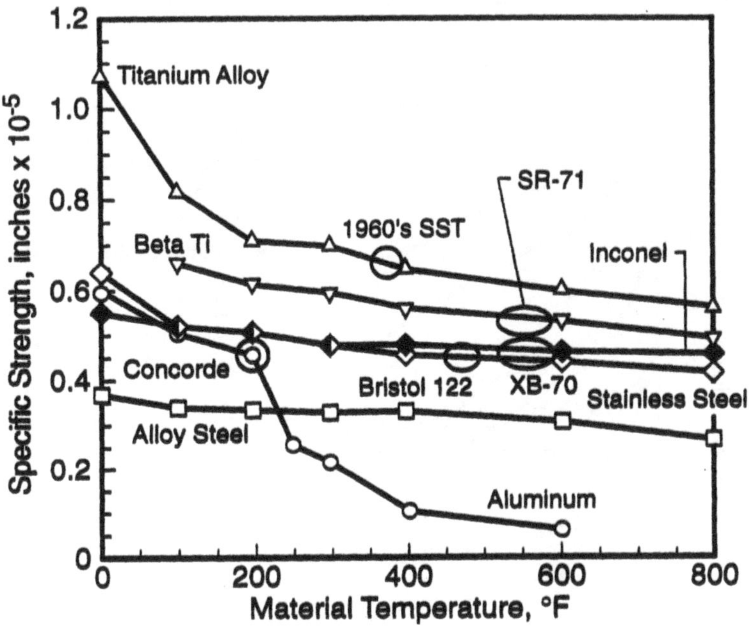

Fig. B.1. Specific strength of structural material as a function of temperature.

temperature below a permissible value; however, the heat so gained in the fuel has not been accounted for further in the propulsion system performance.

It is interesting to examine the capability of some of the newer, projected 1990's materials of the 1960's era, as illustrated in Fig. B.2. Some additional possibilities have been demonstrated, beyond the capability of carbon-carbon materials, for silicon carbide-carbon (SiC / C) and silicon carbide-silicon carbide (SiC / SiC) materials up to 3900°F (1700°C) in oxidizing and combustor applications, at least over tens of hours of operation.

In this connection, it is worthwhile considering some historical data on structural weight per unit surface area as a function of design surface temperature, as shown in Fig. B.3. The figure includes material capability of the 1960s as well as the then-projections for the 1980s, and other estimates based on the use of currently available titanium alloys to replace the Hastalloy and TD nickel alloys of the 1960s. The three solid lines in Fig. B.3 are for three hot, unprotected structures of aluminum, titanium, and René 41 superalloy. The stair-step broken lines represent 1) titanium

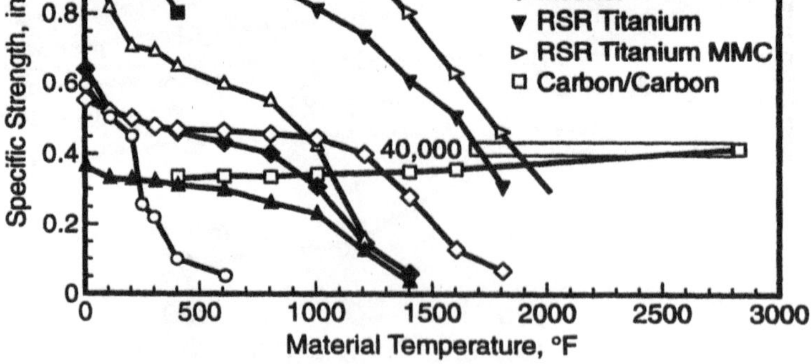

Fig. B.2 Comparison of specific strength of a number of materials as a function of temperature.

TPS (thermal protection system) added to an aluminum thermally protected structure, and 2) titanium, René 41, Hastalloy X, and TD nickel TPS added to a titanium thermally protected structure. Each additional material added, of course, causes an increase in structural weight. These basic curves are for a hypersonic vehicle with about an hour of hypersonic cruise flight capability. Figure B.3 also shows, with open and solid circles, estimates for a research aircraft with a flight at Mach 12 over a test duration of about 5 min. Results from the vehicle sizing program are shown by square symbols for a potential flight vehicle. An estimate for a cruising vehicle with an active energy removal system (for removing the 2.2% of thermal energy that was shown earlier to enter the structure) is also shown (curve X) in the figure. It may be pointed out that the estimates are based variously on the 1960s, 1970s, and 1980s projections, as indicated in Fig. B.4. The estimates in Fig. B.3 also include the use of superalloy and refractory metal shingles. The refractory metals, columbium and molybdenun, are about 1.6 times heavier than steel, whereas

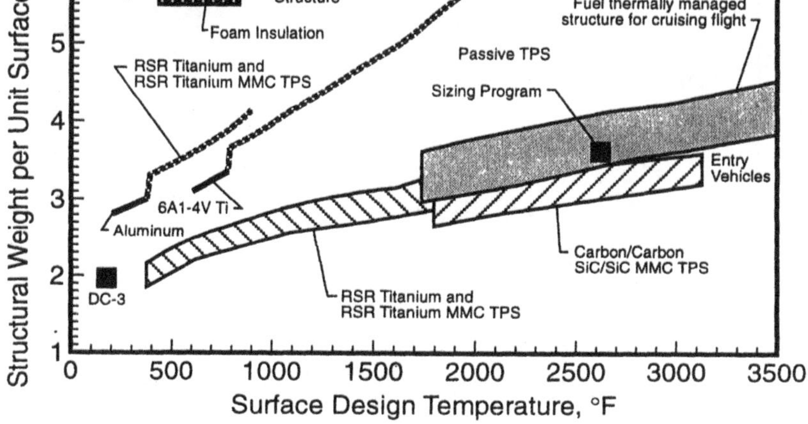

Fig. B.3 Historical data on structural weight per unit surface area as a function of surface temperature.

titanium is about three times heavier. If the refractory metals are replaced by carbon-carbon and RSR titanium alloys, which can be used up to 2100°F or higher, then a considerable savings would be accomplished.

Figure B.4 provides an estimate for the structural weight with the replacement of superalloy and refractory metals. This figure also shows an estimate for a structure incorporating a thermal protection system based on the use of fuel for cooling. An estimate obtained using the vehicle sizing program is also indicated in the figure; the estimate is based on including the benefits from use of titanium shingles in both cruising and entry sections of the flight. The estimated value of 3.5 lbm / ft^2 is utilized in all of the baseline sizing investigations. For additional improvement, the 1990 estimate from the 1960's projection, a value of 3.2 lbm / ft^2 can be utilized.

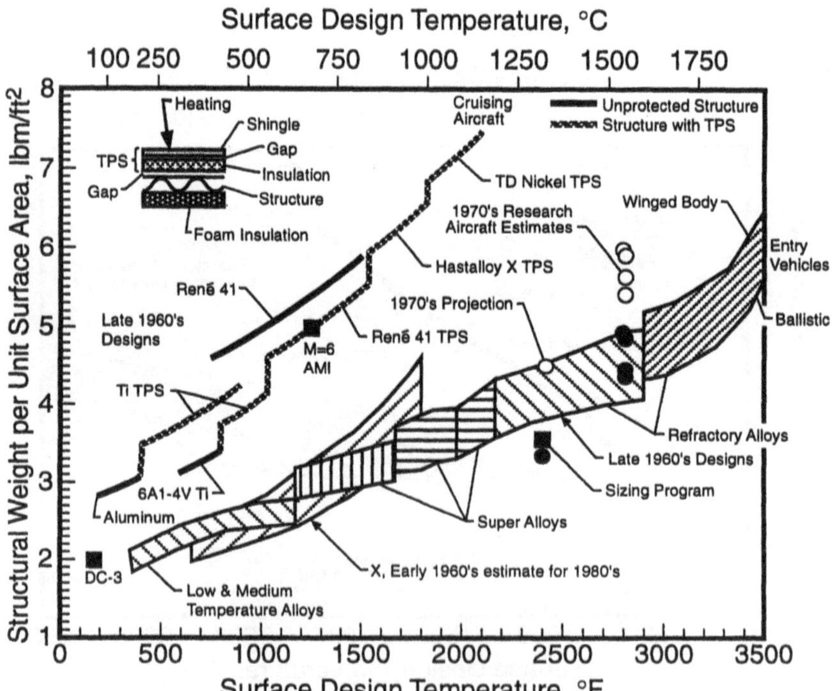

Fig. B.4 Comparison of historical data and projections for protected and unproptected structures.

Appendix C:

Historical Data and Projections on Hypersonic Vehicle Concepts

P. Czysz

The basis for examining historical data and projections in the current case is the same as in the case of materials. In the case of vehicle concepts and the associated geometry, the data available pertain to airplanes of various speed ranges, and also, the single stage to orbit rocket motors of recent times.

A. Propulsion-Propellant Index.

An estimate of the index has been made for a series of combinations of rocket (R) and an airbreather (A) as shown in Table C.1, based on assumed industrial technology capability of the 1978-1988 timeframe for propulsion systems operating with 30% slush hydrogen and triple-point oxygen. In all cases, the payload is assumed to be 30,000 lbm. The four ranges of flight velocity for the four engines chosen are 0 – 3000, (ii) 3000 – 6420, 6000 – 15400, and 15400 – 25600 fps. It can be observed that the index, calculated with respect to ΔV for the airbreather, does not vary by more than ±9% from an all-rocket to an all-airbreather system, whereas the average value for all of the systems considered is 4.01, with a standard deviation value of 0.366.

B. Configuration Size Characteristics

In continuation of the notation introduced in Section IV.B, one may write

$$T = S_p / \mathcal{V}_T^{0.667} = \left(\mathcal{V}_T S_p^{1.5}\right)^{-0.667} = \tau^{-0.667} \quad (C.1)$$

$$S = S_w / \mathcal{V}_T^{0.667} = \omega\tau^{-0.667} \quad (C.2)$$

$$\upsilon = \mathcal{V} / \mathcal{V}_T \tag{C.3}$$

$$G = \omega / \tau\upsilon \tag{C.4}$$

In Eq. (C.1) the planform area S_p is treated as the main parameter rather than the total volume $\mathcal{V}_T$. In the case of low-speed airplanes, the total weight of the vehicle is normally expressed as wing loading times the planform area, whereas, in the case of high-speed vehicles the volume and shape of the vehicle also become important parameters. Thus, in the foregoing notation, the vehicle volume and weight are considered as the primary parameters.

The propellant volume is expressed by the following relation:

$$\upsilon_P = \frac{\mathcal{V}_P}{\mathcal{V}_T} S_p^{-0.0717} \tag{C.5}$$

which corresponds to Eq. (28) with $n_F = 0.0717$. Then Eq. (C.5) may also be written as

$$\upsilon_P = \frac{\mathcal{V}_P}{\tau} S_p^{-1.5717} \tag{C.6}$$

It follows that

$$\frac{\mathcal{V}_P}{\mathcal{V}_T} = \frac{\mathcal{V}_P}{\tau} S_p^{-1.5717} \tag{C.7}$$

It is assumed in the foregoing that the limit for υ_F corresponds to the value when the planform area is equal to 10,000 ft^2.

Next, υ_T is sought to be expressed in terms of $\mathcal{V}_F$ and $\mathcal{V}_{PY}$ in relation to $\mathcal{V}_T$. It is clear that

$$\upsilon_T = \upsilon_P + \upsilon_{PY} \tag{C.8}$$

or,

$$\upsilon_{PY} = \upsilon_T + \upsilon_P \tag{C.9}$$

Considering Mach 8 – 12 hypersonic vehicles of the 1960s, one may assume an average fuel density range of 13 – 8 lbm / ft^3. For relatively dense, nonpassenger-type vehicles, one can estimate the volume based on the mean propellant density. In those vehicles, υ_{PY} may be assumed to have been on the order of 0.01 – 0.15 of the υ_F value. Since passengers as well as space-deployable payloads are of the same density as that of slush hydrogen fuel, one may assume that a suitable estimate for υ of the cargo would be on the same order of magnitude as the amount of hydrogen required. Thus, for a cargo of a given volume, one may write

$$\upsilon_T = \frac{\mathcal{V}_P + \mathcal{V}_{PY}}{\mathcal{V}_P} S_p^{-0.0717} \tag{C.10}$$

This correlation is in the nature of a correction to Eq. (C.8), the latter being the nominal volume relationship. It is then expected that a reasonable, representative value is obtained for $\mathcal{V}_P$ of space launchers with low-density space payloads by subtracting 0.07 from the υ_{PY} of the high-density payload vehicle value. In other words, $\mathcal{V}_P$ of space launchers is larger roughly by an amount equal to 0.07 of the value of $\mathcal{V}_{PY}$ of high-density payload vehicles.

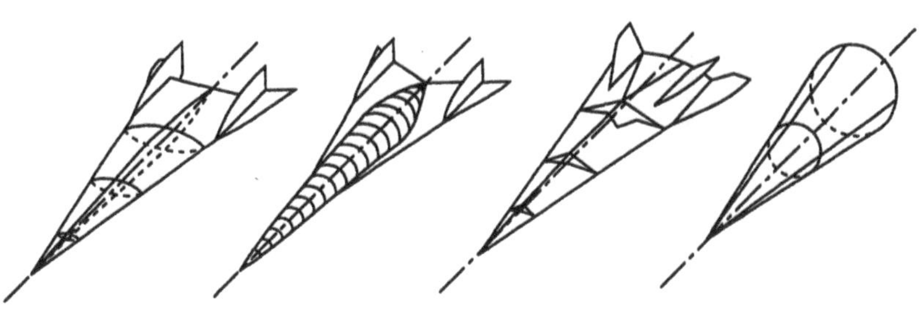

Fig. C.1. Four reference configuration shapes.

Table C.2. Characteristics of reference concept configurations

	Blended body	Wing-body	Waverider	Right circular cone
τ	0.104	0.104	0.104	0.483
ω	2.65	2.76	2.89	3.88
S	4.52	4.52	4.52	1.63
T	11.97	12.49	14.37	6.30
ν	0.471	0.451	0.389	0.467
K_{str}	0.142	0.142	0.142	0.197
βC_{D_0}	0.1021	0.0870	0.0865	0.1397
$(L/D)_{12}$	3.99	3.96	3.92	3.03
G	54.01	58.95	82.58	16.48

3. Characteristics of Reference Concepts

The four reference concept geometries chosen are, the blended body, the wing-body, and the waverider, and the right circular cone as shown in Fig. C.1. The characteristics for those configurations are presented in Table C.2, under the following assumptions:

1) In all cases, the leading-edge sweep angle is 78 deg arc.

2) The value of τ for the nonconical shapes is 0.104 corresponding to a value of L / D equal to 2.5 – 2.7, that is considered adequate for leaving an arbitrary near-Earth orbit and landing in the continental US.

3) The value of τ for the right circular cone is a consequence of its geometry.

It is interesting to observe in Fig. C.2 the relation between *S*, the planform area, and *T*, Eq. (C.1), a parameter similar to τ but based on the total volume of the vehicle, for a number of aircraft

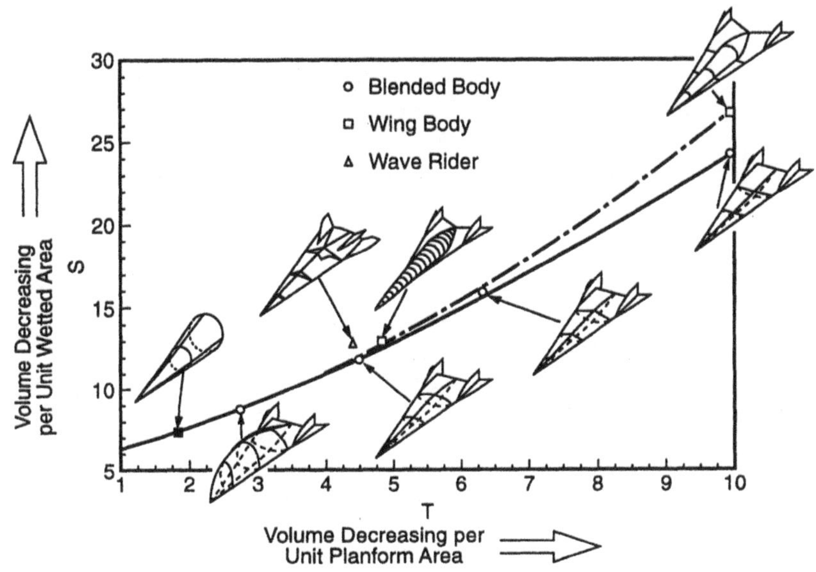

Fig. C.2 Characteristics of reference configuration shapes.

configurations. The figure also shows the parameters for the four reference concept geometries. It is clear that the hypersonic concept geometries are consistent with the aircraft geometries, and one may conclude that hypersonic bodies are shaped with certain trends and features that permit various configuration geometries within the framework of certain fixed trends and features.

The wetted area-to-planform area ratio ω for the four reference concepts is presented as a function of τ in Fig. C.3. Some possible vehicle applications and operational limits are also indicated in the figure. It may be of interest to note that above a value of τ equal to 9.19, the right circular cone has a smaller wetted area (or larger ω) than the other conceptual shapes. Thus, for an SSTO rocket, a cone or a derived shape such as a trapezoidal pyramid becomes the optimal shape.

The curves given in Fig. C.2 can be represented by the following relations:

Blended body:

$$\omega = -64.796\tau^3 + 39.602\tau^2 - 1.627\tau + 2.460$$

Wing-body:

$$\omega = -95.387\tau^3 + 59.897\tau^2 - 5.815\tau + 2.827$$

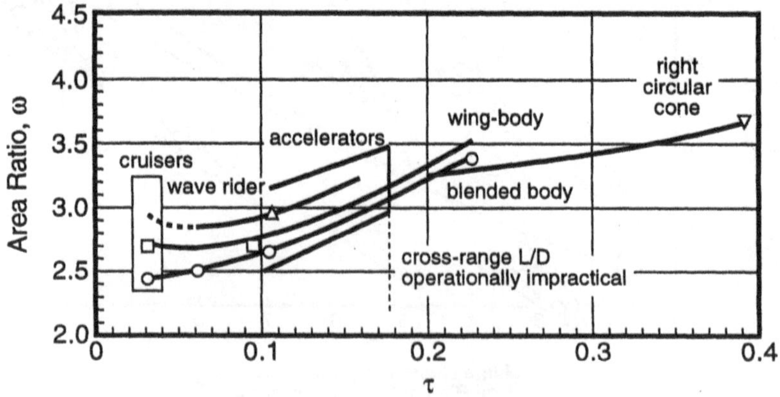

Fig. C.3 Geometry of reference configuration shapes.

Waverider:

$$\omega = -101.22\tau^3 + 63.560\tau^2 - 6.171\tau + 3.000$$

Right circular cone:

$$\omega = 1.913\tau^3 + 1.845\tau^2 + 0.230\tau + 3.123$$

The other geometric parameter of interest, namely, the propellant volume fraction, v_F, includes propulsion considerations. The variation of v_F with respect to T is given for the four selected configurations in Fig. C.4. It may be observed that the values for the cone and the blended body are contiguous, and, also, the winged body has a clear advantage in the case of slender configurations.

The curves given in Fig. C.4 can be represented by the following relations.

Blended body:

$$\nu = -2.811E-5T^3 + 8.227E-4T^2 - 6.867E-3T + 0.4838$$

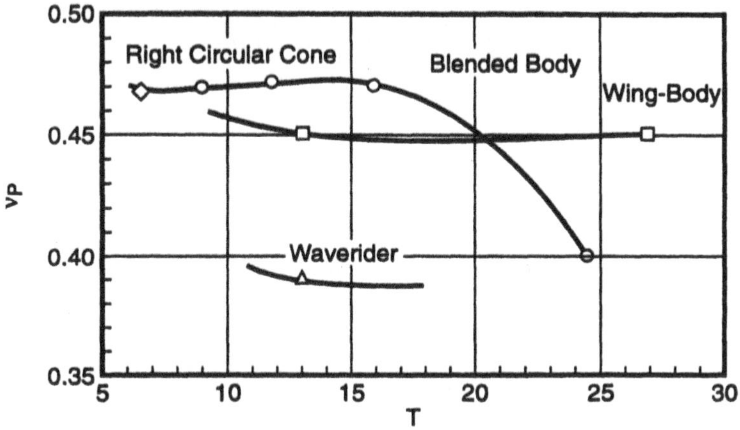

Fig. C.4 Propellant volume fraction as a function of T for the reference configuration shapes.

Wing-body:

$$\nu = -2.811E-6T^3 + 5.172E-4T^2 - 1.114E-2T + 0.5241$$

Waverider:

$$\nu = -1.931E-6T^3 + 2.944E-4T^2 - 8.241E-3T + 0.4512$$

Right circular cone:

$$\nu = -7.971E-4T^3 + 1.917E-2T^2 - 1.514E-3T + 0.8065$$

Finally, one needs a means of determining the drag under different hypersonic flight conditions. One of the correlations obtained based on hypersonic test data yields the total drag under accelerating, cruise, and glide conditions as a function of zero-lift drag:

$$(\beta C_D)_{accel} = 1.1\beta C_{DO} \tag{C.11}$$

$$(\beta C_D)_{cruise} = 1.5\beta C_{DO} \tag{C.12}$$

and

$$(\beta C_D)_{glide} = 2.0\beta C_{DO} \tag{C.13}$$

where $\beta = \left(M^2 - 1\right)^{1/2}$, and

$$\beta C_{DO} = 0.05772\exp(0.4759K_D) \tag{C.14}$$

and

$$K_D = \omega^{0.75} / T^{0.5} = \omega^{0.75} / \tau^{0.6667} \tag{C.15}$$

The correlation is illustrated in Fig. C.5.

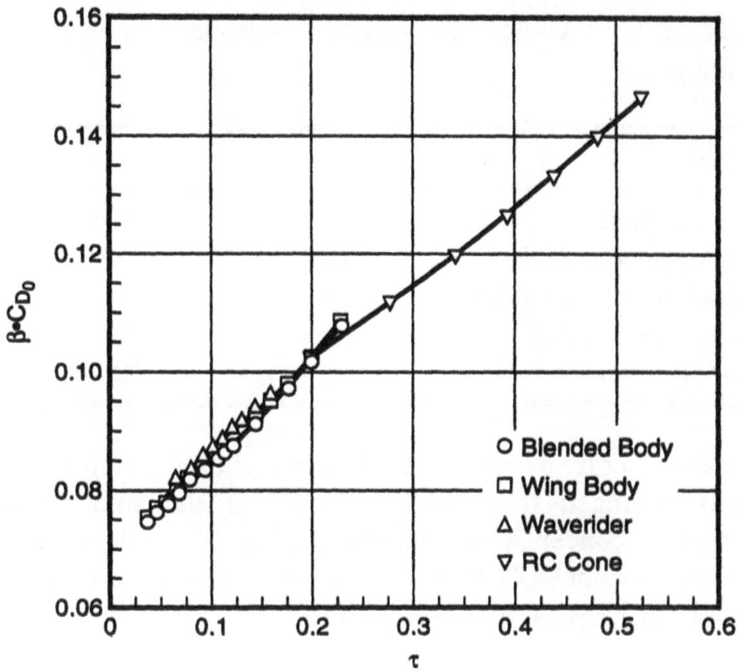

Fig. C.5 Correlation of drag coefficeint factor for the reference configuration shapes.

The maximum value of L / D at a flight Mach number of 12 is taken as

$$(L/D)_{M=12,max} = (\beta C_{DO})^{-0.5636} \qquad (C.16)$$

Another useful correlation for the four reference configuration concepts is that shown in Fig. C.6, where βC_{DO} is shown as a function of $S = S_W / v^{0.667} = \omega / \tau^{0.667}$. It can be observed that the drag coefficient for the four reference configurations shows a smooth variation from the stout configuration to one that is leaner. The drag is the least for the wing body and the blended body.

An interesting variation of the correlation shown in Fig. C.6 for the four configuration concepts is one that relates the weight ratio to the trajectory-averaged thrust-to-drag ratio. To set up such a correlation for a vehicle, one needs to assume parameters of the following type.

Parameters	Example
Type of vehicle	SSTO
T/D at a specific flight Mach number	3.2 at Mach 12
Propulsion-propellant index	4.09
Vehicle weight ratio	2.70
Trajectory-averaged (D/D)	3.20

Then one can set up a correlation of the weight ratio with respect to the thrust-averaged thrust-to-drag ratio for the four reference configuration concepts, as shown in Fig. C.7, assuming the propulsion concept $AAAR_{22.2}$. The zero-lift drag coefficient is found assuming a parabolic drag polar.

In addition to the data provided in Figs. C.5 – C.7, other data used in subsequent analyses are provided in Fig. C.8.

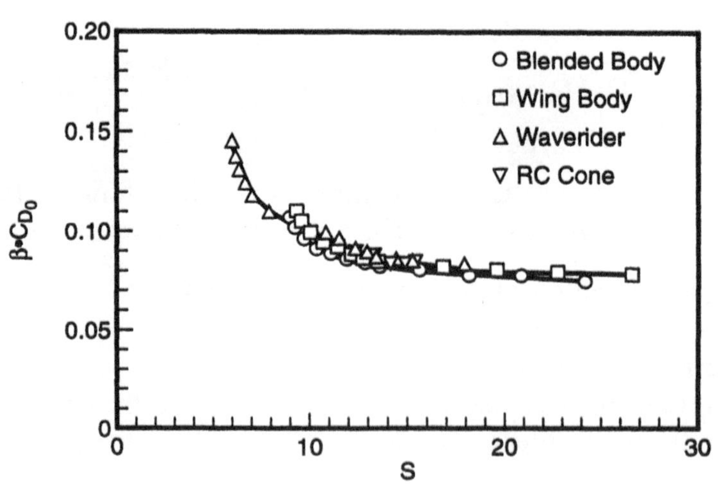

Fig. C.6 Correlation of drag coefficient factor with shape factor, S, for the reference configuration shapes.

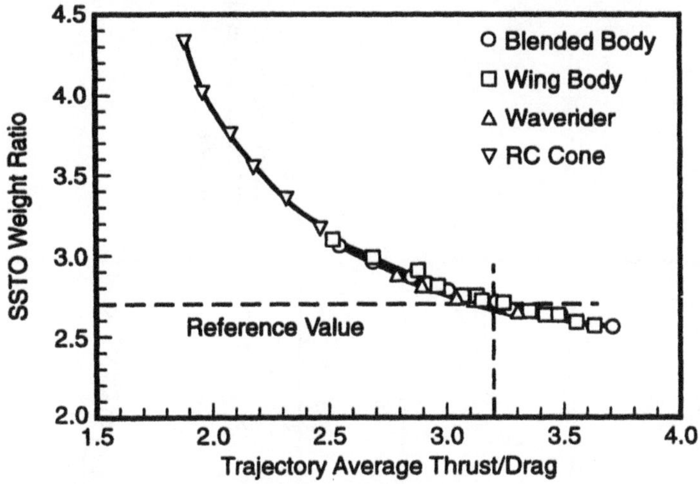

Fig. C.7 Correlation of weight ratio as a function of trajectory-averaged thrust-to-drag ratio for the reference configuration shapes.

D. Parameters of the Propulsion-Propellant Index for the Reference Concepts

The propulsion-propellant index, I_{PP}, defined by Eq. (27), involves the common factor of two parameters, the payload W_{PY} and the structure weight W_{ST}, as can be seen from Eqs. (25) and (26). In each case of the reference concepts, one can consider a specific value of payload, and set up relations as follows for the two parameters. Considering W_{PY}, one can write.

$$\frac{W_{PY}}{W_E} = C_1 \exp(C_2 \tau) \qquad (C.17)$$

For a 7-ton payload, the constants C_1 and C_2 have the following values based on the representative values in Fig. C.9.

Configuration	C_1	C_2
Blended body	0.121	2.1414
Wing-body	0.124	2.1046
Waverider	0.127	2.1002
Right circular cone	0.118	1.2871

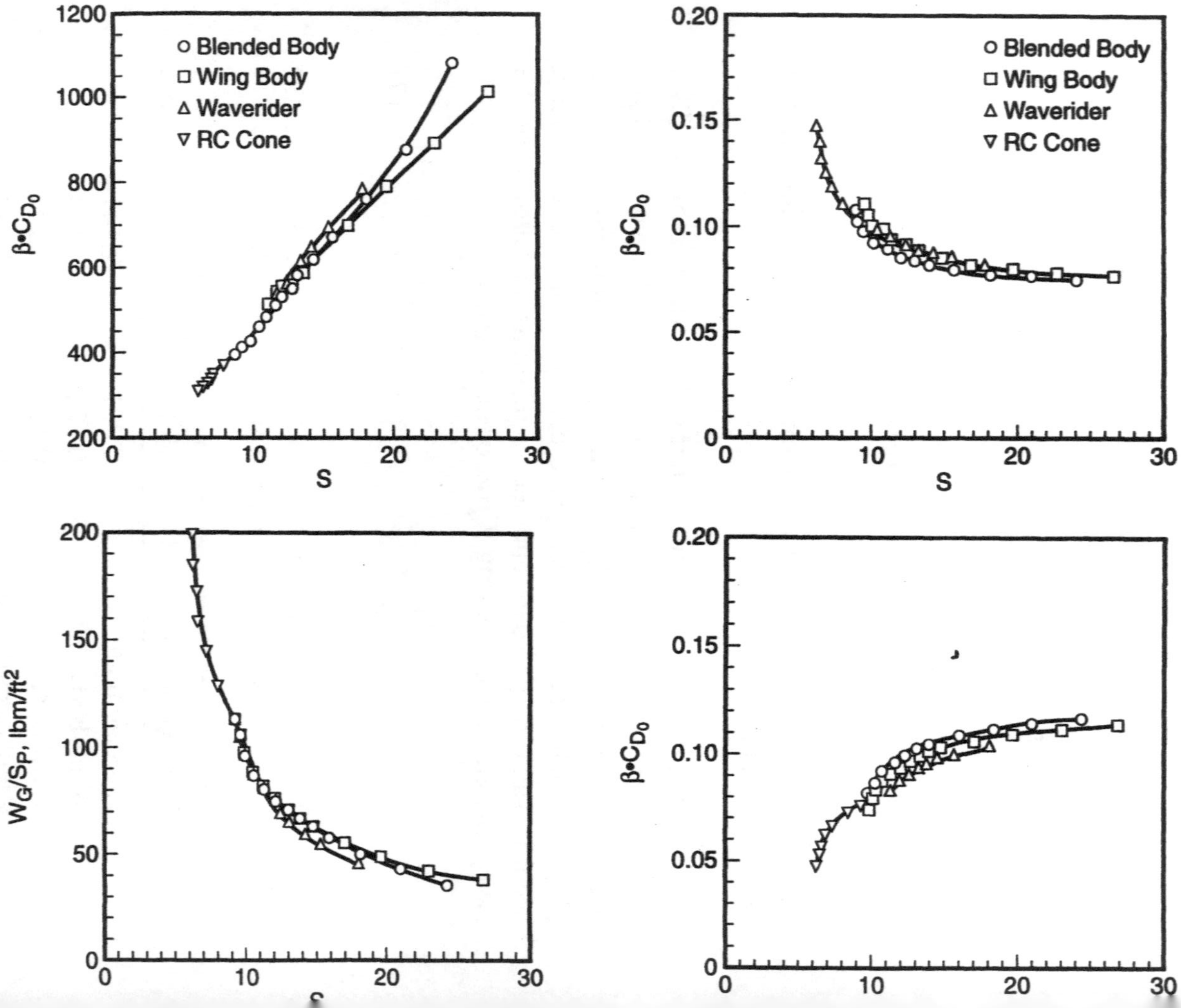

Fig. C.8 Drag characteristics of reference configuration shapes.

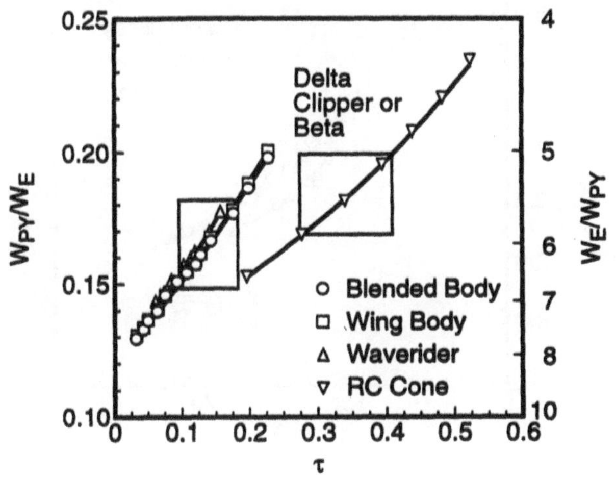

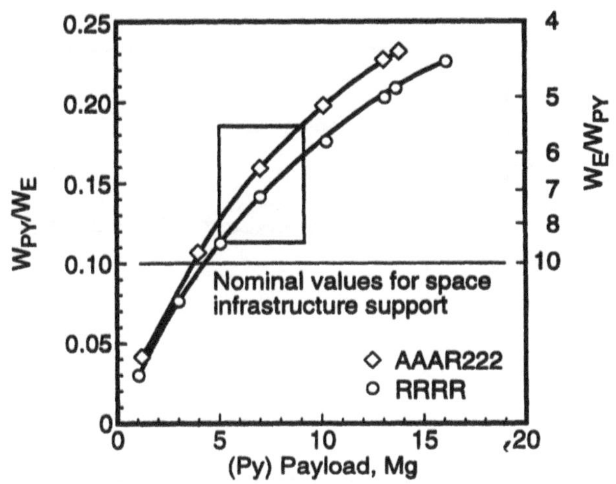

Fig. C.9. Example of a 7-ton payload vehicles with different reference configuration shapes and propulsion systems.

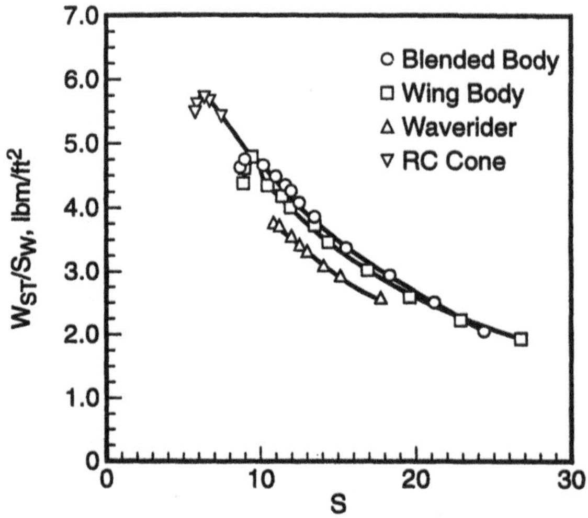
7.0
6.0
5.0
4.0
3.0
2.0
1.0
0
W_{ST}/S_W, lbm/ft^2
0
10
20
30
S
Blended Body
Wing Body
Waverider
RC Cone

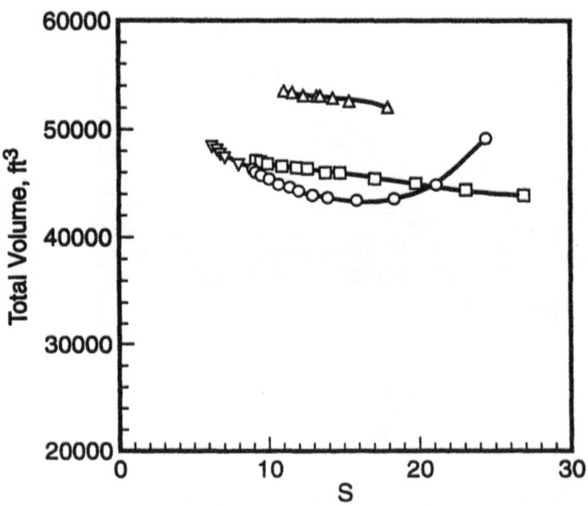
60000
50000
40000
30000
20000
Total Volume, ft^3
0
10
20
30
S

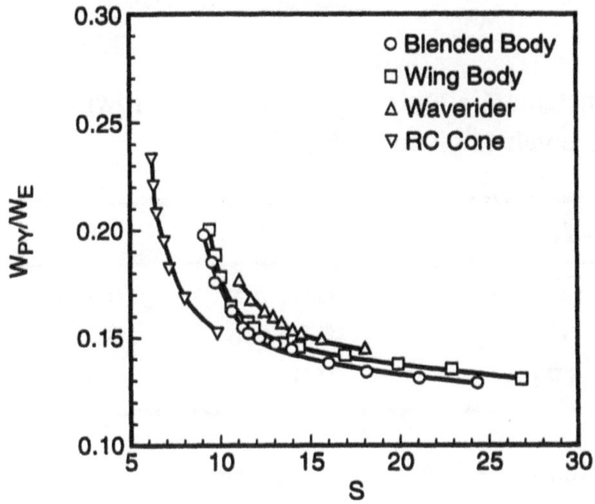
Blended Body
Wing Body
Waverider
RC Cone
W_{PY}/W_E
0.30
0.25
0.20
0.15
0.10
5
10
15
20
25
30
S

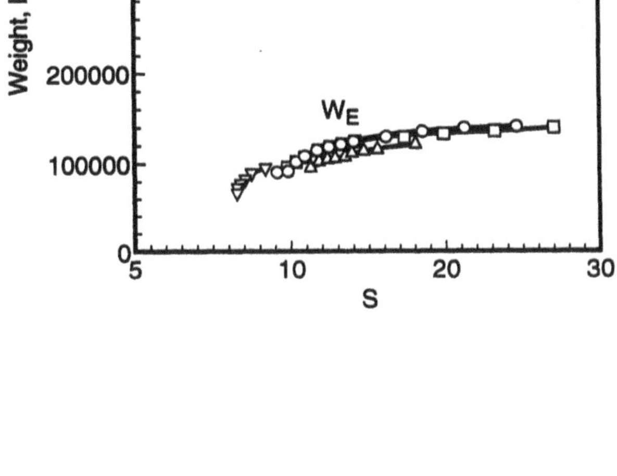
W_G
W_E
Weight, lbm
500000
400000
300000
200000
100000
0
5
10
20
30
S

For a value of $\tau = 0.10$ in the case of a blended-body configuration, one can write

$$\frac{W_{PY}}{W_E} = C_3 \exp\left[C_4 \ln\left(W_{PY}\right)\right] \qquad \text{(C.18)}$$

with the constants C_3 and C_4 given by the following, based on the representative values given in Fig. C.9.

Value	C_3	C_4
median	0.03508	0.7113
$AAAR_{22.2}$	0.03786	0.7069
RRRR	0.03208	0.7173

Then for different payloads, one can write

$$\frac{W_{PY} / W_E}{\left(W_{PY} / W_E\right)_{7\text{ton}}} = \frac{\exp\left[C_4 \ln\left(W_{PY}\right)\right]}{3.001} \qquad \text{(C.19)}$$

Considering the variation of both τ and W_{PY}, one can write, combining Eqs. (C.18) and (C.19),

$$\frac{W_{PY}}{W_E} = C_1 \exp\left(C_2 \tau\right) \frac{\exp\left[C_4 \ln\left(W_{PY}\right)\right]}{3.001} \qquad \text{(C.20)}$$

Next, considering W_{ST}, one can define a correlation as follows:

$$\frac{W_{ST}}{W_E} = K_{ST}\, S_p^{0.138} \left(1 - \frac{W_{EN} + W_{SE}}{W_E}\right) \qquad \text{(C.21)}$$

The system weight $\left(W_{EN} + W_{SE}\right)$ is defined in Eq. (19). The term K_{ST} may be written as

$$K_{ST} = 0.228\tau^{0.206} \pm 0.03 \qquad \text{(C.22)}$$

and is shown in Fig. C.10 as a function of τ. The figure includes examples of some aircraft, and the results obtained using the sizing program.

Several observations can be made from Fig. C.10:

1) The advantage of the right circular cone is volume at the expense of increased drag. For a vertical takeoff and landing rocket the drag is less important than for an airbreather, for which the thrust-to-drag ratio is critical.

2) The payload fraction can be increased with the size of the vehicle, whereas vehicle slenderness causes a reduction in the payload fraction. It is possible that smaller and larger payload vehicles may have the same empty weight. For example, a 1.5-ton payload may require an empty weight equal to 22 times the payload, whereas a 3.5-ton payload vehicle may require an empty weight of 10 times the payload.

It may be recognized that such vehicle weight scaling may be particularly significant in regard to demonstrators for vehicles with various missions. Thus, in the case of a cruise vehicle, a demonstrator and the cruise vehicle have the same performance, but are not required to have the same range, and scale with size in regard to various weight parameters. However, in the case of an

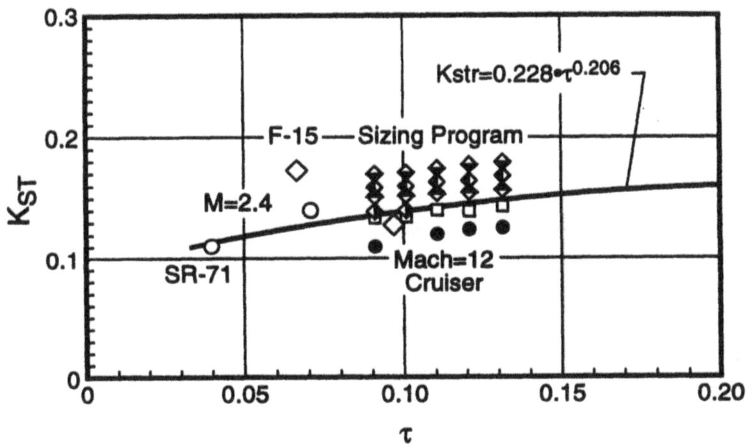

Fig. C.10 Some estimates from the sizing program by use with Eqs. (C.21) and (C.22).

accelerating vehicle, a demonstrator should be expected to fulfill both performance and range, and, therefore, do not scale with size. In fact, there is a penalty for the small size of the demonstrator.

It is also significant to note that advances in structural materials have to be sufficiently large before small vehicles can be attempted. In fact, a large vehicle with low strength materials may have a greater possibility of convergence than a smaller vehicle based on the use of materials with insufficiently large structural strength.

E. Estimated Values of I_{PP} and Vehicle Size

Some reference values of I_{PP} based on the projections of data from the 1960s and 1970s are as follows:

$$I_{PP} = 10\frac{\rho_P}{(R_W - 1)\mu} = 10\frac{\omega}{0.228\tau^{1.206}\upsilon}\,\frac{1 + (W_{PY}/W_{WE})}{S_p^{0.7097}} \tag{C.23a}$$

$$I_{PP} = 10\frac{8.35}{1.95 \cdot 3.5} = 10\frac{2.9}{0.228 \cdot 0.11 \cdot 0.41}\,\frac{1.14}{S_p^{0.7007}} \tag{C.23b}$$

$$I_{PP} = 12.2 = \frac{3220}{S_p^{0.7007}} \quad \text{or} \quad S_p = 2580\,\text{ft}^2 \tag{C.23c}$$

F. Propellant Volume

Based on Eqs. (1) and (12), one may seek a relation between the payload and the propellant volume required for the mission with given concept configuration, and propulsion-propellant system. The sizing program results show that the volume of the propellant is nearly constant for all of the

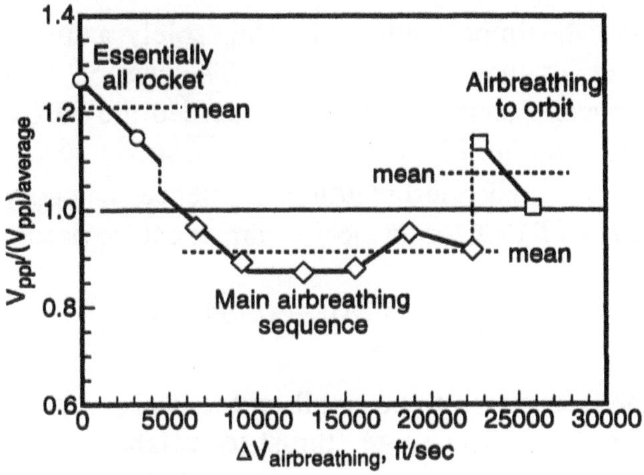

Fig. C.11 Influence on volume of propellant for ΔV obtained with airbreathing propulsion.

propulsion-propellant schemes of Table C.1. Thus, one may write

$$\mathcal{V}_P = \left(\mathcal{V}_P\right)_{W_{PY}=0} + \frac{\partial \mathcal{V}_P}{\partial W_{PY}} W_{PY} \qquad \text{(C.24)}$$

For the 10 different propulsion systems investigated with the sizing program, the average propellant volume was found to be 41,621 ft^3 for a 30,000-lb payload. The variation for the different propulsion systems is shown in Fig. C.11. The figure shows the mean volume ratio for three propulsion categories, namely, the essentially rocket category, the main airbreathing sequence category, and the airbreathing to near-orbital velocity category.

For category 1, one may assume

$$\mathcal{V}_P = 20{,}591 + 0.989 W_{PY} \qquad \text{(C.25)}$$

It may be feasible to incorporate an airbreather in this category up to a low speed of 4500 ft / s. The mass of an airbreathing engine with heavy turbomachinery is an issue in its use. Another issue is the occurrence of a pinch point at transonic speeds for an SSTO vehicle, and the impossibility of using solely a turbomachinery-based engine in that speed regime. The use of a rocket-based combined-cycle system is indicated provided the RBCC system is also utilized at higher flight speeds.

For the main airbreathing sequence category 2, it is assumed that an RBCC or an ejector ramjet can be used, and

$$\mathcal{V}_P = 15{,}731 + 0.734 W_{PY} \tag{C.26}$$

The volume is assumed to be small on the basis of the high thrust-to-weight ratio. The high thrust-to-weight ratio RBCC is applicable over the flight velocity range of 5000 – 22,000 ft / s.

Last, in the case of category 3, it is assumed that

$$\mathcal{V}_P = 21{,}575 + 0.763 W_{PY} \tag{C.27}$$

The effective I_{SP} can be expected to drop for an airbreather at flight speeds in excess of 22,000 ft / s. Thus there arises a hypersonic pinch point in this case, and a consequent increase in vehicle volume, further accentuated by the use of a low-density fuel.

Generalizing the foregoing, one may write

$$\mathcal{V}_P = C_7 + C_8 W_{PY} \tag{C.28}$$

Then, using Eq. (C.7), it follows that

$$S_P = \left(\frac{\mathcal{V}_P}{\upsilon_P \tau}\right)^{1/1.5717} = \left(\frac{\mathcal{V}_P}{\upsilon_P \tau}\right)^{0.63625} \left(\frac{C_7 + C_8 W_{PY}}{\upsilon_P \tau}\right)^{0.63625} \tag{C.29}$$

Thus a relation is established among S_P and $\mathcal{V}_P$ on one hand, and S_P and W_{PY} on the other, with $\left(\upsilon_P \tau\right)$ appearing as a parameter set in each. For υ_P, one can write, as stated in the text after Eq. (C.10), the following:

$$\upsilon_P = \upsilon_P \text{ (for high-density payload vehicle)} - 0.07 \tag{C.30}$$

G. Propulsion-Propellant Index

It is possible to establish a relation between a change in τ, relative to a reference value of τ, and the change in the I_{PP}, the propulsion-propellant index, as follows.

A reference vehicle is considered with a value of $\tau = \tau_{ref}$, $\beta C_{DO} = (\beta C_{DO})_{ref}$, and $T/D = T/D_{ref}$, and with a propulsion-propellant system giving $I_{SP} = (I_{SP})_{ref}$, and, hence, $I_{SPE} = (I_{SPE})_{ref}$ based on $(T/D)_{ref}$. If τ is changed, one can write for the new vehicle configuration

$$\frac{T}{D} = \left(\frac{T}{D}\right)_{ref} \frac{(\beta C_{DO})_{ref}}{\beta C_{DO}} \tag{C.31}$$

where βC_{DO} is the value with the changed value of τ.

Similarly, considering the weight ratio, R_W, one can write for the new vehicle

$$R_W = (R_W)_{ref} \exp\left[\frac{I_{SPE(ref)}}{I_{SP}}\right] \tag{C.32}$$

Hence, I_{PP} for the new vehicle can be expressed as follows:

$$I_{PP} = (I_{PP})_{ref} \frac{(R_W - 1)_{ref}}{R_W - 1} \tag{C.33}$$

References

[1]Johnson, D., "Beyond the X-30: Incorporating Mission Capability," AIAA 91-5078, Orlando, FL, 1991.

[2]Plokhikh, V. P., "Sensitivity Analysis of SSTO Reusable Vehicle Parameters," IAF-89-223, Malaga, 1989.

[3]Schindel, L., "Design of High Performance Ramjet or Scramjet Powered Vehicles," AIAA 89-0379, Reno, NV, 1989.

[4]Frederick, J., Sutton, R., and Martens, R., "Turbine Engine Selection Procedure," *Proceedings of the 3rd International Symposium on Airbreathing Engines*, DGLR Fachbuch Nr. 6, 1976.

[5]Czysz, P., "Rocket-Based Combined Cycles Offer Additional Options," *Proceedings of the 11th International Symposium on Airbreathing Engines*, Tokyo, 1993.

[6]Czysz, P., and Murthy, S. N. B., "Energy Analysis of High-Speed Flight Systems," in *High-Speed Flight Propulsion Systems*, Progress in Aeronautics and Astronautics, Vol. 137, American Institute of Aeronautics and Astronautics, Washington D.C., 1991.

[7]Curran, E. T., "The Potential and Practicality of High Speed Combined Cycle Engines," in *Hypersonic Combined Cycle Propulsion*, AGARD Conference Proceedings, AGARD No. 479, Neuilly-sur-Saine, 1990, pp. K 1-9.

[8]Maita, M., Mosi, T., and Okhami, Y., "System Studies on SSTO Space Plane," AIAA 91-5012, Orlando, FL, December 1991.

[9]Nguyen, H. N., and Norton, J. E., "Performance Assessment of a Reusable, Winged, Variable-Stage Launch System," IAF 92-0864, Washington, D.C., 1992.

[10]Staubacher, W., and Wimbauer, J., "Design Sensitivities of Airbreathing Vehicles," AIAA 93-5099, Munich, 1993.

[11]Kerrebrock, J. L., "Some Readily Quantifiable Aspects of Scramjet Engine Performance," *Journal of Propulsion and Power*, Vol. 8, No. 5, Sept.-Oct. 1992.

[12]Billig, F. S., "Propulsion System from Take-Off to High-Speed Flight," in *High-Speed Flight Propulsion Systems*, Progress in Astronautics and Aeronautics, Vol. 137, American Institute of Aeronautics and Astronautics, Washington, D.C., 1991.

[13]Vandenkerkhove, J., and Czysz, P., "Launcher Vehicle Sizing Program," included in this volume.

[14]Billig, F. S., and Van Wie, D., "Efficiency Parameters for Inlets Operating at Hypersonic Speeds," *Proceedings of the Eighth International Symposium on Airbreathing Engines*, Cincinnati, OH, 1987.

[15]Kutschenreuter, P. H., Subramanian, S. V., Gaeta, Jr. R. J., Hickey, P. K., and Davis, J. A., "A Design Approach to High Mach Number Scramjet Performance," AIAA 92-4248, Hilton Head, NC, 1992.

[16]Builder, C. H., "On the Thermodynamic Spectrum of Airbreathing Engines," AIAA 64-243, 1964.

[17]Curran, E. T., Leingang, J., Carreiro, L., and Petters, D., "Further Studies of Kinetic Energy Methods in High Speed Ramjet Cycle Analysis," AIAA 92-3807, Nashville, TN, 1992.

[18]Riggins, D. W., and McClinton, C. R., "Thrust Modeling for Hypersonic Engines," AIAA 95-6081, Chattanooga, TN, 1995.

[19]Heiser, W. H., and Pratt, D. T., *Hypersonic Airbreathing Propulsion*, AIAA Education Series, American Institute of Aeronautics and Astronautics, Washington, DC, 1994.

[20]Swithenbank, J., Eames, I. W., Chin, S. B., Ewan, B. C. R., Yang, Z., Cao, J., and Zhao, X., "Turbulent Mixing in Supersonic Combustion Systems," in *High Speed Flight Propulsion Systems*, Progress in Astronautics and Aeronautics, Vol. 137, American Institute of Aeronautics and Astronautics, Washington, DC, 1991.

[21]Harney, D. J., "Similarity of Nonequilibrium-Equilibrium in Supersonic Flows," Report No. FDM-TM-67-1, Air Force Flight Dynamics Laboratory, Wright-Patterson Air Force Base, 1967.

[22]Hankey, W. L., Neumann, R. D., and Flynn, E. H., "Design Procedures for Computing Aerodynamic Heating at Hypersonic Speeds," Report No. ASD TR 62-71, Air Force Flight Dynamics Laboratory, Wright-Patterson Air Force Base, 1962.

[23]Sunakawa, M., and Sakamoto, A., "Current Status of R&D in Japan on Materials for Space Planes," AIAA 91-5096, Orlando, FL, 1991.

[24]Ronald, T., "Structural Materials for NASP," AIAA 91-5101, Orlando, FL, 1991.

[25]Grallert, H., and Vollimer, K., "Conceptual Design of the Thermal Protection System for the Reference Concept SANGER by Means of Advanced Methods," AIAA 93-5085, Munich, 1993.

[26]Anon., *Hypersonic Research Facilities Study*, Vol. II, Pt. 2, Flight Vehicle Synthesis, Report No. NASA CR 114324, OART-NASA, 1970.

[27]Anon., *Mission Requirements of Lifting Systems. A Study of Engineering Aspects*, Report No. B-947, NAS 9-3562, August 1963.

[28]Anon., *Development of Military Hypersonic Flight Test Experiment*, Report No. NA-78-488, 1979.

[29]Escher, W. J. D., "Synerjet for Earth/Orbit Propulsion: Revisiting the 1966 NASA/Marquardt Composite (Airbreathing/Rocket) Propulsion System Study", SAE 85-1163, Aerospace Vehicle Requirements Conference, Washington, DC, 1985.

[30]Anon., "Air Augmented Rocket Propulsion Concepts," Report No. AFAL-TR-88-004, 1988.

Author Index

PROGRESS IN ASTRONAUTICS AND AERONAUTICS SERIES VOLUMES

***1. Solid Propellant Rocket Research (1960)**
Martin Summerfield
Princeton University

***2. Liquid Rockets and Propellants (1960)**
Loren E. Bollinger
Ohio State University
Martin Goldsmith
The Rand Corp.
Alexis W. Lemmon Jr.
Battelle Memorial Institute

***3. Energy Conversion for Space Power (1961)**
Nathan W. Snyder
Institute for Defense Analyses

***4. Space Power Systems (1961)**
Nathan W. Snyder
Institute for Defense Analyses

***5. Electrostatic Propulsion (1961)**
David B. Langmuir
Space Technology Laboratories, Inc.
Ernst Stuhlinger
NASA George C. Marshall Space Flight Center
J. M. Sellen Jr.
Space Technology Laboratories, Inc.

***6. Detonation and Two-Phase Flow (1962)**
S. S. Penner
California Institute of Technology
F. A. Williams
Harvard University

***7. Hypersonic Flow Research (1962)**
Frederick R. Riddell
AVCO Corp.

***8. Guidance and Control (1962)**
Robert E. Roberson
Consultant
James S. Farrior
Lockheed Missiles and Space Co.

***9. Electric Propulsion Development (1963)**
Ernst Stuhlinger
NASA George C. Marshall Space Flight Center

***10. Technology of Lunar Exploration (1963)**
Clifford I. Cumming
Harold R. Lawrence
Jet Propulsion Laboratory

***11. Power Systems for Space Flight (1963)**
Morris A. Zipkin
Russell N. Edwards
General Electric Co.

***12. Ionization in High-Temperature Gases (1963)**
Kurt E. Shuler, Editor
National Bureau of Standards
John B. Fenn
Associate Editor
Princeton University

***13. Guidance and Control–II (1964)**
Robert C. Langford
General Precision Inc.
Charles J. Mundo
Institute of Naval Studies

***14. Celestial Mechanics and Astrodynamics (1964)**
Victor G. Szebehely
Yale University Observatory

***15. Heterogeneous Combustion (1964)**
Hans G. Wolfhard
Institute for Defense Analyses
Irvin Glassman
Princeton University
Leon Green Jr.
Air Force Systems Command

***16. Space Power Systems Engineering (1966)**
George C. Szego
Institute for Defense Analyses
J. Edward Taylor
TRW Inc.

***17. Methods in Astrodynamics and Celestial Mechanics (1966)**
Raynor L. Duncombe
U.S. Naval Observatory
Victor G. Szebehely
Yale University Observatory

***18. Thermophysics and Temperature Control of Spacecraft and Entry Vehicles (1966)**
Gerhard B. Heller
NASA George C. Marshall Space Flight Center

***19. Communication Satellite Systems Technology (1966)**
Richard B. Marsten
Radio Corporation of America

***20. Thermophysics of Spacecraft and Planetary Bodies: Radiation Properties of Solids and the Electromagnetic Radiation Environment in Space (1967)**
Gerhard B. Heller
NASA George C. Marshall Space Flight Center

*Out of print.

***21. Thermal Design Principles of Spacecraft and Entry Bodies (1969)**
Jerry T. Bevans
TRW Systems

***22. Stratospheric Circulation (1969)**
Willis L. Webb
Atmospheric Sciences Laboratory, White Sands, and University of Texas at El Paso

***23. Thermophysics: Applications to Thermal Design of Spacecraft (1970)**
Jerry T. Bevans
TRW Systems

24. Heat Transfer and Spacecraft Thermal Control (1971)
John W. Lucas
Jet Propulsion Laboratory

25. Communication Satellites for the 70's: Technology (1971)
Nathaniel E. Feldman
The Rand Corp.
Charles M. Kelly
The Aerospace Corp.

26. Communication Satellites for the 70's: Systems (1971)
Nathaniel E. Feldman
The Rand Corp.
Charles M. Kelly
The Aerospace Corp.

27. Thermospheric Circulation (1972)
Willis L. Webb
Atmospheric Sciences Laboratory, White Sands, and University of Texas at El Paso

28. Thermal Characteristics of the Moon (1972)
John W. Lucas
Jet Propulsion Laboratory

***29. Fundamentals of Spacecraft Thermal Design (1972)**
John W. Lucas
Jet Propulsion Laboratory

***30. Solar Activity Observations and Predictions (1972)**
Patrick S. McIntosh
Murray Dryer
Environmental Research Laboratories, National Oceanic and Atmospheric Administration

31. Thermal Control and Radiation (1973)
Chang-Lin Tien
University of California at Berkeley

32. Communications Satellite Systems (1974)
P.L. Bargellini
COMSAT Laboratories

33. Communications Satellite Technology (1974)
P.L. Bargellini
COMSAT Laboratories

***34. Instrumentation for Airbreathing Propulsion (1974)**
Allen E. Fuhs
Naval Postgraduate School
Marshall Kingery
Arnold Engineering Development Center

35. Thermophysics and Spacecraft Thermal Control (1974)
Robert G. Hering
University of Iowa

36. Thermal Pollution Analysis (1975)
Joseph A. Schetz
Virginia Polytechnic Institute
ISBN 0-915928-00-0

***37. Aeroacoustics: Jet and Combustion Noise; Duct Acoustics (1975)**
Henry T. Nagamatsu, Editor
General Electric Research and Development Center
Jack V. O'Keefe, Associate Editor
The Boeing Co.
Ira R. Schwartz, Associate Editor
NASA Ames Research Center
ISBN 0-915928-01-9

***38. Aeroacoustics: Fan, STOL, and Boundary Layer Noise; Sonic Boom; Aeroacoustics Instrumentation (1975)**
Henry T. Nagamatsu, Editor
General Electric Research and Development Center
Jack V. O'Keefe, Associate Editor
The Boeing Co.
Ira R. Schwartz, Associate Editor
NASA Ames Research Center
ISBN 0-915928-02-7

39. Heat Transfer with Thermal Control Applications (1975)
M. Michael Yovanovich
University of Waterloo
ISBN 0-915928-03-5

***40. Aerodynamics of Base Combustion (1976)**
S.N.B. Murthy, Editor
J. R. Osborn, Associate Editor
Purdue University
A.W. Barrows
J. R. Ward,
Associate Editors
Ballistics Research Laboratories
ISBN 0-915928-04-3

---

*Out of print.

41. Communications Satellite Developments: Systems (1976)
Gilbert E. LaVean
Defense Communications Agency
William G. Schmidt
CML Satellite Corp.
ISBN 0-915928-05-1

***42. Communications Satellite Developments: Technology (1976)**
William G. Schmidt
CML Satellite Corp.
Gilbert E. LaVean
Defense Communications Agency
ISBN 0-915928-06-X

***43. Aeroacoustics: Jet Noise, Combustion and Core Engine Noise (1976)**
Ira R. Schwartz, Editor
NASA Ames Research Center
Henry T. Nagamatsu, Associate Editor
General Electric Research and Development Center
Warren C. Strahle, Associate Editor
Georgia Institute of Technology
ISBN 0-915928-07-8

***44. Aeroacoustics: Fan Noise and Control; Duct Acoustics; Rotor Noise (1976)**
Ira R. Schwartz, Editor
NASA Ames Research Center
Henry T. Nagamatsu, Associate Editor
General Electric Research and Development Center
Warren C. Strahle, Associate Editor
Georgia Institute of Technology
ISBN 0-915928-08-6

***45. Aeroacoustics: STOL Noise; Airframe and Airfoil Noise (1976)**
Ira R. Schwartz, Editor
NASA Ames Research Center
Henry T. Nagamatsu, Associate Editor
General Electric Research and Development Center
Warren C. Strahle, Associate Editor
Georgia Institute of Technology
ISBN 0-915928-09-4

***46. Aeroacoustics: Acoustic Wave Propagation; Aircraft Noise Prediction; Aeroacoustic Instrumentation (1976)**
Ira R. Schwartz, Editor
NASA Ames Research Center
Henry T. Nagamatsu, Associate Editor
General Electric Research and Development Center
Warren C. Strahle, Associate Editor
Georgia Institute of Technology.
ISBN 0-915928-10-8

***47. Spacecraft Charging by Magnetospheric Plasmas (1976)**
Alan Rosen
TRW Inc.
ISBN 0-915928-11-6

48. Scientific Investigations on the Skylab Satellite (1976)
Marion I. Kent
Ernst Stuhlinger
NASA George C. Marshall Space Flight Center
Shi-Tsan Wu
University of Alabama
ISBN 0-915928-12-4

49. Radiative Transfer and Thermal Control (1976)
Allie M. Smith
ARO Inc.
ISBN 0-915928-13-2

***50. Exploration of the Outer Solar System (1976)**
Eugene W. Greenstadt
TRW Inc.
Murray Dryer
National Oceanic and Atmospheric Administration
Devrie S. Intriligator
University of Southern California
ISBN 0-915928-14-0

***51. Rarefied Gas Dynamics, Parts I and II(two volumes) (1977)**
J. Leith Potter
ARO Inc.
ISBN 0-915928-15-9

***52. Materials Sciences in Space with Application to Space Processing (1977)**
Leo Steg
General Electric Co.
ISBN 0-915928-16-7

***53. Experimental Diagnostics in Gas Phase Combustion Systems (1977)**
Ben T. Zinn, Editor
Georgia Institute of Technology
Craig T. Bowman, Associate Editor
Stanford University
Daniel L. Hartley, Associate Editor
Sandia Laboratories
Edward W. Price, Associate Editor
Georgia Institute of Technology
James G. Skifstad, Associate Editor
Purdue University
ISBN 0-915928-18-3

*Out of print.

54. Satellite Communication: Future Systems (1977)
David Jarett
TRW Inc.
ISBN 0-915928-18-3

***55. Satellite Communications: Advanced Technologies (1977)**
David Jarett
TRW Inc.
ISBN 0-915928-19-1

***56. Thermophysics of Spacecraft and Outer Planer Entry Probes (1977)**
Allied M. Smith
ARO Inc.
ISBN 0-915928-20-5

***57. Space-Based Manufacturing from Nonterrestrial Materials (1977)**
Gerald K. O'Neill, Editor
Brian O'Leary, Assistant Editor
Princeton University
ISBN 0-915928-21-3

***58. Turbulent Combustion (1978)**
Lawrence A. Kennedy
State University of New York at Buffalo
ISBN 0-915928-22-1

***59. Aerodynamic Heating and Thermal Protection Systems (1978)**
Leroy S. Fletcher
University of Virginia
ISBN 0-915928-23-X

60. Heat Transfer and Thermal Control Systems (1978)
Leroy S. Fletcher
University of Virginia
ISBN 0-915928-24-8

***61. Radiation Energy Conversion in Space (1978)**
Kenneth W. Billman
NASA Ames Research Center
ISBN 0-915928-26-4

***62. Alternative Hydrocarbon Fuels: Combustion and Chemical Kinetics (1978)**
Craig T. Bowman
Stanford University
Jorgen Birkeland
Department of Energy
ISBN 0-915928-25-6

***63. Experimental Diagnostics in Combustion of Solids (1978)**
Thomas L. Boggs
Naval Weapons Center
Ben T. Zinn
Georgia Institute of Technology
ISBN 0-915928-28-0

***64. Outer Planet Entry Heating and Thermal Protection (1979)**
Raymond Viskanta
Purdue University
ISBN 0-915928-29-9

***65. Thermophysics and Thermal Control (1979)**
Raymond Viskanta
Purdue University
ISBN 0-915928-30-2

***66. Interior Ballistics of Guns (1979)**
Herman Krier
University of Illinois at Urbana-Champaign
Martin Summerfield
New York University
ISBN 0-915928-32-9

***67. Remote Sensing of Earth from Space: Role of "Smart Sensors" (1979)**
Roger A. Breckenridge
NASA Langley Research Center
ISBN 0-915928-33-7

68. Injection and Mixing in Turbulent Flow (1980)
Joseph A. Schetz
Virginia Polytechnic Institute and State University
ISBN 0-915928-35-3

***69. Entry Heating and Thermal Protection (1980)**
Walter B. Olstad
NASA Headquarters
ISBN 0-915928-38-8

***70. Heat Transfer, Thermal Control, and Heat Pipes (1980)**
Walter B. Olstad
NASA Headquarters
ISBN 0-915928-39-6

***71. Space Systems and Their Interactions with Earth's Space Environment (1980)**
Henry B. Garrett
Charles P. Pike
Hanscom Air Force Base
ISBN 0-915928-41-8

***72. Viscous Flow Drag Reduction (1980)**
Gary R. Hough
Vought Advanced Technology Center
ISBN 0-915928-44-2

***73. Combustion Experiments in a Zero-Gravity Laboratory (1981)**
Thomas H. Cochran
NASA Lewis Research Center
ISBN 0-915928-48-5

***74. Rarefied Gas Dynamics, Parts I and II (two volumes) (1981)**
Sam S. Fisher
University of Virginia
ISBN 0-915928-51-5

---

*Out of print.

***75. Gasdynamics of Detonations and Explosions (1981)**
J. R. Bowen
University of Wisconsin at Madison
N. Manson
Universite de Poitiers
A. K. Oppenheim
University of California at Berkeley
R. I. Soloukhin
Institute of Heat and Mass Transfer, BSSR Academy of Sciences
ISBN 0-915928-46-9

76. Combustion in Reactive Systems (1981)
J. R. Bowen
University of Wisconsin at Madison
N. Manson
Universite de Poitiers
A. K. Oppenheim
University of California at Berkeley
R. I. Soloukhin
Institute of Heat and Mass Transfer, BSSR Academy of Sciences
ISBN 0-915928-47-7

***77. Aerothermodynamics and Planetary Entry (1981)**
A. L. Crosbie
University of Missouri-Rolla
ISBN 0-915928-52-3

***78. Heat Transfer and Thermal Control (1981)**
A. L. Crosbie
University of Missouri-Rolla
ISBN 0-915928-53-1

***79. Electric Propulsion and Its Applications to Space Missions**
(1981)
Robert C. Finke
NASA Lewis Research Center
ISBN 0-915928-55-8

***80. Aero-Optical Phenomena (1982)**
Keith G. Gilbert
Leonard J. Otten
Air Force Weapons Laboratory
ISBN 0-915928-60-4

***81. Transonic Aerodynamics (1982)**
David Nixon
Nielsen Engineering & Research, Inc.
ISBN 0-915928-65-5

***82. Thermophysics of Atmospheric Entry (1982)**
T. E. Horton
University of Mississippi
ISBN 0-915928-66-3

***83. Spacecraft Radiative Transfer and Temperature Control (1982)**
T. E. Horton
University of Mississippi
ISBN 0-915928-67-1

84. Liquid-Metal Flows and Magneto-hydrodynamics (1983)
H. Branover
Ben-Gurion University of the Negev
P. S. Lykoudis
Purdue University
A. Yakhot
Ben-Gurion University of the Negev
ISBN 0-915928-70-1

***85. Entry Vehicle Heating and Thermal Protection Systems: Space Shuttle, Solar Starprobe, Jupiter Galileo Probe (1983)**
Paul E. Bauer
McDonnell Douglas Astronautics Co.
Howard E. Collicott
The Boeing Co.
ISBN 0-915928-74-4

***86. Spacecraft Thermal Control, Design, and Operation (1983)**
Howard E. Collicott
The Boeing Co.
Paul E. Bauer
McDonnell Douglas Astronautics Co.
ISBN 0-915928-75-2

***87. Shock Waves, Explosions, and Detonations (1983)**
J. R. Bowen
University of Washington
N. Manson
Universite de Poitiers
A. K. Oppenheim
University of California at Berkeley
R. I. Soloukhin
Institute of Heat and Mass Transfer, BSSR Academy of Sciences
ISBN 0-915928-76-0

88. Flames, Lasers, and Reactive Systems (1983)
J. R. Bowen
University of Washington
N. Manson
Universite de Poitiers
A. K. Oppenheim
University of California at Berkeley
R. I. Soloukhin
Institute of Heat and Mass Transfer, BSSR Academy of Sciences
ISBN 0-915928-77-9

***89. Orbit-Raising and Maneuvering Propulsion: Research Status and Needs (1984)**
Leonard H. Caveny
Air Force Office of Scientific Research
ISBN 0-915928-82-5

*Out of print.

90. Fundamental of Solid-Propellant Combustion (1984)
Kenneth K. Kuo
Pennsylvania State University
Martin Summerfield
Princeton Combustion Research Laboratories, Inc.
ISBN 0-915928-84-1

91. Spacecraft Contamination: Sources and Prevention (1984)
J. A. Roux
University of Mississippi
T. D. McCay
NASA Marshall Space Flight Center
ISBN 0-915928-85-X

92. Combustion Diagnostics by Nonintrusive Methods (1984)
T. D. McCay
NASA Marshall Space Flight Center
J. A. Roux
University of Mississippi
ISBN 0-915928-86-8

93. The INTELSAT Global Satellite System (1984)
Joel Alper
COMSAT Corp.
Joseph Pelton
INTELSAT
ISBN 0-915928-90-6

94. Dynamics of Shock Waves, Explosions, and Detonations (1984)
J. R. Bowen
University of Washington
N. Manson
Universite de Poitiers
A. K. Oppenheim
University of California at Berkeley
R. I. Soloukhin
Institute of Heat and Mass Transfer, BSSR Academy of Sciences
ISBN 0-915928-91-4

95. Dynamics of Flames and Reactive Systems (1984)
J. R. Bowen
University of Washington
N. Manson
Universite de Poitiers
A. K. Oppenheim
University of California at Berkeley
R. I. Soloukhin
Institute of Heat and Mass Transfer, BSSR Academy of Sciences
ISBN 0-915928-92-2

96. Thermal Design of Aeroassisted Orbital Transfer Vehicles (1985)
H. F. Nelson
University of Missouri-Rolla
ISBN 0-915928-94-9

97. Monitoring Earth's Ocean, Land, and Atmosphere from Space–Sensors, Systems, and Applications (1985)
Abraham Schnapf
Aerospace Systems Engineering
ISBN 0-915928-98-1

98. Thrust and Drag: Its Prediction and Verification (1985)
Eugene E. Covert
Massachusetts Institute of Technology
C. R. James
Vought Corp.
William F. Kimzey
Sverdrup Technology AEDC Group
George K. Richey
U.S. Air Force
Eugene C. Rooney
U.S. Navy Department of Defense
ISBN 0-930403-00-2

99. Space Stations and Space Platforms–Concepts, Design, Infrastructure, and Uses (1985)
Ivan Bekey
Daniel Herman
NASA Headquarters
ISBN 0-930403-01-0

100. Single- and Multi-Phase Flows in an Electromagnetic Field: Energy, Metallurgical, and Solar Applications (1985)
Herman Branover
Ben-Gurion University of the Negev
Paul S. Lykoudis
Purdue University
Michael Mond
Ben-Gurion University of the Negev
ISBN 0-930403-04-5

101. MHD Energy Conversion: Physiotechnical Problems (1986)
V. A. Kirillin
A. E. Sheyndlin
Soviet Academy of Sciences
ISBN 0-930403-05-3

102. Numerical Methods for Engine-Airframe Integration (1986)
S.N.B. Murthy
Purdue University
Gerald C. Paynter
Boeing Airplane Co.
ISBN 0-930403-09-6

103. Thermophysical Aspects of Re-Entry Flows (1986)
James N. Moss
NASA Langley Research Center
Carl D. Scott
NASA Johnson Space Center
ISBN 0-930430-10-X

*Out of print.

***104. Tactical Missile Aerodynamics (1986)**
M. J. Hemsch
PRC Kentron, Inc.
J. N. Nielson
NASA Ames Research Center
ISBN 0-930403-13-4

105. Dynamics of Reactive Systems Part I: Flames and Configurations; Part II: Modeling and Heterogeneous Combustion (1986)
J. R. Bowen
University of Washington
J.-C. Leyer
Universite de Poitiers
R. I. Soloukhin
Institute of Heat and Mass Transfer, BSSR Academy of Sciences
ISBN 0-930403-14-2

106. Dynamics of Explosions (1986)
J. R. Bowen
University of Washington
J.-C. Leyer
Universite de Poitiers
R. I. Soloukhin
Institute of Heat and Mass Transfer, BSSR Academy of Sciences
ISBN 0-930403-15-0

***107. Spacecraft Dielectric Material Properties and Spacecraft Charging (1986)**
A.R. Frederickson
U.S. Air Force Rome Air Development Center
D. B. Cotts
SRI International
J. A. Wall
U.S. Air Force Rome Air Development Center
F. L. Bouquet
Jet Propulsion Laboratory, California Institute of Technology
ISBN 0-930403-17-7

***108. Opportunities for Academic Research in a Low-Gravity Environment (1986)**
George A. Hazelrigg
National Science Foundation
Joseph M. Reynolds
Louisiana State University
ISBN 0-930403-18-5

109. Gun Propulsion Technology (1988)
Ludwig Stiefel
U.S. Army Armament Research, Development and Engineering Center
ISBN 0-930403-20-7

110. Commercial Opportunities in Space (1988)
F. Shahrokhi
K. E. Harwell
University of Tennessee Space Institute
C. C. Chao
National Cheng Kung University
ISBN 0-930403-39-8

111. Liquid-Metal Flows: Magnetohydrodynamics and Application (1988)
Herman Branover, Michael Mond, and Yeshajahu Unger
Ben-Gurion University of the Negev
ISBN 0-930403-43-6

112. Current Trends in Turbulence Research (1988)
Herman Branover, Micheal Mond, and Yeshajahu Unger
Ben-Gurion University of the Negev
ISBN 0-930403-44-4

113. Dynamics of Reactive Systems Part I: Flames; Part II: Heterogeneous Combustion and Applications (1988)
A. L. Kuhl
R & D Associates
J. R. Bowen
University of Washington
J.-C. Leyer
Universite de Poitiers
A. Borisov
USSR Academy of Sciences
ISBN 0-930403-46-0

114. Dynamics of Explosions (1988)
A. L. Kuhl
R & D Associates
J. R. Bowen
University of Washington
J.-C. Leyer
Universite de Poitiers
A. Borisov
USSR Academy of Sciences
ISBN 0-930403-47-9

115. Machine Intelligence and Autonomy for Aerospace (1988)
E. Heer
Heer Associates, Inc.
H. Lum
NASA Ames Research Center
ISBN 0-930403-48-7

116. Rarefied Gas Dynamics: Space Related Studies (1989)
E. P. Muntz
University of Southern California
D. P. Weaver
U.S. Air Force Astronautics Laboratory (AFSC)
D. H. Campbell
University of Dayton Research Institute
ISBN 0-930403-53-3

---

*Out of print.

117. Rarefied Gas Dynamics: Physical Phenomena (1989)
E. P. Muntz
University of Southernn California
D. P. Weaver
U.S. Air Force Astronautics Laboratory (AFSC)
D. H. Campbell
University of Dayton Research Institute
ISBN 0-930403-54-1

118. Rarefied Gas Dynamics: Theoretical and Computational Techniques (1989)
E.P. Muntz
University of Southernn California
D. P. Weaver
U.S. Air Force Astronautics Laboratory (AFSC)
D. H. Campbell
University of Dayton Research Institute
ISBN 0-930403-55-X

119. Test and Evaluation of the Tactical Missile (1989)
Emil J. Eichblatt Jr.
Pacific Missile Test Center
ISBN 0-930403-56-8

120. Unsteady transonic Aerodynamics (1989)
David Nixon
Nielsen Engineering & Research, Inc.
ISBN 0-930403-52-5

121. Orbital Debris from Upper-Stage Breakup (1989)
Joseph P. Loftus Jr.
NASA Johnson Space Center
ISBN 0-930403-58-4

122. Thermal-Hydraulics for Space Power, Propulsion and Thermal Management System Design (1989)
William J. Krotiuk
General Electric Co.
ISBN 0-930403-64-9

123. Viscous Drag Reduction in Boundary Layers (1990)
Dennis M. Bushnell
Jerry N. Hefner
NASA Langley Research Center
ISBN 0-930403-66-5

***124. Tactical and Strategic Missile Guidance (1990)**
Paul Zarchan
Charles Stark Draper Laboratory, Inc.
ISBN 0-930403-68-1

125. Applied Computational Aerodynamics (1990)
P. A. Henne
Douglas Aircraft Company
ISBN 0-930403-69-X

126. Space Commercialization: Launch Vehicles and Programs (1990)
F. Shahrokhi
University of Tennessee Space Institute
J. S. Greenberg
Princeton Synergetics Inc.
T. Al-Saud
Ministry of Defense and Aviation Kingdom of Saudi Arabia
ISBN 0-930403-75-4

127. Space Commercialization: Platforms and Processing (1990)
F. Shahrokhi
University of Tennessee Space Institute
G. Hazelrigg
National Science Foundation
R. Bayuzick
Vanderbilt University
ISBN 0-930403-76-2

128. Space Commercialization: Satellite Technology (1990)
F. Shahrokhi
University of Tennessee Space Institute
N. Jasentuliyana
United Nations
N. Tarabzouni
King Abulaziz City for Science and Technology
ISBN 0-930403-77-0

129. Mechanics and Control of Large Flexible Structures (1990)
John L. Junkins
Texas A&M University
ISBN 0-930403-73-8

130. Low-Gravity Fluid Dynamics and Transport Phenomena (1990)
Jean N. Koster
Robert L. Sani
University of Colorado at Boulder
ISBN 0-930403-74-6

131. Dynamics of Deflagrations and Reactive Systems: Flames (1991)
A. L. Kuhl
Lawrence Livermore National Laboratory
J.-C. Leyer
Universite de Poitiers
A. A. Borisov
USSR Academy of Sciences
W. A. Sirignano
University of California
ISBN 0-930403-95-9

---
*Out of print.

132. Dynamics of Deflagrations and Reactive Systems: Heterogeneous Combustion (1991)
A. L. Kuhl
Lawrence Livermore National Laboratory
J.-C. Leyer
Universite de Poitiers
A. A. Borisov
USSR Academy of Sciences
W. A. Sirignano
University of California
ISBN 0-930403-96-7

133. Dynamics of Detonations and Explosions: Detonations (1991)
A. L. Kuhl
Lawrence Livermore National Laboratory
J.-C. Leyer
Universite de Poitiers
A. A. Borisov
USSR Academy of Sciences
W. A. Sirignano
University of California
ISBN 0-930403-97-5

134. Dynamics of Detonations and Explosions: Explosion Phenomena (1991)
A. L. Kuhl
Lawrence Livermore National Laboratory
J.-C. Leyer
Universite de Poitiers
A. A. Borisov
USSR Academy of Sciences
W. A. Sirignano
University of California
ISBN 0-930403-98-3

135. Numerical Approaches to Combustion Modeling (1991)
Elaine S. Oran
Jay P. Boris
Naval Research Laboratory
ISBN 1-56347-004-7

136. Aerospace Software Engineering (1991)
Christine Anderson
U.S. Air Force Wright Laboratory
Merlin Dorfman
Lockheed Missiles & Space Company, Inc.
ISBN 1-56347-005-0

137. High-Speed Flight Propulsion Systems (1991)
S.N.B. Murthy
Purdue University
E.T. Curran
Wright Laboratory
ISBN 1-56347-011-X

138. Propagation of Intensive Laser Radiation in Clouds (1992)
O. A. Volkovitsky
Yu. S. Sedenov
L. P. Semenov
Institute of Experimental Meteorology
ISBN 1-56347-020-9

139. Gun Muzzle Blast and Flash (1992)
Gunter Klingenburg
Fraunhofer-Institut fur Kurzzeitdynamik, Ernst-Mach-Institut (EMI)
Joseph M. Heimerl
U.S. Army Ballistic Research Laboratory (BRL)
ISBN 1-56347-012-8

140. Thermal Structures and Materials for High-Speed Flight (1992)
Earl. A. Thornton
University of Virginia
ISBN 1-56347-017-9

141. Tactical Missile Aerodynamics: General Topics (1992)
Michael J. Hemsch
Lockheed Engineering & Sciences Company
ISBN 1-56347-015-2

142. Tactical Missile Aerodynamics: Prediction Methodology (1992)
Michael R. Mendenhall
Nielsen Engineering & Research, Inc.
ISBN 1-56347-016-0

143. Nonsteady Burning and Combustion Stability of Solid Propellants (1992)
Luigi De Luca
Politecnico di Milano
Edward W. Price
Georgia Institute of Technology
Martin Summerfield
Princeton Combustion Research Laboratories, Inc.
ISBN 1-56347-014-4

144. Space Economics (1992)
Joel S. Greenberg
Princeton Synergetics, Inc.
Henry R. Hertzfeld
HRH Associates
ISBN 1-56347-042-X

145. Mars: Past, Present, and Future (1992)
E. Brian Pritchard
NASA Langley Research Center
ISBN 1-56347-043-8

146. Computational Nonlinear Mechanics in Aerospace Engineering (1992)
Satya N. Atluri
Georgia Institute of Technology
ISBN 1-56347-044-6

147. Modern Engineering for Design of Liquid-Propellant Rocket Engines (1992)
Dieter K. Huzel
David H. Huang
ISBN 1-56347-013-6

*Out of print.

148. Metallurgical Technologies, Energy Conversion, and Magneto-hydrodynamic Flows (1993)
Herman Branover
Yeshajahu Unger
Ben-Gurion University of the Negev
ISBN 1-56347-019-5

149. Advances in Turbulence Studies (1993)
Herman Branover
Yeshajahu Unger
Ben-Gurion University of the Negev
ISBN 1-56347-018-7

150. Structural Optimization: Status and Promise (1993)
Manohar P. Kamat
Georgia Institute of Technology
ISBN 1-56347-56-X

151. Dynamics of Gaseous Combustion (1993)
A. L. Kuhl
Lawrence Livermore National Laboratory
J.-C. Leyer
Universite de Poitiers
A. A. Borisov
USSR Academy of Sciences
W. A. Sirignano

152. Dynamics of Heterogeneous Gaseous Combustion and Reacting Systems (1993)
A. L. Kuhl
Lawrence Livermore National Laboratory
J.-C. Leyer
Universite de Poitiers
A. A. Borisov
USSR Academy of Sciences
W. A. Sirignano
University of California
ISBN 1-56347-058-6

153. Dynamic Aspects of Detonations (1993)
A. L. Kuhl
Lawrence Livermore National Laboratory
J.-C. Leyer
Universite de Poitiers
A. A. Borisov
USSR Academy of Sciences
W. A. Sirignano
University of California
ISBN 1-56347-057-8

154. Dynamic Aspects of Explosion Phenomena (1993)
A. L. Kuhl
Lawrence Livermore National Laboratory
J.-C. Leyer
Universite de Poitiers
A. A. Borisov
USSR Academy of Sciences
W. A. Sirignano
University of California
ISBN 1-56347-059-4

155. Tactical Missile Warheads (1993)
Joseph Carleone
Aerojet General Corporation
ISBN 1-56347-067-5

156. Toward a Science of Command, Control, and Communications (1993)
Carl R. Jones
Naval Postgraduate School
ISBN 1-56347-068-3

157. Tactical and Strategic Missile Guidance Second Edition (1994)
Paul Zarchan
Charles Stark Draper Laboratory, Inc.
ISBN 1-56347-077-2

158. Rarefied Gas Dynamics: Experimental Techniques and Physical Systems (1994)
Bernie D. Shizgal
University of British Columbia
David P. Weaver
Phillips Laboratory
ISBN 1-56347-079-9

159. Rarefied Gas Dynamics: Theory and Simulations (1994)
Bernie D. Shizgal
University of British Columbia
David P. Weaver
Phillips Laboratory
ISBN 1-56347-080-2

160. Rarefied Gas Dynamics: Space Sciences and Engineering (1994)
Bernie D. Shizgal
University of British Columbia
David P. Weaver
Phillips Laboratory
ISBN 1-56347-081-0

161. Teleoperation and Robotics in Space (1994)
Steven B. Skaar
University of Notre Dame
Carl F. Ruoff
Jet Propulsion Laboratory, California Institute of Technology
ISBN 1-56347-095-0

162. Progress in Turbulence Research (1994)
Herman Branover
Yeshajahu Unger
Ben-Gurion University of the Negev
ISBN 1-56347-099-3

*Out of print.

163. Global Positioning System: Theory and Applications Volume I (1996)
Bradford W. Parkinson
Stanford University
James J. Spilker Jr.
Stanford Telecom
Penina Axelrad,
Associate Editor
University of Colorado
Per Enge,
Associate Editor
Stanford University
ISBN 1-56347-107-8

164. Global Positioning System: Theory and Applications Volume II (1996)
Bradford W. Parkinson
Stanford University
James J. Spilker Jr.
Stanford Telecom
Penina Axelrad,
Associate Editor
University of Colorado
Per Enge,
Associate Editor
Stanford University
ISBN 1-56347-106-X

165. Developments In High-Speed Vehicle Propulsion Systems (1996)
S.N.B. Murthy
Purdue University
E. T. Curran
Wright Laboratory
ISBN 1-56347-176-0

*Out of print.

www.ingramcontent.com/pod-product-compliance
Lightning Source LLC
Chambersburg PA
CBHW051255200926
51612CB00001B/53
9781563471766